WITHDRAWN

PROBLEMS IN ORGANIC REACTION MECHANISMS

Problems in Organic Reaction Mechanisms

Hermann Höver

*Union Rheinische Braunkohlen Kraftstoff A.G.
Wesseling, Germany*

WILEY-INTERSCIENCE
a division of John Wiley & Sons Ltd
London   New York   Sydney   Toronto

Copyright © 1970 John Wiley & Sons Ltd.
All Rights Reserved. No part of this publication may be reproduced, stored in a retrieval system, or transmitted, in any form or by any means, electronic, mechanical photocopying, recording or otherwise, without the prior written permission of the Copyright owner.

Library of Congress catalog card number 70-108591

ISBN 0 471 41637 1

Printed in Northern Ireland at The Universities Press, Belfast

Preface

The deduction of reaction mechanisms, when starting materials and reaction conditions are known, is an efficient method of studying organic chemistry if the student is assisted in the interpretation of the facts. This requirement is met in a seminar, where the problems can be discussed by professor and students together.

Since most textbooks which contain collections of problems do not provide the student with the opportunity to check results, this volume attempts to use a method which is similar to seminar teaching. The book is divided into two principal parts. The first part contains the problems, and in the second part they are thoroughly discussed. Most of the problems deal with aliphatic addition, elimination, and substitution reactions. Some aromatic substitutions, and radical and photochemical reactions, are discussed in Chapters 7–9. Examples have been chosen of such a degree of difficulty that the reader has still to work at them after having read the book only once. Since more than one product is usually formed in a chemical reaction, products are presented with their percent yields if available. The interpretation of the individual problems is based essentially on the authors' discussions in the chemical literature. Long-term investigations which have appeared in the literature during several years have been followed to the latest possible date. However, some important papers published in 1968 and 1969 could not be included, or have been incorporated only as references.

Many of the interpretations have been reviewed by the authors, to whom I express my gratitude. In numerous instances additional facts are given, analogies have been mentioned, and alternative mechanisms have been suggested. If possible, it has been shown why a reaction proceeds just by the path which it actually takes. In order to close the gap between current textbooks and this treatise, brief introductions have been added to the individual chapters.

Although most of the problems are taken from the literature of the last ten years, the book does not claim to represent a cross-section of current organic chemistry. It should be considered rather as a collection of interesting problems, some of which could perhaps be replaced by more instructive ones; any suggestions which would contribute to making this book more representative, more stimulating, and more instructive would be appreciated.

The suggested mechanisms are in general not purely speculative. The isolation of intermediates, kinetic and stereochemical studies, tracer studies, and other methods make these mechanisms appear to be at least very likely. In other examples only the starting materials and the products are known; in these cases the intermediate, unknown transformations can only be postulated on the basis of analogies.

The justification for mechanistic speculation, even for the student with little experience, is self-evident, provided that such speculation is considered as a means of penetrating intellectually into the course of the reaction. This intellectual approach leads to the points of attack for the actual physical measurements and investigations of the mechanisms. A well known example, namely Kekulé's interpretation of the aromatic properties of benzene, illustrates this point: his interpretation, although not exactly correct, was a tremendous stimulation for subsequent investigations.

Acknowledgement is made to the following for permission to use material which appeared in their publications: The American Chemical Society for the diagrams on pages 158, 173, 237–239, and 241, and the table on page 159; Elsevier Publishing Company for diagrams on pages 372 and 379, and the table on page 378; Academic Press Inc. for formulae on pages 319–321, and tables on pages 329 and 331.

I am very grateful to Professor Criegee, Prof. H. C. Brown, Professor Grob, and Professor Wittig. I am particularly indebted to Dr. J. Schantl and Dr. Joe Baker, of the University of Wisconsin, who revised the chapter on photochemistry, and to Dr. M. Schlosser of the University of Heidelberg and many others who read portions of the manuscript. It is a pleasure to thank Mrs. Buchholtz, of Union Kraftstoff, Wesseling, who typed the manuscript.

HERMANN HÖVER

Contents

PART 1 STATEMENT OF THE PROBLEMS

1. Addition Reactions 3
2. Cycloadditions 13
3. Valence Isomerizations 21
4. β-Eliminations 23
5. Fragmentations 27
6. Aliphatic Substitution 29
7. Aromatic Substitution 45
8. Radical Reactions 47
9. Photochemistry 50
10. Miscellaneous Reactions 53

PART 2 DISCUSSION OF THE PROBLEMS

1. Addition Reactions 59

   I. Electrophilic additions to the C=C bond 59
   II. Nucleophilic additions to the C=C bond 78
   III. Additions to the C≡C bond 88
   IV. Nucleophilic additions to the C=O bond 95
   V. Electrophilic additions to the C=O bond 110
   VI. Electrophilic additions to the C≡N bond 117
   VII. Miscellaneous additions 121

2. Cycloadditions 129

   I. 1,1-Additions 130
   II. 1,2-Additions 153
   III. 1,3-Dipolar additions 165
   IV. Diels–Alder reactions 170
   V. Trimerization of acetylenes 182

3. Valence Isomerizations **189**
   I. Diallylic systems and monovinyl-cyclopropanes and -cyclobutanes 192
   II. Conjugated polyenes 195

4. β-Eliminations **202**
   I. Saytzeff and Hofmann rules 203
   II. Thermal *cis*-eliminations 207
   III. Miscellaneous eliminations 213

5. Fragmentation **223**
   I. Fragmentation of γ-amino halides 223
   II. Fragmentation of α-amino ketoximes 226
       Beckmann fission 228
   III. Miscellaneous fragmentations 230
   IV. α-Fragmentations 234

6. Aliphatic Substitution **236**
   I. S_N1 substitution 236
   II. Carbonium ions 241
   III. Participation of neighboring carbon in carbonium ion reactions
        (non-classical carbonium ions) 265
        The cyclopropylcarbinyl system 265
        The norbornyl system 276
   IV. S_N2 substitution 284
       (a) S_N2 substitution at the saturated carbon atom . 284
       (b) S_N2 substitution at the carbonyl carbon atom . 298
   V. S_Ni substitution 308
   VI. S_N2 substitution at hetero atoms 311
       (a) S_N2 substitution at oxygen 311
       (b) S_N2 substitution at bivalent sulfur . . . 313
       (c) S_N2 substitution at chlorine 315
   VII. S_E1 substitution 317
   VIII. S_E2 and S_Ei substitution 326
   IX. Carbanions 328
       Miscellaneous reactions 332, 350
       Ylides 340
       Ambident ions 347
       Birch reductions 358

7. Aromatic Substitutions **371**

    I. Electrophilic aromatic substitution 371
        (a) *Reaction intermediates* 371
        (b) *Substituent effects and Hammett equation* . . . 373
        (c) *Transition states*. 374
        (d) *ortho/para ratio* 375
        (e) *Modifications of the Hammett equation* . . . 376
    II. Nucleophilic aromatic substitution 383

8. Radical Reactions **388**

    I. Radical substitution 389
    II. Radical addition 391
    III. Autoxidation. 395
    IV. Decomposition of hydroperoxides 396
    V. Radical mechanisms in lead tetraacetate oxidations . . 399

9. Photochemistry **405**

    Theoretical background 405
        Absorption 405
        Deactivation 407
    Examples 413

10. Miscellaneous Reactions **423**

Index **445**

Part 1

Statement of the Problems

Unless stated otherwise, the reader should try to elucidate the correct structures of the products with the aid of the total formulae, and to derive reasonable reaction mechanisms; in a number of cases the spectroscopic data will be useful. Proposing and justifying reaction mechanisms should be considered as the principal exercises. The solution of the problems is facilitated by designation of the reactions with capital letters denoting a principal step: A = an addition reaction; C = a cycloaddition; V = valence isomerization; E = elimination; F = fragmentation; S = substitution; AS = aromatic substitution; R = radical reaction; P = photoreaction; M = miscellaneous reactions.

The starting materials, most important intermediates, and products are numbered to facilitate their recognition in Part 2. The intermediates in brackets have not been isolated but are transformed into more stable products under the reaction conditions.

1

Addition Reactions

I. ELECTROPHILIC ADDITIONS TO THE C=C BOND

A-1. Addition of DBr to indene

THIS yields 81% of *cis*-adduct and 19% of *trans*-adduct. The reaction may proceed by three mechanisms:

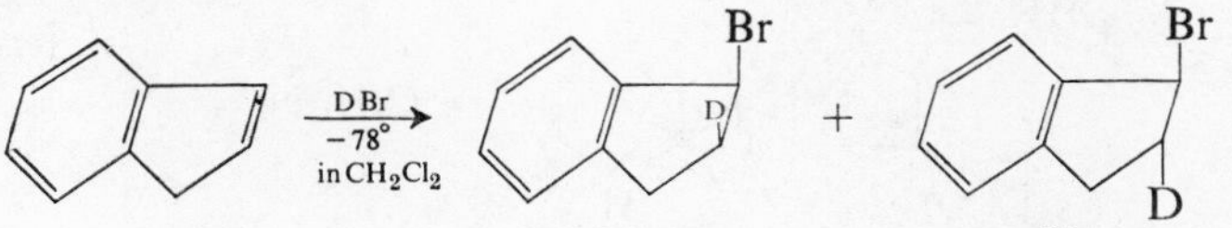

$$\text{indene} \xrightarrow[\text{in CH}_2\text{Cl}_2]{\text{DBr}, -78°} \text{(81\%) cis-Br,D adduct} + \text{19\% trans-Br,D adduct}$$

(a) *via* a π-complex;
(b) *via* a free solvated carbonium ion; or
(c) *by* intermediate formation of a tight ion-pair

Which mechanism best explains the results?

A-2. Addition of bromine to 3-benzamidocyclohexene

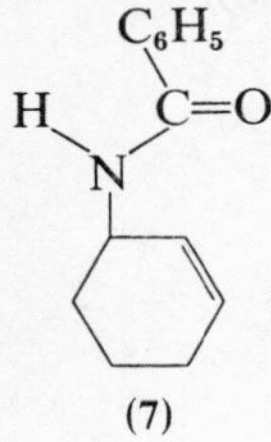

$$(7) \xrightarrow[\text{CH}_3\text{OH}]{\text{Br}_2} [C_{13}H_{15}BrNO]^+ Br^- + C_{13}H_{15}Br_2NO + C_{13}H_{15}Br_2NO$$
$$\quad\quad\quad\quad\quad\quad\quad (9) \quad\quad\quad\quad\quad (10a) \quad\quad\quad\quad (10b)$$
$$\quad\quad\quad\quad\quad\quad\quad\; \downarrow H_2O \quad 60\%$$
$$\quad\quad\quad\quad\quad\quad\quad C_{13}H_{16}BrNO_2$$
$$\quad\quad\quad\quad\quad\quad\quad\quad\; (11)$$

(7) is 3-benzamidocyclohexene (with C$_6$H$_5$–C(=O)–NH– group)

A-3. Acylation of olefins

$$H_2C=CH-CH_3 + H_3CCOCl \xrightarrow{AlCl_3} C_5H_8O + [C_7H_{10}O_2]$$
$$\quad\quad\quad\quad\quad\quad\quad\quad\quad\quad\quad\quad\quad (15) \quad\quad \text{not isolated}$$
$$\quad\quad\quad\quad\quad\quad\quad\quad\quad\quad\quad\quad\quad\quad\quad\quad\quad (17)$$
$$\quad\quad\quad\quad\quad\quad\quad\quad\quad\quad\quad\quad\quad\quad\quad\quad\quad \downarrow HCl$$
$$\quad\quad\quad\quad\quad\quad\quad\quad\quad\quad\quad\quad\quad [C_7H_9O]^+Cl^- \; (18)$$

A-4. Ozonization of olefins

indene $\xrightarrow[NH_3]{O_3}$ [C$_9$H$_9$NO$_2$] $\xrightarrow{NH_3}$ C$_9$H$_7$N + NH$_2$OH
                            not isolated   (35)
                               (33)   62%

A-5. Addition of bromine to 1,2,3,4-tetraphenyl-1,3-butadiene

$$\underset{H}{\overset{H_5C_6}{C}}=\underset{C_6H_5}{\overset{H_5C_6}{C}}-\underset{H}{\overset{C_6H_5}{C}}=\underset{}{\overset{C_6H_5}{C}} + Br_2 \longrightarrow C_{28}H_{20}$$

(40) 70%

A-6. Addition of acrylonitrile to cyclopentanone pyrrolidine enamine

cyclopentanone + pyrrolidine $\longrightarrow$ C$_9$H$_{15}$N

C$_9$H$_{15}$N + H$_2$C=CH—CN → C$_{12}$H$_{18}$N$_2$ $\xrightarrow{H_2O}$ C$_8$H$_{11}$NO + C$_4$H$_9$N
                                    (42)   (43) 67%

A-7. Rearrangement of 1-(α-methylenefurfuryl)pyrrolidine

2-acetylfuran + pyrrolidine $\xrightarrow{HCl}$ [C$_{10}$H$_{13}$NO] $\longrightarrow$ C$_{10}$H$_{13}$NO
                                                 not isolated   (49)
                                                    (47)

A-8. Formation of cyclobutane derivatives from enamines

pyrrolidine + cyclopentanone $\longrightarrow$ C$_9$H$_{15}$N

The presence of (54) has been   C$_{13}$H$_{19}$NO$_2$
demonstrated by NMR and IR   (54)
                                              ↑
                                     storage at
                                     room temp.

C$_9$H$_{15}$N + HC≡C—COOCH$_3$ $\xrightarrow{<35°}$ C$_{13}$H$_{19}$NO$_2$ $\xrightarrow{\Delta}$ C$_{13}$H$_{19}$NO$_2$
                                                       (52)   (53) 65%

Characteristic IR absorption of (53): 1666 cm^{-1} Hydrol. ↓
                                      1599
                                      1514   C$_9$H$_{12}$O$_3$
Characteristic IR absorption of (54): 1680 cm^{-1} (55)
                                      1549

A-9. Alkylation of enamines with alkyl halides

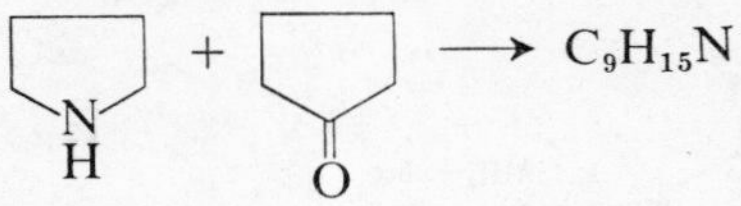

[pyrrolidine] + [cyclopentanone] → $C_9H_{15}N$

$C_9H_{15}N + Br-CH_2-CH=CH-CH_3 \rightarrow C_{13}H_{21}N + [C_{13}H_{22}N]^+Br^-$

↓ Hydrol. ↓ Hydrol.

$C_9H_{14}O$ $C_9H_{14}O$
(57) 80%   (56) 20%

Characteristic IR absorption of (56): 1730 cm^{-1}
1630–40
998
910

Characteristic IR absorption of (57): 1730 cm^{-1}
1630–40
965

A-10. Addition of hydrogen iodide to 3,5,5-trimethyl-1,2-cyclohexanedione 2,3-monoenol

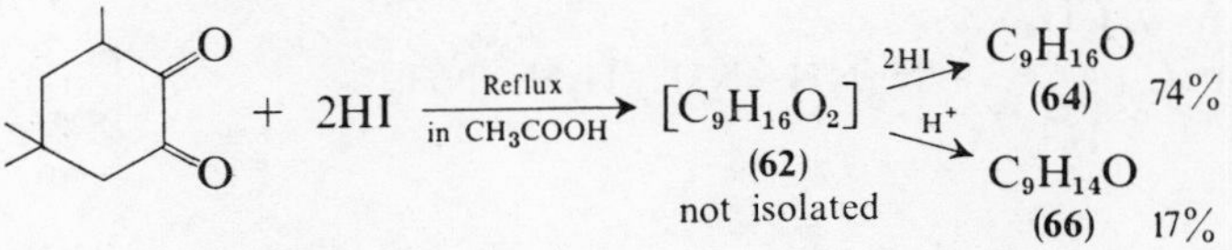

[3,5,5-trimethyl-1,2-cyclohexanedione 2,3-monoenol] + 2HI $\xrightarrow[\text{in CH}_3\text{COOH}]{\text{Reflux}}$ $[C_9H_{16}O_2]$ $\underset{H^+}{\overset{2HI}{\rightleftarrows}}$

$C_9H_{16}O$ (64) 74%

(62) not isolated

$C_9H_{14}O$ (66) 17%

Characteristic IR absorption of (64): 1708 cm^{-1}
Characteristic IR absorption of (66): 1680 cm^{-1}
1640

II. NUCLEOPHILIC ADDITIONS TO THE C=C BOND

A-11. Addition of OH$^-$ to trans-γ-bromodypnone

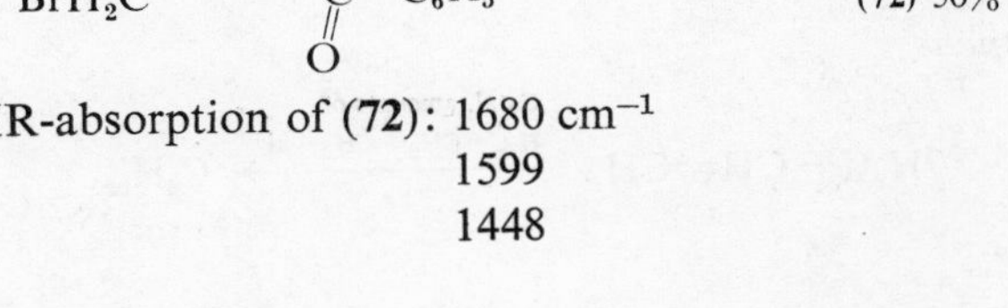

$\begin{array}{c} C_6H_5 \\ BrH_2C \end{array}\!\!>\!\!C\!\!=\!\!C\!\!<\!\!\begin{array}{c} H \\ C(=O)-C_6H_5 \end{array}$ $\xrightarrow{\text{NaOH-C}_2\text{H}_5\text{OH}}$ $C_{16}H_{14}O_2$
(72) 50%

Characteristic IR-absorption of (72): 1680 cm^{-1}
1599
1448

Addition of lithium aluminum hydride to double bonds

A-12. (*i*) *To tropone*

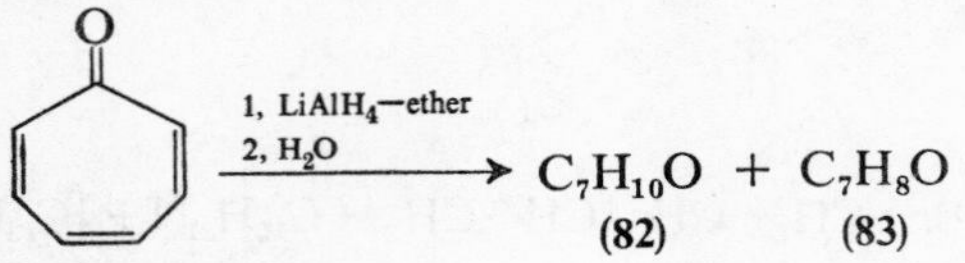

$$\text{tropone} \xrightarrow[\text{2, } H_2O]{\text{1, LiAlH}_4-\text{ether}} C_7H_{10}O + C_7H_8O$$
$$\quad\quad\quad\quad\quad\quad\quad\quad (82)\quad\quad (83)$$

A-13. (*ii*) *To norbornadien-7-ol*

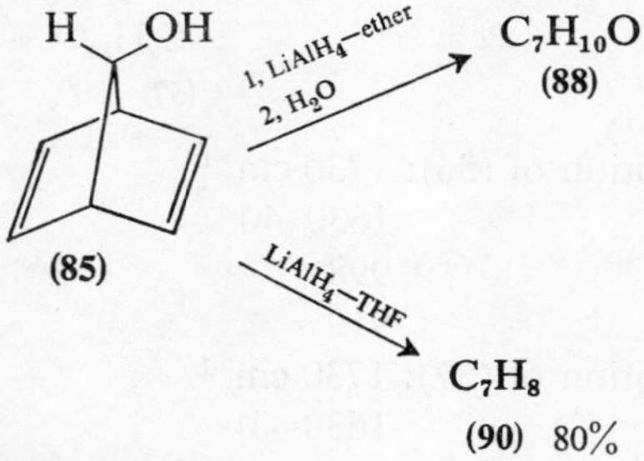

$$(85) \xrightarrow[\text{2, } H_2O]{\text{1, LiAlH}_4-\text{ether}} C_7H_{10}O \;(88)$$

$$(85) \xrightarrow{\text{LiAlH}_4-\text{THF}} C_7H_8 \;(90)\; 80\%$$

Addition of sodium hydrazide to olefins

A-14. (*i*) *To β-methylstyrene*

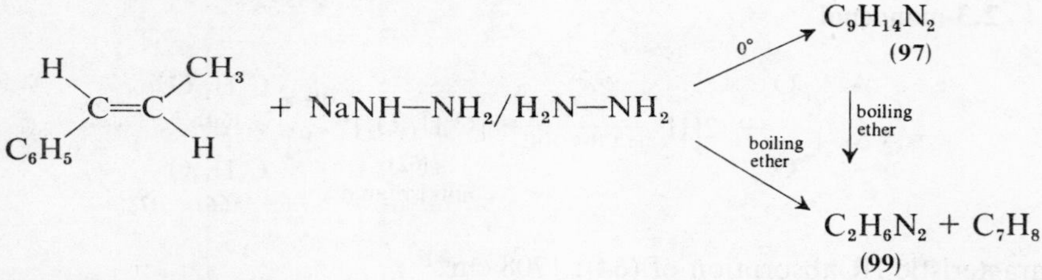

$$\underset{C_6H_5}{\overset{H}{>}}C=C\underset{H}{\overset{CH_3}{<}} + \text{NaNH}-\text{NH}_2/\text{H}_2\text{N}-\text{NH}_2 \xrightarrow{0°} C_9H_{14}N_2 \;(97)$$

$$\xrightarrow[\text{ether}]{\text{boiling}} C_2H_6N_2 + C_7H_8 \;(99)$$

(with boiling ether branch from (97))

A-15. (*ii*) *To trans-stilbene*

The reaction of *trans*-stilbene with sodium hydrazide leads to 90% of 1,2-diphenylethane. Propose a reaction mechanism.

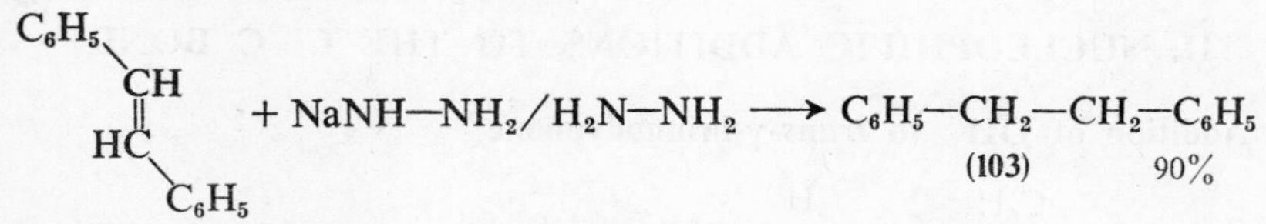

$$\underset{C_6H_5}{\overset{C_6H_5}{>}}\text{CH}=\text{CH} + \text{NaNH}-\text{NH}_2/\text{H}_2\text{N}-\text{NH}_2 \longrightarrow C_6H_5-CH_2-CH_2-C_6H_5$$
$$\quad\quad\quad\quad\quad\quad\quad\quad\quad\quad\quad\quad\quad\quad\quad\quad\quad (103)\quad\quad\quad 90\%$$

Addition of aluminum alkyls and hydrides to olefins

A-16. (*i*) *To propene*

$$2H_3C-CH=CH_2 \xrightarrow{\text{Catal. amounts of } H_3C-CH_2-CH_2-\text{al}} C_6H_{12} \;(105)$$

A-17. (ii) Isomerizations

$$H_3C-CH_2-CH=CH-CH_2-CH_3 + al\text{-}H \rightarrow C_6H_{13}al$$
(107)

$$al\text{-}H = (iso\text{-}C_4H_9)_2Al\text{-}H$$

III. ADDITIONS TO THE TRIPLE BOND

A-18. Addition of bromine to 1-penten-4-yne

Which product is obtained in the bromination of 1-penten-4-yne with 1 mol. of bromine?

$$H_2C=CH-CH_2-C\equiv CH + Br_2 \xrightarrow{\text{in } CCl_4} \text{?}$$

A-19. Addition of malononitrile to dimethyl acetylenedicarboxylate

[pyridinium]
CH_3COO^-
+ $H_2C\begin{smallmatrix}CN\\CN\end{smallmatrix}$ + 2 $\begin{smallmatrix}COOCH_3\\C\\\|\\C\\COOCH_3\end{smallmatrix}$ $\longrightarrow$ $[C_{14}H_{10}N_2O_7]$ + [pyridinium] + CH_3OH
not isolated CH_3COO^-
(111)

(111) + [pyridinium] $\longrightarrow$ $[C_5H_6N]^+ [C_{14}H_9N_2O_7]^- + CH_3COOH$
CH_3COO^- (112) 60%

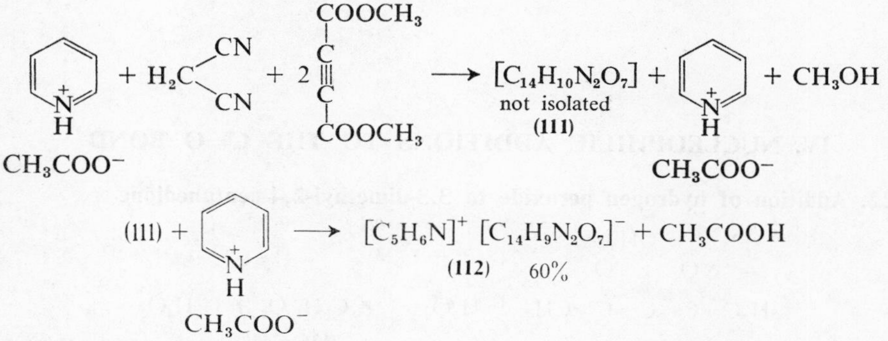

Characteristic absorption of (112):

IR	2198 cm^{-1}	NMR	0.94—2.1 τ	(6) multiplet
	1742		6.20	(3) singlet
	1715		6.40	(3) singlet
			6.41	(3) singlet

A-20. Addition of a phenacylidenetriphenylphosphorane to dimethyl acetylenedicarboxylate

$$C_6H_5-\underset{\underset{O}{\|}}{C}-CH=P(C_6H_5)_3 + H_3COOC-C\equiv C-COOCH_3 \begin{array}{c}\xrightarrow{CH_3OH} C_{32}H_{27}O_5P \text{ (116)}\\ \xrightarrow{CH_3CN} C_{32}H_{27}O_5P \text{ (118)}\end{array}$$

[See p. 8 for spectra.]

Characteristic absorptions of:

(116) IR 1715 cm^{-1} NMR 2—3 τ (20) multiplet
       1608   4.81 (1) singlet
       1587   6.67 (3) singlet
       1511   6.97 (3) singlet

(118) IR 1724 cm^{-1} NMR 2—3 τ (20) multiplet
       1639   4.39 (1) broad singlet
                              6.27 (3) singlet
                              6.84 (3) singlet

A-21. Addition of triphenylaluminum to diphenylacetylene

$$C_6H_5-C\equiv C-C_6H_5 + (C_6H_5)_3Al \xrightarrow{200°} C_{26}H_{19}Al \xrightarrow{H_2O} C_{20}H_{16}$$
$$\text{(124)} \qquad \text{(123)} \qquad \text{(127)}$$

IV. NUCLEOPHILIC ADDITIONS TO THE C=O BOND

A-22. Addition of hydrogen peroxide to 3,3-dimethyl-2,4-pentanedione

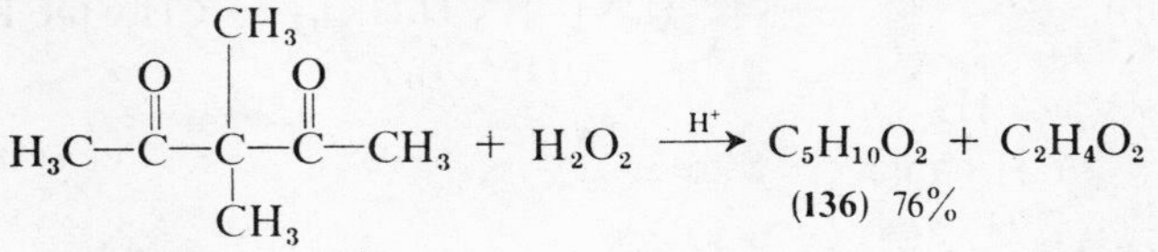

$$\underset{(133)}{H_3C-\underset{\parallel}{\underset{O}{C}}-\underset{\underset{CH_3}{|}}{\overset{\overset{CH_3}{|}}{C}}-\underset{\parallel}{\underset{O}{C}}-CH_3} + H_2O_2 \xrightarrow{H^+} \underset{(136)\ 76\%}{C_5H_{10}O_2} + C_2H_4O_2$$

A-23. Addition of LiAlH$_4$ to tropolone methyl ether

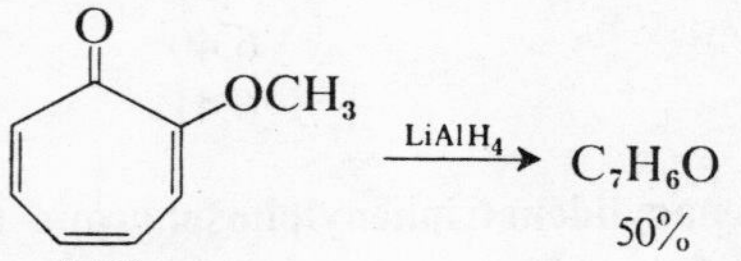

(142) $\xrightarrow{\text{LiAlH}_4}$ C$_7$H$_6$O 50%

A-24. Addition of phosphorus tris(dimethylamide) to benzaldehyde

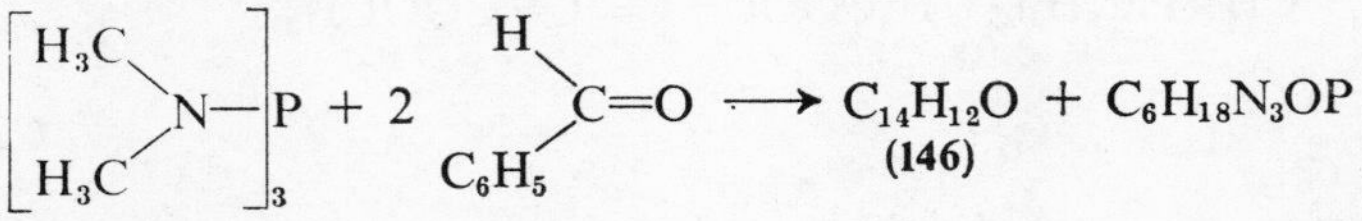

$$\left[\underset{H_3C}{\overset{H_3C}{\diagdown}}N-\right]_3 P + 2\ \underset{C_6H_5}{\overset{H}{\diagdown}}C=O \longrightarrow \underset{(146)}{C_{14}H_{12}O} + C_6H_{18}N_3OP$$
(144)

A-25. Additions of trimethyl phosphite

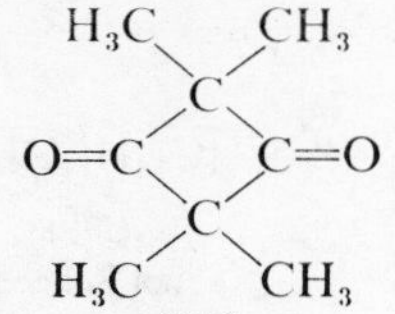

$$\text{(152)} \xrightarrow{\text{P(OCH}_3)_3} C_{11}H_{21}O_5P$$
(155) 47%

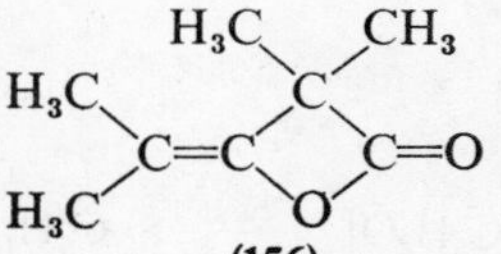

$$\text{(156)} \xrightarrow{\text{P(OCH}_3)_3} C_{11}H_{21}O_5P \xrightarrow{\text{CH}_3\text{OH/HCl}} C_9H_{16}O_3 + C_3H_9OP$$
(159) 67%

Characteristic IR and NMR absorption of:

(155)	IR	1750 cm^{-1}	NMR	6.37 τ	(3) singlet
		1685		6.45	(6) doublet
(159)	IR	1730 cm^{-1}		8.73	(6) singlet
		1660		8.87	(6) doublet

A-26. Addition of diazoethane to an azine

$$C_6H_5\text{—}\underset{O}{\overset{}{C}}\text{—CH=N—N=CH—CH}_3 + CH_3\text{—CH=}\ddot{N}\text{—}\ddot{N} \longrightarrow C_{12}H_{14}N_2O$$
(163)   (168) 16%

$$\xrightarrow{H_3O^+} C_{10}H_{10}N_2 + C_2H_4O$$
(170)

$$C_{10}H_{10}N_2 + C_2H_4O \rightarrow C_{22}H_{22}N_4 + H_2O$$
(170)   (171)

Characteristic IR and UV absorption of (168):

IR   3310 cm^{-1} UV   298 mμ (log ε = 4.36)
       1645   344 mμ (log ε = 3.95)

A-27. Reaction of 1,3-diphenyl-3-phospholin 1-oxide with phenyl isocyanate

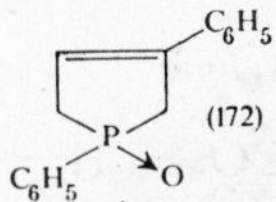

$$2\ C_6H_5\text{—N=C=O} \xrightarrow[\text{catalyst}]{\text{(172)}} C_{13}H_{10}N_2$$
(173)   (176)

Clemmensen reduction

A-28. (*i*) *Of 2,2'-diacetylbiphenyl*

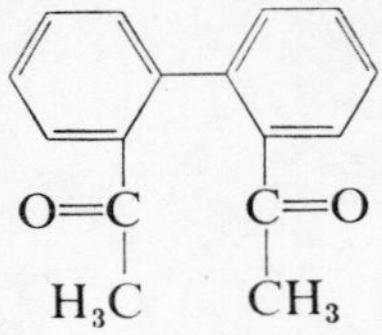

(177) $\xrightarrow{\text{Zn-HCl}}$ [$C_{16}H_{16}O_2$] $\xrightarrow{H^+}$ [$C_{16}H_{14}O$] $\xrightarrow{\text{Zn-HCl}}$ $C_{16}H_{14}$
(178) not isolated (179) not isolated (180)

A-29. (*ii*) *Of 2,4-pentanedione*

$H_3C-\underset{\underset{O}{\|}}{C}-CH_2-\underset{\underset{O}{\|}}{C}-CH_3$ $\xrightarrow{\text{Zn/HCl}}$ [$C_5H_{10}O_2$] $\xrightarrow{H^+}$ [C_5H_8O] $\xrightarrow{\text{Zn/HCl}}$ $C_5H_{10}O$
(182) not isolated (183) not isolated (184)

$C_5H_{10}O$ $\xrightarrow{\text{Zn-HCl}}$ C_5H_{12}
(184)   (185)

V. ELECTROPHILIC ADDITIONS TO THE C=O BOND

Addition of PCl_5

A-30. (*i*) *To 4-methyl-4-(trichloromethyl)-2,5-cyclohexadienone*

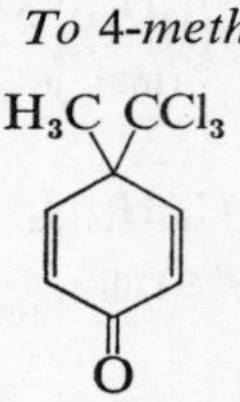

(192) + PCl_5 [PCl_4^+ PCl_6^-] $\longrightarrow$ $C_8H_6Cl_4$ + $POCl_3$ + HCl
(195)

A-31. (*ii*) *To 3,4-dimethyl-4-(trichloromethyl)-2,5-cyclohexadienone*

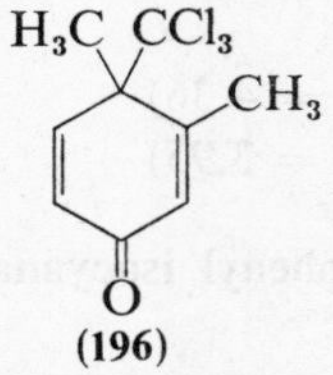

(196) + PCl_5 [PCl_4^+ PCl_6^-] $\longrightarrow$ $C_9H_8Cl_4$ + $POCl_3$ + HCl
(200) 88%

A-32. (*iii*) *To 1,2-dibenzoylethylene*

$C_6H_5-CO-CH=CH-CO-C_6H_5$ + $PCl_5[PCl_4^+PCl_6^-]$ $\longrightarrow$
(201)
$C_{16}H_{11}ClO$ + $POCl_3$ + HCl
(202)

Addition Reactions

A-33. *Meerwein–Ponndorf–Verley–Oppenauer reaction*

$$C_6H_5-\underset{O}{\overset{\|}{C}}-C_6H_5 + Al\left(O-\underset{CH_3}{\overset{CH_3}{\underset{|}{CH}}}\right)_3 \longrightarrow$$

$$C_{13}H_{12} + 2\,H_3C-CO-CH_3 + \underset{\text{unstable}}{[C_3H_7-O-Al=O]}$$
(203)

VI. ELECTROPHILIC ADDITIONS TO THE C≡N BOND

Reaction of tertiary alcohols with CH_3CN in the presence of concentrated sulfuric acid

A-34. (*i*) *2-(1-Cyclopentenyl)-2-methylpropanol*

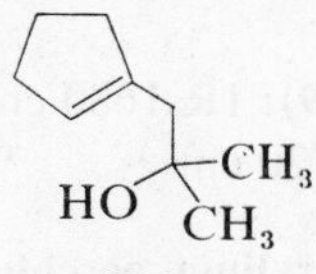

<chemical structure> $\xrightarrow[CH_3CN]{H_2SO_4}$ $C_{11}H_{19}NO$ (206) 63%

(204)

Conc. H_2SO_4 is added dropwise to a cold mixture of (204) and CH_3CN
Characteristic IR absorption of (206): 1666 cm^{-1}

A-35. (*ii*) *2-(2-Hydroxycyclopentyl)-2-methylpropanol*

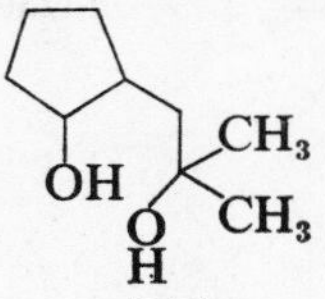

<chemical structure> $\xrightarrow[H_2SO_4]{CH_3CN}$ $C_{11}H_{17}N$ (207) 76%

(209)

Characteristic IR and UV absorption of (207):

    IR 1667 cm^{-1} UV 263 mμ (log ε 3.63)
          1600

A-36. Addition of phenethyl chloride to benzonitrile

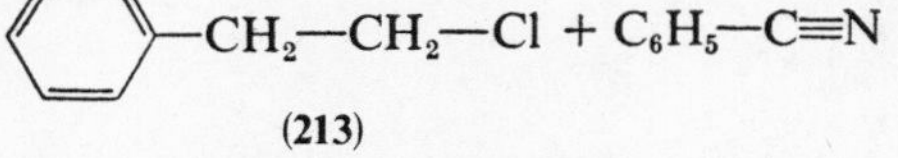

<chemical structure>—CH_2—CH_2—Cl + C_6H_5—C≡N $\xrightarrow{SbCl_5}$ $C_{15}H_{13}N$ + $HSbCl_6$
(213)   (214)

A-37. Addition of nitrilium salts to nitriles

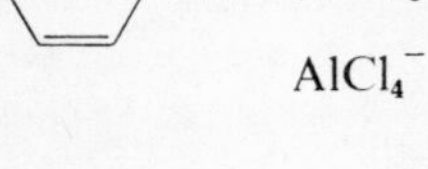

<chemical structure>—$\overset{+}{N}$=C—C_6H_5 + C_6H_5—C≡N $\longrightarrow$ $[C_{20}H_{15}N_2]^+$ $AlCl_4^-$ $\xrightarrow{NaOH}$ $C_{20}H_{14}N_2$
$AlCl_4^-$ not isolated   (215)

VII. MISCELLANEOUS ADDITIONS

A-38. Addition of 2,3-dimethyl-2-butene to hexafluorothioacetone

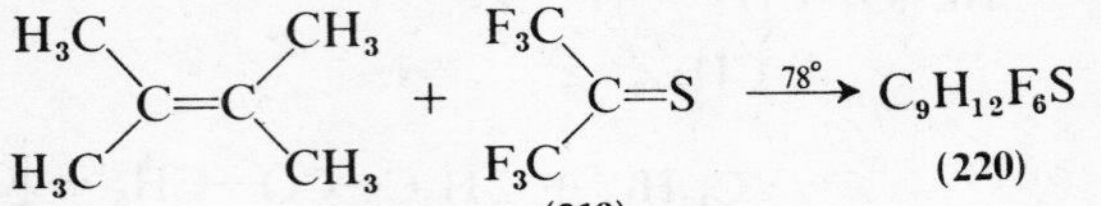

$$\underset{(219)}{(H_3C)_2C=C(CH_3)_2} + \underset{}{(F_3C)_2C=S} \xrightarrow{-78°} \underset{(220)}{C_9H_{12}F_6S}$$

A-39. Hydride addition to 2,4,6-trimethylpyrylium perchlorate

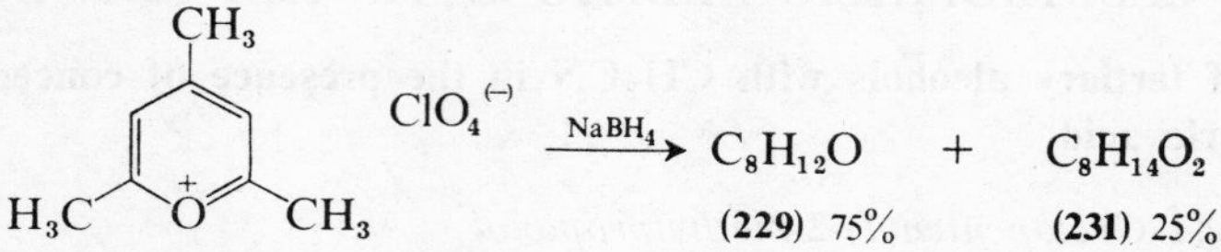

$$\underset{(227)}{[2,4,6\text{-trimethylpyrylium}]^+ ClO_4^{(-)}} \xrightarrow{NaBH_4} \underset{(229)\ 75\%}{C_8H_{12}O} + \underset{(231)\ 25\%}{C_8H_{14}O_2}$$

Characteristic UV and IR absorption of **(229)**: IR 1680 cm^{-1} UV 401 mμ
Characteristic IR absorption of **(231)**: IR 1720 cm^{-1}

A-40. Addition of NH$_3$ to 2,4,6-triphenylpyrylium perchlorate

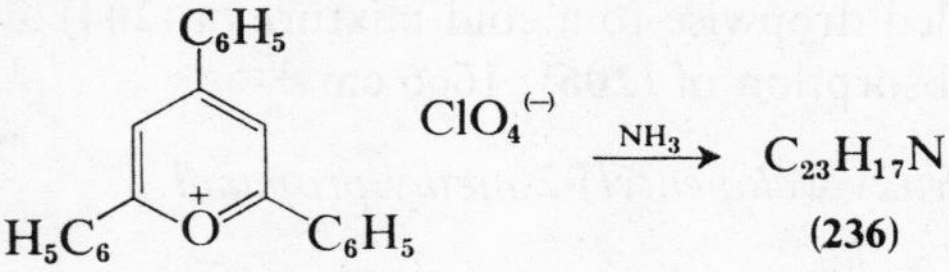

$$\underset{(232)}{[2,4,6\text{-triphenylpyrylium}]^+ ClO_4^{(-)}} \xrightarrow{NH_3} \underset{(236)}{C_{23}H_{17}N}$$

2
Cycloadditions

I. 1,1-ADDITIONS

C-1. Epoxidation of bicycloheptadiene

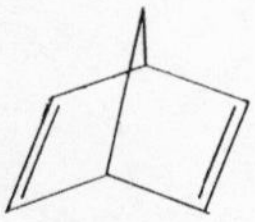

$$\text{[bicycloheptadiene]} + H_3C-\overset{O}{\underset{\|}{C}}-O-O-H \longrightarrow \underbrace{C_7H_8O + C_7H_8O}_{70\%} + CH_3COOH$$
$$\text{(246)} \quad \text{(247)}$$
$$\text{(246):(247)} = 7:3$$

Characteristic IR and NMR spectroscopic data of:

(246) IR 1695 cm^{-1} NMR   0.03 τ   doublet (1)
                                  1.59   quartet (1)
                                  2.8—2.0 multiplet (4)
                                  5.8   singlet (2)

(247) NMR 1.9—1.8 τ multiplet (2)   5.32 τ 2 doublets (1)
            2.8—2.4 multiplet (1)   5.75   doublet (1)
            4.83   singlet (1)   6.39   2 doublets (1)
            4.93   2 doublets (1)

C-2. Rearrangement of α-pinene oxide

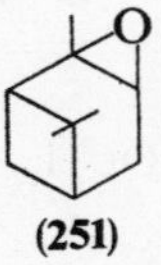

$$\text{(251)} \xrightarrow{ZnBr_2} C_{10}H_{16}O$$
$$\quad\quad\quad\quad\quad\text{(252)}$$

Reactions of methylene:

C-3. (i) With cyclohexene

Four products are obtained in the reaction of cyclohexene with diazomethane. The formation of the various products and the product distribution depend on the presence of additives such as benzophenone or Cu-powder. Explain the results.

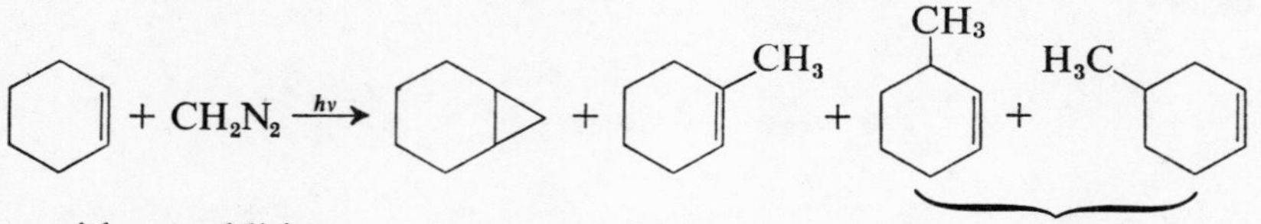

without additives	1	0.24	1.3
+ benzophenone	1	traces	0.42
+ Cu-powder	1	0	0

C-4. (*ii*) *With* cis- *and* trans-*butene*

Predict the configuration of the addition products of the following reactions:

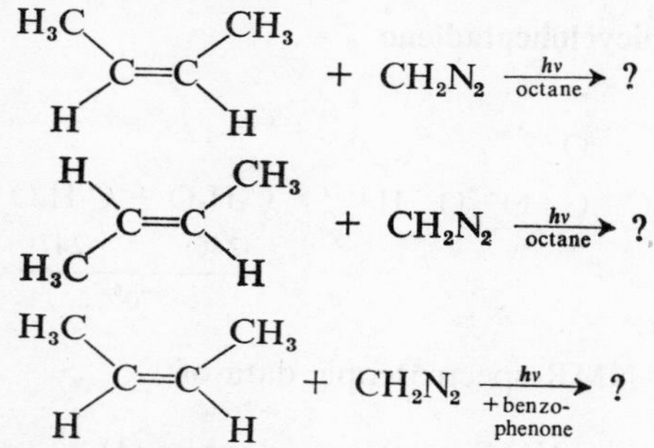

$$\underset{H}{\overset{H_3C}{>}}C=C\underset{H}{\overset{CH_3}{<}} + CH_2N_2 \xrightarrow[octane]{h\nu} ?$$

$$\underset{H_3C}{\overset{H}{>}}C=C\underset{H}{\overset{CH_3}{<}} + CH_2N_2 \xrightarrow[octane]{h\nu} ?$$

$$\underset{H}{\overset{H_3C}{>}}C=C\underset{H}{\overset{CH_3}{<}} + CH_2N_2 \xrightarrow[\substack{+benzo-\\phenone}]{h\nu} ?$$

C-5. Intramolecular addition of a carbene

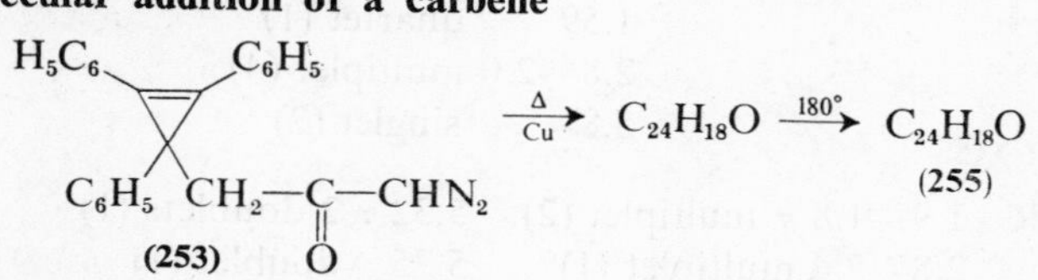

$$\text{(253)} \xrightarrow[Cu]{\Delta} C_{24}H_{18}O \xrightarrow{180°} C_{24}H_{18}O \quad \text{(255)}$$

C-6. Synthesis of cyclopropyl derivatives (i)

$$\text{[pyrrole-Li]} + CH_2Cl_2 + CH_3Li \longrightarrow ?$$

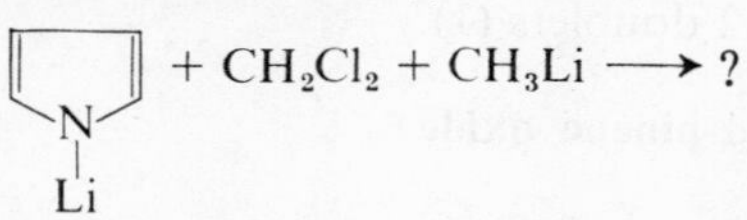

C-7. Synthesis of cyclopropyl derivatives (ii)

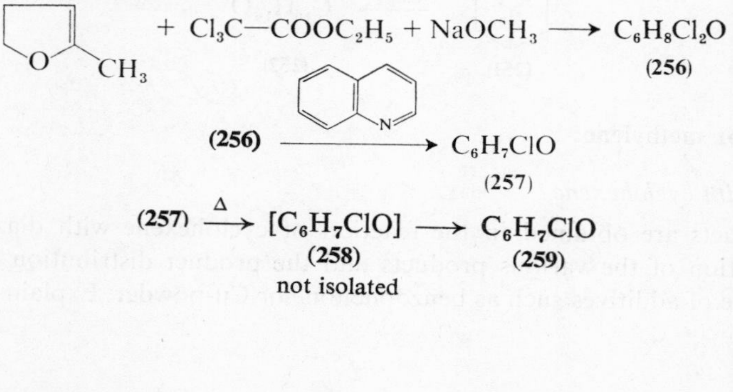

$$\text{[2-methylfuran]} + Cl_3C-COOC_2H_5 + NaOCH_3 \longrightarrow C_6H_8Cl_2O \quad \text{(256)}$$

$$\text{(256)} \xrightarrow{\text{quinoline}} C_6H_7ClO \quad \text{(257)}$$

$$\text{(257)} \xrightarrow{\Delta} [C_6H_7ClO] \longrightarrow C_6H_7ClO$$
$$\quad\quad\quad\quad\text{(258)} \quad\quad\quad \text{(259)}$$
$$\quad\quad\quad\text{not isolated}$$

C-8. Addition of difluorocarbene

$$F_3C-C\equiv C-CF_3 + CF_2 \text{ (from } (CF_3)_3PF_2) \xrightarrow[\text{gas phase}]{100°} C_5F_8$$
(260)

(260) $\xrightarrow[100°]{CF_2}$ C_6F_{10}
(261)

(261) $\xrightarrow{300°}$ C_6F_{10}
(262) 95%

(262) $\xrightarrow{350°}$ C_6F_{10}
(263)

C-9. Transannular insertion of a carbene

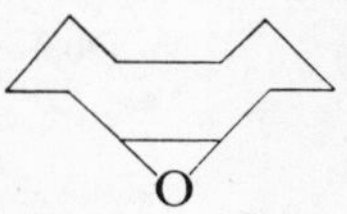

(264) $\xrightarrow[\text{2, H}_2\text{O}]{\text{1, LiN(C}_2\text{H}_5)_2}$ $C_{10}H_{18}O$ + $C_{10}H_{18}O$ + $C_{10}H_{18}O$
(265) 83%   (266) 9%   (267) 8%

C-10. Addition to isocyanides

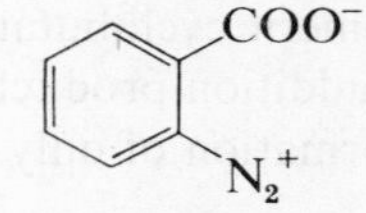

(270) + $:\bar{C}\equiv\overset{+}{N}-C_6H_5$ $\longrightarrow$ $C_{14}H_9NO_2$
(274)

C-11. Addition to nucleophilic carbenes

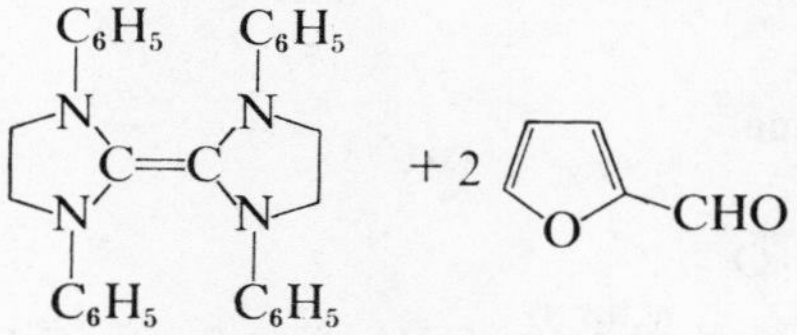

+ 2 furfural-CHO $\longrightarrow$ 2 $C_{20}H_{18}N_2O_2$
(280)

C-12. Insertion of a nitrene into a C–H bond

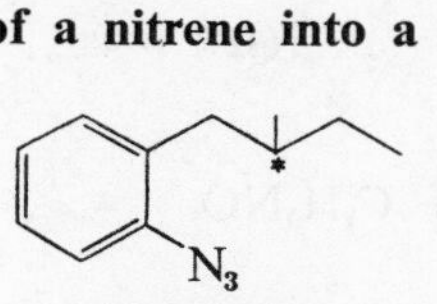

(282) $\longrightarrow$ $C_{11}H_{15}N$ + $C_{11}H_{15}N$
(284)   (285)

Optical activity is partly retained in (284) and completely retained in (285).

C-13. Insertion of positive nitrogen into a C–H bond

[Structure (286): 4-bromo-7-(2-methylpropan-2-yl)indanone oxime-like structure with C(CH₃)₂CH₃ group, Br, and C=N–OH]

$\xrightarrow{\text{polyphosphoric acid}}$ C$_{13}$H$_{14}$BrN + C$_{13}$H$_{16}$BrNO + C$_{13}$H$_{16}$BrNO

(289) 75%   (287) 16%   (288) 4%

C-14. Insertion of positive oxygen into a C–H bond

[Structure (297): cyclohexane with gem-dimethyl and C(CH₃)(O–OH) group]

$\xrightarrow[\text{2, H}_2\text{O}]{\text{1, C}_6\text{H}_5\text{–SO}_2\text{Cl, pyridine}}$ C$_9$H$_{18}$O$_2$ + C$_9$H$_{18}$O$_2$ + C$_9$H$_{16}$O

(300)   (301)   (303) 5–10%

(303) is formed in the first step as a by-product.

II. 1,2-ADDITIONS

C-15. Addition of F$_2$C=CCl$_2$ to dienes

Addition of F$_2$C=CCl$_2$ to isoprene leads to two isomeric cyclobutane derivatives in the ratio 7:1 out of four theoretically possible 1,2-addition products. What are the structures of the products. Try to explain: a, the formation of only two isomers; b, the isomer ratio.

F$_2$C=CCl$_2$ + H$_2$C=C(CH$_3$)–CH=CH$_2$ ⟶ C$_7$H$_8$Cl$_2$F$_2$ + C$_7$H$_8$Cl$_2$F$_2$

(307)   (312)   (313)
                                            7   :   1

C-16. Addition of methylenesulfene

H$_3$C–SO$_2$Cl + H$_2$C=C(N–morpholino)$_2$ $\xrightarrow[\text{THF}]{(C_2H_5)_3N}$ C$_{11}$H$_{20}$N$_2$O$_4$S + (C$_2$H$_5$)$_3$\overset{+}{N}–H Cl$^-$

              (324)   (327) 48%

↓ C$_6$H$_6$

C$_7$H$_{11}$NO$_3$S + (C$_2$H$_5$)$_3$$\overset{+}{N}$H Cl$^-$ + C$_4$H$_9$NO

(326) 79%

Characteristic NMR absorption of (326): 4.38 τ (2) singlet, 5.3 τ (1) singlet
Characteristic NMR absorption of (327): 2.94 (3) singlet, 4.23 (1) singlet

Cycloadditions

C-17. Addition of hexafluoroacetone to ethoxyacetylene

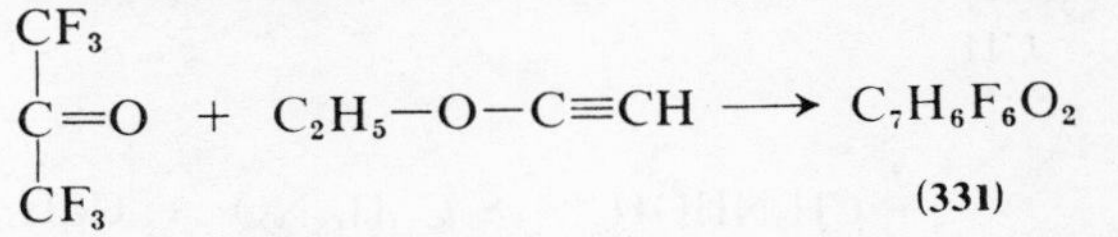

$$\underset{(328)}{\underset{CF_3}{\overset{CF_3}{C}}=O} + \underset{(329)}{C_2H_5-O-C\equiv CH} \longrightarrow \underset{(331)}{C_7H_6F_6O_2}$$

C-18. Cyclodimerization of acetylenes and allenes

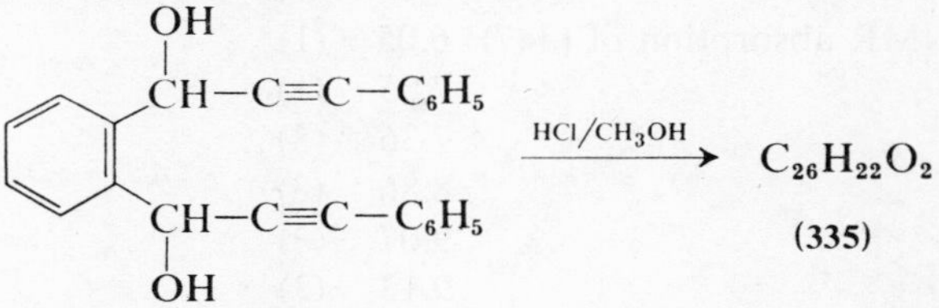

(333): 1,2-bis[CH(OH)-C≡C-C₆H₅]benzene $\xrightarrow{HCl/CH_3OH}$ $C_{26}H_{22}O_2$ (335)

Characteristic UV absorption of **(335)**: 230, 272, 283, 295, 307, 320 mμ

III. 1,3-DIPOLAR ADDITIONS

C-19. Addition of N^2-phenylhydrazonoyl chloride

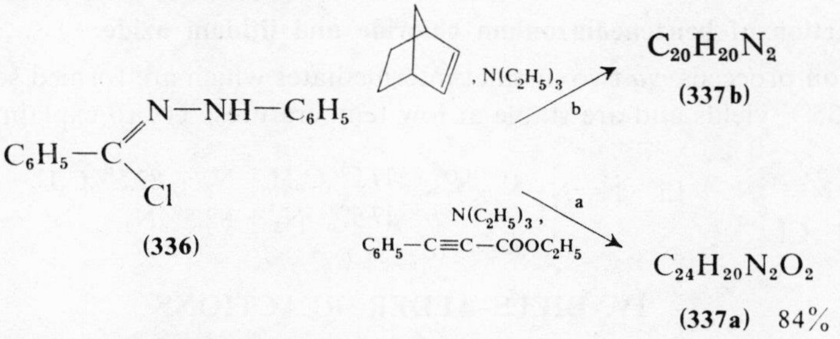

$$\underset{(336)}{C_6H_5-\underset{Cl}{\overset{N-NH-C_6H_5}{C}}}$$

$\xrightarrow[b]{\text{norbornene}, N(C_2H_5)_3}$ $C_{20}H_{20}N_2$ (337b)

$\xrightarrow[a]{N(C_2H_5)_3, \; C_6H_5-C\equiv C-COOC_2H_5}$ $C_{24}H_{20}N_2O_2$ (337a) 84%

C-20. Diphenyldiazomethane and propiolic esters

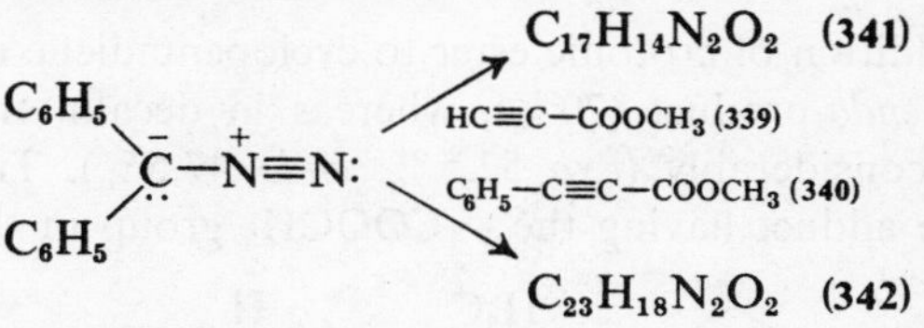

$$\underset{C_6H_5}{\overset{C_6H_5}{\diagdown}}\overset{\bar{}}{\underset{..}{C}}-\overset{+}{N}\equiv N:$$

$\xrightarrow{HC\equiv C-COOCH_3 \; (339)}$ $C_{17}H_{14}N_2O_2$ (341)

$\xrightarrow{C_6H_5-C\equiv C-COOCH_3 \; (340)}$ $C_{23}H_{18}N_2O_2$ (342)

C-21. Addition of nitrones to C=C bonds (i)

$$\begin{array}{c} C_6H_5-CHO \\ + \\ C_6H_5NHOH \end{array} \longrightarrow [C_{13}H_{11}NO] \xrightarrow[60°]{C_6H_5-CH=CH_2} C_{21}H_{19}NO$$

Not isolated   (344) 92–99%

C-22. Addition of nitrones to C=C bonds (ii)

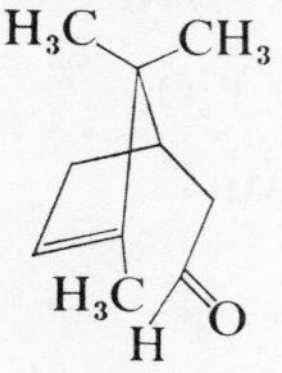

(345) + CH₃NHOH ⟶ $C_{11}H_{19}NO$ + $C_{11}H_{19}NO$
                                (347)   (348)

Characteristic NMR absorption of (347): 6.05 τ (1)
　　　　　　　　　　　　　　　　　　　　7.05　(1)
　　　　　　　　　　　　　　　　　　　　7.36　(3)
　　　　　　　　　　　　　　　　　　　　8.86　(3)
　　　　　　　　　　　　　　　　　　　　9.07　(3)
　　　　　　　　　　　　　　　　　　　　9.13　(3)

Absorption of (348) is analogous, except at 6.05 τ where no absorption is observed.

C-23. Decomposition of diazoacetic ester in benzonitrile

$$N_2HC-COOC_2H_5 \xrightarrow[145°]{C_6H_5-CN} C_{11}H_{11}NO_2$$
                                                                 (350)

C-24. Reaction of benzenediazonium chloride and lithium azide

The reaction proceeds *via* two distinct intermediates which are formed severally in 65% and 35% yields and are stable at low temperatures. Try to explain.

$C_6H_5-\overset{+}{N}\equiv{}^{15}N: + Li^+ \; {}^-\ddot{N}=\ddot{N}-\ddot{N}\; \xrightarrow{H_2O}$ 　17.5% $C_6H_5-N_3$ + 82.5% $C_6H_5-{}^{15}N_3$
　　　　　Cl⁻　　　　　　　　　　　　　　　　　　　　17.5% $^{15}N_2$ + 82.5% N_2

IV. DIELS–ALDER REACTIONS

C-25. Methyl crotonate and cyclopentadiene

The Diels–Alder addition of crotonic ester to cyclopentadiene in acetic acid yields predominantly an *endo*-product (70%), whereas in decalin the amount of *exo*-product increases considerably (*exo* 52.5%, *endo* 47.5%). Try to explain. The *endo*-product is the adduct having the —COOCH₃ group in the *endo*-position.

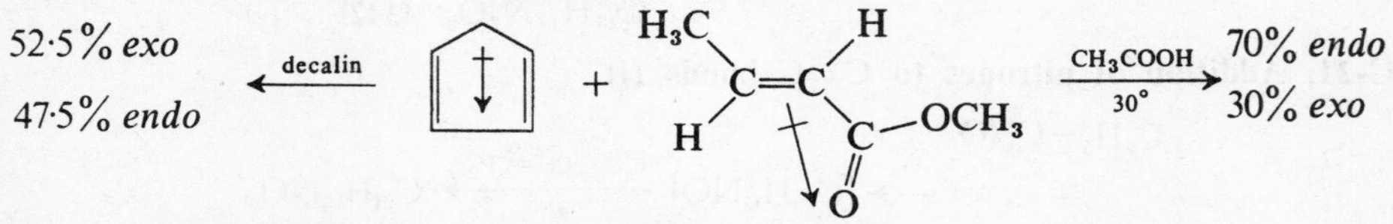

52·5% exo　　　　　　　　　　　　　　　　　　　　　　　　　　　70% endo
　　　　　⟵ decalin　　[cyclopentadiene] ↕ + methyl crotonate　　⟶ CH₃COOH, 30°
47·5% endo　　　　　　　　　　　　　　　　　　　　　　　　　　　30% exo

The crossed arrows indicate the dipole moments.

C-26. 3,5-Dichloro-2-pyrone and maleic anhydride

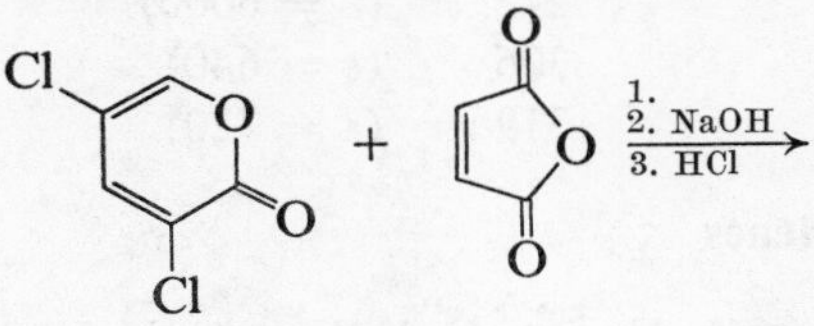

(365) + maleic anhydride $\xrightarrow[3.\ HCl]{\substack{1.\\2.\ NaOH}}$ $C_8H_5ClO_4$ (368)

C-27. Rearrangement of α-dicyclopentadien-1-ol

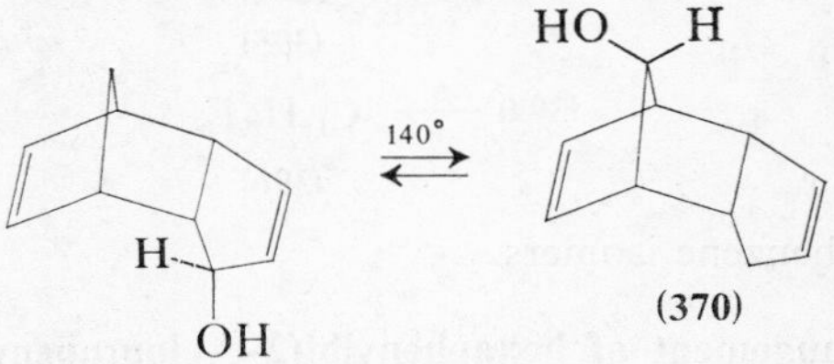

(369) $\underset{}{\overset{140°}{\rightleftarrows}}$ (370)

(369) rearranges stereoselectively to (370). Propose a mechanism.

C-28. Addition of benzyne to 2-pyrone

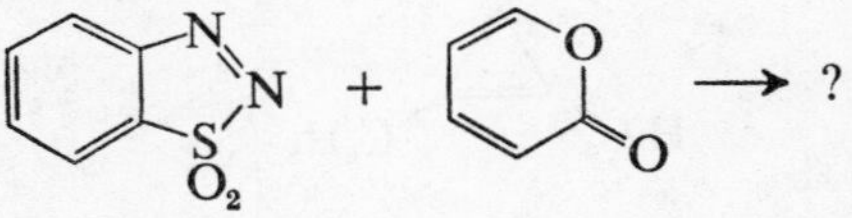

(373) + (375) ⟶ ?

C-29. Addition of hexafluoro-2-butyne to benzene

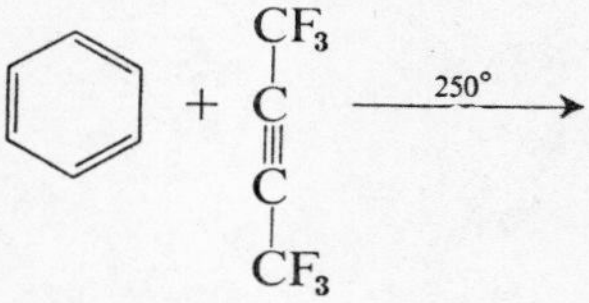

benzene + $\underset{CF_3}{\overset{CF_3}{\underset{\|}{\underset{C}{\overset{C}{\|}}}}}$ $\xrightarrow{250°}$ $C_8H_4F_6$ + $C_{14}H_6F_{12}$ + $C_{14}H_6F_{12}$

(379) 8%   (382)   (383)

↓ +F₃C—C≡C—CF₃   ↓ Oxidn.   ↓ Oxidn.

$C_{10}H_2F_{12}$ $C_{14}H_4F_{12}$ $C_{14}H_4F_{12}$

(380) 2%   (384) 30%   (385) 6%

Characteristic UV absorption of (384): 222 mμ ($\varepsilon = 78000$) (ethanol)
                                     260   ($\varepsilon = 4030$)
                                     308   ($\varepsilon = 4130$)
                                     323   ($\varepsilon = 2580$)

[Contd. on p. 20.]

Characteristic UV absorption of (385): 221 mμ ($\varepsilon = 107000$) (isooctane)
  272   ($\varepsilon = 6000$)
  306   ($\varepsilon = 640$)
  319   ($\varepsilon = 720$)

C-30. Trimerization of acetylenes

$$3(CH_3)_3C-C\equiv C-F \xrightarrow{<0°} C_{18}H_{27}F_3 + C_{18}H_{27}F_3$$
$$\qquad\qquad\qquad\qquad\qquad (393) \qquad (394)$$

$$(393) \xrightarrow{\Delta} C_{18}H_{27}F_3$$
$$\qquad\qquad (395)$$

$$(394) \xrightarrow{\Delta} C_{18}H_{27}F_3$$
$$\qquad\qquad (396)$$

All the products are benzene isomers.

C-31. Thermal rearrangement of hexaphenylbi(2-cyclopropenyl)

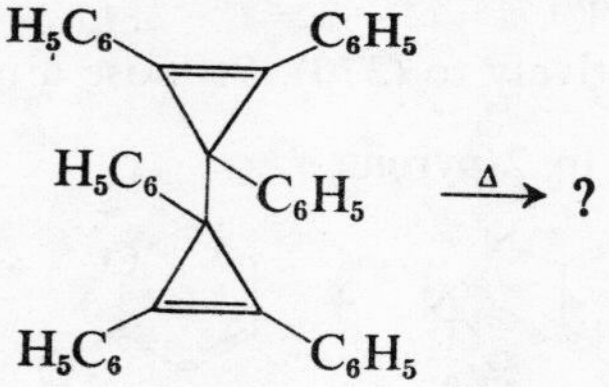

3

Valence Isomerizations

I. DIALLYLIC SYSTEMS AND MONOVINYL-CYCLOPROPANES AND -CYCLOBUTANES

V-1. Rearrangement of 1,1-dicyclopropylethylene

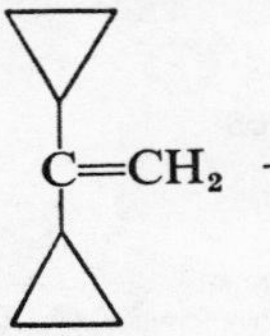

$$\text{(dicyclopropyl)}C=CH_2 \xrightarrow{400°} C_8H_{12} + C_8H_{12}$$
$$\qquad\qquad\qquad\quad 35\% \quad\; 61\%$$
$$\qquad\qquad\qquad\quad (413) \quad (414)$$

V-2. Rearrangement of bicyclo[3.2.0]hept-2-en-6-yl acetate

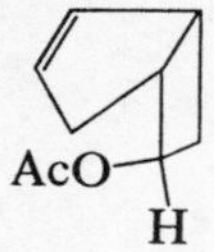

$$(415) \xrightarrow[\text{in decalin}]{300°} C_9H_{12}O_2$$
$$\qquad\qquad\qquad (416)$$

V-3. Thermal isomerization of thujone

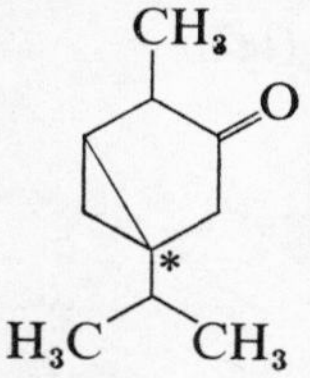

$$(419) \xrightarrow{\Delta} C_{10}H_{16}O + C_{10}H_{16}O + C_{10}H_{16}O$$
$$\qquad\qquad\quad (421) \qquad\quad (422) \qquad\quad (423)$$
$$\qquad\qquad\text{Retention of} \;\; \text{Optically inactive} \;\; \text{Retention of}$$
$$\qquad\qquad\text{configuration} \qquad\qquad\qquad\qquad \text{configuration}$$

Characteristic NMR absorption of:

(421)		(422)		(423)	
3.02 (1)	singlet	3.3 (1)	multiplet	2.93 (1)	doublet
7.8—7.96 (2)	doublet	7.9 (2)	multiplet	8.0—7.7 (2)	multiplet
8.28 (3)	singlet	8.3 (3)	singlet	8.28 (3)	two doublets
8.85 (3)	singlet	9.0—9.1 (2)	multiplet	9.0 (3)	singlet
9.15—9.0 (6)	two doublets	9.1—9.0 (6)	doublet	9.2—9.09	doublet (3)
9.0—9.2 (1)	multiplet			9.42—9.34	doublet (3)
				9.4—9.0 (1)	multiplet

II. CONJUGATED POLYENES

V-4. Isomerization of 3-phenyl-3-cyclobutene-1,2-dione

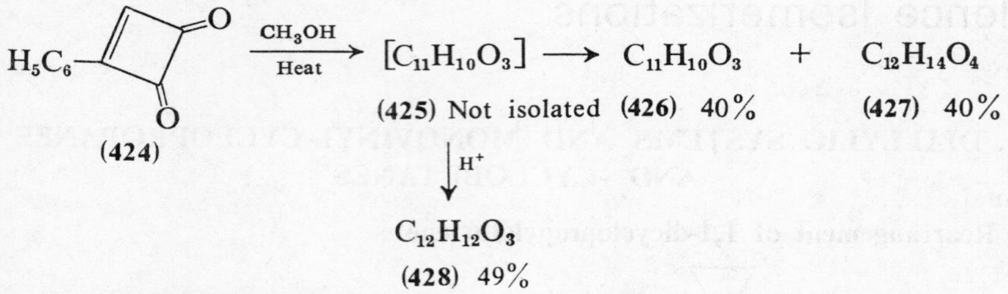

$$H_5C_6-\text{(cyclobutenedione)} \xrightarrow[\text{Heat}]{CH_3OH} [C_{11}H_{10}O_3] \longrightarrow C_{11}H_{10}O_3 + C_{12}H_{14}O_4$$

(424)   (425) Not isolated   (426) 40%   (427) 40%

$$\downarrow H^+$$

$$C_{12}H_{12}O_3$$

(428) 49%

V-5. Isomerization of cyclooctatriene derivatives

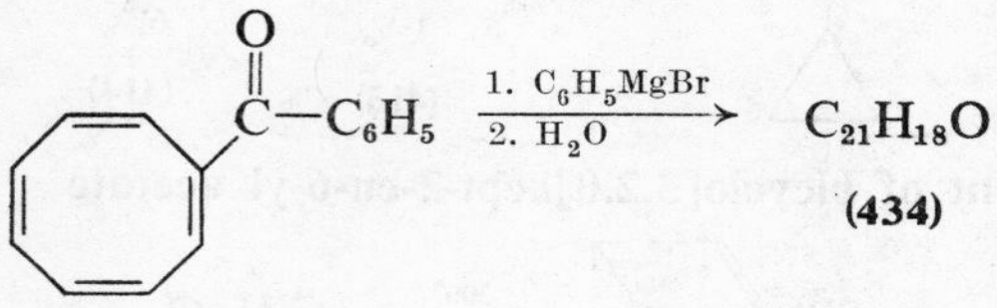

$$\text{(cyclooctatriene)}-\overset{O}{\underset{}{C}}-C_6H_5 \xrightarrow[\text{2. } H_2O]{\text{1. } C_6H_5MgBr} C_{21}H_{18}O$$

(432a)   (434)

V-6. Rearrangement of 1,6-epoxy(10)annulene

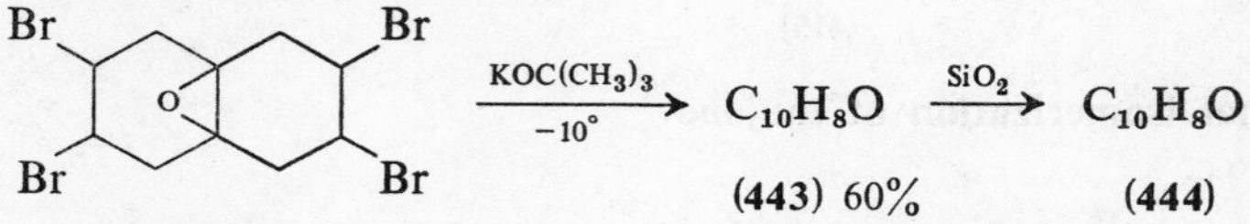

$$\text{(tetrabromoepoxide)} \xrightarrow[-10°]{KOC(CH_3)_3} C_{10}H_8O \xrightarrow{SiO_2} C_{10}H_8O$$

(441)   (443) 60%   (444)

4

β-Eliminations

E-1. Decomposition of tetraalkylammonium hydroxides

Thermal decomposition of the ammonium hydroxides (**449**) and (**452**) yields two olefins in each case. Which olefin is formed predominantly in each reaction? Justify your decision.

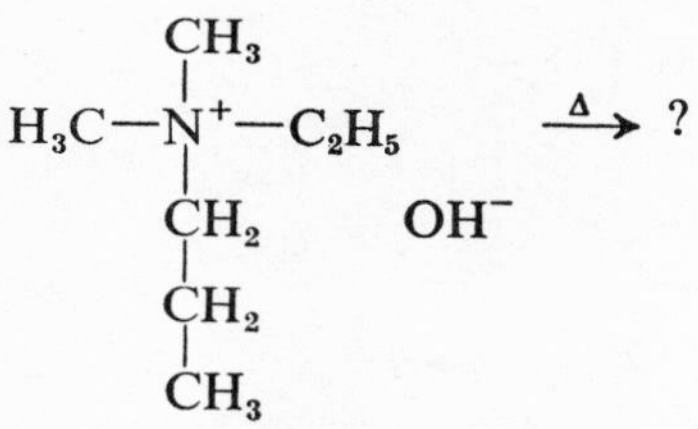

$$H_3C-\overset{\overset{CH_3}{|}}{\underset{\underset{\underset{CH_3}{|}}{\underset{CH_2}{|}}{\underset{CH_2}{|}}}{N^+}}-C_2H_5 \quad OH^- \quad \xrightarrow{\Delta} \quad ?$$

(**449**)

$$H_3C-\overset{\overset{CH_3}{|}}{\underset{\underset{\underset{CH_3}{|}}{C}}{\underset{CH_3}{|}}}{N^+}-C_2H_5 \quad OH^- \quad \xrightarrow{\Delta} \quad ?$$

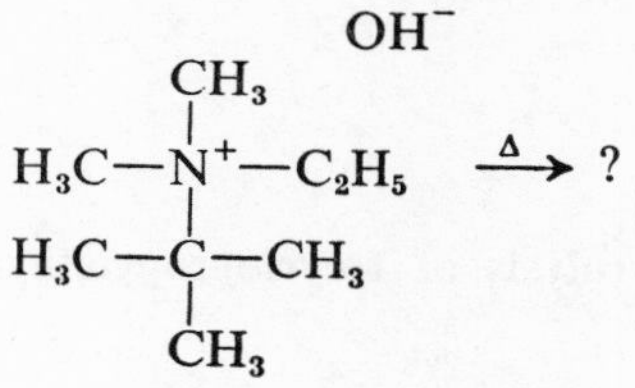

(**452**)

E-2. Reaction of *trans*-2-phenylcyclohexyl *p*-toluenesulfonate with potassium hydroxide

In the reaction of (**453**) with KOH in methanol, 64% of (**454**) is obtained if X = $^+$N(CH$_3$)$_3$, and 20% of (**454**), as well as 53% of (**455**), if X = OTs. Explain this.

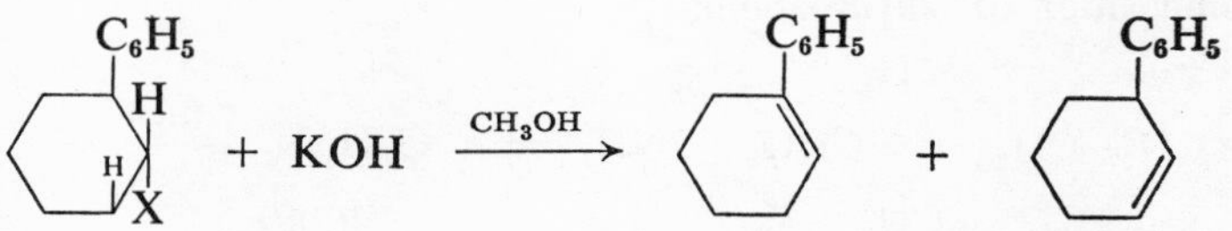

(**453**) + KOH $\xrightarrow{CH_3OH}$ (**454**) + (**455**)

X = $^+$N(CH$_3$)$_3$ 64%   2%
X = *p*-CH$_3$C$_6$H$_4$SO$_2$—O—   20%   53%

E-3. Thermal elimination from 3-ethyl-6-phenyl-5-hexen-3-ol

$$C_6H_5CH=CH-CH_2-C(C_2H_5)_2-OD \xrightarrow{500°}$$

(**456**)

E-4. Rearrangements of (+)-3-hydroxymethyl-4-carene

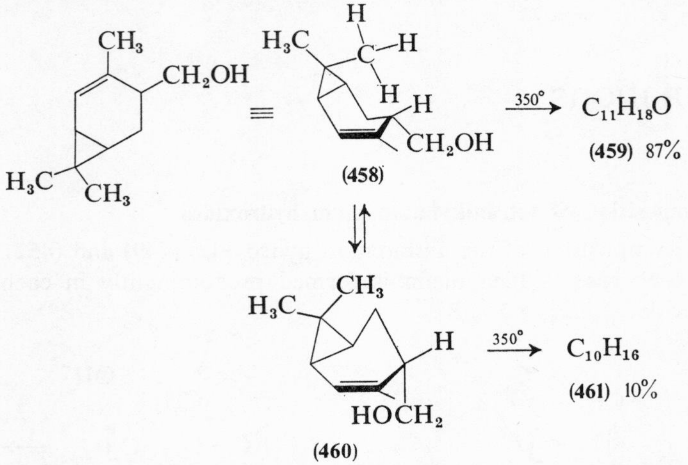

$$\text{(458)} \xrightarrow{350°} C_{11}H_{18}O \quad \text{(459) 87\%}$$

$$\updownarrow$$

$$\text{(460)} \xrightarrow{350°} C_{10}H_{16} \quad \text{(461) 10\%}$$

E-5. Pyrolysis of 1-cyclopropylethyl acetate

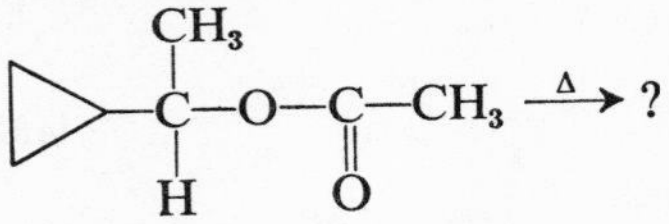

cyclopropyl–CH(CH$_3$)–O–C(=O)–CH$_3$ $\xrightarrow{\Delta}$?

E-6. Cope degradation of amine oxides

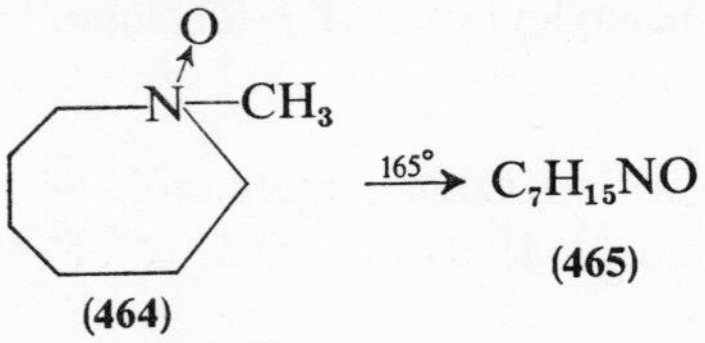

$$\text{(464)} \xrightarrow{165°} C_7H_{15}NO \quad \text{(465)}$$

E-7. Decomposition of "hydroamides"

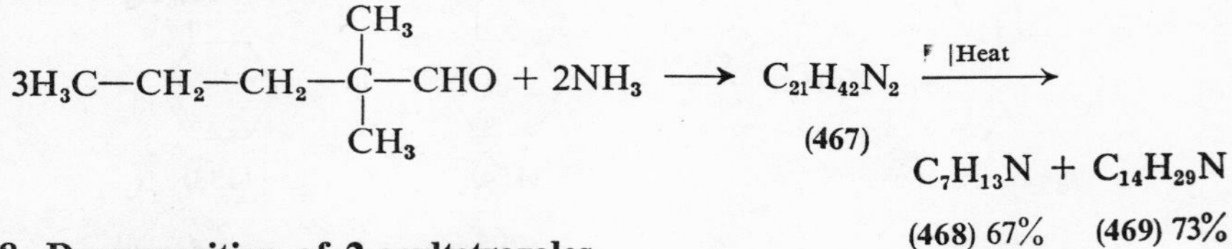

$$3\,H_3C-CH_2-CH_2-C(CH_3)_2-CHO + 2\,NH_3 \longrightarrow C_{21}H_{42}N_2 \xrightarrow{F\ |\text{Heat}} C_7H_{13}N + C_{14}H_{29}N$$

(467)   (468) 67%   (469) 73%

E-8. Decomposition of 2-acyltetrazoles

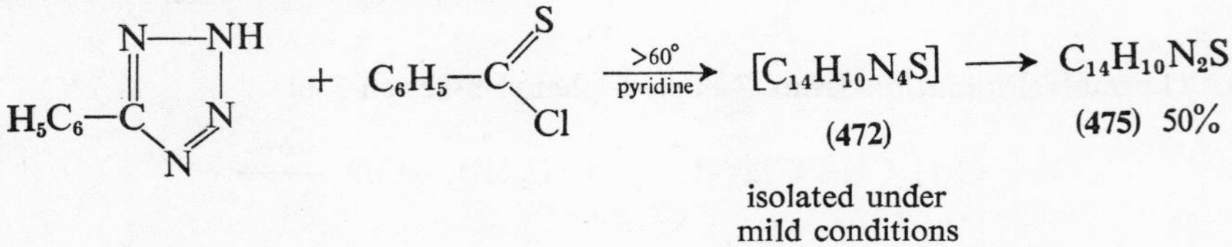

$$H_5C_6\text{-tetrazole} + C_6H_5-C(=S)Cl \xrightarrow[\text{pyridine}]{>60°} [C_{14}H_{10}N_4S] \longrightarrow C_{14}H_{10}N_2S$$

(472)   (475) 50%

isolated under mild conditions

β-Eliminations

E-9. Reaction of 1,1-dimethylpyrrolidinium bromide with phenyllithium

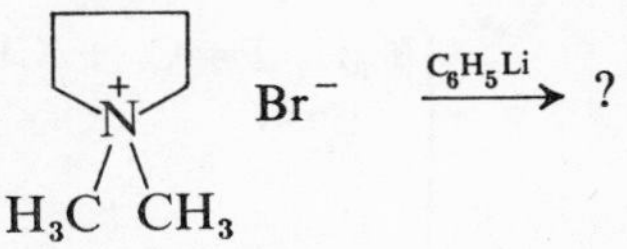

$$\text{(pyrrolidinium with N}^+\text{(CH}_3\text{)}_2\text{)} \; Br^- \xrightarrow{C_6H_5Li} ?$$

(476)

E-10. Decomposition of diallylammonium salts by base

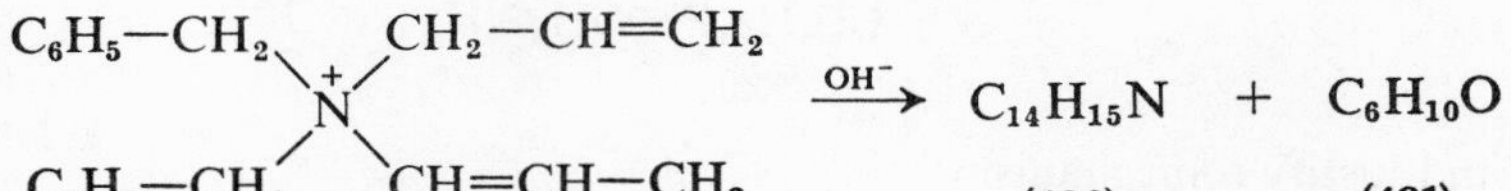

$$\begin{array}{c} C_6H_5-CH_2 \\ C_6H_5-CH_2 \end{array} N^+ \begin{array}{c} CH_2-CH=CH_2 \\ CH=CH-CH_3 \end{array} \xrightarrow{OH^-} C_{14}H_{15}N + C_6H_{10}O$$

(479)   (480)   (481)

E-11. Reaction of (trichloromethyl)oxirane with methyllithium

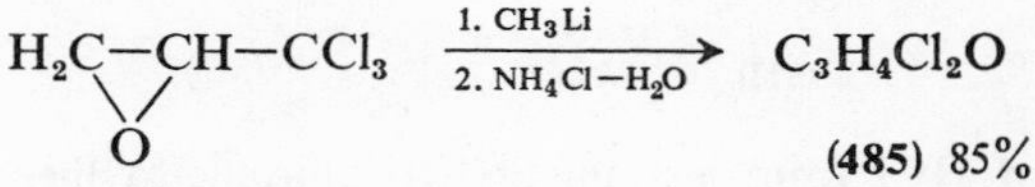

$$H_2C\underset{O}{-}CH-CCl_3 \xrightarrow[2.\; NH_4Cl-H_2O]{1.\; CH_3Li} C_3H_4Cl_2O$$

(482)   (485) 85%

E-12. Rearrangement of 4-phenyl-1,2-diazabicyclo[3.2.0]hept-2-en-6-one

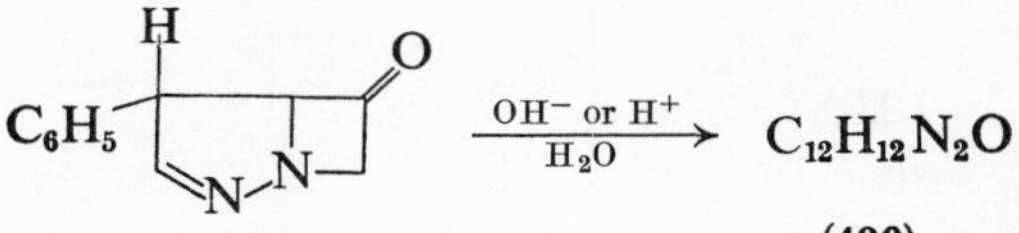

$$\xrightarrow[H_2O]{OH^- \text{ or } H^+} C_{12}H_{12}N_2O$$

(489)   (490)

E-13. Ring cleavage of 2-methyl-5-phenylisoxazolium chloride

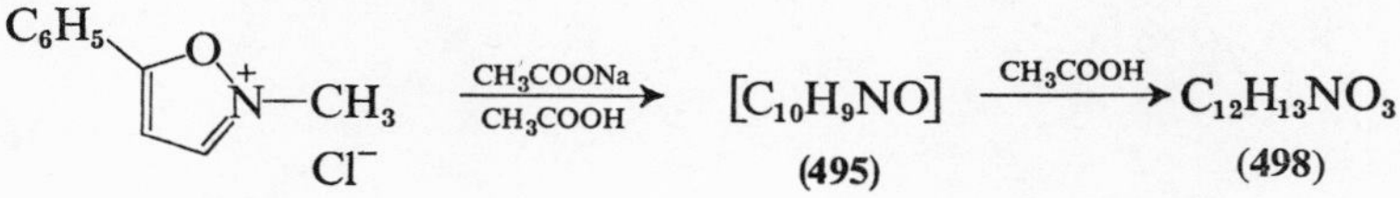

$$\text{(isoxazolium)} \; Cl^- \xrightarrow{CH_3COONa / CH_3COOH} [C_{10}H_9NO] \xrightarrow{CH_3COOH} C_{12}H_{13}NO_3$$

(494)   (495)   (498)

The presence of (495) has been demonstrated by IR spectroscopy
Characteristic absorption of (498): IR 3472 cm^{-1} 1634 cm^{-1}
                         1770   1515
                         1678   1190
                UV 269 mμ ($\varepsilon = 24\,000$)

E-14. Reaction of phosphonium salts with base

$(C_6H_5)_3P^+\!-\!CH_2Cl \xrightarrow{OH^-}$
$\begin{cases} (C_6H_5)_3P\!=\!O + CH_3Cl \quad 50\text{–}70\% \\ \quad (503) \\ (C_6H_5)_3P + CH_2O \quad\quad 8\% \\ \quad (507) \\ (C_6H_5)_2\overset{O}{\underset{\|}{P}}\!-\!CH_2C_6H_5 \quad 10\% \\ \quad (505) \end{cases}$

(501)

Propose and justify a mechanism.

E-15. Elimination of sulfur from 1,1-dichlorocyclopropa[b][1]benzothiopyran

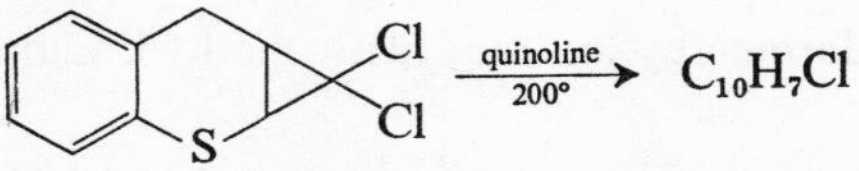

(511) $\xrightarrow[200°]{\text{quinoline}}$ $C_{10}H_7Cl$ (513)

E-16. Elimination of HCl from pentamethyl(trichloromethyl)benzene

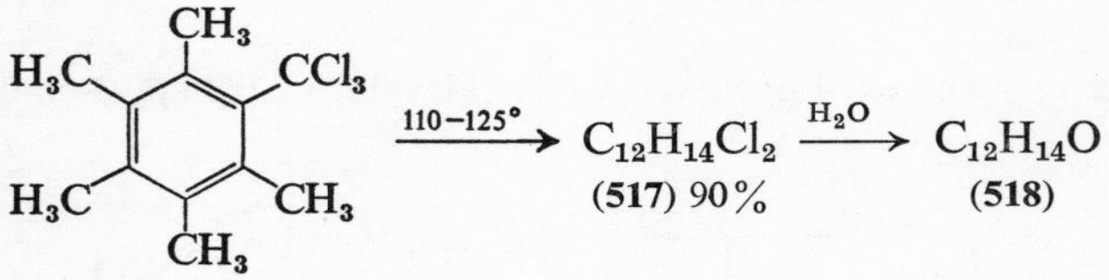

(518) $\xrightarrow{110\text{–}125°}$ $C_{12}H_{14}Cl_2$ (517) 90% $\xrightarrow{H_2O}$ $C_{12}H_{14}O$ (518)

5
Fragmentations

F-1. Fragmentation of 3-bromo-*N*,*N*-dimethyladamantan-1-amine

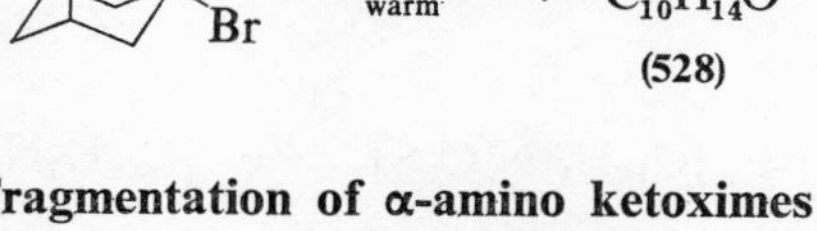

3-bromo-N,N-dimethyladamantan-1-amine $\xrightarrow[\text{warm}]{80\% \text{ ethanol}}$ $C_{10}H_{14}O$ (528)   (528) forms a semicarbazone

F-2. Fragmentation of α-amino ketoximes

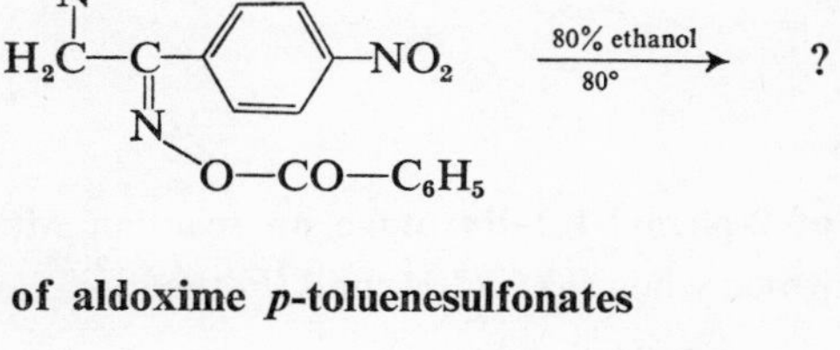

$\xrightarrow[80°]{80\% \text{ ethanol}}$?

F-3. Fragmentation of aldoxime *p*-toluenesulfonates

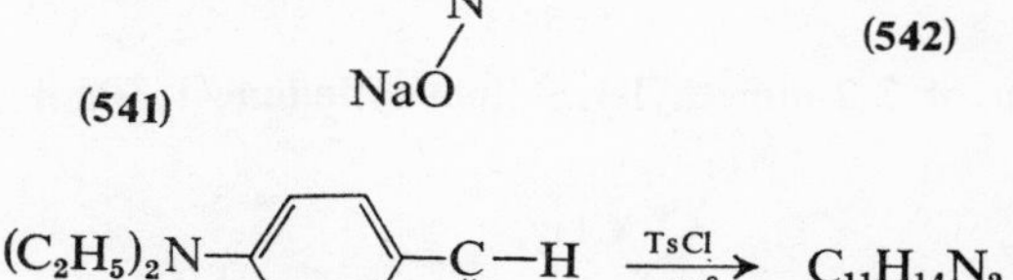

$(C_2H_5)_2N$—C₆H₄—CH=N—ONa (541) $\xrightarrow[-30°]{\text{TsCl}}$ $C_{11}H_{14}N_2$ (542)

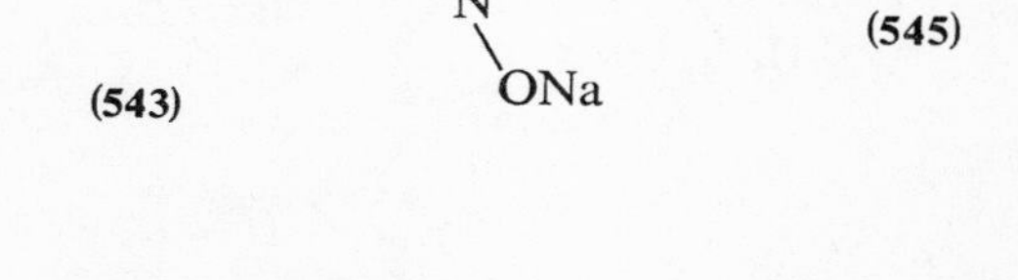

$(C_2H_5)_2N$—C₆H₄—CH=N—ONa (543) $\xrightarrow[-30°]{\text{TsCl}}$ $C_{11}H_{14}N_2$ (545)

F-4. Fragmentation of 5-(p-tosyloxyimino)-1-decalone

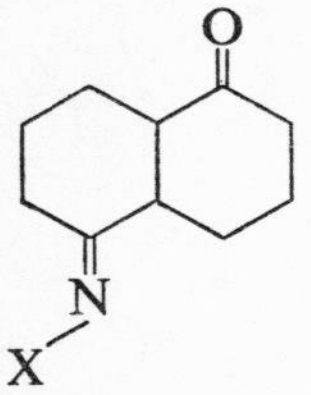

(546) $\xrightarrow[X=OTs]{80\% C_2H_5OH}$ $C_{10}H_{15}NO_2$ (547)

(546) $\xrightarrow[X=OTs]{NaOCH_3}$ $C_{10}H_{13}NO$ (548) 68–85%

(546) $\xrightarrow[X=OH]{N-NaOH - TsCl}$ $C_{10}H_{15}NO_2$ (549) 60%

(549) has characteristic IR absorption at 3500 and 2240—2260 cm^{-1}.

F-5. Fragmentation of 5-nitroso-2-phenylpyrimidine-4,6-diamine

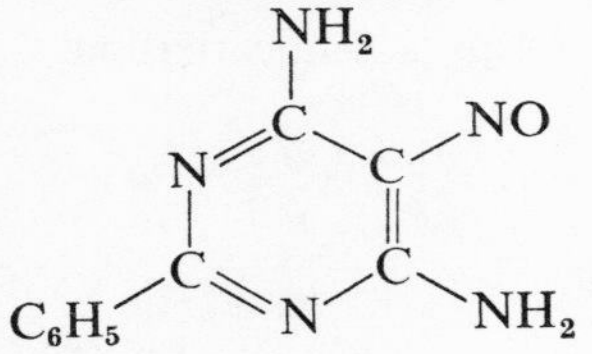

(550) $\xrightarrow[\text{in refluxing pyridine}]{TsCl}$ $[C_{16}H_{13}N_5O_3S]$ ⟶ $[C_{10}H_7N_5]$

Not isolated (551) Not isolated (552)

⟶ $C_{10}H_7N_5$

30% (553)

F-6. Fragmentation of 2-phenyl-1,3-dioxolane on reaction with butyllithium

What products are formed when (a) LiC$_4$H$_9$ and (b) the dioxolane is used in excess?

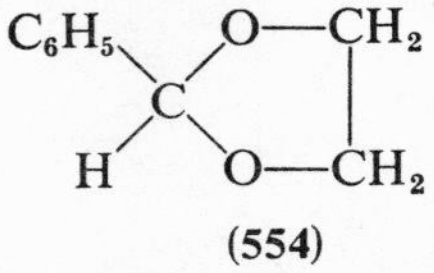

(554) $\xrightarrow{LiC_4H_9\text{-}n}$?

F-7. Fragmentation of 2,2-dimethyl-1,3-diphenylindane-1,3-diol

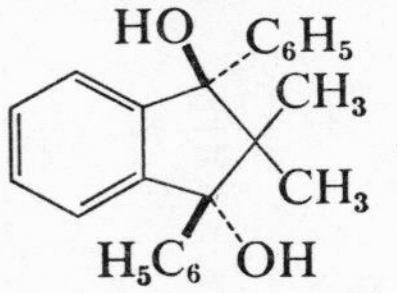

(556) $\xrightarrow[150-160°]{KHSO_4}$ $C_{23}H_{20}O$ (557)

6

Aliphatic Substitution

I. $S_N 1$ SUBSTITUTION
II. CARBONIUM IONS

S-1. Solvolysis of 5,6-dibromo-5,6-dihydrodibenzo[a,e]cyclooctene

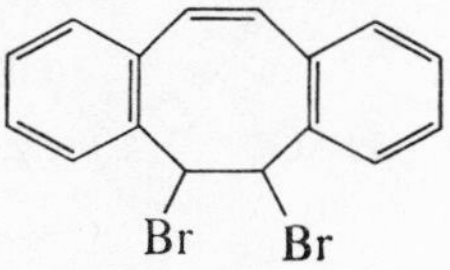

(568) $\xrightarrow{\text{1, CH}_3\text{OH}-\text{K}_2\text{CO}_3}_{\text{2, H}^+}$ $C_{16}H_{12}O$ (571)

S-2. Rearrangement of 5,11-epoxy-10,11-dihydro-5-methoxy-10-phenyl-[$5H$]dibenzo[a,d]cyclohepten-10-ol

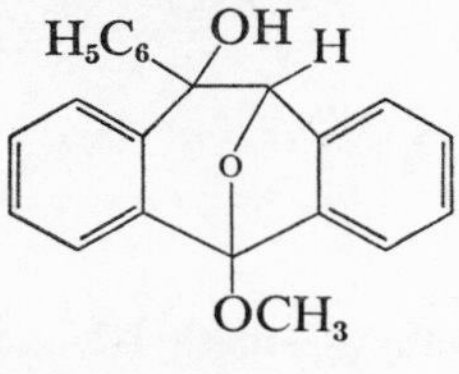

(573) $\xrightarrow{\text{H}_3\text{O}^+}$ $C_{20}H_{14}O$ (574)

S-3. Rearrangement of 8,9-dihydro-α,α-diphenyl-9-([$1H$]bicyclo[6.1.0]-nonene)methanol

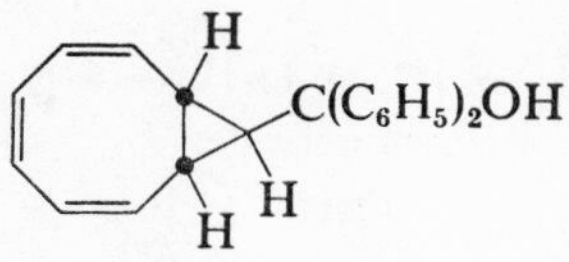

(575) $\xrightarrow{\text{HBF}_4}$ $C_{22}H_{18}$ (577)

S-4. $S_N 1$ reactions of allyl halides

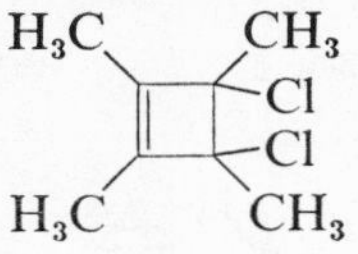

(580) $\xrightarrow[\text{NaHCO}_3 \atop \text{room temp.}]{\text{H}_2\text{O}_2-\text{THF}-}$ $C_8 H_{14} O_5$ (585)

S-5. Reaction of (n-butylthio)acetaldehyde dimethyl acetal with 1,2-ethanedithiol

$$n\text{-}C_4H_9\text{-}S\text{-}CH_2\text{-}CH(OCH_3)_2 + HS\text{-}CH_2\text{-}CH_2\text{-}SH \xrightarrow{H^+} C_8H_{16}S_3$$
$$\text{(589)} \qquad\qquad\qquad \text{(590)} \qquad\qquad \text{(594) 82\%}$$

S-6. Rearrangements of 3,4-epoxy-2,2,4-trimethylpentyl isobutyrate

$$\underset{\substack{\text{CH}_3 \\ | \\ \text{H}_3\text{C}-\text{C}-\text{CH}-\text{C}-\text{CH}_2-\text{O}-\text{C}-\text{CH}-\text{CH}_3 \\ \diagdown\!\!\diagup \quad | \qquad\qquad \| \quad | \\ \text{O} \quad\;\; \text{CH}_3 \qquad\;\; \text{O} \;\; \text{CH}_3}}{} \xrightarrow{\text{HCl} \atop \text{ether}} \begin{array}{l} C_{12}H_{22}O_3 + C_{12}H_{22}O_3 \\ \text{(597) 5\%} \quad\;\; \text{(598) 5\%} \\ + C_{12}H_{22}O_3 + C_{12}H_{22}O_3 \\ \text{(600) 70\%} \quad\;\; \text{(589) 20\%} \end{array}$$

$$C_{12}H_{22}O_3 \longrightarrow C_{12}H_{22}O_3; \; C_{12}H_{22}O_3 \longrightarrow C_{12}H_{22}O_3$$
$$\text{(598)} \qquad\;\;\; \text{(599)} \qquad\;\;\; \text{(599)} \qquad\;\;\; \text{(600)}$$

Saponification of (597) → $C_8H_{16}O_2$ + $C_4H_8O_2$
$\qquad\qquad\qquad\qquad\;\;\;$ (597a)

Saponification of (598) and (599) → $C_8H_{16}O_2$ + $C_4H_8O_2$
$\qquad\qquad\qquad\qquad\qquad\qquad\;\;$ (598a)

Saponification of (600) → $C_8H_{16}O_2$ + $C_4H_8O_2$
$\qquad\qquad\qquad\qquad\;\;\;$ (600a)

S-7. Rearrangement of ethyl 1-(2-formylethyl)-3-methyl-2-oxocyclohexanecarboxylate

[structure (611): cyclohexanone with H₃C, CHO-CH₂-CH₂- and -COOC₂H₅ substituents] $\xrightarrow{H^+}$ $[C_{13}H_{19}O_3]^+$ $\longrightarrow$ $C_{13}H_{16}O_2$
$\qquad\qquad\qquad\qquad\qquad\qquad\qquad\qquad\qquad$ not isolated
$\qquad\qquad$ (611) $\qquad\qquad\qquad\qquad\qquad$ (613) $\qquad\qquad$ (616) 3%

$\qquad\qquad\qquad\qquad\qquad\qquad\qquad\qquad\downarrow \qquad\searrow$

$\qquad\qquad\qquad\qquad\qquad\qquad\qquad C_{13}H_{18}O_3 \qquad C_{13}H_{18}O_3 + C_{13}H_{18}O_3$
$\qquad\qquad\qquad\qquad\qquad\qquad\;\;$ (620) 5% $\qquad\;\;$ (614) $\qquad\quad$ (615)
$\qquad\qquad\qquad\qquad\qquad\qquad\qquad\qquad\qquad\qquad\;\underbrace{\qquad\qquad\qquad}_{43\%}$

Characteristic absorption of (620): IR 1728, 1667 cm^{-1}
$\qquad\qquad\qquad\qquad\qquad\qquad$ UV 252 mμ, $\varepsilon = 12000$

Aliphatic Substitution

S-8. Rearrangement of 1,5,5-trimethylbicyclo[2.1.1]hexane-6-carboxylic acid

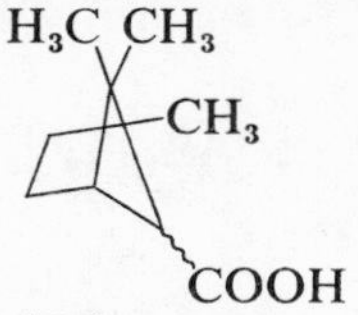

(623) $\xrightarrow{H_2SO_4}$ $C_{10}H_{16}O_2$ + $C_{10}H_{16}O_2$
(627) 50%   (628) 20%

Characteristic IR absorption of (627): 1770 cm^{-1}
Characteristic IR absorption of (628): 3500

S-9. Decomposition of *o*-nitrobenzoyldiazomethane

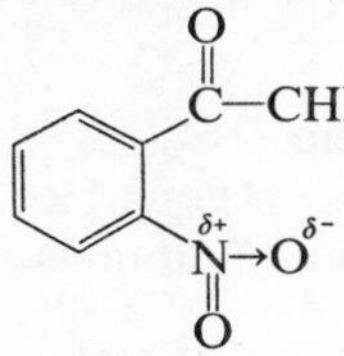

(632) $\xrightarrow[\text{or dil. } H_2SO_4]{\text{AcOH}}$ $C_8H_5NO_3$
(635) 24%

S-10. Reaction of *o*-nitrobenzhydrol with thionyl chloride

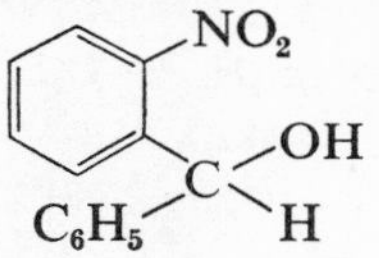

(642) $\xrightarrow[CHCl_3]{SOCl_2}$ $C_{13}H_8ClNO$
(643)

III. PARTICIPATION OF NEIGHBORING CARBON IN CARBONIUM ION REACTIONS

S-11. Solvolysis of 1,4-dihydro-3,5-dimethoxybenzyl *p*-toluenesulfonate

$C_{15}H_{18}O_5S$ $\xleftarrow[\text{AcOH}]{Ac_2O}$ (657) $\xrightarrow{\text{pyridine}}$ $C_9H_{12}O_2$
(662)   (659)

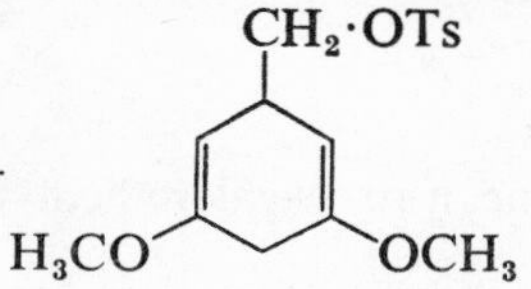

S-12. Rearrangement of 2-chloro-4,5-dihydro-1-benzothiepin-5-ol

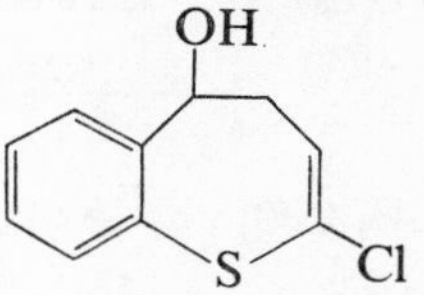

(663) $\xrightarrow{H^+}$ $C_{10}H_8OS$ $\xrightarrow[\text{2, HCl}]{\text{1, NaOH}}$ $C_{10}H_{10}O_2S$
(665)   (665a)

S-13. Solvolysis of (2,3-diphenyl-2-cyclopropenyl)methyl *p*-toluenesulfonate

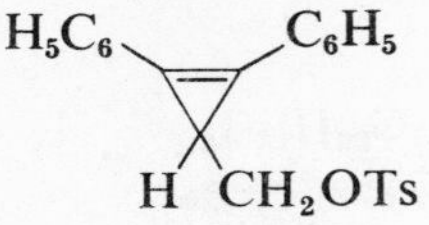

$$\text{(666)} \xrightarrow[\text{CH}_3\text{CN}]{\text{H}_2\text{O}} \text{C}_{16}\text{H}_{14}\text{O} + \text{C}_{16}\text{H}_{14}\text{O} + \text{C}_{16}\text{H}_{14}\text{O}$$

Starting material (666): cyclopropene with H_5C_6 and C_6H_5 on the double bond carbons, and H, CH$_2$OTs on the sp^3 carbon.

Products: (668a) 50–60%, (669) 5%, (670) 20–25%

Characteristic absorption of (668a): NMR 5.2 τ (1) quartet
6.5 (1) singlet
7.4 (2) multiplet

Characteristic absorption of (669): IR 2750, 1680 cm^{-1}
UV 275 mμ

Characteristic absorption of (670): IR 1655 cm^{-1}
NMR 3.4 τ (1) quartet
8.0 τ (3) doublet

S-14. Reaction of 7-chlorobicycloheptadiene with NaBH$_4$

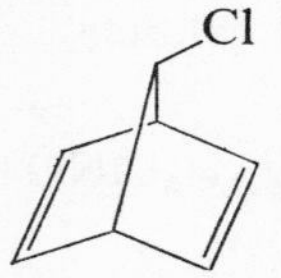

$$\text{(673)} \xrightarrow[\text{65\% diglyme}]{\text{NaBH}_4} \text{C}_7\text{H}_8 + \text{C}_7\text{H}_8$$

(682) 12%   (683) 83%

S-15. Reaction of 7-chlorobicycloheptadiene with NaCN

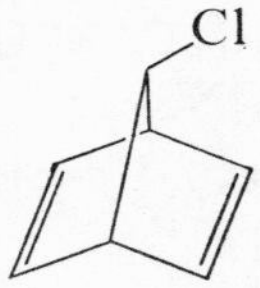

$$\text{(673)} \xrightarrow[\text{C}_2\text{H}_5\text{OH}]{\text{NaCN–H}_2\text{O–}} \text{C}_8\text{H}_7\text{N}$$

(699)

S-16. Reaction of norbornadiene with *N*-chlorodiethylamine

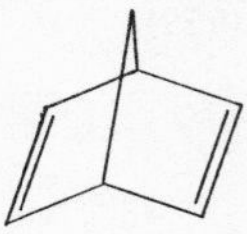

$$\text{norbornadiene} + (\text{C}_2\text{H}_5)_2\text{NCl} \xrightarrow[\text{CH}_3\text{COOH}]{\text{H}_2\text{SO}_4-} \text{C}_{11}\text{H}_{18}\text{ClN} + \text{C}_{11}\text{H}_{18}\text{ClN}$$

(720)   (2 stereoisomers)
                (721)

$$+ \text{C}_9\text{H}_{11}\text{ClO}_2 + \text{C}_9\text{H}_{11}\text{ClO}_2$$

(722)   (2 stereoisomers)
                (723) and (724)

S-17. Reaction of 6,6-dimethylbicyclo[2.1.1]hexan-2-amine with HNO_2

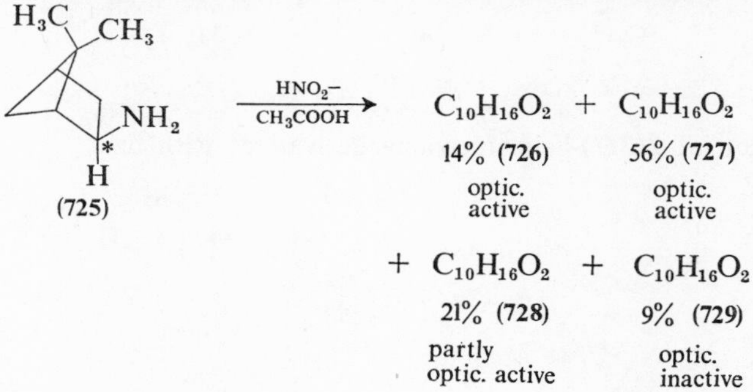

(725) $\xrightarrow[CH_3COOH]{HNO_2^-}$ $C_{10}H_{16}O_2$ + $C_{10}H_{16}O_2$
14% (726)   56% (727)
optic. active   optic. active

+ $C_{10}H_{16}O_2$ + $C_{10}H_{16}O_2$
21% (728)   9% (729)
partly optic. active   optic. inactive

IV. S_N2 SUBSTITUTION

S-18. Reaction of cyclic carbonates with KSCN

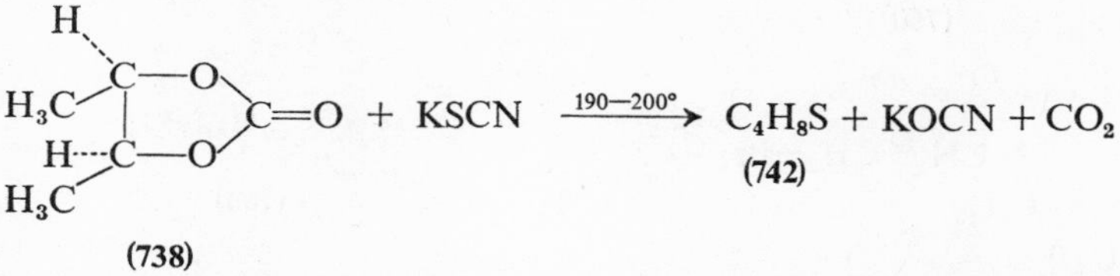

(738) + KSCN $\xrightarrow{190-200°}$ C_4H_8S + KOCN + CO_2
                                  (742)

S-19. Reaction of trans-1,2-dimethyloxirane with potassium methyl xanthate

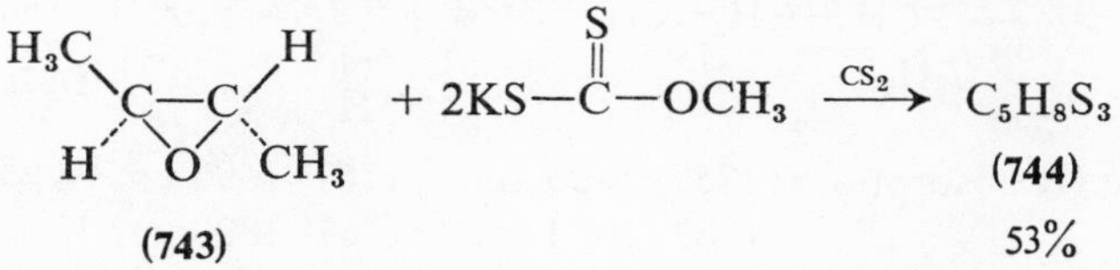

(743) + 2KS—C(=S)—OCH$_3$ $\xrightarrow{CS_2}$ $C_5H_8S_3$
                                    (744) 53%

S-20. Reaction of epichlorohydrin with cyclopentadienylsodium

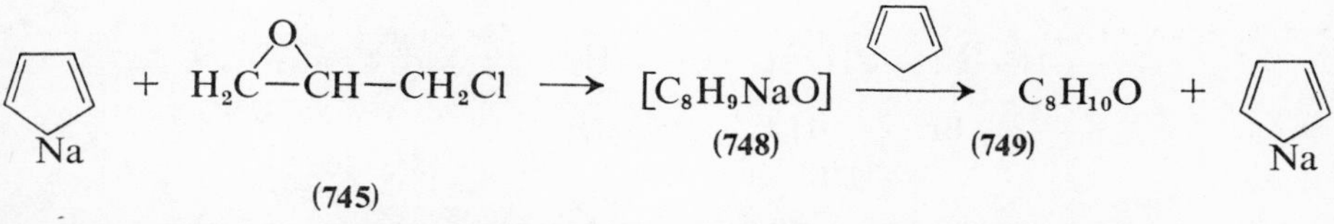

$\underset{Na}{\bigcirc}$ + $H_2C(-O-)CH-CH_2Cl$ → $[C_8H_9NaO]$ → $C_8H_{10}O$ + $\underset{Na}{\bigcirc}$
           (745)   (748)   (749)

S-21. Reaction of 1-[(phenylthio)carbonyl]aziridine with sodium iodide

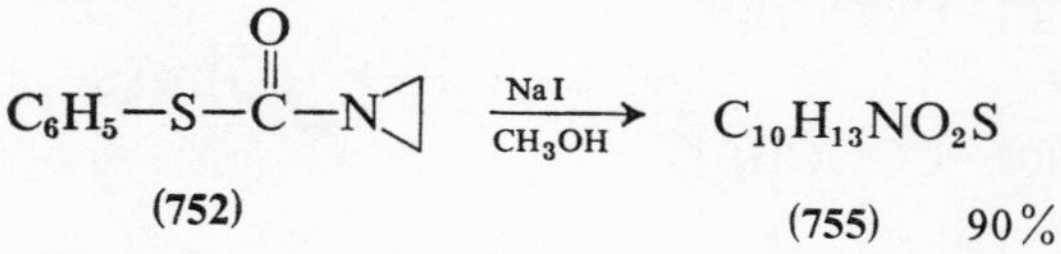

C_6H_5—S—C(=O)—N◁ $\xrightarrow[CH_3OH]{NaI}$ $C_{10}H_{13}NO_2S$
(752)   (755)   90%

S-22. Reaction of 1-(arylazo)aziridines with sodium iodide

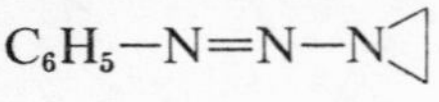

$C_6H_5-N=N-N\triangleleft$ $\xrightarrow{NaI}$ $C_8H_9N_3$
(756)   (757)

S-23. Reaction of 2(3H)-benzofuranone derivatives with base

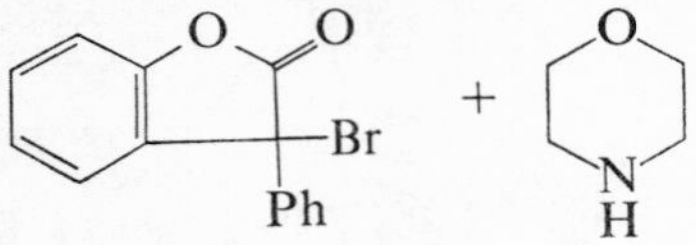

[benzofuranone with Br and Ph] + [morpholine] → $C_{18}H_{17}NO_3$
(758)   (759) 76%

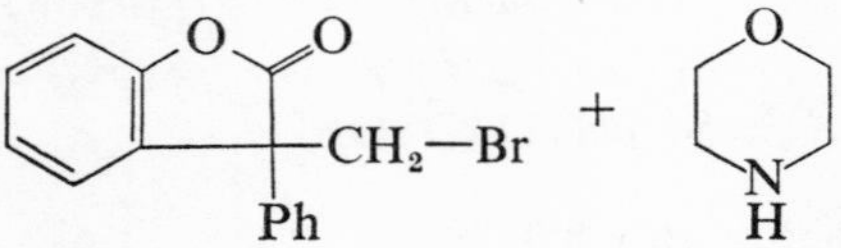

[benzofuranone with CH_2-Br and Ph] + [morpholine] → $C_{19}H_{19}NO_3$
(760)   (761) 81%

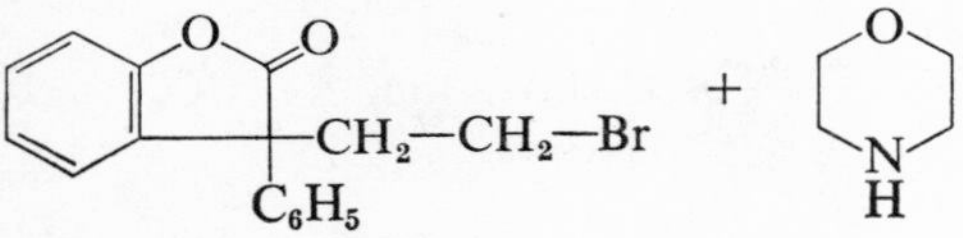

[benzofuranone with CH_2-CH_2-Br and C_6H_5] + [morpholine] → $C_{20}H_{21}NO_3$ + $C_{20}H_{21}NO_3$
(762)   (764) 92%   (765)

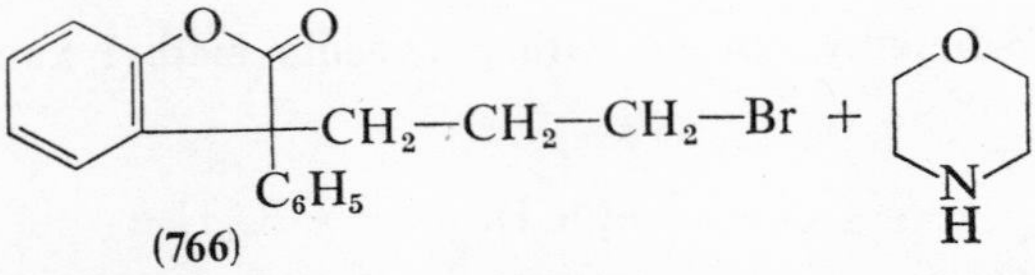

[benzofuranone with CH_2-CH_2-CH_2-Br and C_6H_5] + [morpholine] → $C_{21}H_{23}NO_3$
(766)   (767) 85%

Characteristic IR absorptions: (759) 1800 cm^{-1}; (761) 1690 cm^{-1}; (765) 1690 cm^{-1}; (767) 1800 cm^{-1}.

S-24. Solvolysis of γ,δ-dibromo ketones

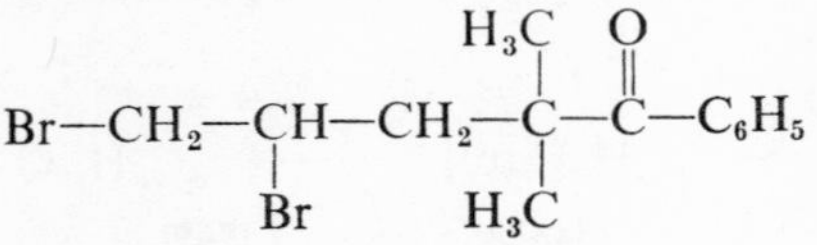

$Br-CH_2-CH(Br)-CH_2-C(CH_3)(CH_3)-C(=O)-C_6H_5$ $\xrightarrow[\text{dioxane-}H_2O]{10\% NaOH}$ $C_{13}H_{16}O_2$
(771)   (773)

S-25. Reaction of dimethyl 2,2'-dichloro-3,3'-thiodipropionate with Na₂S

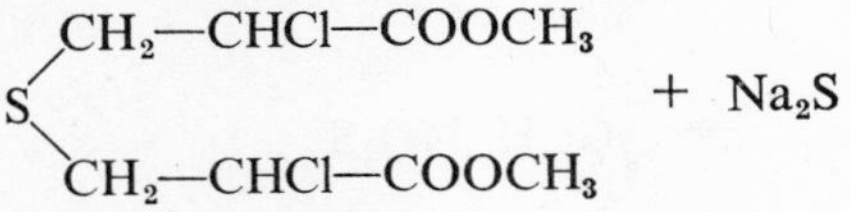

S(CH$_2$—CHCl—COOCH$_3$)$_2$ + Na$_2$S → $C_8H_{12}O_4S_2$ + $C_8H_{12}O_4S_2$
(774)   (776) ~20%   (777) ~50%

Aliphatic Substitution

S-26. Arbusow–Michaelis reaction

$$(C_2H_5O)_3P + HC(F)(Br)-COOC_2H_5 \longrightarrow C_8H_{16}FO_5P$$
(781) 83%

S-27. Perkow reaction

$$(C_2H_5O)_3P + Cl_2HC-CHO \longrightarrow C_6H_{12}ClO_4P$$
(782)

S-28. Thermolysis of piperidinomethyl thioacetate

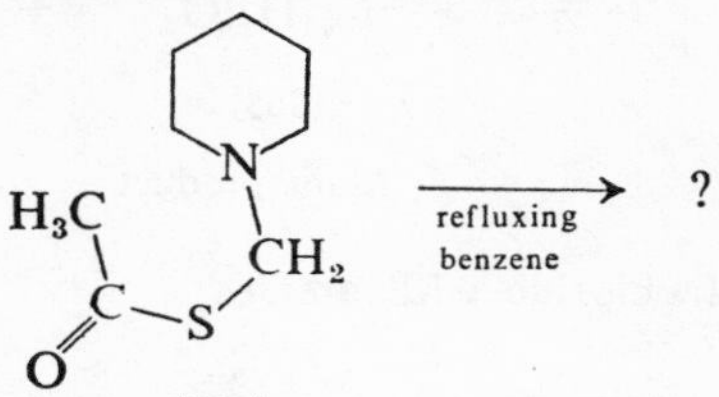

(786) $\xrightarrow{\text{refluxing benzene}}$?

S-29. Cyclizations of ethyl 6-oxo-5-phenylheptanoate

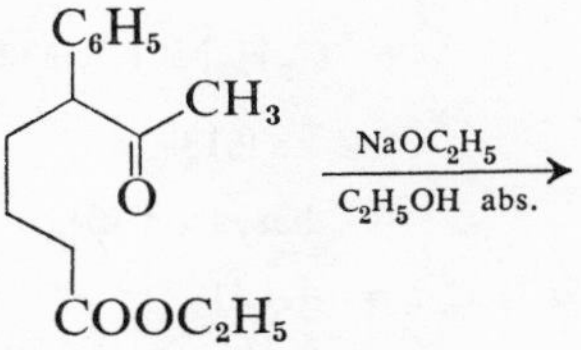

(787) $\xrightarrow[\text{C}_2\text{H}_5\text{OH abs.}]{\text{NaOC}_2\text{H}_5}$ $\left. \begin{array}{c} C_{15}H_{18}O_2 \\ (793) \\ \\ C_{15}H_{18}O_2 \\ (794) \end{array} \right\}$ 38% + $C_{13}H_{14}O_2$ (791)

Characteristic IR absorption of (793): 1730 cm^{-1}
Characteristic IR absorption of (794): 1710
Characteristic absorption of (791);   1740 (red colour with FeCl$_3$) 2.9—2.76 τ (5)
　　　　　　　　　　　　　　　　　　1705　　　　　　　　　　　　　　　　　7.8 τ (3)
　　　　　　　　　　　　　　　　　　1605
　　　　　　　　　　　　　　　　　　1595
　　　　　　　　　　　　　　　　　　1500
　　　　　　　　　　　　　　　　　　760
　　　　　　　　　　　　　　　　　　700

S-30. Reaction of 4-oxocyclohexyl benzoate with KOC(CH$_3$)$_3$

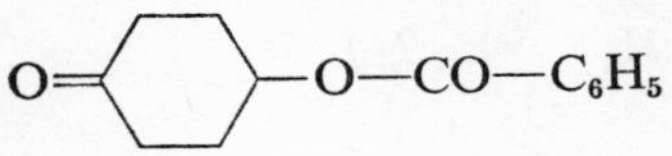

(794) $\xrightarrow[\text{2, HCl}]{\text{1, (CH}_3\text{)}_3\text{COK}}$ $C_{13}H_{14}O_3$ [Spectra on p. 36]
　　　　　　　　　　　　　　　(798)

Characteristic absorption of (**798**): IR 1720 cm^{-1} NMR −0.98 τ (1)
                                           1685   2.1   (2)
                            UV 244 mμ (ε = 16 700)   7.52   (3)
                                                                 8.42   (3)
                                                                 9.12   (2)

S-31. Cleavage of β-diketones

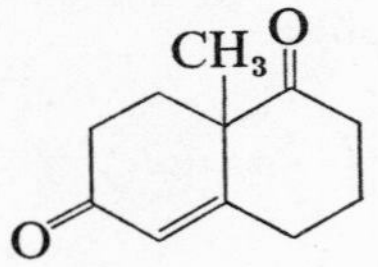

(**802**) $\xrightarrow{\text{NaOCH}_3}$ $C_{11}H_{14}O_2$ (**805**) Main product + $C_{12}H_{18}O_3$ (**804**) 10%

S-32. Reaction of oxalyl chloride with amides

C_6H_5—CH_2—C(=O)—NH_2 + Cl—C(=O)—C(=O)—Cl → $C_{10}H_7NO_3$ $\xrightarrow{\Delta}$

(**811**) 72%

$C_9H_7NO_2$ + CO
(**813**)
50% based on (**811**)

Characteristic absorption of (**811**): UV 240 mμ (log ε = 4.12)
                                           330   = 4.16
                       of (**813**): IR 2250 cm^{-1}

S-33. Preparation of cyanic esters

C_2H_5—O—C(=S)—Cl + NaN_3 $\xrightarrow{20°}$ C_3H_5NO
                                              (**816**) 60%

V. S_Ni SUBSTITUTION

S-34. Pyrolysis of thiocarbonates

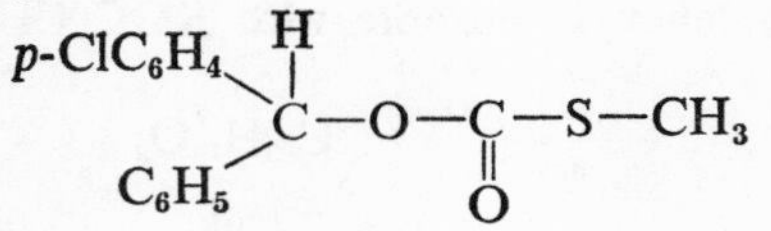

p-ClC$_6$H$_4$\
        C(H)(C$_6$H$_5$)—O—C(=O)—S—CH$_3$ $\xrightarrow{\Delta}$?

(**823**)

VI. S_N2 SUBSTITUTION AT HETEROATOMS

S-35. Reaction of peroxides with phosphines

In the reaction of benzoyl peroxide with optically active phosphine (826) in CH_3CN, 88% of racemic phosphine oxide is obtained, whereas in petroleum ether 49% of racemic phosphine oxide is formed. Explain this.

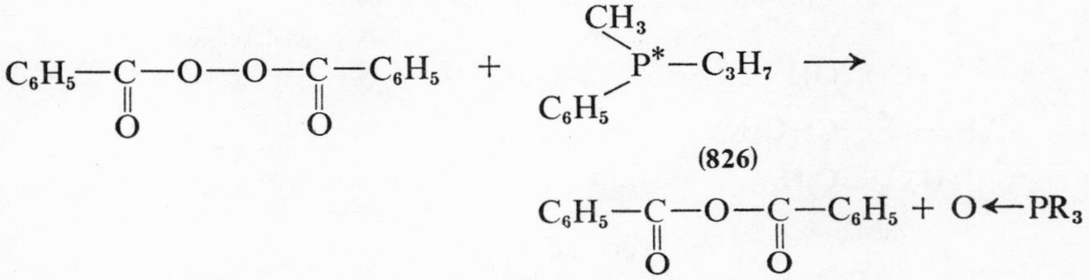

$$C_6H_5-\underset{\underset{O}{\|}}{C}-O-O-\underset{\underset{O}{\|}}{C}-C_6H_5 + \underset{C_6H_5}{\overset{CH_3}{>}}P^*-C_3H_7 \longrightarrow$$

(826)

$$C_6H_5-\underset{\underset{O}{\|}}{C}-O-\underset{\underset{O}{\|}}{C}-C_6H_5 + O \leftarrow PR_3$$

S-36. Reaction of a disulfide with a phosphine

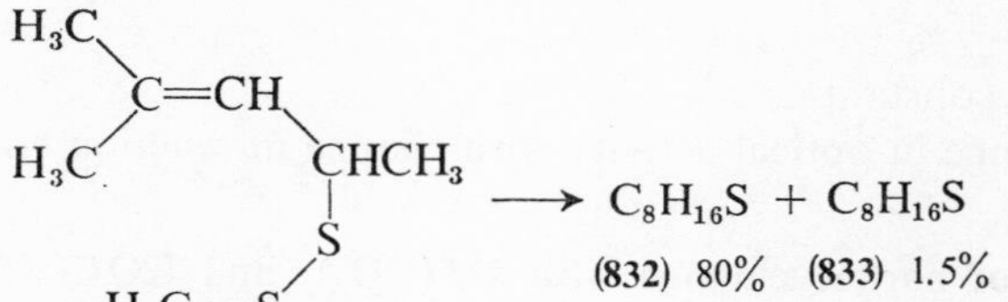

$$\underset{H_3C}{\overset{H_3C}{>}}C=CH\underset{\underset{H_5C_2-S}{}}{\overset{CHCH_3}{\diagdown S \diagup}} \longrightarrow C_8H_{16}S + C_8H_{16}S$$

(829)   (832) 80%   (833) 1.5%

S-37. Reaction of an α-halogenoacetamide with a phosphine

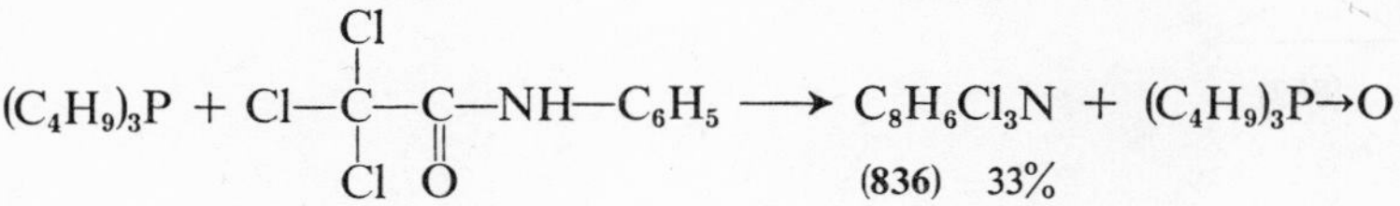

$$(C_4H_9)_3P + Cl-\underset{\underset{Cl}{|}}{\overset{\overset{Cl}{|}}{C}}-\underset{\underset{O}{\|}}{C}-NH-C_6H_5 \longrightarrow C_8H_6Cl_3N + (C_4H_9)_3P \to O$$

(836) 33%

S-38. Reaction of a chloro compound with a phosphite ester

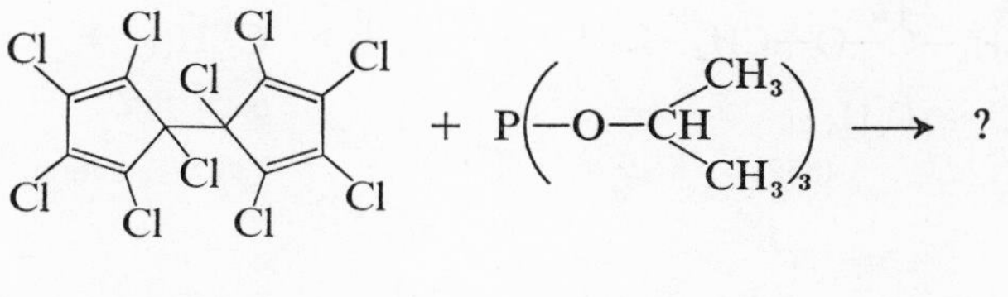

(837) + $P\left(-O-CH\underset{CH_3}{\overset{CH_3}{<}}\right)_3 \longrightarrow$?

VII. S_E1 SUBSTITUTION

S-39. Carbon as leaving group

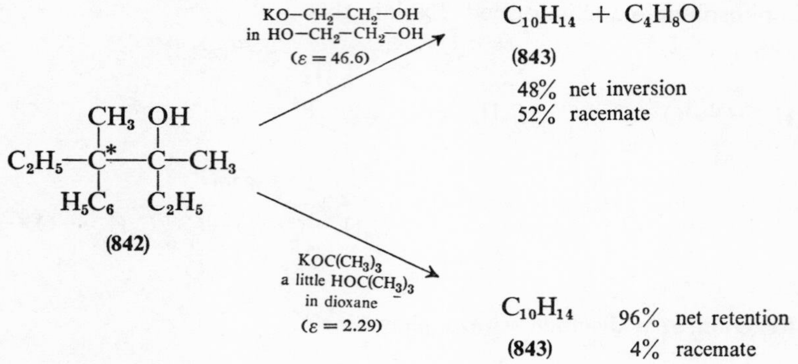

$$C_2H_5-\overset{*}{\underset{H_5C_6}{\underset{|}{C}}}(CH_3)-\underset{C_2H_5}{\underset{|}{C}}(OH)-CH_3$$
(842)

$\xrightarrow{\begin{array}{c}KO-CH_2-CH_2-OH\\ \text{in } HO-CH_2-CH_2-OH\\ (\varepsilon = 46.6)\end{array}}$ $C_{10}H_{14} + C_4H_8O$
(843)
48% net inversion
52% racemate

$\xrightarrow{\begin{array}{c}KOC(CH_3)_3\\ \text{a little } HOC(CH_3)_3\\ \text{in dioxane}\\ (\varepsilon = 2.29)\end{array}}$ $C_{10}H_{14}$ 96% net retention
(843) 4% racemate

ε = dielectric constant

Explain the change in optical activity with change in reaction conditions.

S-40. Reaction of nortricyclanone with $OS(CD_3)_2$ and $KOC(CH_3)_3$

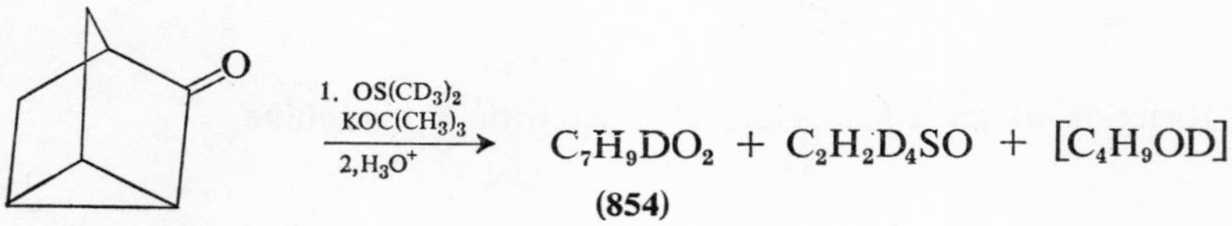

(852) $\xrightarrow[2, H_3O^+]{\begin{array}{c}1.\ OS(CD_3)_2\\ KOC(CH_3)_3\end{array}}$ $C_7H_9DO_2 + C_2H_2D_4SO + [C_4H_9OD]$
(854)

S-41. Oxygen as leaving group

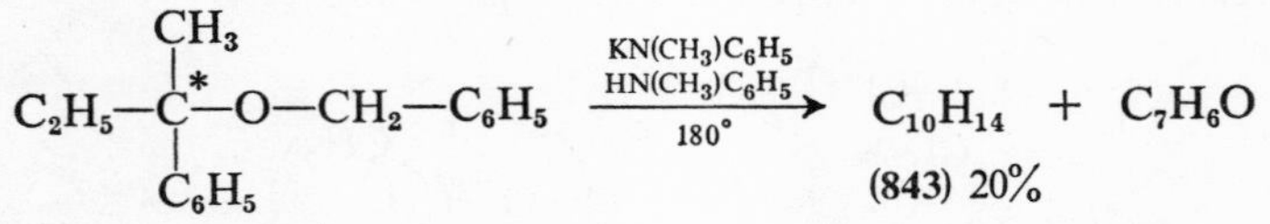

$$C_2H_5-\overset{*}{\underset{C_6H_5}{\underset{|}{C}}}(CH_3)-O-CH_2-C_6H_5$$
(855)

$\xrightarrow[180°]{\begin{array}{c}KN(CH_3)C_6H_5\\ HN(CH_3)C_6H_5\end{array}}$ $C_{10}H_{14} + C_7H_6O$
(843) 20%
29% retention

Propose a mechanism, which explains retention of configuration and racemization in the product.

S-42. Wittig rearrangement

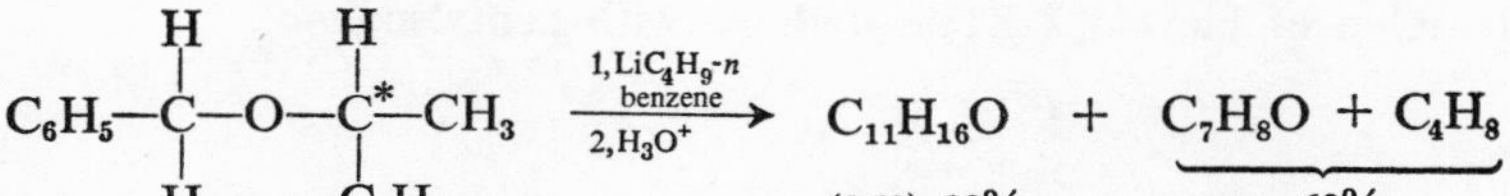

$$C_6H_5-\underset{H}{\overset{H}{C}}-O-\underset{C_2H_5}{\overset{*}{C}}-CH_3 \xrightarrow[2, H_3O^+]{1, LiC_4H_9\text{-}n \\ \text{benzene}} C_{11}H_{16}O + \underbrace{C_7H_8O + C_4H_8}$$

(858)   (860) 38%   62%
                           partial retention

Propose a mechanism, which explains retention of configuration and racemization in the product.

S-43. Meisenheimer rearrangement of amine oxides

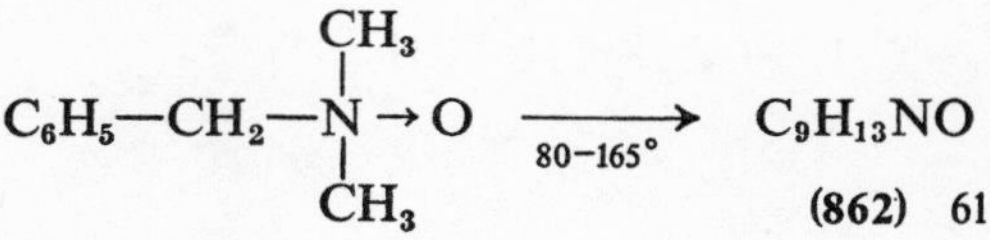

$$C_6H_5-CH_2-\underset{CH_3}{\overset{CH_3}{N}}\rightarrow O \xrightarrow{80-165°} C_9H_{13}NO$$

(861)   (862) 61%

VIII. S_E2 AND S_Ei SUBSTITUTION

S-44. Reaction of *trans*-4-methylcyclohexylmercuric chloride with bromine

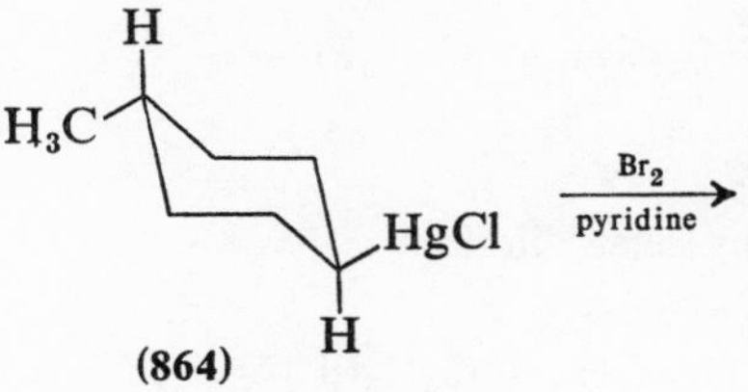

trans-4-methylcyclohexyl-HgCl (864) $\xrightarrow[\text{pyridine}]{Br_2}$ (865)

Complete retention of configuration

Explain the retention of configuration.

S-45. Reaction of diphenylmercury with acetic acid and tributylphosphine

$$(C_6H_5)_2Hg + 2CH_3COOH \longrightarrow (CH_3COO)_2Hg + 2C_6H_6$$

$$(CH_3COO)_2Hg + (n\text{-}C_4H_9)_3P \longrightarrow (n\text{-}C_4H_9)_3P \longrightarrow O + (CH_3CO)_2O + Hg$$
                                                                                                                                                             80%

Propose and justify a mechanism.

IX. CARBANIONS

S-46. Reaction of bicyclo[2.2.1]heptadiene with pentylsodium

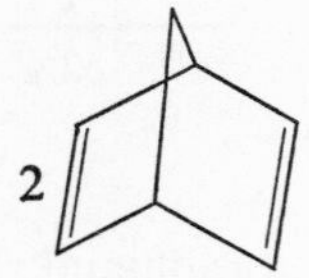

2 [bicyclo[2.2.1]heptadiene] $\xrightarrow[\text{3, }H_3O^+]{\text{1, 2 }n\text{-}C_5H_{11}Na \atop \text{2, }CO_2}$ $C_{12}H_{12}O_4$ + $2C_2H_2$

(871)

S-47. Reaction of 6-phenyl-1-hexene with butylpotassium

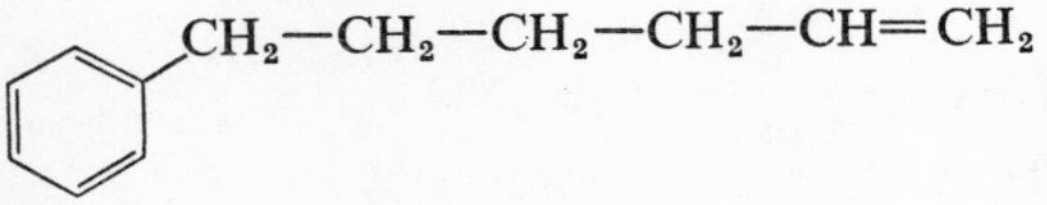

[phenyl]—CH_2—CH_2—CH_2—CH_2—$CH=CH_2$ $\xrightarrow[\text{2, }H_3O^+]{\text{1, }n\text{-}C_4H_9K}$

(872)

$C_{12}H_{16}$ + $C_{12}H_{14}$ + Isomeric phenyl-hexenes
(877)   (876)
9.4% cis   10.6%   14%
48.0% trans

S-48. Reaction of phenyl vinyl ether with phenyllithium

C_6H_5—O—CH=CH_2 + C_6H_5Li $\longrightarrow$?
(878)

S-49. Cleavage of vinyl sulfides by phenyllithium

C_6H_5—S—CH=CH_2 + C_6H_5Li $\longrightarrow$?
(879)

S-50. Unusual reaction of a vinyl sulfide with butyllithium

CH_3—[CH_2]$_8$—$\underset{\underset{CH_3}{|}}{\overset{\overset{CH_3}{|}}{C}}$—S—CH=$CH_2$ + n-C_4H_9Li $\longrightarrow$

$C_{12}H_{24}$ + C_2H_3LiS + C_4H_{10}
(882) 64%

S-51. Ramberg–Bäcklund reaction

$$CH_3-CH_2-SO_2-CHBr-CH_3 \xrightarrow[50°]{\text{NaOH 40\% dioxane–H}_2\text{O}} ?$$
(884)

S-52. Neber rearrangement

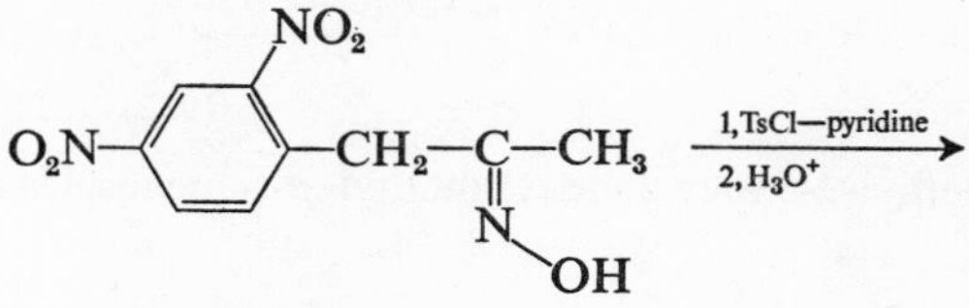

$$O_2N-\underset{NO_2}{\underset{|}{C_6H_3}}-CH_2-\underset{\underset{OH}{\overset{|}{N}}}{\overset{\|}{C}}-CH_3 \xrightarrow[2, H_3O^+]{1, TsCl-pyridine} C_9H_9N_3O_5$$
(893)

S-53. Wittig reaction. (i)

$$[(C_6H_5)_3P-CH_3]^+Br^- \xrightarrow[2, (C_6H_5)_2C=O]{1, C_6H_5Li} ?$$
(902)

S-54. Wittig reaction. (ii) Reaction of (methoxycarbonylmethylene)triphenylphosphorane with benzaldehyde

In the reaction of benzaldehyde with this ylide, methyl cinnamate is obtained with a *trans:cis* ratio of 84:16. Write transition states and intermediates for the reaction and explain the stereoselectivity.

$$C_6H_5-CHO + (C_6H_5)_3P=CHCOOCH_3 \longrightarrow$$

$$(C_6H_5)_3PO + \underset{H}{\overset{C_6H_5}{>}}C=C\underset{COOCH_3}{\overset{H}{<}} + \underset{H}{\overset{C_6H_5}{>}}C=C\underset{H}{\overset{COOCH_3}{<}}$$

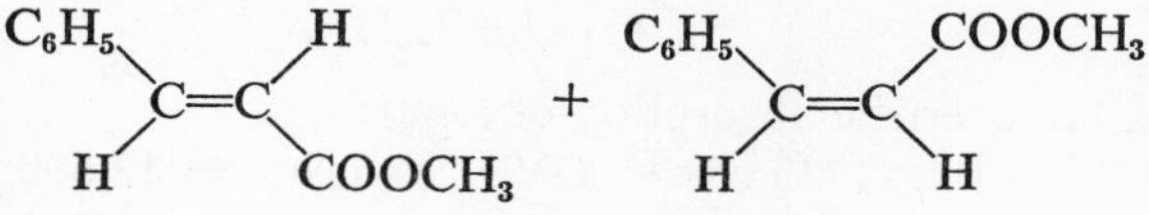

84 : 16

S-55. Reaction of a sulfur ylide with an aldehyde

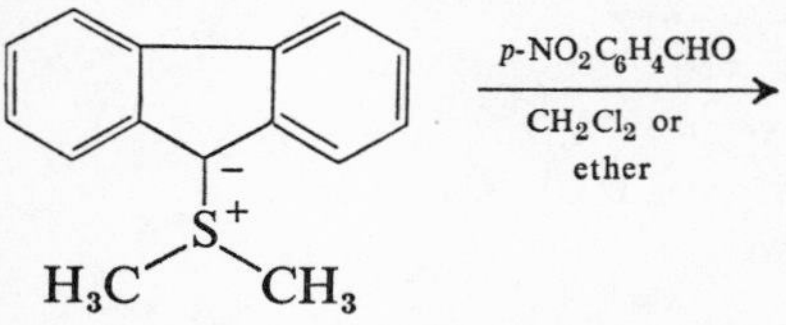

9-(dimethylsulfonio)fluorenide ylide (916) $\xrightarrow[\text{CH}_2\text{Cl}_2 \text{ or ether}]{p\text{-NO}_2\text{C}_6\text{H}_4\text{CHO}}$ $C_{20}H_{13}NO_3$ + C_2H_6S + $C_{22}H_{19}NO_3S$

(918)   (920)

40% in CH$_2$Cl$_2$ 25% in CH$_2$Cl$_2$
8% in ether   83% in ether

Characteristic IR absorption of (918): 850 cm^{-1}
Characteristic IR absorption of (920): 3450

S-56. Reaction of 2-chlorovinyl methyl ketone with ethyl acetoacetate

$$H_3CCOCH_2COOC_2H_5 + H_3CCOCH=CHCl \xrightarrow[K_2CO_3]{C_6H_6} C_{10}H_{12}O_3 + C_{14}H_{16}O_4$$
(925) (924) (927) (929)

S-57. Substitution of a β-keto carbonyl compound at the γ-carbon atom

$$C_6H_5CH_2CH_2COCHNa-CHO \xrightarrow[3,\ NH_3]{1,\ KNH_2;\ 2,\ C_6H_5COOCH_3} C_{18}H_{15}NO$$
(930) (932) 72%

S-58. Reaction of (1-methyl-2-oxocyclohexyl)methyl p-toluenesulfonate with base

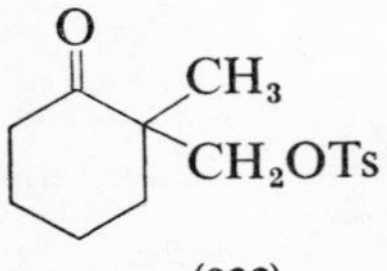

(933) $\xrightarrow[CH_3OH]{NaOH}$ $C_8H_{12}O$ + $C_8H_{12}O$
(934) (935)

S-59. A condensation and rearrangement similar to the abnormal Chapman rearrangement

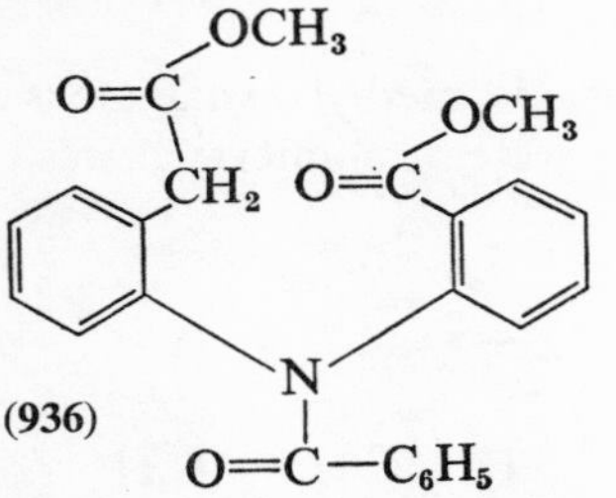

(936) $\xrightarrow[\text{in benzene}]{\text{Excess of } CH_3ONa}$ $C_{23}H_{17}NO_4$
(940) 72%

Characteristic absorption of (940):
　　IR　1735 cm⁻¹　UV　268 mμ ε = 12 600　NMR　−1.7 τ　(1)
　　　　1650　　　　　　320　　　　　11 300　　　　　1.67—3.5 (13)
　　　　1630　　　　　　　　　　　　　　　　　　　　 6.37　(3)

S-60. Reaction of bicyclo[4.2.0]octa-1,3,5-trien-7-ol and -7-one with base

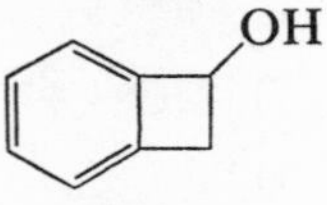

(945) $\xrightarrow{OH^-}$ C_8H_8O
(947) 72%

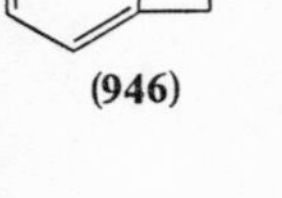

(946) $\xrightarrow{OH^-}$ $C_8H_8O_2$ + $C_8H_8O_2$
(948) 40%　(949) 40%

S-61. Reaction of 3-hydroxycyclobutanone with base

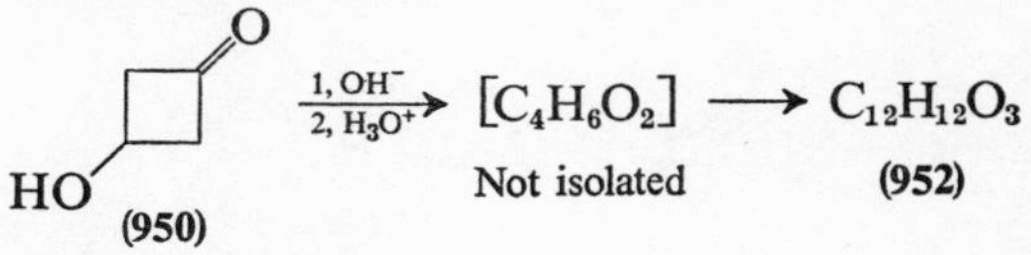

(950) $\xrightarrow{\text{1, OH}^-}_{\text{2, H}_3\text{O}^+}$ [$C_4H_6O_2$] $\longrightarrow$ $C_{12}H_{12}O_3$ (952)
Not isolated

S-62. Reaction of a cyclobutanone derivative with Grignard reagents

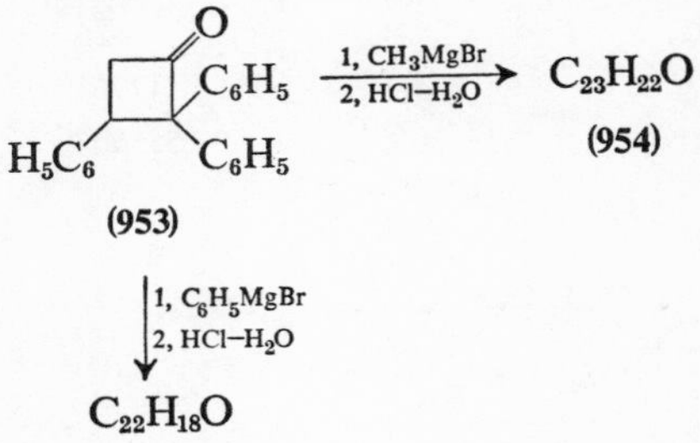

(953) $\xrightarrow{\text{1, CH}_3\text{MgBr}}_{\text{2, HCl–H}_2\text{O}}$ $C_{23}H_{22}O$ (954)

$\downarrow$ 1, C_6H_5MgBr
  2, HCl–H$_2$O

$C_{22}H_{18}O$ (957)

S-63. Birch reduction. (i)

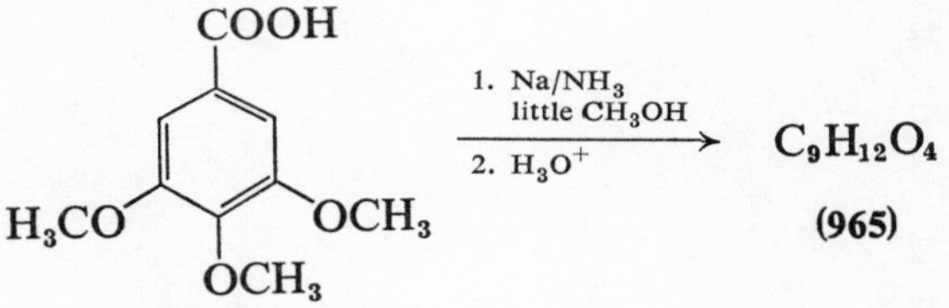

(962) $\xrightarrow[\text{2. H}_3\text{O}^+]{\text{1. Na/NH}_3 \text{ little CH}_3\text{OH}}$ $C_9H_{12}O_4$ (965)

S-64. Birch reduction. (ii)

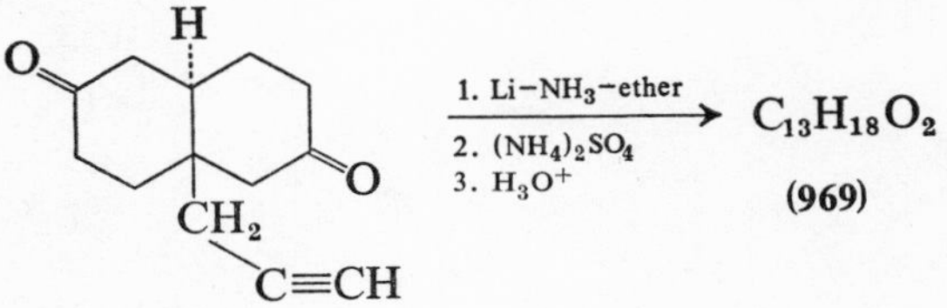

(966) $\xrightarrow[\text{3. H}_3\text{O}^+]{\text{1. Li–NH}_3\text{–ether} \atop \text{2. (NH}_4\text{)}_2\text{SO}_4}$ $C_{13}H_{18}O_2$ (969)

S-65. Birch reduction. (iii)

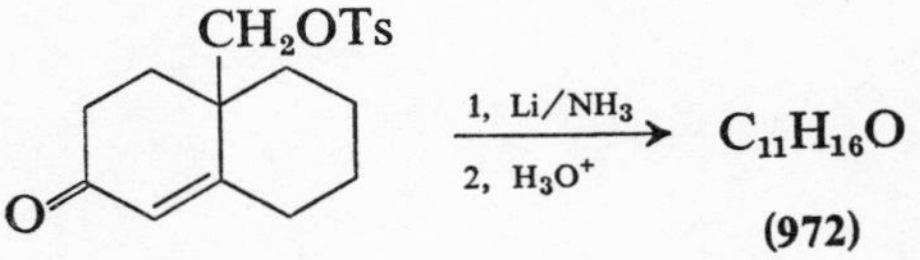

(970) $\xrightarrow{\text{1, Li/NH}_3}_{\text{2, H}_3\text{O}^+}$ $C_{11}H_{16}O$ (972)

S-66. Reaction of 2-chloro-2,2-diphenylacetanilide with NaH

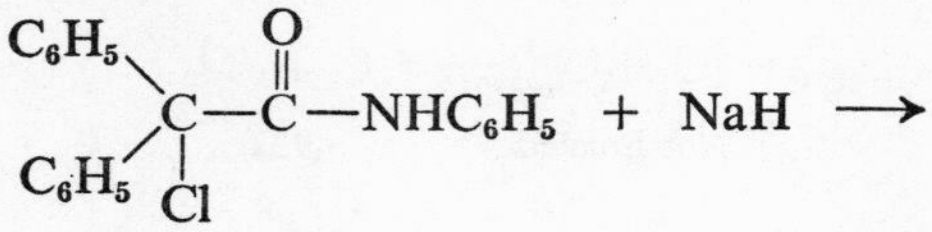

$$\underset{(973)}{\underset{C_6H_5}{\overset{C_6H_5}{>}}\underset{Cl}{C}-\overset{O}{\overset{\|}{C}}-NHC_6H_5} + NaH \longrightarrow$$

$$\underset{\underset{26-30\%}{(974)}}{C_{20}H_{15}NO} + \underset{\underset{55-60\%}{(975)}}{C_{20}H_{15}NO} + \underset{\underset{3-5\%}{(976)}}{C_{20}H_{15}NO}$$

7
Aromatic Substitution

I. ELECTROPHILIC AROMATIC SUBSTITUTION

AS-1. Nitration

Explain why in the nitration of compound (**976d**) with HNO_3–H_2SO_4 predominantly *para*-nitration occurs, whereas with N_2O_5 substitution takes place in predominantly the *ortho*-position.

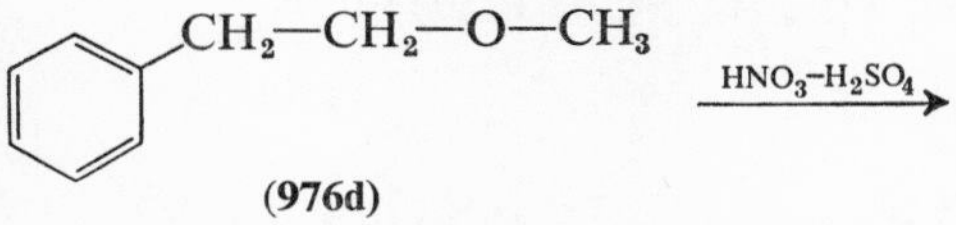

[Structure of 976d: benzene with CH_2–CH_2–O–CH_3 substituent]

$\xrightarrow{HNO_3-H_2SO_4}$

para-isomer (CH_2–CH_2–O–CH_3 with NO_2 para): 62.4%
+ ortho-isomer: 28.9%
+ meta-isomer: 8.7%

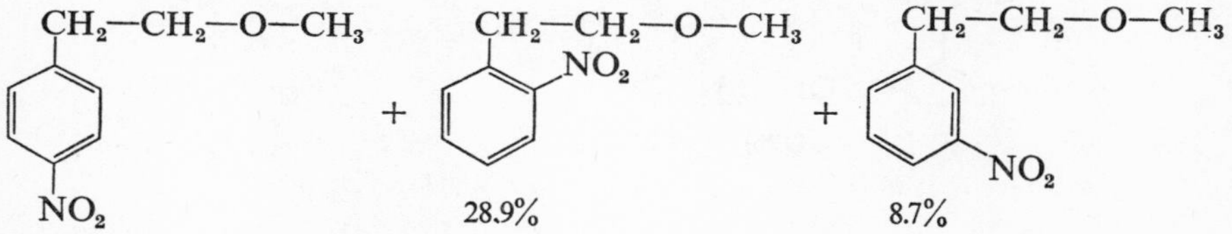

(**976d**) + N_2O_5 ⟶ *para* 29.8%, *ortho* 66.0%, and *meta* 4.2%

AS-2. Partial rate factors

Calculate the partial rate factors k/k_0 for the chlorination of toluene in the *meta*- and the *para*-position with elemental chlorine in CH_3COOH at 25°.

The chlorination yields *ortho* 59.78%, *meta* 0.48%, and *para* 39.74%. The ratio of the substitution rates of benzene to toluene is 1:344, independently of the site of substitution.

AS-3. Electrophilic substitution by OH^+

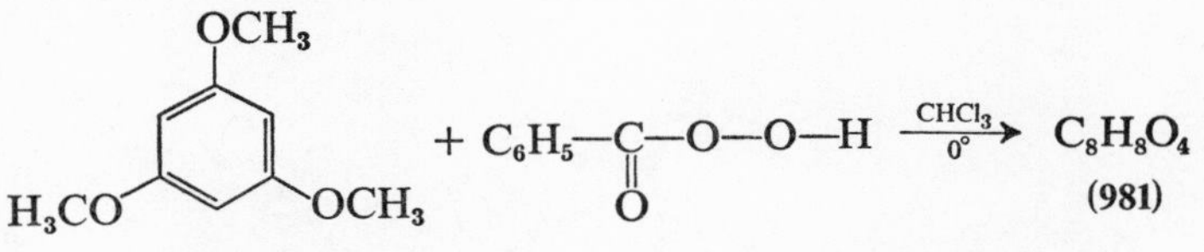

[Structure 979: 1,3,5-trimethoxybenzene with OCH_3, H_3CO, OCH_3] + C_6H_5–C(=O)–O–O–H $\xrightarrow[0°]{CHCl_3}$ $C_8H_8O_4$ (**981**)

AS-4. Reaction of 2,4,6-trimethylphenoxide with chloramine

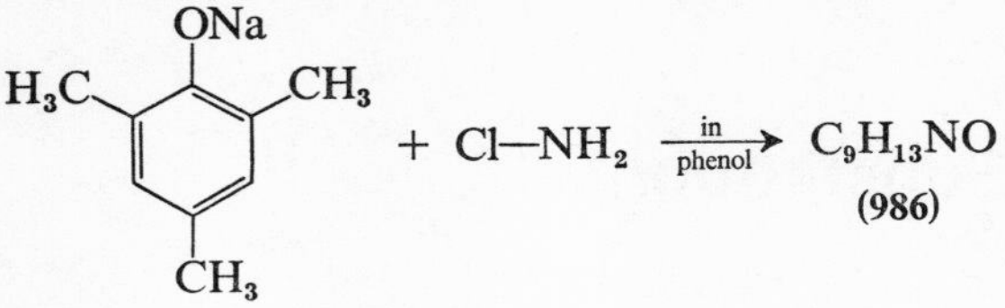

2,4,6-trimethylphenoxide sodium salt + Cl—NH$_2$ $\xrightarrow[\text{phenol}]{\text{in}}$ C$_9$H$_{13}$NO

(986)

AS-5. Reaction of pyridine oxide with acetic anhydride

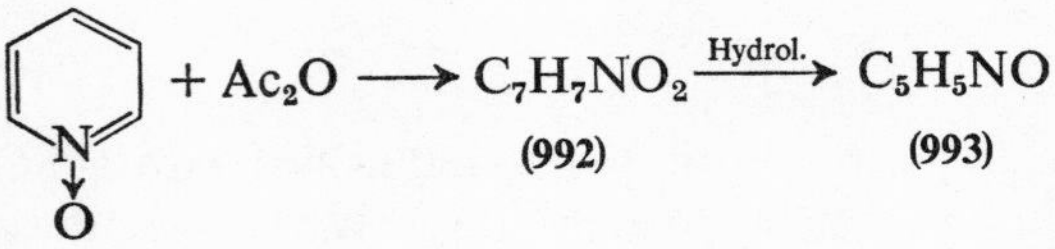

Pyridine N-oxide + Ac$_2$O $\longrightarrow$ C$_7$H$_7$NO$_2$ $\xrightarrow{\text{Hydrol.}}$ C$_5$H$_5$NO

(992)   (993)

II. NUCLEOPHILIC AROMATIC SUBSTITUTION

AS-6. Smiles rearrangement

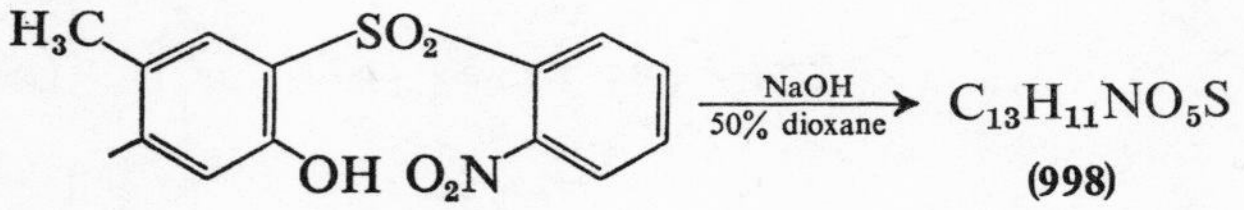

(996) $\xrightarrow[\text{50\% dioxane}]{\text{NaOH}}$ C$_{13}$H$_{11}$NO$_5$S

(998)

8
Radical Reactions

I. RADICAL SUBSTITUTION

R-1. Chlorination of *n*-heptane

The reaction of *n*-heptane with elemental chlorine at 20° under irradiation yields a mixture of primary and secondary monochlorides. The ratio of primary to secondary chlorides corresponds to a chlorination rate ratio of 1:3.3. Among the secondary hydrogen atoms no difference in the rate of chlorination is observed. If, however, *n*-heptane is chlorinated with *tert*-butyl hypochlorite at 0° under irradiation, the following relative rates are observed:

Position	1	2	3	4
Relative rate	1.1	11.3	8.5	7.1

tert-Butyl hypochlorite reacts by the following mechanism:

$$(CH_3)_3C-O\cdot + HR \longrightarrow (CH_3)_3COH + R\cdot$$
$$R\cdot + (CH_3)_3C-OCl \longrightarrow RCl + (CH_3)_3C-O\cdot$$

Chlorination with elemental chlorine in benzene solution at 20° under irradiation leads to the following relative rates:

Position	1	2	3
Relative rate	1.9	8.6	9.1

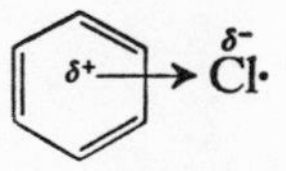

(1001)

Chlorine atoms form π-complexes with benzene, which presumably are the actual chlorination agents **(1001)**.

Explain how the relative rates of chlorination depend on the position of the CH_2 group in the hydrocarbon chain and on the chlorinating agent.

II. RADICAL ADDITION

R-2. Addition of CCl$_4$ to 1-heptyne

$$H_3C-[CH_2]_4-C\equiv CH + CCl_4 \xrightarrow{(C_6H_5COO)_2} C_8H_{12}Cl_2 + C_8H_{12}Cl_4 + C_7H_{12}Cl_2$$
(1001)   (1007) 20%   (1002) 40%   (1003) 6%

R-3. Addition of oxygen to 1,2,4,6-tetraphenylthiabenzene

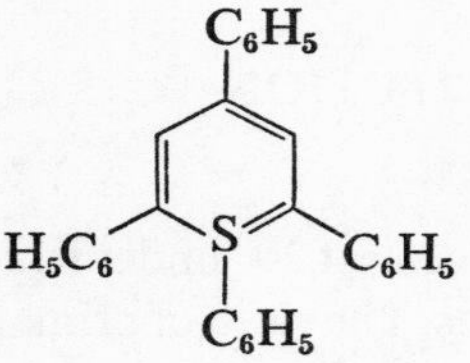

(1012) $\xrightarrow[\text{2, HCl}]{\text{1, O}_2}$ C$_{23}$H$_{16}$O$_2$ + C$_6$H$_6$S
        (1014)

(1012) structure: 1,2,4,6-tetraphenylthiabenzene with C$_6$H$_5$ groups at positions 2, 4, 6 and C$_6$H$_5$ on S (position 1), H$_5$C$_6$ at position 2.

III. OXIDATION

R-4. Autoxidation of 2,4-dimethylpentane and 2,4-pentanediol

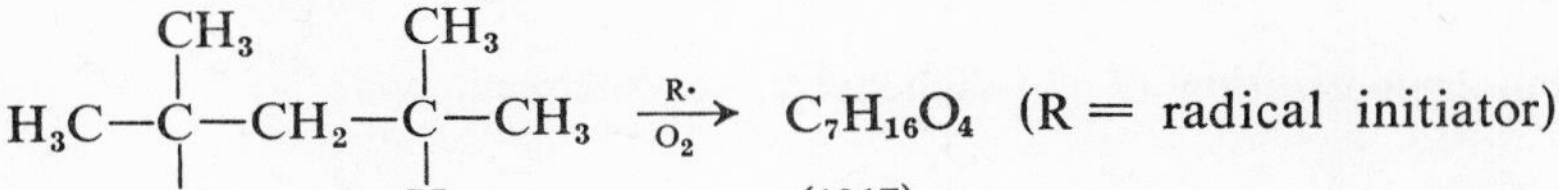

$$H_3C-\underset{H}{\overset{CH_3}{C}}-CH_2-\underset{H}{\overset{CH_3}{C}}-CH_3 \xrightarrow[O_2]{R\cdot} C_7H_{16}O_4$$
(1016) → (1017)   (R = radical initiator)

$$H_3C-\underset{H}{\overset{OH}{C}}-CH_2-\underset{H}{\overset{OH}{C}}-CH_3 \xrightarrow[O_2]{R\cdot} C_5H_{10}O_2$$
(1021)

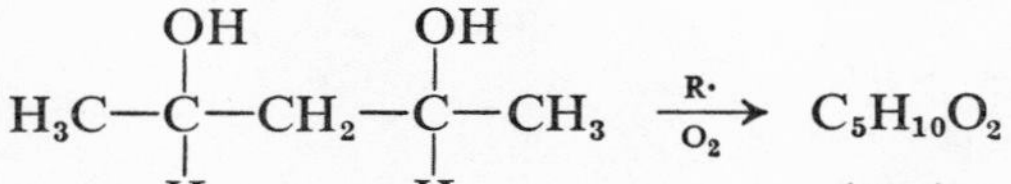

R-5. Reaction of *tert*-butyl hydroperoxide with KOH and *o*-phthalonitrile

The reaction

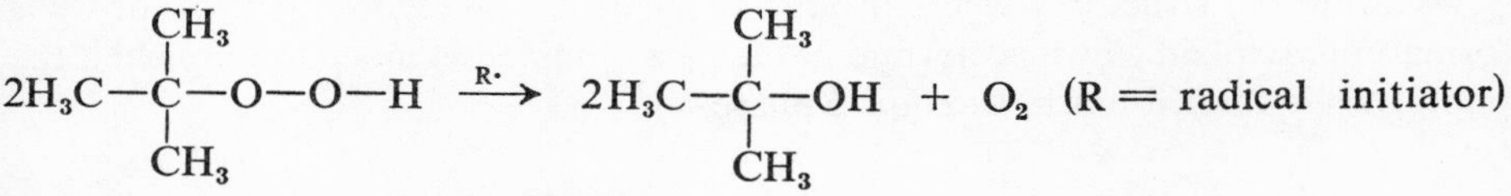

$$2H_3C-\underset{CH_3}{\overset{CH_3}{C}}-O-O-H \xrightarrow{R\cdot} 2H_3C-\underset{CH_3}{\overset{CH_3}{C}}-OH + O_2 \quad (R = \text{radical initiator})$$

is a radical chain reaction. Propose a mechanism.

The radical initiator is formed in accord with the following equation:

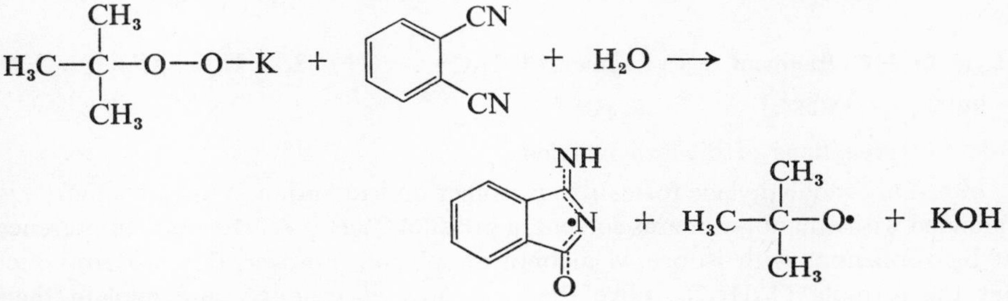

$$H_3C-\underset{CH_3}{\overset{CH_3}{C}}-O-O-K + \underset{CN}{\overset{CN}{\bigcirc}} + H_2O \longrightarrow$$

$$\underset{O}{\overset{NH}{\bigcirc}}\!\!\!\dot{N} + H_3C-\underset{CH_3}{\overset{CH_3}{C}}-O\cdot + KOH$$

Propose a mechanism.

R-6. Oxidation of primary alcohols by lead tetraacetate

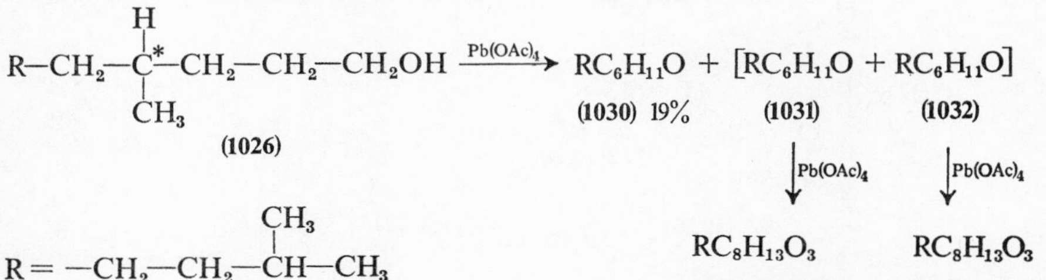

$$R-CH_2-\underset{CH_3}{\overset{H}{\overset{|}{C^*}}}-CH_2-CH_2-CH_2OH \xrightarrow{Pb(OAc)_4} RC_6H_{11}O + [RC_6H_{11}O + RC_6H_{11}O]$$
(1026) (1030) 19%   (1031)   (1032)

$$\downarrow Pb(OAc)_4 \quad \downarrow Pb(OAc)_4$$

$$R = -CH_2-CH_2-\underset{}{\overset{CH_3}{\overset{|}{CH}}}-CH_3 \qquad RC_8H_{13}O_3 \qquad RC_8H_{13}O_3$$
(1033) 33.3%   (1034) 19.7%

9
Photochemistry

P-1. Photoisomerization of 4,4-diphenyl-2,5-cyclohexadienone

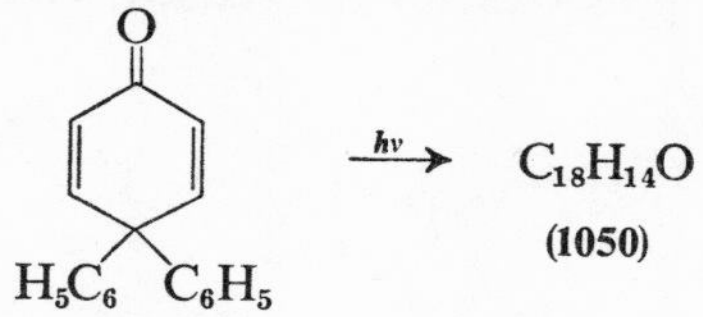

$$(1045) \xrightarrow{h\nu} C_{18}H_{14}O \quad (1050)$$

P-2. Photoreactions of 2-pentanone

Gas-phase photolysis of 2-pentanone yields a C_2 and a C_3 fragment, and an isomer $C_5H_{10}O$. Other major products are biacetyl and *n*-hexane.

What are the structures of the products? What is a likely mechanism for their formation? Explain why biacetyl and hexane are obtained in much lower yield if the photolysis is carried out in the liquid phase.

$$H_3C-\underset{\underset{O}{\|}}{C}-CH_2-CH_2-CH_3 \xrightarrow[\substack{\text{room temp.}\\\text{gas phase}}]{h\nu}$$
(1052)

$$C_5H_{10}O + C_2 \text{ fragment} + C_3 \text{ fragment} + H_3C-\underset{\underset{O}{\|}}{C}-\underset{\underset{O}{\|}}{C}-CH_3 + H_3C-[CH_2]_4-CH_3$$
(1055)   (1055a)   (1055b)

P-3. Photoreactions of dibenzoylethylene

trans-Dibenzoylethylene forms its *cis*-isomer on irradiation. When irradiation is extended with ethanol–water as solvent, a product $C_{18}H_{18}O_3$ is formed. In presence of benzophenone with isopropyl alcohol as solvent, however, the photo-product has the formula $C_{16}H_{14}O_2$. Give the structures of products and explain their formation.

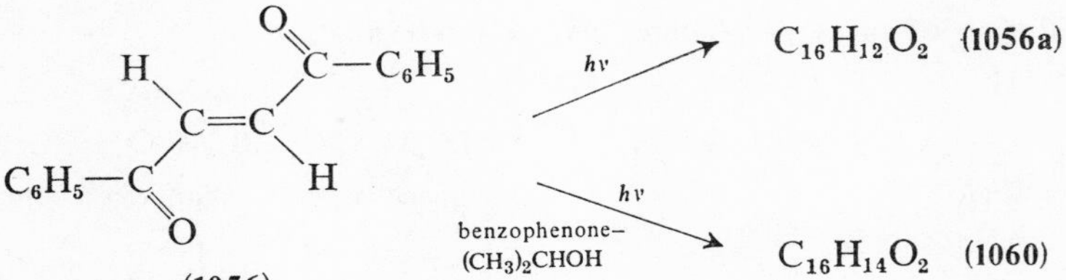

(1056) $\xrightarrow{h\nu}$ $C_{16}H_{12}O_2$ (1056a)

(1056) $\xrightarrow[\text{benzophenone–}(CH_3)_2CHOH]{h\nu}$ $C_{16}H_{14}O_2$ (1060)

(1056a) $\xrightarrow[C_2H_5OH-H_2O]{h\nu}$ $C_{18}H_{18}O_3$ (1059)

P-4. Photorearrangement of epoxy ketones

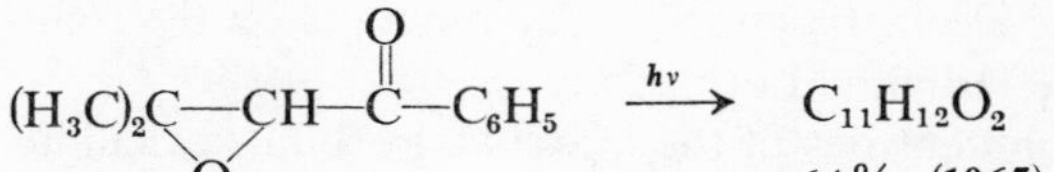

$$(H_3C)_2C-\underset{O}{CH}-\overset{O}{\overset{\|}{C}}-C_6H_5 \quad \xrightarrow{h\nu} \quad C_{11}H_{12}O_2$$
(1061)   64% (1065)

$$\underset{H_5C_6}{\overset{H_3C}{>}}C-\underset{O}{CH}-\overset{O}{\overset{\|}{C}}-CH_3 \quad \xrightarrow{h\nu} \quad C_{11}H_{12}O_2$$
(1066)   (1070)

P-5. Photoreactions of stilbene

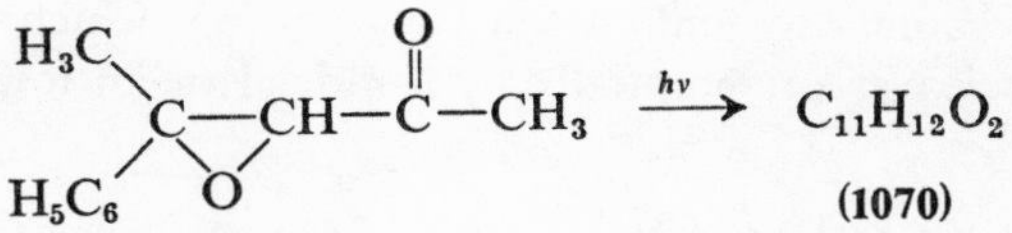

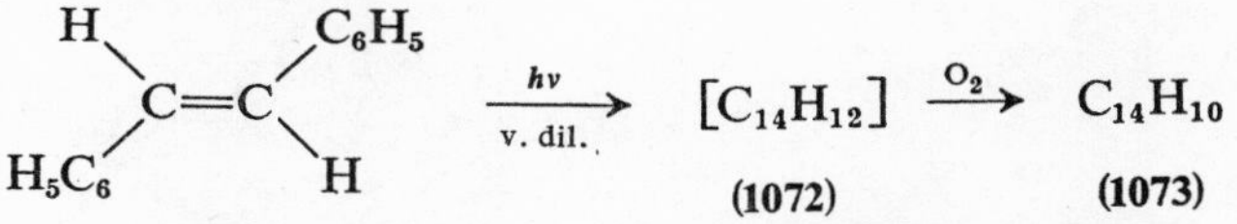

$$\underset{H_5C_6}{\overset{H}{>}}C=C\underset{H}{\overset{C_6H_5}{<}} \quad \xrightarrow[\text{v. dil.}]{h\nu} \quad [C_{14}H_{12}] \quad \xrightarrow{O_2} \quad C_{14}H_{10}$$
(1071)   (1072)   (1073)
      not isolated

$$\underset{H_5C_6}{\overset{H}{>}}C=C\underset{H}{\overset{C_6H_5}{<}} \quad \xrightarrow[\text{saturated solution}]{h\nu} \quad C_{28}H_{24}$$
       (1074)

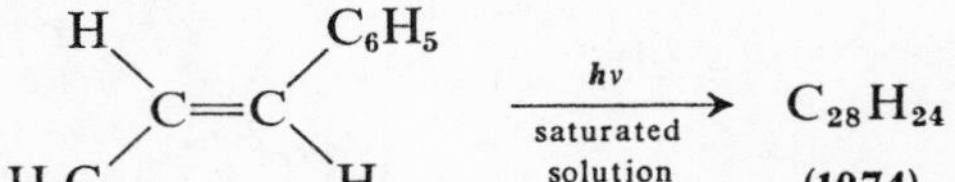

P-6. Substituent effects in excited aromatic compounds

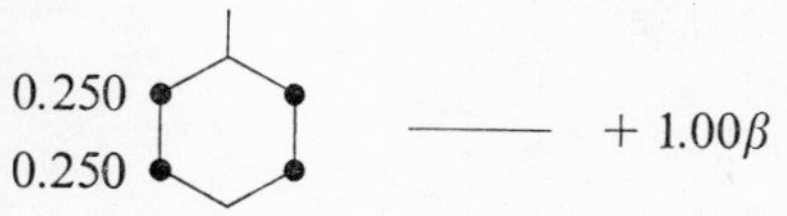

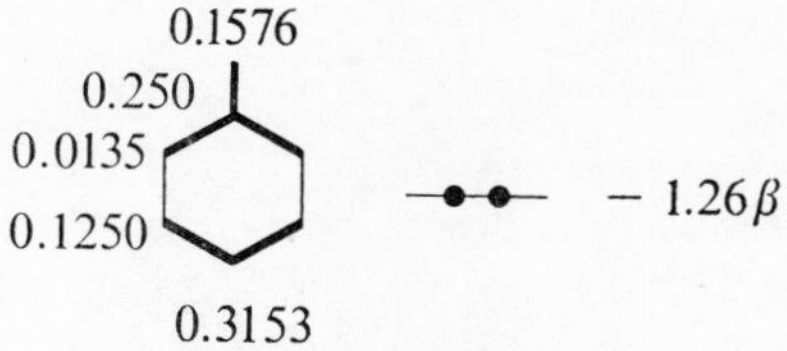

0.250, 0.250 — toluene-like — $+1.00\beta$

0.1576, 0.250, 0.0135, 0.1250, 0.3153 — -1.26β

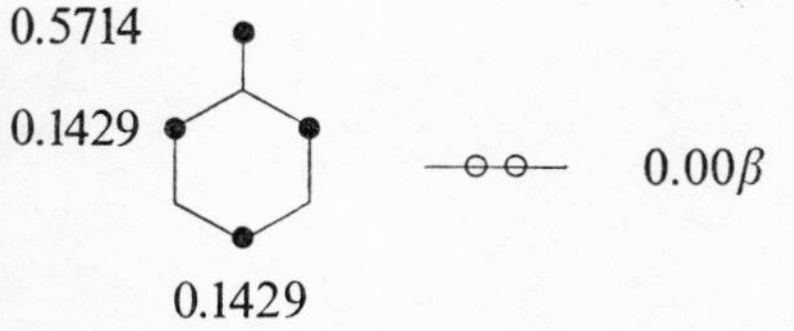

0.5714, 0.1429, 0.1429 — 0.00β

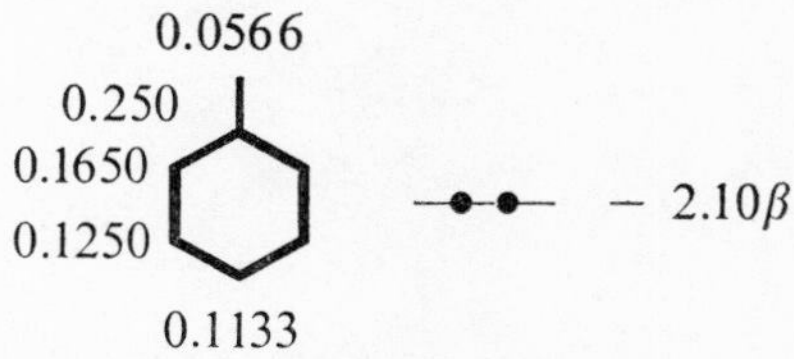

0.0566, 0.250, 0.1650, 0.1250, 0.1133 — -2.10β

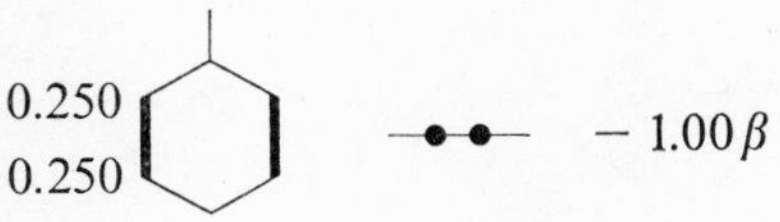

0.250, 0.250 — -1.00β

The Figure on p. 51 shows the five lowest orbitals of the seven benzyl orbitals, indicated by reinforced lines and dots. The numbers at the individual carbon atoms represent the electron densities. The numbers on the right-hand side are the energies in β of the individual orbitals.

From the electron densities of the individual orbitals calculate the total electron densities for the benzyl anion in the *meta-* and the *para-*position for the ground state and the first excited state. Demonstrate, by assuming that the effect of the OCH_3 group corresponds to that of the CH_2^- group, which of the compounds **(1075)** and **(1076)** cleaves preferentially by a radical or an ionic mechanism.

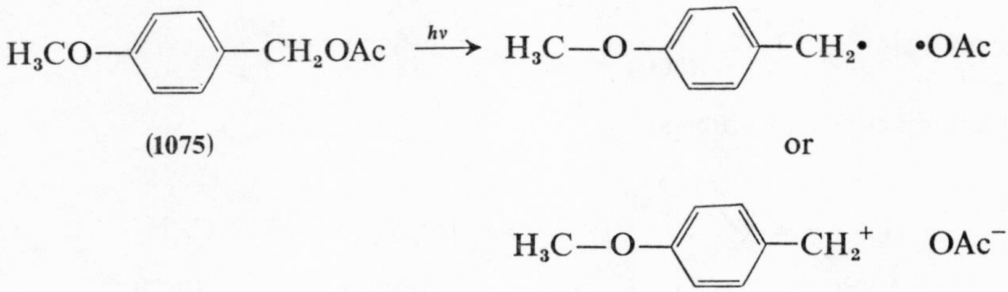

H_3CO—⟨❬❭⟩—CH_2OAc $\xrightarrow{h\nu}$ H_3C—O—⟨❬❭⟩—$CH_2\bullet$ $\bullet OAc$

(1075)   or

H_3C—O—⟨❬❭⟩—CH_2^+ OAc^-

⟨❬❭⟩—CH_2OAc $\xrightarrow{h\nu}$ ⟨❬❭⟩—$CH_2\bullet$ $\bullet OAc$
H_3CO H_3CO or

⟨❬❭⟩—CH_2^+ OAc^-
H_3CO

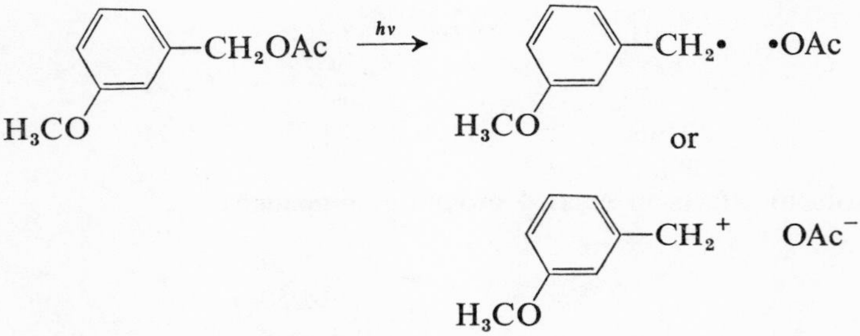

10

Miscellaneous Reactions

M-1. Acid-catalyzed rearrangement of *N*-allylaniline

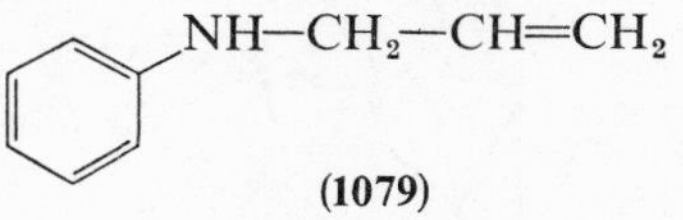

$$\text{C}_6\text{H}_5\text{NH}-\text{CH}_2-\text{CH}=\text{CH}_2 \xrightarrow{\text{HCl}, 180°} \text{C}_9\text{H}_{11}\text{N} + \text{C}_9\text{H}_9\text{N}$$

(1079)   (1081) ~20%   (1082) ~12%

$$\text{C}_9\text{H}_{11}\text{N} \xrightarrow{\text{HCl}, \Delta} \text{C}_9\text{H}_9\text{N} + \text{H}_2$$

M-2. Base-catalyzed rearrangement of 3-amino-1,2,4-benzotriazine 1-oxide

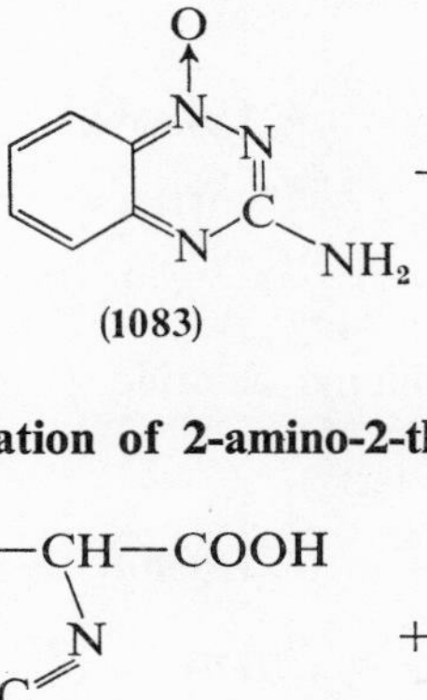

(1083) $\xrightarrow{\text{OH}^-}$ $\text{C}_7\text{H}_6\text{N}_4\text{O}$ $\xrightarrow{\text{OH}^-}$ $\text{C}_6\text{H}_5\text{N}_3$

(1085) 16%   (1086) 37%

M-3. Benzoylation of 2-amino-2-thiazoline-4-carboxylic acid in a basic medium

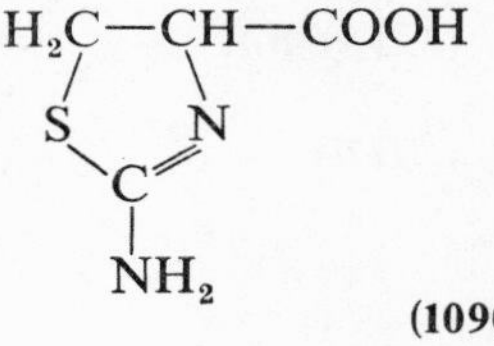

(1090) + $\text{C}_6\text{H}_5\text{COCl}$ $\xrightarrow{1,\ \text{OH}^-;\ 2,\ \text{H}_3\text{O}^+}$ $\text{C}_{11}\text{H}_{12}\text{N}_2\text{O}_4\text{S}$

(1096)

M-4. Reaction of oximes with nitrous acid. (i)

$$\text{H}_3\text{C}-\underset{\text{O}}{\overset{\|}{\text{C}}}-\underset{\text{CH}_3}{\text{C}}=\text{N}-\text{OH} \xrightarrow{\text{HNO}_2} (1099)$$

(1097)

M-5. Reaction of oximes with nitrous acid. (ii)

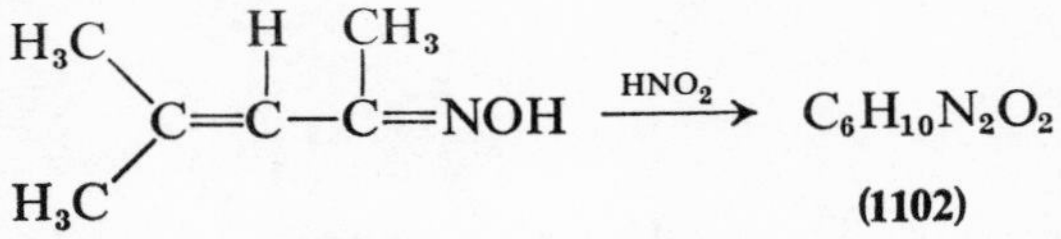

$$\underset{H_3C}{\overset{H_3C}{>}}C=\underset{H}{\overset{}{C}}-\underset{CH_3}{\overset{}{C}}=NOH \xrightarrow{HNO_2} C_6H_{10}N_2O_2$$

(1100) → (1102)

M-6. Base-catalyzed acyloin rearrangement

Propose and justify a mechanism for the reaction:

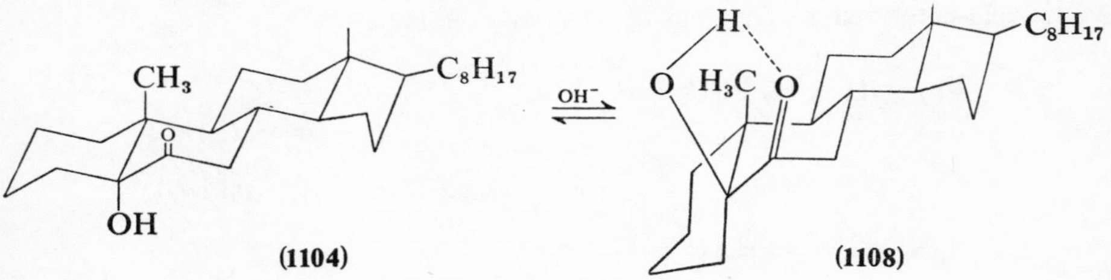

(1104) $\xrightarrow{OH^-}$ (1108)

M-7. Pummerer rearrangement

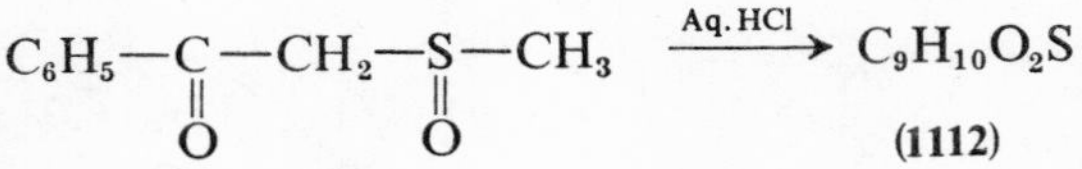

$$C_6H_5-\underset{O}{\overset{\|}{C}}-CH_2-\underset{O}{\overset{\|}{S}}-CH_3 \xrightarrow{Aq.\ HCl} C_9H_{10}O_2S$$

(1109) → (1112)

M-8. Reaction of o-benzoylbenzoic acid with thionyl chloride

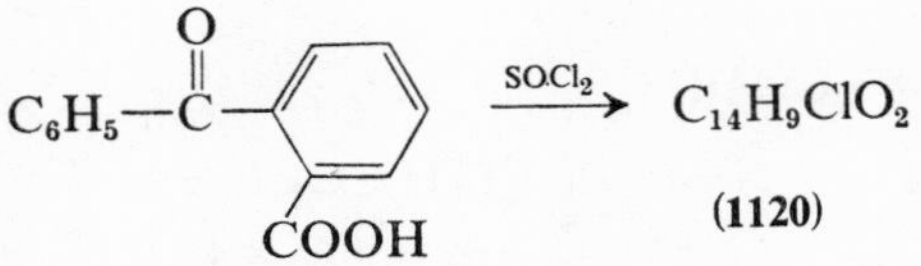

(1118) $\xrightarrow{SOCl_2}$ $C_{14}H_9ClO_2$ (1120)

M-9. Rearrangement of O-glycylsalicylamide

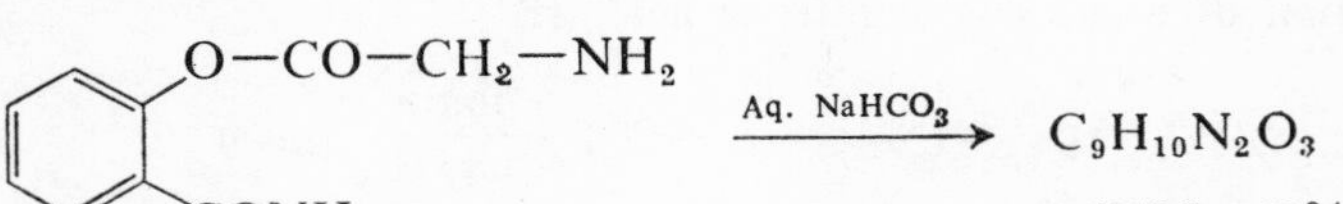

(1123) $\xrightarrow{Aq.\ NaHCO_3}$ $C_9H_{10}N_2O_3$ (1124) 48%

M-10. Cyclopropenylium cations

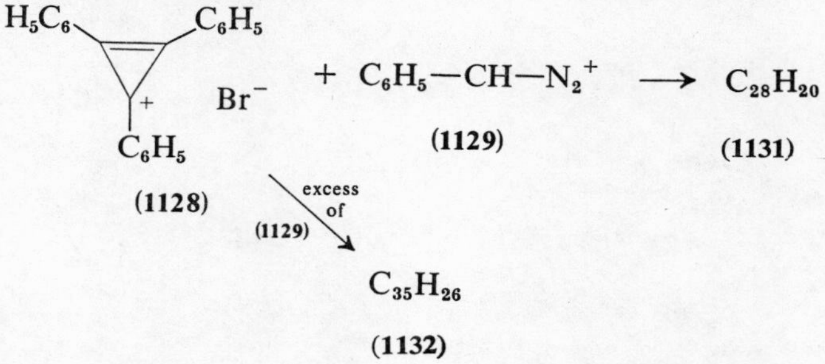

H_5C_6–△(+)(C_6H_5, C_6H_5) Br^- + C_6H_5–CH–N_2^+ → $C_{28}H_{20}$

(1128)   (1129)   (1131)

↓ excess of (1129)

$C_{35}H_{26}$

(1132)

Part 2

Discussion of the Problems

1

Addition Reactions

A CHARACTERISTIC property of unsaturated bonds is their ability to undergo addition reactions. Ionic additions to the C=C bond are usually initiated by attack of an electrophile. Nucleophilic attack occurs only if the double bond is sufficiently polarized. On the other hand, additions to the triple bond are in general initiated by nucleophilic attack. Many additions to the C=O bond are catalyzed by acids and thus start with the addition of an electrophile.

The various mechanisms by which polar additions take place are discussed below in detail in their bearing on the individual examples. Radical additions are presented separately in Chapter 8.

I. ELECTROPHILIC ADDITIONS TO THE C=C BOND

A-1, A-2. Stereochemistry of the hydrogen halide and halogen addition.

In the addition of hydrogen halides to many olefins, the first step is assumed to consist of interaction of an approaching proton with the π-electron cloud of the double bond whereby a π-complex (1) is formed*. π-Complexes are commonly

(1)   (2)

described by formulae such as (1) or (2)†. If in the next step the π-complex is attacked by the halide ion, *trans*-addition products are obtained, since the nucleophile can approach the π-complex only from the rear.

Stereospecific *trans*-additions are the reactions of HI with tiglic acid and angelic acid[1] and the addition of HCl or HBr to 1,2-dimethylcyclopentene or 1,2-dimethylcyclohexene[2]. Whether in these additions π-complexes are really responsible for the stereospecificity is still a matter of controversy.

---
* π-Complexes are also formed in electrophilic aromatic substitution.
† The arrow in (1) indicates that the bonding electrons are furnished by the double bond. Nonclassical formulations such as (2) are discussed on p. 61.

HCl + [1,2-dimethylcyclohexene] → [π-complex with H⁺...Cl⁻] → [product: H, H₃C, CH₃, Cl on cyclohexane]

In many cases, the π-complex is converted into a carbonium ion (**4**), by formation of a σ-bond between the proton and one specific carbon atom of the double bond, at a faster rate than the π-complex is attacked by the halide ion.

$$\text{C=C} + 3\text{HY} \rightleftarrows \text{[C=C}\rightarrow\text{H}\cdots\text{Y}(\text{YH})_2\text{]} \longrightarrow \text{[C=C}\rightarrow\text{H}^+ + \text{Y}^-(\text{YH})_2\text{]}$$

(3)   (1)

[Scheme: carbonium ion (4) with H⁺ and Y⁻ leads to cis and trans products]

equal amounts of *cis* and *trans*

trans

The rate law of the addition reaction in non-ionizing solvents indicates that two additional molecules of HX are involved, which probably serve to pull the halide ion away from the proton as indicated in (**3**).

Rate determinations measure the slowest step of a reaction. The influence of the additional HX molecules on the reaction rate therefore shows that transformation (**3**) → (**1**) must be the slowest, *i.e.* the rate-determining step of the reaction. Addition of the halide ion to the ion (**4**) leads to a 1:1 mixture of *cis*- and *trans*-adducts. In the case of an unsymmetrically substituted olefin, for example, isobutene, formation of the more highly substituted, more stable carbonium ion is favored; consequently, the nucleophile adds to the more highly substituted carbon atom (Markownikoff's rule.)

A-1. *Addition of DBr to indene*[3]

Recently, several addition reactions of DBr to aryl-substituted olefins have been shown to yield predominantly *cis*-addition products[3]. Indene, for example, yields 81% of *cis*-adduct at −78° in CH_2Cl_2.

$$\text{indene} \xrightarrow[\text{in CH}_2\text{Cl}_2]{\underset{-78°}{\text{DBr}}} \text{(Br, D cis adduct)} \ (81\%) \ + \ \text{(Br, D trans adduct)} \ 19\%$$

These results have been explained by the direct formation of tight ion-pairs, without preceding occurrence of π-complexes. It is reasonable to assume, that the formation of a tight ion-pair is especially favored over a π-complex in the case of such olefins, in which the developing positive charge is stabilized by suitable substituents. The anion of the tight ion-pair is thought to be located on the side of the olefin where the proton is already attached to one of the olefinic carbon atoms. It may either add at this side to the electrophilic centre, yielding a *cis*-adduct, or migrate to the opposite side of the carbonium ion, yielding a *trans*-adduct. With increasing lifetime of the carbonium ion, the probability of *trans*-addition increases. Accordingly, a slight increase in the amount of *trans*-adduct has been observed with rising polarity of the solvent.

Polar addition of Cl_2, Br_2, or I_2 to simple olefins yields *trans*-addition products. The stereospecificity has been interpreted, as in many cases of hydrogen halide addition, by the primary formation of π-complexes. The reaction begins with the addition of Br^+ or another halogen cation. Since the carbon–halogen bonds (5) are less stable than the C–H bond in classical carbonium ions, and since π-complex formation depends not only on the electron-donating properties of the olefinic double bond, but also on the back-donation of p-electrons from the electrophile to the antibonding orbital of the double bond, π-complex formation with Br^+ or I^+ (6) is much more favorable than in the case of attack by a proton[3].

$$\underset{(5)}{\overset{Br}{\underset{|}{>}C-\overset{+}{C}<}} \quad \underset{(6)}{\overset{Br^+}{\underset{\updownarrow}{>}C=C<}} \text{ or } >\overset{Br}{\underset{+}{C\cdots C}}<$$

The mechanism of addition thus depends on the type of halogen atom as well as on the substituents at the olefinic carbon atoms. For example, simple olefins yield *trans*-adducts with Cl_2; but styrene, with a phenyl group which stabilizes the developing positive charge of the classical carbonium ion, yields *cis*- and *trans*-adducts. Thus Cl_2, which forms a stronger C–X bond than Br_2 takes an intermediate position, forming other classical cations in the form of tight ion-pairs or π-complexes.

$$>\!\!\underset{C}{\overset{C}{\|}}\!\!< \ + \ Br_2 \ \rightleftarrows \ >\!\!\underset{C}{\overset{C}{\|}}\!\!< \!\!\rightarrow Br-Br \ \xrightarrow{\text{Slow}} \ >\!\!\underset{C}{\overset{C}{\|}}\!\!< \ \rightleftarrows \ Br^+ \ \xrightarrow{Br^-} \ \underset{Br-\overset{|}{C}-}{\overset{-\overset{|}{C}-Br}{}} \ \textit{trans}$$

In solvents of low polarity, formation of a bromonium ion, which involves charge separation (the solvent has, of course, an even stronger effect in the case of classical cations), is rendered more difficult; here it has been found that the rate of bromination depends, not only on a term of second order, but also on one of third order, (first order in olefin and second order in halogen). It is likely that the formation of the π-complex is accelerated by another Br_2 molecule, which stabilizes the separating Br^- by formation of a trihalide ion Br_3^-. The same applies to I_2.

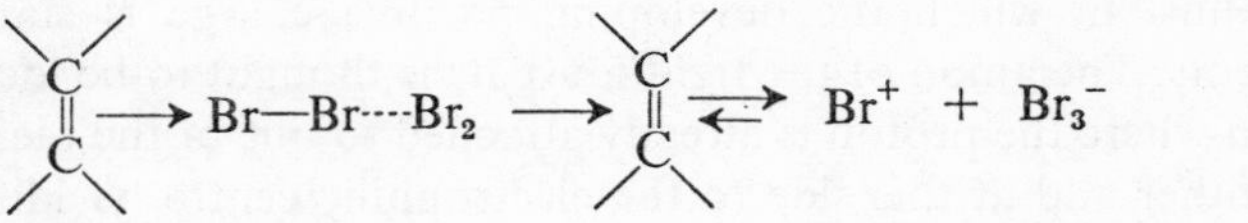

$$\underset{C}{\overset{C}{\parallel}} \longrightarrow Br-Br\cdots Br_2 \longrightarrow \underset{C}{\overset{C}{|}} \rightleftarrows Br^+ + Br_3^-$$

A-2. *Addition of bromine to 3-benzamidocyclohexene* (7)[4]

Bromination of 3-benzamidocyclohexene (7) leads to a bromonium ion (8), which is attacked from the rear by the adjacent carbonyl-oxygen instead of by Br^-. The resulting oxazolinium salt (9) has been isolated in better than 60%

yield. With increasing bromide ion concentration (for example, on addition of LiBr), the yields of bromo-compounds (10a) and (10b) increase at the expense of (9). To some extent, the solvent also reacts with (8). The formation of (12) by nucleophilic attack by the carbonyl-oxygen of (10b) does not occur.

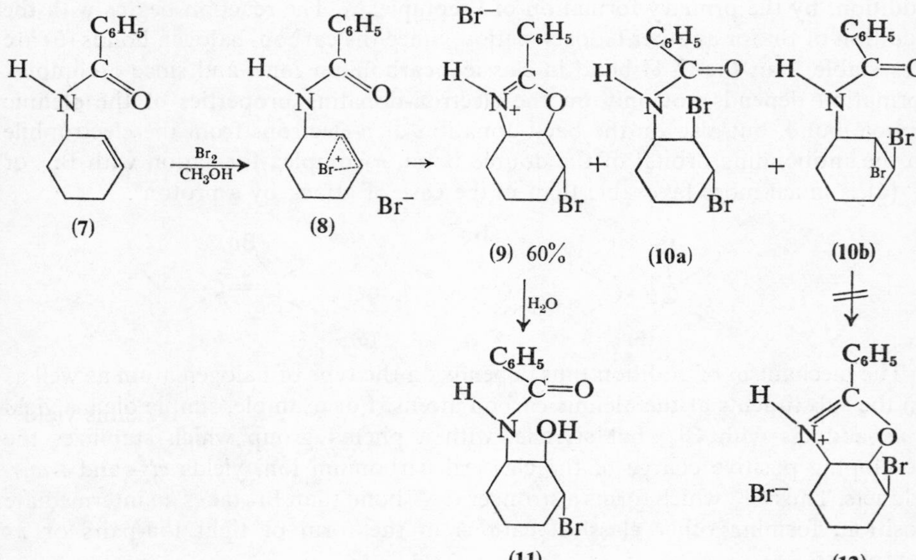

A-3. Acylation of olefins

Friedel–Crafts catalysts such as $AlCl_3$, $ZnCl_2$, and $SnCl_4$ form acylcarbonium salts with acyl halides*. These salts are present in polar solvents as solvated ions, in non-polar solvents as tight ion-pairs or strongly polarized complexes. In the presence of olefins, the carbonium ion attacks the double bond in analogy to Friedel–Crafts reactions of aromatic compounds. The addition obeys Markownikoff's rule.

A-3. *Addition of acetyl chloride to propene*[6]

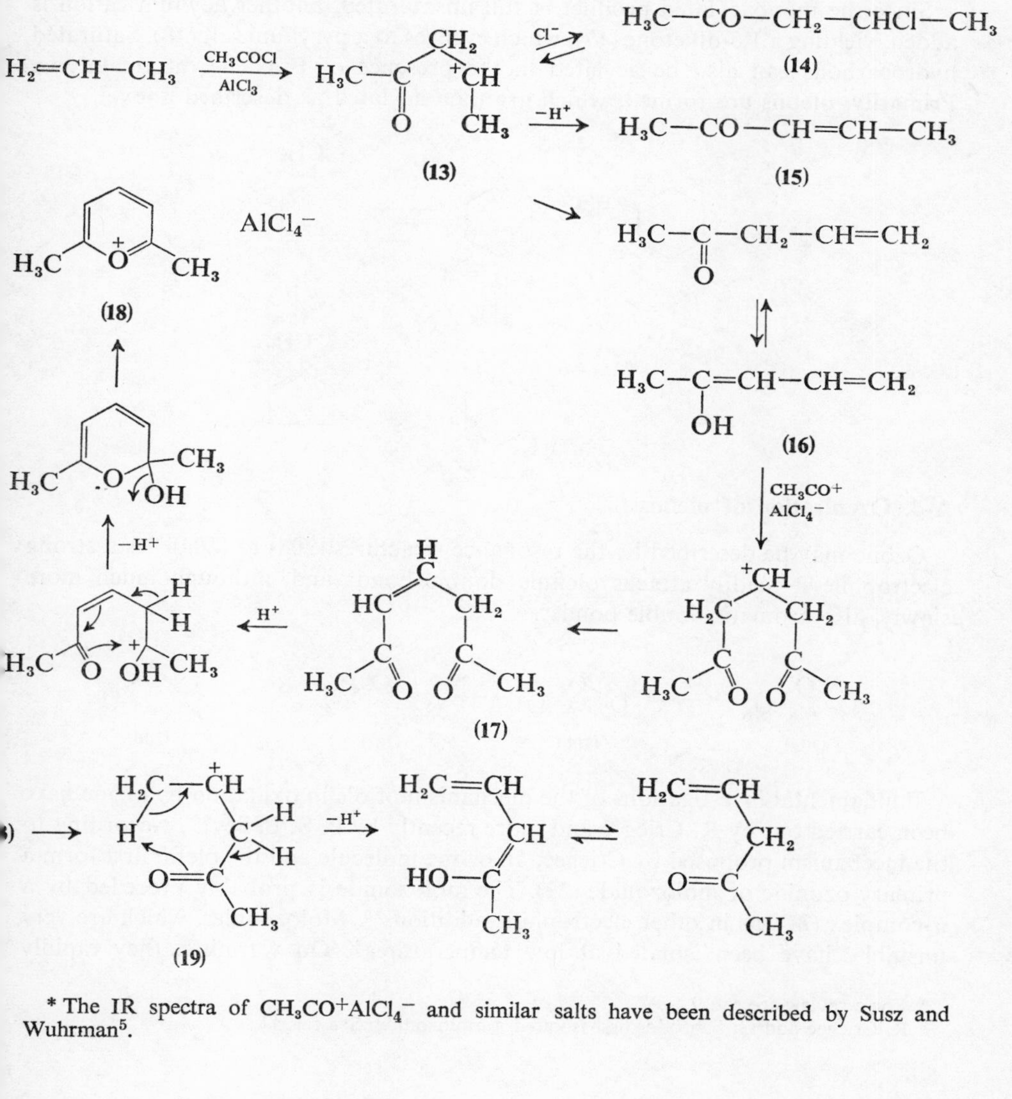

---

* The IR spectra of $CH_3CO^+AlCl_4^-$ and similar salts have been described by Susz and Wuhrman[5].

The β-keto carbonium ion (**13**), which is formed in the first step of the reaction of acetyl chloride with propene, is stabilized by addition of a chloride ion (**14**) or by loss of a proton, (**15**) and (**16**). At low temperatures, (**14**) may be isolated; at higher temperatures only (**15**) and (**16**) are obtained. If the reaction is carried out under conditions that do not permit equilibration of the isomers, the non-conjugated isomer (**16**) is formed predominantly. The faster elimination of a terminal proton may be explained by an energetically favorable cyclic transition state (**19**), in which the carbonyl group assists the proton abstraction.

Since the monoacylated product is still unsaturated, another acylium cation is added, yielding a 1,5-diketone (**17**), which cyclizes to a pyrylium salt (**18**). Saturated hydrocarbons can also be acylated in the presence of Friedel–Crafts catalysts. Primarily, olefins are formed, which are then acylated as described above[7].

A-4. Ozonization of olefins

Ozone may be described by the resonance structures (**20a**) to (**20d**). As a strong electrophile it readily attacks olefinic double bonds and, although much more slowly, also aromatic double bonds[8].

Fundamental investigations of the mechanism of olefin oxidation by ozone have been carried out by R. Criegee and more recently by P. S. Bailey[9]*. According to the mechanism proposed by Criegee, an ozone molecule and the olefin first form a primary ozonide or molozonide (**22**). The molozonide is probably preceded by a π-complex (**21**), as in other electrophilic additions[10]. Molozonides, which are very unstable, have been isolated at low temperatures†. On warming, they rapidly

---

* For a review see ref. 9.
† R. Criegee and G. Schröder first isolated a molozonide. See ref. 11.

decompose to a carbonyl compound and a peroxidic zwitterion (23), which is converted into more stable products in several ways.

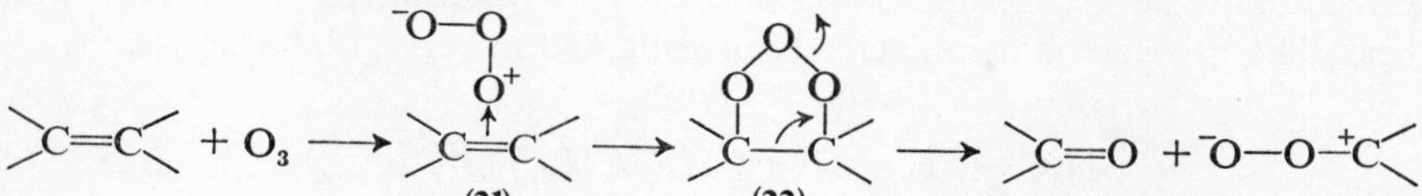

The zwitterion (23) may react with solvent molecules, for example, with methanol, to form an ether hydroperoxide (24). In inert solvents it may react with the carbonyl compound which arises from the decomposition of (22), to yield the ozonide (25). In general, (25) is only formed with carbonyl groups of aldehydes. In those cases, in which ketones are formed from (22), the zwitterion (23) yields exclusively dimeric, trimeric, or polymeric peroxides (26) or (27). By addition of an aldehyde, for example, of formaldehyde, the zwitterion can be trapped.

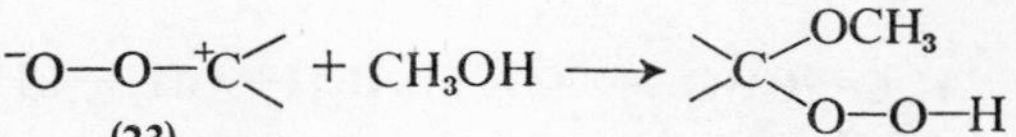

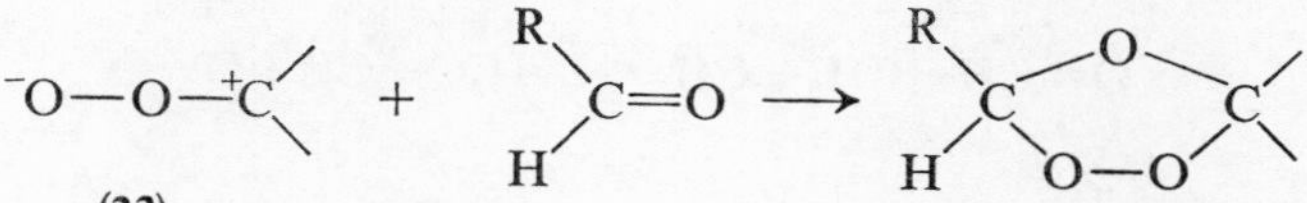

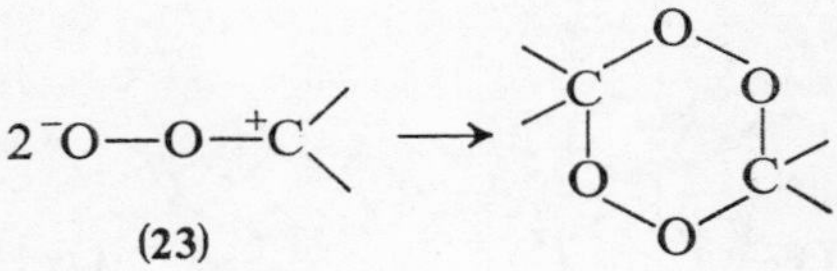

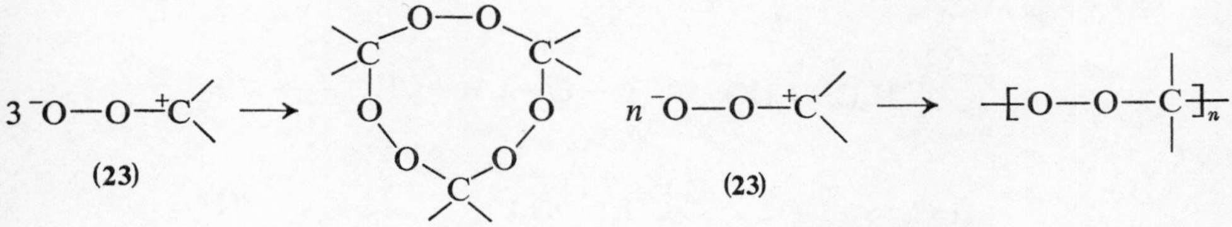

Ozonization of an unsymmetrical olefin such as 2-pentene should yield three normal ozonides (28), (29), and (30). Recently it was shown that these ozonides are actually formed[12]. If steric interactions of the substituents are insignificant, the ozonides (28), (29), and (30) should consist of approximately equal amounts of cis- and trans-isomers (a and b). It has been found, however[13], that the ratios of cis- to trans-ozonides are often significantly different from 1, depending on the geometry of the starting olefin. In order to accommodate these results, it has been assumed that the Criegee mechanism is only one possible reaction path among others. The problem is being actively investigated in several laboratories[13a].

$$H_3C-CH=CH-CH_2-CH_3 + O_3 \longrightarrow$$

$$H_3C-CHO + {}^-O-O-{}^+CH-CH_2-CH_3 +$$

$$O=CH-CH_2-CH_3 + H_3C-{}^+CH-O-O^-$$

$$H_3C-CHO + {}^-O-O-{}^+CH-CH_2-CH_3 \longrightarrow$$

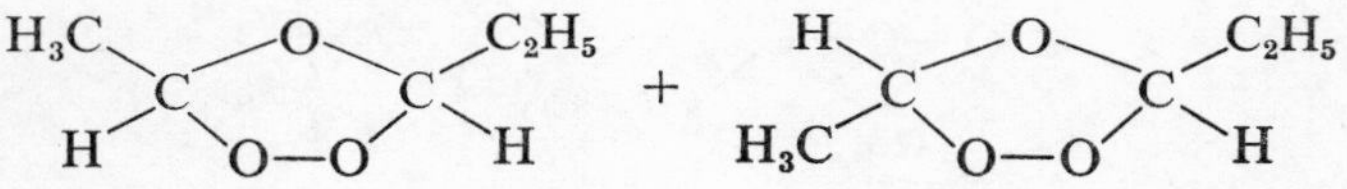

+

(28a)   (28b)

$$H_3C-CHO + {}^-O-O-{}^+CH-CH_3 \longrightarrow$$

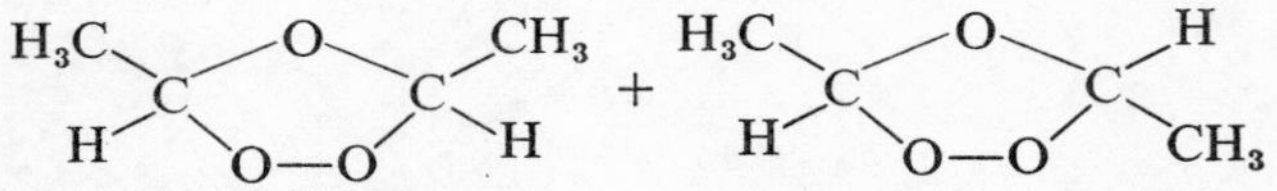

(29a)   (29b)

$$H_5C_2-CHO + {}^-O-O-{}^+CH-C_2H_5 \longrightarrow$$

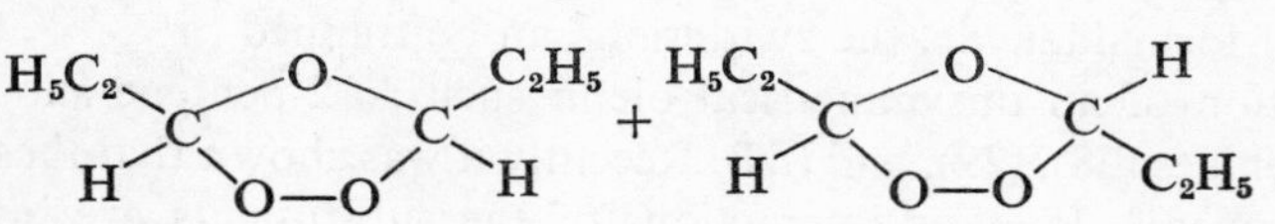

(30a)   (30b)

A-4. Ozonization of indene in liquid ammonia[14]

$$\text{(indene)} \xrightarrow{O_3} \text{(31)} \xrightarrow{NH_3} \text{(32)} \longrightarrow$$

(33) $\xleftarrow{NH_3}$ → (intermediate) $+ \text{HO}\leftarrow{}^+NH_3$ →

(34) $+ NH_2OH \xrightarrow{-H_2O}$ (35) 62%

The products obtained in the ozonization of indene in liquid ammonia can be satisfactorily interpreted by the Criegee mechanism. Zwitterion (31) yields the hydroperoxide (32) by reaction with NH_3. NH_3 reduces (33) to intermediate (34), which by loss of water forms isoquinoline (35). Nucleophilic attack of NH_3 at an oxygen atom of (33) presumably takes place in the direction of the O–O axis as indicated.

A-5. Addition of Cl_2 and Br_2 to butadiene

Addition of chlorine to conjugated dienes may occur in the 1,2- or the 1,4-position. In many cases the 1,2-adduct is preferentially formed at room temperature. With rising temperature, however, the proportion of 1,4-adduct increases. If reaction conditions are maintained that allow equilibration of the 1,2- and 1,4-adducts, for example, if $ZnCl_2$ is added, the 1,4-adduct predominates even at room temperature. Consequently, the faster formation of the 1,2-adduct in the absence of catalyst is a kinetically controlled reaction, *i.e.* the free activation energy of 1,2-addition is lower than that of 1,4-addition. However, the 1,4-adduct is thermodynamically more stable than the 1,2-adduct. Sufficiently long reaction times or the presence of a catalyst leads to the thermodynamic equilibrium, in which the 1,4-adduct predominates (cf. Figure 1, p. 68).

In the first step of the addition of Cl_2 or Br_2 to butadiene, a carbonium ion (38) is formed, but this may be preceded by formation of a π-complex (36) or (37)*. Since, however, the 1,4-addition yields *trans*-1,4-dihalogeno-2-butene[15] and (37) would lead to a *cis*-1,4-adduct, the classical cation (38) is probably formed before addition of Br^- occurs (see also the addition of Br_2 to isolated double bonds). The preferential formation of the *trans*-adduct from (38) instead of a 1,4-*cis*-adduct may be

* For π-complexes formed in other electrophilic addition reactions see examples A-1 to A-4. A 1,4-bromonium ion and a 1,4-chloronium ion have been prepared from 2'-bromo- and 2'-chloro-2-biphenylylamine as shown on p. 68.

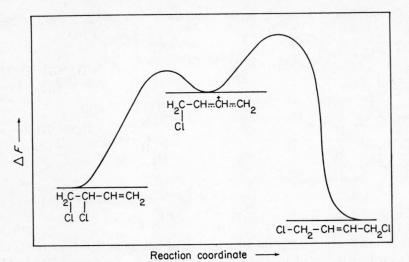

Figure 1.

explained by the weaker steric and electrostatic interactions between the halogen atoms in the former.

$$H_2C=CH-CH=CH_2 \xrightarrow{Br_2} \left[\underset{(36)}{H_2\overset{+}{C}\!=\!=\!CH-CH=CH_2} \text{ or } \underset{(37)}{\begin{array}{c}Br\\ C\!=\!=\!C\\ \overset{+}{\parallel}\\ C\!=\!=\!C\end{array}} \right] Br^-$$

$$\longrightarrow \underset{(38)}{H_2C-CH\!=\!=\!CH\!=\!=\!CH_2}\underset{Br}{|}$$

[structure: 2-amino-2'-bromobiphenyl] $\xrightarrow{HNO_2}$ [structure: dibenzofuran-like cation with Br⁺]

A-5. *Addition of bromine to 1,2,3,4-tetraphenyl-1,3-butadiene*[16]

If Br_2 is added to the tetraphenylbutadiene, the mesomeric cation (39) is formed. In (39), C-4 is in a favorable position for electrophilic substitution at the *cis*-phenyl ring, which takes place faster than the addition of Br^-. Loss of HBr leads to product (40).

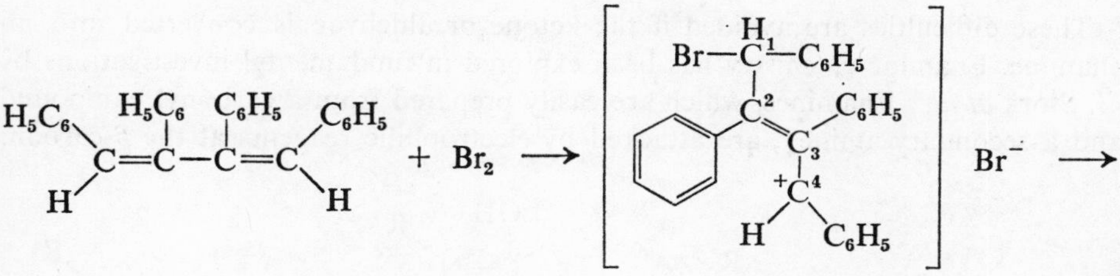

Addition Reactions

[Reaction scheme (39): H₅C₆-CH=C(C₆H₅)-C(C₆H₅)=CH-C₆H₅ + Br₂ → brominated intermediate → Br⁻ →]

[Structure (40): 70%]

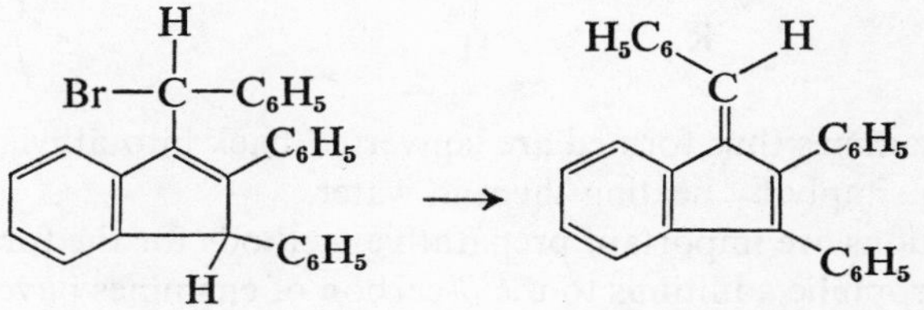

An analogous reaction[17] has been observed in the chlorination of tetraphenyl-butadiene with $SOCl_2$.*

A-6 & A-7. Alkylation of enamines with electrophilic olefins

α-Alkylation of a ketone can be achieved by converting the ketone into an enolate anion by a strong base. The enolate is then alkylated by an alkyl halide at the α-carbon atom if proper conditions are observed. In a competing reaction, alkylation at the enolate oxygen may occur†. The strongly basic reaction medium often causes additional undesirable reactions, for example, self-condensation of the ketone or polyalkylation, owing to fast proton exchange between already alkylated enolate and unchanged enolate.

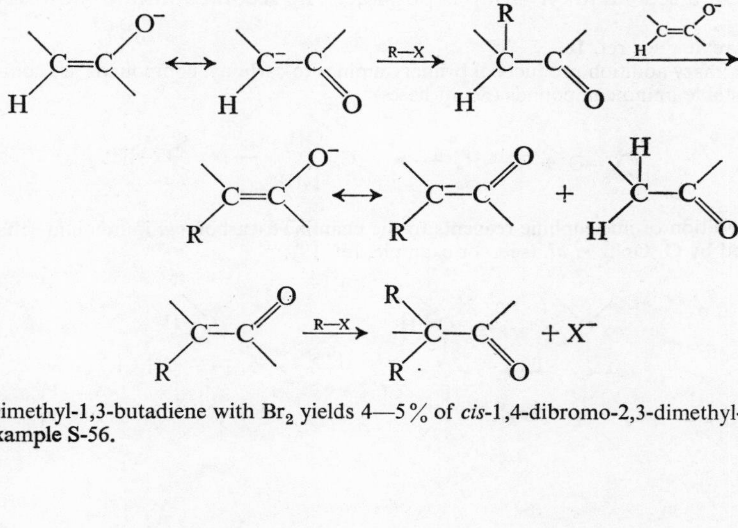

[Reaction scheme showing enolate resonance, alkylation with R-X, proton exchange, and second alkylation to give dialkylated ketone + X⁻]

---
* 2,3-Dimethyl-1,3-butadiene with Br_2 yields 4—5% of cis-1,4-dibromo-2,3-dimethyl-2-butene.
† See example S-56.

These difficulties are avoided if the ketone or aldehyde is converted into an enamine. Enamine chemistry has been explored in fundamental investigations by G. Stork et al.* Enamines, which are easily prepared from a carbonyl compound and a secondary amine†, are attacked by electrophilic reagents at the β-carbon.

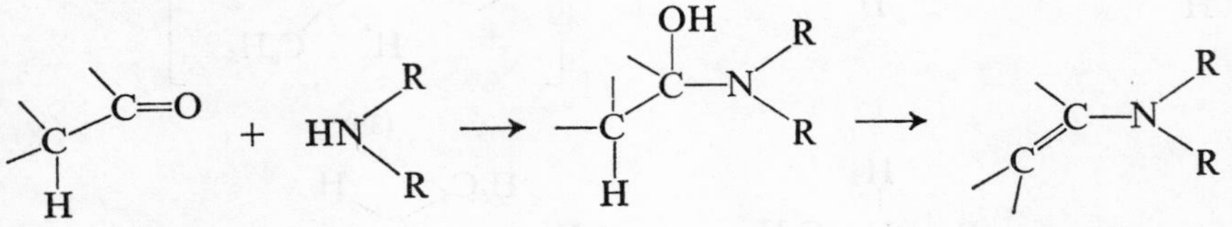

The alkylated enamines thus formed are converted back into alkylated ketones by hydrolysis, often simply by heating them in water.

Enamine reactions are important preparative methods for the formation of C–C bonds. The electrophilic additions to the β-carbon of enamines have been classified into three groups‡:

(a) reactions with electrophilic olefins, for example, with acrylonitrile;
(b) alkylations with alkyl halides; and
(c) acylations.

In the reaction of enamines with electrophilic olefins as well as with alkyl halides, N- or C-alkylation may occur. In contrast to the alkylation with alkyl halides, however, attack of electrophilic olefins on the nitrogen atom does not interfere with the course of the reaction, since the ammonium salt formation is reversible.

A-6. *Addition of acrylonitrile to cyclopentanone pyrrolidine enamine*

Irreversible proton migration in (**41**) shifts the equilibrium completely to the side of the β-carbon addition product. Since the reaction product is again an enamine, addition of a second alkyl group is possible. The second addition, however, occurs

---

* For a review see ref. 18.

† The primary addition products of primary amines to carbonyl compounds are converted into the more stable imino compounds (Schiff bases):

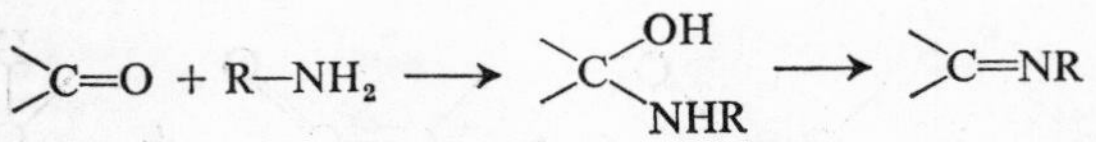

‡ The addition of nucleophilic reagents to the enamine α-carbon *via* immonium salts has been investigated by G. Opitz et al. (see, for example, ref. 19):

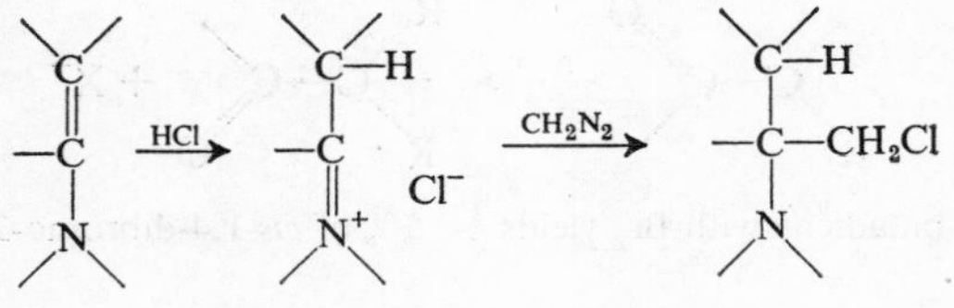

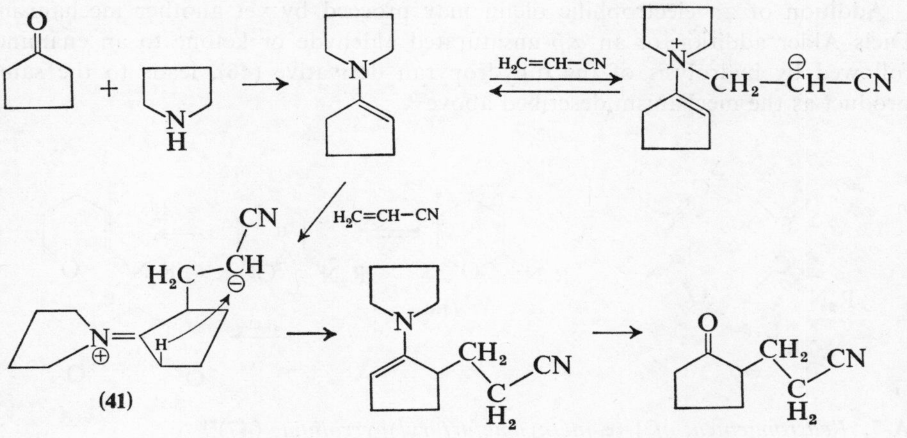

much more slowly than the primary one. The rate retardation is probably caused by the steric interaction between the first substituent and the N-alkyl group (cf. **44**). The second addition necessitates coplanarity in the transition state, because the C–N bond has partial double bond character. If R^1 is different from H, steric interaction between it and the N-alkyl group renders coplanarity more difficult and leads to a higher transition-state energy.

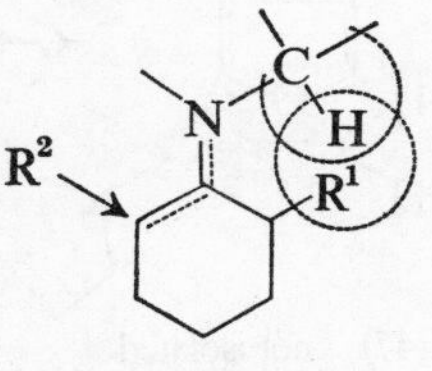

Also for steric reasons an unsymmetrically substituted ketone yields the less substituted enamine. Alkylation of this enamine leads to a ketone, which is substituted at the α- and α'-carbon atoms (**45**), in contrast to the alkylation of enolates which takes place predominantly at the side which is already substituted.

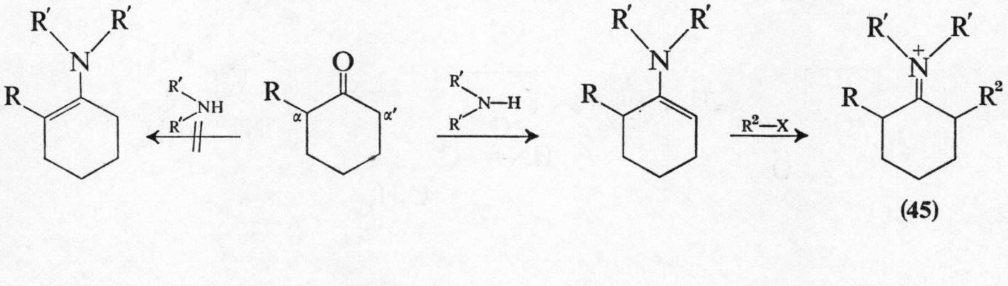

Addition of an electrophilic olefin may proceed by yet another mechanism: Diels–Alder addition of an α,β-unsaturated aldehyde or ketone to an enamine, followed by hydrolysis of the dihydropyran derivative (46), leads to the same product as the mechanism described above[20].

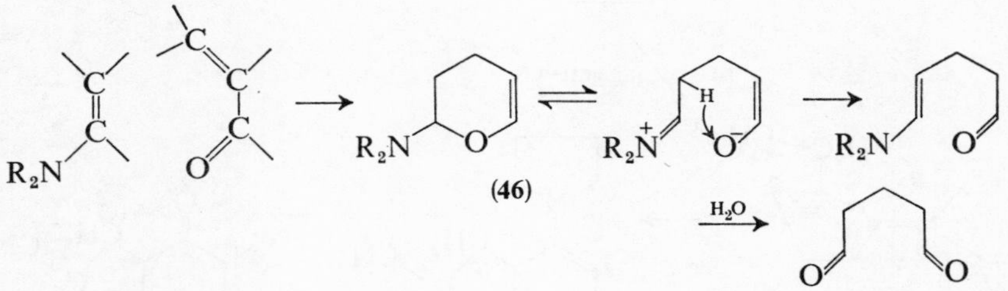

A-7. *Rearrangement of 1-(α-methylenefurfuryl)pyrrolidine (47)*[21]

The enamine (47) is obtained from α-acetylfuran and pyrrolidine. Although the aromatic furan system usually reacts as a nucleophile with acids, PCl_5, and other electrophiles with subsequent ring opening, it may react as an electrophile in sterically favorable cases.

The negative enamine carbon atom of (47) attacks the α-position of the furan ring; the product (48) rearranges to *o*-pyrrolidinophenol (49). Nucleophiles and electrophiles attack the furan system preferentially at the α-carbon.

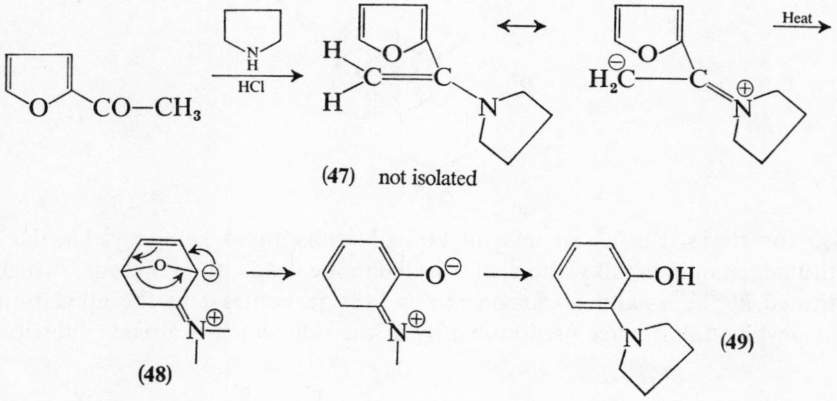

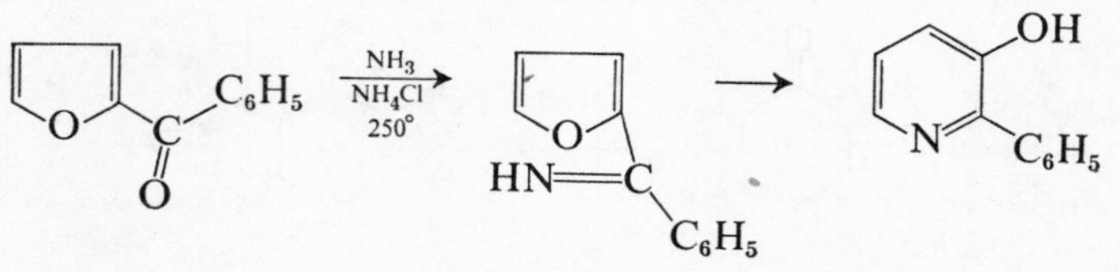

A similar rearrangement has been observed in the reaction of α-benzoylfuran with ammonia as shown at the foot of p. 72.

Acetylfuran itself rearranges to some extent, presumably *via* its enol.

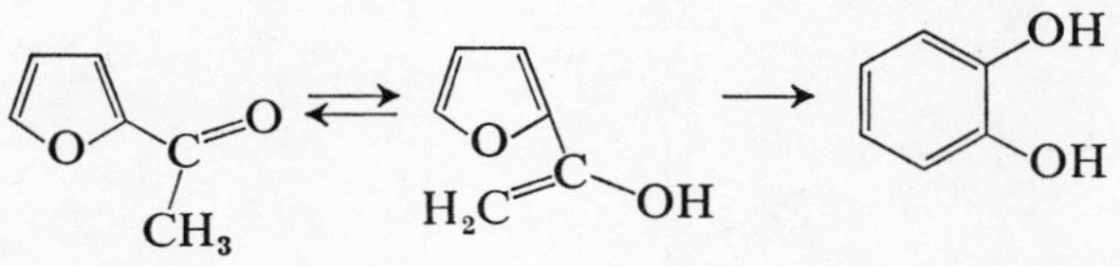

A-8. Formation of cyclobutane derivatives from enamines[22]

Intermediate (**41**) is stabilized by a proton shift across a six-membered cyclic transition state. In aldehyde enamines such a proton is not available. Intermediate (**50**), which is analogous to (**41**), can rearrange to a more stable product only by migration of a proton from the β-carbon atom *via* a four-membered cyclic transition state or by formation of a cyclobutane derivative. Formation of cyclobutanes takes place exclusively if the β-proton is replaced by an alkyl group as in (**51**).

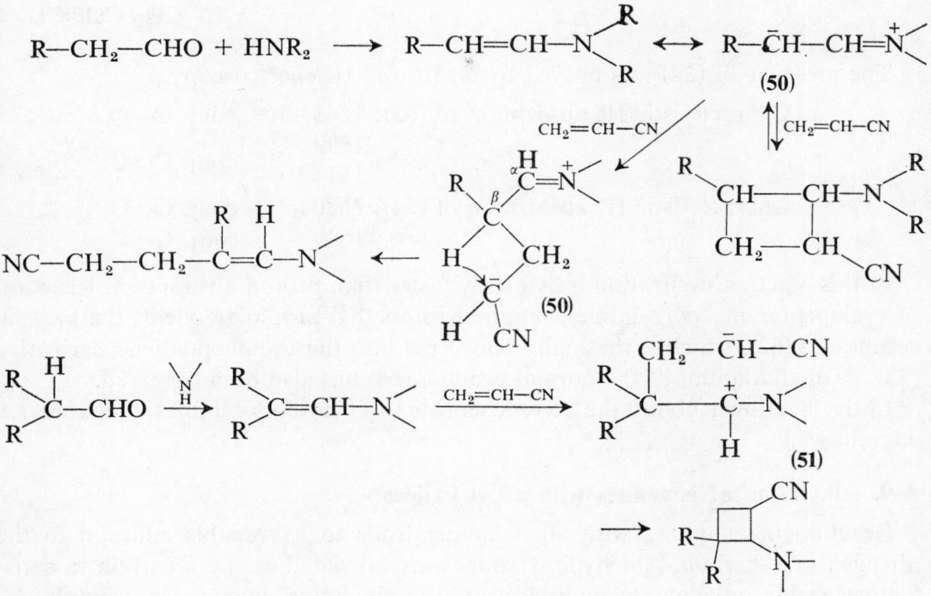

Four-membered rings are readily formed in the reaction of enamines with acetylenes.

Addition of methyl propiolate to cyclopentanone–pyrrolidine enamine[22]

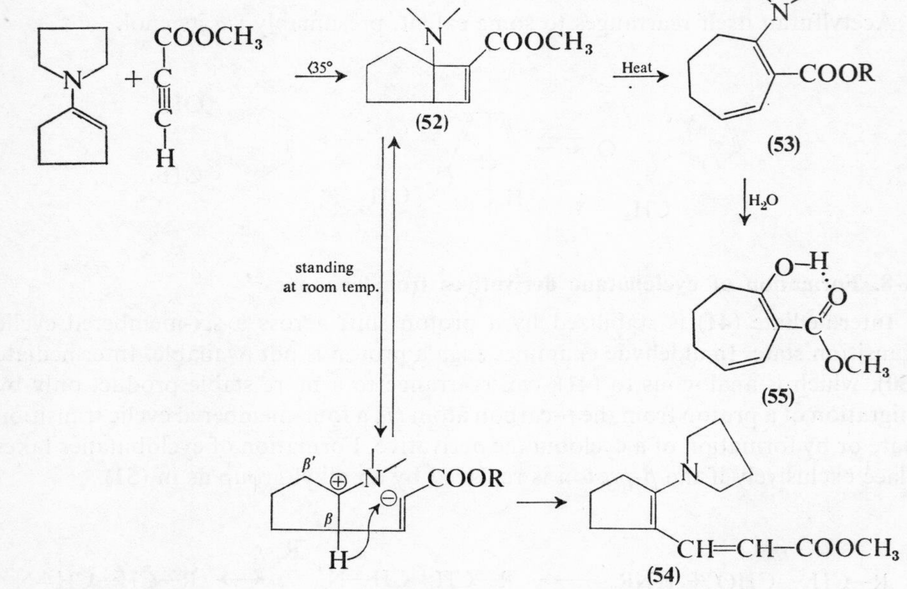

The presence of (54) was proved by NMR and IR spectroscopy.

Characteristic IR absorption of (53): 1666 cm^{-1} conj. C=O
1599
1514 } conj. C=C

Characteristic IR absorption of (54): 1680 conj. C=O
1549 conj. C=C

In this reaction cyclization is evidently faster than proton abstraction. Reaction of cyclopentanone–pyrrolidine enamine with methyl propiolate yields the bicyclic compound (52), which is thermally converted into the cycloheptadiene derivative (53). A small amount of the normal product (54) has also been observed.

Little is known about the stereochemistry of the cyclobutanes prepared *via* enamines[23].

A-9. Alkylation of enamines with alkyl halides

Reaction of enamines with alkyl halides leads to irreversible addition to the nitrogen or β-carbon. Aldehyde enamines are alkylated at the β-carbon in satisfactory yields only by strongly electrophilic alkylating agents, for example, by benzyl halides or α-halo-nitriles and -ketones.

Ketone enamines, however, are also alkylated by methyl, ethyl, or propyl iodide. In some cases, rearrangement of the *N*-addition product leads to the *C*-addition product. For example, in the reaction of cyclopentanone–pyrrolidine enamine with

2-butenyl bromide, 20% of the rearranged product (**56**) and 80% of the C-alkylation product (**57**) are formed[24].

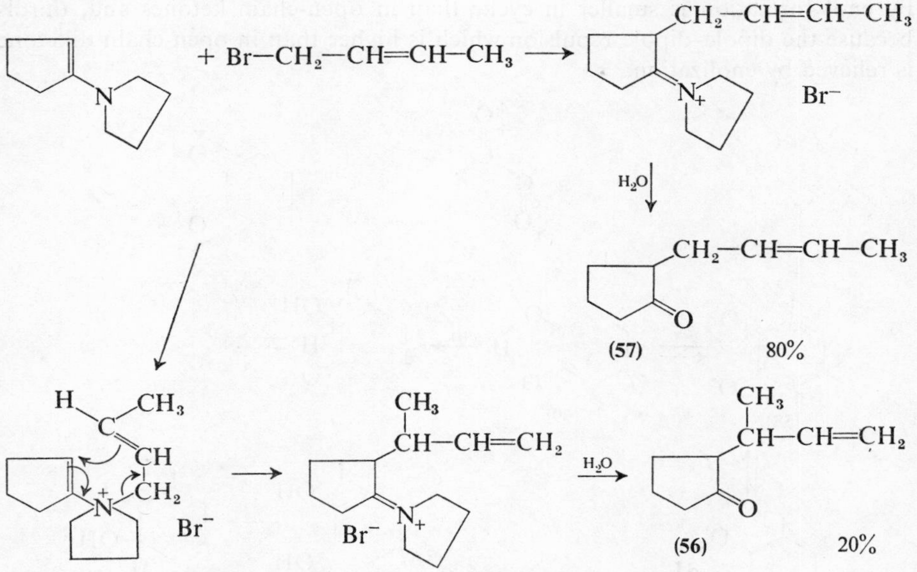

Characteristic IR absorption of (**56**): 1730 cm^{-1} C=O
1630—40 C=C
998 C—H^1
910 C(H^2)(H^2), C=C(H^1)(H^2)(H^2)

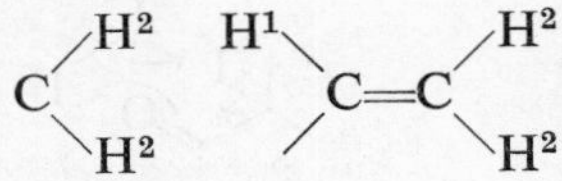

Characteristic IR absorption of (**57**): 1730 C=O
1630—40 C=C
965 (H)C=C(H)

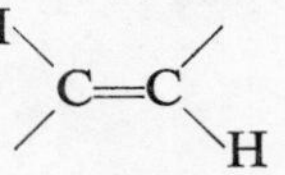

Ketone enamines are acylated at the β-carbon in good yields. A method for the preparation of long-chain dicarboxylic acids is based on this reaction[25].

A-10. Addition of hydrogen iodide to 3,5,5-trimethyl-1,2-cyclohexanedione 2,3-monoenol (59)[26]

Hydrogen iodide reduces α-ketols and α-diketones at reflux temperature to monoketones. The reduction of α-diketones probably proceeds *via* α-ketols.

In the first step of the reduction of (**58**), I$^-$ probably adds to the enol double bond of (**59**), which is in equilibrium with (**58**).

3,5,5-Trimethyl-1,2-cyclohexanedione (**58**) is almost completely enolized, like the unsubstituted 1,2-cyclohexanedione. The enol is favoured because an α,β unsaturated carbonyl system is formed, and also because loss of entropy caused by enol formation is smaller in cyclic than in open-chain ketones and, thirdly, because the dipole–dipole repulsion which is higher than in open-chain diketones is relieved by enolization.

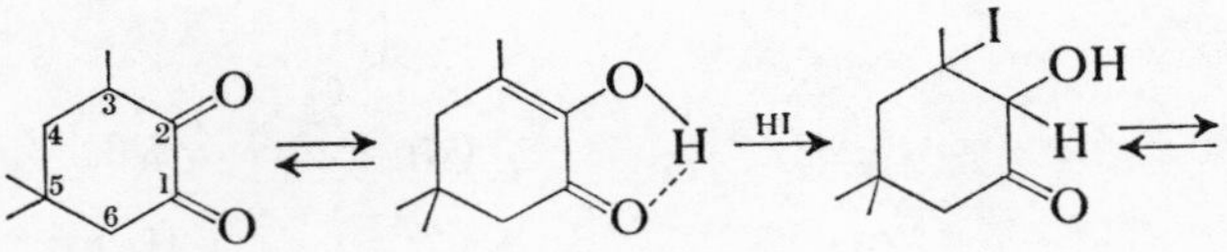

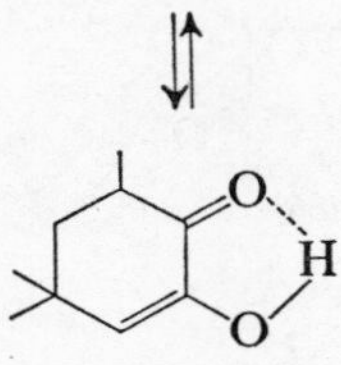

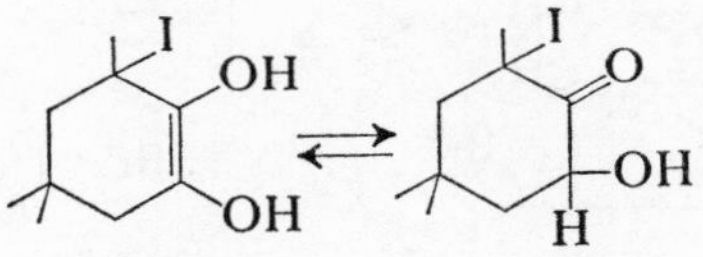

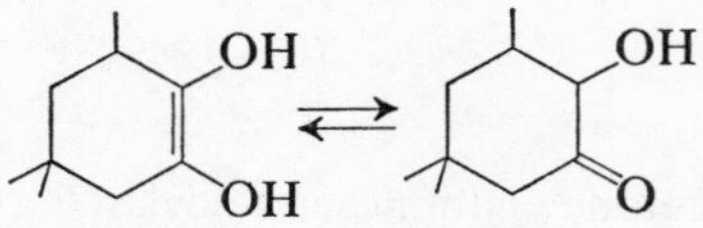

Characteristic IR absorption of **(64)**: 1708 cm^{-1} C=O
Characteristic IR absorption of **(66)**: 1680   C=O
                                         1640   C=C

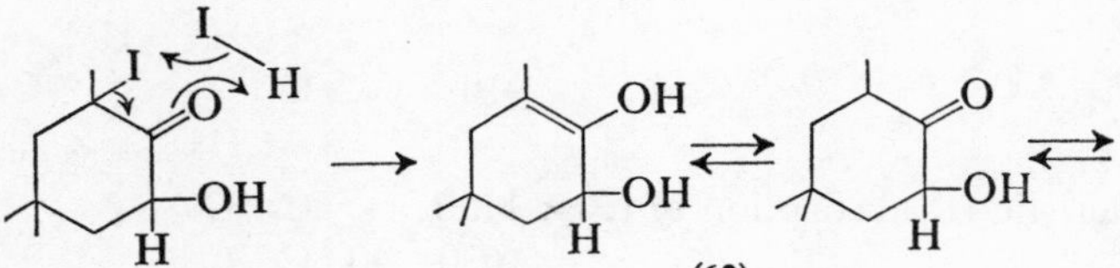

As a consequence of the inductive effect of the 3-methyl group, enol **(59)** is favored over enol **(60)***. An excess of HI reduces **(61)**, possibly by a concerted mechanism, to the α-ketol **(62)**[27].

* For reactions of ambident ions see p. 347.

Addition Reactions

A second possible mechanism starts with direct addition of HI to one of the C=O groups:

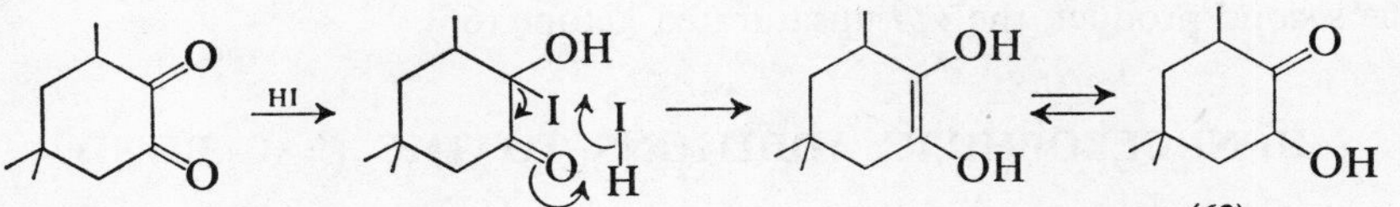

[Scheme showing (58) → intermediate → (62)]

[Second scheme showing analogous addition → (62)]

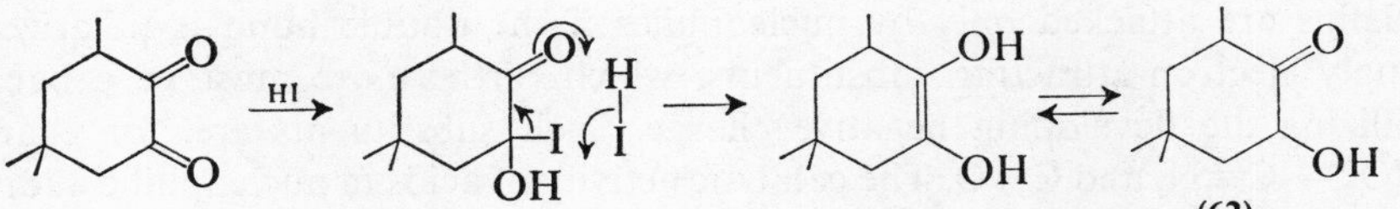

Reduction of the α-ketol (62) to ketone (64) takes place, according to the authors, *via* an allylic carbonium ion (63). Since (63) is more stable than the allylic

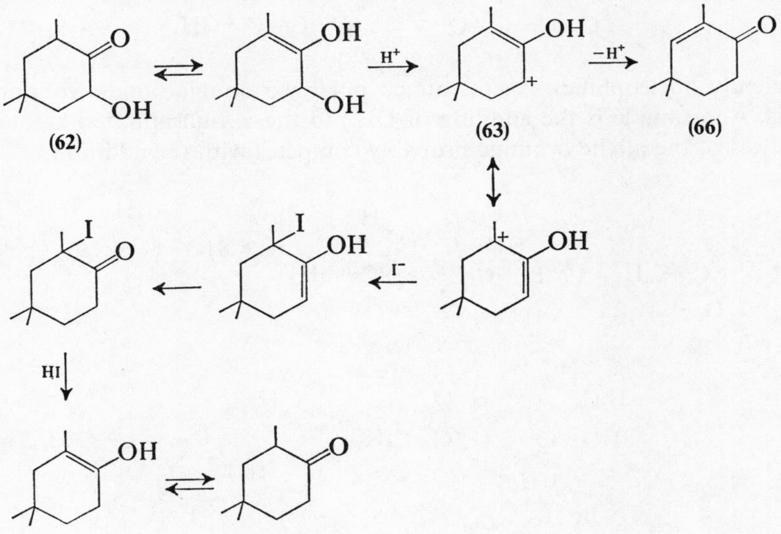

[Scheme: (62) ⇌ enediol → (63) → (66); (63) resonance structure → I⁻ addition → iodo intermediate → (with HI) → enol ⇌ (64)]

[Scheme: (62) ⇌ tautomer ⇌ enediol → H⁺ → (65) resonance structures]

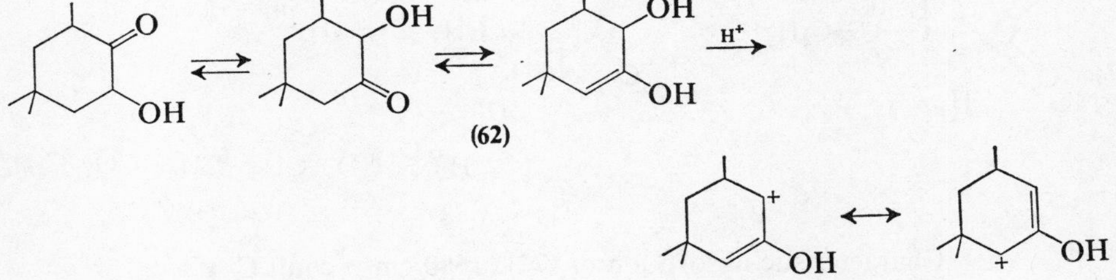

carbonium ion **(65)** [one of the resonance structures of **(63)** is a tertiary carbonium ion], ketone **(64)** should be predominantly formed. Loss of a proton from **(63)** leads to the second product, the α,β-unsaturated ketone **(66)**.

II. NUCLEOPHILIC ADDITIONS TO THE C=C BOND

A-11. Addition of OH⁻ to *trans*-γ-bromodypnone (67)[28]

Olefins are attacked only by nucleophiles if the double bond is polarized by strongly electron-attracting substituents, which furthermore must be capable of stabilizing the developing negative charge. Such substituents are, for example, $-NO_2$, $-C \equiv N$, and $C=O$. The catalytic activity of acids in nucleophilic additions is due to protonation of the electron-attracting group, whose effect is thus considerably increased.

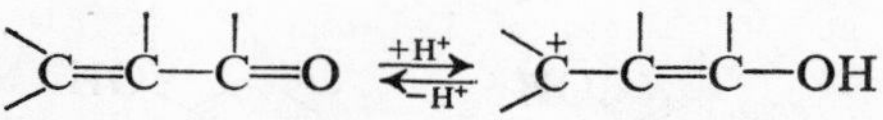

$$\text{>C=C-C=O} \underset{-H^+}{\overset{+H^+}{\rightleftharpoons}} \text{>}\overset{+}{\text{C}}\text{-C=C-OH}$$

Sufficiently nucleophilic reagents attack polarized double bonds without acid-catalysis. An example is the addition of OH⁻ to the α,β-unsaturated ketone **(67)**. (Hydrolysis of the allylic bromine probably competes with the addition.)

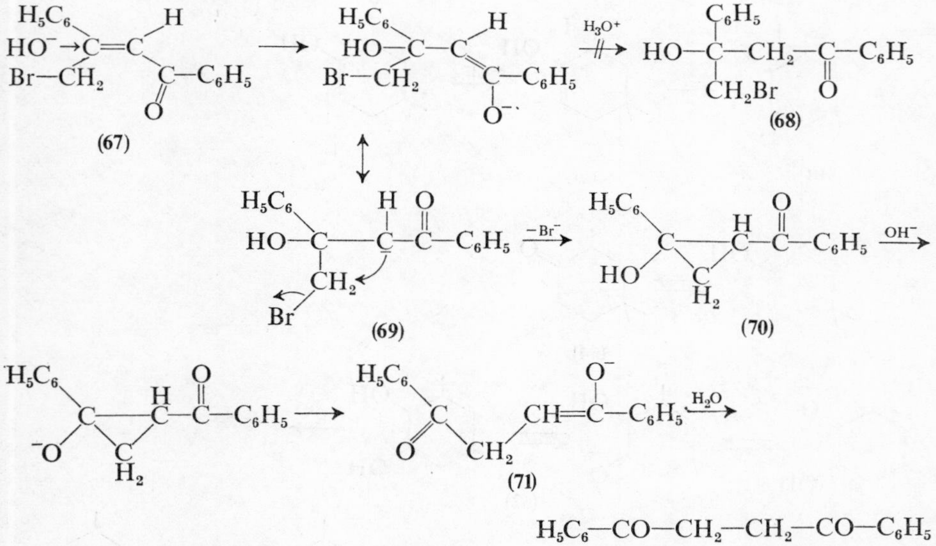

$$H_5C_6-CO-CH_2-CH_2-CO-C_6H_5$$
(72) ~50%

Characteristic absorption of **(72)**: 1680 cm⁻¹ conj. C=O
1599
1448 } arom. C=C

In this reaction dibenzoylethane (**72**) is obtained in approximately 50% yield instead of the normal addition product (**68**). The reaction probably proceeds by conversion of enolate (**69**) into the cyclopropyl derivative (**70**), in analogy with the Favorsky rearrangement*. Step (**70**) → (**71**) conforms to a retroaldol condensation. In (**71**), however, the carbonyl fragments remain linked to each other by a methylene group†.

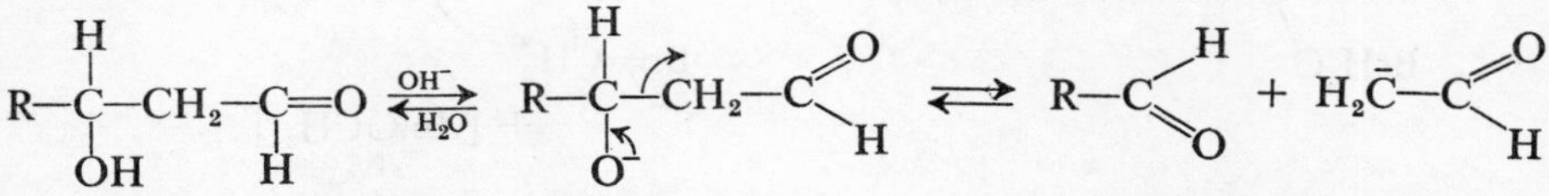

$$R-\overset{H}{\underset{OH}{C}}-CH_2-\overset{H}{C}=O \underset{H_2O}{\overset{OH^-}{\rightleftarrows}} R-\overset{H}{\underset{O^-}{C}}-CH_2-C\overset{O}{\underset{H}{}} \rightleftarrows R-\overset{H}{C}\overset{O}{\underset{O}{}} + H_2\bar{C}-C\overset{O}{\underset{H}{}}$$

A related rearrangement has been observed in the reaction of epoxide (**73**) with Zn or other metals[30].

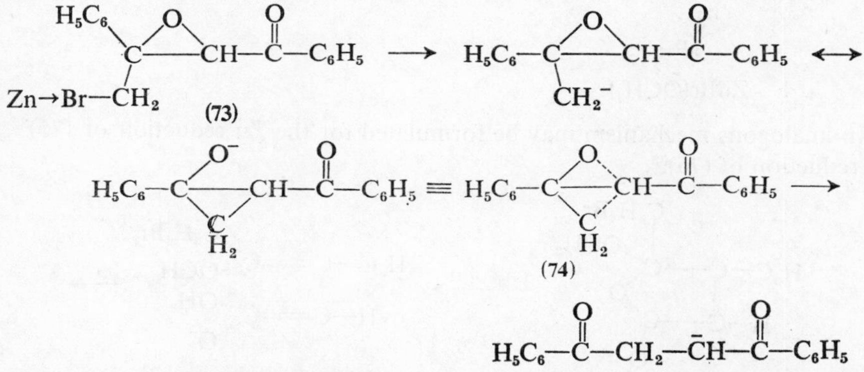

(**73**) → ... (**74**) → $H_5C_6-\overset{O}{\overset{\|}{C}}-CH_2-\bar{C}H-\overset{O}{\overset{\|}{C}}-C_6H_5$

---

* In some cases of the Favorsky rearrangement a substituted cyclopropanone is formed as an intermediate:

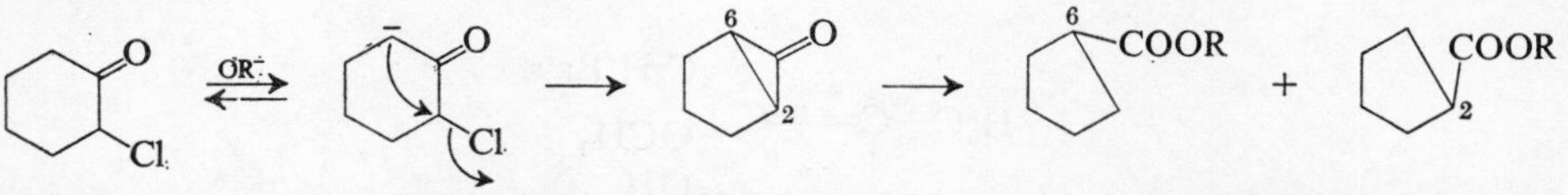

Or the rearrangement may occur by a different mechanism, very similar to the benzilic acid rearrangement:

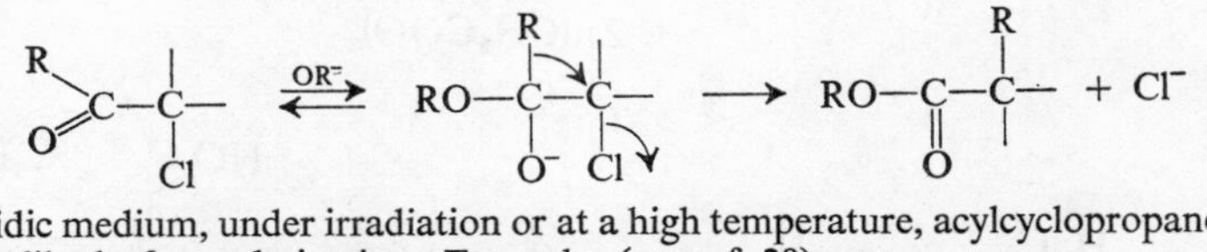

$$\underset{Cl}{\overset{R}{\underset{}{\overset{}{O=C-C-}}}} \overset{OR^-}{\rightleftarrows} RO-\overset{R}{\underset{O^-}{C}}-\overset{}{\underset{Cl}{C}}- \rightarrow RO-\overset{R}{\underset{O}{C}}-\overset{}{\underset{}{C}}- + Cl^-$$

† In an acidic medium, under irradiation or at a high temperature, acylcyclopropane derivatives rearrange to dihydrofuran derivatives, Examples (see ref. 29) are:

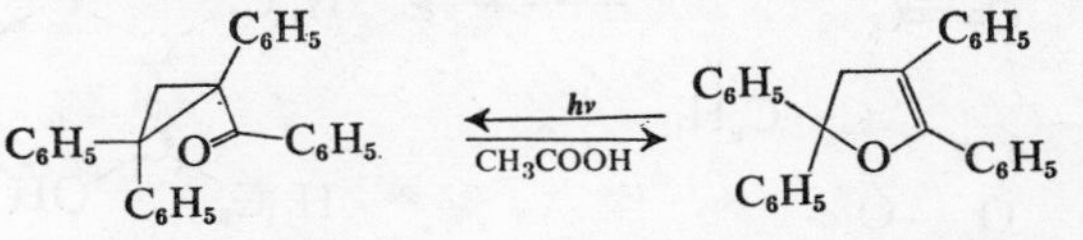

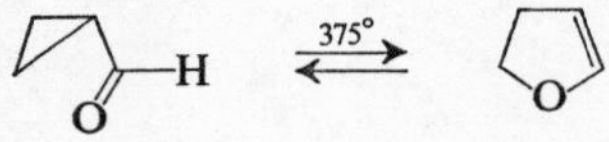

A reaction mechanism which avoids the highly strained transition state (74) has been proposed by R. E. Lutz et al.[28,31]

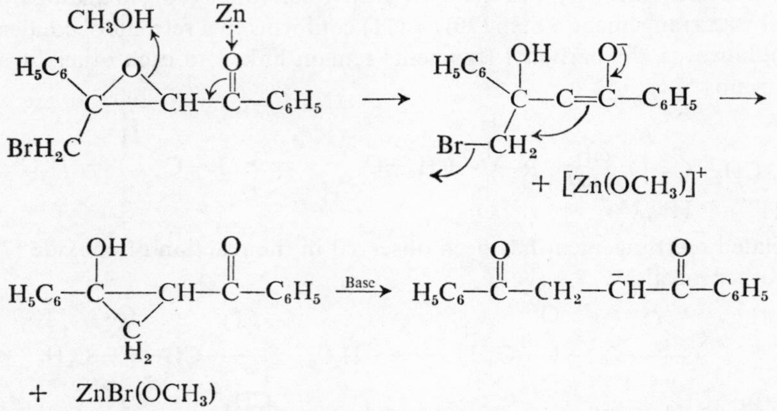

[Scheme showing: epoxide with CH₃OH and Zn → intermediate with OH, C=C, Br-CH₂ and [Zn(OCH₃)]⁺ → cyclopropane intermediate with OH → (Base) → H₅C₆-C(O)-CH₂-CH⁻-C(O)-C₆H₅ + ZnBr(OCH₃)]

An analogous mechanism may be formulated for the Zn reduction of (75)[32] or the reduction of (76)[33].

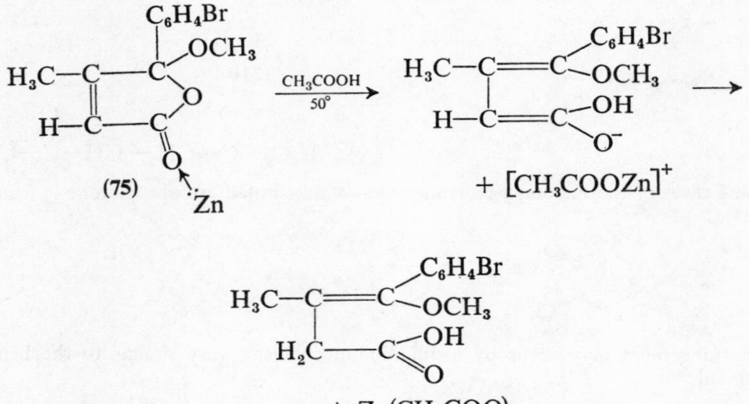

[Scheme for (75): four-membered ring with C₆H₄Br, OCH₃, O, Zn → CH₃COOH, 50° → open chain enol form with [CH₃COOZn]⁺ → final product + Zn(CH₃COO)₂]

[Scheme for (76): H₁₁C₉-C(=C-H)-C(=O)-... with H₁₁C₉ groups → Zn-CH₃COOH → diol product with HO, C₉H₁₁, H₁₁C₉, OH]

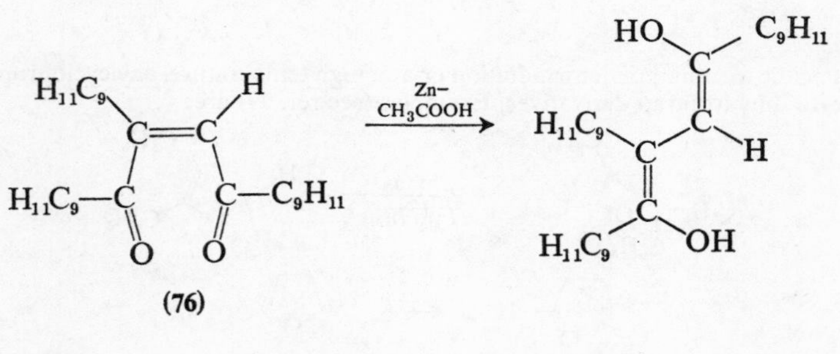

A-12 & A-13. Addition of lithium aluminum hydride to double bonds

In addition of the hydride ion to double bonds, complex hydrides such as $NaAlH_4$ or $NaBH_4$ serve as hydrogen-transfer agents. The nature of the attacking species and the rate of reduction are strongly dependent on the solvent[34]. In the first step, H^- adds to the centre of lowest electron density. The negative charge, which is formed during the addition at a carbon atom or a hetero atom of the double bond, is stabilized by complex-formation with neutral AlH_3. The newly formed complex anion continues to donate H^-, until it is converted into tetraalkylaluminate (77) or tetraalkoxyaluminate (78).

$$\text{>C=C<} + [AlH_4]^- \longrightarrow \text{>C(H)-C<}^- \xrightarrow{AlH_3} \text{>C(H)-C(AlH_3)<} \longrightarrow$$

$$[Al(-\overset{|}{\underset{|}{C}}-\overset{|}{\underset{|}{C}}-)_4]^- \quad (77)$$

$$\text{>C=O} + [AlH_4]^- \longrightarrow \text{>C(H)-O}^- \xrightarrow{AlH_3} \text{>C(H)-O-AlH_3}^- \longrightarrow [Al(OR)_4]^- \quad (78)$$

Esters are reduced to aldehydes by nucleophilic substitution in the first reaction step.

$$\underset{\underset{H^-}{\uparrow}}{R-\overset{\overset{O}{\|}}{C}-OR'} \longrightarrow R'-O^- + R-C\overset{\nearrow O}{\underset{\searrow H}{}}$$

A-12. *Addition of* $LiAlH_4$ *to tropone* (79)[35]

The particular arrangement of the double bonds in (79) leads to 1,8-addition of $LiAlH_4$ and Grignard reagents. Both reagents also add to the ends of α,β-unsaturated carbonyl compounds, yielding β-substituted ketones. Complex (80), which on hydrolysis affords (83), is presumably in equilibrium with (81). The C=O group in (81) is reduced by a second molecule of $LiAlH_4$ to give the alcohol (82). (See p. 82.)

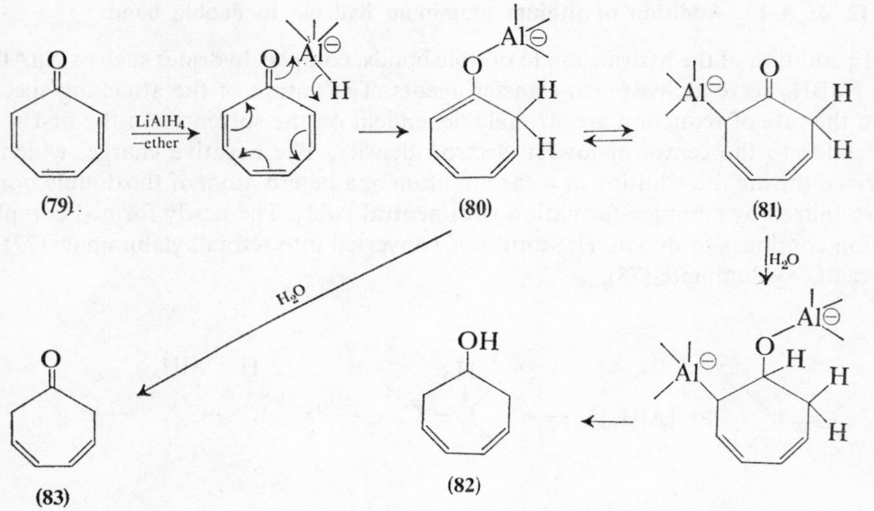

A-13. Addition of LiAlH₄ to norbornadien-7-ol (85)[36]

At temperatures around 100° the C=C double bond of α-olefins can be smoothly reduced[37] by LiAlH₄. At room temperature, however, C=C double bonds are attacked by LiAlH₄ only if the developing negative charge is stabilized by conjugation, as, for example, in the reduction of dimethylfulvene (84), or in cases where a particularly favorable steric arrangement exists between the double bond and the attacking hydride ion. Such a special case is the reduction of norbornadien-7-ol (85). The strong base LiAlH₄ attacks primarily the OH group, forming an O–Al bond (86). In (86) H⁻ can readily migrate to the double bond. That this steric arrangement is a precondition for migration of H⁻ is evident in the inertness of *anti*-norbornen-7-ol (91) and norbornene-*syn*-7-methanol (92), whereas the *syn*-alcohol (93) is reduced. In (87) the neutral aluminum coordinates with the negative carbon atom. If the reaction is carried out in a better coordinating solvent, for example, in tetrahydrofuran, the aluminum coordinates with the solvent instead of the negative carbon atom (89). In that case, the reaction takes a different course: rearrangement of (89) leads to cycloheptatriene (90) in 80% yield.

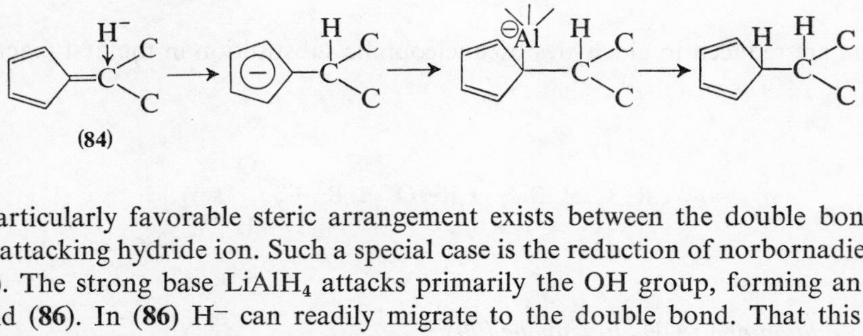

(85) (86)

(87) (88)

(89) (90) 80%

(91) (92) (93)

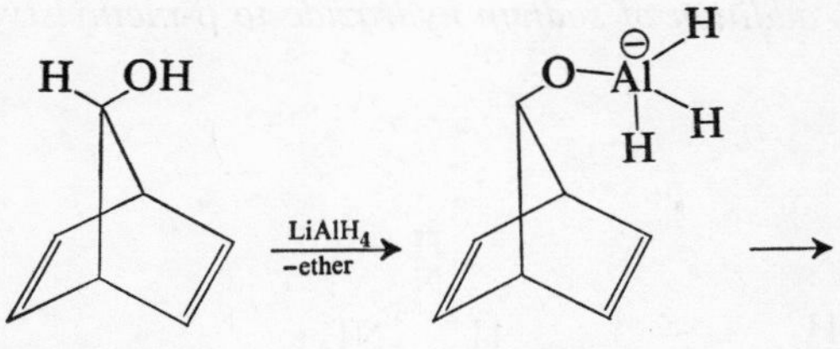

A-14 & A-15. Nucleophilic addition of sodium hydrazide to olefins*

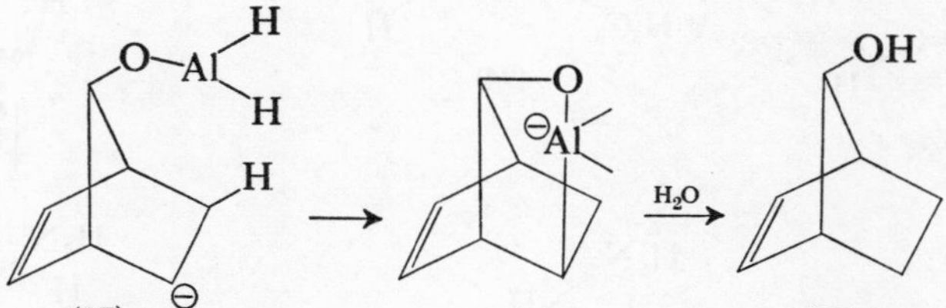

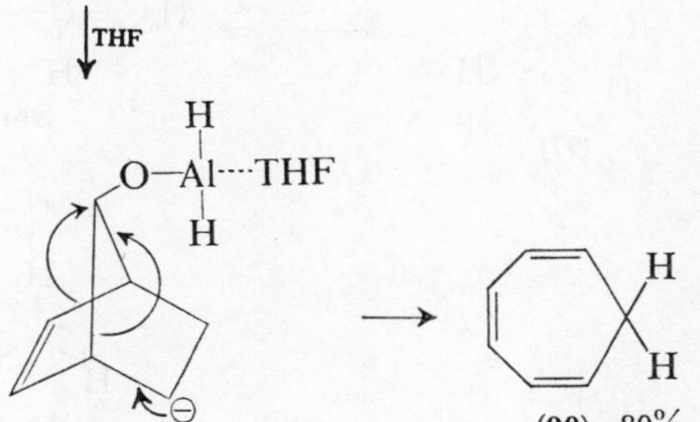

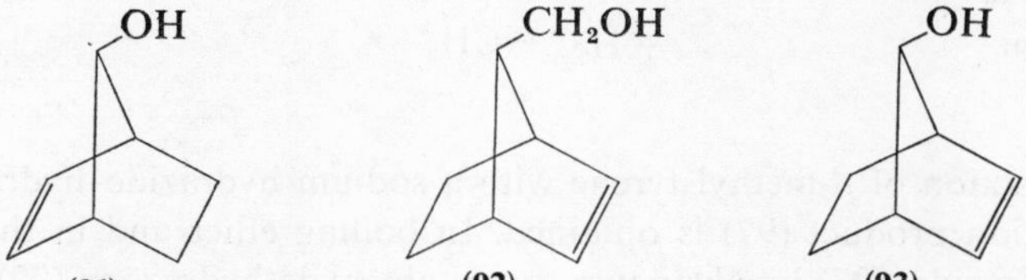

The very strong nucleophile sodium hydrazide is capable of many interesting reactions. Although it does not attack isolated double bonds, it readily reacts with aryl-substituted and other conjugated double bonds even at 0°.

* For reviews of reactions of sodium hydrazide see ref. 38.

A-14. *Nucleophilic addition of sodium hydrazide to β-methylstyrene*

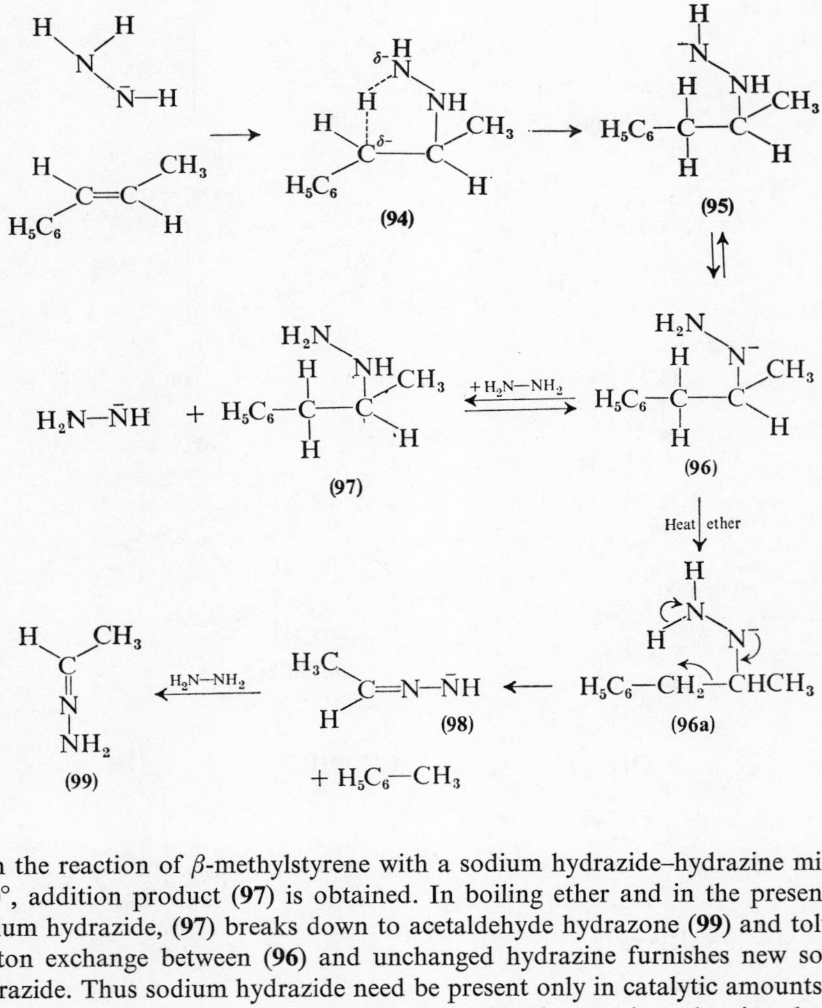

In the reaction of β-methylstyrene with a sodium hydrazide–hydrazine mixture at 0°, addition product **(97)** is obtained. In boiling ether and in the presence of sodium hydrazide, **(97)** breaks down to acetaldehyde hydrazone **(99)** and toluene. Proton exchange between **(96)** and unchanged hydrazine furnishes new sodium hydrazide. Thus sodium hydrazide need be present only in catalytic amounts.

Styrene does not polymerize anionically during the reaction, showing that free carbanions are not formed. For this reason it has been assumed that the negative carbon atom in **(94)** forms a hydrogen bond with the NH_2 group. Hydrazines that are disubstituted at one N atom do not add to olefin and this indicates that hydrogen bonding is essential for the addition to take place. Another explanation for the lack of anionic polymerization would be formation of a tight ion pair.

The complex **(94)** forms an intermediate **(96)** *via* **(95)***. The cleavage of **(97)** in boiling ether to **(98)** and toluene is characteristic of aryl-substituted olefins. It only occurs, however, if each N atom of **(97)** carries at least one hydrogen atom.

Decomposition according to the mechanism below can be excluded, since anion **(100)** is stable.

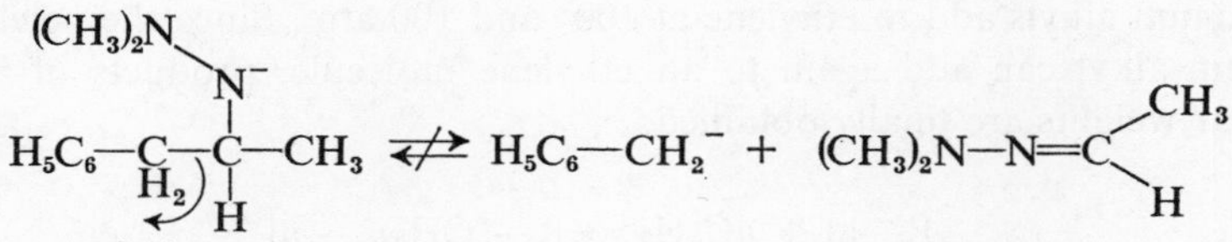

$$H_5C_6-\underset{H_2}{C}-\underset{\underset{N^-}{|}}{\underset{|}{C}}-CH_3 \rightleftarrows H_5C_6-CH_2^- + (CH_3)_2N-N=C\begin{smallmatrix}CH_3\\H\end{smallmatrix}$$
$$(CH_3)_2N$$
(100)

A-15. *Reduction of* trans-*stilbene with sodium hydrazide*

In certain cases the reaction of aryl-substituted olefins with sodium hydrazide leads to the reduction of the C=C double bond. This proceeds very readily in the presence of free hydrazine by way of **(101)** and **(102)**. Reduction predominates over

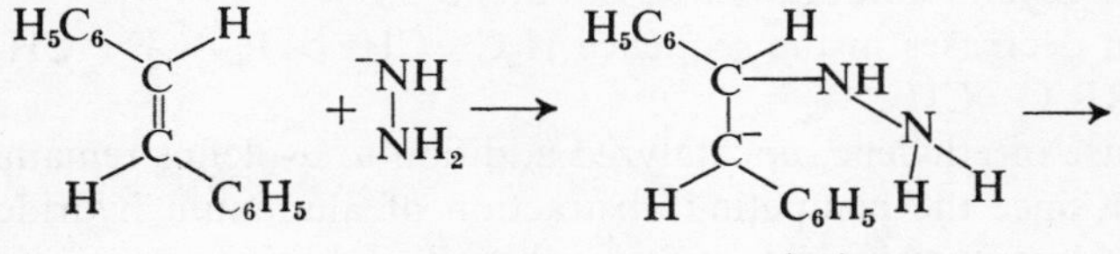

(101)

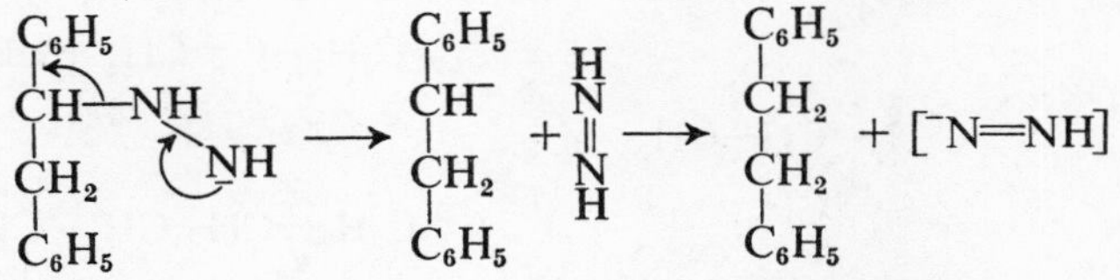

(102) → → **(103)** 90% + [⁻N=NH]

* The equilibrium **(95)** ⇌ **(96)** was confirmed by reactions of sodium methylhydrazide with pyridine and with styrene: the former gave the *N,N'*-substituted hydrazine **(100a)** but the latter gave compound **(101a)** with both substituents on the same nitrogen atom:

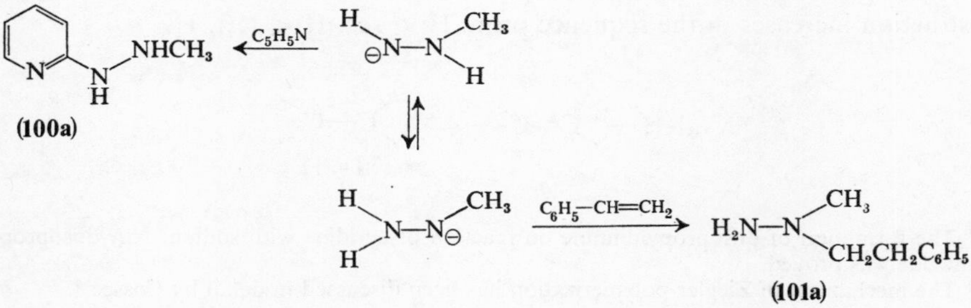

(100a) ←——C_5H_5N——

(101a)

cleavage in those cases where loss of diimine* is accompanied by formation of a resonance-stabilized carbanion. For example, *trans*-stilbene is reduced to bibenzyl (**103**) in better than 90% yield even at 0°.

A-16 & A-17. Addition of aluminum alkyls and hydrides to olefins

Aluminum alkyls add to ethylene at 100° and 100 atm. Since the newly formed aluminum alkyl can add again to an ethylene molecule, products of very high molecular weights are finally obtained.

$$\text{R—al} + n\text{C}_2\text{H}_4 \rightarrow \text{R—(CH}_2)_{2n}\text{—al}$$

The rate of reaction is enormously accelerated if catalytic amounts of compounds of transition elements are added, for example, $TiCl_4$. With these "Ziegler" catalysts ethylene is polymerized at room temperature without application of pressure†. Aluminum alkyls add not only to ethylene but also to α-olefins, internal olefins, and in the case of trimethylaluminum even to isobutene[40].

The rate of addition decreases in the sequence $H_2C{=}CH_2$ > R—CH=CH$_2$ > R—CH=CH—R > RR′C=CH$_2$.

In contrast to the case of ethylene, uncatalyzed addition to α-olefins remains at the stage of the dimer since the competing abstraction of aluminum hydride or monoalkyl- or dialkyl-aluminum hydride is faster than the addition.

$$\begin{array}{c}
\text{C}_3\text{H}_7\text{—CH—CH}_2\text{—al} \\
| \\
\text{CH}_3
\end{array}
\underset{i}{\overset{\nearrow}{\underset{\searrow}{}}}
\begin{array}{c}
\text{C}_3\text{H}_7\text{—C=CH}_2 + \text{al—H} \\
| \\
\text{CH}_3 \\
\\
\text{C}_3\text{H}_7\text{—CH—CH}_2\text{—CH—CH}_2\text{—al} \\
|| \\
\text{CH}_3\text{CH}_3
\end{array}$$

i, +CH$_3$—CH=CH$_2$

Aluminum hydrides add faster to double bonds than do aluminum alkyls. The reversible addition obeys Markownikow's rule and yields *cis*-adducts. Presumably the hydrogen and the aluminum atom add simultaneously. The ease of al–H abstraction increases in the sequence prim. H < sec. H < tert. H.

$$\text{al–H} + \text{\textbackslash C=C/} \rightleftarrows \text{\textbackslash C—C/} \atop \text{alH}$$

---

* The formation of diisopropyldiimine on reaction of acridine with sodium *N,N*-diisopropylhydrazide was proved.

† The mechanism of Ziegler polymerization has been discussed in detail by Cossee[39].

A-16. *Addition of* n-*propylaluminum to propene*

$$H_3C-CH=CH_2 + H_3C-CH_2-CH_2-al \rightleftarrows C_3H_7-\underset{\underset{CH_3}{|}}{CH}-CH_2-al$$
(104)

$$C_3H_7-\underset{\underset{CH_3}{|}}{CH}-CH_2-al \rightleftarrows C_3H_7-\underset{\underset{CH_3}{|}}{C}=CH_2 + al-H$$
(104)   (105)

$$al-H + H_3C-CH=CH_2 \rightleftarrows H_3C-CH_2-CH_2-al$$

In the first step of the dimerization of propene, addition of propylaluminum, which is present in catalytic amounts, to propene takes place. In addition product **(104)** a tertiary hydrogen atom is available, and this is readily abstracted as aluminum hydride. Addition of aluminum hydride to propene then regenerates propylaluminum. Propene, whose aluminum compound does not possess a tertiary hydrogen atom, thus liberates 2-methyl-1-pentene **(105)** from (2-methylpentyl)aluminum[41]*.

Much industrial research has been directed to oligomerization† reactions of olefins. By the use of suitable catalysts specific dimerizations of numerous olefins and dienes may be achieved.

A-17. *Isomerization of internal aluminum alkyls*

$$H_3C-CH_2-CH=CH-CH_2-CH_3 + {>}al-H \rightleftarrows$$
$$H_3C-CH_2-\underset{\underset{al{<}}{|}}{CH}-CH_2-CH_2-CH_3 \rightleftarrows$$
(106)

$$H_3C-CH=CH-CH_2-CH_2-CH_3 + {>}al-H \rightleftarrows$$
$$H_3C-\underset{\underset{al{<}}{|}}{CH}-CH_2-CH_2-CH_2-CH_3 \rightleftarrows$$

$$H_2C=CH-CH_2-CH_2-CH_2-CH_3 + {>}al-H \rightleftarrows$$
$${>}al-CH_2-CH_2-CH_2-CH_2-CH_2-CH_3$$
(107)

$${>}al-H \equiv (C_4H_9)_2Al-H$$

* For a review of aluminum alkyls see ref. 42.
† The term oligomerization denotes formation of addition products of low molecular weight by dimerization, trimerization, or tetramerization.

If diisobutylaluminum hydride is added to an internal olefin, for example to 3-hexene, and if the reaction mixture is heated for several hours, the primary aluminum alkyl (**107**) is obtained instead of the expected addition product (**106**). Presumably by a sequence of addition and elimination steps an equilibrium is obtained between all possible olefins. Since the primary aluminum alkyl is thermodynamically the most stable one*, it is obtained as the main product. This reaction, like polymerization reactions, is accelerated by transition-metal salts[43].

The uncatalyzed aluminum alkyl isomerization corresponds to the alkylboron isomerization which has been explored by H. C. Brown in exemplary investigations[44].

III. ADDITIONS TO THE C≡C BOND

A-18. Addition of bromine to 1-penten-4-yne

Halogens react with C=C double bonds by electrophilic addition†. One might expect the acetylene bond, which possesses two pairs of π-electrons, to have an even more pronounced tendency to react with electrophiles. Numerous investigations have shown, however, that the triple bond is less reactive towards electrophiles, but is attacked much faster by nucleophiles, than the C=C double bond. For example, alcohols and amines readily add to acetylenes, whereas addition of hydrogen halides is relatively slow[45].

This unexpected behavior is probably caused by the high electron density in the centre of the triple bond. The C–C distance in acetylene is by 0.14 Å shorter than in ethylene. Since the overlap integrals of π-bonds increase greatly with decreasing C–C distance, whereas those of σ-bonds vary only little, the mobility (polarizability) of the π-electrons decreases greatly. As a consequence the positive nuclear charge on the carbon atoms is less shielded externally, leaving them more accessible to nucleophiles. This also explains the acidity of the acetylenic hydrogen atoms‡.

Owing to its lower polarizability the acetylene bond is less capable than the double bond of forming π-complexes. Provided that the bromination of acetylene is an electrophilic addition§, the slower addition of Br_2 to the triple bond of 1-penten-4-yne is explicable[46].

$$CH_2=CH-CH_2-C≡CH \xrightarrow[CCl_4]{Br_2} CH_2-CH-CH_2-C≡CH$$
$$\phantom{CH_2=CH-CH_2-C≡CH \xrightarrow[CCl_4]{Br_2} } ||$$
$$\phantom{CH_2=CH-CH_2-C≡CH \xrightarrow[CCl_4]{Br_2} } BrBr$$

(**108**) 90%

---

* See p. 329.
† See p. 59.
‡ The acidity of the acetylenic hydrogen atom also follows from the considerable s-character of the acetylenic C–H bond (see p. 329).
§ Addition of halogen to the triple bond yields *trans*-adducts. Birch reduction of the triple bond also gives *trans*-products, but hydrogenation with Pt, Pd, or Raney Ni catalysts yields *cis*-olefins.

However, the possibility that the bromination is a nucleophilic addition cannot be definitely excluded. Supporting evidence is the increasing rate of addition on going from mono- to poly-ynes. Even C=C double bonds are brominated by a nucleophilic mechanism in the case of α,β-unsaturated ketones.

Since the addition of bromine is accelerated in the latter case by HBr much more than would be expected for the acid strength of HBr, attack by Br_3^- on the conjugated double bond is probably the first reaction:

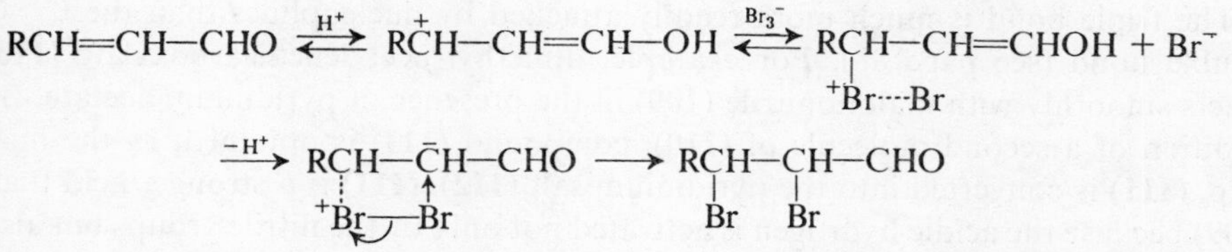

$$RCH=CH-CHO \underset{}{\overset{H^+}{\rightleftarrows}} R\overset{+}{C}H-CH=CH-OH \underset{}{\overset{Br_3^-}{\rightleftarrows}} \underset{\overset{\uparrow}{{}^+Br\cdots Br}}{RCH-CH=CHOH} + Br^-$$

$$\xrightarrow{-H^+} \underset{\overset{\uparrow}{{}^+Br\!-\!Br}}{RCH-\bar{C}H-CHO} \longrightarrow \underset{\underset{Br}{|}\underset{Br}{|}}{RCH-CH-CHO}$$

A-19. Addition of malononitrile (109) to dimethyl acetylenedicarboxylate (110)[47]

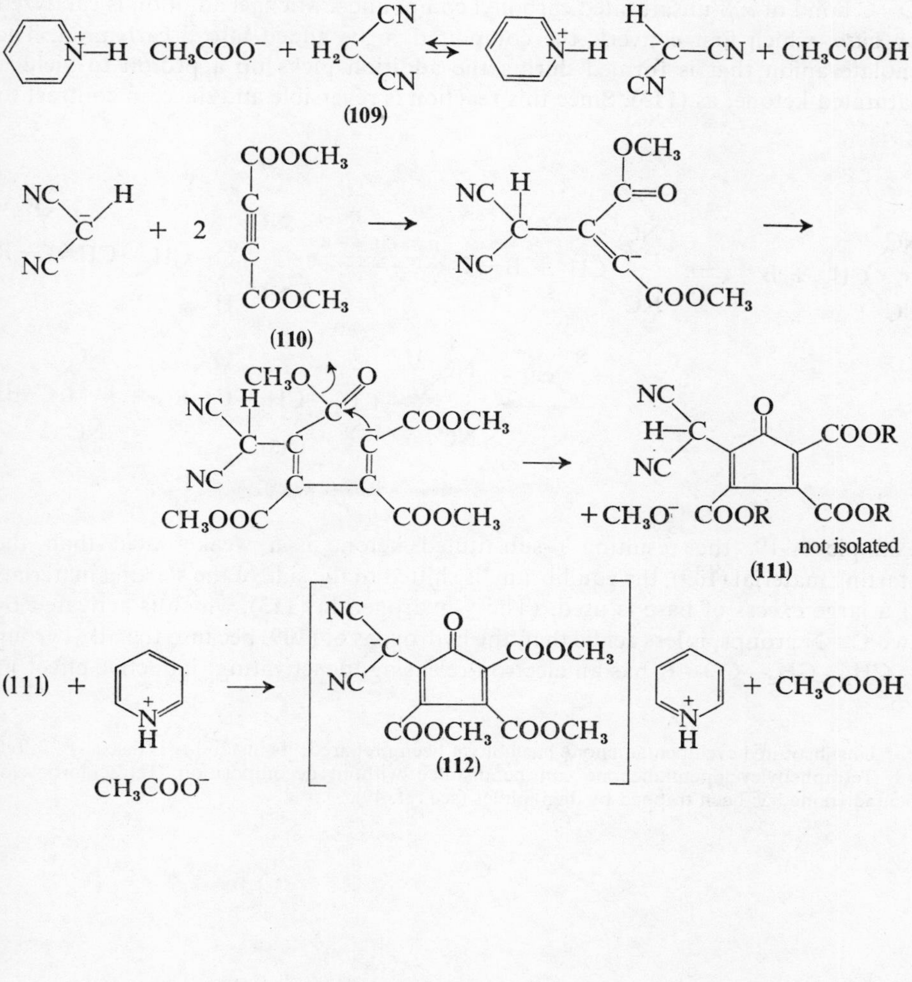

Characteristic IR and NMR absorption of (**112**):

2198 cm^{-1}	C≡N	0.94—2.1 τ (6) multiplet, protonated pyridine
1742	5-ring ketone	6.20 (3) singlet, OCH$_3$
1715	C=O conj. ester	6.40 (3) singlet, OCH$_3$
		6.41 (3) singlet, OCH$_3$

The triple bond is much more readily attacked by nucleophiles than the C=C double bond (see page 88). For example, dimethyl acetylenedicarboxylate (**110**) reacts smoothly with malononitrile (**109**) in the presence of pyridinium acetate. By addition of a second molecule of (**110**), compound (**111**) is obtained. In the final step, (**111**) is converted into the pyridinium salt (**112**). (**111**) is a stronger acid than (**109**) because the acidic hydrogen is activated not only by the nitrile groups but also by the electron-attracting cyclopentadienone group*.

The reaction resembles the Michael addition, in which carbanions add to the C=C bond of α,β-unsaturated carbonyl compounds. Michael addition is catalyzed by base, which first converts the compound to be added into a carbanion. The enolate anion that is formed during the addition picks up a proton to yield a saturated ketone, as (**113**). Since this reaction is reversible and since, in contrast to

example A-19, the resulting β-substituted ketone is a weaker acid than the starting material (**109**), the equilibrium is shifted to the side of the starting materials if a large excess of base is used. (The γ-hydrogen in (**113**), which is activated by two C≡N groups, is less acidic than the hydrogens of (**109**) because the alkyl group —CH$_2$—CH$_2$—CO—R has an electron-releasing, deactivating, inductive effect.)

---

* Unsubstituted cyclopentadienone has not yet been prepared; its instability is discussed in ref. 48. Tetraphenylcyclopentadienone can be distilled without decomposition. Tetrachlorocyclopentadienone has been trapped by dienophiles (see ref. 49).

A-20. Addition of phenacylidenetriphenylphosphorane (114) to dimethyl acetylenedicarboxylate (110)[50]

$$H_5C_6-\underset{\underset{O}{\|}}{C}-CH=P(C_6H_5)_3$$

(114)

$$\updownarrow \quad + H_3COOC-C\equiv C-COOCH_3 \longrightarrow$$

(110)

$$H_5C_6-\underset{\underset{O}{\|}}{C}-\overset{-}{C}H-\overset{+}{P}(C_6H_5)_3$$

$$H_5C_6-\underset{\underset{O}{\|}}{C}-CH-\underset{\overset{+}{P}(C_6H_5)_3}{\overset{COOR}{\overset{|}{C}}}\overset{H}{\underset{COOR}{\overset{|}{C^-}}} \xrightarrow{CH_3OH} H_5C_6-\underset{\underset{O}{\|}}{C}-CH-\underset{\overset{+}{P}(C_6H_5)_3}{\overset{COOR}{\overset{|}{C}}}\overset{H}{\underset{COOR}{\overset{\|}{C}}}$$

(115)

↓ Ether

$$-H^+ \downarrow$$

$$H_5C_6-\underset{\underset{O}{\|}}{C}-\underset{P(C_6H_5)_3}{\overset{COOR}{\overset{|}{C}}}-\underset{COOR}{\overset{H}{\overset{\|}{C}}}$$

(116)

(117): cyclobutene structure with $H_5C_6-C(=O)-CH-C(COOR)=C(COOR)-P(C_6H_5)_3$

$$\longrightarrow H_5C_6-\underset{\underset{O}{\|}}{C}-CH=\underset{\underset{P(C_6H_5)_3}{|}}{\overset{COOR}{\overset{|}{C}}}\underset{}{\overset{}{C}-COOR}$$

(118)

Characteristic absorption of (116):

IR		NMR
1715 cm⁻¹	C=O conj. ester	2—3 τ (20) multiplet, aromatic
1608	C=C olefin or aromatic	4.81 (1) singlet, vinyl
1587	⎫ C=C aromatic	6.67 (3) singlet, OCH₃
1511	⎭	6.97 (3) singlet, OCH₃

Characteristic absorption of (118):

IR		NMR
1724 cm⁻¹	C=O conj. ester	2—3 τ (20) multiplet, aromatic
1639	(C₆H₅)₃P=C—COOR	4.39 (1) broad singlet, H—C=C
		6.27 (3) singlet, OCH₃
		6.84 (3) singlet, OCH₃

Phosphorus ylides such as phenacylidenetriphenylphosphorane (114) react with (110) by addition of the nucleophilic ylide-carbon to the triple bond. If the reaction is carried out in a protic (proton-donating) solvent, the *trans*-addition product (116) is obtained. In aprotic solvents, however, the intermediate primarily formed (**115**), cyclizes to intermediate (**117**), which finally rearranges to (**118**)*.

Phosphoranes (pentavalent phosphorus compounds) have a trigonal-bipyramidal geometry (**119**) (sp^3d) with angles of approximately 120° in the plane and 90°

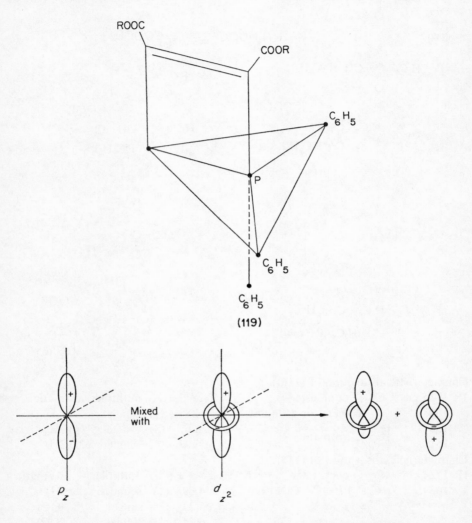

between plane and apex of the pyramid. The bonds are formed by sp^2-hybridization of two p-orbitals of the phosphorus atom with the s-orbital and hybridization of the third p-orbital with a d-orbital. The first mixing leads to the bonds in the plane, the

* For reactions of ylides with carbonyl compounds see example S-53.

latter to the bonds perpendicular to the plane. By further slight mixing, five identical hybrids are obtained*.

The four-membered ring in (119) is arranged perpendicularly to the basal plane of the bipyramid. Since the angle at the phosphorus atom can adopt a minimum value of 71°, resulting in release of angle strain at the carbon atoms, the phosphorane ring is less strained than the cyclobutene ring (120). In accordance with

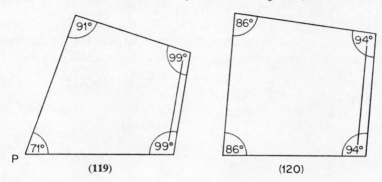

these considerations, formation of compound (122) *via* cyclobutene (121) has not been observed.

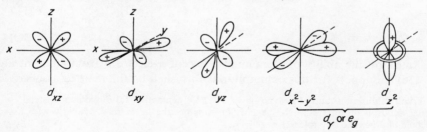

* Shapes of the five *d*-orbitals are as illustrated (from ref. 66).

A-21. Addition of triphenylaluminum (123) to diphenylacetylene (124)[51]

Aluminum alkyls and aryls add under appropriate conditions to disubstituted acetylenes, for example to tolane (124). These addition reactions, for example, of dialkylaluminum hydride yield *cis*-addition products*.

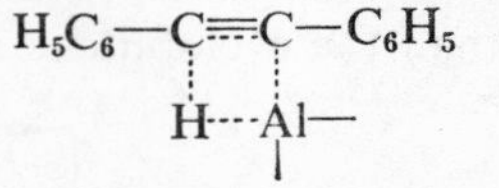

$$H_5C_6-C\equiv C-C_6H_5$$
$$H\cdots Al-$$

The reaction of triphenylaluminum (123) with tolane (124) at 200° yields an aluminole (127), a representative of a new class of heterocyclic compounds[53]. The

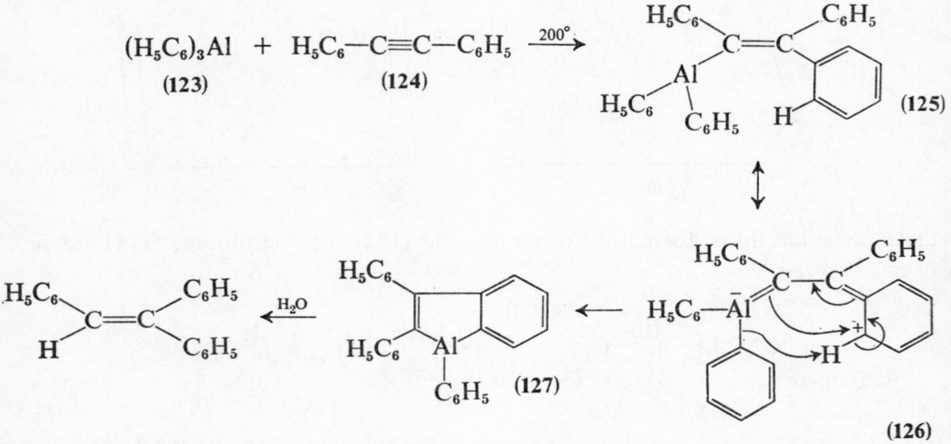

$$(H_5C_6)_3Al + H_5C_6-C\equiv C-C_6H_5 \xrightarrow{200°}$$
(123) (124)

intermediate (125) can be isolated, like the analogous Li compound (128) in the reaction of phenyllithium with diphenylacetylene. The compound (125) is converted into (127) by elimination of benzene. The ring formation is probably

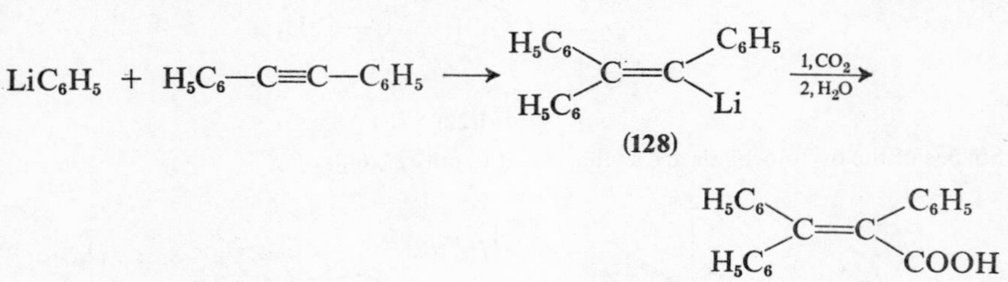

$$LiC_6H_5 + H_5C_6-C\equiv C-C_6H_5 \longrightarrow$$

(128) $\xrightarrow{1, CO_2}{2, H_2O}$

facilitated by the ability of aluminum to accept π-electrons from adjacent atoms (cf. 126)[54]. Support for this assumption is furnished by the marked association of

---

* Triethylaluminum and 3-hexyne give 1,1,2,3,4-pentaethyl-1,3-butadiene instead of the normal addition product 3-ethyl-3-hexene. Presumably the latter reacts rapidly with a second molecule of 3-hexyne (see ref. 52).

alkenylaluminum compounds, which is probably caused by strong polarization of the double bond[55].

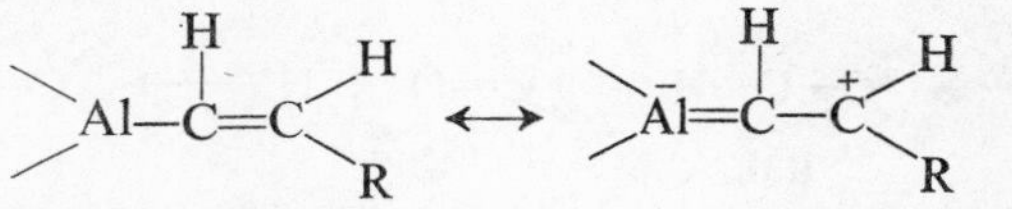

$$\text{Al–C=C} \quad \leftrightarrow \quad \text{Al=C–C}^+$$

An alternative mechanism is the electrophilic or radical addition of the aluminum atom to the benzene ring **(128a)**[56].

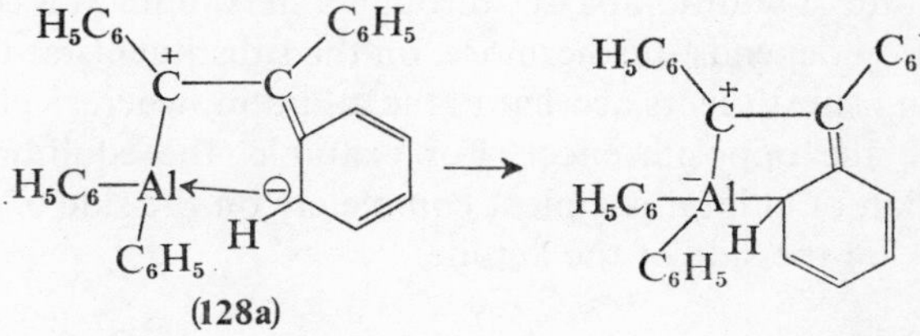

(128a)

Analogous reactions are the preparation of 5-phenyldibenzaluminole **(129)** and the heterocyclic compound **(130)**.

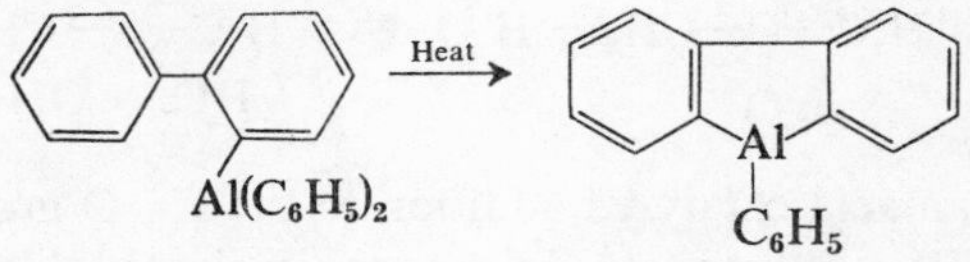

(129)

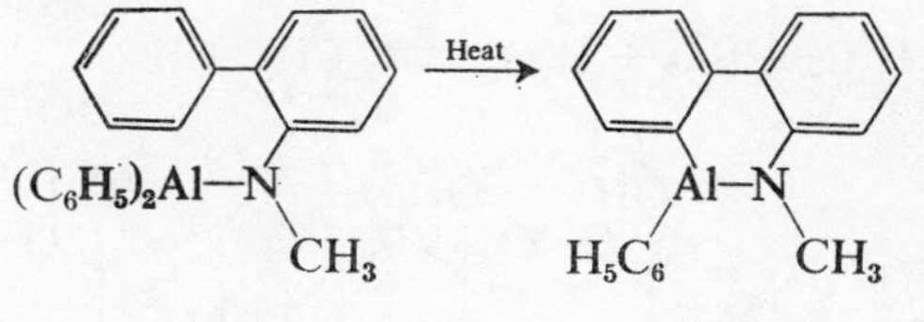

(130)

IV. NUCLEOPHILIC ADDITIONS TO THE C=O BOND

A-22. Addition of hydrogen peroxide to aldehydes and ketones

The strongly polarized carbonyl double bond may be attacked by electrophilic reagents at the oxygen atom or by nucleophiles at the carbon atom. Nucleophilic additions to the carbonyl-carbon are catalyzed by acids. Since many of the addition reactions are catalyzed both by H$^+$ and by Lewis acids (for example, the hydration of carbonyl compounds) one speaks of general acid-catalysis in contrast to the specific acid-catalysis caused exclusively by H$^+$. Addition of a proton to the oxygen of the carbonyl group yields the conjugate acid **(131)**. By hydrogen bonding to the oxygen the acid may form a complex (**132**). In both cases the reactivity of the carbonyl-carbon atom towards nucleophiles is markedly increased.

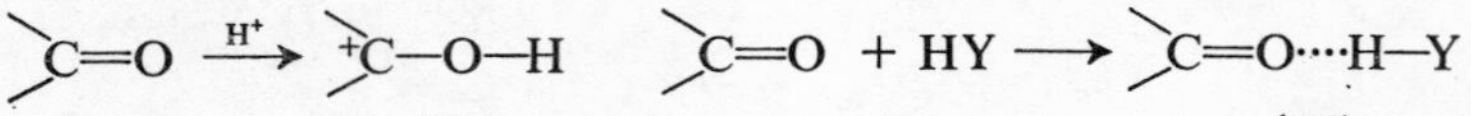

$$\ce{>C=O} \xrightarrow{H^+} \ce{>{\overset{+}{C}}-O-H} \qquad \ce{>C=O + HY -> >C=O\cdots H-Y}$$
$$\text{(131)} \hspace{5cm} \text{(132)}$$

If the concentration of the attacking nucleophile is reduced by the addition of the catalyst, for example, by formation of an ammonium salt if an amine is used as nucleophile, the rate of addition passes through a maximum at a certain pH value. The rate of addition depends, furthermore, on the substituents at the C=O group. Electron-attracting substituents accelerate the addition, whereas electron-donating substituents have the opposite effect. For example, the equilibrium of the uncatalyzed hydration of chloral is almost completely on the side of the hydrate, but with acetone it is on the side of the ketone.

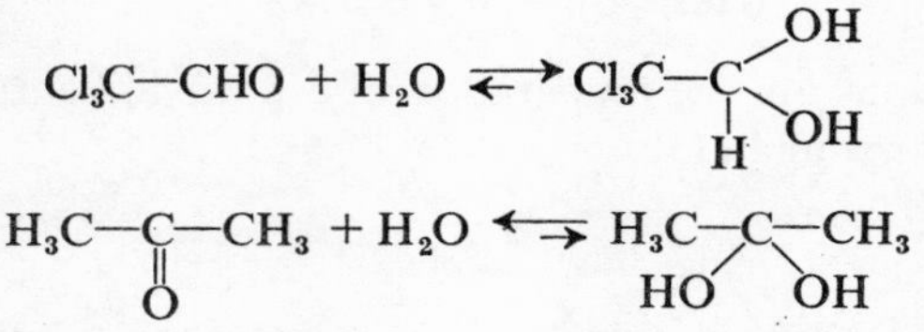

$$\ce{Cl_3C-CHO + H_2O <=> Cl_3C-\underset{H}{\overset{OH}{C}}-OH}$$

$$\ce{H_3C-\underset{O}{\overset{||}{C}}-CH_3 + H_2O <=> H_3C-\underset{HO\ \ OH}{C}-CH_3}$$

An example of an acid-catalyzed addition to the C=O group is the reaction of H_2O_2 with aldehydes and ketones*. In the case of simple aldehydes, the α-hydroxy

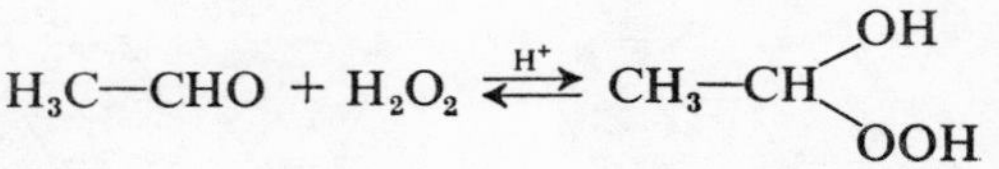

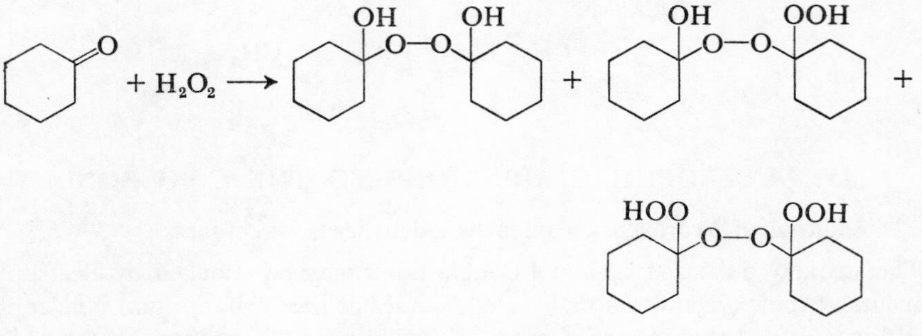

$$\ce{H_3C-CHO + H_2O_2 <=>[H^+] CH_3-CH{<}^{OH}_{OOH}}$$

[cyclohexanone] + H_2O_2 → [bis(1-hydroxycyclohexyl) peroxide] + [1-hydroxy-1'-hydroperoxy dicyclohexyl peroxide] +

[bis(1-hydroperoxycyclohexyl) peroxide]

hydroperoxides initially formed can be isolated; with ketones one usually obtains only secondary products.

If H_2O_2 is added to cyclic ketones in the presence of strong acids, lactones are

---

* For a review see ref. 57.

obtained according to the Baeyer–Villiger reaction)*. Peresters are formed as intermediates, which break down to tight ion pairs. By analogy with the Criegee rearrangement†, and the ozone cleavage of double bonds‡, a cation with positive oxygen is considered to be formed. Migration of an alkyl or aryl group to the electron-deficient oxygen and subsequent abstraction of a proton lead to the lactone.

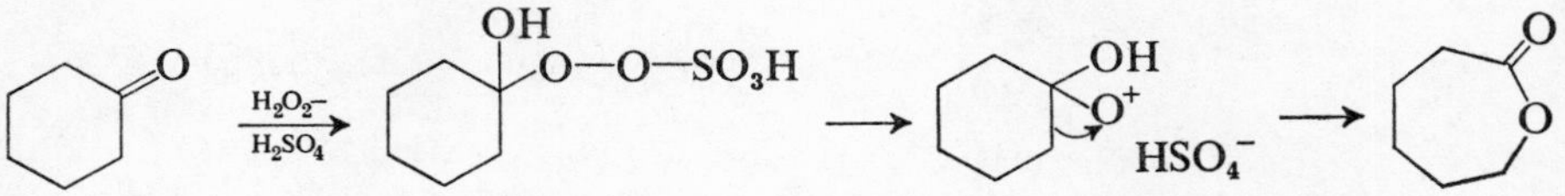

A-22. *Addition of* H_2O_2 *to* 3,3-*dimethyl*-2,4-*pentanedione* **(133)**[59]

The reaction of 3,3-dimethyl-2,4-pentanedione (**133**) with H_2O_2 in the presence of acids yields pivalic acid (**136**) in 76% yield. The reaction probably passes through

(133) **(134)**

$\xrightarrow{-H^+}$ $(CH_3)_3C{-}COOH + CH_3COOH$

(135) **(136)** 76%

(137)

(138)

the cyclic hydroperoxide (**134**). Analogous hydroperoxides (**137**) and (**138**) have been isolated by A. Rieche *et al.*[60]. In an acidic medium, (**134**) probably yields the

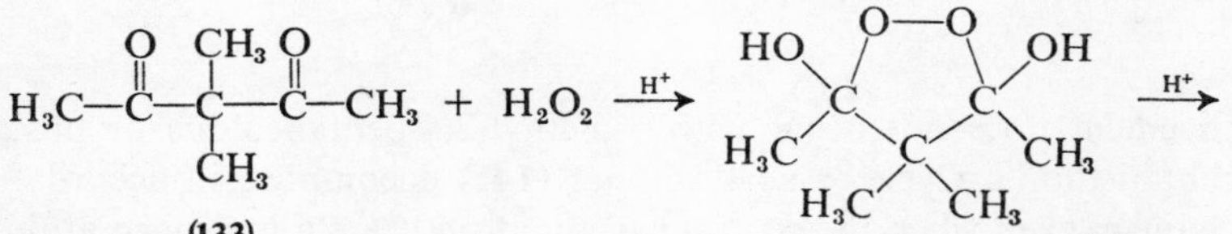

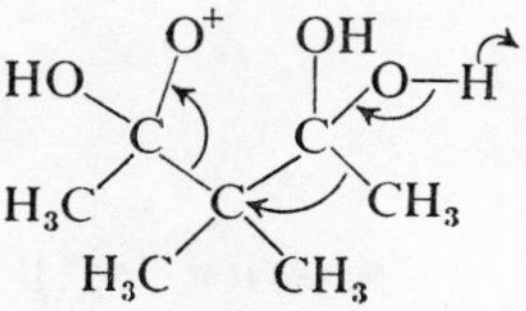

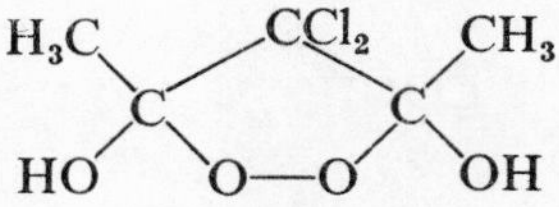

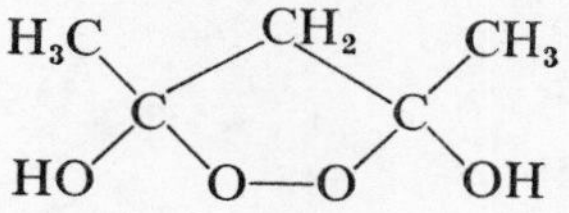

* Rearrangement of open-chain ketones to esters has rarely been observed. An example is the rearrangement of benzylideneacetone in reaction with peracetic acid[58]:

$$C_6H_5{-}CH{=}CH{-}CO{-}CH_3 \rightarrow C_6H_5{-}CH{=}CH{-}O{-}CO{-}CH_3$$

† See example C-14.
‡ See example A-4.

cation (135). By analogy with the Baeyer–Villiger reaction, migration of an alkyl group to the positive oxygen could occur, leading to the compounds (139) and (140). The rearrangement that actually takes place is presumably favored by the formation of a relatively stable tertiary carbonium ion as intermediate.

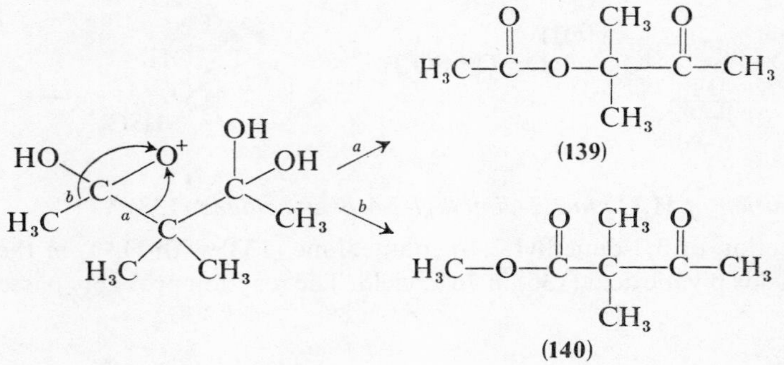

The rearrangement does not occur with 3-methyl-2,4-pentanedione or unsubstituted 2,4-pentanedione. Cyclic α-acyl ketones (141) decompose in accordance with this mechanism even when the central carbon atom bears a hydrogen atom[61].

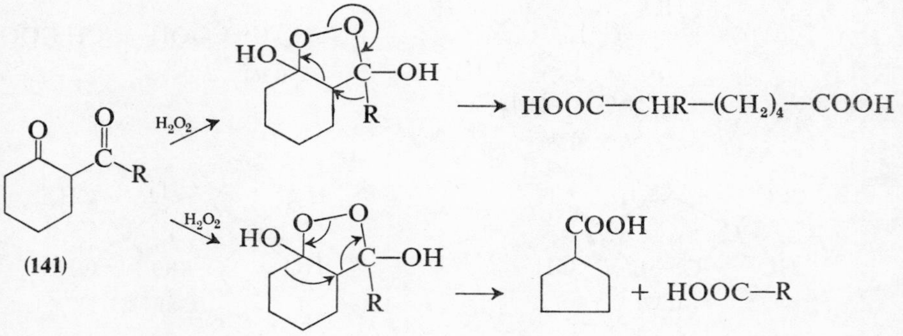

A-23. Addition of LiAlH₄ to tropolone methyl ether (142)[62]

Reaction of tropolone methyl ether (142) with LiAlH$_4$ yields 50% of benzaldehyde. The ring contraction is probably initiated by addition of H⁻ to the C=O double bond*. Evidence has been presented, that the norcaradiene (143) is formed as an intermediate[63].

* See examples A-12 and A-13.

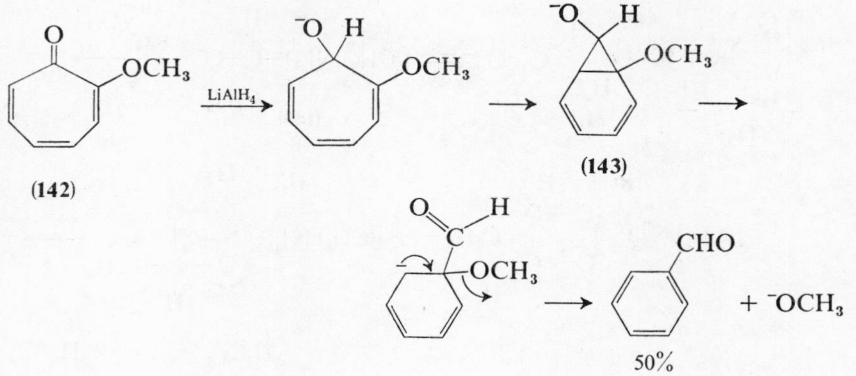

A-24 & A-25. Addition reactions of trivalent phosphorus compounds

Compounds of trivalent phosphorus react as nucleophiles, but in certain cases also as electrophiles, depending on the nature of the substituents at the phosphorus atom.

The relatively high electrophilic reactivity of trivalent phosphorus is due to the large radius of this atom compared with nitrogen, and the high polarizability of its electron cloud. Electrophilic reactions usually lead to substitution at the phosphorus atom.

$$R^1R^2P\text{—}Cl + RO^- \rightarrow R^1R^2P\text{—}OR + Cl^-$$

The weakly basic trialkylphosphines react as nucleophiles with alkyl halides, yielding phosphonium salts. Trialkyl phosphites with alkyl halides give trialkoxy-alkylphosphonium salts (Arbusow reaction)*.

$$R_3P + R'Br \rightarrow R_3R'P^+ \ Br^-$$
$$(RO)_3P + R'Br \rightarrow (RO)_3R'P^+ \ Br^- \rightarrow (RO)_2R'P\text{=}O + RBr$$

A-24. *Addition of phosphorous tris(dimethylamide)* **(144)** *to benzaldehyde*[64]

Trivalent phosphorus compounds of exceptionally high nucleophilicity are the phosphorous triamides. If benzaldehyde is added to (144), the phosphorus atom attacks the carbonyl-carbon, forming adduct (145), which can be isolated. With an excess of aldehyde, good yields of *trans*- and *cis*-epoxides (146) are obtained (2-pyridinecarbaldehyde gives 90% of epoxide).

For this reaction a mechanism has been proposed which resembles the mechanism of the formation of azines (148) from phosphorus ylides (147) and diazo compounds[65] or of cyclopropanes from phosphorus ylides and epoxides[66].

* See p. 295.

$$[\underset{H_3C}{\overset{H_3C}{N}}-]_3 P + \underset{H_5C_6}{\overset{H}{\diagdown}}C=O \rightleftharpoons [(CH_3)_2N]_3\overset{+}{P}-\underset{C_6H_5}{\overset{H}{C}}-O^- \xrightarrow{\overset{H}{\underset{H_5C_6}{\diagdown}}C=O}$$

(144)   (145)

$$[(CH_3)_2N]_3\overset{+}{P}\underset{O-\underset{H}{\overset{|}{C}}-C_6H_5}{\overset{\overset{R}{\overset{|}{C}}-O}{\diagup\diagdown}} \longrightarrow [(CH_3)_2N]_3P\underset{O-\underset{H}{\overset{|}{C}}-C_6H_5}{\overset{\overset{H_5C_6}{\overset{|}{C}}-O}{\diagup\diagdown}} \longrightarrow$$

$$[(CH_3)_2N]_3PO + \underset{H}{\overset{H_5C_6}{\diagdown}}C\underset{O}{\overset{\diagup\diagdown}{\quad}}C\underset{C_6H_5}{\overset{H}{\diagup}}$$

(146)

$$\underset{R}{\overset{R}{\diagdown}}\overset{-}{C}-\overset{+}{P}(C_6H_5)_3 + \overset{+}{N}=N-\overset{R'}{\underset{R'}{\overset{|}{C}}} \longrightarrow R-\underset{N=N}{\overset{\overset{R}{\overset{|}{C}}-\overset{+}{P}(C_6H_5)_3}{\diagdown}}\overset{R'}{\underset{R'}{\overset{-}{C}}} \longrightarrow$$

(147)

$$\underset{R}{\overset{R}{\diagdown}}C=N-N=C\underset{R'}{\overset{R'}{\diagup}}$$

(148)

$$R-\underset{H}{\overset{\overset{O}{\diagup\diagdown}}{C}}CH_2 + (H_5C_6)_3P=\overset{H}{\underset{}{C}}-COOR' \longrightarrow$$

$$H_2C\underset{CHR-O^-}{\overset{\overset{COOR'}{\underset{|}{CH}-\overset{+}{P}(C_6H_5)_3}}{\diagup}} \longrightarrow R-CH-\overset{CH_2}{\overset{|}{CH}}-COOR' + (H_5C_6)_3PO$$

The formation of epoxides can be initiated in principle also by nucleophilic attack of the phosphorus atom at the carbonyl-oxygen, yielding (**149**) *etc.*

Such an addition has been observed, for example, in the reaction of chloranil (**150**) with a trialkyl phosphite[67].

Since in (**149**) the carbanion is not stabilized by conjugation, as in (**151**), this mechanism is not very likely.

[Scheme showing: (R₂N)₃P + O=CHR' → (R₂N)₃P⁺—O—C⁻HR' + H(R')C=O →]
(149)

[Scheme: (R₂N)₃P⁺—O—C(R')(H) / ⁻O—C(H)(R') → cyclic (R₂N)₃P—O—C(R')(H) / O—C(H)(R')]

[Scheme: tetrachloro-benzoquinone (150) + (RO)₃P → tetrachloro-phenoxide with O—P⁺(OR)₃ substituent (151)]

A-25. *Addition of trimethyl phosphite to 2,2,4,4-tetramethyl-1,3-cyclobutanedione* **(152)** *and 3-hydroxy-2,2,4-trimethyl-3-pentenoic acid lactone* **(156)**[68]

The formulae for these reactions are shown on pp. 102–103.

Trimethyl phosphite reacts with **(152)** and **(156)** by nucleophilic attack at the C=O carbon atom. Compounds of trivalent phosphorus, like other nucleophiles, react with C=C double bonds only if these are sufficiently polarized by electron-attracting groups. For example, tributylphosphine readily adds to benzylidene-malononitrile[69].

[Scheme: H₅C₆—CH=C(CN)₂ + Bu₃P → H₅C₆—CH(—)—C(CN)₂ with ⁺P(Bu)₃ on central carbon]

During the rearrangements of the adducts **(153)** and **(157)** to the intermediates **(154)** and **(158)**, the C–C bond in **(153)** and the C–O bond in **(157)** are not completely broken because such a mechanism would lead to a common intermediate **(160)** and consequently to analogous products from **(152)** and **(156)**. It is reasonable to assume, therefore, that the ring expansions **(153)** → **(154)** and **(157)** → **(158)** proceed by a synchronous mechanism.

Migration of a CH₃O group in **(154)** or **(158)** leads to the products **(155)** and **(159)**. The driving force of the reactions is probably strain relief by opening of the four-membered ring and neutralization of the charges developing on P and O. Stabilization of **(153)** and **(157)** by an Arbusow reaction*, which would lead to the products **(161)** and **(162)**, has not been observed.

* See p. 295.

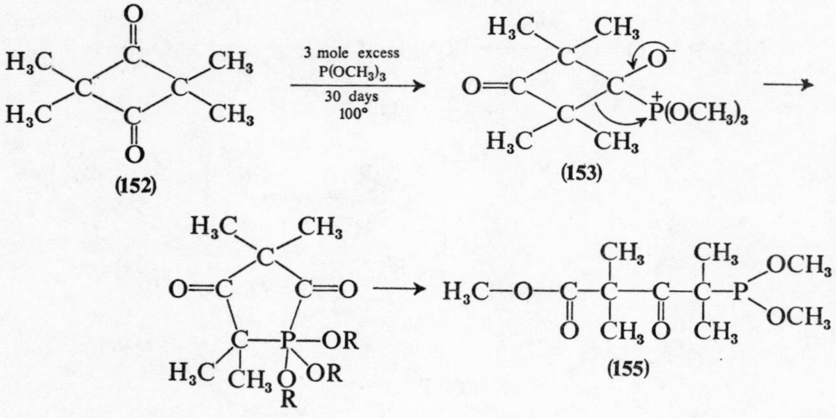

(152) (153) (154) (155)

(156) (157) (158) (159) 67%

Characteristic absorption of (**155**):

IR	1750 cm^{-1}	C=O (COOR)	NMR 6.37 τ(3)	singlet,	$COOCH_3$
	1685	C=O	6.45 (6)	doublet,	$P(OCH_3)_2$
			8.73 (6)	singlet,	$ROOCC(CH_3)_2$
			8.87 (6)	doublet,	$CO{\cdot}C(CH_3)_2P$

Characteristic absorption of (**159**):

IR 1730 cm^{-1} C=O (COOR)

1660 C=O

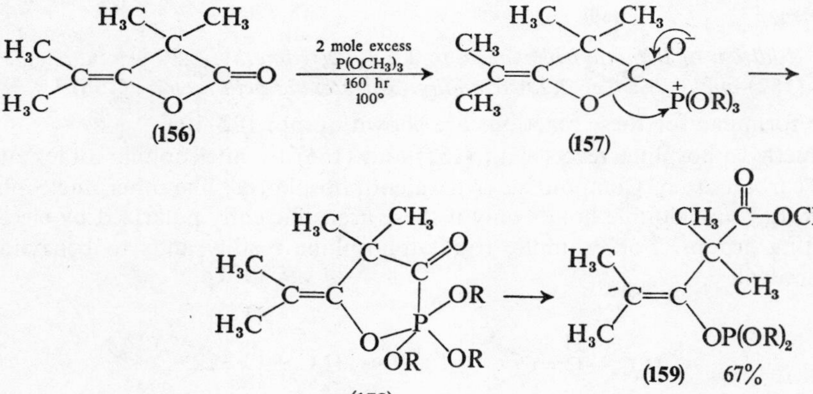

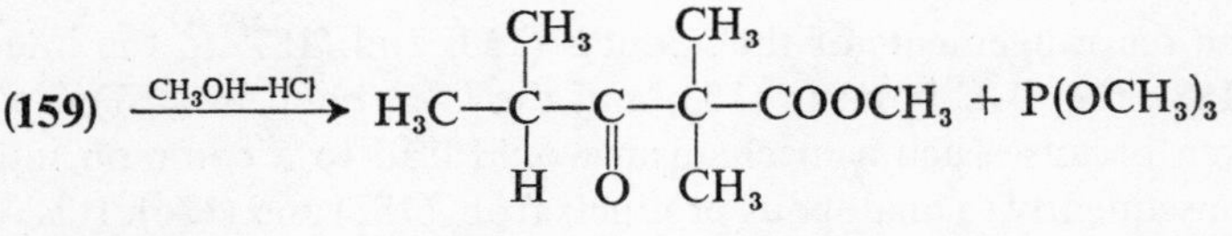

(153) (157) (160)

(153) (161)

(157) (162)

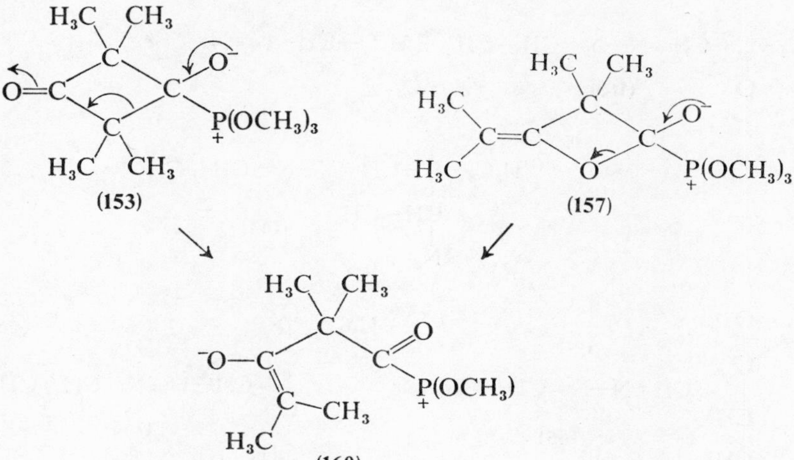

A-26. Addition of diazoethane to an azine (163)[70] (see p. 104)

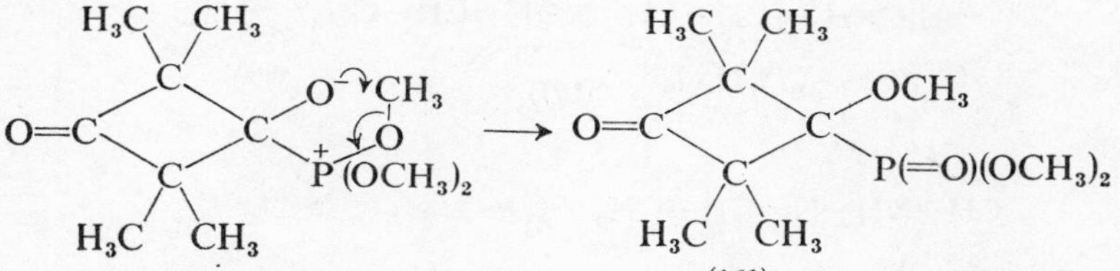

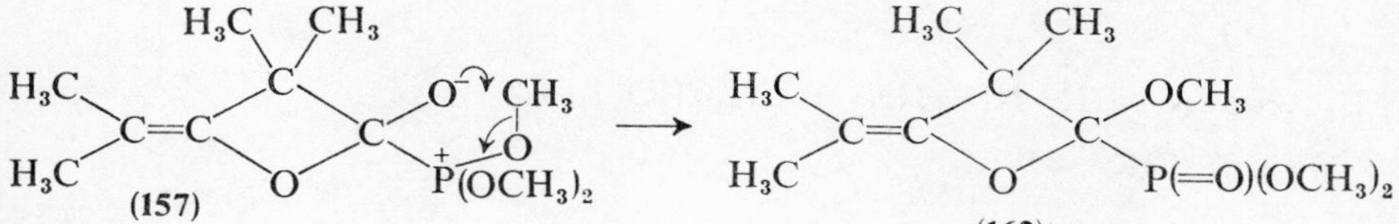

In 1-ethylidene-2-phenacylidenehydrazine (**163**) both the C═O group and the C═N double bond can, in principle, react with diazoethane. As a consequence of the higher electrophilicity of the C═O carbon, nucleophilic attack of diazoethane at the C═O group is favored. (In contrast to carbenes, which are electrophiles, aliphatic diazo-compounds react as nucleophiles.)

In the reaction of (**163**) with diazoethane[70], intermediate (**164**) is formed, which can be converted into a more stable product in three ways: (1) by phenyl migration to the carbon atom, from which the diazonium nitrogen separates; (2) by migration of the —CH═N—N═CH—CH_3 group; or (3) by formation of an epoxide.

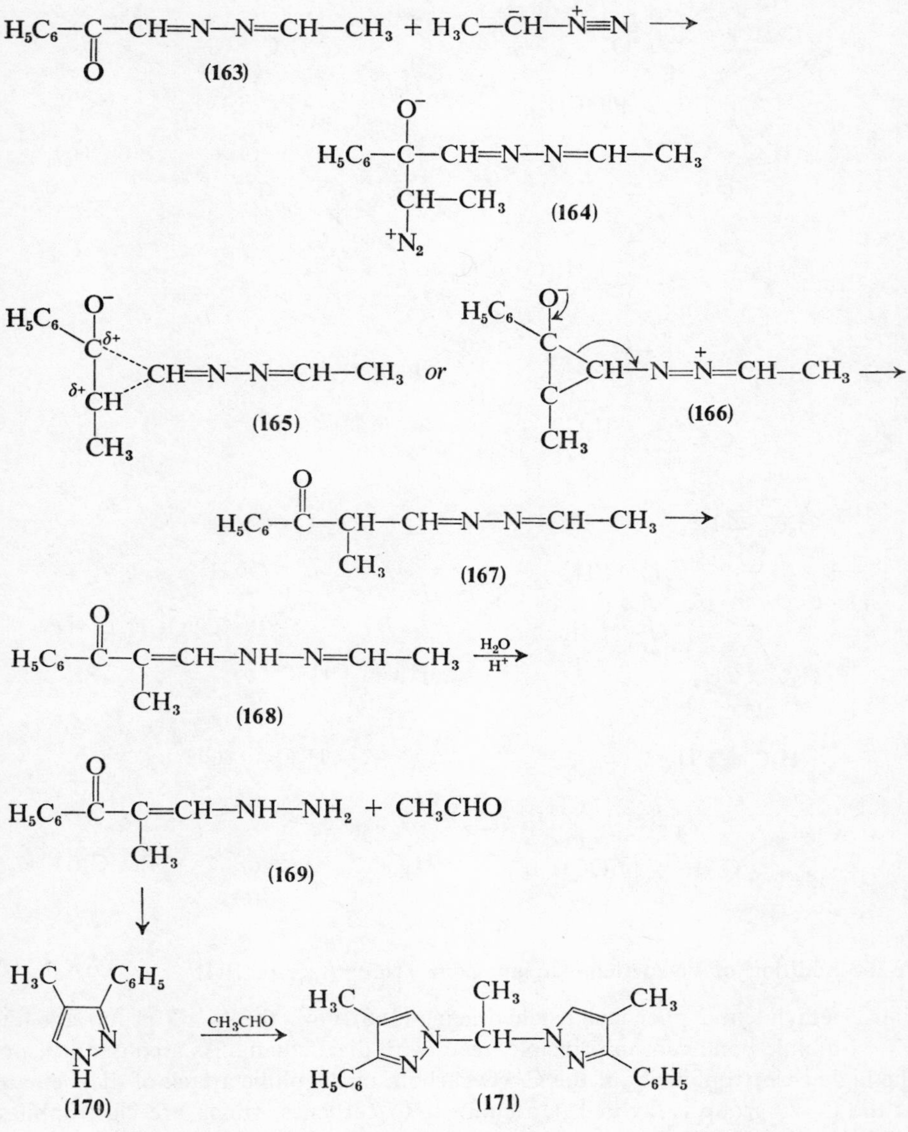

$$H_5C_6-\underset{O}{\underset{\|}{C}}-CH=N-N=CH-CH_3 + H_3C-\bar{C}H-\overset{+}{N}\equiv N \longrightarrow$$
(163)

$$H_5C_6-\underset{\underset{+N_2}{\overset{|}{CH-CH_3}}}{\overset{O^-}{\overset{|}{C}}}-CH=N-N=CH-CH_3$$
(164)

$$H_5C_6\underset{\underset{CH_3}{\overset{|}{\overset{\delta+}{CH}}}}{\overset{O^-}{\overset{\delta+}{C}}}\cdots CH=N-N=CH-CH_3 \quad or \quad H_5C_6\underset{\underset{CH_3}{\overset{|}{C}}}{\overset{O^-}{\overset{\curvearrowleft}{C}}}\underset{}{\overset{}{\underset{CH}{\triangle}}}N=\overset{+}{N}=CH-CH_3 \longrightarrow$$
(165) (166)

$$H_5C_6-\underset{O}{\underset{\|}{C}}-\underset{CH_3}{\underset{|}{CH}}-CH=N-N=CH-CH_3 \longrightarrow$$
(167)

$$H_5C_6-\underset{O}{\underset{\|}{C}}-\underset{CH_3}{\underset{|}{C}}=CH-NH-N=CH-CH_3 \xrightarrow[H^+]{H_2O}$$
(168)

$$H_5C_6-\underset{O}{\underset{\|}{C}}-\underset{CH_3}{\underset{|}{C}}=CH-NH-NH_2 + CH_3CHO$$
(169)

↓

(170) →[CH₃CHO] (171)

Characteristic absorption of (**168**): UV 298 mμ (ln ε 4.36) ⎫
344   (ln ε 3.95) ⎭ conj.

IR   3310 cm⁻¹ —NH—
       1645   —N=CH—

The rate of rearrangement of the substituents at the carbonyl group increases with increasing nucleophilicity of the migrating group. In the pinacol rearrangement it has been shown that the rate increases in the following series from left to right[71]: p-chlorophenyl < phenyl < p-tolyl < p-methoxyphenyl. If the C=O group is substituted by strongly electron-attracting substituents, i.e., if they are of low nucleophilicity, rearrangement does not occur and epoxide formation is observed instead. Examples are the reactions of diazomethane with chloral and with o-nitrobenzaldehyde.

$$Cl_3CCHO \xrightarrow{CH_2N_2} Cl_3C-\underset{H}{\underset{|}{C}}-\overset{+CH_2}{\overset{|}{C}}-O^- \longrightarrow \underset{}{\overset{O}{\triangle}}-CCl_3$$

In the case of **(164)** the —CH=N—N=CH—CH$_3$ group actually migrates, probably across transition state **(165)** or intermediate **(166)**. This rearranges to **(167)**, which is converted into the more stable system of conjugated double bonds **(168)** (usually an enamine suffers the opposite conversion into the more stable imine structure —CH=C—NH— → —CH$_2$—C=N—). **(168)**, which can be isolated, is hydrolyzed by acids to the hydrazine derivative **(169)**, which cyclizes to product **(170)**. Two molecules of **(170)** with one molecule of acetaldehyde, which is formed during hydrolysis of **(168)**, yield product **(171)**.

A-27. Reaction of 1,3-diphenyl-3-phospholin 1-oxide (172) with phenyl isocyanate (173)[72]

Additions to isocyanic esters are initiated by nucleophilic attack at the carbonyl-carbon atom. In the reaction of **(172)** with **(173)** one would therefore expect the oxygen of **(172)** to add to the C=O carbon of **(173)**, as shown on p. 106. This assumption was confirmed by determination of the ρ value of the addition*.

$$\lg k - \lg k_0 = \rho\sigma$$

In the Hammett equation above, k represents the rate of addition to phenyl isocyanates which are substituted in the *para*-position of the aromatic ring. k_0 is the rate of addition to unsubstituted phenyl isocyanate. σ expresses the resonance and inductive effect of a substituent on the reaction rate and ρ is a parameter that depends on the type of reaction but is independent of the substituents†. ρ is determined by evaluating the slope of the best straight line through the points which are obtained by plotting $\lg k/k_0$ or simply $\lg k$ against σ. Determining the

---

* For a discussion of the Hammett equation see p. 373.
† For tables of σ and ρ values see ref. 73.

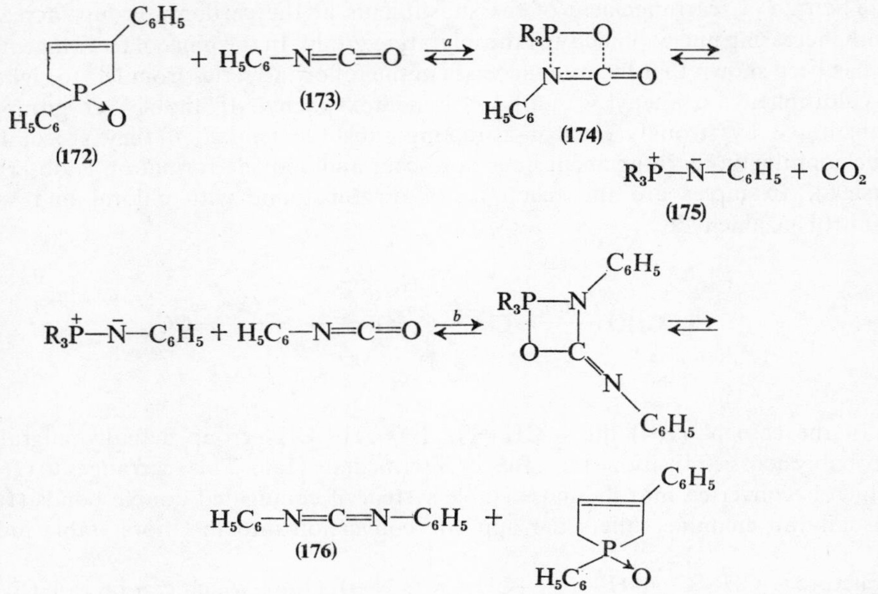

kinetics of the reaction of (172) with (173) has shown that the reaction is of the first order in both (172) and (173). The ρ value obtained is similar to the ρ values of other nucleophilic additions to isocyanic esters. Electron-donating substituents, which increase the electron density at the carbonyl-carbon, reduce the reaction rate. Furthermore, it was shown that the polarity of the P–O bond has considerable influence on the reaction: the activity of the phospholin oxide increases with increasing polarity of this bond.[74]

Reaction step *a* must be rate-determining. If *b* were rate-determining, the reaction would be of the second order in (173). In that case, a large excess of (172) should prevent step *b*, since (173) would be consumed before it could react. However, if step *a* is rate-determining, the phosphorimide (175) can react with available (173) to give the product. The investigations have shown that (176) is indeed formed, and in good yield, in the presence of an excess of the oxide (172). Formation of the imide (175), whose existence as an intermediate is supported by the findings of Staudinger and Meyer[75], may be best explained as involving a four-membered, cyclic transition state. Supporting evidence is the high activation entropy of the reaction, which is characteristic for a relatively rigid transition state, and, secondly, the strong influence of *ortho*-substituents on the benzene ring of (173) on the reaction rate: *ortho*-substituents obviously have a steric effect on the formation of the four-membered ring.

A-28 & A-29. Clemmensen reductions

By Clemmensen reduction, which is usually carried out with amalgamated Zn in 20—40% HCl, carbonyl groups of aldehydes and ketones are reduced to CH_2 groups. Other reaction products may be alcohols, olefins, and rearrangement products. A satisfactory interpretation of the principal reaction and of the side reactions has been furnished by D. Staschewski[76].

Before the reaction mechanism is discussed, the so-called "nascent hydrogen" which is assumed to be active in reductions with Zn/HCl and other metal/acid couples must be characterized. Although radicals may be formed during the reaction, the nascent hydrogen presumably does not consist of hydrogen atoms, since the latter would also reduce olefins or would polymerize them: however, neither reaction has been observed. The protons that reach the metal surface *via* H_3O^+ ions are discharged there, but, instead of being reduced to H atoms, they form bonds with the metal surface by filling their empty *s*-orbitals with electrons from the electron gas of the metal. With increasing electron pressure of the metal, *i.e.* with increasing reduction potential, the hydrogen that is bound to the metal surface increasingly adopts hydride character.

$$M\text{—}H \xrightarrow{\text{Increase of red. potential}} M\blacktriangleleft H$$

(Charge compensation is provided by the process $Zn \rightarrow Zn^{2+}$.)

In principle, separation of a hydride ion from the metal surface can be caused by an approaching proton (*a*) or by an approaching conjugate acid (*b*), resulting in formation of H_2 or an alcohol. Reduction of carbonyl compounds is started by

$$(a) \quad M\blacktriangleleft H + H^+ \longrightarrow M^+ + H_2$$
$$(b) \quad M\blacktriangleleft H + {}^+CR_2OH \longrightarrow M^+ + CHR_2OH$$

addition of the carbonyl-carbon atom to the metal surface which is covered in part by hydrogen atoms. Instead of the free carbonyl compound, the conjugate acid may actually add to the surface. A necessary condition, however, is sufficient electrophilicity of the carbonyl-carbon; this is not great enough in carboxylic acids or esters, which are therefore not reduced by Zn/HCl.

$$M + {}^+CR_2OH \longrightarrow MCR_2OH$$

The carbonyl group may separate from the metal surface by attack of a proton, yielding an alcohol.

$$M\blacktriangleleft CR_2OH + H^+ \longrightarrow M^+ + CHR_2OH$$

With numerous metal/acid couples, alcohol formation according to this mechanism actually takes place.

One may ask why reduction with amalgamated Zn leads to the methylene stage. The cause has to be looked for in the high overvoltage at the amalgamated Zn. At the metal surface, the electrokinetic processes that permit the separation of H⁻ ions by approaching protons or other cations are obstructed by mechanisms that are not exactly understood. As a consequence of the overvoltage a relatively high hydride density exists at the metal surface. Attack of a H⁻ in combination with the electron pressure of the metal thus causes it to replace the OH group of the molecule which is bound to the surface:

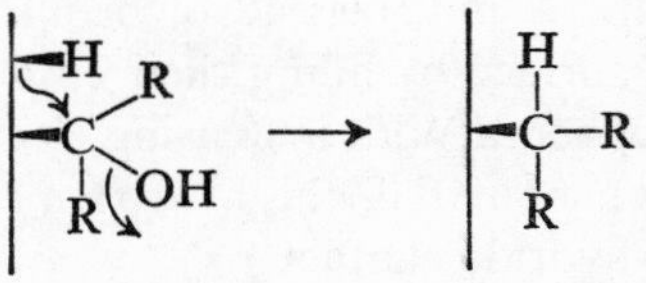

In the slowest step of the reaction (here overvoltage is also effective) the hydrocarbon is finally abstracted as an anion by an attacking proton.

A second mechanism that results in replacement of the OH group is the migration of H, Alkyl, or Aryl, with their bonding electrons, to the position of the leaving OH group. In these rearrangements, olefins are obtained. The formation of olefins predominates if conjugated double bonds can be formed:

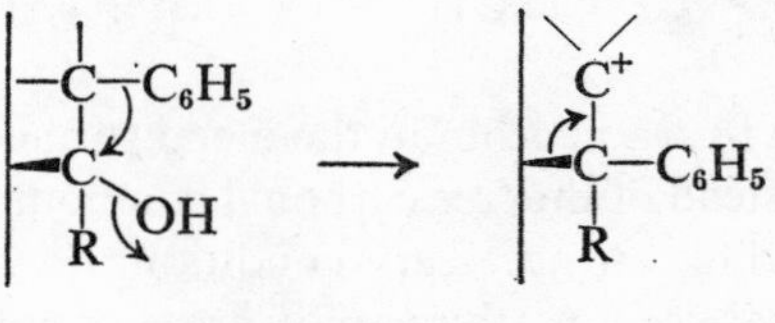

The cation, which is adsorbed at the metal surface, passes through the stage of a radical during the reduction. If on the metal surface enough short-lived radical intermediates are present, dimerization occurs, which results in the formation of pinacols. Pinacol formation is favored by resonance stabilization of the radicals. Bonding between two radicals is particularly favorable if they are in the same molecule, as in the examples A-28 and A-29. The neutral diol readily separates from the surface. The acid present subsequently causes rearrangement.

A-28. *Reduction of 2,2'-diacetylbiphenyl* **(177)**[77]

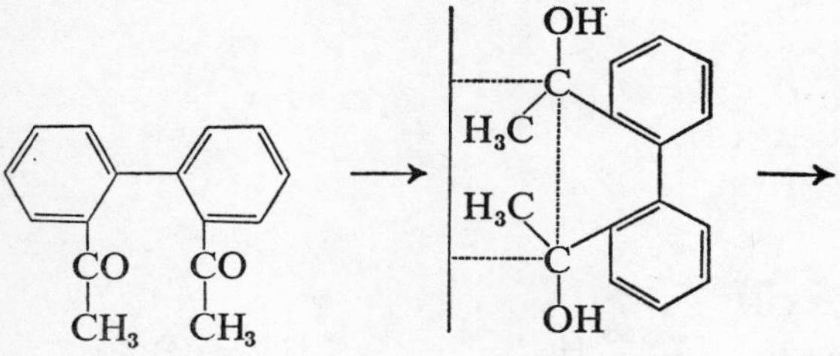

A-29. *Reduction of 2,4-pentanedione* **(181)**[78]

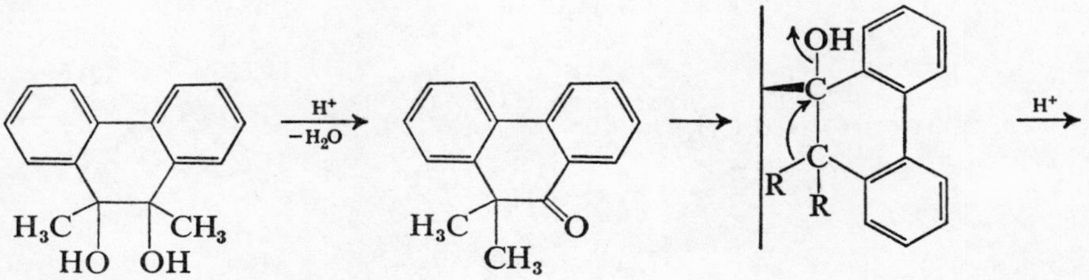

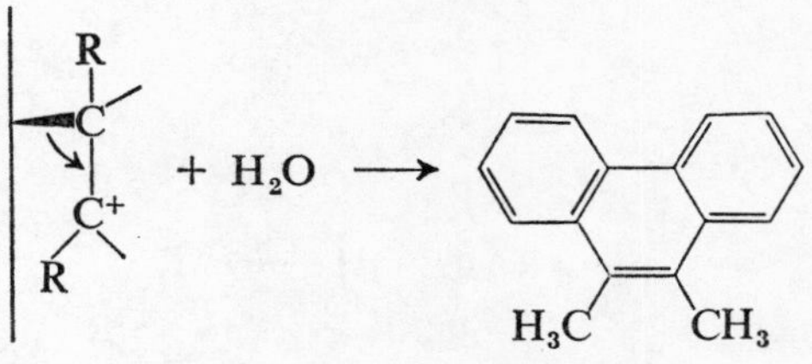

In example A-28, phenanthrone (**179**) is formed from 2,2'-diacetylbiphenyl (**177**), whereas an α,β-unsaturated ketone (**183**) is obtained from 2,4-pentanedione (**181**) in example A-29. Products (**179**) and (**183**) are further reduced to 9,10-dimethylphenanthrene (**180**) and 3-methyl-2-butanone (**184**). If (**184**) is exposed long enough to the reducing agent, it is converted into 2-methylbutane in the way outlined above.

Alcohols can be reduced to saturated hydrocarbons if they form carbonium ions in the strongly acidic medium of the Clemmensen reduction. Ketones with electronegative α-substituents such as OH, OR, SR, NH_2, NR_2, Cl, or Br are also reduced to saturated hydrocarbons, as shown schematically on p. 110.

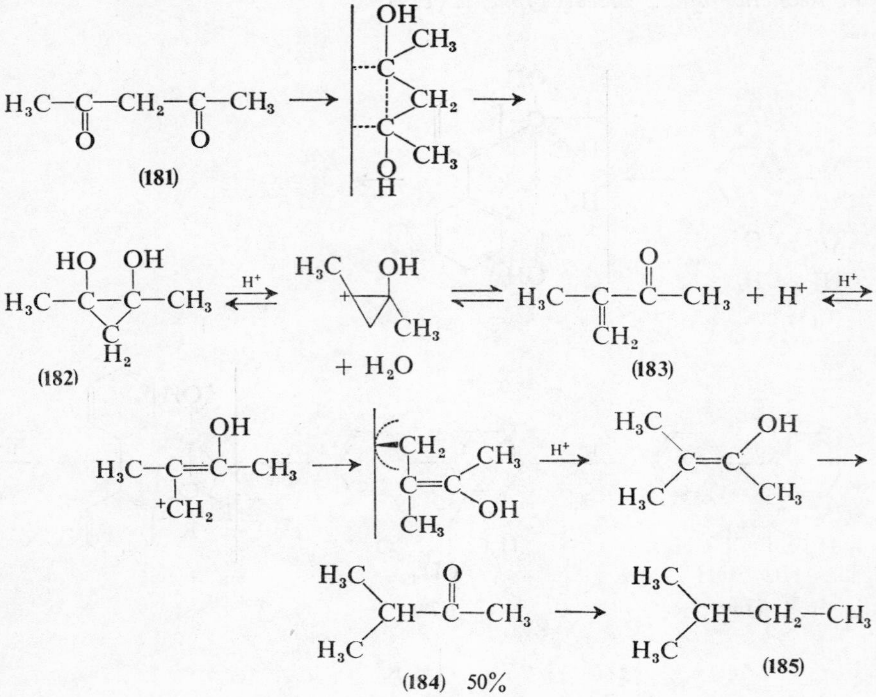

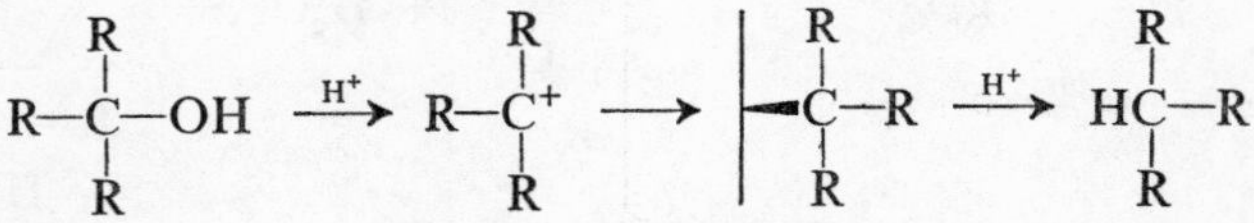

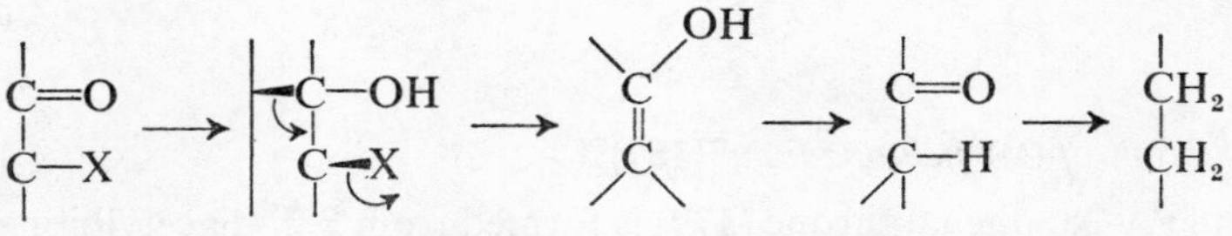

V. ELECTROPHILIC ADDITIONS TO THE C=O BOND

A-30 to A-32. Addition of PCl$_5$ to the C=O bond

We have seen that additions to the carbon atom of the C=O bond are facilitated by preceding addition of a proton to the carbonyl-oxygen.* Attack at the oxygen may occur also by other electrophilic reagents, for example, by PCl$_5$.

---
* See p. 95.

Addition Reactions

PCl$_5$, which is present as PCl$_4^+$PCl$_6^-$ in the solid state,[79] is probably at least partly dissociated into PCl$_4^+$ and PCl$_6^-$ in polar solvents. According to the mechanism proposed by M. S. Newman and L. L. Woods Jr.[80] PCl$_4^+$ attacks the carbonyl-oxygen in the first step of the reaction. Addition product (186) either decomposes to

$$2\text{PCl}_5 \rightleftarrows \text{PCl}_4^+ \text{ PCl}_6^-$$

$$\text{R}-\text{CO}-\text{CH}_3 + \text{PCl}_4^+ \rightleftarrows \text{R}-\overset{+}{\underset{\text{OPCl}_4}{\text{C}}}-\text{CH}_3 \longrightarrow \text{R}-\overset{+}{\underset{\text{Cl}}{\text{C}}}-\text{CH}_3 + \text{POCl}_3$$

(186)   (187)

$$\xrightarrow{\text{Cl}^-} \text{R}-\underset{\text{Cl}}{\overset{\text{Cl}}{\text{C}}}-\text{CH}_3 \quad (189)$$

$$\text{R}-\underset{\text{CH}_3}{\overset{\text{Cl}}{\text{C}^+}} \text{POCl}_4^- \longleftarrow \text{R}-\underset{\text{OPCl}_4}{\overset{\text{Cl}}{\text{C}}}-\text{CH}_3 \longrightarrow \text{R}-\underset{\text{O}-\text{P}-\text{Cl}}{\overset{\text{Cl}}{\text{C}}}\text{CH}_2 \longrightarrow \text{R}-\underset{\text{Cl}}{\text{C}}=\text{CH}_2$$

(191)   (188)   Cl$_3$ (190)

$$\downarrow -\text{POCl}_3 \qquad\qquad\qquad\qquad\qquad + \text{POCl}_3 + \text{HCl}$$

(189)

the chloro carbonium ion (187) or it forms the intermediate (188) by addition of Cl$^-$ which is furnished by PCl$_6^-$; (188) can be converted into (187) by loss of POCl$_4^-$ or it can form a chloro olefin (190). Chloro olefins are often observed in reactions of aldehydes and ketones with PCl$_5$. (188) can alternatively dissociate to a tight ion pair (191), which yields the dichloride (189). The latter is also formed from (187) by addition of Cl$^-$.

The formation of chloro olefins seems to involve chloro carbonium ions only in those cases in which the positive charge is sufficiently stabilized. Otherwise the chloro olefins are formed by path (188) → (190). For example, in the reaction of 4-phenyl-2-butanone and 5-phenyl-2-pentanone with PCl$_5$ in CH$_2$Cl$_2$ at 25°, 2-chloro-4-phenyl-1-butene and 2-chloro-5-phenyl-1-pentene are obtained as principal products[81]. No indane or tetrahydronaphthalene derivative is observed—this

$$\text{H}_5\text{C}_6-\text{CH}_2-\text{CH}_2-\underset{\text{O}}{\overset{\|}{\text{C}}}-\text{CH}_3 \xrightarrow{\text{PCl}_5} \text{H}_5\text{C}_6-\text{CH}_2-\text{CH}_2-\underset{\text{Cl}}{\text{C}}=\text{CH}_2$$

$$\text{H}_5\text{C}_6-\text{CH}_2-\text{CH}_2-\text{CH}_2-\underset{\text{O}}{\overset{\|}{\text{C}}}-\text{CH}_3 \xrightarrow{\text{PCl}_5} \text{H}_5\text{C}_6-\text{CH}_2-\text{CH}_2-\text{CH}_2-\underset{\text{Cl}}{\text{C}}=\text{CH}_2$$

formation would be expected from carbonium intermediates. However, reaction of benzylidenedeoxybenzoin with PCl$_5$ in CH$_2$Cl$_2$ affords 1-chloro-2,3-diphenyl-indene[82].

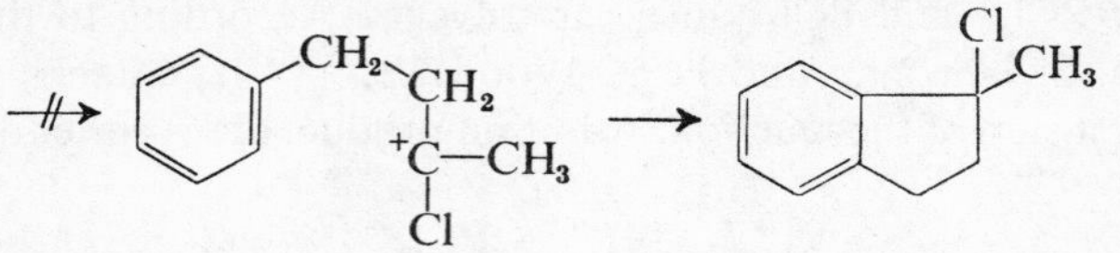

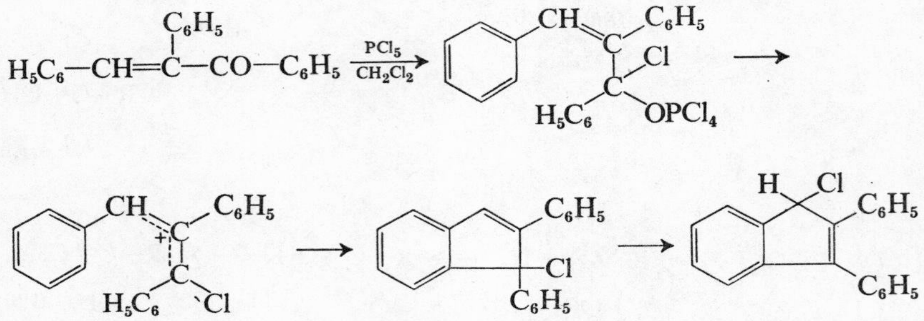

Furthermore, the mode of attack of PCl$_5$ on the C=O group seems to depend, in particular, on the solvent. Dissociation of PCl$_5$ to PCl$_4^+$ PCl$_6^-$ probably takes place only in strongly polar solvents. In less polar solvents such as CH$_2$Cl$_2$ or CCl$_4$, PCl$_5$ is not dissociated and attacks directly as monomer or dimer, yielding intermediates of type (**188**), which collapse to chloro olefin (**190**) or chloro carbonium ion (**187**).

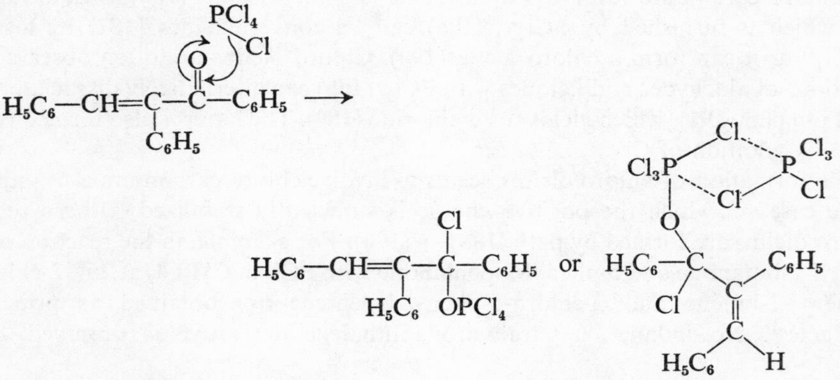

A-30. *Reaction of PCl$_5$ with 4-methyl-4-(trichloromethyl)-2,5-cyclohexadienone* (**192**)[83]

In example A-30, the chloro carbonium ion (**193**) is converted into a stable product by migration of a CH$_3$ group, *via* transition state (**194**) and with subsequent loss of a proton, yielding (**195**).

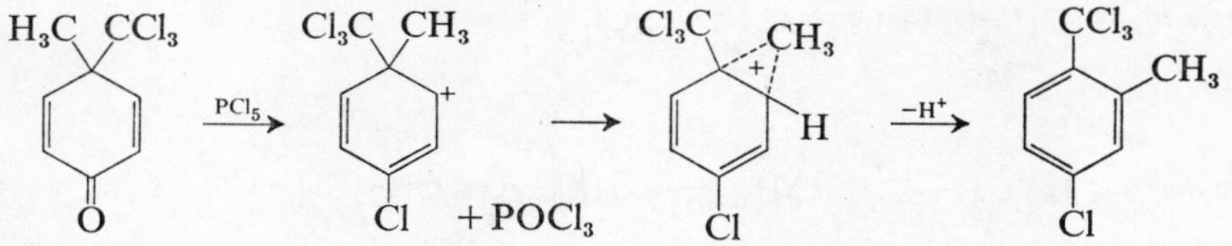

Addition Reactions

[Scheme: (192) → (193) + POCl₃ → (194) → (195), with PCl₅ and −H⁺ steps]

A-31. *Reaction of 3,4-dimethyl-4-(trichloromethyl)-2,5-cyclohexadienone* **(196)** *with* PCl_5[84]

In example A-31, 3,4-dimethyl-4-(trichloromethyl)-2,5-cyclohexadienone **(196)** is converted into a chloro carbonium ion **(197)**. Since in **(197)** migration of an alkyl

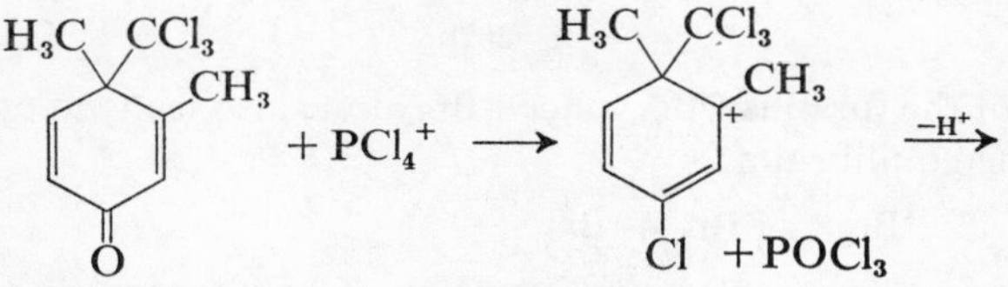

[Scheme: (196) + PCl_4^+ → (197) + $POCl_3$ → $-H^+$ → (198) → (199) → (200) 88%]

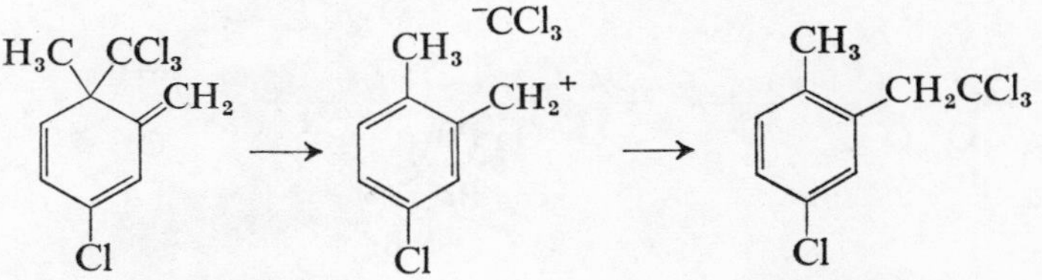

group does not lead to aromatization, the very reactive intermediate **(198)** is probably formed. This is then converted into 4-chloro-1-methyl-2-(2,2,2-trichloroethyl)benzene **(200)** by 1,3-migration of the CCl_3 group. The preferential migration of the CCl_3 group indicates that cleavage of the C–CCl_3 bond and the bond formation with the electrophilic exomethylene-carbon do not occur simultanously, because in that case migration of the more nucleophilic CH_3 group would be favored (see A-30). Presumably a tight ion pair **(199)** is formed first, which is converted into the product **(200)**.

A-32. *Reaction of 1,2-dibenzoylethylene* (*1,4-diphenyl-2-butene-1,4-dione*) **(201)** *with* PCl_5[85] (see formulae on p. 114)

In example A-32, dibenzoylethylene **(201)** cyclizes in the presence of PCl_5 to 3-chloro-2,5-diphenylfuran **(202)**. Cyclization occurs only if the C=C bond bears at least one hydrogen atom. With PBr_5, however, furan formation from *cis*-1,4-diketones is observed even if there is no such hydrogen atom. Possibly the agent

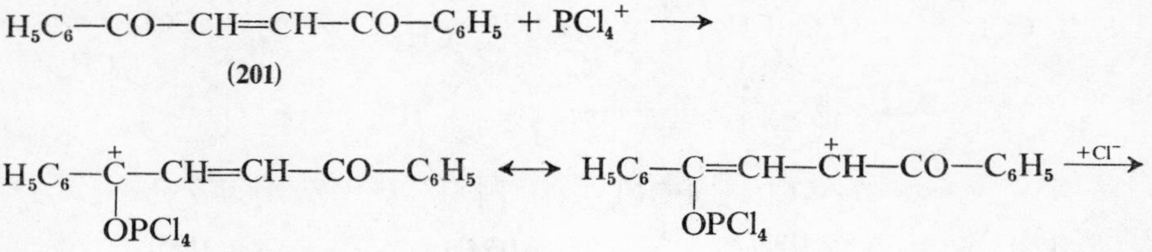

$H_5C_6-CO-CH=CH-CO-C_6H_5 + PCl_4^+ \longrightarrow$
(201)

$H_5C_6-\overset{+}{C}-CH=CH-CO-C_6H_5 \longleftrightarrow H_5C_6-\overset{+}{C}=CH-\overset{+}{C}H-CO-C_6H_5 \xrightarrow{+Cl^-}$
$\quad\quad |\!\!\!\!\!OPCl_4 \quad\quad\quad\quad\quad\quad\quad\quad\quad\quad\quad |\!\!\!\!\!OPCl_4$

[structure diagram showing cyclization to furan] $\xrightarrow{25-40°}$ [furan product] $+ POCl_3 + HCl$

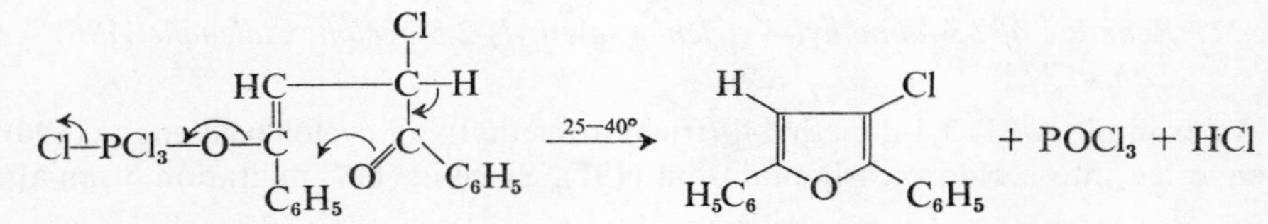

(202)

actually causing formation of the furan is PBr_3, since PBr_3 alone also leads to cyclization. PBr_3 is formed in the equilibrium:

$$PBr_5 \rightleftharpoons PBr_3 + Br_2$$

The reaction may be formulated in analogy to Newman (a) or according to a mechanism proposed by C. F. Wilcox Jr.[86] and J. B. Conant[87] (b).

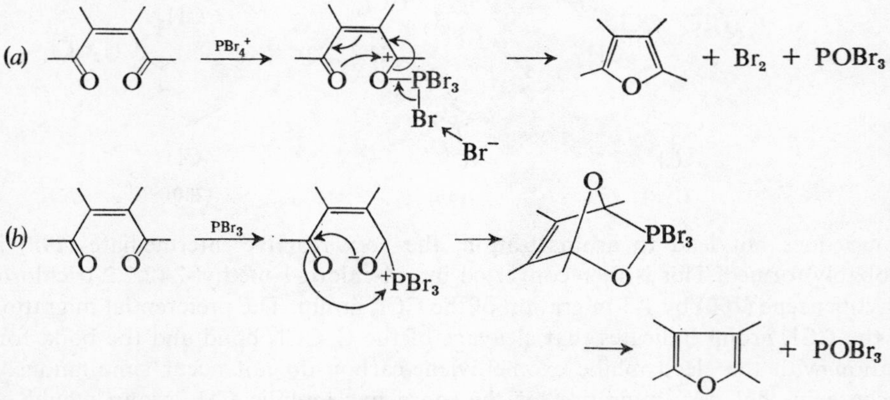

(a) [mechanism scheme via PBr_4^+] $+ Br_2 + POBr_3$

(b) [mechanism scheme via PBr_3] $+ POBr_3$

A-33. Meerwein–Ponndorf–Verley–Oppenauer reaction

The equilibrium between aldehydes or ketones and the corresponding alcohols in the presence of metal alkoxides is used for the preparative reduction of carbonyl compounds (Meerwein–Ponndorf–Verley reaction)[88] or for the oxidation of alcohols (Oppenauer reaction)[89]. The reaction is initiated by addition of the

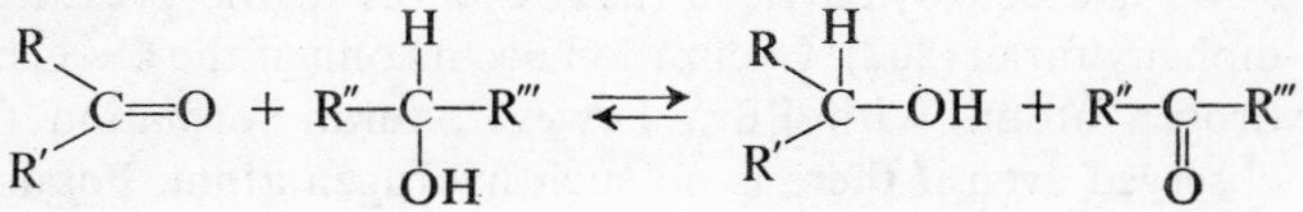

$\underset{R'}{\overset{R}{>}}C=O + R''-\underset{OH}{\overset{H}{\underset{|}{C}}}-R''' \rightleftharpoons \underset{R'}{\overset{R}{>}}C-OH + R''-\underset{O}{\overset{H}{\underset{\|}{C}}}-R'''$

aluminum alkoxide to the oxygen atom of the carbonyl group. In the, presumably, rate-determining step a hydride ion from the alkoxide group is transferred to the carbonyl-carbon atom[90]. Supporting evidence for this mechanism is the effect of substituents at the C=O group on the reaction rate; chloroacetone reacts faster than acetone, because the α-chlorine atoms increase the electrophilicity of the carbonyl-carbon by its $-I$ effect[91].

$$\begin{array}{c}R\\R'\end{array}\!\!\!>\!\!C\!=\!O \quad\rightleftarrows\quad \begin{array}{c}R\\R'\end{array}\!\!\!>\!\!\overset{+}{C}\!-\!O\!\!\!\diagdown\!\!\!\!\!\begin{array}{c}\\al^-\end{array} \quad\rightleftarrows\quad \begin{array}{c}R\\R'\end{array}\!\!\!>\!\!\overset{H}{\underset{|}{C}}\!-\!O\!\!\!\diagdown\!\!\!\!\!\begin{array}{c}\\al^-\end{array} \quad\rightleftarrows\quad \begin{array}{c}R\\R'\end{array}\!\!\!>\!\!CH\!-\!O\!-\!al$$

$$\begin{array}{c}R''\\R'''\end{array}\!\!\!>\!\!CH\!-\!O\!-\!al \qquad\qquad \begin{array}{c}R''\\R'''\end{array}\!\!\!>\!\!\overset{H}{\underset{|}{C}}\!-\!O \qquad\qquad \begin{array}{c}R''\\R'''\end{array}\!\!\!>\!\!\overset{+}{C}\!-\!O \qquad\qquad +\;\begin{array}{c}R''\\R'''\end{array}\!\!\!>\!\!C\!=\!O$$

In order to reduce aldehydes or ketones on a preparative scale, one usually uses aluminum isopropoxide in an excess of 2-propanol. The alcohol of the aluminum alkoxide that is formed by reduction of the carbonyl compound is continuously displaced by 2-propanol. Theoretically, only catalytic amounts of aluminum

$$\begin{array}{c}R\\R'\end{array}\!\!\!>\!\!C\!=\!O \;+\; al\!-\!O\!-\!CH\!\!\!<\!\!\begin{array}{c}CH_3\\CH_3\end{array} \quad\rightleftarrows\quad \begin{array}{c}R\\R'\end{array}\!\!\!>\!\!\overset{H}{\underset{|}{C}}\!-\!O\!-\!al \;+\; \overset{CH_3}{\underset{CH_3}{\overset{|}{C}\!=\!O}}$$

$$\begin{array}{c}R\\R'\end{array}\!\!\!>\!\!\overset{H}{\underset{|}{C}}\!-\!O\!-\!al \;+\; H\!-\!\overset{CH_3}{\underset{CH_3}{\overset{|}{C}\!-\!OH}} \quad\rightleftarrows\quad \begin{array}{c}R\\R'\end{array}\!\!\!>\!\!\overset{H}{\underset{|}{C}}\!-\!OH \;+\; al\!-\!O\!-\!CH\!\!\!<\!\!\begin{array}{c}CH_3\\CH_3\end{array}$$

alkoxide are necessary for the reduction; in order to shorten the reaction time, however, one uses at least a two- or three-fold excess.

A suitable ketone for the preparative oxidation of alcohols is acetone. Since, in order to be oxidized, the alcohol must be present as aluminum alkoxide, so one either prepares the aluminum alkoxide first or adds a stoichiometric amount of aluminum *tert*-butoxide, which exchanges butyl alcohol for the alcohol to be oxidized[92]. Adkins *et al.* have determined the oxidation potentials of numerous

$$R''R'''CHOH + Bu^t\!-\!O\!-\!al \rightleftarrows Bu^tOH + R''R'''CHOal$$
$$RR'C\!=\!O + R''R'''CH\!-\!O\!-\!al \rightleftarrows RR'CH\!-\!O\!-\!al + R''R'''C\!=\!O$$

ketones[93]. Ketones that are particularly useful for the oxidation have been found to be cyclohexanone and *p*-benzoquinone, as well as the inexpensive acetone. Instead of aluminum alkoxides, other metal alkoxides have also been used, for example, Mg, Sn, Zr, and alkali-metal alkoxides.

A-33. *Reduction of benzophenone*

Meerwein *et al.* observed that reduction of benzophenone yields, not only benzhydrol, but also 7% of diphenylmethane (**203**). Benzhydrol has been reduced under Meerwein–Ponndorf–Verley conditions to diphenylmethane in 28% yield[94]. Recently it has been shown that reduction to hydrocarbons takes place in very good yields if the C=O group is substituted by two aryl groups[95].

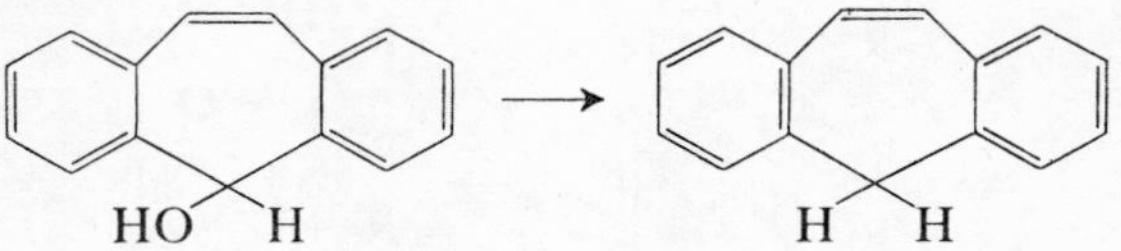

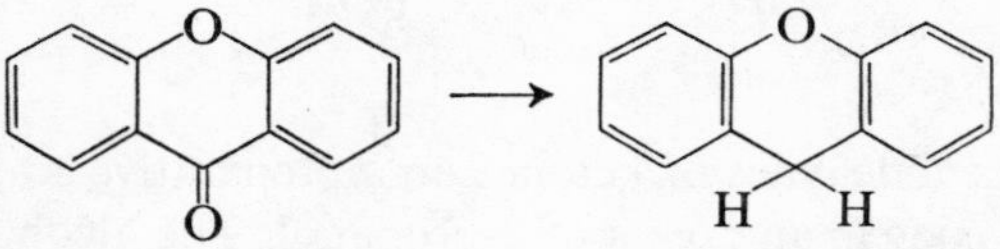

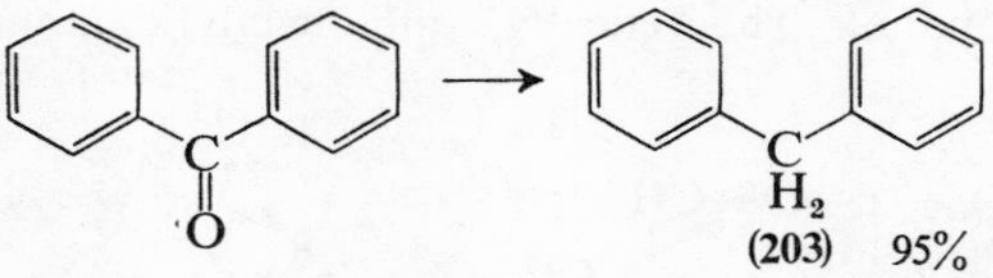

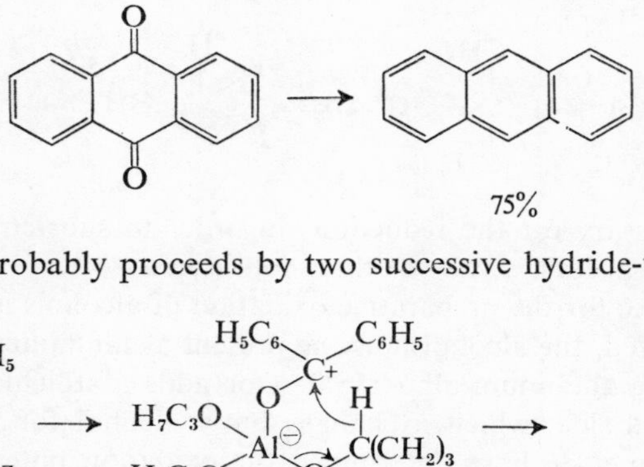

The reduction probably proceeds by two successive hydride-transfer reactions.

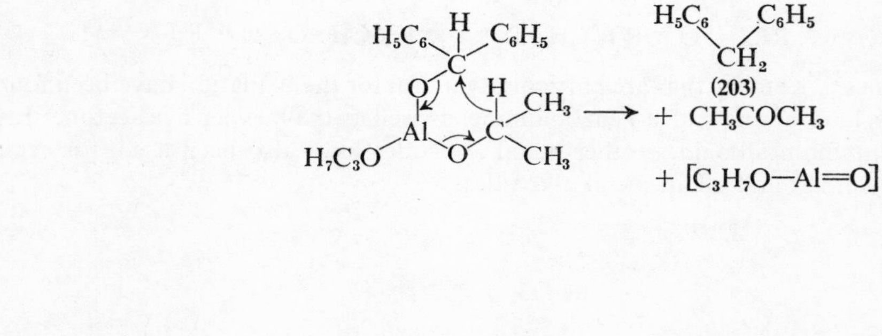

Another hydride-transfer reaction is the Cannizzaro reaction, which leads in basic
Other hydride-transfer reactions are the Cannizzaro reaction, which leads in basic
medium to alcohol and acid by disproportionation of two molecules of aldehyde.

$$R-\overset{H}{\underset{O}{C}} + OH^- \longrightarrow R-\overset{OH}{\underset{O^-}{\overset{|}{C}}}-H \quad \overset{R}{\underset{H}{\overset{|}{C}}}=O \xrightarrow{\text{Slow}}$$

$$R-\overset{OH}{\underset{O}{C}} + H-\overset{R}{\underset{H}{\overset{|}{C}}}-O^- \longrightarrow \begin{array}{c} R-COO^- \\ + \\ R-CH_2OH \end{array}$$

In the Tischtschenko reaction ester and alcohol are formed by reaction of two aldehyde molecules with an alkoxide ion.

$$R-\overset{H}{\underset{O}{C}} + {}^-O-R \longrightarrow R-\overset{OR}{\underset{O^-}{\overset{|}{C}}}-H \quad \overset{R}{\underset{R}{\overset{|}{C}}}=O \xrightarrow{\text{Slow}} R-\overset{}{\underset{O}{\overset{\|}{C}}}-OR + R-\overset{R}{\underset{O^-}{\overset{|}{C}}}-H$$

VI. ELECTROPHILIC ADDITIONS TO THE C≡N BOND

A-34 & A-35. Reaction of tertiary alcohols with CH$_3$CN in the presence of concentrated sulfuric acid[96]

Branched olefins and secondary or tertiary alcohols form carbonium ions[97] in concentrated sulfuric acid.* In the presence of nitriles, N-substituted amides are

$$\underset{H_3C}{\overset{H_3C}{>}}C=CH_2 + H_2SO_4 \longrightarrow$$

$$[(CH_3)_3C]^+ \ {}^-O-SO_3H \xrightarrow{CH_3CN} [H_3C-\overset{+}{C}=N-C(CH_3)_3]\ SO_4H^- \xrightarrow{H_2O}$$

$$H_3C-\underset{OH}{\overset{|}{C}}=N-C(CH_3)_3 \longrightarrow H_3C-\underset{O}{\overset{\|}{C}}-NH-C(CH_3)_3$$

obtained by addition of the carbonium ion to the nitrile-nitrogen atom and subsequent reaction with water.

* Other strongly acidic non-aqueous media for formation of carbonium ions are HF/SbF$_5$, HClO$_4$/CH$_3$COOH/(CH$_3$CO)$_2$O, and HBF$_4$/CH$_3$COOH/(CH$_3$CO)$_2$O.

A-34. (*i*) 3-(1-*Cyclopentenyl*)-2-*methyl*-2-*propanol* (**204**)

If the *N*-alkyl group of the addition product is substituted in the 3- or 4-position by OH or SH, or if a double bond is located in position 2 or 3, cyclization can yield a variety of heterocyclic systems[98]. For example, the alcohol (**204**) yields 63% of a dihydrooxazine (**206**), if 96% sulfuric acid is added dropwise to its cold solution in an excess of acetonitrile.

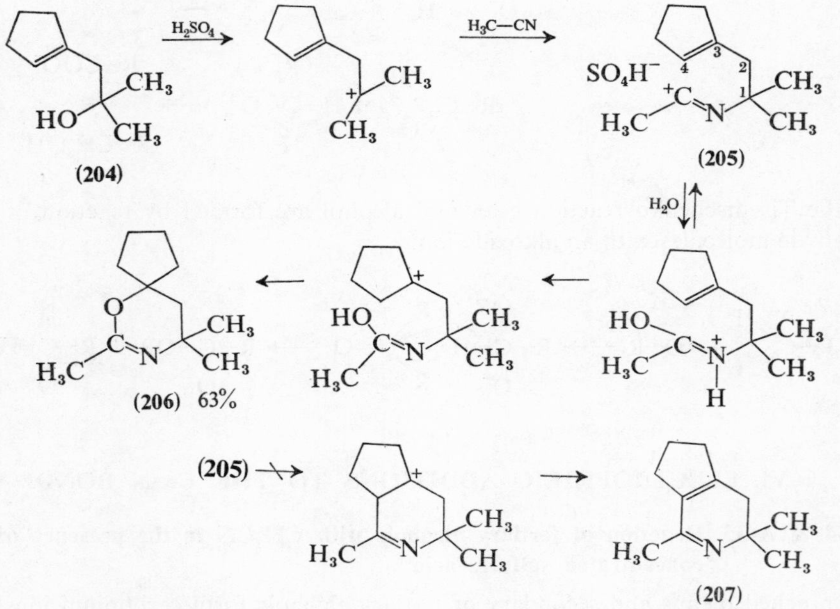

Characteristic IR absorption of (**206**): 1666 cm^{-1} —O—C=N—

The isomeric oxazine (**208**), which would be formed by prior addition of a proton to the double bond, has not been observed.

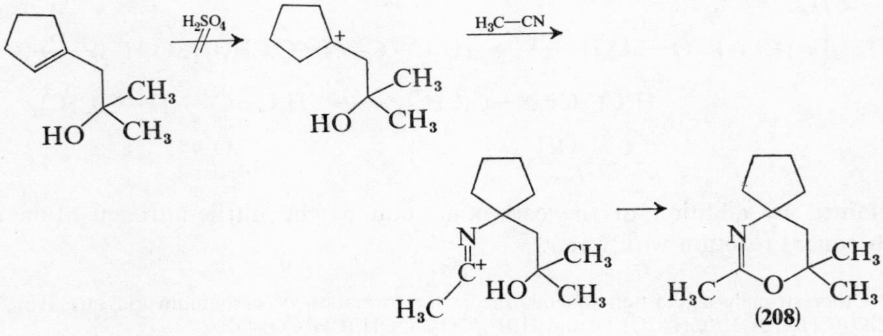

Formation of 4,5,6,7-tetrahydro-1,3,3-trimethyl-3*H*-cyclopenta[*c*]pyridine (**207**) from intermediate (**205**), which should be favored in the absence of water, has not been observed. If (**204**) was added to a mixture of acetonitrile and concentrated sulfuric acid in order to lower the initial water activity only polymerization took place.

However employment of a starting material (**209**), which did not possess a double bond but could be converted into an olefin *in situ* during the reaction, led to the desired product (**207**).[99]

A-35. (*ii*) *3-(2-Hydroxycyclopentyl)-2-methyl-2-propanol* (**209**)

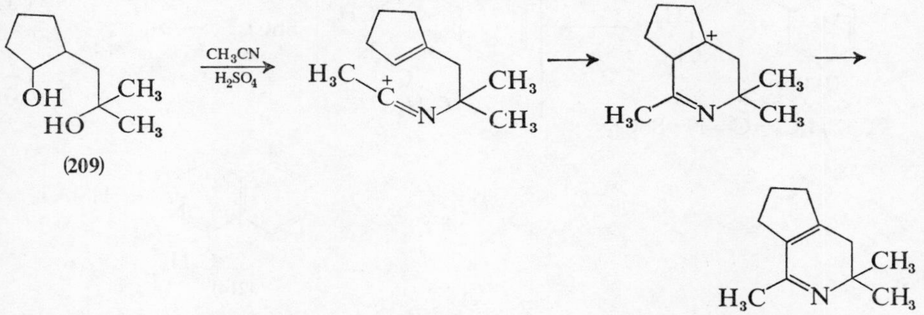

Characteristic UV absorption of (**207**): 263mμ log ε = 3.63 conj. system
Characteristic IR absorption of (**207**): 1667 cm^{-1} C=N conj.
1600 C=C conj.

The product (**207**) was also obtained from 3-(3-cyclopentenyl)-2-methyl-2-propanol (**210**) by adding it to a mixture of CH_3CN and concentrated H_2SO_4. If H_2SO_4 was added to the alcohol and CH_3CN, as in example A-34, only the oxazine (**206**) was obtained.

A-36. Addition of phenethyl chloride to benzonitrile in the presence of $SbCl_5$[100]

With strong complexing agents such as $SbCl_5$, primary alkyl halides can add to nitriles, yielding nitrilium salts (**211**). Actually the carbonium ion adds, not to the

$$R-C\equiv N + R'-Cl + SbCl_5 \rightleftarrows R-C^+=N-R'SbCl_6^-$$
(**211**)

free nitrile, but to a complex (**212**) of nitrile and metal halide.

$$R-C\equiv N + SbCl_5 \rightleftarrows R-C\equiv N\rightarrow SbCl_5$$
(**212**)

The rate-determining step of the reaction is presumably heterolysis of the carbon–halogen bond, which is caused by the complexing agent and results in a carbonium ion. The rate increases enormously on passing from primary to tertiary halides.[101] The formation of carbonium ions from primary, secondary, and tertiary alkyl fluorides and SbF_5 has been directly observed with the aid of NMR spectroscopy*[102].

The mesomeric nitrilium cation, which yields amides with water, cyclizes to heterocyclic compounds by electrophilic substitution if an aromatic nucleus is available. An example is the reaction of phenethyl chloride (213) with benzonitrile and $SbCl_5$.

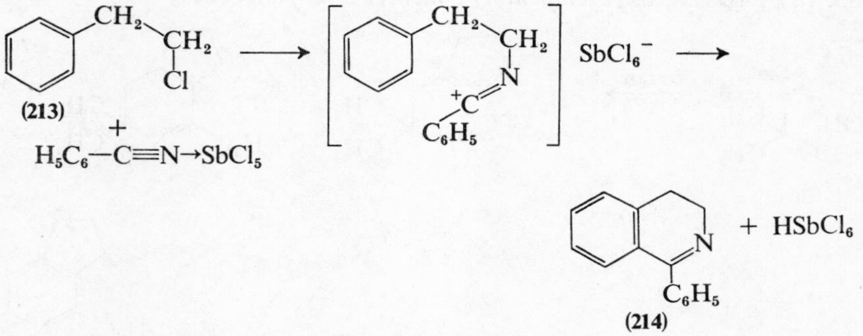

A-37. Formation of quinazolines by addition of nitrilium salts to nitriles[103]

By addition of a second nitrile molecule, nitrilium salts may form quinazolines (215).

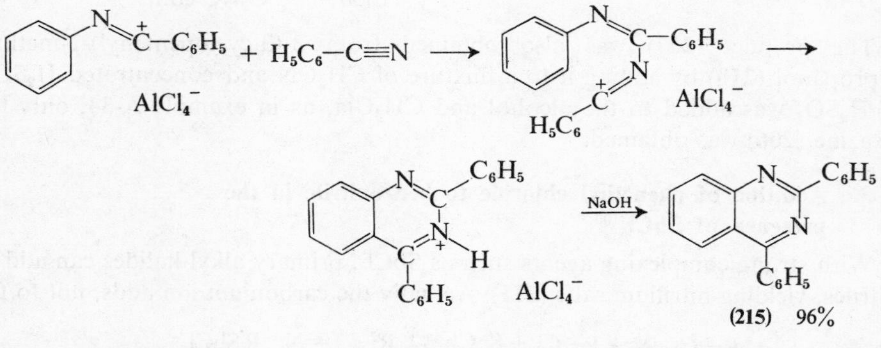

Alkylation of nitriles by oxonium salts (216), which have been shown by Meerwein to be extremely strong alkylation agents[104], also leads to nitrilium salts.

$$R-CN + [R_3'O]^+SbCl_6^- \rightarrow [R-C^+=N-R']SbCl_6^- + R_2'O$$
(216)

* See example A-34.

Other methods for the preparation of nitrilium salts are the reaction of N-substituted imidoyl chlorides (217) with $SbCl_5$ and of nitriles with diazonium fluoroborates.

$$R-CCl=N-R + SbCl_5 \rightarrow [R-C^+=N-R]SbCl_6^-$$
(217)

$$R-C\equiv N + RN_2^+BF_4^- \rightarrow [R-C^+=N-R]BF_4^- + N_2$$

Very stable salts (218) are obtained by reaction of nitrile–metal halide complexes (212) with acid chlorides.

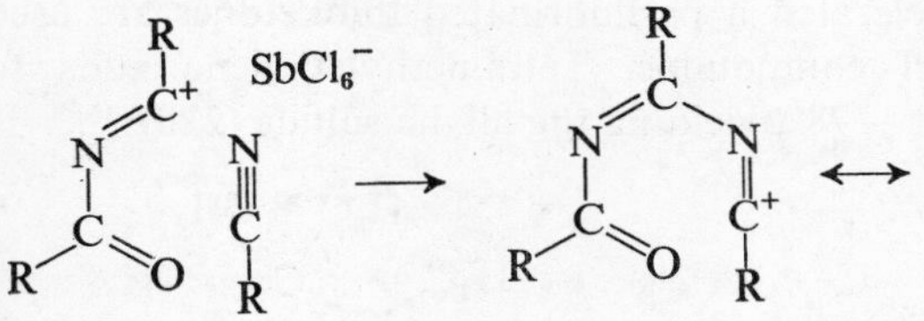

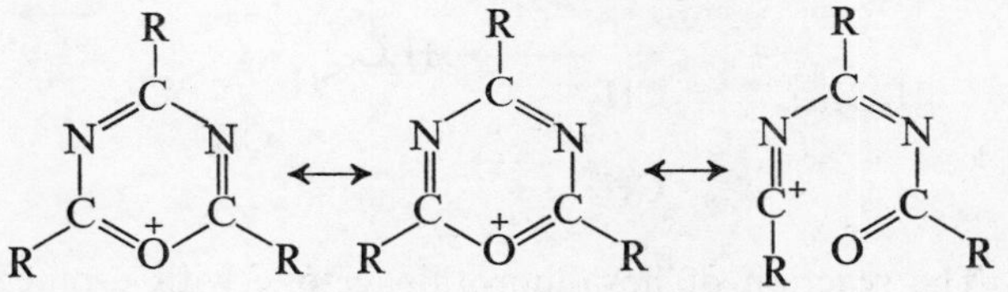

(218)

VII. MISCELLANEOUS ADDITIONS

A-38. ene-Addition of 2,3-dimethyl-2-butene to hexafluorothioacetone (219)[105]

Olefins containing allylic hydrogen add to C=C and C=O bonds at temperatures above 150° by "ene-addition"*.

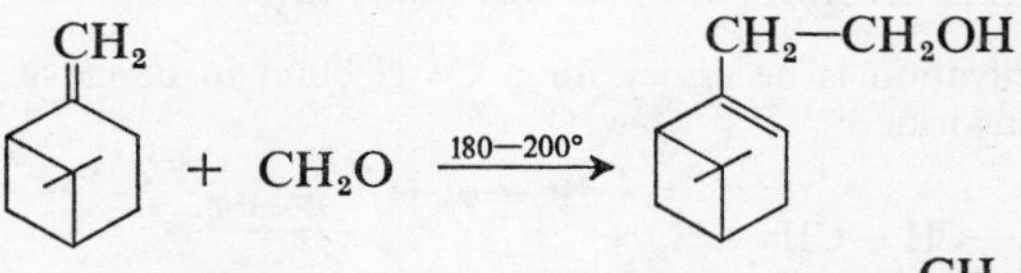

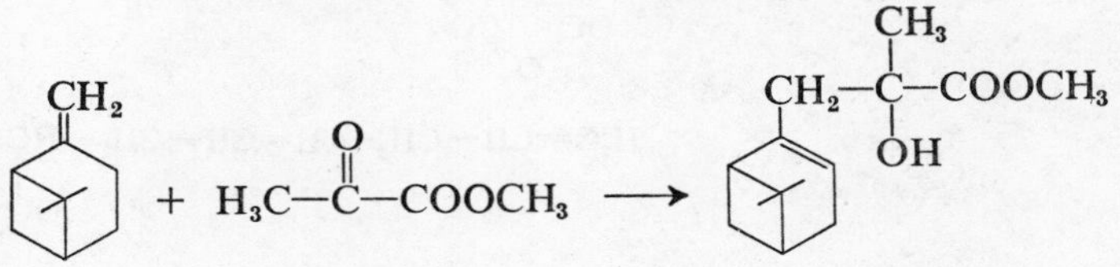

* See p. 122.

The reaction is assumed to proceed by a synchronous, ionic mechanism through a cyclic transition state. In the reverse reaction, β-hydroxy olefins can be cleaved to olefins and carbonyl compounds by heat†.

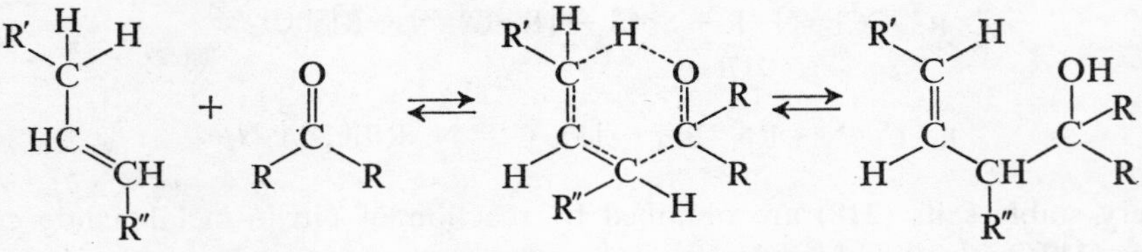

The stereochemistry of ene-addition has been investigated by J. A. Berson et al.[107] ene-Addition is very greatly accelerated if perfluorinated thioketones are used instead of unsubstituted carbonyl compounds. Tetramethylethylene adds to hexafluorothioacetone (219) even at −78°, yielding the allylic sulfide (220)[105].

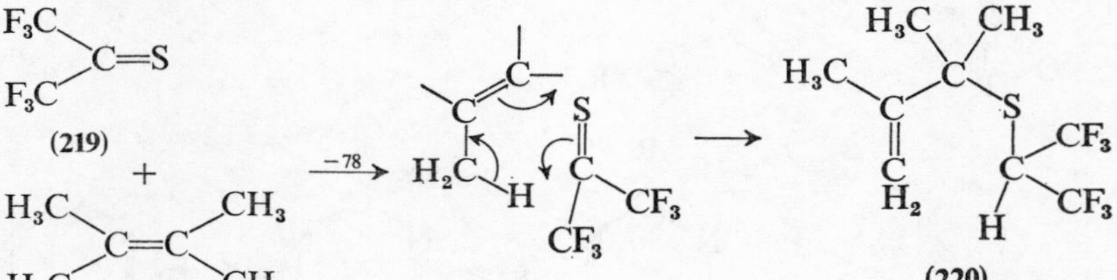

The reaction of hexafluorothioacetone with α-pinene (221) provides evidence that the addition takes place by a synchronous mechanism (222). A stepwise mechanism (223) should lead to a product (224). It is assumed that the allylic hydrogen of the olefin adds as a proton to the C=S carbon atom. The C=S bond seems to be polarized oppositely to the polarization of normal thioketones (dipole moment of thiobenzophenone in benzene at 20° = 3.40 D; that of benzophenone = 2.5—3.0 D)[108] as a consequence of the strong electron-attracting effect of the fluorine atoms. This polarization is also indicated by the reactions of hexafluorothioacetone with 2,2,2-trifluoroethanethiol (225) and with HCl[109].

---

* [From p. 121] Activation is necessary for a C=C bond to be capable of addition, as, for example, in maleic anhydride.[106]

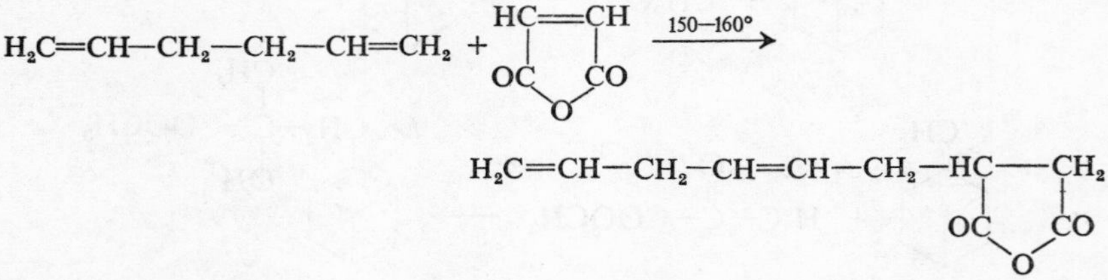

$$H_2C=CH-CH_2-CH_2-CH=CH_2 + \text{(maleic anhydride)} \xrightarrow{150-160°}$$

$$H_2C=CH-CH_2-CH=CH-CH_2-HC\underset{O}{\overset{}{\underset{OC\ \ CO}{\diagdown\diagup}}}CH_2$$

† See example E-3.

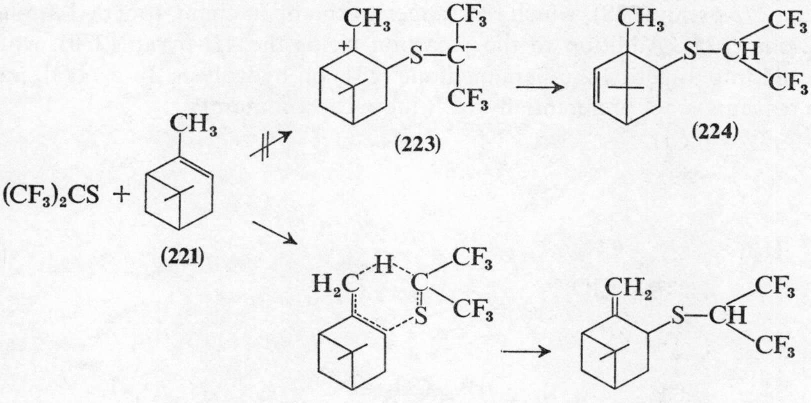

$$F_3C-CH_2-SH + F_3C-CS-CF_3 \xrightarrow{CsF} F_3C-CH_2-S-S-CH(CF_3)_2$$
(225)

$$HCl + 2F_3C-CS-CF_3 \longrightarrow (CF_3)_2CH-S-S-CCl(CF_3)_2$$

The C=S bond of perfluorinated thioketones is also a very reactive dienophile; for example, with butadiene the compound (226) is obtained[110].

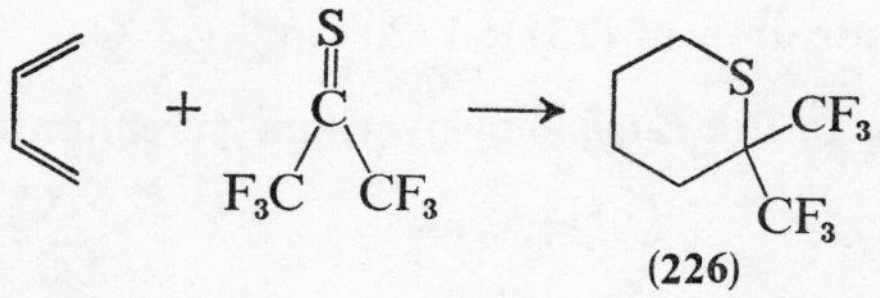

A-39. Hydride addition to 2,4,6-trimethylpyrylium perchlorate (227)*

Reduction of functional groups with complex metal hydrides such as LiAlH$_4$ and NaBH$_4$ takes place by transfer of a hydride anion to the center of lowest electron density†.

In the reduction of 2,4,6-trimethylpyrylium perchlorate (227) with NaBH$_4$, H⁻ adds either to the α- or the γ-carbon atom. Addition to the α-carbon leads to the

---

* For pyrylium salts see ref. 111. For reduction of pyridinium salts by NaBH$_4$ see ref. 112. For reduction of quinolizinium salts by LiAlH$_4$ or NaBH$_4$ see ref. 113.
† See example A-12.

unstable 2H-pyran (**228**), which rearranges to the open-chain 4-methyl-3,5-hepta-dien-2-one (**229**). Addition to the γ-carbon yields the 4H-pyran (**230**), which is converted into 4-methyl-2,6-heptanedione (**231**) on hydrolysis. In general, nucleophilic reagents react preferentially with the α-carbon atom[114].

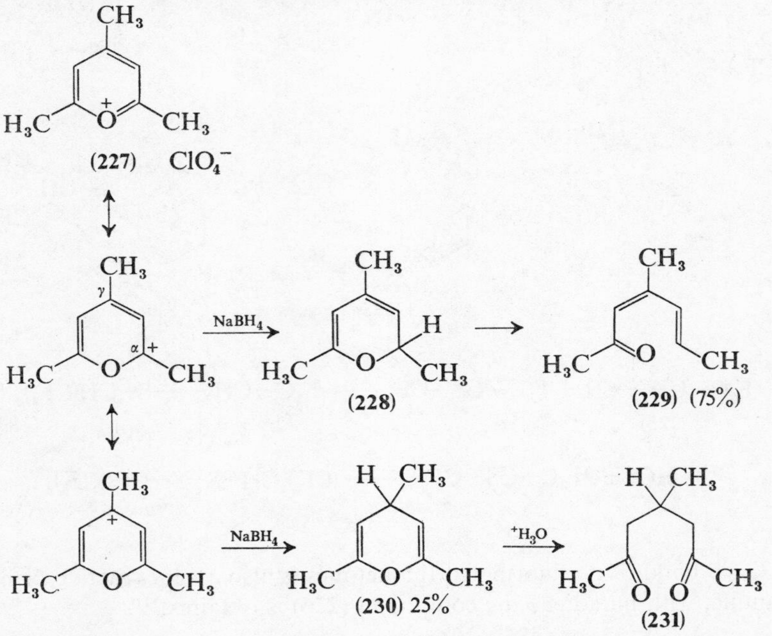

Characteristic absorption of (**229**): 401mμ conj. system; 1680 cm^{-1} C=O conj.
Characteristic IR absorption of (**231**): 1720 cm^{-1} C=O

A-40. Addition of NH_3 to 2,4,6-triphenylpyrylium perchlorate (**232**)[115]

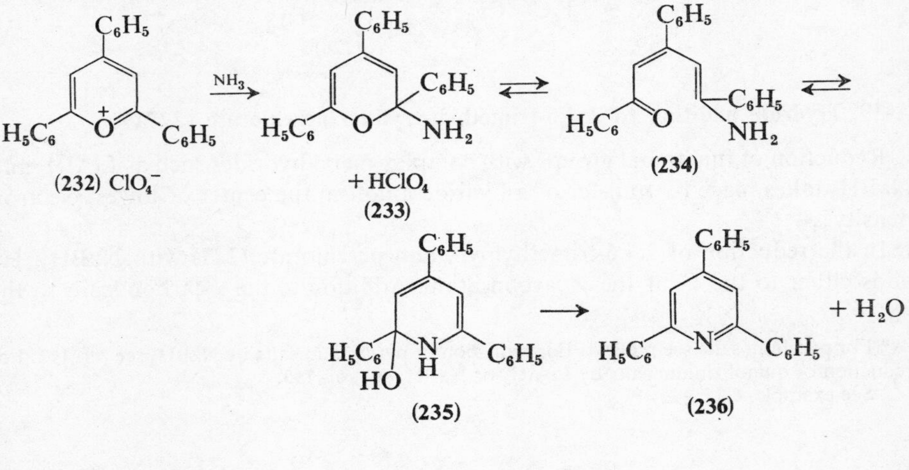

The nucleophilic NH_3 adds to the 2-position of pyrylium salt (232). Stabilization of the positive charge at the α-carbon by conjugation with suitable substituents facilitates the addition. The product (233) is in equilibrium with (234) and (235). Loss of water leads to triphenylpyridine (236). Pseudobases such as (233) can be isolated[116].

REFERENCES

[1] W. G. Young, R. T. Dillon, and H. J. Lucas, *J. Amer. Chem. Soc.*, **51**, 2528 (1929).
[2] G. S. Hammond and T. D. Nevitt, *J. Amer. Chem. Soc.*, **76**, 4121 (1954); G. S. Hammond and C. H. Collins, *ibid*, **82**, 4323 (1960).
[3] M. J. S. Dewar and R. C. Fahey, *Angew. Chem. Intern. Ed. Engl.* **3**, 245 (1964).
[4] L. Goodman, S. Winstein, and R. Boschan, *J. Amer. Chem. Soc.*, **80**, 4312 (1958); L. Goodman and S. Winstein, *ibid.*, **79**, 4788 (1957); M. Bergmann, F. Dreyer, and F. Radt, *Chem. Ber.*, **54**, 2139 (1921); G. Kresze, G. Schulz, and H. Zimmer, *Tetrahedron*, **18**, 675 (1962).
[5] B. P. Susz and J. J. Wuhrmann, Helv. Chim. Acta, **40**, 971 (1957).
[6] A. T. Balaban, D. Farcasiu, and C. D. Nenitzescu, *Tetrahedron*, **18**, 1075 (1962). For a review see: C. D. Nenitzescu and A. T. Balaban, "Friedel-Crafts and Related Reactions," ed. by G. A. Olah, Vol. III, p. 1033, Interscience Publishers, New York, 1964.
[7] H. Pines and J. M. Mavity, "The Chemistry of Petroleum Hydrocarbons," Vol. III, ed. by B. T. Brooks, C. E. Boord, S. S. Kurtz, and L. Schmerling, p. 9, Reinhold Publishing Corporation, New York, 1955.
[8] R. Criegee and H. Höver, *Chem. Ber.*, **93**, 2521 (1960).
[9] P. S. Bailey, *Chem. Rev.*, **58**, 925 (1958).
[10] P. S. Bailey, S. B. Mainthia, and C. J. Abshire, *J. Amer. Chem. Soc.*, **82**, 6136 (1960).
[11] *Chem. Ber.*, **93**, 689 (1960); see also P. S. Bailey, J. A. Thompson, and B. A. Shoulders, *J. Amer. Chem. Soc.*, **88**, 4098 (1966); F. L. Greenwood, *J. Org. Chem.*, **29**, 1321 (1964); L. J. Durham and F. L. Greenwood, *Chem. Commun.*, **16**, 843 (1967).
[12] L. D. Loan, R. W. Murray, and P. R. Story, *J. Amer. Chem. Soc.*, **87**, 737 (1965).
[13] R. W. Murray, R. D. Youssefyeh, and P. R. Story, *J. Amer. Chem. Soc.*, **88**, 3143, 3144, 3655 (1966); F. L. Greenwood, *ibid.*, p. 3146.
[13a] C. E. Bishop and P. R. Story, *J. Amer. Chem. Soc.*, **90**, 1905 (1968); P. R. Story, C. E. Bishop, J. R. Burgess, R. W. Murray, and R. D. Youssefyeh, *ibid.*, p. 1907; E. E. Erickson, R. T. Hansen, and J. Harkins, *J. Amer. Chem. Soc.*, **90**, 6777 (1968).
[14] M. I. Fremery and E. K. Fields, *J. Org. Chem.*, **29**, 2240 (1964).
[15] K. Mislow and H. M. Hellmann, *J. Amer. Chem. Soc.*, **73**, 244 (1951); K. Mislow, *ibid.*, **75**, 2512 (1953).
[16] L. I. Smith and H. H. Hoehn, *J. Amer. Chem. Soc.*, **63**, 1184 (1941); A. Orechoff, *Chem. Ber.*, **47**, 89 (1914).
[17] O. J. Sweeting and J. R. Johnson, *J. Amer. Chem. Soc.*, **68**, 1057 (1946).
[18] G. Stork, A. Brizzolara, H. Landesman, J. Szmuszkoviez, and R. Terell, *J. Amer. Chem. Soc.*, **85**, 207 (1963).
[19] G. Opitz, A. Griesinger, and H. W. Schubert, *Ann. Chem.*, **665**, 91 (1963).
[20] G. Opitz and I. Löschmann, *Angew. Chem.* **72**, 523 (1960); G. Opitz and H. Holtmann, *Ann. Chem.*, **684**, 79 (1965).
[21] L. Birkofer and G. Daum, *Chem. Ber.*, **95**, 183 (1962).
[22] K. C. Brannock, R. D. Burpitt, V. W. Goodlett, and J. G. Thweatt, *J. Org. Chem.*, **29**, 818 (1964).
[23] K. C. Brannock, A. Bell, R. D. Burpitt, and C. A. Kelly, *J. Org. Chem.*, **29**, 801 (1964).
[24] G. Opitz, H. Mildenberger, and H. Suhr, *Ann. Chem.*, **649**, 47 (1961).
[25] S. Hünig and E. Lücke, *Chem. Ber.*, **92**, 652 (1959); S. Hünig and W. Lendle, *ibid.*, **93**, 913 (1960).

[26] W. Reusch and R. Lemahieu, *J. Amer. Chem. Soc.*, **86**, 3068 (1964).
[27] M. S. Newman, *J. Amer. Chem. Soc.*, **73**, 4993 (1951).
[28] R. E. Lutz, L. T. Slade, and P. A. Zoretic, *J. Org. Chem.*, **28**, 1358 (1963).
[29] D. W. Boykon, Jr., and R. E. Lutz, *J. Amer. Chem. Soc.*, **86**, 5046 (1964); C. L. Wilson, *ibid.*, **69**, 3002 (1947).
[30] H. H. Wasserman, N. E. Aubrey, and H. E. Zimmerman, *J. Amer. Chem. Soc.*, **75**, 96 (1953).
[31] R. E. Lutz, personal communication.
[32] R. E. Lutz, *J. Amer. Chem. Soc.*, **56**, 1378 (1934).
[33] R. E. Lutz and C.-K. Dien, *J. Org. Chem.*, **23**, 1861 (1958).
[34] H. C. Brown and K. Ischikawa, *J. Amer. Chem. Soc.*, **83**, 4372 (1961); H. C. Brown, E. J. Mead, and B. C. Subba Rao, *ibid.*, **77**, 6201 (1955); N. L. Paddock, *Nature*, **167**, 1070 (1951); N. G. Gaylord, *Reduction with Complex Metal Hydrides*, Interscience Publishers, New York, 1956.
[35] O. L. Chapman, D. J. Pasto, and A. A. Griswold, *J. Amer. Chem. Soc.*, **84**, 1213 (1962).
[36] B. Franzus and E. J. Snyder, *J. Amer. Chem. Soc.*, **87**, 3423 (1965); **85**, 3902 (1963); P. R. Story, *J. Org. Chem.*, **26**, 287 (1961).
[37] K. Ziegler, H. Gellert, H. Martin, K. Nagi, and J. Schneider, *Ann. Chem.*, **589**, 91 (1954).
[38] Th. Kauffmann, *Angew. Chem. Intern. Ed. Engl.* 3, 342 (1964); see also Th. Kauffmann, K. Lötzsch, and D. Wolf, *Chem. Ber.* **99**, 3148 (1966); Th. Kauffmann, L. Bán, H. Hacker, S. M. Hage, J. Hansen, H. Henkler, Ch. Kosel, K. Lötzsch, H. Müller, E. Rauch, W. Schoeneck, J. Schulz, J. Sobel, S. Spaude, R. Weber, D. Wolf, and H. Zengel in *Neuere Methoden der präparativen organischen Chemie*, ed. W. Foerst, Vol. IV, p. 62, Verlag Chemie, Weinheim/Bergstrasse, 1966.
[39] P. Cossee, *J. Catalysis*, 3, 80, 89, 99 (1964).
[40] W. Pfohl, *Ann. Chem.*, **629**, 207 (1960).
[41] K. Ziegler, H. G. Gellert, K. Zosel, E. Holzkamp, J. Schneider, M. Söll, and W. R. Kroll, *Ann. Chem.*, **629**, 121 (1960).
[42] K. Ziegler, *Organometallic Chemistry*, ed. H. Zeiss, p. 194, Reinhold Publ. Corp., 1960.
[43] H. Höver, H. Mergard, and F. Korte, *Ann. Chem.*, **685**, 89 (1965).
[44] H. C. Brown, *Hydroboration*, W. A. Benjamin Inc., New York, 1962.
[45] F. Bohlmann, *Angew. Chem.*, **69**, 82 (1957).
[46] C. Prevost, P. Souchay, and Mlle. J. Chauvelier, *Bull. Soc. Chim. France*, **18**, 714 (1951).
[47] E. LeGoff and R. B. LaCount, *J. Org. Chem.*, **29**, 423 (1964).
[48] E. W. Garbisch Jr. and R. F. Sprecher, *J. Amer. Chem. Soc.*, **88**, 3435 (1966).
[49] W. H. Dietsche, *Tetrahedron Letters*, **1966**, 201.
[50] J. B. Hendrickson, C. Hall, R. Rees, and J. F. Templeton, *J. Org. Chem.*, **30**, 3312 (1965).
[51] J. J. Eisch and W. C. Kaska, *J. Amer. Chem. Soc.*, **84**, 1501 (1962); **88**, 2213 (1966); J. J. Eisch and S. M. E. Healy, *ibid.*, **86**, 4221 (1964).
[52] G. Wilke and H. Müller, *Ann. Chem.*, **629**, 222 (1960).
[53] G. Wilke and H. Müller, *Chem. Ber.*, **89**, 444 (1956).
[54] A. W. Laubengayer, *Special Publication No.* 15, p. 150, The Chemical Society, London, 1961.
[55] Ref. 52, p. 224.
[56] J. J. Eisch and W. C. Kaska, *J. Amer. Chem. Soc.*, **88**, 2976 (1966).
[57] A. Rieche, *Angew. Chem.*, **70**, 251 (1958).
[58] J. E. Leffler, *Chem. Rev.*, **45**, 410 (1949).
[59] G. B. Payne, *J. Org. Chem.*, **26**, 4793 (1961).
[60] A. Rieche, Lecture at the Max Planck Institut für Kohleforschung, Mülheim/Ruhr, 1966.
[61] S. I. Zavialov, L. P. Vinogradova, and G. V. Kondratieva, *Tetrahedron*, **20**, 2745 (1964).
[62] J. W. Cook, R. A. Raphael, and A. J. Scott, *J. Chem. Soc.*, **1952**, 4416.
[63] W. von E. Doering and D. B. Denney, *J. Amer. Chem. Soc.*, **77**, 4619 (1955).
[64] V. Mark, *J. Amer. Chem. Soc.*, **85**, 1884 (1963); *Org. Syn.*, **46**, 31, 42 (1966).
[65] G. Wittig and M. Schlosser, *Tetrahedron*, **18**, 1023 (1962).
[66] D. B. Denney, J. J. Vill, and M. J. Boskin, *J. Amer. Chem. Soc.*, **84**, 3944 (1962).
[67] F. Ramirez, E. H. Chen, and S. Dershowitz, *J. Amer. Chem. Soc.*, **81**, 4338 (1959).
[68] W. G. Bentrude and E. R. Witt, *J. Amer. Chem. Soc.*, **85**, 2522 (1963).
[69] L. Horner and K. Klüpfel, *Ann. Chem.*, **591**, 69 (1955); Z. Rappoport and S. Gertler, *J. Chem. Soc.*, **1964**, 1360.

Addition Reactions   127

[70] P. Yates, D. G. Farnum, and D. W. Wiley, *Tetrahedron*, **18**, 881 (1962).
[71] *Molecular Rearrangements*, ed. P. de Mayo, Vol. I, p. 22, Interscience Publishers, Inc., New York, 1963.
[72] T. W. Campbell, J. J. Monagle, and V. S. Foldi, *J. Amer. Chem. Soc.*, **84**, 3673 (1962); J. J. Monagle, T. W. Campbell, and H. F. McShane, Jr., *ibid.*, p. 4288.
[73] H. H. Jaffé, *Chem. Rev.*, **53**, 191 (1953).
[74] J. J. Monagle, *J. Org. Chem.*, **27**, 3851 (1962).
[75] H. Staudinger and J. Meyer, *Helv. Chim. Acta*, **2**, 640 (1919).
[76] D. Staschewski, *Angew. Chem.*, **71**, 726 (1959).
[77] D. M. Hall, J. E. Ladbury, M. S. Lesslie, and E. F. Turner, *J. Chem. Soc.*, **1956**, 3475.
[78] N. J. Cusack and B. R. Davis, *Chem. and Ind. (London)*, **1964**, 1426.
[79] D. Clark, H. M. Powell, and A. F. Wells, *J. Chem. Soc.*, **1942**, 642.
[80] M. S. Newman and L. L. Woods, Jr., *J. Amer. Chem. Soc.*, **81**, 4300 (1959).
[81] M. S. Newman, G. Fraenkel, and W. N. Kirn, *J. Org. Chem.*, **28**, 1851 (1963).
[82] M. S. Newman and G. Kangars, *J. Org. Chem.*, **30**, 3105 (1965).
[83] K. von Auwers and W. Julicher, *Chem. Ber.*, **55**, 2167 (1922).
[84] M. S. Newman and L. L. Woods, Jr., *J. Org. Chem.*, **23**, 1236 (1958).
[85] R. E. Lutz and F. N. Wilder, *J. Amer. Chem. Soc.*, **56**, 2145 (1934); R. E. Lutz and W. J. Welstead, *ibid.*, **85**, 755 (1963).
[86] C. F. Wilcox, Jr., and M. P. Stevens, *J. Amer. Chem. Soc.*, **84**, 1258 (1962).
[87] J. B. Conant and A. A. Cooke, *J. Amer. Chem. Soc.*, **42**, 830 (1920).
[88] A. L. Wilds, *Org. Reactions*, Vol. II, p. 178, ed. R. Adams, J. Wiley and Sons, New York, 1946.
[89] C. Djerassi, *Org. Reactions*, Vol. VI, p. 207, ed. R. Adams, J. Wiley and Sons, New York, 1951.
[90] R. B. Woodward, N. L. Wendler, and F. J. Brutschy, Jr., *J. Amer. Chem. Soc.*, **67**, 1425 (1945).
[91] B. J. Yager and C. K. Hancock, *J. Org. Chem.*, **30**, 1174 (1965).
[92] A. Lauchenauer and H. Schinz, *Helv. Chim. Acta*, **32**, 1265 (1949).
[93] Ref. 89, p. 228.
[94] H. Meerwein, B. von Bock, B. Kirschnick, W. Lenz, and A. Migge, *J. Prakt. Chem.*, **147**, 211 (1936).
[95] R. D. Hoffsommer, B. Taub, and N. L. Wendler, *Chem. and Ind. (London)*, **1964**, 482.
[96] J. J. Ritter and P. P. Minieri, *J. Amer. Chem. Soc.*, **70**, 4045 (1948); F. R. Benson and J. J. Ritter, *ibid.*, **71**, 4128 (1949); E. J. Tillmanns and J. J. Ritter, *ibid.*, **79**, 839 (1957); A. I. Meyers and W. Y. Libano, *J. Org. Chem.*, **26**, 1682 (1961); A. I. Meyers, *J. Org. Chem.*, **25**, 1147 (1960).
[97] N. C. Deno and J. J. Houser, *J. Amer. Chem. Soc.*, **86**, 1741 (1964); N. C. Deno, D. B. Boyd, J. D. Hodge, C. U. Pittman, Jr., and J. O. Turner, *ibid.*, p. 1745; N. C. Deno and C. U. Pittman, Jr., *ibid.*, p. 1744.
[98] L. M. Trefonas, J. Schneller, and A. I. Meyers, *Tetrahedron Letters*, **1961**, 785; A. I. Meyers, J. Schneller, and N. K. Ralhan, *J. Org. Chem.*, **28**, 2944 (1963).
[99] A. I. Meyers and J. J. Ritter, *J. Org. Chem.*, **23**, 1918 (1958).
[100] M. Lora-Tamayo, R. Madronero, and G. G. Munoz, *Chem. Ber.*, **93**, 289 (1960); M. Lora-Tamayo, R. Madronero, G. G. Munoz, and M. Stud, *ibid.*, **94**, 199 (1961).
[101] H. Meerwein, P. Laasch, R. Mersch, and J. Spille, *Chem. Ber.*, **89**, 209 (1956).
[102] G. A. Olah, W. S. Tolgyesi, S. J. Kuhn, M. E. Moffatt, I. J. Bastien, and E. B. Baker, *J. Amer. Chem. Soc.*, **85**, 1328 (1963); G. A. Olah, M. B. Comisarow, C. A. Cupas, and C. U. Pittman Jr., *ibid.*, **87**, 2997 (1965); G. A. Olah, E. B. Baker, J. C. Evans, W. S. Tolgyesi, J. S. McIntyre, and J. J. Bastien, *ibid.*, **86**, 1360 (1964).
[103] H. Meerwein, P. Laasch, R. Mersch, and J. Nentwig, *Chem. Ber.*, **89**, 224 (1956).
[104] R. Criegee, *Angew. Chem.*, **78**, 347 (1966).
[105] W. J. Middleton, E. G. Howard, and W. H. Sharkey, *J. Amer. Chem. Soc.*, **83**, 2589 (1961).
[106] K. Alder, F. Pascher, and A. Schmitz, *Chem. Ber.*, **76**, 27 (1943); R. T. Arnold and J. S. Showell, *J. Amer. Chem. Soc.*, **79**, 419 (1957).
[107] J. A. Berson, R. G. Wall, and H. D. Perlmutter, *J. Amer. Chem. Soc.*, **88**, 187 (1966).
[108] C. E. Hunter and J. R. Partington, *J. Chem. Soc.*, **1933**, 87.
[109] W. J. Middleton, *J. Org. Chem.*, **30**, 1395 (1965); W. J. Middleton and W. H. Sharkey, *ibid.*, p. 1384.
[110] W. J. Middleton, *J. Org. Chem.*, **30**, 1390 (1965).

[111] A. T. Balaban, G. Mihai, and C. D. Nenitzescu, *Tetrahedron*, **18,** 257 (1962).
[112] P. S. Anderson, W. E. Krueger, and R. E. Lyle, *Tetrahedron Letters*, **1965,** 4011.
[113] T. Miyadera and Y. Kishida, *Tetrahedron Letters*, ibid., p. 905.
[114] J. Kontecky, *Coll. Czech. Chem. Commun.*, **24,** 1608 (1959); A. T. Balaban and C. D. Nenitzescu, *Chem. Ber.*, **93,** 599 (1960).
[115] K. Dimroth in *Neuere Methoden der Präparativen Organischen Chemie*, Vol. III, p. 239, Verlag Chemie, Weinheim/Bergstraße, 1961; K. Dimroth, K. Wolf, and H. Kroke, *Ann. Chem.*, **678,** 183 (1964).
[116] J. A. Berson, *J. Amer. Chem. Soc.*, **74,** 358 (1952); H. R. Hensel, *Ann. Chem.*, **611,** 97 (1958).

2

Cycloadditions

SINCE cycloadditions have received considerable attention during recent years, they are treated separately in the present Chapter.

Addition-reactions, which lead to cyclic compounds, may be divided in four groups:

1. In 1,1-additions one reactive center forms two bonds with an olefin. The products are three-membered rings. To this group belong: the epoxidation of olefins, the addition of carbenes and carbenoids to unsaturated compounds, and reactions which result in unstable, cyclic intermediates as in the bromination of olefins, which proceeds *via* a cyclic bromonium ion*. (Additions to "stable carbenes" such as isocyanides and fulminic acid and its derivatives are also 1,1-additions, although they do not usually lead to cyclic products. An exception is discussed in example C-10, page 142.) The reverse reaction is α-elimination (see page 137).

2. In 1,2-additions, two adjacent reaction centers form four-membered rings by addition to a second unsaturated molecule.

3. In 1,3-additions a zwitterion adds to an unsaturated bond, yielding a five-membered ring.

4. A well-known 1,4-addition is the Diels–Alder reaction which leads to six-membered rings.

A detailed classification of cycloadditions has been given by R. Huisgen, R. Grashey, and J. Sauer[1] and is shown in the Table on p. 130. The numbers on the left-hand side of the numerical equations indicate the relative positions of the reactive centers in the two participating molecules (for example, 4 indicates 1,4-addition). The number on the right-hand side indicates the number of atoms contained in the new ring of the cyclic product.

The same authors formulated six rules, which characterize cycloadditions:

1. The product of a cycloaddition is the sum of the components. Cycloadditions are not accompanied by elimination of small molecules.
2. Cycloadditions do no involve the breaking of σ-bonds.
3. The number of σ-bonds increases in cycloadditions.
4. Cycloadditions of more than two reactants are multi-step processes; only the

---

* See example A-2.

last step, leading to the ring structure, is strictly speaking a cycloaddition.

5. Intramolecular cycloadditions are possible if the necessary functional groups are present in the same molecule.

6. The term cycloaddition should only be used if the cyclic product has a finite life time. The ring structure must be an intermediate, not merely a transition state.

Classification of cycloadditions	
Combinations	Examples
2 + 1 = 3	Olefins + carbenes
2 + 2 = 4	Cyclobutanes from two unsaturated molecules
3 + 2 = 5	1,3-Dipolar addition
4 + 1 = 5	Dienes + SO_2
3 + 3 = 6	Dimerization of 1,3-dipoles
4 + 2 = 6	Diels–Alder addition
4 + 3 = 7	Anthracene + ozone?
4 + 4 = 8	Photodimerization of anthracene
6 + 2 = 8	4HC≡CH → cyclooctatetraene?
8 + 4 = 12	Addition of butadiene dimer to butadiene during trimerization of butadiene*

I. 1,1-ADDITIONS

C-1 & C-2. Epoxidations

Three mechanisms have been proposed for the epoxidation of olefins with peracids: (1) Synchronous addition of oxygen to the double bond without preceding heterolysis of the peracid[2] (mechanism 1). (2) Cleavage of the peracid into $RCOO^-$ and OH^+, followed by addition of OH^+ to the double bond; presumably a π-complex is formed first, which is converted into the epoxide by loss of a proton (mechanism 2); this mechanism should be favored in strongly polar solvents, for

Mechanism 1   Mechanism 2

example, in HCOOH. (3) Recently a third mechanism has been proposed[3] that is related to the formation of epoxides by addition of the ozone zwitterion† to

---

* The trimerization of butadiene and similar reactions have been described by Wilke *et al.*[1a]
† See example A-4.

certain olefins, for example, to tetracyanoethylene. By synchronous 1,3-addition of zwitterion (238) to the olefin, a hydroxyperoxide (239) is formed, which decomposes to a carboxylic acid and an epoxide.* The same mechanism has been proposed for the epoxidation of tetracyanoethylene by the ion (240). All three mechanisms

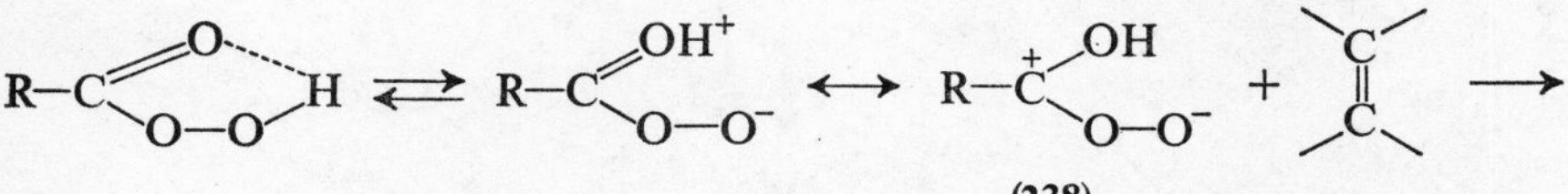

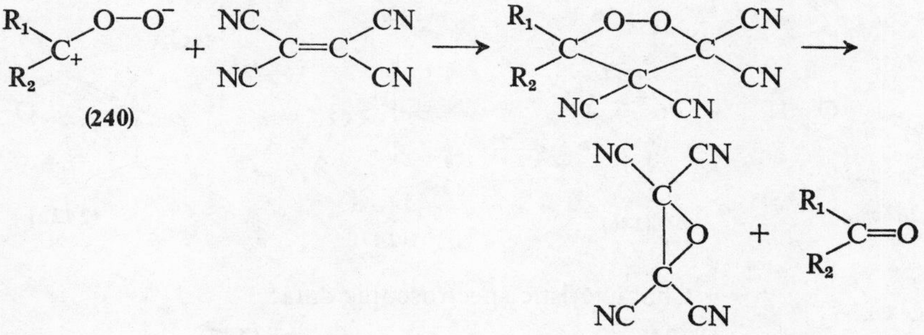

explain the stereospecificity of the epoxidation by which *cis*-olefins are oxidized to *cis*-epoxides and *trans*-olefins to *trans*-epoxides.

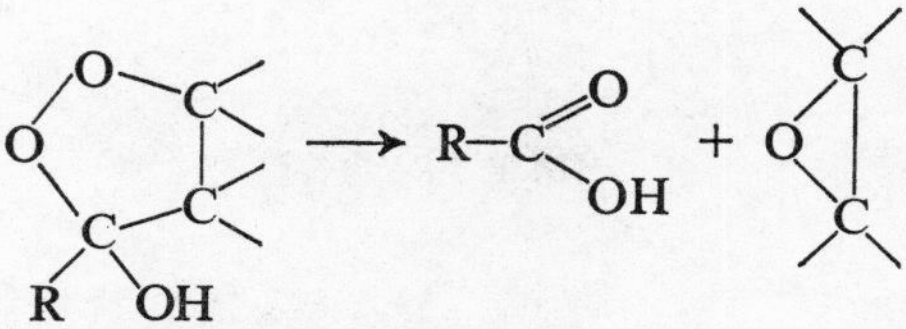

C-1. Epoxidation of bicycloheptadiene[4] (see formulae on pp. 132—133)

The epoxidation of bicycloheptadiene presumably leads first to the epoxide (241), which is cleaved to the equilibrating cations (242a), (242b), and (242c)†. Possibly the cations are formed directly by addition of OH^+ to one carbon atom of the double bond via a π-complex (243), since the epoxide (241) could not be detected. Cation (242) may rearrange to more stable products in different ways. In analogy to the rearrangement of norbornene epoxide (244) in the presence of acid catalysts, (242c) could rearrange to 1,2-dihydrobenzaldehyde (245). This rearrangement has not

---
* But see Bingham et al.[3a]
† See non-classical carbonium ions (p. 265).

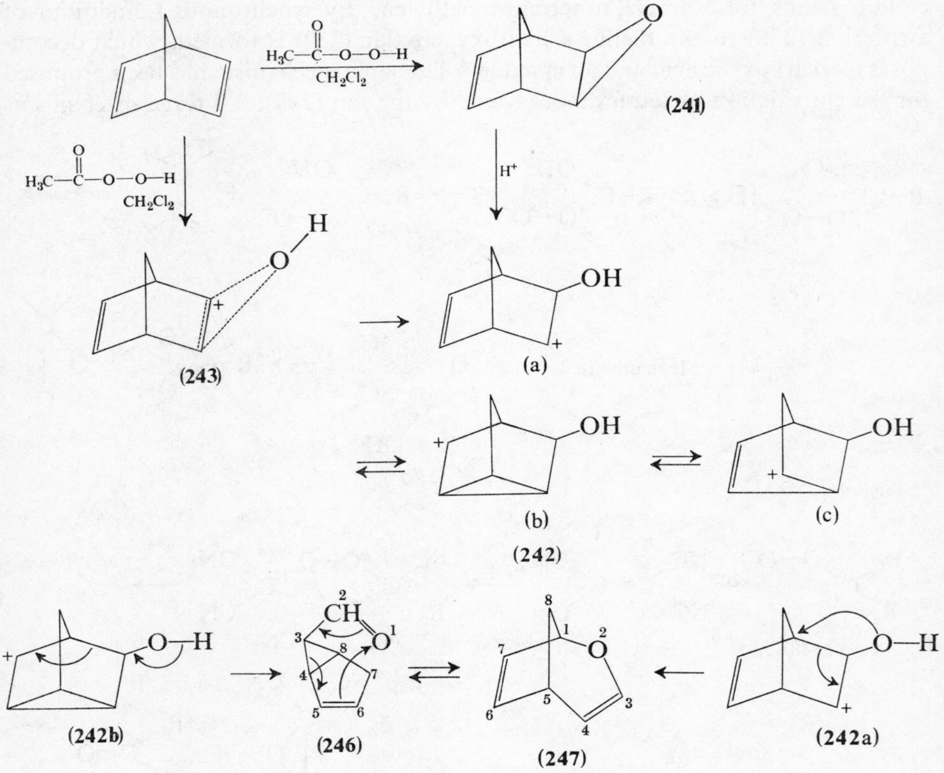

Characteristic spectroscopic data:

of (246)			of (247)		
1695 cm^{-1}		—CHO	1.9—1.8 τ	(2) multiplet	H$_8$
0.03 τ	(1) doublet	H$_2$	2.8—2.4	(1) multiplet	H$_5$
1.59	(1) quartet	H$_3$	4.83	(1) singlet	H$_1$
2.8—2.0	(4) multiplet	H$_4$, H$_7$, H$_8$	4.93	(1) 2 doublets	H$_4$
5.8	(2) singlet	H$_5$, H$_6$	5.32	(1) 2 doublets	H$_7$
			5.75	(1) doublet	H$_3$
			6.39	(1) 2 doublets	H$_6$

been observed. Instead of the compound (245), an aldehyde (246) in equilibrium with (247) has been isolated in 70% yield. The *endo*-structure of (246) has been established by lactonization by iodine. Rey and Dreiding[5] have shown that (246) is in equilibrium with its valence isomer (247) at room temperature [(246):(247) = 7:3]. The equilibrium is shifted completely to the side of the aldehyde in reactions involving the aldehydic carbonyl group. The conversion of

(244)

(242c) **(245)**

(246)

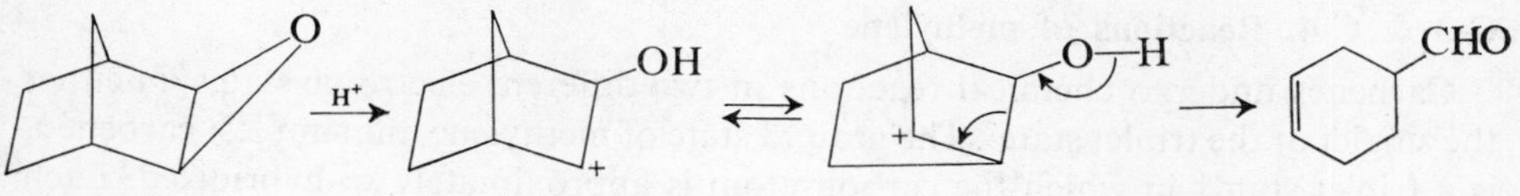

(**246**) into (**247**) corresponds to the Cope rearrangement* and is therefore called an oxy-Cope rearrangement. Like (**246**), which rearranges to (**247**), the *cis*-isomer of compound (**248**) equilibrates with the dihydrooxepin (**249**) already at 50°. However, above 400°, (**248**) rearranges irreversibly to (**250**)[5a].

(249) **(248)** **(250)**

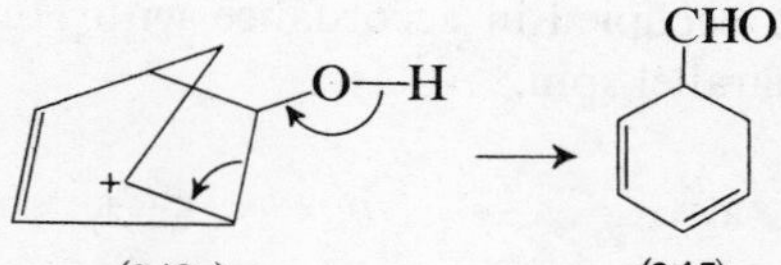

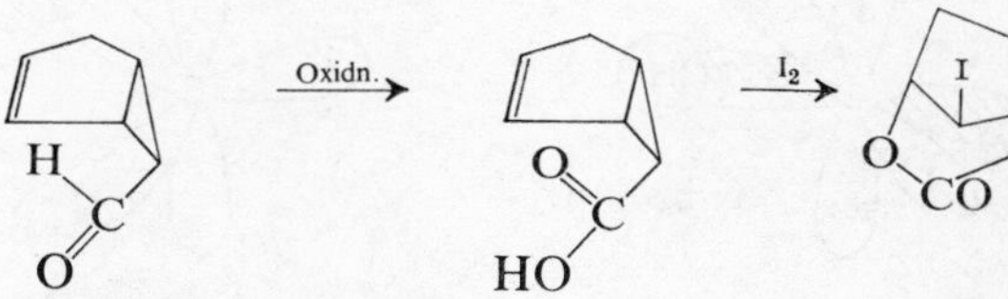

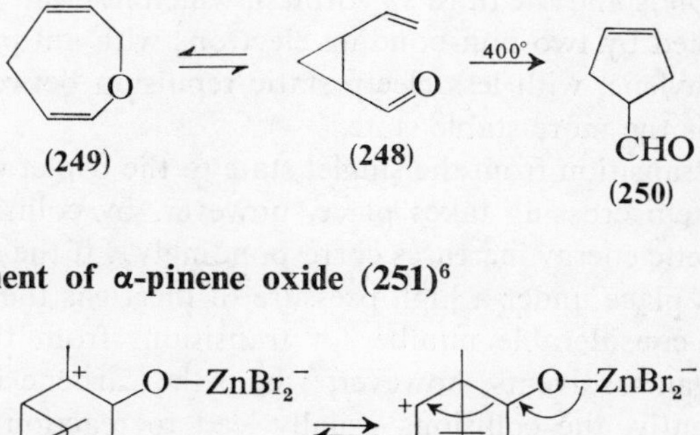

C-2. Rearrangement of α-pinene oxide (251)[6]

(251) **(252)**

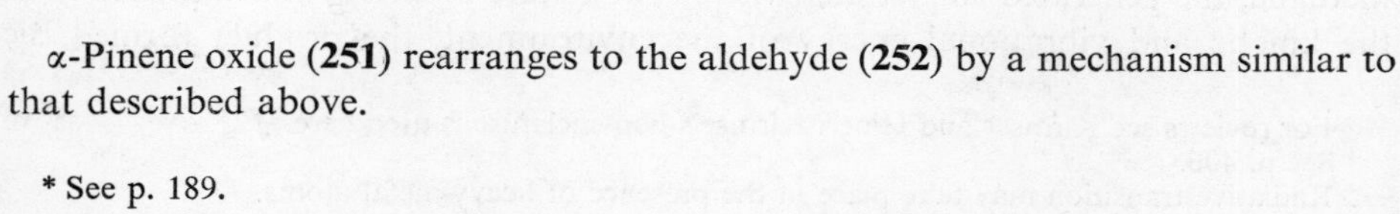

α-Pinene oxide (**251**) rearranges to the aldehyde (**252**) by a mechanism similar to that described above.

* See p. 189.

C-3 & C-4. Reactions of methylene

Carbenes undergo chemical reactions in two different electronic states, in either the singlet or the triplet state[7]. The ground state of methylene, the simplest carbene*, is a triplet state† in which the carbon atom is approximately sp-hybridized. Each of the sp-orbitals is used for a C–H bond. The molecule is linear and the two degenerate p-orbitals are occupied in accordance with Hund's rule, each by one electron; the pair have parallel spin.

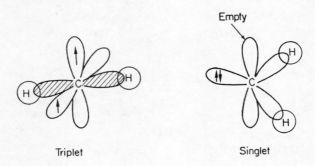

Triplet   Singlet

Most methods of generating methylene, for example by photolysis of diazomethane or ketene or by thermal decomposition of diazirine or diazomethane, lead to singlet methylene.

In the singlet state the methylene carbon is sp^2-hybridized. Two sp^2-orbitals are used for the C–H bonds and the third sp^2-orbital, which has more s-character than the others, is occupied by two non-bonding electrons with antiparallel spins. The triplet state of methylene, with less electrostatic repulsion between the two non-bonding electrons, is the more stable state.

Simple radiative transition from the singlet state to the triplet ground state does not occur; intersystem crossing takes place, however, by collision with another molecule whose kinetic energy increases correspondingly‡. If the decomposition of diazomethane takes place under a high pressure of inert gas the great number of collisions causes a considerable number of transitions from the singlet to the triplet state. In organic solvents, however, where the carbene intermediate also collides very frequently, the collisions usually lead to reaction with the solvent before transition to the triplet state occurs.

Since by some methods of preparation, for example by photolysis of diazomethane, the generated singlet methylene is in a state of energy much higher than the kinetic and vibrational energy of the environment, the product formed by

---

* For reviews see Kirmse[8] and Hine[9]. Kirmse's nomenclature is used here.
† See p. 406.
‡ Radiative transition may take place in the presence of heavy-metal atoms.

collision with the "hot" singlet methylene often rearranges in a secondary reaction. For example, *cis*-dimethylcyclopropane, which is formed primarily by reaction of singlet methylene with *cis*-2-butene, rearranges partly to *trans*-dimethylcyclopropane.

C-3. (*i*) *Reactions of methylene with cyclohexene*[10]

Triplet methylene may be generated exclusively if diazomethane is decomposed photolytically in a solvent in the presence of a sensitizer, for example, benzophenone. A sensitizer is a molecule, having a relatively low-lying first excited singlet state, which is rapidly converted into the triplet state by collision. Since the triplet state of the sensitizer is of somewhat higher energy than the triplet state of diazomethane, the triplet energy is transfered directly to the diazomethane by collision†.

$(C_6H_5)_2C=O \xrightarrow[3130 \text{ Å}]{h\nu} (C_6H_5)_2C=O^*$ (singlet)

$(C_6H_5)_2C=O^*$ (singlet) $\longrightarrow (C_6H_5)_2C=O^*$ (triplet)

$(C_6H_5)_2C=O^*$ (triplet) $+ CH_2N_2 \longrightarrow (C_6H_5)_2C=O + CH_2N_2^*$ (triplet)

$CH_2N_2^*$ (triplet) $\longrightarrow CH_2$ (triplet) $+ N_2$

Singlet and triplet methylene differ characteristically in their chemical reactivities. Singlet methylene adds stereospecifically to olefins, yielding cyclopropyl derivatives of the same configuration as the olefin. (Note the analogy to epoxidation.) It is inserted into C–H bonds, presumably by a three-centre mechanism. The selectivity between primary, secondary, and tertiary C–H bonds is only insignificant. If insertion takes place into the C–H bond of an asymmetric C atom, the optical activity is preserved.

Triplet methylene, however, does not add stereospecifically to double bonds. This is to be expected, since as a consequence of the diradical character of triplet methylene one bond is formed first, thus permitting rotation around the C–C axis of the olefin. The insertion into C–H bonds is considerably less pronounced, since triplet methylene is a more stable carbene. Insertion into the C–H bond of an asymmetric carbon leads to a racemic product.

† See "Photochemistry" on p. 410.

$$-\overset{|}{\underset{|}{C}}-H + \uparrow\uparrow CH_2 \longrightarrow -\overset{|}{\underset{|}{C}}\bullet + \bullet CH_3 \longrightarrow -\overset{|}{\underset{|}{C}}-CH_3$$

$$\underset{R}{\overset{R}{>}}C=C\underset{R}{\overset{R}{<}} + \uparrow\uparrow CH_2 \longrightarrow \underset{}{>}C\overset{\bullet CH_2}{\underset{\bullet}{C}}C< \longrightarrow \underset{R}{\overset{R}{\triangle}}\underset{}{\overset{R}{}} + \underset{}{\overset{R}{\triangle}}\underset{}{\overset{R}{}}$$

Cyclopropyl derivatives may be prepared by a third method. If CH_2I_2 is treated with a Zn/Cu couple or if a diazo compound is heated with copper powder in the presence of an olefin, formal addition of CH_2 to the double bond takes place. The nature of the reactive intermediate is not exactly known; it differs, however, from singlet and triplet methylene. For example, it is not capable of insertion into a C–H bond.

Cf. reactions of methylene with cyclohexene:

cyclohexene + CH_2N_2 $\xrightarrow{h\nu}$	bicyclic	+	methylcyclohexane	+	methylcyclohexene	+	methylcyclohexene
Without additives	1	0.24	1.3	
+ Benzophenone	1	Trace	0.42	
+ Cu powder	1	0	0	

C-4. (ii) *Addition of methylene to* cis- *and* trans-*2-butene*[10]

$$\underset{H_3C}{\overset{H}{>}}C=C\underset{H}{\overset{CH_3}{<}} + CH_2N_2 \xrightarrow[Octane]{h\nu} \text{cis-dimethylcyclopropane}$$

$$\underset{H}{\overset{H_3C}{>}}C=C\underset{H}{\overset{CH_3}{<}} + CH_2N_2 \xrightarrow[Octane]{h\nu} \text{trans-dimethylcyclopropane}$$

$$\underset{H}{\overset{H_3C}{>}}C=C\underset{H}{\overset{CH_3}{<}} + CH_2N_2 \xrightarrow[+\text{Benzo-phenone}]{h\nu} \text{trans} + \text{cis}$$
$$\qquad\qquad\qquad\qquad\qquad\qquad\qquad 2:1$$

C-5. Intramolecular carbene addition by thermal decomposition of 3-diazo-1-(1,2,3-triphenylcyclopropenyl)-2-propanone (253)[11]

Intramolecular cycloaddition of carbenes to double bonds is a very useful method of preparing highly strained polycyclic systems[12]. Attempts to synthesize the most highly condensed system, tetrahedrane (p. 137, top), by this route have, however, so far failed[13].

Tetrahedrane

If 3-diazo-1-(1,2,3-triphenylcyclopropenyl)-2-propanone (253) is heated in the presence of copper powder, a strained compound (254) is obtained; this yields the triphenylphenol (255), when heated to 180°, presumably by cleavage of the 1,6 and 4,5 carbon-carbon bonds.

(253) → Δ/Cu → (254) → 180° → (255)

C-6 & C-7. Synthesis of cyclopropyl derivatives by reaction of polyhalomethanes with olefins in the presence of strong bases (α-elimination)

If $CHCl_3$, $CHBr_3$, or methylene dihalides are treated with strong bases in the presence of olefins, cyclopropyl derivatives are obtained.[14] This reaction has been thoroughly studied for $CHCl_3$[9]. It has been shown by Hine that exchange of H in $CHCl_3$ for deuterium in a basic medium is faster than any other possible reaction (for example, S_N2 substitution of a halide ion by a base molecule). It follows that, by reaction of $CHCl_3$ with base, a trichloromethyl carbanion is primarily formed (*a*), which either reacts with base to yield CO and formate (*b*), or decomposes to dichlorocarbene and chloride (*c*). Combination of (*a*) and (*c*) represents α-elimination.

(*a*) $HCCl_3 + OH^- \rightleftarrows CCl_3^- + H_2O$

(*b*) $CCl_3^- + H_2O \xrightarrow{Slow} \bar{C}Cl_2-\overset{+}{O}H_2 + Cl^-$

$\bar{C}Cl_2-\overset{+}{O}H_2 \xrightarrow[H_2O]{OH^-} CO$ and $HCOO^-$

(*c*) $CCl_3^- \xrightarrow{Slow} CCl_2 + Cl^-$

(*d*) $CCl_2 \xrightarrow[H_2O]{OH^-} CO$ and $HCOO^-$

The electrophilic dichlorocarbene reacts with base to yield also CO and $HCOO^-$ (*d*). The actual formation of dichlorocarbene from $CHCl_3$ has been demonstrated by addition of Cl^- ions to the reaction mixture: addition of chloride ions retards the transformations of CCl_3^{-} [15]. Since Cl^- is believed not to have an effect on reaction (b), the products should arise at least partially from CCl_2.

The reaction of halomethanes with olefins in the presence of strong bases, which takes place in aprotic solvents (in proton-donating solvents solvolysis of the halomethanes competes with cyclopropyl formation), is a stereospecific cycloaddition reaction. Monohalocarbenes yield two isomers:

$$\underset{H}{\overset{R}{>}}C=C\underset{H}{\overset{R}{<}} \xrightarrow[\text{base}]{CH_2Cl_2} \underset{H}{\overset{R\;\;\;R}{\triangle}}{Cl} + \underset{Cl}{\overset{R\;\;\;R}{\triangle}}{H}$$

Cyclopropyl formation has long been considered to take place exclusively by carbene addition to the olefin. Recently it has been shown[16] that cyclopropyl derivatives can also be formed by direct interaction of a halo carbanion with an olefin*. Cyclopropyl formation can therefore no longer be considered as evidence

$$CBrCl_3 + CH_3Li \longrightarrow CCl_3Li + CH_3Br + CH_3CCl_3 + LiBr$$

$$\bigcirc\!=\; + \; CCl_3Li \longrightarrow \bigcirc\!\!\!\!\diagdown\!\!\!\!\overset{Cl}{\underset{Cl}{C}}\cdots\bar{Cl}\cdots\overset{+}{Li} \longrightarrow \bigcirc\!\!\!\!\triangleleft\!\!\overset{Cl}{_{Cl}} + LiCl$$

for the presence of carbenes. The best evidence for formation of carbenes is insertion into C–H bonds. In most α-eliminations no insertion is observed. The same applies to the decomposition of trichloroacetic acid and ethyl trichloroacetate by base, which also proceed through carbanions:

$$CCl_3COO^- \longrightarrow CO_2 + CCl_3^-$$

$$CCl_3\text{--}COOR + {}^-OR \longrightarrow Cl_3C\text{--}\underset{\underset{OR}{|}}{\overset{\overset{O^-}{|}}{C}}\text{--}OR \longrightarrow CCl_3^- + RO\text{--}\overset{\overset{O}{\|}}{C}\text{--}OR$$

If insertion occurs, it differs characteristically from the three-centre insertion of singlet methylene. In the latter case, the optical activity of the molecule into whose C–H bond insertion occurs is preserved, whereas in insertions that result from α-elimination racemic products are obtained.

The bicyclic products, which are formed in examples C-6 and C-7, either by carbene or by carbanion addition, rearrange to more stable products.

* This mechanism probably applies also to addition of CH to olefins on reaction of CH_2I_2 with Zn/Cu.

C-6. *Reaction of 1-lithiopyrrole with methylene chloride and methyllithium*[17]

$$CH_3Li + CH_2Cl_2 \longrightarrow CHCl_2Li \longrightarrow :CHCl + LiCl$$

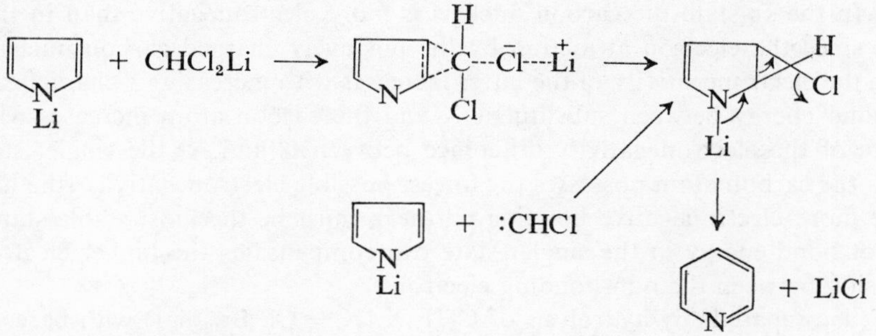

C-7. *Reaction of 2,3-dihydro-5-methylfuran with ethyl trichloroacetate and sodium methoxide*[18]

$$Cl_3C-COOC_2H_5 + NaOCH_3 \longrightarrow Cl_3C\underset{OR}{\overset{\overset{O^-}{|}}{C}}-OR \longrightarrow CCl_3^- \longrightarrow CCl_2 + Cl^-$$

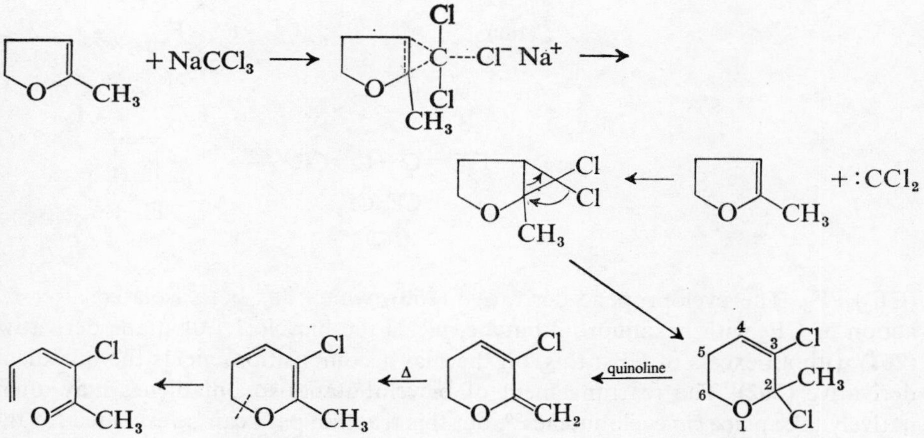

(259)   (258)   (257)   (256)

In example C-7, quinoline does not attack position 5 of compound (**256**), but abstracts a proton from the α-methyl group, yielding (**257**); the intermediates (**256**) and (**257**) could be isolated. When heated, (**257**) rearranges to (**258**), which is transformed into its open-chain valence isomer (**259**).

C-8. Addition of difluorocarbene to hexafluoro-2-butyne[19]

CF_2, the most stable dihalocarbene*, adds stereospecifically to double bonds. Like CCl_2 and CBr_2 but in contrast to CH_2†, it has a singlet ground state. The carbon atom is sp^2-hybridized in the singlet state and sp-hybridized in the triplet state. In the sp-state the carbon nucleus is more electronegative than in the sp^2-state, since the electron attraction by the positively charged carbon nucleus and hence the electronegativity of the latter increases with increasing s-character. Since the bond energy between substituent X and the carbon atom increases with the square of the electronegativity difference between C and X, the singlet state, in which the carbon atom possesses the lowest possible electronegativity (the halogen is the more electronegative bonding partner), must be the most stable state. The gain of bond-energy in the singlet state overcompensates the higher electrostatic repulsion between the non-bonding electrons.

CF_2 is prepared[9] by hydrolysis of CHF_2X (X = Cl, Br, or I) with base, by reaction of CF_3Br with alkyllithiums, or by decomposition of CF_2Cl–COONa or of

$$F_3C-C{\equiv}C-CF_3 + :CF_2 \; [\text{from} \; (CF_3)_3PF_2] \xrightarrow[\text{gas phase}]{100°}$$

(260) → (261) → (263)

(262)

$(CF_3)_3PF_2$. The cyclopropene derivative (260), which has been isolated after reaction of CF_2 with hexafluoro-2-butyne, yields the bicyclo[1.1.0]butane derivative (261) with an excess of CF_2; this, on thermal decomposition, yields the butadiene derivative (262). The rearrangement of bicyclobutanes to butadienes may alternatively take place *via* cyclobutenes[20], but this reaction path can be excluded for the conversion (261) → (262) since perfluoro-2,3-dimethyl-1,3-butadiene (262) is obtained, and none of the 1,3-bis(trifluoromethyl) derivative that would be expected from a cyclobutene.

---

* The stability of dihalocarbenes decreases in the sequence $CF > CCl_2 > CBr_2$. Decisive factors are the differences in electronegativity and ability of the halogen to donate electrons to carbon: $\ddot{X}-\ddot{C}-\ddot{X} \leftrightarrow \overset{+}{X}{=}\ddot{C}^{-}-\ddot{X} \leftrightarrow \ddot{X}-\ddot{C}^{-}{=}\overset{+}{X}$.

† See p. 134.

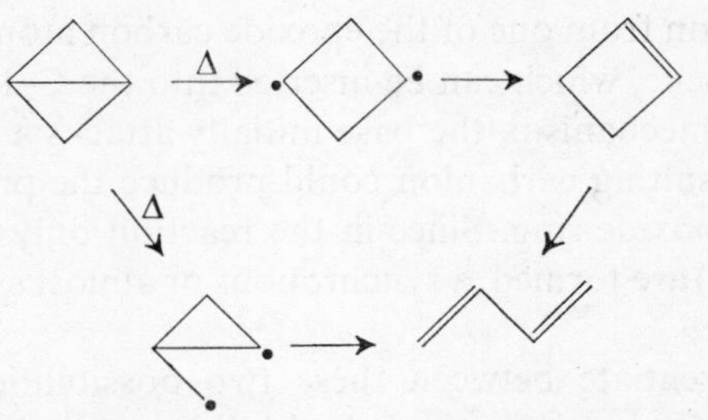

Finally, (262) is converted into (263) at 350°. The pronounced tendency of fluorinated olefins to form four-membered rings has been repeatedly described[21].

C-9. Transannular insertion of carbene by reaction of *cis*-cyclodecene oxide (264) with lithium diethylamide[22]

Transannular reactions are observed between two reaction centers of cyclic compounds, when these centers are separated by several atoms that are, however, in close steric vicinity of one another[23]. Such sterically favorable conformations are frequently encountered in medium-sized rings.

Usually hydrogen migrates between the two reaction centers. However, transannular migrations of other groups have also been observed[24]. Transannular migration of hydrogen can be catalyzed by acid or base or may proceed by a radical mechanism*.

The formation of the products (265), (266), and (267) from *cis*-cyclodecene oxide and lithium diethylamide can be explained by two mechanisms. In one case the

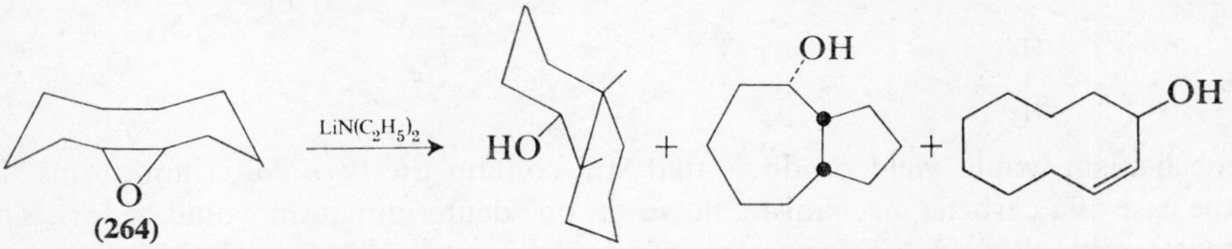

(264)

(265) cis, cis-1-Decalol 83%

(266) endo-cis-Bicyclo[5.3.0.]-decan-2-ol 9%

(267) 2-Cyclo-decen-1-ol 8%

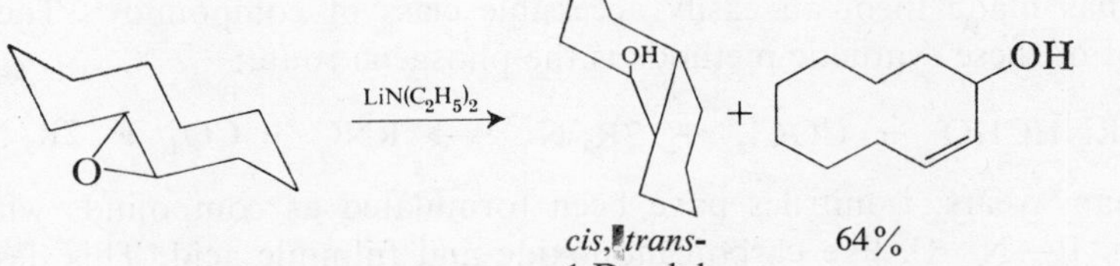

cis, trans-1-Decalol 36%

64%

* For examples see ref. 25.

base abstracts a proton from one of the epoxide carbon atoms. Ring opening of the epoxide yields a carbene, which can be inserted into the C–H bonds at C-6, C-7, or C-10. In the second mechanism, the base initially attacks a hydrogen atom at C-6, C-7, or C-10; the resulting carbanion could produce the products by nucleophilic substitution at the epoxide ring. Since in the reaction only single stereoisomers of (265), (266), and (267) are formed, a synchronous or almost synchronous mechanism must be followed.

In order to differentiate between these two possibilities, the deuterated *cis*-cyclodecene oxide (268) has been treated with lithium diethylamide[26]. A carbanion

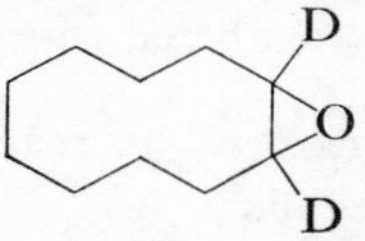

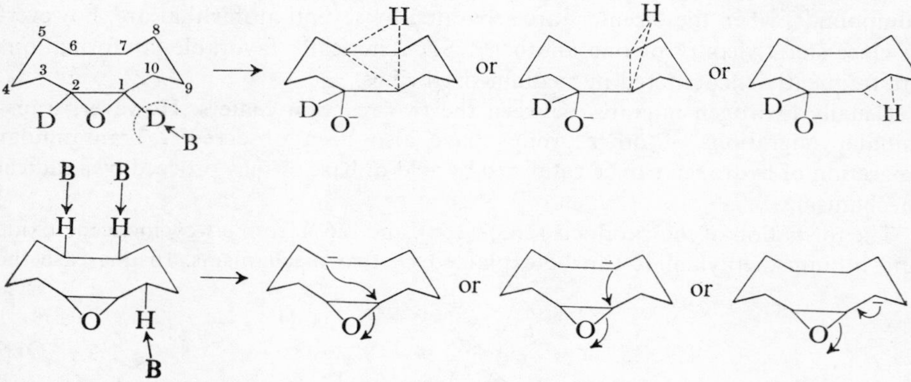

mechanism would yield products that still contain the two deuterium atoms. In the case of a carbene mechanism, however, one deuterium atom would be lost. The reaction resulted in the formation of monodeuterated (265) and (266). Product (267) contained 1.5 deuterium atoms. The result is good evidence that (265) and (266) are formed by way of carbenes, whereas (267) is formed by both mechanisms.

C-10. Addition to isocyanides

The development of new methods for the preparation of isocyanides by I. Ugi et al. has made them an easily accessible class of compound[27]. The most important of these synthetic methods is the phosgene route:

$$\text{RNHCHO} + \text{COCl}_2 + 2\text{R}_3'\text{N} \longrightarrow \text{RNC} + \text{CO}_2 + 2\text{R}_3'\text{NHCl}$$

For many years, isonitriles have been formulated as compounds with bivalent carbon, R—N=C, like carbon monoxide and fulminic acid. This description is supported by their chemical reactions, which consist of 1,1-additions to the terminal carbon, yielding formimide derivatives*. Dipole measurements and determinations of the atomic distances have shown, however, that the C–N distance is very close to that for a triple bond. Consequently, of the mesomeric structures (268) and (269), the former possesses the greater weight.

$$:\bar{C}\equiv\overset{+}{N}R \longleftrightarrow :C=N-R$$

(268)   (269)

Addition of benzenonium-2-carboxylate (271) to phenyl isocyanide (272)[29]

An example of 1,1-addition to an isocyanide is the reaction of compound (271) with phenyl isocyanide (272). The diazonium compound (270) has been used as a starting material for preparation of the very reactive benzyne[30].

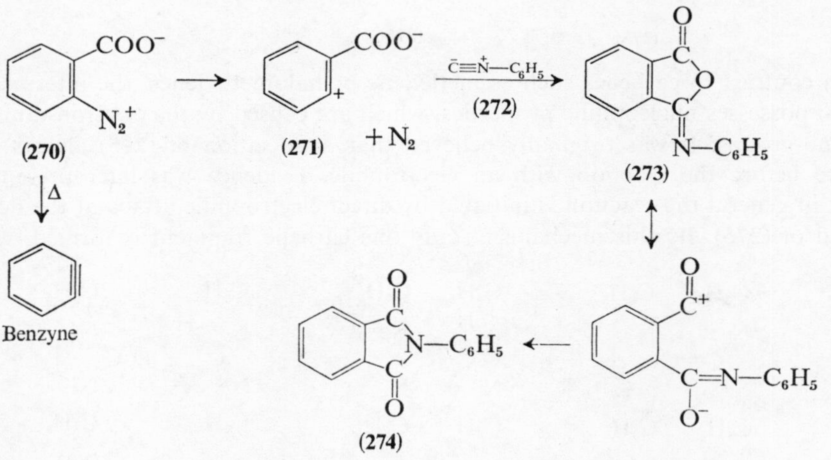

In principle, benzyne can be formed by synchronous or by stepwise loss of CO_2 and N_2. By the reaction of (270) with the nucleophilic (272), it has been demonstrated that nitrogen is lost first. By addition of (272) to the intermediate (271), compound (273) is obtained, which rearranges to the more stable *N*-phenylphthalimide (274). If (270) is decomposed in the presence of $Ni(CO)_4$, phthalic acid anhydride is obtained by an analogous reaction.

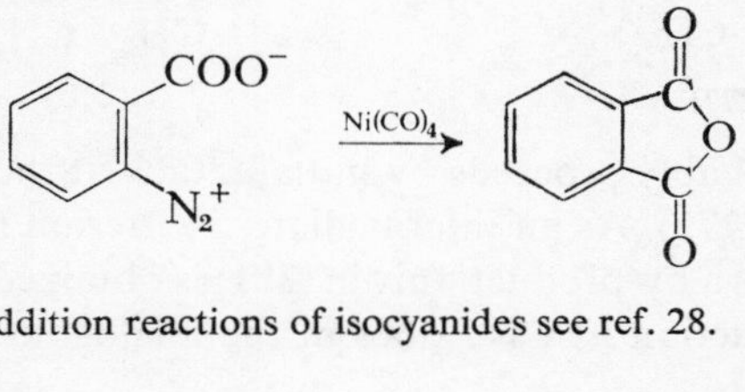

* For a review of such addition reactions of isocyanides see ref. 28.

C-11. Addition to nucleophilic carbenes

"Nucleophilic carbenes" which were first prepared and investigated by H. W. Wanzlick et al. are chemically related to isocyanides[31].

The reaction of 1,1',3,3'-tetraphenylbi-(2-imidazolidinylidene) (**275**) with C-H acidic compounds leads to adducts that are formally 1,1-adducts to the carbene (**276**) formed by dissociation of (**275**).

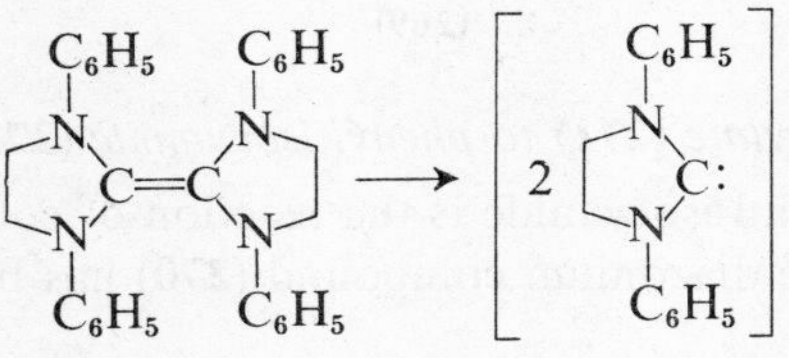

In contrast to carbenes such as methylene or halomethylenes, the intermediate (**276**) possesses nucleophilic properties, which are caused by the electron-donating amino-groups. It was originally believed that dissociation of (**275**) always took place before the reaction with an electrophile. Evidence was later presented[32] that in general the reaction is initiated by direct electrophilic attack at the double bond of (**275**). By this mechanism, only one carbene fragment is formed, which reacts with HX as shown above.

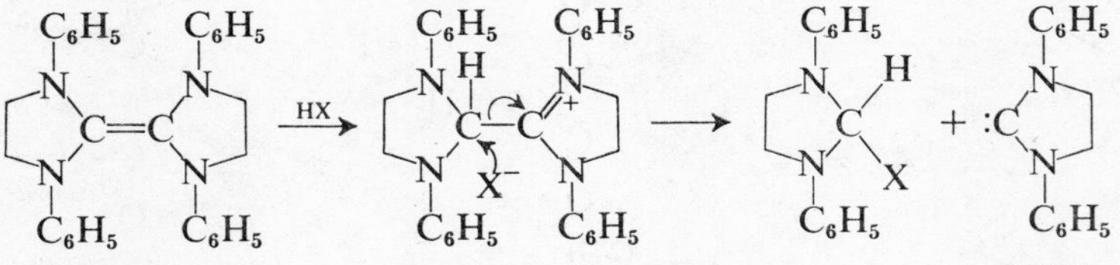

Another class of "carbene dimers" is prepared by reaction of benzothiazolium halides with base[33]; for example, (**277**) yields the "dimer" (**278**).

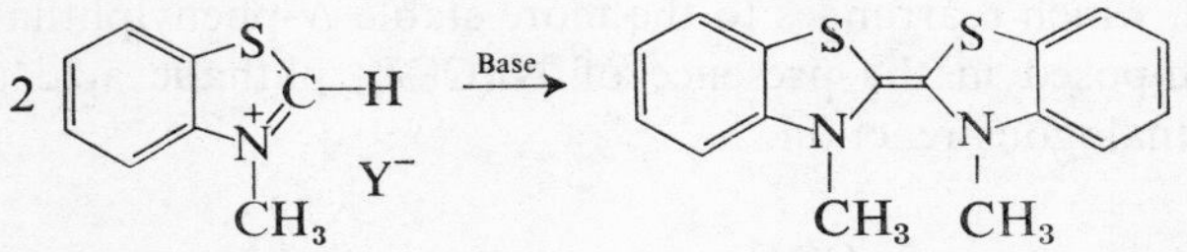

A reaction that probably proceeds by initial attack at the dimer[34] is the addition of furfuraldehyde to (**275**). As an intermediate, zwitterion (**279**) is assumed, which rearranges to (**280**). As a by-product, furoin (**281**) is obtained; evidently, the carbene (**276**) catalyzes the reaction as CN^- does in the analogous benzoin condensation.

Benzoin condensation

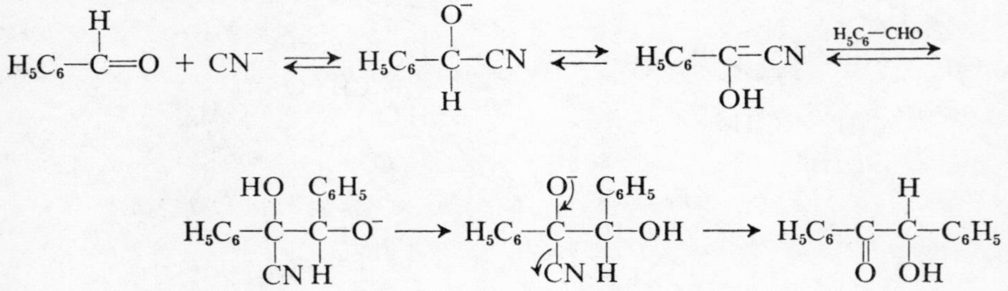

$$H_5C_6-\underset{H}{\overset{H}{C}}=O + CN^- \rightleftarrows H_5C_6-\underset{H}{\overset{O^-}{C}}-CN \rightleftarrows H_5C_6-\underset{OH}{\overset{}{C}}\!\!=\!\!CN \xleftarrow{H_5C_6-CHO}$$

$$H_5C_6-\underset{CN}{\overset{HO}{C}}-\underset{H}{\overset{C_6H_5}{C}}-O^- \longrightarrow H_5C_6-\underset{CN}{\overset{O^-}{C}}-\underset{H}{\overset{C_6H_5}{C}}-OH \longrightarrow H_5C_6-\underset{O}{\overset{}{C}}-\underset{OH}{\overset{H}{C}}-C_6H_5$$

C-12. Insertion of nitrene into a C–H bond

Nitrenes (imenes) are generated thermally or photolytically from azides[35]. The very reactive intermediates are converted into more stable compounds by insertion into C–H bonds, addition to nucleophiles, or abstraction of a hydrogen atom from a vicinal carbon atom.

Thermal decomposition of o-(2-methylbutyl)phenyl azide **(282)**[36]

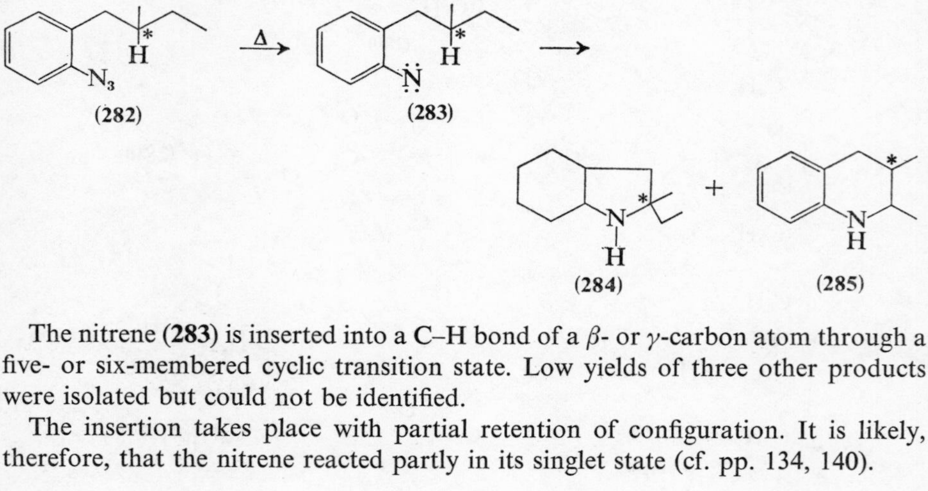

(282) → **(283)** →

(284) + **(285)**

The nitrene **(283)** is inserted into a C–H bond of a β- or γ-carbon atom through a five- or six-membered cyclic transition state. Low yields of three other products were isolated but could not be identified.

The insertion takes place with partial retention of configuration. It is likely, therefore, that the nitrene reacted partly in its singlet state (cf. pp. 134, 140).

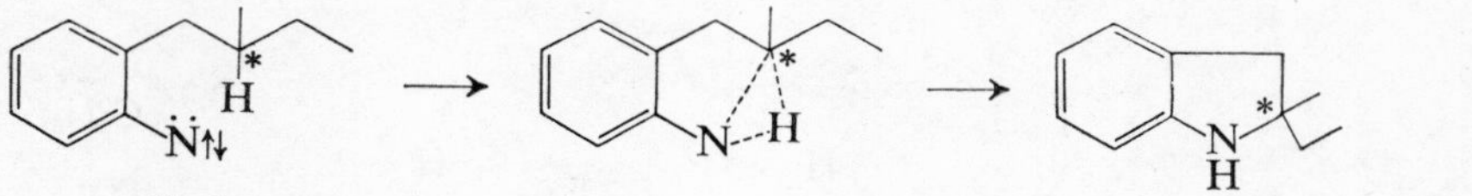

Singlet

Alkylnitrenes are believed to have a triplet ground state[37], like methylene and alkylcarbenes.

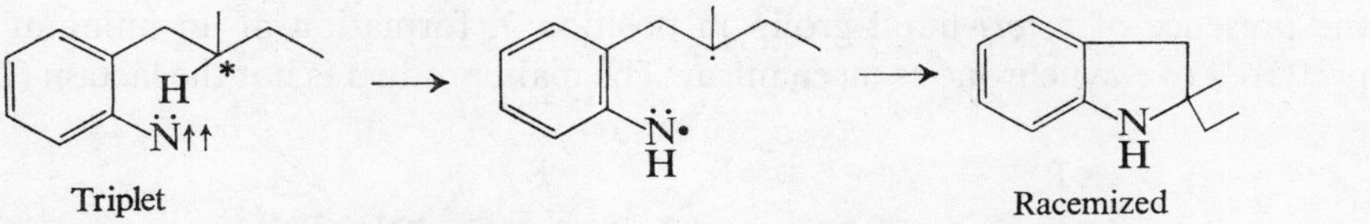

C-13. Insertion of positive nitrogen into a C–H bond

During Beckmann rearrangement, the substituent in the *trans*-position to the oxime ester group migrates to the electron-deficient nitrogen. It is generally agreed that separation of the acid anion from the nitrogen atom and migration of the substituent occur synchronously (see p. 228). If the synchronous rearrangement is sterically hindered, the reaction proceeds in two steps by primary formation of an iminium ion, which rearranges non-stereospecifically in the subsequent step.

Reaction of 4-*bromo*-7-tert-*butylindanone oxime* (**286**) *with polyphosphoric acid*[38]

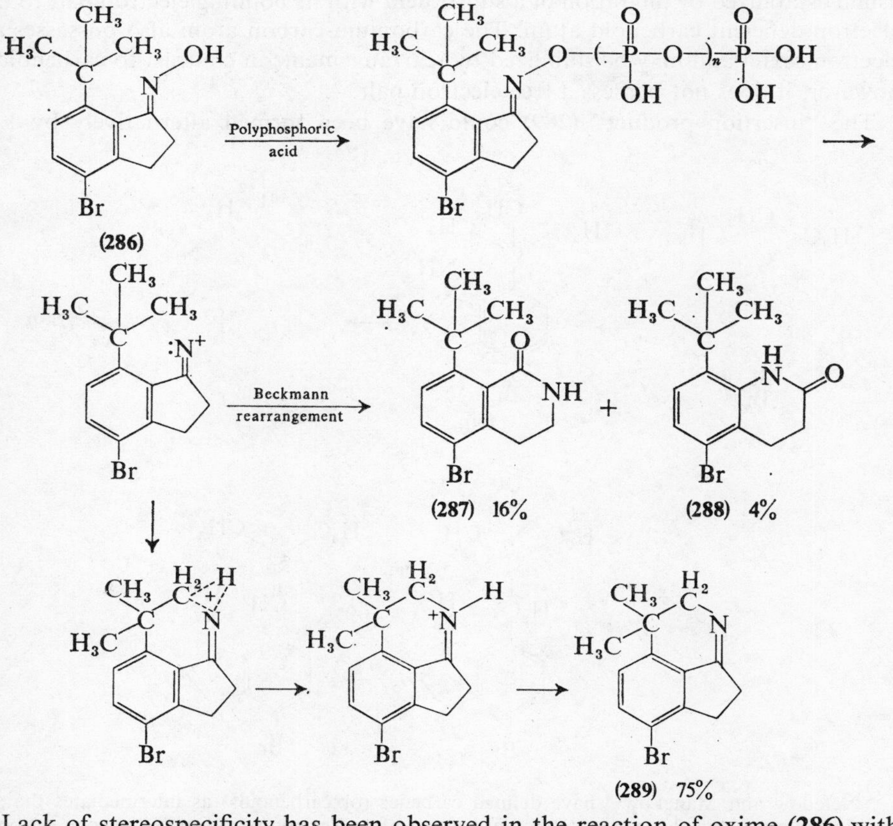

Lack of stereospecificity has been observed in the reaction of oxime (**286**) with polyphosphoric acid. As a consequence of the rigidity of the indanone system and of the presence of a *tert*-butyl group in position 7, formation of an iminium ion is preferred to a synchronous mechanism. The main product is not the lactam (**287**) or

$$\begin{matrix} R \\ \diagdown \\ C: \\ \diagup \\ R \end{matrix} \qquad R-\ddot{N} \qquad \begin{matrix} R \\ \diagdown \\ N:^+ \\ \diagup \\ R \end{matrix} \qquad \begin{matrix} R \\ \diagdown \\ C=N:^+ \\ \diagup \\ R \end{matrix} \qquad R-\ddot{\underset{..}{O}}{}^+$$

(**288**), but the imine (**289**). The authors assume that, in analogy to the insertion of oxygen (see example C-14) or of a singlet carbene, positive nitrogen is inserted into a C–H bond in the close proximity, since bivalent positive nitrogen and univalent positive oxygen possess an electron sextet and free electron pairs as do carbenes and nitrenes*. Reactive intermediates, which possess an electron structure analogous to that of carbenes, are designated carbenoids[40]. Positively charged carbenoids are usually stabilized by migration of a substituent with its bonding electron pair to the electron-deficient carbenoid atom. The carbonium-carbon atom also possesses an electron sextet and may be stabilized by rearrangement; in contrast to carbenoids, however, it does not possess a free electron pair.

The "insertion product" (**289**) could have been formed alternatively by 1,6-

---

* Nefedow and Manekow[39] have defined carbenes (or carbenoids) as intermediates that, in addition to having an electron sextet and at least one free electron pair, form either one sp^2 bond or two sp^3 bonds by their reactions. This definition excludes R_2N^+ and RO^+ since they form only one sp^3 bond; however, if they are inserted into C–H bonds these intermediates exhibit typical carbene properties.

hydride shift. This mechanism is, however, less likely since no neophyl rearrangement of a carbonium ion (**290**) has been observed.

Another possible mechanism is the initial formation of a nitrene (**291**). In this case, the azirine (**292**) would be expected as principal or by-product, but it could not be detected[41]. This mechanism is also rendered unlikely by reaction of the

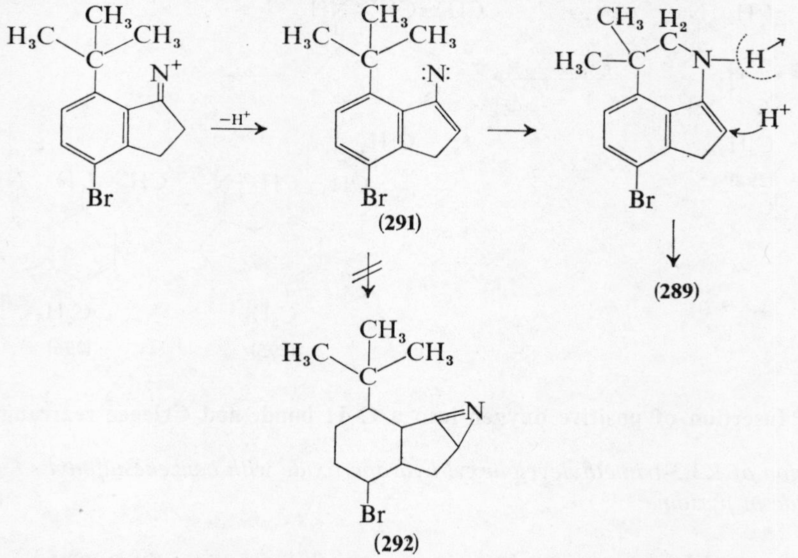

oxime (**286**) with deuterated polyphosphoric acid, which did not yield a monodeuterated product (**289**).

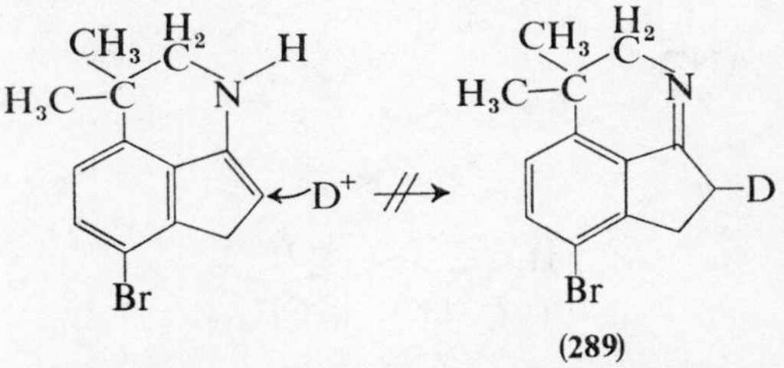

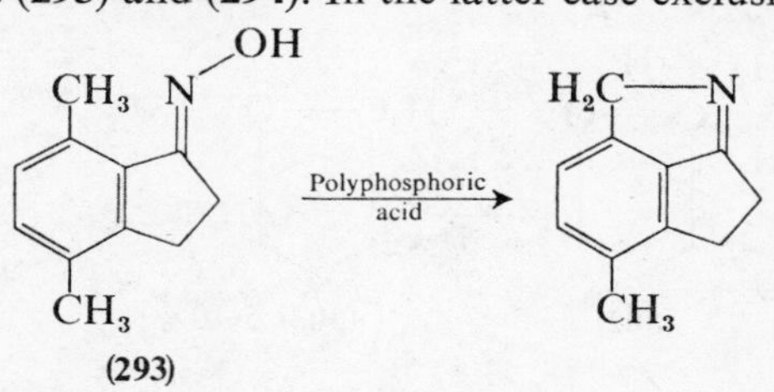

Formal insertion, as in the reaction of (**286**), has also been observed in the reactions of the oximes (**293**) and (**294**). In the latter case exclusive insertion into the α-C–H bond was observed. It is likely that product (**296**) is formed by a 1,5-hydride shift in a carbonium ion mechanism. Possibly the "insertion" leading to (**295**) is also actually a hydride shift *via* the same carbonium ion.

(**294**) (**295**) (**296**)

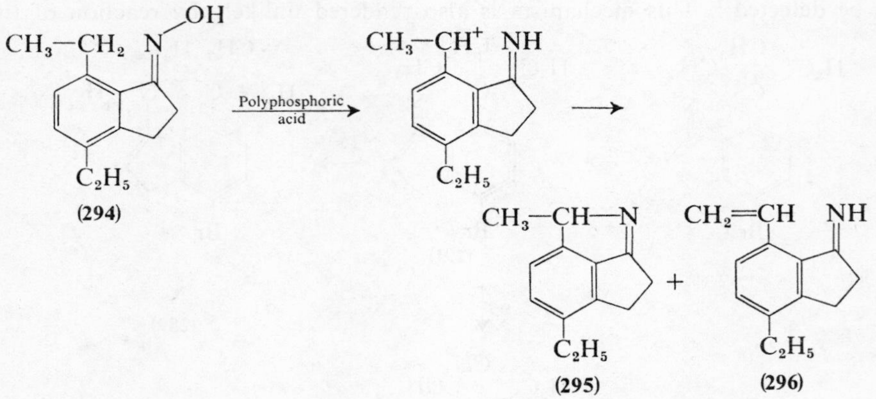

C-14. Insertion of positive oxygen into a C–H bond, and Criegee rearrangement

Reaction of 1,3,3-*trimethylcyclohexyl hydroperoxide with benzenesulfonyl chloride in pyridine*[42,43]

(**297**) (**298**) (**299**) (**300**) (**301**) (**302**) (**303**) 5–10%

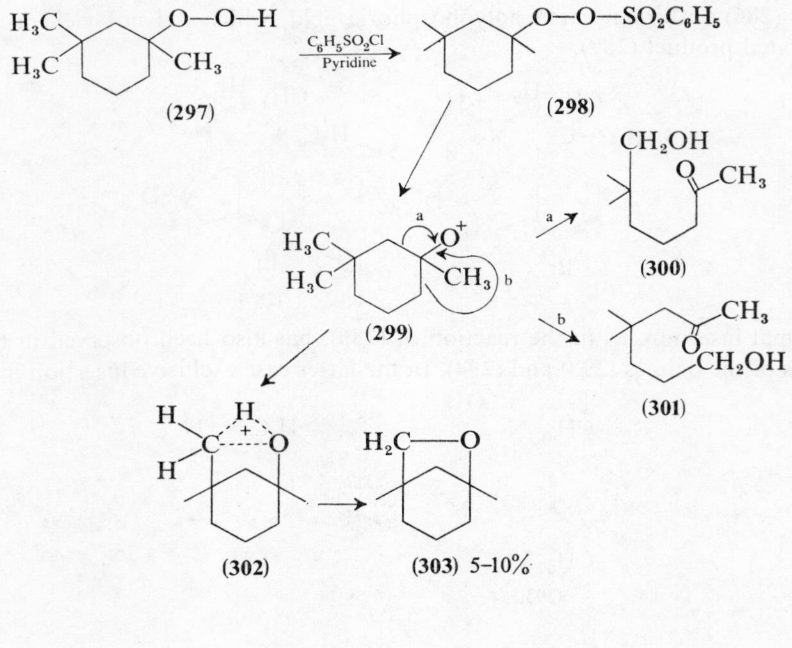

Treatment of the hydroperoxide (**297**) with benzenesulfonyl chloride in pyridine yields a perester (**298**), which forms an oxygen cation (**299**) by heterolysis of the O–O bond. The cation rearranges to the ketones (**300**) and (**301**) (Criegee rearrangement)[44]. Presumably the acid anion does not separate completely but forms a tight ion pair with the cation. In analogy to carbonium ion rearrangements or rearrangements of intermediates with electron-deficient nitrogen, the driving force of the rearrangement is the presence of an electron sextet. The rate of the Criegee rearrangement increases with increasing strength of the acid that is used for esterification of the hydroperoxide and with the polarity of the solvent. Only tertiary hydroperoxides undergo rearrangement: secondary and primary hydroperoxides form carbonyl compounds by loss of an α-proton.

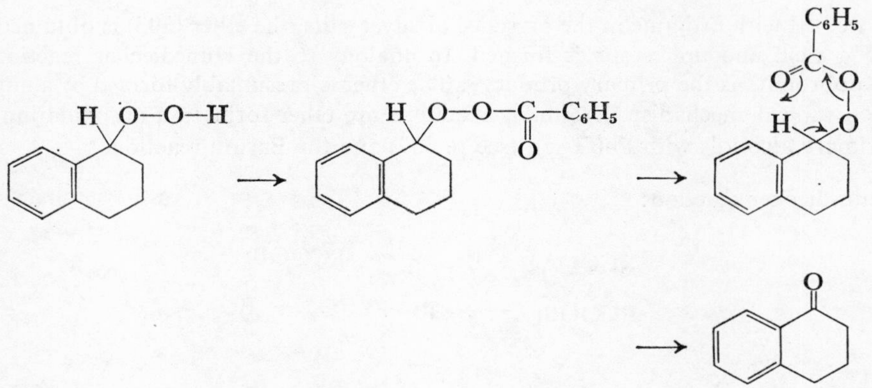

A closely related rearrangement is the Baeyer–Villiger reaction (see p. 97). Tertiary oxygen radicals react like oxygen cations. The stability of the developing carbon radical or cation determines which C–C bond is preferentially broken.

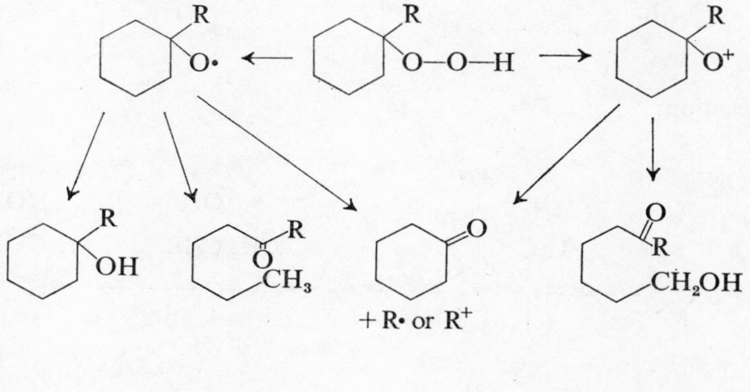

The rearrangement of (297) yields 5–10% of the ether (303) besides the normal rearrangement products. That reaction has been explained by insertion of the univalent positive oxygen (302), which is a carbenoid (see p. 148 and footnote on p. 148), into the sterically favorably arranged C–H bond of the γ-methyl group. An alternative mechanism is a 1,5-hydride migration followed by cyclization. If (297a)

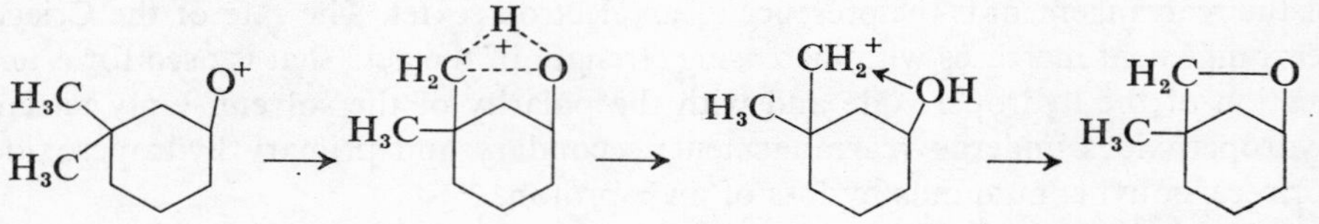

is treated with bromine in the presence of silver salts, the ether (303) is obtained in 75% yield and no ketone is formed. In analogy to the Hunsdiecker reaction a hypobromite is the primary product[43]. The ether is presumably formed by a multistep radical mechanism[45]. Similar reactions are ether formation on oxidation of primary alcohols with $Pb(OAc)_4$ (see p. 339) and the Barton reaction.

Hunsdiecker reaction:

$$RCOOAg + Br_2 \longrightarrow RCOOBr$$

$$RCOOBr \longrightarrow R\cdot + CO_2 + Br\cdot$$

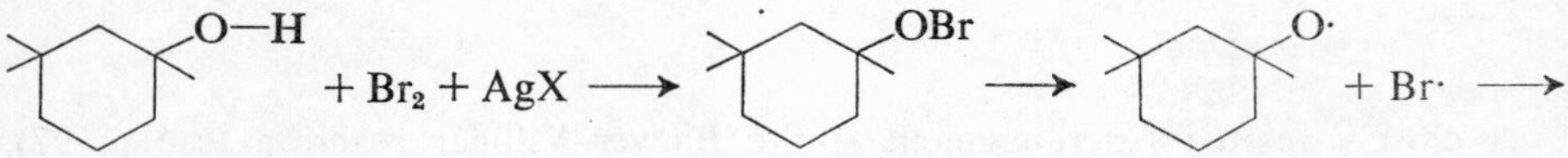

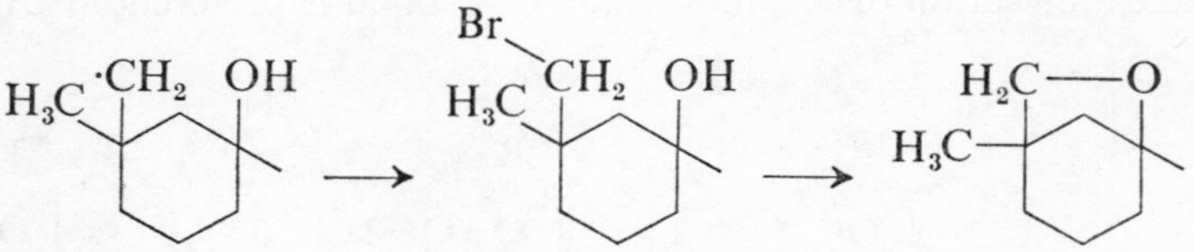

Barton reaction:

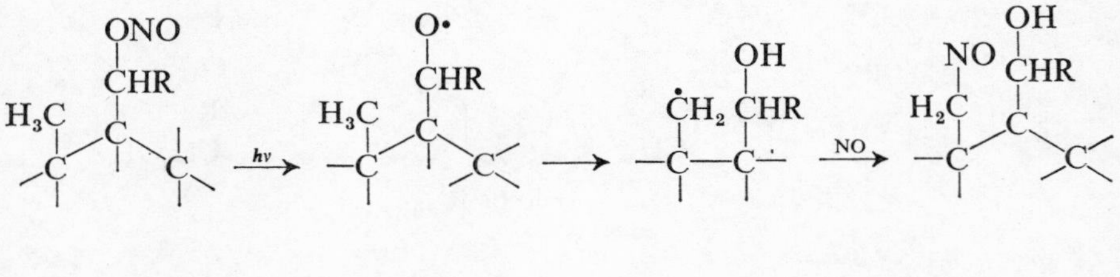

II. 1,2-ADDITIONS

C-15. Addition of 1,1-dichloro-2,2-difluoroethylene to dienes[46]

Formation of four-membered rings may be achieved by thermal, catalytic, or photolytic dimerization of unsaturated compounds*.

If tetrafluoroethylene is heated to 200°, octafluorocyclobutane is obtained. Formation of stereoisomers is not possible in this reaction, but if unsymmetrically substituted monomers are used stereoisomers may be formed.

To some thermal dimerizations an orientation rule applies that restricts the number of isomers. The rule states that the products have those structures which are formed through the most stable diradical intermediate†.

Heating acrylonitrile, for example, leads to the head-to-head products (304). Since in (303) free rotation around the C–C bonds is possible, *cis*- and *trans*-cyclobutane derivatives are obtained. A head-to-tail adduct (306) would be formed

$$2CH_2=CH-CN \longrightarrow \begin{array}{c} H_2C-CH-CN \\ | \\ H_2C-\overset{\cdot}{C}H-CN \end{array} \longrightarrow$$

(303)

(304)

from the less stable intermediate (305).

$$2CH_2=CH-CN \longrightarrow \begin{array}{c} H_2C-\overset{\cdot}{C}H-CN \\ | \\ NC-CH-\overset{\cdot}{C}H_2 \end{array}$$

(305)

(306)

An ionic mechanism is unlikely, since it would also lead to a head-to-tail adduct.

$$2CH_2=CH-CN \longleftrightarrow \overset{+}{C}H_2-CH=C=N^- \longrightarrow \begin{array}{c} H_2C-\bar{C}H-CN \\ | \\ NC-CH-\overset{+}{C}H_2 \end{array}$$

Addition of 1,1-dichloro-2,2-difluoroethylene (307) to isoprene or chloroprene leads to the products (312) and (313)[46]. In accordance with the biradical mechanism,

---

* For a review of the classes of unsaturated compound that can be thermally dimerized see ref. 47.

† The rule fails in the case of ionic dimerization, for instance of ketene. See also Woodward–Hoffmann rules for concerted cycloadditions on pages 158 and 159.

no (314) or (315) is formed, since the diradicals (308) and (309) are better stabilized by chlorine than are (310) and (311) by fluorine.

$$F_2C=CCl_2 + H_2\overset{1}{C}=\overset{2}{\underset{R}{C}}-\overset{3}{C}H=\overset{4}{C}H_2 \longrightarrow \begin{array}{c} F_2C-\overset{\bullet}{C}Cl_2 \\ | \quad | \\ H_2\overset{1}{C}-\overset{2}{\underset{\underset{3}{C}H=\overset{4}{C}H_2}{\overset{\bullet}{C}R}} \end{array} \longrightarrow$$

(307)   R = CH$_3$ or Cl   (308)   (312)

(310)   (311)   (309)   (313)

(312):(313) = 7:1

(314)   (315)   (316)

Comparison of the rate of 1,2-addition of (307) to isoprene and chloroprene with the rate of addition of (307) to butadiene shows that substituents in position 2 have a slight accelerating effect, this probably being due to stabilization of the developing carbon radicals of the transition state by the substituents.

Substituents, however, retard 3,4-additions. In the intermediate (309), the carbon atom that carries the substituent does not have radical character. Retardation of the cycloaddition is probably caused by steric interaction of the substituent with other parts of the molecule when the molecule tries to attain the planar arrangement necessary for cyclization. 1,2-Addition is therefore favored over 3,4-addition in the case of isoprene and chloroprene (see Table 1.) One further point has yet to be clarified, namely, why formation of four-membered rings dominates over the Diels–Alder reaction. An explanation is at hand if the reaction really proceeds by a two-step mechanism *via* an intermediate diradical. Since in the intermediate diradical one electron is in resonance with the allylic double bond, free rotation around the C$_2$–C$_3$ axis is prevented. In the rather rigid intermediate (316), C-4 is too far from C-5 to permit formation of a six-membered ring.

Table 1 shows that the mechanism of the 1,2-addition differs strikingly in several respects from the synchronous Diels–Alder reaction. 1,2-Addition can therefore not take place by a similar synchronous mechanism. In contrast to the Diels–Alder

Table 1. Effect of substituents and stereochemistry on the relative rates of some addition reactions

	Reaction with (307)		
Diene	1,2-Addition	3,4-Addition	Diels–Alder reaction with maleic anhydride
⁄⁀⁄	1	1	1
CH₃-substituted	2.3	0.42	2.25
Cl-substituted	1.2	0.30	0.10
trans		2.3	V. slow
cis		1.65	3.3
cyclopentadiene	0.113	0.113	1350

reaction the influence of polar groups is only insignificant. The typical blocking effect of terminal *cis*-substituents on the Diels–Alder reaction is not observed. Finally, cyclic dienes react very slowly, whereas in the Diels–Alder reaction they react 10³ times faster than open-chain butadienes.

In the cycloaddition of (307) to 2,4-hexadiene, Bartlett *et al.* have demonstrated a clear-cut two step mechanism *via* biradical intermediates. The products, summarized in Table 2, can only be explained by biradical intermediates with free rotation around the C_1–C_2 axis, as formulated on p. 156.

Table 2. Reaction of $F_2C=CCl_2$ with stereoisomeric 2,4-hexadienes

	(317)	(318)	(319)	(320)
trans-trans	84.2	15.8	0	0
trans-cis	44.2	13.7	34.2	7.9
cis-cis	0	0	75.9	24.1

Some cycloadditions seem to proceed by a dipolar two-step mechanism. In contrast to the addition of 1,1-dichloro-2,2-difluoroethylene to 1,4-dimethylbutadiene, these cycloadditions may proceed stereospecifically or lead to isomerized products, depending on the steric crowding in the transition state[48].

The ratio of the two isomers is 1:2.5

Cycloadditions

The *cis*- and *trans*-isomers **(321)** [bis(trifluoromethyl)]-maleonitrile and -fumaronitrile react rapidly with electron-rich unsubstituted vinyl compounds, even at room temperature, yielding cyclobutane derivates. *tert*-Butyl vinyl sulfide **(322)** reacts with each isomer stereospecifically. However, reaction of the *trans*-nitrile

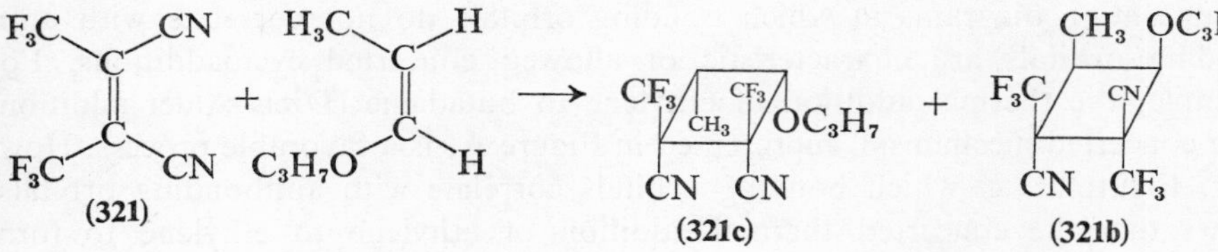

with propyl *cis*-propenyl ether **(321a)** leads to two products, in one of which the CF_3 groups are *cis* to one another. Analogously the reaction of *cis*-**(321)** with **(321a)** yields a product with the CF_3 groups in *trans*-position to one another. The products and the dependence of rate on the polarity of the solvent indicate that these reactions involve dipolar intermediates. Bond a is formed first, since only in this case is the positive charge stabilized by X. Formation of bond a is rate-determining. Bond b is formed in a subsequent ready collapse of the internal ion pair. When there is strong steric crowding, rotation occurs around d, before b is formed. Thus rotation occurs, before ring closure to the expected products **(321d)** and **(321e)** takes place.

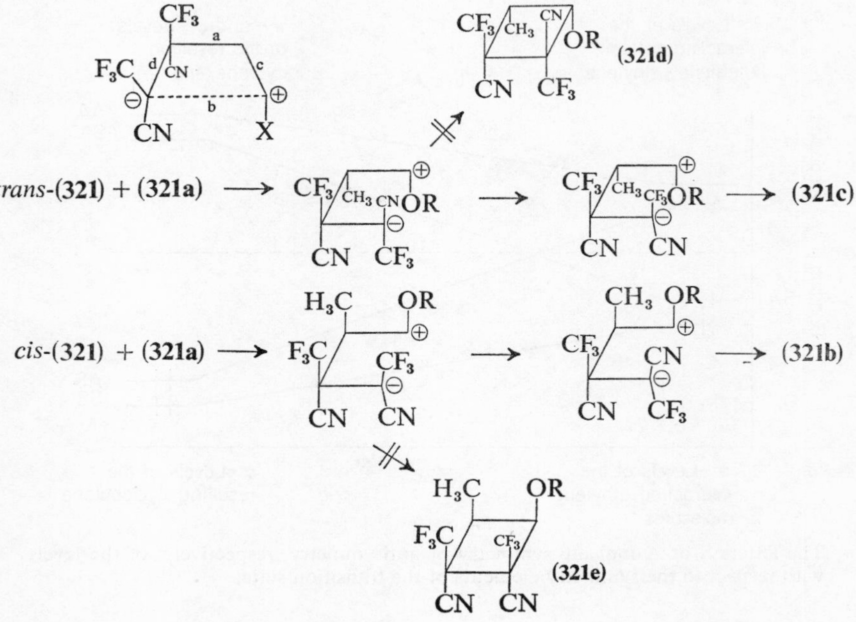

Hoffmann and Woodward[49] showed how to correlate the molecular orbitals of the interacting starting materials involved in cycloaddition reactions with the molecular orbitals of the product by classifying the levels with respect to the symmetry elements of the transition state, and this led to predictions as to the feasibility and mechanism of cycloaddition reactions.

Correlation diagrams in which bonding orbitals do not correlate with antibonding orbitals are characteristic of allowed concerted cycloadditions. For example, the thermal addition of ethylene to butadiene (Diels–Alder addition) by a concerted mechanism, represented in Figure A, is a favorable process. However, Figure B, in which bonding orbitals correlate with antibonding orbitals, shows that the concerted thermal addition of ethylene to ethylene to form

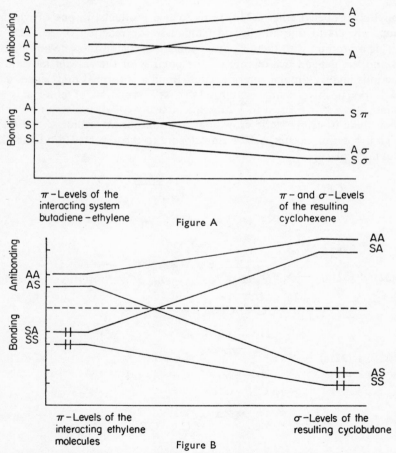

Figure A

Figure B

The letters S or A indicate symmetry or antisymmetry, respectively, of the levels with respect to the symmetry elements of the transition state.

cyclobutane is a highly unfavorable process. The situation is reversed in photochemical reactions.

Table 3 summarizes allowed concerted cycloaddition reactions. Cycloadditions that do not follow these rules, for example, the addition C-15, proceed by a multi-step mechanism. In Table 3 the numbers below m, n, and p are the

Table 3. Summary of concerted cycloadditions (for the significance of the m, n, p entries see the text).

Number of π-bonds disappearing from the starting materials	Number of σ-bonds newly formed in the product	Thermal reaction				Photolytic reaction			
		m		n		m		n	
2	2	4		2		2		2	
		6		4		4		4	
		8		2		6		2	
		m	p	p		m	p	p	
3	3	2	2	2		4	2	2	
		2	4	4					
		6	2	2					
		m	n	p	p	m	n	p	p
4	4	4	2	2	2	2	2	2	2

numbers of π-electrons of the molecules p, m, and n that participate in the cyclizations. For example, an entry 2 indicates that an isolated double bond is involved in the reaction. The synchronous Diels–Alder reaction is characterized by m = 4 and n = 2. A synchronous dimerization of two molecules with isolated double bonds is designated by m = 2 and n = 2; according to Table 3, this reaction preferentially follows a synchronous mechanism if it is effected by light. 6,2 is a system, whose one component has an unsaturated system of three conjugated double bonds and the second component only one isolated double bond. Where p appears twice, a component with two unsaturated systems is involved, which are independent of one another: bicycloheptadiene, for example, would be designated p,p; it may react with one partner m or two partners m and n. A m,p,p (2,2,2) reaction is the addition of tetracyanoethylene to bicycloheptadiene (see p. 160).

C-16. Addition of methylenesulfene to 1,1-dimorpholinoethylene[50]

Like some cyclobutanes, heterocyclic four-membered rings can be synthesized by thermal dimerization. A well-known example is the dimerization of ketene[51]. These additions proceed by a polar two-step mechanism. Interesting examples are the additions of sulfenes* to unsaturated compounds containing electron-rich double bonds, such as enamines and enediamines.

* The cyclodimerization of sulfenes has been described by Opitz et al.[51a]

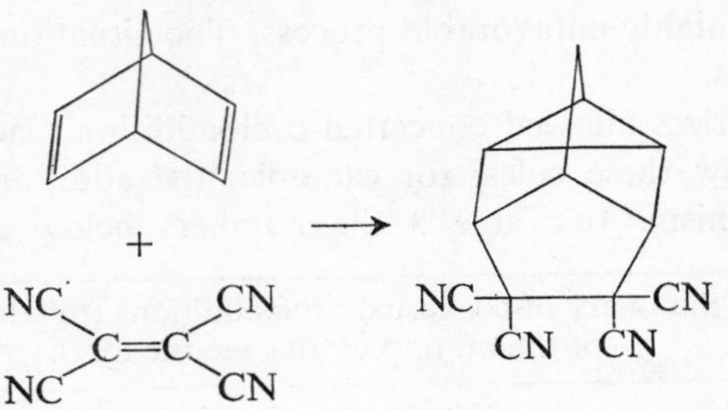

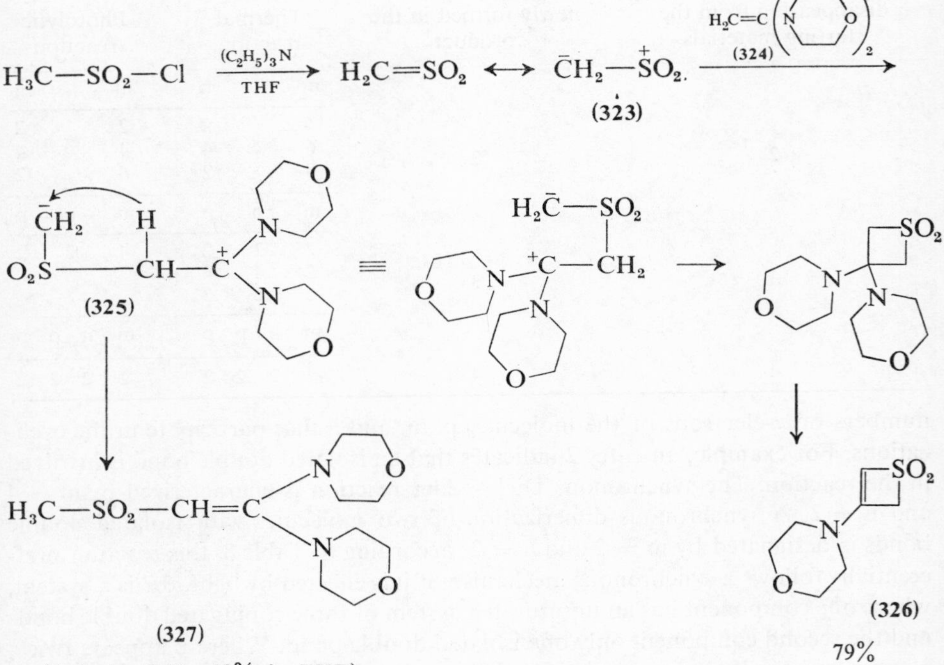

$$H_3C-SO_2-Cl \xrightarrow[THF]{(C_2H_5)_3N} H_2C=SO_2 \leftrightarrow \bar{C}H_2-\overset{+}{S}O_2 \xrightarrow{(324)}$$
(323)

where (324) = $H_2C=C(N(CH_2CH_2)_2O)_2$

(325) → (intermediate with morpholine groups) → (326) 79%

(327) (little in C$_6$H$_6$, 48% in THF): $H_3C-SO_2-CH=C$(morpholine)$_2$

(326): morpholine-substituted thietane-1,1-dioxide

Characteristic NMR absorption of (326): 4.38 τ (2) singlet —CH$_2$—

5.3 (1) singlet $\overset{H}{\diagdown}C=C\diagup$

Characteristic NMR absorption of (327): 2.94 (3) singlet CH$_3$

4.23 (1) singlet $\overset{H}{\diagdown}C=C\diagup$

The very reactive ylide-like sulfenes, which it has not yet been possible to isolate, react smoothly at low temperatures with the enediamine (324). Sulfenes are prepared by reaction of tertiary amines with sulfonyl chlorides that have α-hydrogen atoms. The formation of four-membered rings is facilitated by activation of the α-hydrogen as in α-toluenesulfonyl chloride. The presence of sulfenes as intermediates has been demonstrated[52] by the reaction of sulfonyl chlorides with amines in the presence of deuterated alcohol*. One obtains monodeuterated

$$H_3C-SO_2Cl + CH_3OD \xrightarrow{R_3N} H_2CD-SO_2-OCH_3 + R_3NH^+Cl^-$$

$$H_3C-SO_2Cl \xrightarrow[CH_3OD]{\not\longrightarrow} H_2CD-SO_2Cl + HCD_2-SO_2Cl + D_3C-SO_2Cl$$

sulfonic esters. The absence of di- and tri-deuterio ester precludes H–D exchange as a mechanism for incorporation of deuterium in the ester but is consistent with a sulfene intermediate.

In example C-16[50], the sulfene (323) is attacked by the nucleophilic (324). The resulting zwitterion (325) either cyclizes to the thieten (326) or is converted into the acyclic product (327). In polar solvents, (327) is formed predominantly.

C-17. Addition of hexafluoroacetone to ethoxyacetylene[54]

$$CF_3-\underset{\underset{(328)}{}}{\overset{\overset{CF_3}{|}}{C}}=O + \underset{(329)}{HC{\equiv}C-OC_2H_5} \longrightarrow \underset{(330)}{F_3C-\overset{\overset{CF_3}{|}}{\underset{\underset{HC=C-OC_2H_5}{|}}{C}}-O} \longrightarrow$$

$$\underset{(331)}{\overset{F_3C}{\underset{F_3C}{\diagdown}}C=CH-COOC_2H_5}$$

Hexafluoroacetone (328) adds to ethoxyacetylene (329) even below 0°. The resulting oxetene (330) rearranges at 70° to the acrylic ester (331). If the reaction is carried out at room temperature, (331) is obtained directly. Thermal addition of perfluorocarbonyl compounds to olefins usually leads to alcohols as illustrated in example A-38 (p. 121). Only vinyl ethers are known to give oxetanes with fluoro ketones under thermal conditions[55]. Photoexcited fluoro ketones, however, do react by cycloaddition to fluoro olefins and olefins to give oxetanes[56]. Perfluorinated thioacetone also adds to vinyl ethers[57], yielding thietanes such as (332).

$$F_3C-\underset{S}{\overset{\parallel}{C}}-CF_3 + H_3CO-CH=CH_2 \longrightarrow \underset{(332)}{H_3CO-\underset{S-C(CF_3)_2}{\overset{|}{CH}}-CH_2}$$

In the presence of BF_3, α,β-unsaturated carboxylic esters are obtained by addition of aldehydes or ketones to ethoxyacetylenes or sulfides[58]. This reaction is thought to

---
* Other mechanisms are discussed in ref. 53.

proceed through an oxetene intermediate.

$$\underset{R}{\overset{R'}{\diagdown}}C=O\cdots BF_3 + HC\equiv C-OC_2H_5 \longrightarrow \left[\begin{array}{c} R' \\ \underset{R}{\diagdown}\overset{\diagup}{C}-CH \\ R\overset{\diagup}{\underset{BF_3}{O}}\overset{\diagdown}{\underset{}{C}}\overset{\diagdown}{OC_2H_5} \end{array}\right] \longrightarrow$$

$$\underset{R}{\overset{R'}{\diagdown}}C=CH-COOC_2H_5 + BF_3$$

In many reactions, however, acetylenes that still bear a hydrogen atom at one of the acetylene-carbon atoms do not participate in cyclization but react by addition of the relatively acidic C–H bond to a carbonyl group. The best known example is the addition of unsubstituted acetylene to carbonyl compounds in the ethynylation reaction[59]. The reaction, which is carried out under pressure and in the presence of heavy-metal acetylides, yields 1-alkyn-3-ols.

$$HC\equiv CH + RR'CO \xrightarrow{\text{Catalyst}} HOCRR'-C\equiv CH$$

Lithium and sodium acetylides add to carbonyl compounds at atmospheric pressure[60].

Another reaction of this type is the Mannich synthesis, in which by reaction of phenyl- or vinyl-acetylenes with formaldehyde and a secondary amine, 1-alkyn-3-amines are obtained[61]. As in other Mannich reactions, a necessary condition is

$$C_6H_5-C\equiv CH + CH_2O + HNR_2 \longrightarrow C_6H_5-C\equiv C-CH_2-NR_2 + H_2O$$

sufficient acidity of the acetylenic hydrogen and, because of the electron-donating effect of the alkoxy-group, alkoxyacetylenes do not react in this way.

C-18. Acid-catalyzed cyclodimerization of acetylenes and allenes

The intramolecular cyclodimerization of the diacetylene (**333**) is catalyzed by HCl[62]. An allene (**334**) is probably formed as an intermediate.

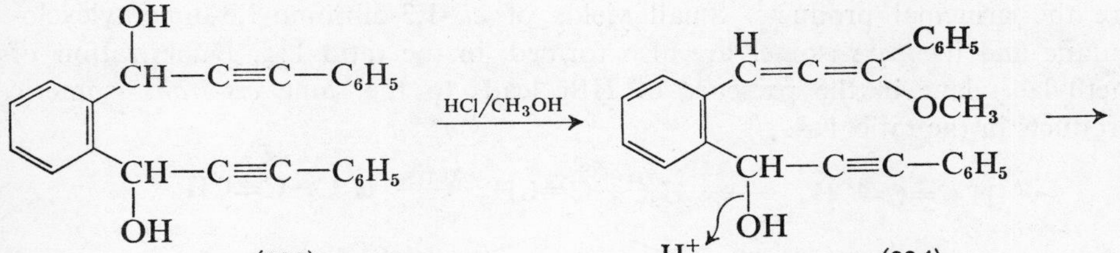

Characteristic UV bands of (**335**), all assignable to the naphthalene nucleus are at 230, 272, 283, 295, 307, and 320 mμ.

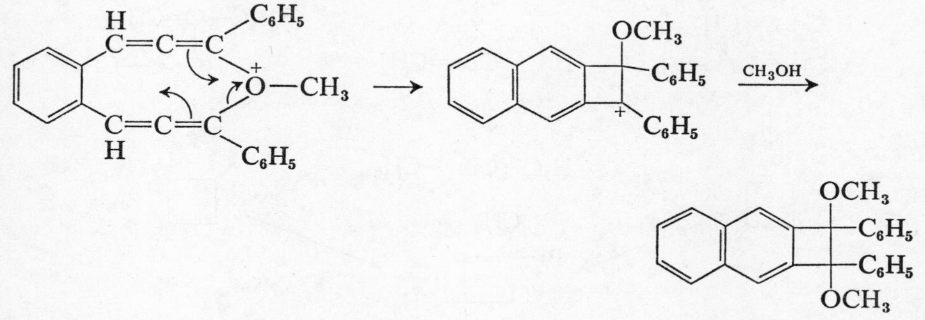

Formation of allenes on treatment of a 1-alkyn-3-ol with thionyl chloride has been demonstrated in a similar reaction[63].

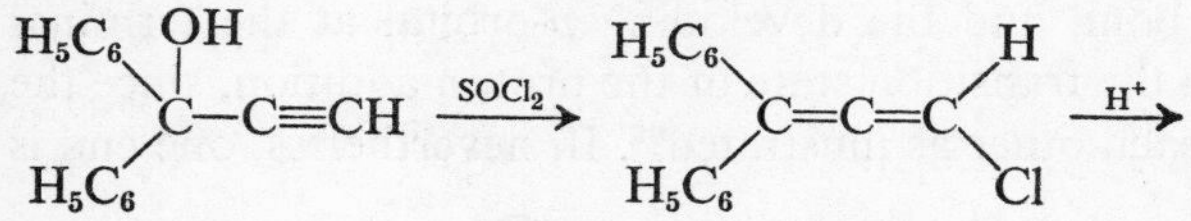

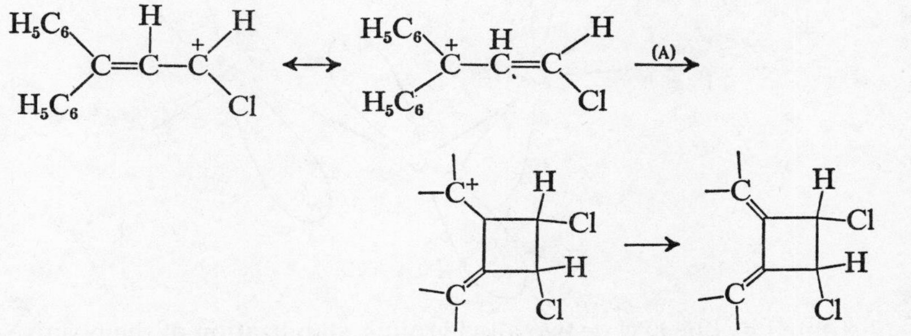

Unsubstituted allene can also be cyclodimerized in the presence of mineral acids*[64]. If allene is treated with HBr, 2-bromopropene and 2,2-dibromopropane

* Thermal cyclodimerization has been known for many years [65].

are the principal products. Small yields of *cis*-1,3-dibromo-1,3-dimethylcyclobutane and its *trans*-isomer are also formed, in the ratio 1:5. Dimerization of methylacetylene in the presence of HBr leads to the same *cis–trans* isomeric products in the ratio 1:4.

$$H_2C=C=CH_2 \xrightarrow{H^+} H_3C-\overset{+}{C}=CH_2 \xleftarrow{H^+} H_3C-C\equiv CH$$

$$\begin{array}{c} CH_2=C=CH_2 \\ H_3C-C=CH_2 \\ H_2\overset{+}{C}-C=CH_2 \end{array} \quad \begin{array}{c} \downarrow Br^- \\ H_3C-CBr=CH_2 \\ \downarrow HBr \\ H_3C-CBr_2-CH_3 \end{array} \quad \begin{array}{c} H_3C-C\equiv CH \\ H_3C-C=CH_2 \\ HC=\overset{+}{C}-CH_3 \end{array}$$

The proton attacks the terminal carbon of the allene molecule. If one side is substituted by two alkyl substituents or by one aryl substituent, the proton adds to the central carbon. One would expect addition to the central carbon to be favored in all cases since that gives an allylic cation; evidently, however, interaction between the *p*-orbitals of the double bond and the developing *p*-orbital at the terminal carbon atom is not possible in the transition state of the proton addition, since the orbitals are perpendicular to each other as illustrated[66]. If, nevertheless, one end is

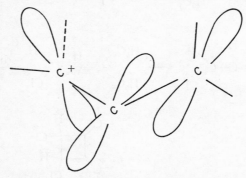

substituted by one aryl or two alkyl groups, stabilization of the positive charge that develops at the terminal carbon is sufficient to cause preferential addition to the central C atom. It remains obscure why in the addition to unsubstituted allene the vinyl cation is favored over the primary carbonium ion[67].

III. 1,3-DIPOLAR ADDITIONS

The chemistry of 1,3-dipolar additions has been reported in a series of excellent papers by R. Huisgen et al. The cycloaddition permits the preparation of numerous five-membered ring systems[68]. The 1,3-dipoles can be divided into a main group with internal octet stabilization and a smaller group without such stabilization. "Internal octet stabilization" means that the 1,3-dipoles can be described by resonance structures in which all participating atoms possess a complete electron octet.

Examples of the first group are:

diazoalkanes $\quad \overset{+}{\underset{..}{N}}=\underset{..}{N}-\overset{\diagup}{\underset{\diagdown}{C}} \longleftrightarrow \underset{..}{N}\equiv\overset{+}{N}-\overset{\diagup}{\underset{\diagdown}{\underset{..}{C}}}$

nitro compounds $\quad :\overset{+}{\underset{..}{O}}-\overset{}{\underset{..}{N}}-\overset{-}{\underset{..}{\underset{|}{O}}}: \longleftrightarrow \overset{}{\underset{..}{O}}=\overset{+}{\underset{|}{N}}-\overset{-}{\underset{..}{O}}:$

ozone $\quad :\overset{+}{\underset{..}{O}}-\overset{}{\underset{..}{O}}-\overset{-}{\underset{..}{O}}: \longleftrightarrow \overset{}{\underset{..}{O}}=\overset{+}{\underset{}{O}}-\overset{}{\underset{..}{O}}:^{-}$

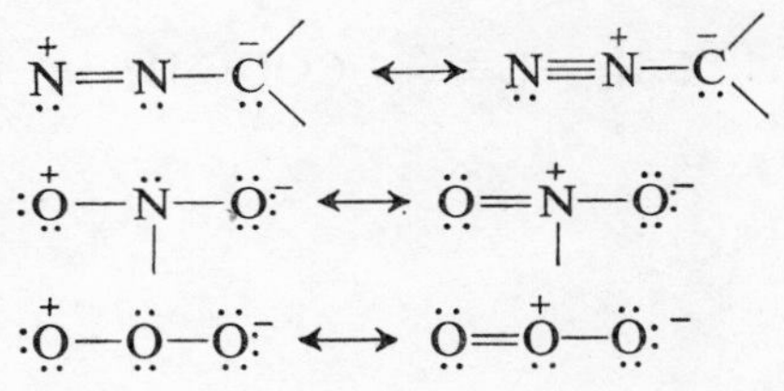

Examples of the second group are:

keto carbenes $\quad -\overset{+}{\underset{}{C}}=\overset{|}{\underset{}{C}}-\overset{-}{\underset{..}{O}}:^{-} \longleftrightarrow -\overset{..}{\underset{}{C}}-\overset{|}{\underset{}{C}}=\overset{}{\underset{..}{O}}$

vinylcarbenes $\quad -\overset{+}{\underset{}{C}}=\overset{}{\underset{|}{C}}-\overset{-}{\underset{..}{\underset{\diagdown}{C}}} \longleftrightarrow -\overset{..}{\underset{}{C}}-\overset{}{\underset{|}{C}}=\overset{\diagup}{\underset{\diagdown}{C}}$

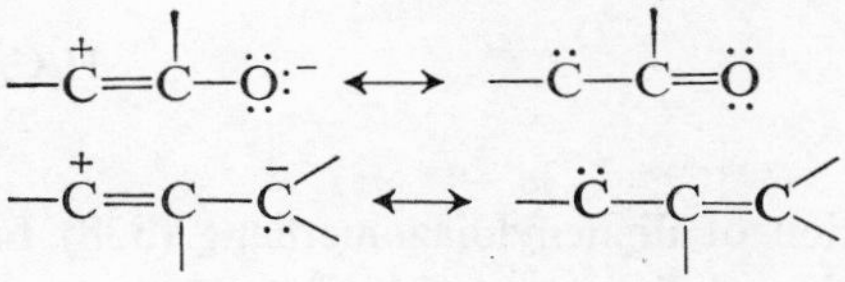

The 1,3-dipolar addition very probably takes place by a concerted process, like other cycloadditions. Evidence is the stereoselectivity, the lack of solvent-dependence, and the strongly negative activation entropies, which last are characteristic of rigid, highly ordered transition states (see p. 170).

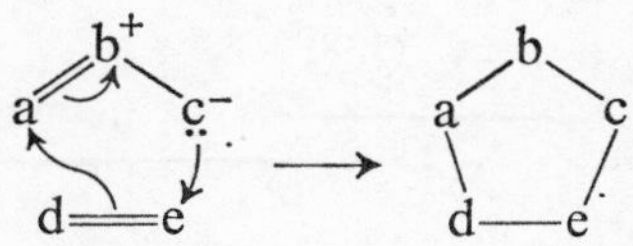

In a multicentre reaction the σ-bonds do not have to be formed at exactly the same rate; the formation of one bond may be further advanced in the transition state than that of the other, so that partial charges appear in the transition state. Stabilization of these charges by suitable substituents lowers the energy of the transition state and accelerates the addition. Thus, whereas normal olefins react only slowly, olefins containing phenyl, C=O or C=N substituents react faster by several powers of ten.

C-19. Reaction of N^2-phenylhydrazonoyl chloride (336) with triethylamine[69]

Diphenylnitrimine (336a), formed as intermediate when the hydrazonoyl chloride (336) is treated with a base, affords the diazole (337a) and the diazoline (337b) in the reactions illustrated.

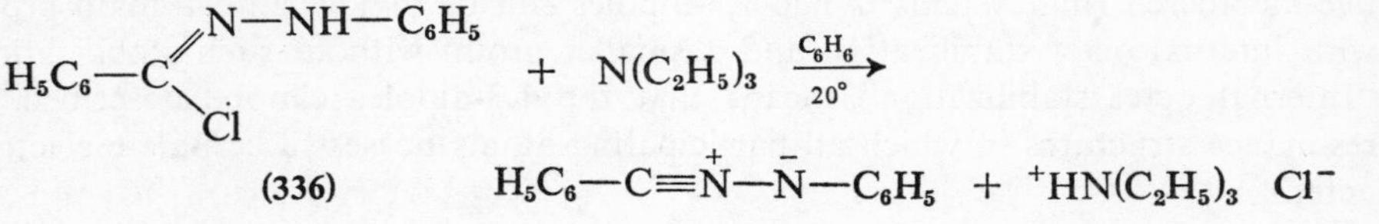

$$H_5C_6-C\begin{smallmatrix}N-NH-C_6H_5\\ \\ Cl\end{smallmatrix} + N(C_2H_5)_3 \xrightarrow[20°]{C_6H_6}$$

(336)

$$H_5C_6-C\equiv\overset{+}{N}-\overset{-}{N}-C_6H_5 + {}^+HN(C_2H_5)_3 \ Cl^-$$

(336a)

(336a) + $C_6H_5-C\equiv C-COOC_2H_5 \longrightarrow$

[structure of (337a) with C_6H_5, N, N, C_6H_5, H_5C_2OOC, C_6H_5]

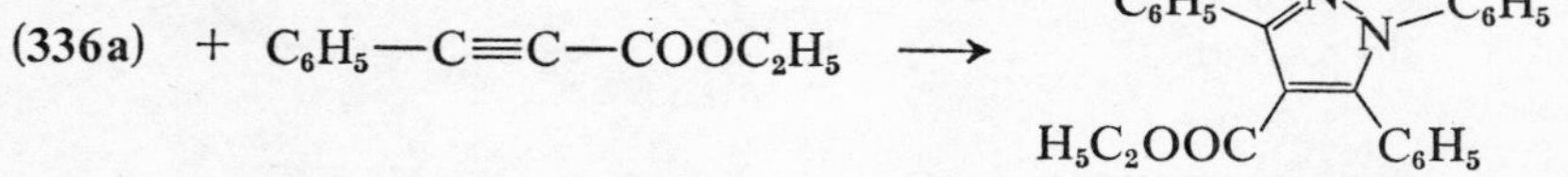

(337a) (84%)

(336a) + Norbornene $\longrightarrow$

[structure of (337b) with C_6H_5, N, N, H_5C_6 and norbornene fused ring]

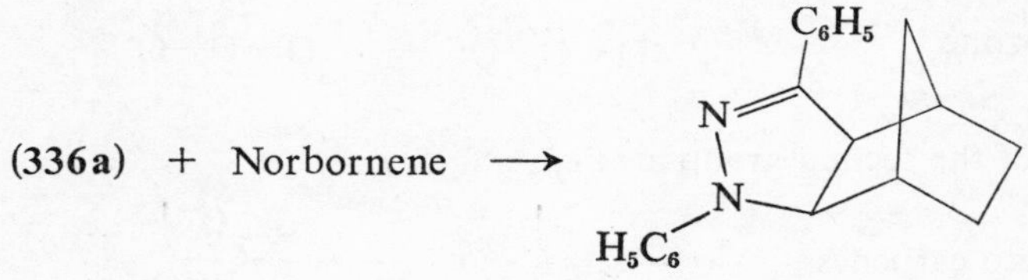

(337b)

C-20. Addition of diphenyldiazomethane (338) to propiolic (339) and phenylpropiolic esters (340)[70]

The orientation of addition may be affected by the strength of the developing sigma bonds and by steric factors.

An example is the reaction of (338) with (339) and (340). Reaction of diphenyl-

$HC\equiv C-COOCH_3 \qquad (C_6H_5)_2\overset{-}{C}-\overset{+}{N}\equiv N \qquad C_6H_5-C\equiv C-COOCH_3$

(339) \qquad\qquad (338) \qquad\qquad (340)

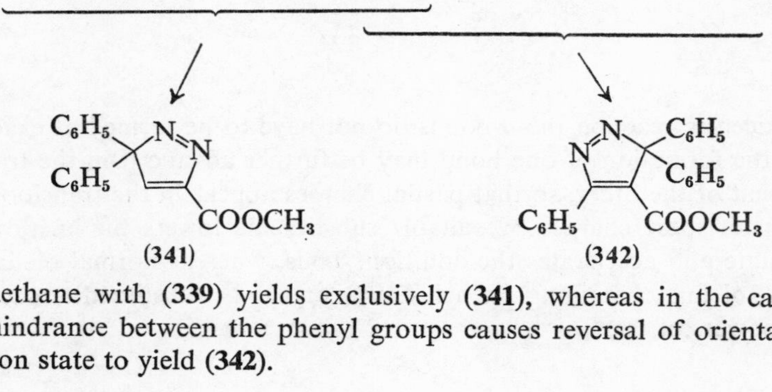

[structure (341): C_6H_5, C_6H_5, N, N, COOCH$_3$]

(341)

[structure (342): N, N, C_6H_5, C_6H_5, C_6H_5, COOCH$_3$]

(342)

diazomethane with (339) yields exclusively (341), whereas in the case of (340) steric hindrance between the phenyl groups causes reversal of orientation in the transition state to yield (342).

C-21 & C-22. Addition of nitrones to C=C bonds

Hydroxylamine yields oximes with carbonyl compounds. N-Substituted hydroxylamines, however, form nitrones* with aldehydes. The hydrates, as (**343**) formed in the latter reaction as intermediates, have been isolated in several cases[71]. Ketones do not react in the same simple way.

C-21. (i) Reaction of benzaldehyde with N-phenylhydroxylamine[72]

Nitrones add to C=C bonds, yielding isoxazolidine derivatives.

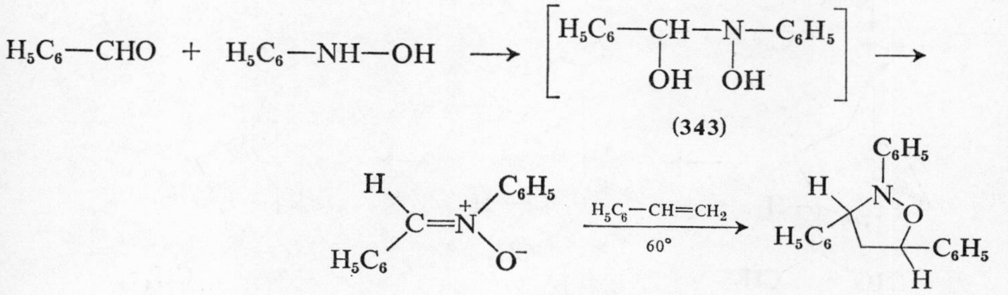

$$H_5C_6-CHO + H_5C_6-NH-OH \longrightarrow \left[\begin{array}{c} H_5C_6-CH-N-C_6H_5 \\ | \quad \quad | \\ OH \quad OH \end{array} \right] \longrightarrow$$

(**343**)

[nitrone structure] $\xrightarrow{H_5C_6-CH=CH_2, 60°}$ [isoxazolidine structure] (**344**) 92-99%

C-22. (ii) Reaction of N-methylhydroxylamine with 2-(2,2,3-trimethyl-3-cyclopentenyl)acetaldehyde (345)

See p. 168 for the formulae.

In the reaction of (**345**) with N-methylhydroxylamine[73] the 1,3-dipole (**346**) formed may add to the ring double bond in two directions, yielding the tricyclic isoxazolidines (**347**) and (**348**).

Characteristic NMR absorption of (**347**):
- 6.05 (1) H on C-5
- 7.05 (1) H on C-3
- 7.36 (3) N–CH$_3$
- 8.86 (3) CH$_3$ at C-4
- 9.07 τ (3) ⎫ gem. CH$_3$ at C-7
- 9.13 τ (3) ⎭

Characteristic NMR absorption of (**348**):
- 7.05 (1) H on C-5
- 7.36 (3) N–CH$_3$
- 8.86 (3) CH$_3$ at C-3
- 9.07 τ (3) ⎫ CH$_3$ at C-2
- 9.13 τ (3) ⎭

no absorption at 6.05

* For a review see ref. 70a.

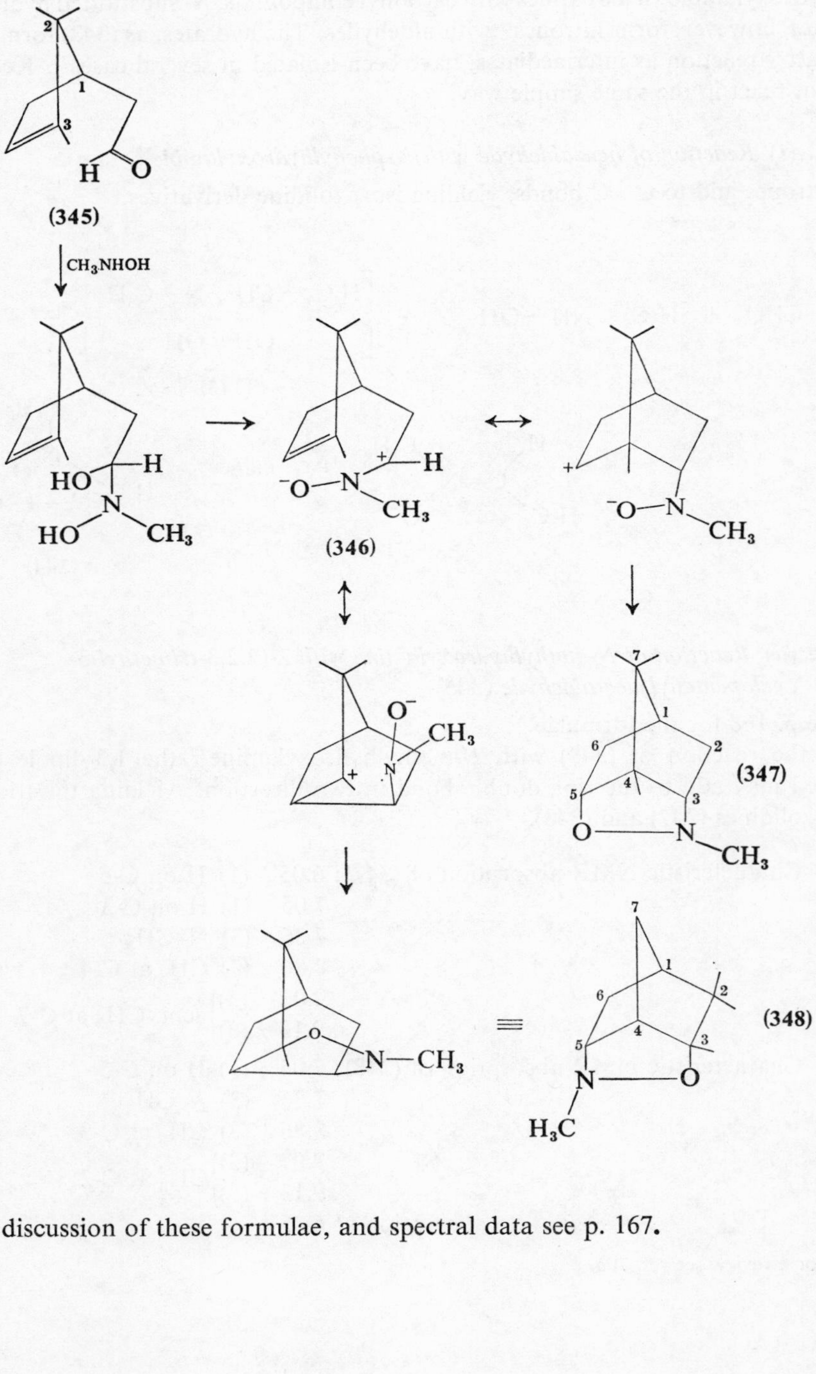

For discussion of these formulae, and spectral data see p. 167.

C-23. Decomposition of diazoacetic ester in benzonitrile[74]

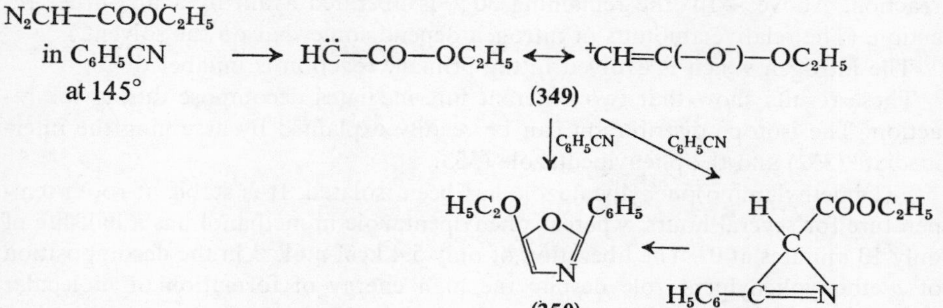

$N_2CH—COOC_2H_5$ in C_6H_5CN at 145° ⟶ $H\ddot{C}—CO—OC_2H_5$ ⟷ $^+CH=C(—O^-)—OC_2H_5$
(349)

$\downarrow C_6H_5CN \quad \searrow C_6H_5CN$

(350) oxazole ← azirine with H_5C_2O, C_6H_5, O, N / H, $COOC_2H_5$, $H_5C_6—C=N$

The carbene ester (349) formed on loss of nitrogen from the diazo ester adds to benzonitrile yielding the oxazole (350). An azirine derivative may be an intermediate[75]. Wolff rearrangement of (349) is not observed, since the ethoxy group has no tendency to migrate.

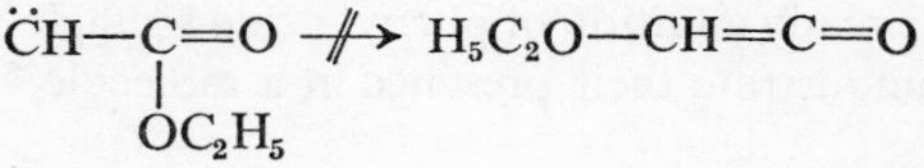

$\ddot{C}H—C=O$ with OC_2H_5 $\not\rightarrow$ $H_5C_2O—CH=C=O$

C-24. Reaction of benzenediazonium chloride with lithium azide[76]

Unsubstituted pentazole (351) which consists of five nitrogen and one hydrogen atom is as yet unknown. Aryl-substituted pentazoles, however, have recently been prepared.

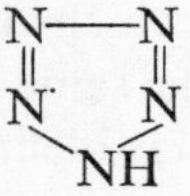

(351)

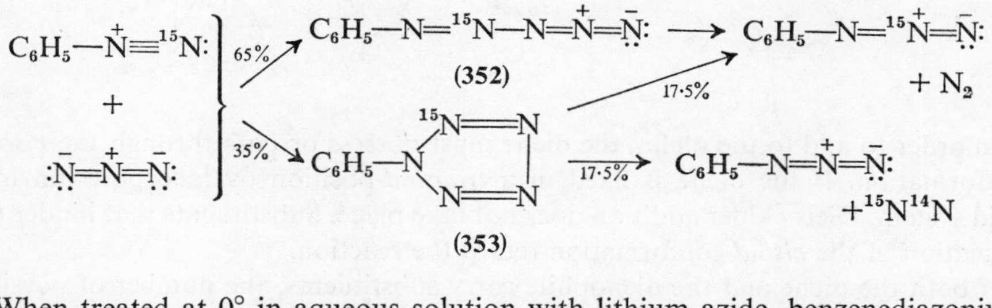

$C_6H_5—\overset{+}{N}=^{15}N:$ + $:\overset{-}{N}=\overset{+}{N}=\overset{-}{N}:$

65% → $C_6H_5—N=^{15}N—N=\overset{+}{N}=\overset{-}{N}:$ (352) → $C_6H_5—N=^{15}\overset{+}{N}=\overset{-}{N}:$ + N_2

35% → $C_6H_5—N$ with $^{15}N=N$ / $N=N$ (353) —17.5%→ $C_6H_5—N=\overset{+}{N}=\overset{-}{N}:$ + $^{15}N^{14}N$

17.5%

When treated at 0° in aqueous solution with lithium azide, benzenediazonium chloride labeled at the β-nitrogen atom affords phenyl azide containing 85% of the label and nitrogen containing 17.5% of the label.

In CH_3OH at $-25°$ 70% of the expected nitrogen is evolved in a first-order reaction. Above $-10°$ the remaining 30% is liberated again by a first-order reaction. (The relative amounts of nitrogen depend somewhat on the solvent.)

The nitrogen which is evolved in the primary reaction is unlabeled[77].

These results show that two different intermediates decompose during the reaction. The isotope distribution can be readily explained by assuming the intermediate (352) and the phenylpentazole (353).

p-(Dimethylamino)phenylpentazole has been isolated. It is stable at room temperature for several hours, whereas phenylpentazole in methanol has a half-life of only 13 minutes at $0°$. The liberation of only 5.4 kcal mole^{-1} in the decomposition of p-ethoxyphenylpentazole despite the high energy of formation of molecular nitrogen (226 kcal mole^{-1}) indicates the aromatic character of the pentazole ring.

Sandmeyer substitution is not observed in the reaction of phenyldiazonium chloride with lithium azide. Azides also add to C=C double bonds, forming five-

$$\langle\bigcirc\rangle-\overset{+}{N}\equiv{}^{15}N + N_3^- \;\not\rightarrow\; \langle\bigcirc\rangle-N_3 + N\equiv{}^{15}N$$

membered rings. The ready addition of phenyl azide to angle-strained double bonds has been used to demonstrate their presence in a molecule[78].

IV. DIELS–ALDER REACTIONS[79]

Dienes and olefins (dienophiles) form six-membered rings by Diels–Alder addition. The stereospecificity and the significant negative activation entropies of the cycloaddition indicate, in analogy to the 1,3-dipolar addition, that the reaction occurs by a synchronous, multi-center mechanism.

cis-*Rule*. A cis-disubstituted olefin leads to a cyclic product in which the substituents are also in cis-position. trans-Olefins yield trans-adducts. The same applies to the diene component. In general, trans-olefins react faster than cis-olefins.

In order to add to the olefin, the diene must possess or pass through the *cisoid*-conformation. If the diene is fixed in a *transoid*-position by incorporation in a rigid system, Diels–Alder addition does not take place. Substituents that hinder the formation of the *cisoid*-conformation retard the reaction.

If both the diene and the dienophile carry substituents, the number of possible isomers increases. For example, two structural isomers are obtained by addition of dienophile (355) to diene (354). In product (356), X and R are in neighboring positions, whereas in (357) they are relatively in the 1,3-position. In general, the isomer with neighboring substituents predominates. Furthermore, X and R may be *cis* or *trans* to each other. If X is in a *trans*-position in the diene, as in (354), X and R are in *cis*-positions in the product. If, however, X is in a *cis*-position in the diene (358), the *trans*-adducts (359) and (360) are obtained. The latter case is of minor importance, however, since dienes with terminal *cis*-substituents react only under forcing reaction conditions if at all. If X is located at the 2-position of the

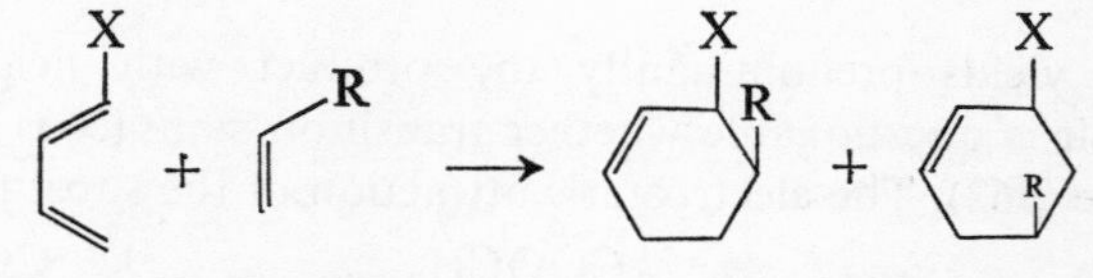

(354) (355) (356) Predominant (357)

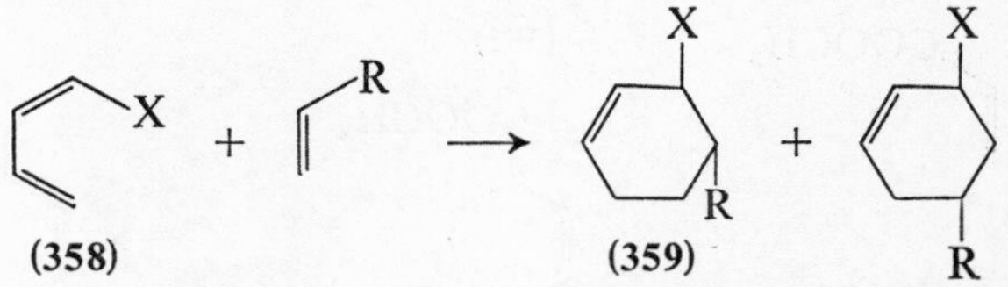

(358) (359) Predominant (360)

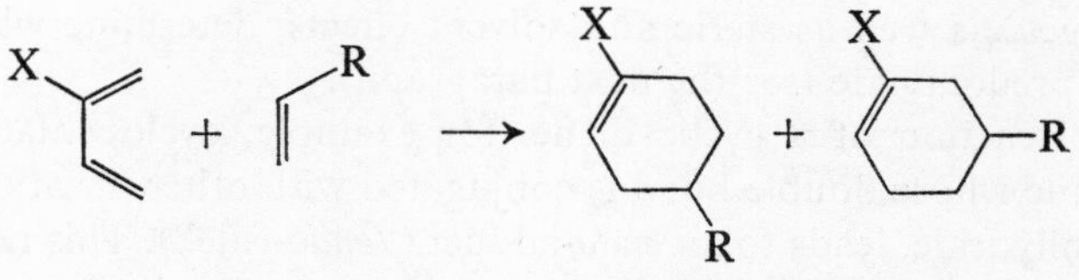

(Predominant)

diene, the 1,4-adduct predominates. It has been proposed that the orientations can be understood if bond formation in multi-center mechanisms does not proceed completely synchronously. If the formation of one bond runs somewhat ahead of the other, the transition state has some ionic or diradical character. Maximum resonance stabilization of the transition state should, therefore, determine the outcome of the addition reaction[80]. However, methyl 2,4-pentadienoate, like

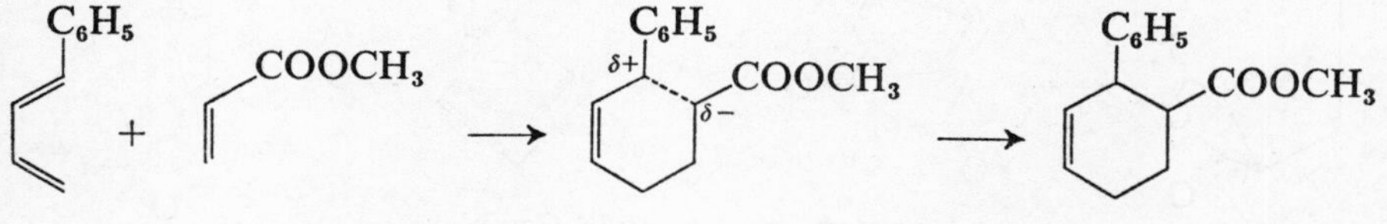

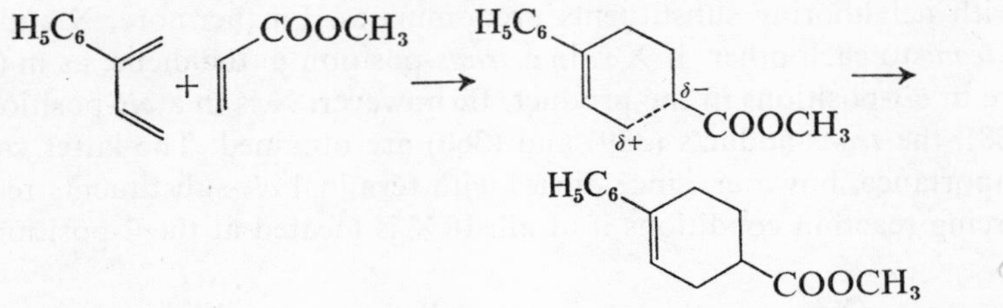

1-phenylbutadiene, yields predominantly the product with neighboring substituents, although it is questionable whether transition state (**361**) is more stable than transition state (**362**). The electrostatic attraction of the substituents carrying

(**361**)

(**362**)

opposite charges, as well as steric and solvent effects, determine whether *cis*- or *trans*-adducts predominate (see the next paragraph).

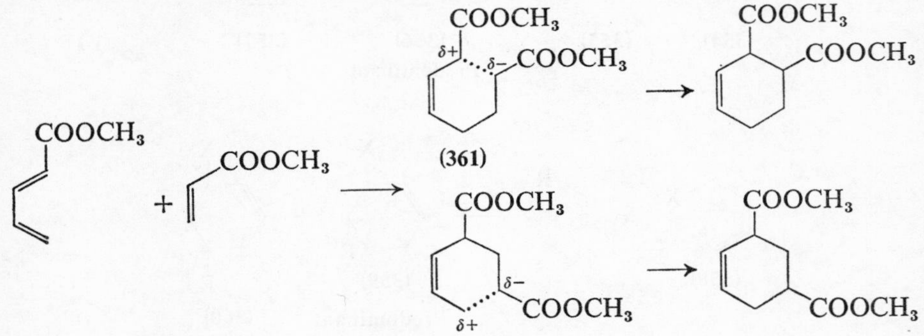

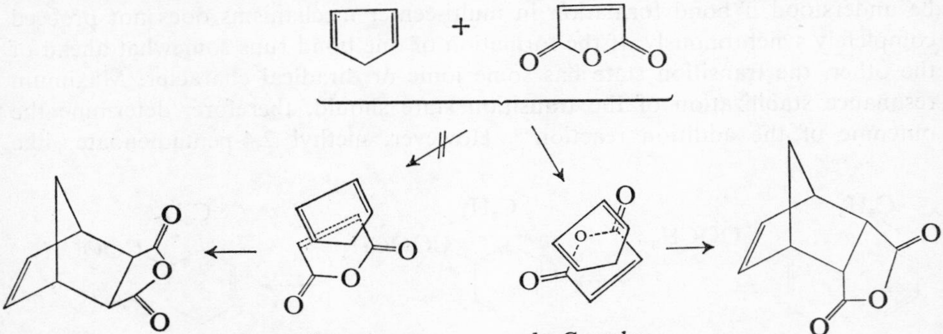

endo-*Rule*. Reaction of a cyclic diene, for example, cyclopentadiene, with a cyclic dienophile whose double bond is conjugated with other unsaturated groups, as in maleic anhydride, leads to an *endo*-adduct ("*endo*-rule"). This result has been rationalized by the theory that a charge-transfer complex is formed by parallel arrangement of the components, which allows maximum interaction between all π-electrons in the diene and the dienophile. The *endo*-rule is obeyed not only in the

exo-Complex *endo*-Complex

reaction of cyclic dienes with cyclic dienophiles, but also in numerous cases where cyclic dienes react with open-chain mono- and di-substituted dienophiles or open-chain dienes react with cyclic dienophiles. There is at present no quantitative theory that satisfactorily explains the stereoselectivities of individual reactions.

Recently, Hoffmann and Woodward[80a] suggested that symmetry-controlled secondary orbital interactions may determine whether an *exo*- or an *endo*-addition product is formed. Consider the addition of butadiene to itself. Inspection of Figure C, which represents symmetry-allowed mixing of unoccupied with occupied levels, shows that, in the *endo*-transition state, energy-lowering secondary orbital inter-

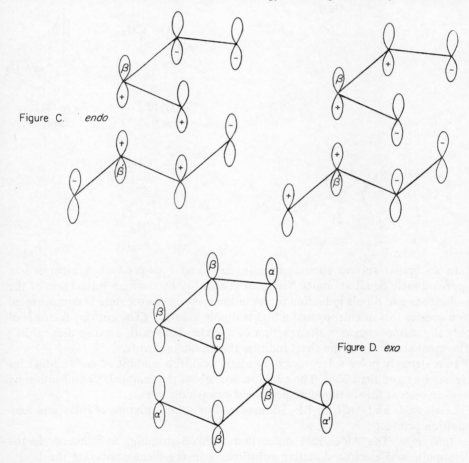

Figure C. *endo*

Figure D. *exo*

actions between β and β' are possible, whereas such interactions cannot take place in the *exo*-transition state (Figure D). The course of addition is probably determined by the interaction of several relatively weak effects, as is shown by the reaction of a crotonic ester with cyclopentadiene.

C-25. Reaction of methyl crotonate with cyclopentadiene[81]

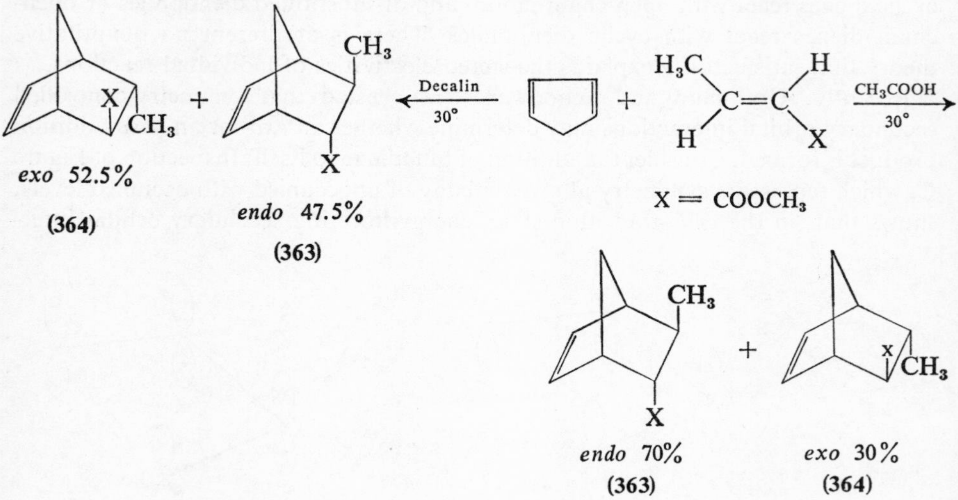
CH₃
X
CH₃
X
Decalin
30°
H₃C
H
C=C
H
X
CH₃COOH
30°
X = COOCH₃
CH₃
X
X
CH₃
exo 52.5%
(364)
endo 47.5%
(363)
endo 70%
(363)
exo 30%
(364)

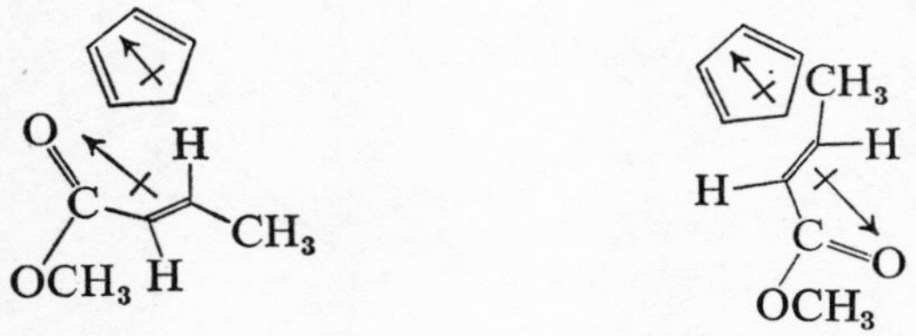
O
H
C
CH₃
OCH₃
H
CH₃
H
H
C
O
OCH₃

endo-Complex *exo*-Complex

In an apolar solvent such as decalin, *exo*- and *endo*-products are formed in approximately equal amounts. The gain in stability by stronger interaction of the π-electrons and dipole induction forces in the *endo*-transition state is compensated by a greater loss of entropy and a higher dipole moment. (The entropy factor is of only slight importance in the reaction of a cyclic diene with a cyclic dienophile.) The crossed arrows in the chart indicate the dipole moments.

In a strongly polar solvent such as acetic acid, the amount of *endo*-adduct increases by more than 20%. The addition now obeys the *endo*-rule. Stabilization by a polar solvent favors the more polar *endo*-transition state.

Lewis acids and $AlCl_3$ or BF_3 adducts lead to a 95:5 mixture of *endo*- and *exo*-addition product[82].

Alder rule. The Alder rule states that electron-attracting substituents in the dienophile and electron-donating substituents in the diene accelerate the Diels–Alder addition[83]. The accelerating effect of Lewis acids is probably due to their ability to co-ordinate with substituents in the dienophile, causing a further decrease of electron density in the dienophile.

Tetracyanoethylene with its strongly electron-attracting substituents reacts with 9,10-dimethylanthracene immeasurably fast[84]. However, no reaction takes place with hexachlorocyclopentadiene, but this diene reacts readily with dienophiles that carry electron-donating substituents. These results clearly indicate that the Alder rule can be reversed. Dienophiles with electron-donating substituents readily react with electron-deficient dienes: they do not react, however, with electron-rich dienes[85].

C-26. Reaction of 3,5-dichloro-2-pyrone with maleic anhydride[86]

Chlorinated 2-pyrones behave like hexachlorocyclopentadiene. The rate of addition to maleic anhydride decreases with increasing chlorination. Whereas the

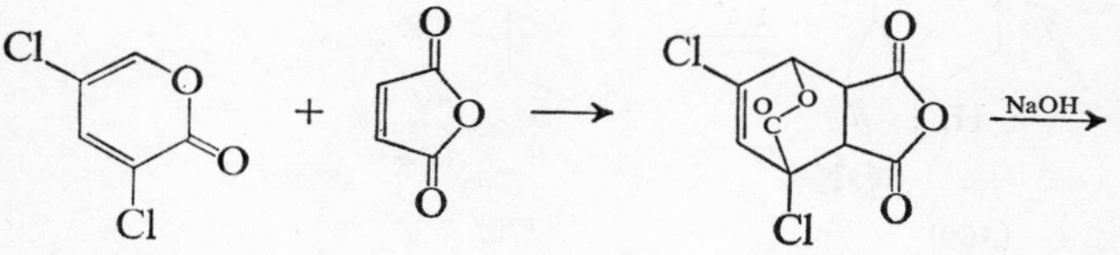

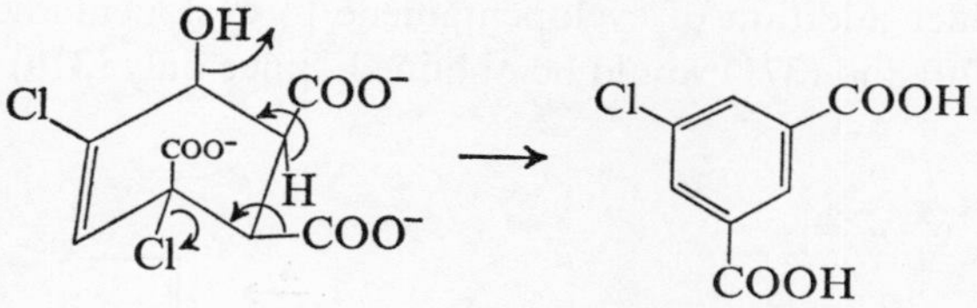

lactone (365) still reacts with maleic anhydride under conditions that allow the trapping of 65% of the adduct (366), the addition to perchlorinated 2-pyrone is possible only under conditions that lead to decomposition of the primary addition product. If the product (366) is treated with diluted sodium hydroxide solution, 5-chloroisophthalic acid (368) is obtained by loss of water from, and decarboxylative dehalogenation of, the intermediate (367).

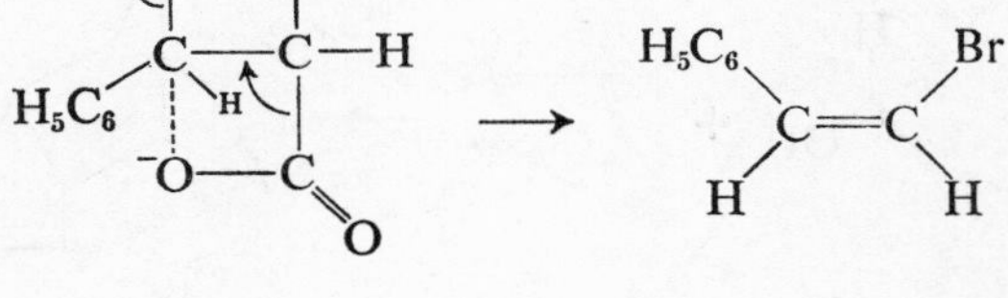

The elimination of CO_2 together with a halide ion, leading to an olefin, may take place stereospecifically in apolar solvents. By synchronous *trans*-elimination a *cis*-olefin is obtained (a). In polar solvents mixtures of *cis*- and *trans*-olefins are formed by stepwise elimination (b)[87].

(b) $H_5C_6-\underset{Br}{CH}-\underset{Br}{CH}-COO^- \longrightarrow H_5C_6-\overset{+}{CH}-\underset{Br}{CH}-COO^- \longrightarrow$

$H_5C_6-CH=CHBr + CO_2$

C-27. Rearrangement of α-dicyclopentadien-1-ol (369)[88]

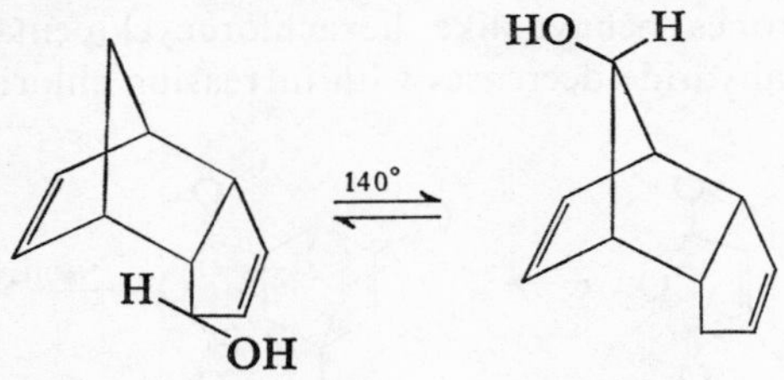

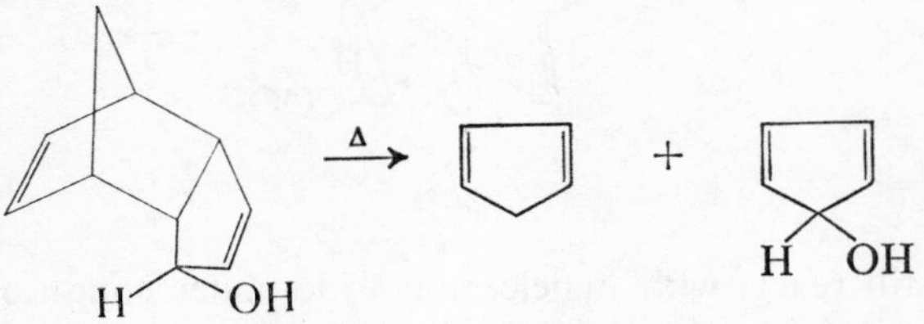

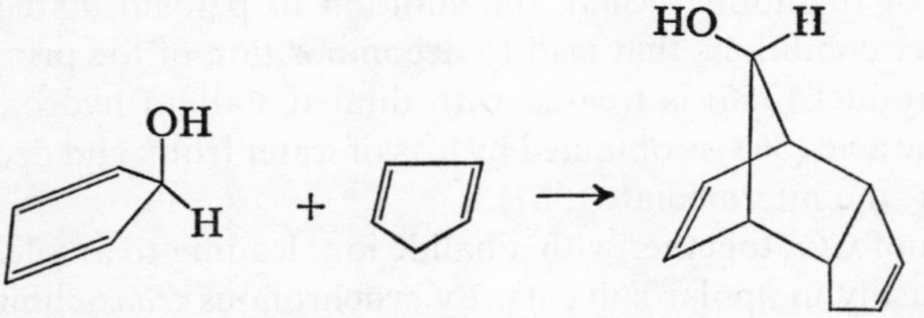

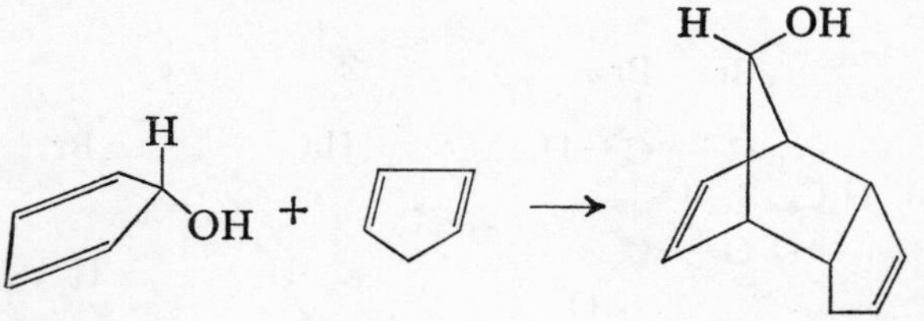

If the alcohol (369) is heated at 140°, stereoselective rearrangement to (370) occurs. If in the first step there were to be retrodiene reaction, followed by Diels–Alder addition of cyclopentadiene to cyclopentadien-1-ol, two isomeric products (370) and (371) would be obtained. Since only (370) is formed, it has been assumed that only the 3a,4-bond is broken and that in the second step bond-formation occurs between C-2 and C-6. On the principle of microscopic reversibility the postulated stepwise mechanism of the rearrangement has led to the conclusion that Diels–Alder additions in general follow a two-step mechanism. It seems, however more likely that the reaction (369) → (370) is actually a special case of the Cope rearrangement (see p. 189), during which the 2,6-bond is formed simultaneously with cleavage of the divinyl-substituted 3a,4-bond.

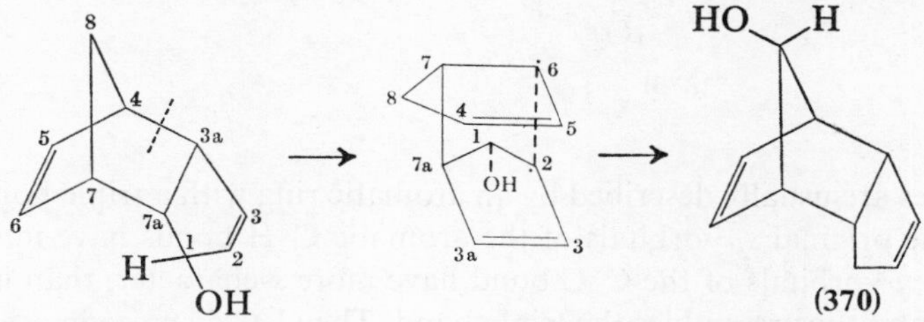

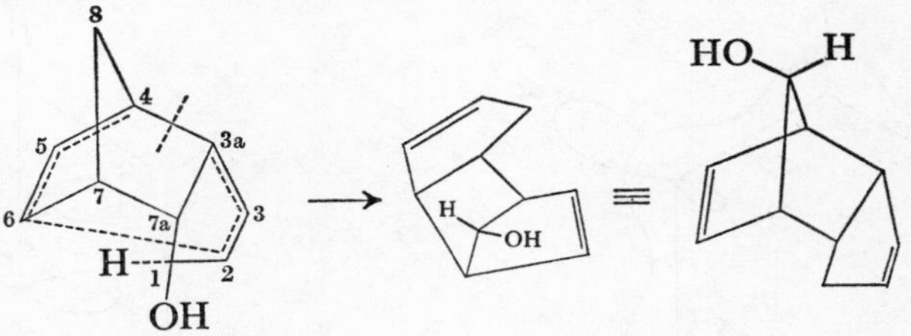

C-28. Addition of benzyne to 2-pyrone[89]

Arynes are particularly reactive dienophiles. The simplest representative is benzyne (dehydrobenzene), which is generated as a short-lived intermediate on reaction of a halobenzene with strong base[90].

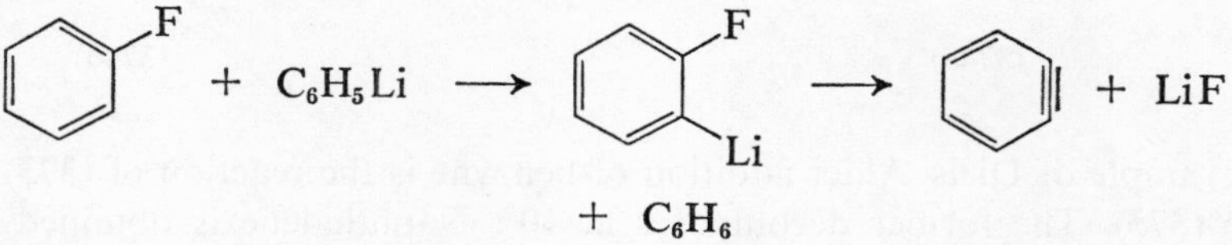

Other methods of generation are the thermal decomposition of the compounds (372) and (373)[91].

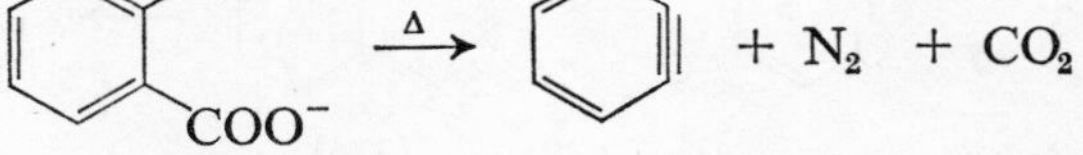

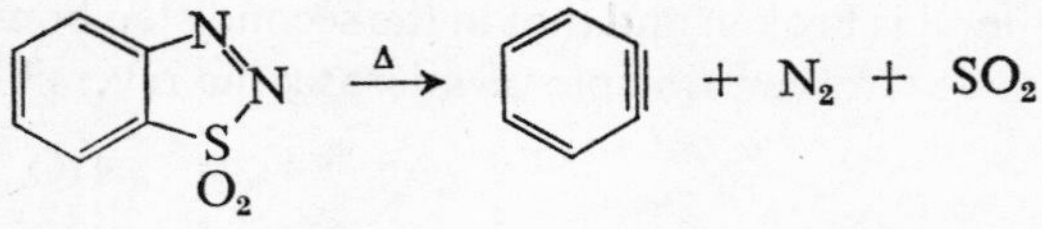

(373) → benzene + N₂ + SO₂

Arynes are usually described by an aromatic ring with a triple bond. In the aryne bond the original sp^2-orbitals of the aromatic C–H bonds have more p-character, but the sp^2-orbitals of the C–C bond have more s-character, than in benzene. The aryne bond thus resembles the triple bond. The C–C distance in arynes is probably smaller than in benzene. The structure of benzyne would be best represented by a formulation between structure (374a) with the original sp^2-orbitals and structure (374b) which contains a triple bond, causing strong distortion of the six-membered ring[92].

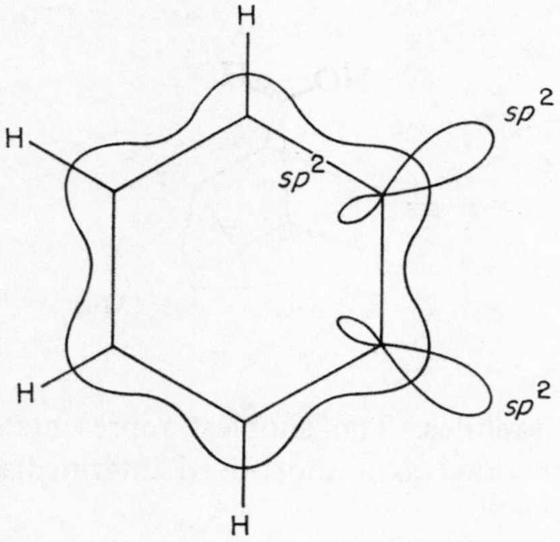

(374a) 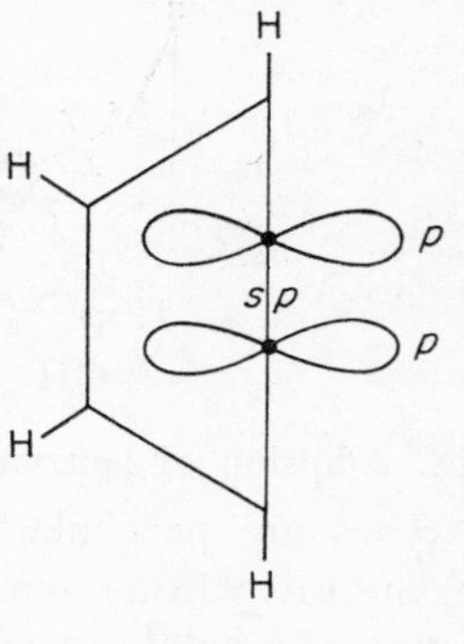   (374b)

An example of Diels–Alder addition of benzyne is the reaction of (373) with 2-pyrone (375). The former decomposes at 10°. Naphthalene is obtained by way of intermediate (376), in 36% yield.

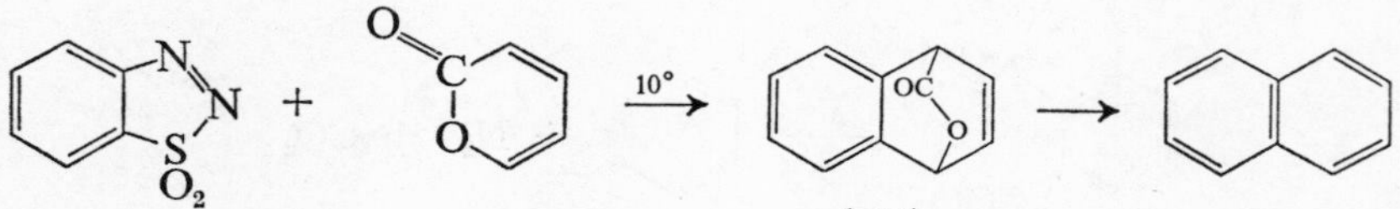

(373) + (375) →10° (376) → naphthalene

C-29. Addition of hexafluoro-2-butyne to benzene[93]

Hexafluoro-2-butyne (377) reacts with benzene by 1,4-addition at elevated temperatures. This is the only known Diels–Alder addition to unsubstituted benzene. The photoaddition of maleic anhydride to benzene, described by Bryce-Smith and Angus[94] is presumably initiated by 1,2-addition.

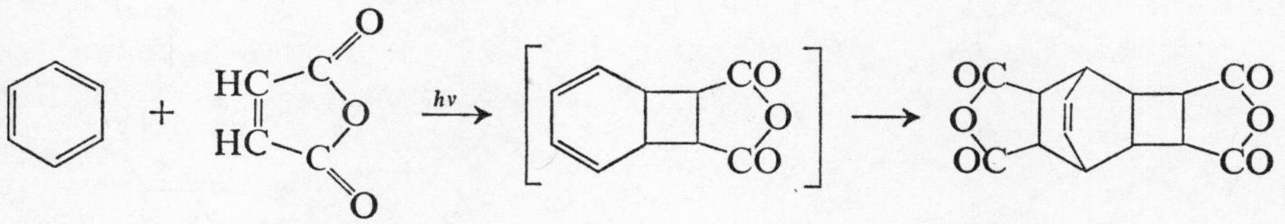

The primary addition product (378), which could not be isolated, yields o-bis-(trifluoromethyl)benzene (379) by retrodiene cleavage. Further reactions are shown in the chart on p. 180.

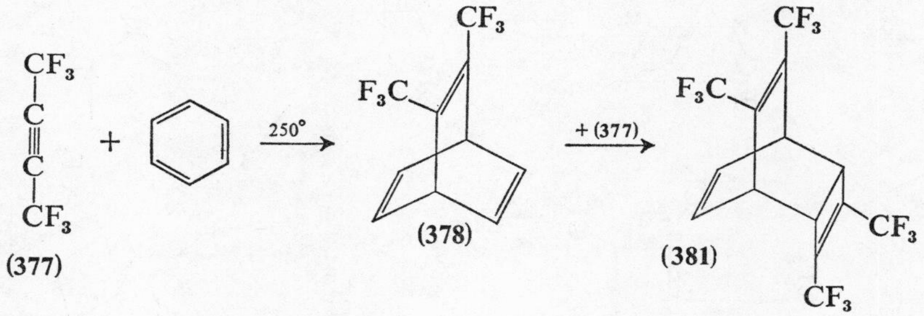

The butyne (377) adds again to this product (379), yielding 1,2,4,5-tetrakis(trifluoromethyl)benzene (380). The formation of 2,3,6,7- (384) and 1,4,6,7-tetrakis-(trifluoromethyl)naphthalene (385) is assumed to involve a common intermediate (381). This is obtained by 1,2-addition of (377) to (378) and can rearrange by path a or by path b. The resulting 9,10-dihydronaphthalenes (382) and (383) are

Characteristic UV absorption of (384):

222 mμ	ε = 78000 (in ethanol)	
260	4030	
308	4130	naphthalene nucleus
323	2580	
Characteristic UV absorption of (385):		
221	107000 in isooctane	
272	6000	
306	640	
319	720	

dehydrogenated to (384) and (385). A related reaction[95] is the pyrolysis of an adduct (386), where X = COOCH$_3$.

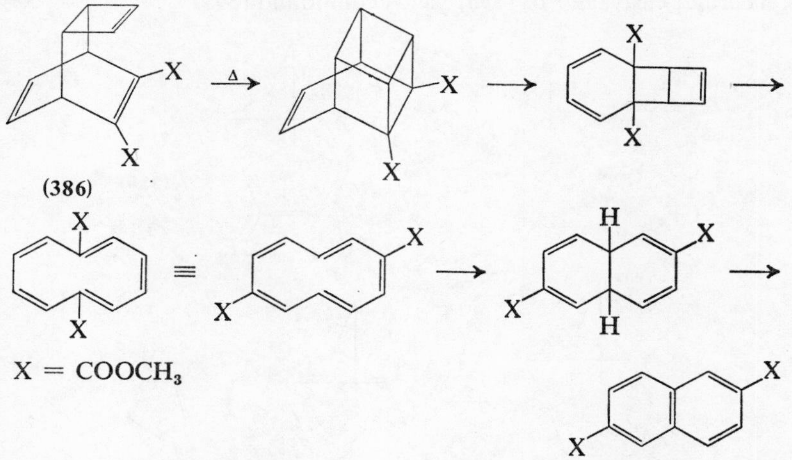

X = COOCH$_3$

An alternative mechanism for the formation of (382) through intermediate (387) has been proposed[93], which requires the formation of a carbene (388) which, however, is rather unlikely.

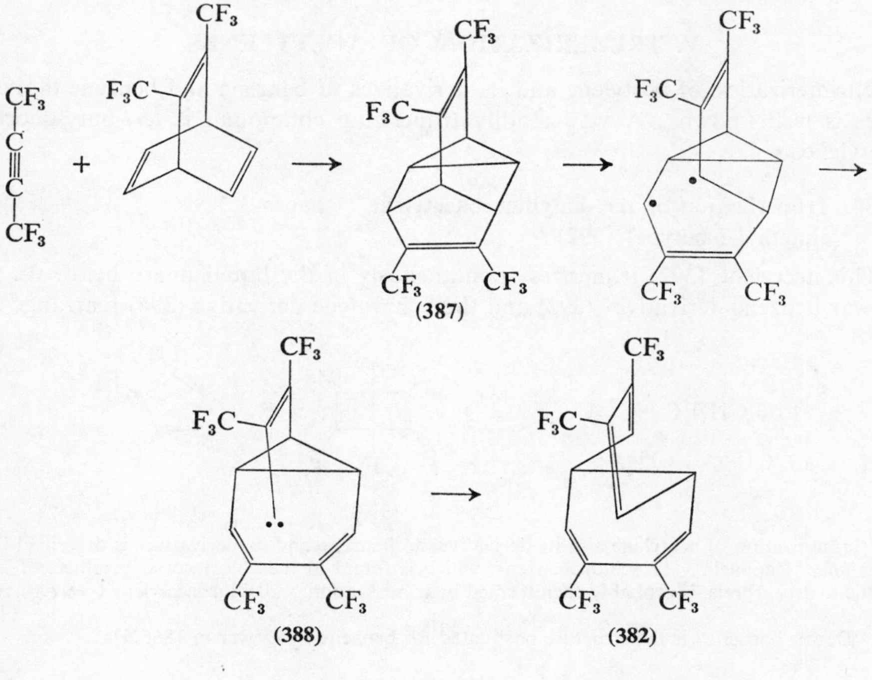

Compounds analogous to (387) are known; for example, (389) is obtained from (377) and bicycloheptadiene, and bicycloheptadiene forms a dimer (390) among others. Thermal cleavage[96] of (390) yields compound (391).

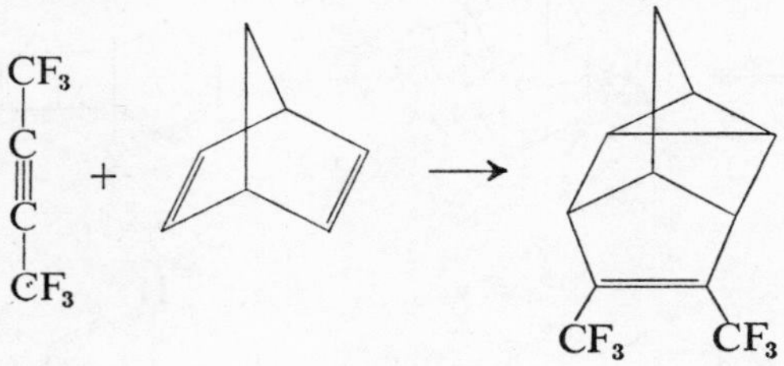

V. TRIMERIZATION OF ACETYLENES

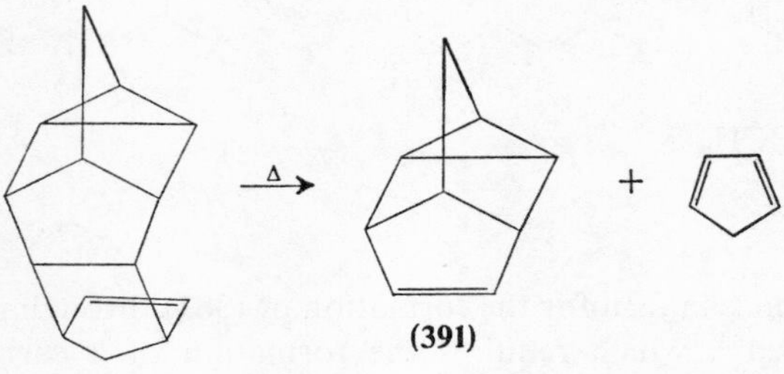

The trimerization of acetylene and its derivatives to benzene and benzene derivatives is well known*. A very readily trimerizing compound is *tert*-butylfluoroacetylene.

C-30. Trimerization of *tert*-butylfluoroacetylene (1-fluoro-3,3-dimethyl-1-butyne) (392)[99]

This acetylene (392) trimerizes spontaneously in the liquid phase below 0°. A Dewar benzene derivative† (393) and the benzvalene derivative (394) rearrange to

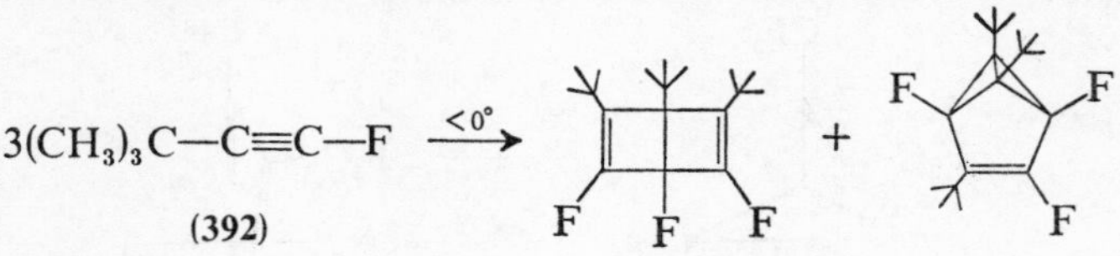

$$3(CH_3)_3C-C\equiv C-F \xrightarrow{<0°}$$

(392)   (393)   (394)

* Trimerization of acetylene and its derivatives to benzene and its derivatives is described by, *inter alios*, Raphael[97]. Hexachlorobenzene, which is found in the chlorinolysis products of aliphatic hydrocarbons, is probably then formed by trimerization of dichloroacetylene (see, *e.g.*, ref. 98).

† "Dewar benzene" is the structure postulated for benzene by Dewar in 1866/67.

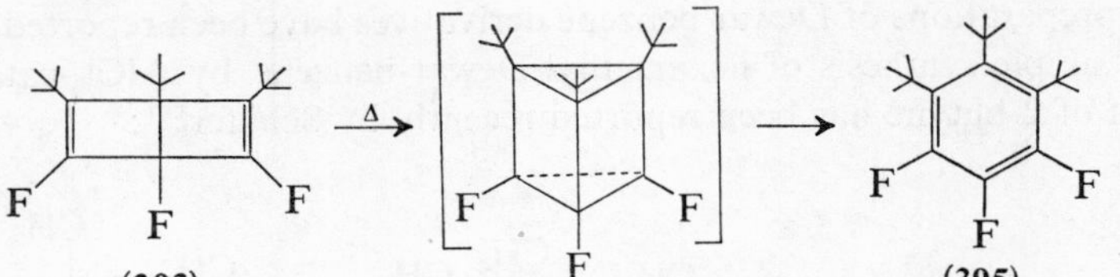

(395) and (396), respectively, on being warmed.

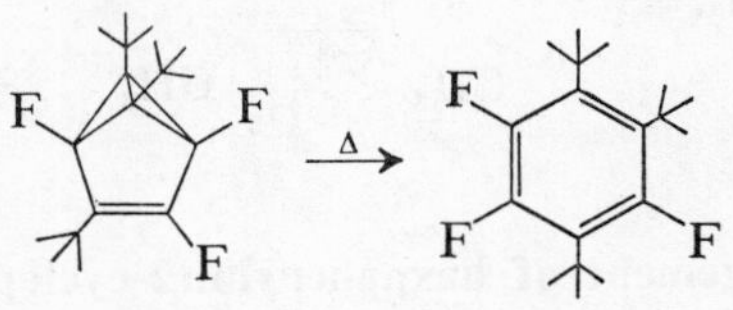

The thermal stability of the Dewar benzene is lower than that of the benzvalene derivative. In general, the stability of the non-aromatic valence isomers of benzene increases with decreasing number of double bonds. The compounds (393) and (394) may be formed by the following mechanism:

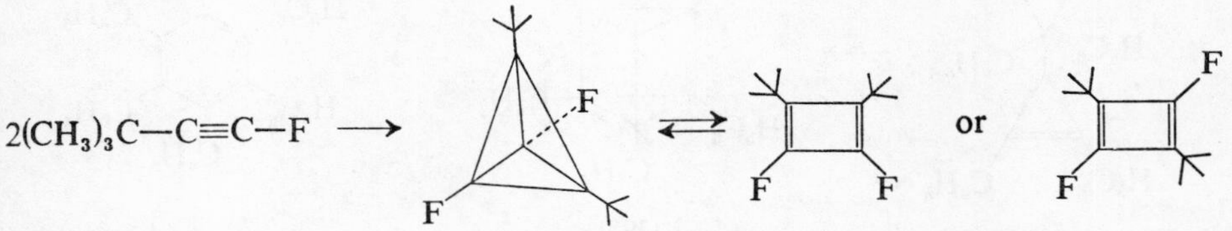

Cyclobutadiene* has a triplet ground state† and thus exhibits diradical character:

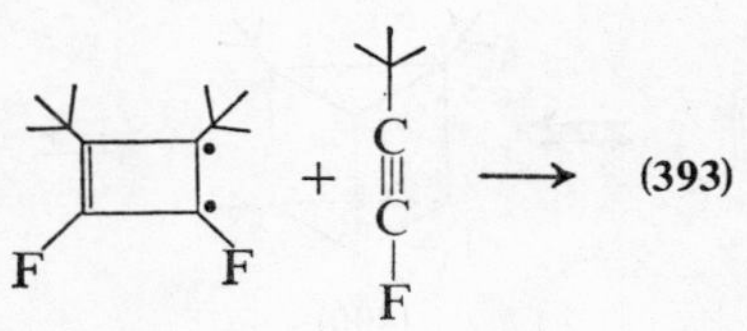

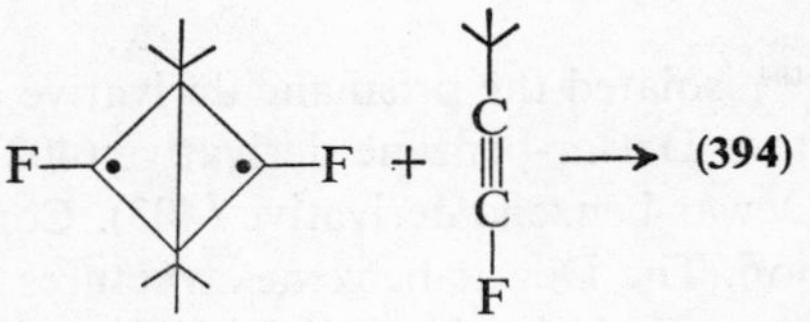

The mechanism illustrated is supported by the trapping of dimeric intermediates.

---
* The first stable substituted cyclobutadiene was reported by Gompper et al.[99a]
† But see Dewar et al.[99b]

Several preparations of Dewar benzene derivatives have been reported recently*. A very simple synthesis of hexamethyl-Dewar-benzene by $AlCl_3$-catalyzed trimerization of 2-butyne has been reported recently by Schäfer[101]:

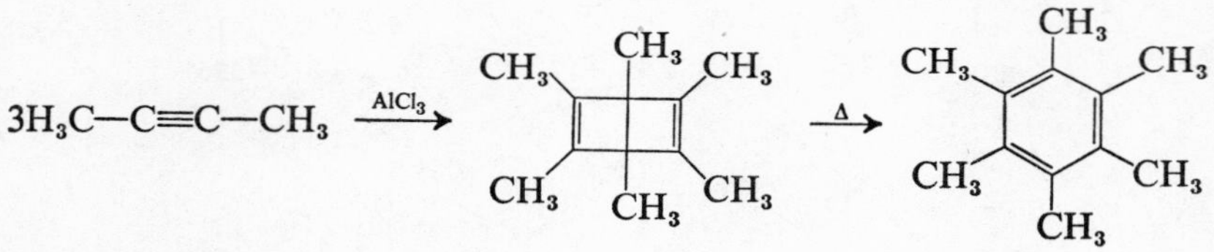

C-31. Thermal rearrangement of hexaphenylbi(2-cyclopropenyl)[102]

Ladenburg prism-structures† such as (397) and (398) have been postulated as intermediates in the thermal rearrangement of (393) and of bi(triphenylcyclopropenyl)[102]. The photoisomerization of (399) is believed[103] to involve formation of a prismane derivative (400).

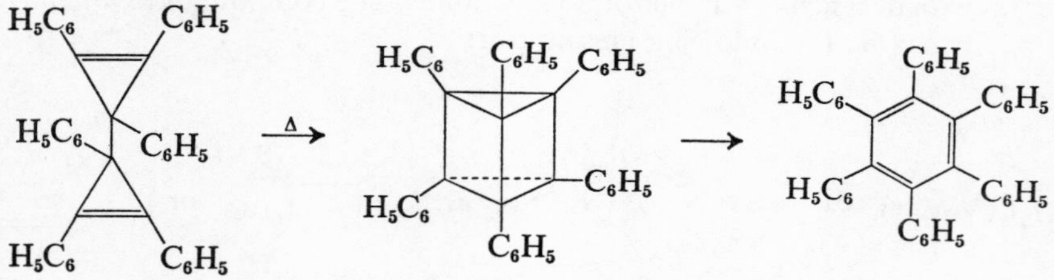

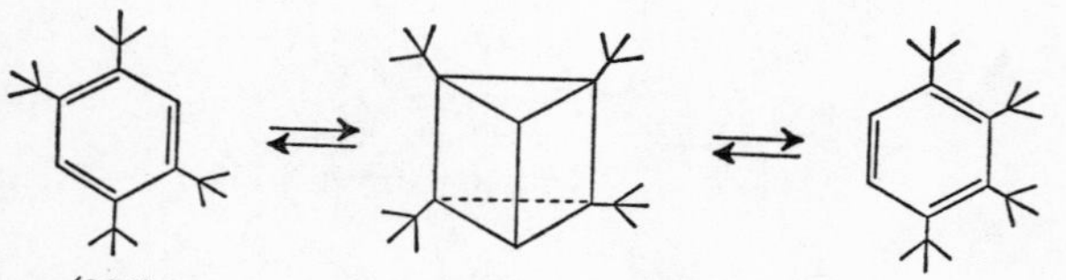

Criegee and Askani[104] isolated the prismane derivative (401). In the presence of acids, this rearranges to a Dewar-benzene derivative (402), but heating (401) with pyridine leads to the Dewar-benzene derivative (403). Conversely, (401) is formed from (402) by irradiation. The Dewar-benzene structures (402) and (403) are converted into aromatic benzene derivatives (404), (405), and (406) on being heated.

* For a survey see Viehe[99] in a footnote, also van Tamelen[100] and Ward et al.[100a]
† Such prism formulae were put forward as possible formulae for benzene and its derivatives by Ladenburg in 1869.

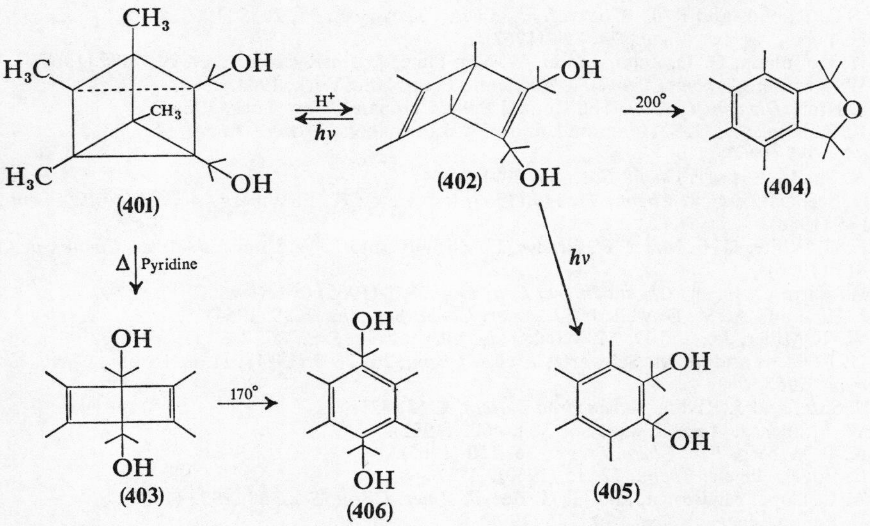

Irradiation of 1,2,4- and 1,3,5-tri-*tert*-butylbenzene yields[105] a mixture of the Ladenburg prism derivative (**407**), the Dewar-benzene derivative (**408**), and the benzvalene derivative (**409**).

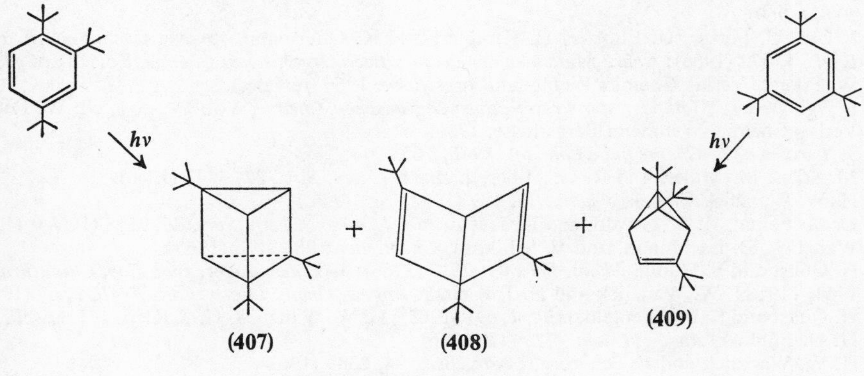

REFERENCES

[1] R. Huisgen, R. Grashey, and J. Sauer, *The Chemistry of the Alkenes*, ed. S. Patai, p. 750, Interscience, Inc., New York, 1964.

[1a] G. Wilke, B. Bogdanovic, P. Hardt, P. Heimbach, W. Keim, M. Kröner, W. Oberkirch, L. Tanaka, E. Steinbrücke, D. Walter, and H. Zimmermann, *Angew. Chem. Intern. Edit. Engl.*, **5**, 151 (1966).

[2] P. D. Bartlett, *Rec. Chem. Progr.*, **11**, 47 (1950).

[3] H. Kwart and D. M. Hoffman, *J. Org. Chem.*, **30**, 419 (1966).

[3a] K. D. Bingham, G. D. Meakins, and G. H. Witham, *Chem. Commun.*, **1966**, 445.

[4] J. Meinwald, S. S. Labana, and M. S. Chadha, *J. Amer. Chem. Soc.*, **85**, 582 (1963).

[5] M. Rey and A. S. Dreiding, *Helv. Chim. Acta*, **48**, 1985 (1965).

[5a] S. J. Rhoads and R. D. Cockroft, *J. Amer. Chem. Soc.*, **91**, 2815 (1969).
[6] E. Vogel, *Angew. Chem.*, **74**, 829 (1962).
[7] R. Hoffmann, G. D. Zeiss, and G. W. Van Dine, *J. Amer. Chem. Soc.*, **90**, 1485 (1968).
[8] W. Kirmse, *Carbene Chemistry*, Academic Press, New York, 1964.
[9] J. Hine, *Divalent Carbon*, The Ronald Press Company, New York, 1964.
[10] K. R. Kopecky, G. S. Hammond, and P. A. Leermakers, *J. Amer. Chem. Soc.*, **83**, 2397 (1961); **84**, 1015 (1962).
[11] A. Small, *J. Amer. Chem. Soc.*, **86**, 2091 (1964).
[12] D. Seebach, *Angew. Chem.*, **77**, 119 (1965) (a review); K. B. Wiberg, *Adv. Alicyclic Chem.*, **2**, 185 (1968).
[13] E. H. White, G E. Maier, R. Graeve, U. Zirngibl, and E. W. Friend, *J. Amer. Chem. Soc.*, **88**, 611 (1966).
[14] W. Kirmse, *Angew. Chem., Intern. Edit. Engl.* **4**, 1 (1965) (a review).
[15] J. Hine and A. M. Dowell, Jr., *J. Amer. Chem. Soc.*, **76**, 2688 (1954).
[16] W. T. Miller, Jr., and D. M. Whalen, *J. Amer. Chem. Soc.*, **86**, 2089 (1964).
[17] G. L. Closs and G. M. Schwartz, *J. Org. Chem.*, **26**, 2609 (1961); H. E. Dobbs, *Chem. Commun.*, **1965**, 56.
[18] S. Sarel and J. Rivlin, *Tetrahedron Letters*, **1965**, 821.
[19] W. Mahler, *J. Amer. Chem. Soc.*, **84**, 4600 (1962).
[20] K. B. Wiberg, *Rec. Chem. Progr.*, **26**, 150 (1965).
[21] E. Vogel, *Angew. Chem.*, **72**, 15 (1960).
[22] A. C. Cope, M. Brown, and H. H. Lee, *J. Amer. Chem. Soc.*, **80**, 2855 (1958).
[23] V. Prelog, *Angew. Chem.*, **70**, 145 (1958).
[24] E. S. Gould, *Mechanismus und Struktur in der Organischen Chemie*, Verlag Chemie GmbH, Weinheim/Bergstraße, 1962, p. 724.
[25] A. C. Cope, R. S. Bly, M. M. Martin, and R. C. Petterson, *J. Amer. Chem. Soc.*, **87**, 3111, 3119 (1965); W. R. Roth, *Ann. Chem.*, **671**, 25 (1964).
[26] A. C. Cope, G. A. Berchtold, P. E. Peterson, and S. H. Sharman, *J. Amer. Chem. Soc.*, **82**, 6370 (1960).
[27] I. Ugi, U. Fetzer, U. Eholzer, H. Knupfer, and K. Offermann, *Angew. Chem. Intern. Edit. Engl.*, **4**, 472 (1965); *Neure Methoden der Präparativen Organischen Chemie*, Vol. IV, p. 37, ed. W. Foerst, Verlag Chemie, Weinheim/Bergstrasse, 1966 (reviews).
[28] I. Ugi, *Neuere Methoden der Präparativen Organischen Chemie*, Vol. IV, p. 1, ed. W. Foerst, Verlag Chemie, Weinheim/Bergstrasse, 1966.
[29] S. Yaroslavsky, *Chem. Ind. (London)*, **1965**, 765.
[30] M. Stiles, M. Miller, and R. G. Miller, *J. Amer. Chem. Soc.*, **82**, 3802 (1960).
[31] H. W. Wanzlick, *Angew. Chem.*, **74**, 129 (1962).
[32] D. M. Lemal, R. A. Lovald, and K. I. Kawano, *J. Amer. Chem. Soc.*, **86**, 2518 (1964); H. W. Wanzlick, B. Lachmann, and E. Schikora, *Chem. Ber.*, **98**, 3170 (1965).
[33] H. Quast and S. Hünig, *Chem. Ber.*, **99**, 2017 (1966); J.-J. Vorsauger, *Bull. Soc. Chim. France*, **1964**, 119; H. W. Wanzlick and H.-J. Kleiner, *Angew. Chem. Intern. Edit. Engl.*, **3**, 65 (1964); H. Quast and E. Frankenfeld, *ibid.*, **4**, 691 (1965); H. W. Wanzlick, H.-J. Kleiner, I. Lasch, and H. U. Füldner, *Angew. Chem.*, **77**, 115 (1966).
[34] H. W. Wanzlick and E. Schikora, *Chem. Ber.*, **94**, 2389 (1961).
[35] L. Horner and A. Christmann, *Angew. Chem.*, **75**, 707 (1963) (review); R. A. Abramovitch and B. A. Davis, *Chem. Rev.*, **64**, 149 (1964).
[36] G. Smolinsky and B. I. Feuer, *J. Amer. Chem. Soc.*, **86**, 3085 (1964).
[37] E. Wasserman, G. Smolinsky, and W. A. Yager, *J. Amer. Chem. Soc.* **86**, 3166 (1964).
[38] P. T. Lansbury, J. G. Colson, and N. R. Mancuso, *J. Amer. Chem. Soc.*, **86**, 5225 (1964); P. T. Lansbury and N. R. Mancuso, *ibid.*, **88**, 1205 (1966).
[39] O. M. Nefedow and M. N. Manakow, *Angew. Chem.*, **78**, 1039 (1966).
[40] M. E. Volpin, Y. D. Koreshkov, V. G. Dulova, and D. N. Kursanov, *Tetrahedron*, **18**, 112 (1962) (review).
[41] G. Smolinsky, *J. Org. Chem.*, **27**, 3557 (1962).
[42] E. J. Corey and R. W. White, *J. Amer. Chem. Soc.*, **80**, 6686 (1958).
[43] R. A. Sneen and N. P. Matheny, *J. Amer. Chem. Soc.*, **86**, 3096, 5503 (1964).

Cycloadditions

[44] R. Criegee, *Ann. Chem.*, **560**, 130 (1948).
[45] G. Smolinsky and B. I. Feuer, *J. Org. Chem.*, **30**, 3216 (1965).
[46] P. D. Bartlett, L. K. Montgomery, and B. Seidel, *J. Amer. Chem. Soc.*, **86**, 616 (1964); L. K. Montgomery, K. Schueller, and P. D. Bartlett, *ibid.*, p. 622; P. D. Bartlett and L. K. Montgomery, *ibid.*, p. 628.
[47] J. D. Roberts and C. M. Sharts, *Organic Reactions*, Vol. 12, p. 17. ed. R. Adams, Interscience, New York, 1962.
[48] S. Proskow, H. E. Simmons, and T. L. Cairns, *J. Amer. Chem. Soc.*, **85**, 2341 (1963); **88**, 5254 (1966).
[49] R. Hoffmann and R. B. Woodward, *J. Amer. Chem. Soc.*, **87**, 2046, 2048, 4388; *Accounts of Chem. Res.*, **1**, 17 (1968); B. Gill, *Quart. Rev. (London)*, **22**, 338 (1968); D. Seebach, *Fortschr. Chem. Forsch.*, **11**, 177 (1969).
[50] R. H. Hasek, R. H. Meen, and J. C. Martin, *J. Org. Chem.*, **30**, 1495 (1965).
[51] See, for example: J. E. Baldwin and J. D. Roberts, *J. Amer. Chem. Soc.*, **85**, 2444 (1963); H. H. Wasserman and E. V. Dehmlow, *ibid.*, **84**, 3786 (1962); J. Bregman and S. H. Bauer, *ibid.*, **77**, 1955 (1955); R. H. Hasek, R. D. Clark, E. V. Elam, and J. C. Martin, *J. Org. Chem.*, **27**, 60 (1962); G. Schwarzenbach and K. Lutz, *Helv. Chim. Acta*, **23**, 1151, 1155 (1940); G. Schwarzenbach and E. Felder, *ibid.*, **27**, 1044 (1944); R. B. Woodward and E. R. Blout, *J. Amer. Chem. Soc.*, **65**, 562 (1943); R. B. Woodward and G. Small, *ibid.*, **72**, 1297 (1950).
[51a] G. Opitz and H. R. Mohl, *Angew. Chem.*, **81**, 36 (1969).
[52] W. E. Truce, R. W. Campbell and J. R. Norell, *J. Amer. Chem. Soc.* **88**, 3599 (1966); J. F. King and T. Durst, *J. Amer. Chem. Soc.*, **87**, 5684 (1965).
[53] W. E. Truce and J. R. Norell, *J. Amer. Chem. Soc.*, **85**, 3231 (1963).
[54] W. J. Middleton, *J. Org. Chem.*, **30**, 1307 (1965).
[55] D. C. England, *J. Amer. Chem. Soc.*, **83**, 2205 (1961).
[56] J. F. Harris and D. D. Coffman, *J. Amer. Chem. Soc.*, **84**, 1553 (1962); E. W. Cook and B. F. Landrum, *J. Heterocyclic Chem.*, **2**, 327 (1965).
[57] W. J. Middleton, *J. Org. Chem.*, **30**, 1395 (1965).
[58] H. Vieregge, H. J. T. Bos, and F. J. Arens, *Rec. trav. Chim.*, **78**, 664 (1959).
[59] W. Reppe et al., *Ann. Chem.*, **596**, 1 (1955).
[60] W. Ried, *Angew. Chem.*, **76**, 933, 973 (1964); *Neuere Methoden der Präparativen Organischen Chemie*, Vol. IV, p. 88, ed. W. Foerst, Verlag Chemie, Weinheim/Bergstrasse, 1966 (reviews).
[61] C. Mannich and F. T. Chang, *Chem. Ber.*, **66**, 418 (1933).
[62] M. P. Cava, B. Hwang, and J. P. Van Meter, *J. Amer. Chem. Soc.*, **85**, 4031 (1963).
[63] P. D. Landor and S. R. Landor, *Proc. Chem. Soc.*, **1962**, 77.
[64] K. Griesbaum, W. Naegele, and G. G. Wanders, *J. Amer. Chem. Soc.*, **87**, 3151 (1965); K. Griesbaum, *ibid.*, **86**, 2301 (1964); *Angew. Chem. Intern. Edit. Engl.*, **5**, 933 (1966).
[65] J. D. Roberts and C. M. Sharts, *Organic Reactions*, Vol. 12, p. 23, 1962.
[66] T. L. Jacobs and R. N. Johnson, *J. Amer. Chem. Soc.*, **82**, 6397 (1960).
[67] P. E. Peterson and J. E. Duddey, *J. Amer. Chem. Soc.*, **85**, 2865 (1963).
[68] R. Huisgen, *Angew. Chem.*, **75**, 604, 742 (1963) (review).
[69] R. Huisgen, M. Seidel, G. Wallbillich, and K. Knupfer, *Tetrahedron*, **17**, 3 (1962); R. Huisgen, H. Knupfer, R. Sustmann, G. Wallbillich, and V. Weberndörfer, *Chem. Ber.*, **100**, 1580 (1967).
[70] E. Buchner and M. Fritsch, *Chem. Ber.*, **26**, 256 (1893); E. Buchner and W. Behaghel, *ibid.*, **27**, 869 (1894); R. Hüttel, J. Riedl, H. Martin, and K. Franke, *ibid.*, **93**, 1425 (1960).
[70a] J. Hamer and A. Macaluso, *Chem. Rev.*, **64**, 473 (1964).
[71] J. Schreiber and H. Wolf, *Ann. Chem.* **357**, 25, footnote 2 (1907).
[72] I. Brüning, R. Grashey, H. Hanck, R. Huisgen, and H. Seill, *Org. Synth.*, **46**, 127 (1966).
[73] N. A. LeBel, G. M. J. Slusarczuk, and L. A. Spurlock, *J. Amer. Chem. Soc.*, **84**, 4360 (1962).
[74] R. Huisgen, H. König, G. Binsch, and H. J. Sturm, *Angew. Chem.*, **73**, 368 (1961).
[75] R. Huisgen, R. Sustmann, and K. Bunge, *Tetrahedron Letters*, **1966**, 3603.
[76] K. Clusius and H. Hürzeler, *Helv. Chim. Acta*, **37**, 798 (1954); I. Ugi in *Advances in Heterocyclic Chemistry*, Vol. III, p. 373, ed. A. R. Katritzky, Academic Press, New York, 1964; R. Huisgen, *Angew. Chem.*, **72**, 359 (1960).
[77] I. Ugi, R. Huisgen, K. Clusius, and M. Vecchi, *Angew. Chem.*, **68**, 753 (1956); R. Huisgen and I. Ugi, *Chem. Ber.*, **90**, 2914 (1957).

[78] K. Alder and G. Stein, *Ann. Chem.* **485,** 211 (1931); **501,** 1 (1933).
[79] Reviews: J. Sauer, *Angew. Chem.* **79,** 76 (1967); R. Huisgen, R. Grashey, and J. Sauer, *The Chemistry of Alkenes,* ed. S. Patai, Interscience Publishers, New York, 1964, p. 878; J. G. Martin and R. K. Hill, *Chem. Rev.,* **62,** 537 (1962); Yu. A. Titov, *Russian Chem. Rev.,* English transl., **1962,** 267.
[80] A. Wasserman, *Diels-Alder Reaction,* Elsevier, Amsterdam, 1965.
[80a] R. Hoffmann and R. B. Woodward, *J. Amer. Chem. Soc.,* **87,** 4388 (1965).
[81] J. A. Berson, Z. Hamlet, and W. A. Mueller, *J. Amer. Chem. Soc.,* **84,** 297 (1962).
[82] J. Sauer and J. Kredel, *Angew. Chem., Intern. Edit., Engl.,* **4,** 989 (1965); *Tetrahedron Letters,* **1966,** 731; T. Inukai and T. Kojima, *J. Org. Chem.,* **31,** 2032 (1966).
[83] J. Sauer, H. Wiest, and A. Mielert, *Z. Naturforsch.,* **17b,** 203 (1962); J. Sauer, D. Lang, and A. Mielert, *Angew. Chem.,* **74,** 352 (1962).
[84] J. Sauer, H. Wiest, and A. Mielert, *Chem. Ber.,* **97,** 3183 (1964).
[85] J. Sauer and H. Wiest, *Angew. Chem.,* **74,** 353 (1962).
[86] G. Märkl, *Chem. Ber.,* **96,** 1441 (1963).
[87] S. J. Cristol and W. P. Norris, *J. Amer. Chem. Soc.,* **75,** 632, 2645 (1953).
[88] R. B. Woodward and T. J. Katz, *Tetrahedron,* **5,** 70 (1959); *Tetrahedron Letters,* No. 5, 19 (1959).
[89] G. Wittig and R. W. Hoffmann, *Chem. Ber.,* **95,** 2718, 2729 (1962).
[90] G. Wittig, *Angew. Chem.,* **69,** 245 (1957); *Angew. Chem. Intern. Edit., Engl.,* **4,** 731 (1965); R. Huisgen and J. Sauer, *Angew. Chem.,* **72,** 91 (1960).
[91] M. Stiles and R. G. Miller, *J. Amer. Chem. Soc.,* **82,** 3802 (1960).
[92] R. Huisgen and J. Sauer, *Angew. Chemie.,* **72,** 107 (1960).
[93] C. G. Krespan, B. C. McKusick, and T. L. Cairns, *J. Amer. Chem. Soc.,* **83,** 3428 (1961).
[94] D. Bryce-Smith and H. J. F. Angus, *Proc. Chem. Soc.,* **1959,** 326; D. Bryce-Smith, *Pure Appl. Chem.,* **16,** 1 (1968).
[95] M. Avram, C. D. Nenitzescu, and E. Marica, *Chem. Ber.,* **90,** 1863 (1957).
[96] L. G. Cannell, *Tetrahedron Letters,* **1966,** 5967.
[97] R. A. Raphael, *Acetylenic Compounds in Organic Synthesis,* Butterworths, London, 1964, p. 159.
[98] H. Höver, *Chem. Ber.,* **100,** 456 (1967).
[99] H. G. Viehe, R. Merenyi, J. F. M. Oth, and P. Valange, *Angew. Chem. Intern. Edit., Engl.,* **3,** 746 (1964); H. G. Viehe, R. Merenyi, J. F. M. Oth, J. R. Senders, and P. Valange, *ibid.,* p. 755; H. G. Viehe, *Angew. Chem.,* **77,** 768 (1965).
[99a] R. Gompper and G. Seybold, *Angew. Chem.,* **80,** 804 (1968).
[99b] M. J. S. Dewar and G. J. Gleicher, *J. Amer. Chem. Soc.,* **87,** 3255 (1965).
[100] E. E. van Tamelen, *Angew. Chem. Intern. Edit., Engl.,* **4,** 738 (1965).
[100a] H. R. Ward and J. S. Wishnote, *J. Amer. Chem. Soc.,* **90,** 1085 (1968).
[101] W. Schäfer, *Angew. Chem.,* **78,** 716 (1966).
[102] R. Breslow and P. Gal, *J. Amer. Chem. Soc.,* **81,** 4747 (1959); R. Breslow, P. Gal, H. W. Chang, and L. J. Altman, *ibid.,* **87,** 5139 (1965); R. Breslow, *Molecular Rearrangements,* Vol. I, ed. P. de Mayo, Interscience Publishers, Inc. New York, 1963, p. 243.
[103] E. M. Arnett and J. M. Bollinger, *Tetrahedron Letters* **1964,** 3803.
[104] R. Criegee and R. Askani, *Angew. Chem.,* **78,** 494 (1966).
[105] K. E. Wilzbach and L. Kaplan, *J. Amer. Chem. Soc.,* **87,** 4004 (1965).

3

Valence Isomerizations

VALENCE isomerizations are thermal, unimolecular rearrangements, in which reorganizations, usually concerted, of π- and σ-electrons take place, accompanied by changes of distances and bond angles.

Univalent substituents at the carbon atoms involved in the rearrangement do not change their positions. Thus the rearrangement of 1,3-pentadiene is not a valence isomerization.

The term valence tautomerism is used for fast valence isomerizations; degenerate valence isomerizations are rearrangements in which starting material and product are the same compound. An example is bullvalene, which is briefly discussed below.

Although a clear-cut definition of valence isomerization is not available at present, the unsaturated compounds that are considered to undergo valence isomerization may be classified into two main groups:

1. Diallylic systems and monovinyl-cyclopropanes or -cyclobutanes.

(The number of σ- and π-bonds does not change during rearrangement.)

These rearrangements, often referred to as Cope rearrangements, take place either by path a or by path b. Path b is also followed in the thermal isomerization of monovinyl-cyclopropanes and -cyclobutanes. If A or A′ is a heteroatom, path a represents the well-known Claisen rearrangement.

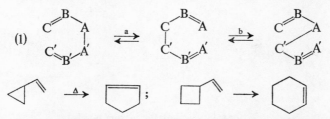

Diallylic systems undergoing rearrangement occupy preferentially a quasi-chair conformation in the transition state. When, for steric reasons, the quasi-chair conformation cannot be formed, isomerization *via* the quasi-boat conformation occurs (see p. 190).

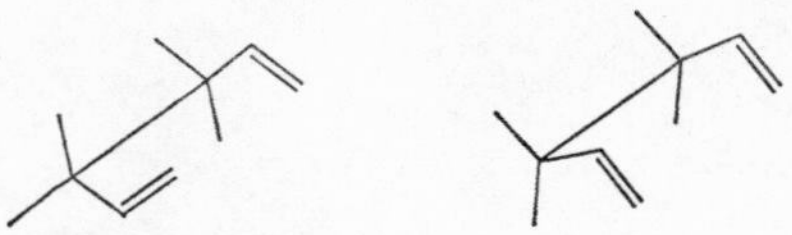

cis-Divinylcyclobutane rearranges at 120° *via* the quasi-boat conformation **(410)**, yielding *cis-cis*-1,5-cyclooctadiene. Conformations **(411)** and **(412)** would lead to the highly strained *cis-trans*- and *trans-trans*-cyclooctadienes. Recently it has been reported that heating *cis*-divinylcyclobutane at 90° affords *cis-trans*-cyclooctadiene in 3% yield, together with 97% of *cis-cis*-cyclooctadiene[1].

cis quasi-boat

(410)

(411) *cis* quasi-chair

(412) *cis* quasi-boat

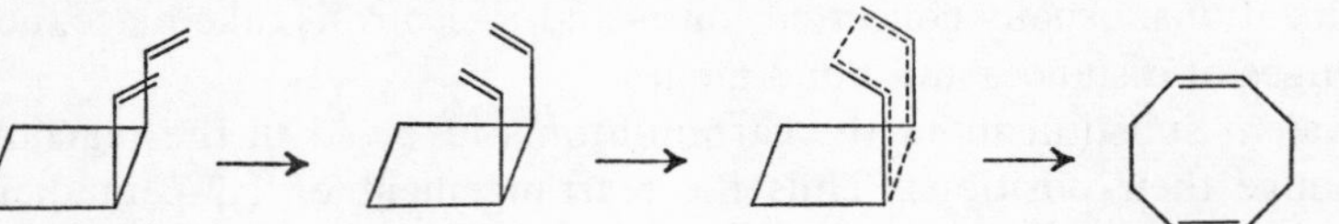

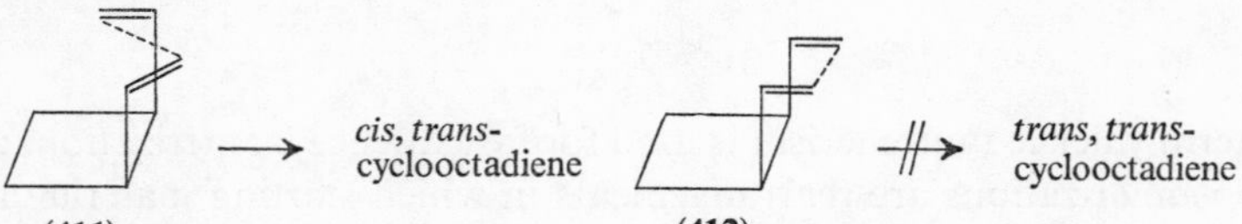

trans-Divinylcyclobutane yields vinylcyclohexene and 1,5-cyclooctadiene at 180–200°; since the vinyl groups are too far from each other to permit cyclization by a synchronous mechanism, the rearrangement takes place by a diradical mechanism (see p. 154). This valence isomerization is one of the relatively rare cases that occur by a diradical stepwise mechanism[1a].

trans

trans

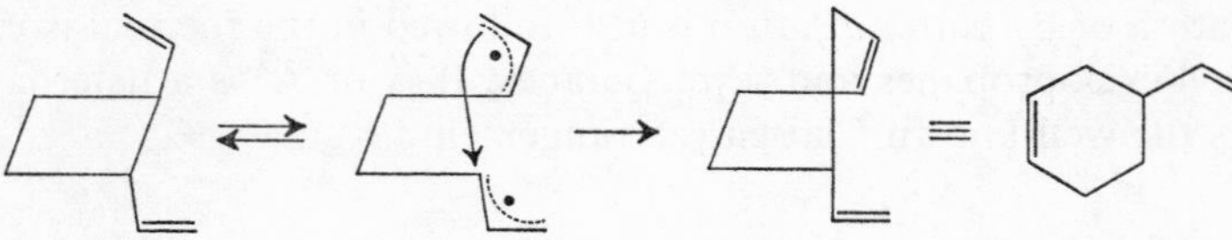

Molecules with fluctuating bonds.

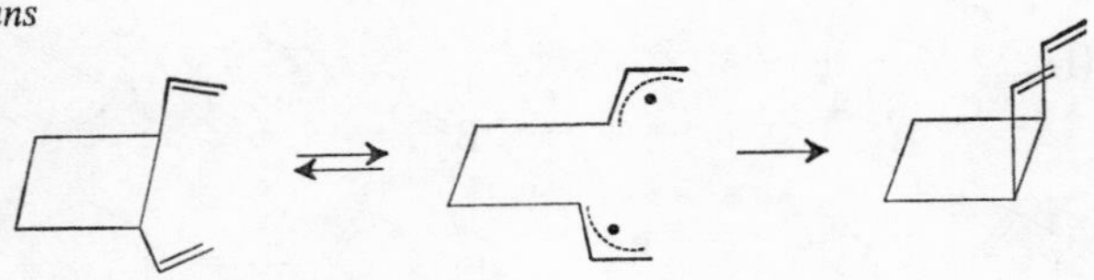

Some unsaturated compounds rearrange into each other at such a high rate that the individual isomers cannot be detected by chemical means. For these cases, NMR spectroscopy is very useful since the magnetic environment of the hydrogen atoms changes with the valence isomerization.

The first representative of these interesting molecules with fluctuating bonds was bullvalene, whose properties have been predicted by Doering and Roth[2], and which has been synthesized by Schröder[3]. In bullvalene each carbon atom forms a bond with every other carbon atom by valence isomerization. The total number of possible isomers is approximately 1.2 millions. Since the individual isomers are

Bullvalene

separated by energy barriers, albeit very low ones, and have therefore a real existence, they are different from resonance structures.

Rearrangements of group 1 are examples of sigmatropic reactions[3a]. A sigmatropic change of order (i,j) has been defined by Hoffmann and Woodward[3a] as the migration of a σ-bond, flanked by one or more π-electron systems, to a new position whose termini are $i-1$ and $j-1$ atoms removed from the original bonded loci, in an uncatalyzed intramolecular process.

Thus the Cope and the Claisen rearrangement are sigmatropic reactions of order (3,3). Sigmatropic reactions like cycloadditions (see pp. 158 and 159), and electrocyclic reactions, which are discussed in the following section, are determined by orbital symmetry relationships.

2. Conjugated polyenes (electrocyclic reactions).

According to Woodward and Hoffmann[3b] the stereochemistry of valence isomers of conjugated polyenes is determined by the symmetry of the highest occupied π-orbital of the open-chain isomer. Overlap between the positive terminal lobes

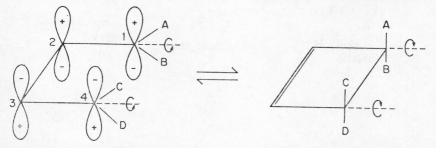

of butadiene is only possible by conrotatory rotation around the C_1—C_2 and C_3—C_4 axis. All conjugated open-chain systems with $4n$ π-electrons behave analogously because they possess the same symmetry. Compounds with $(4n + 2)$ π-electrons, however, undergo cyclization by disrotatory rotation.

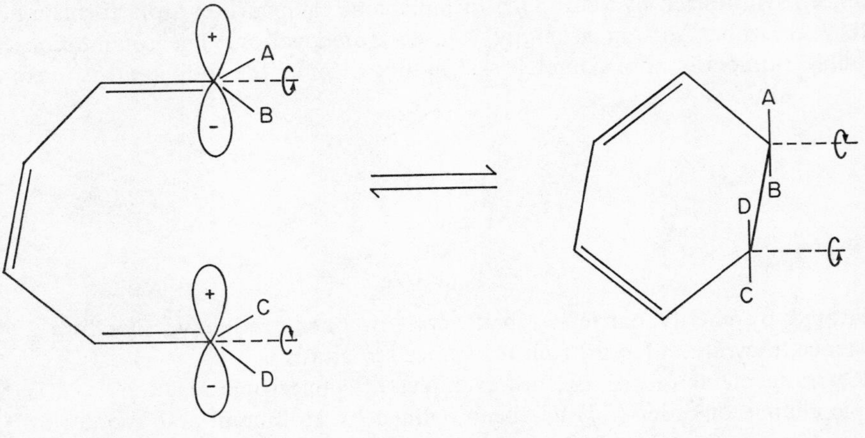

I. DIALLYLIC SYSTEMS AND MONOVINYL-CYCLOPROPANES AND -CYCLOBUTANES

V-1. Rearrangement of 1,1-dicyclopropylethylene[4]

1,1-Dicyclopropylethylene isomerizes by path b of the Cope rearrangement.

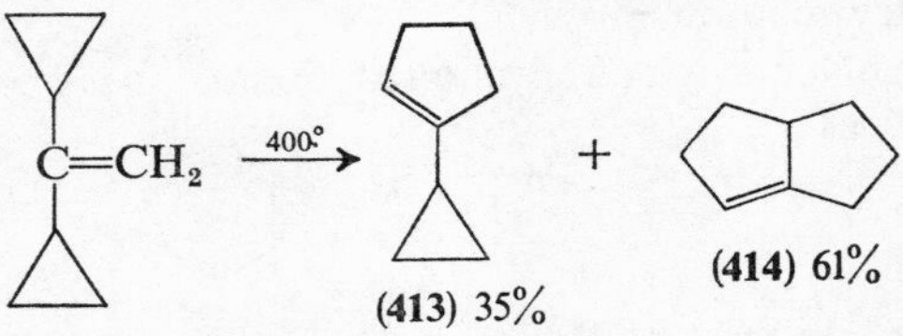

(413) 35%   (414) 61%

V-2. Rearrangement of bicyclo[3.2.0]hept-2-en-6-yl acetate[5]

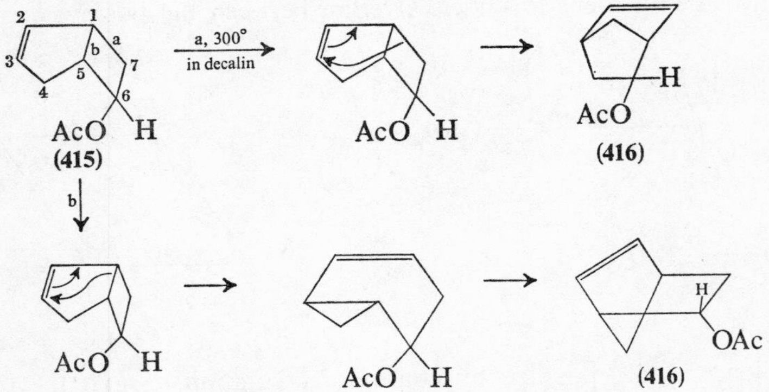

(415) → (416) → (416)

Two mechanisms have been proposed. In both cases the vinylcyclobutane system isomerizes by cleavage of bond a or of bond b. A third mechanism, which begins with the cleavage of (415) to cyclopentadiene and vinyl acetate, followed by Diels–Alder addition, can be only of minor importance, since in the rearrangement an *exo:endo* ratio of 9.4:1 is obtained, whereas Diels–Alder reaction of cyclopentadiene with vinyl acetate leads to an *exo-endo* ratio of 0.43:1.

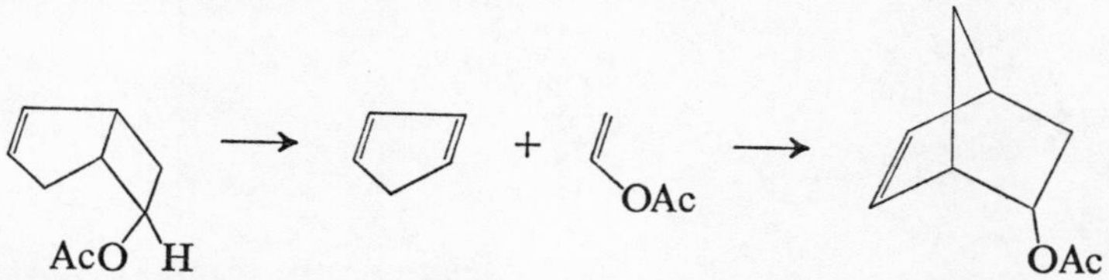

The rearrangement has been studied recently with the deuterated compound (415)[6]. The rearranged product is the bicycloheptene (417). This result shows that, during rearrangement, ring opening of the C-1/C-7 bond occurs, accompanied by 180° rotation about the C_6–C_7 axis.

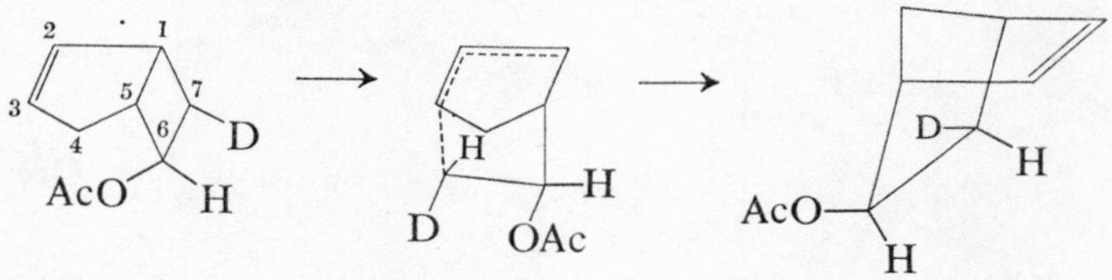

Mechanism b would lead to compound (418).

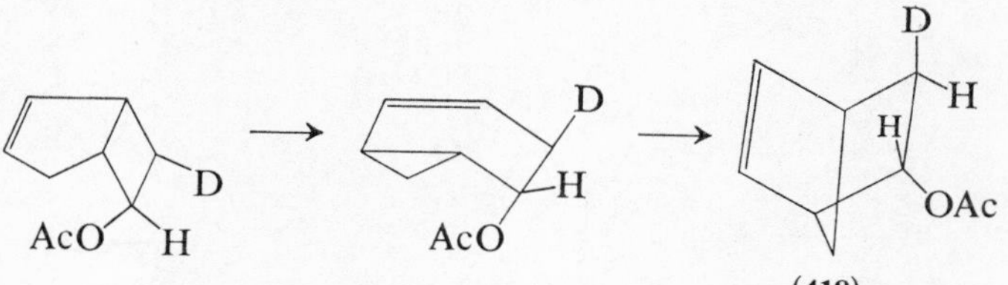

In contrast to the thermal rearrangements of tetrafluorobicyclo[3.2.0]heptene[7] and other bicyclo[3.2.0]heptene systems, no cycloheptadienyl acetate is formed.

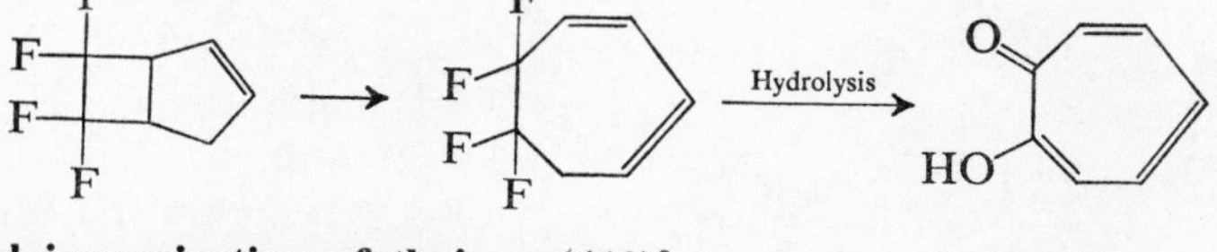

V-3. Thermal isomerization of thujone (419)[8]

It has recently been shown[9] that α-thujene isomerizes at 200°. It is likely, therefore, that thujone isomerizes through its enol forms (420 and 420a). The reaction presumably represents a case of radical isomerization. The products (**421**) and (**423**) are formed with retention of configuration.

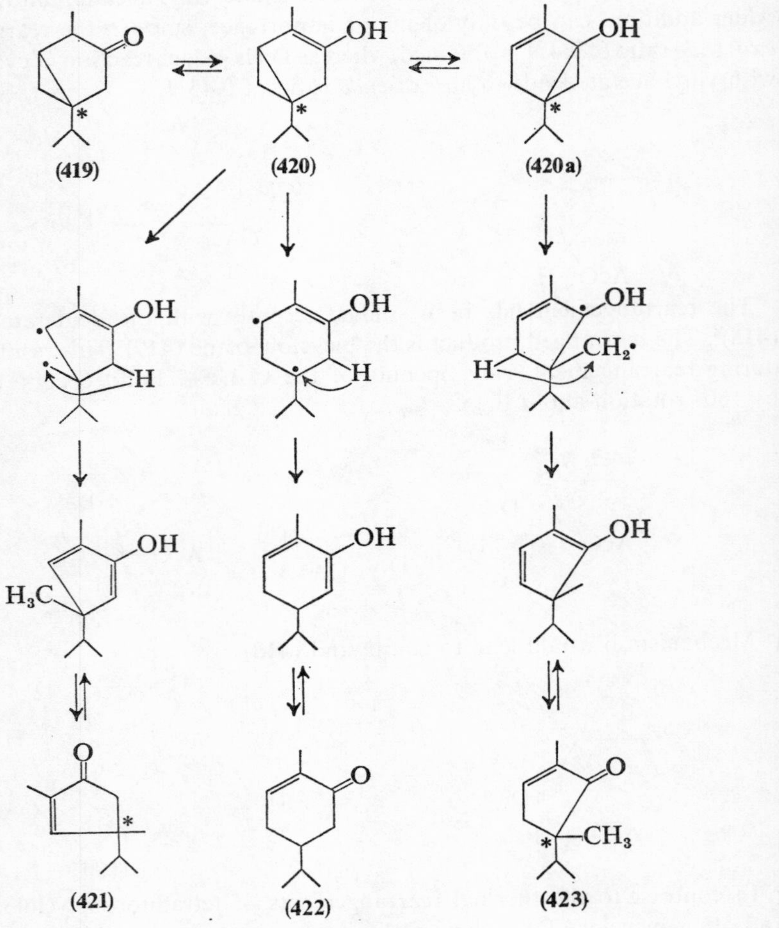

NMR signals (τ)

(**421**)

3.02(1) singlet =C—H
7.80–7.96(2) doublet —C(O)—CH$_2$—
8.28(3) singlet =C—CH$_3$
8.85 (3) singlet CR$_2$—CH$_3$
9.15–9.0(6) 2 doublets —C(CH$_3$)$_2$
9.2—9.0(1) multiplet —CR$_2$—H

(**422**)

3.3(1) multiplet =C—H
7.9 (2) multiplet ⎫ ⎰ =C—CH$_2$— &
9.1–9.0(2) multiplet ⎭ ⎱ —C(O)—CH$_2$—
8.3(3) singlet —CH$_3$
9.1–9.0(6) doublet —C(CH$_3$)$_2$

(423)

2.93(1) doublet =C—H
8.0–7.7(2) multiplet =C—CH$_2$—
8.28(3) 2 doublets =C—CH$_3$
9.0(3) singlet —CR$_2$—CH$_3$
9.2–9.09(3) doublet —CH(CH_3)$_2$
9.42–9.34(3)
9.4–9.0(1) multiplet —CH(CH$_3$)$_2$

II. CONJUGATED POLYENES

V-4. Isomerization of 3-phenyl-3-cyclobutene-1,2-dione (424) in hot CH$_3$OH[10]

The simplest case of valence isomerization yielding a conjugated polyene is the ring opening of cyclobutene and its derivatives (see p. 196).

3-Phenyl-3-cyclobutene-1,2-dione (424) is prepared by cycloaddition of phenylacetylene to chlorotrifluoroethylene, followed by hydrolysis. Despite its considerable ring strain it is more stable than *o*-benzoquinone. Attempts to prepare phenylcyclobutadienediol by hydrogenation were unsuccessful. The dione (424) is attacked by CH$_3$OH at elevated temperatures at the C=O groups. The products (426) and (427) are presumably formed through the common intermediate (425). In the presence of catalytic amounts of acid, (428) is obtained in 49% yield. Formation of the vinylketene (429) seems to be an uncatalyzed thermal reaction. Ketenes have been frequently postulated in ring-opening reactions[11]. The mechanism illustrated is supported by the reaction of dichlorophenylcyclobutene with deuterioacetic acid as shown[12]. At a reaction temperature of 150°, the ester (427)

$$\text{H}_5\text{C}_6-\overset{\text{Cl}}{\underset{\overset{|}{\text{H}}\ \overset{|}{\text{Cl}}}{\square}}=\text{O} + \text{CH}_3\text{COOD, then H}_2\text{O} \longrightarrow \text{H}_5\text{C}_6-\overset{\text{Cl}\quad\text{D}}{\underset{\overset{|}{\text{C}-\text{Cl}}\\ \overset{|}{\text{H}}}{\text{C}}}\overset{}{\diagdown}\text{COOH}$$

predominates, owing either to a change in the ratio k_B/k_C or to the direct formation *via* the diketene (430).

V-5. Isomerizations of cyclooctatriene derivatives

At 80—100° 85% of cyclooctatriene (431) is in equilibrium with 15% of bicyclo-[4.2.0]octa-2,4-diene (432). Both isomers can be isolated[13] below 30°.

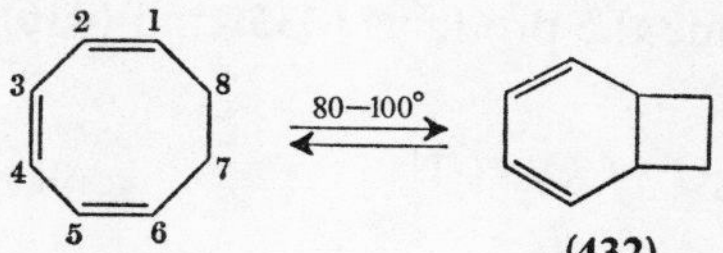

Cyclooctatriene can, however, isomerize by another path. If it is substituted in positions 7 and 8 by unsaturated substituents, all-*trans*-octatetraene derivatives are formed by irreversible ring-opening.

Reaction of benzoylcyclooctatetraene (**432a**) *with phenylmagnesium bromide*[14]

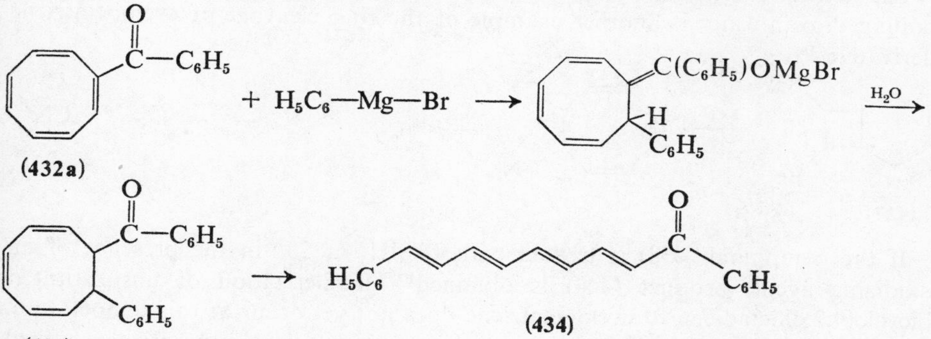

In the reaction of benzoylcyclooctatetraene (**432a**) with phenylmagnesium bromide, 20% of the open-chain ketone (**434**) are obtained; this is very likely formed by way of intermediate (**433**). The driving force for the ring opening is formation of a more stable conjugated system and release of steric strain between the substituents in positions 7 and 8.

That steric interactions may suffice to cause cleavage of the cyclooctatriene ring is demonstrated by the reaction of dipotassiocyclooctatetraene with benzophenone[15].

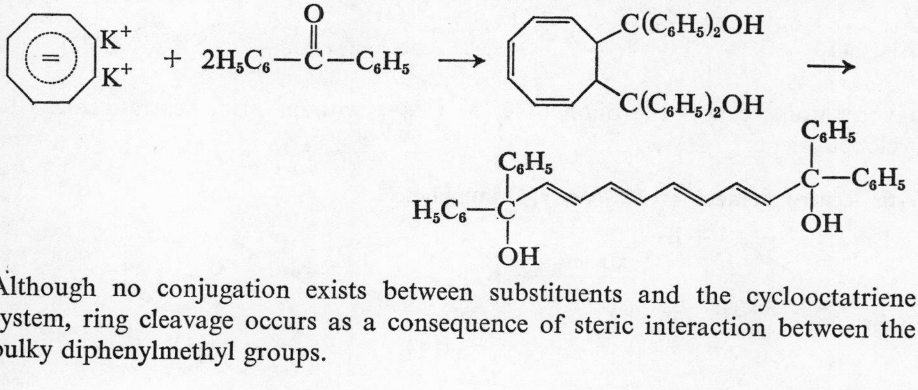

Although no conjugation exists between substituents and the cyclooctatriene system, ring cleavage occurs as a consequence of steric interaction between the bulky diphenylmethyl groups.

On reaction with acetone the products (**435**) and (**436**) are obtained.

(**435**) (**436**)

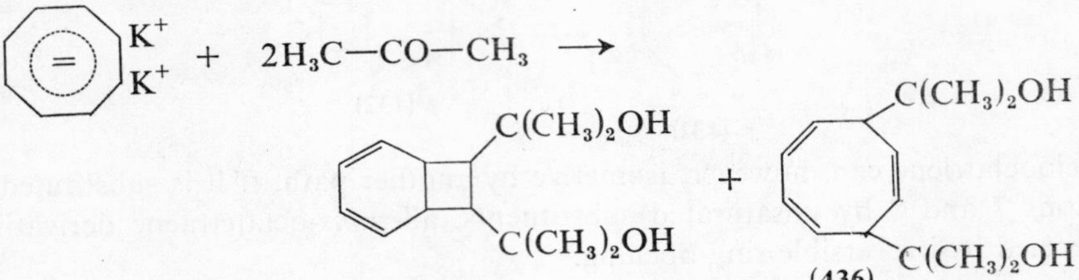

The reaction of 7,8-dibromobicyclo[4.2.0]octa-2,4-diene (**437**) with KCN in boiling dioxan/water is another example of the ring cleavage of cyclooctatriene derivatives[16].

(**437**)

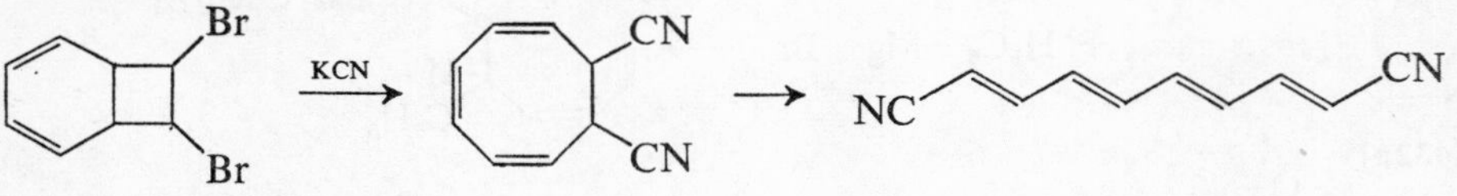

If the compound (**438**) is treated with $LiAlH_4$ at 25° in the presence of an oxidizing agent, product (**440**) is obtained[17]. Isomerization of unsubstituted bicyclo[4.2.0]octadiene to cyclooctatriene does not yet occur at this temperature. It is therefore likely that electrostatic repulsion between the two negatively charged

(**438**) (**439**) (**440**)

oxygen atoms causes the diion (**439**) to cleave without prior rearrangement to cyclooctatriene.

V-6. Rearrangement of 1,6-epoxy[10]annulene[18]

(**441**) (**442**) (**443**) 60%

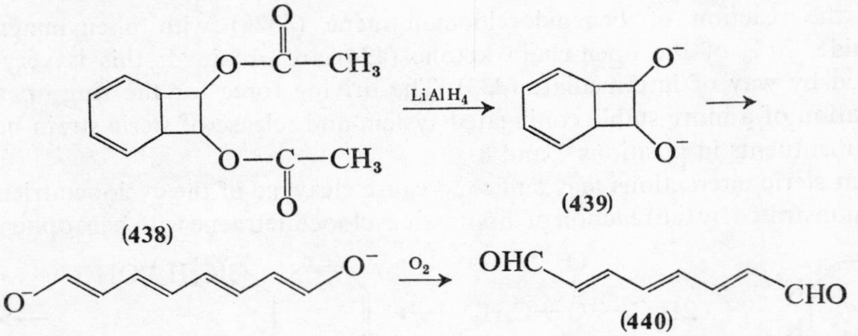

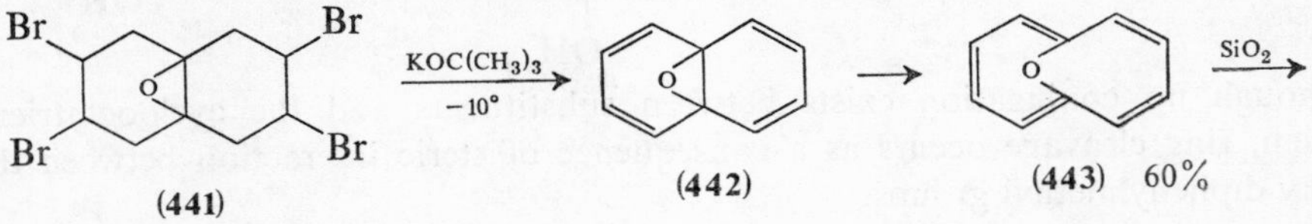

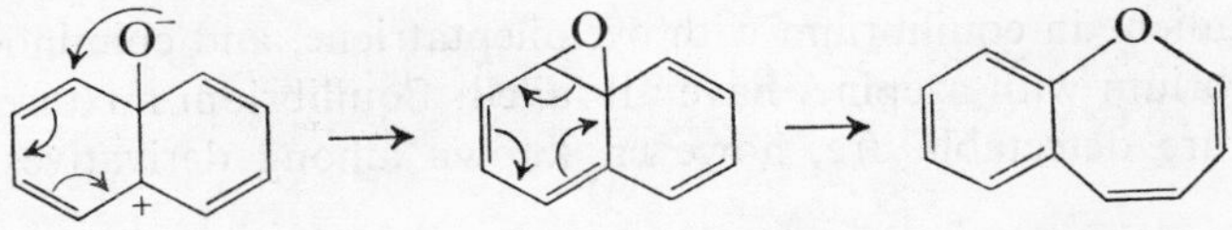

(444)

Dehydrobromination of the tetrabromo compound (441) with potassium *tert*-butoxide at $-10°$ yields, presumably by valence isomerization of intermediate (442), 1,6-epoxy[10]annulene (443) in 60% yield. This rearranges to 1-benzoxepin (444) on silica or alumina[19]. The rearrangement mechanism proposed is supported by the oxepin–epoxycyclohexadiene equilibrium[20], although in a later paper the authors[21] questioned this mechanism.

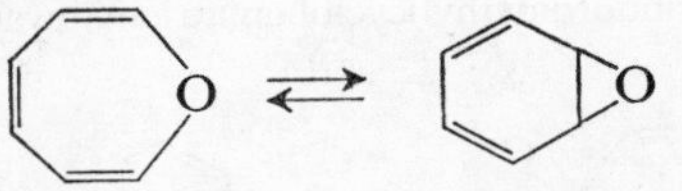

Oxepin and benzene oxide are in very rapid equilibrium at elevated temperatures, but the NMR spectrum of the individual compounds can be observed, if the mixture is frozen at $-127°$.

1,6-Epiminocyclodecapentaene (445) can be prepared analogously to (443).

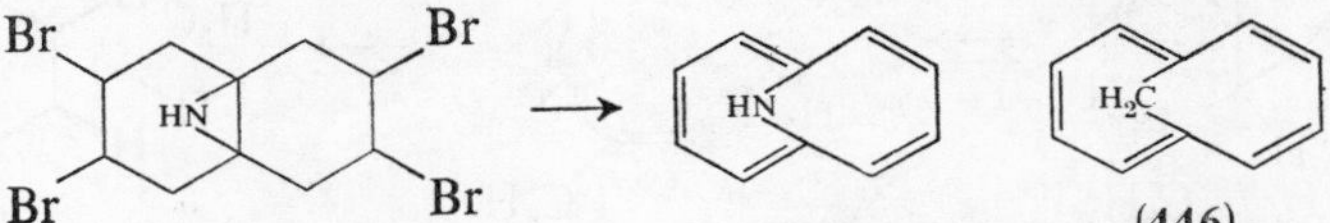

(445)   (446)

Both compounds are analogous to 1,6-methylenecyclodecapentaene (bicyclo-[4.4.1]undecapentaene) (446) which was first prepared by Vogel *et al.*[22] In accordance with Hückel's $4n + 2$ rule the compound (446; $n = 2$) possesses aromatic character; the conjugated 10-electron system is kept in plane by the methylene bridge. Cyclodecapentaene itself, which is not planar because of steric interaction between the hydrogen atoms in positions 1 and 6, is not aromatic[22]. The hydrocarbon (446) can be substituted by electrophiles and does not react with maleic anhydride in boiling benzene.

Oxepin is a heterocyclic analogue of cycloheptatriene. Azepine (447) has recently been prepared[23]. Although oxepin and epoxycyclohexadiene are present in comparable amounts of both in the equilibrium mixture at room temperature, attempts

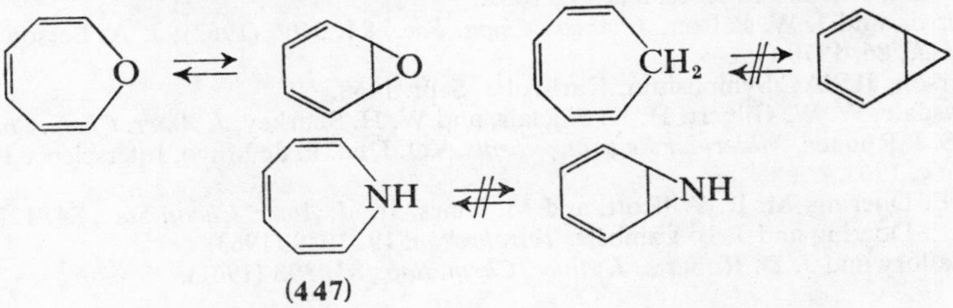

(447)

to find norcaradiene in equilibrium with cycloheptatriene, and epiminocyclohexadiene in equilibrium with azepine, have all failed. Equilibrium mixtures in which both isomers are detectable are, however, known among derivatives of cycloheptatriene*.

* For a review see ref. 24. Also Vogel et al.[25] discussed the equilibria:

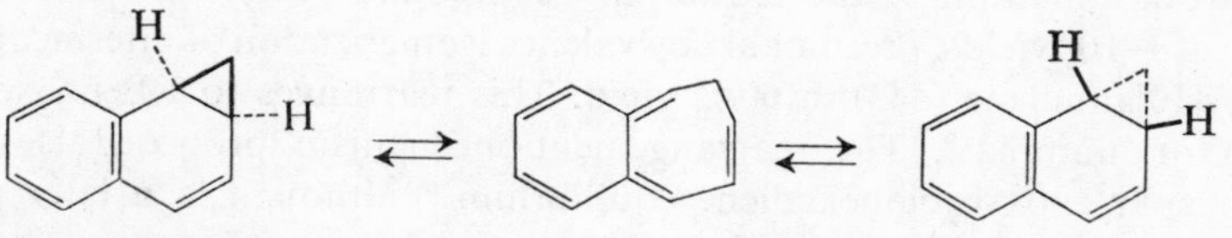

with particular reference to discrimination between the two hydrogen atoms of the methylene group. Ciganek[26] studied 1-(trifluoromethyl)cycloheptatrienyl cyanide, and Maier[27] investigated a diazepine derivative.

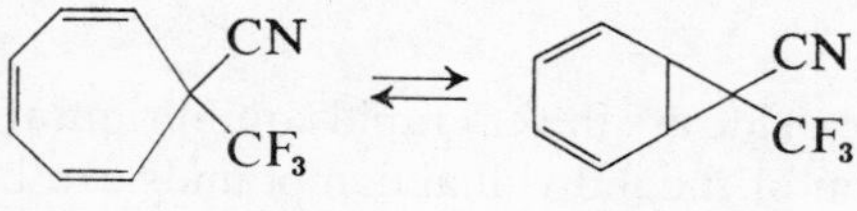

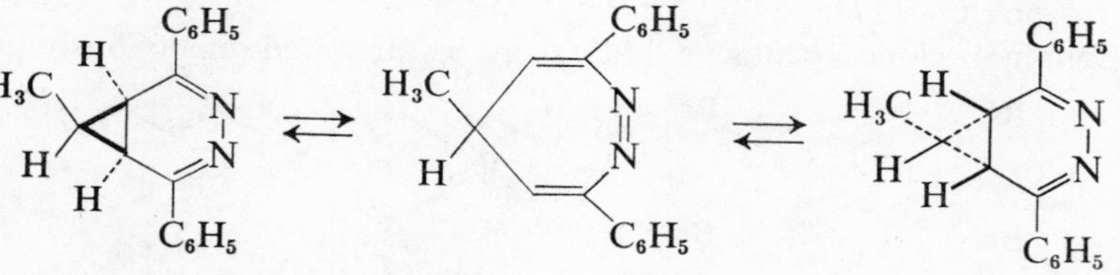

REFERENCES

[1] P. Heimbach, W. Brenner, H. Hey, K. Ploner, and R. Traunmüller, IUPAC Symposium, Karlsruhe, Sept. 1968.
[1a] G. S. Hammond and C. D. DeBoer, *J. Amer. Chem. Soc.*, **86**, 899 (1964).
[2] W. von E. Doering and W. R. Roth, *Tetrahedron*, **19**, 715 (1963).
[3] G. Schröder, J. F. M. Oth, and R. Merenyi, *Angew. Chem., Intern. Edit. Engl.*, **4**, 752 (1965).
[3a] R. B. Woodward and R. Hoffmann, *J. Amer. Chem. Soc.*, **87**, 2511 (1965); H.-J. Hansen, B. Sutter, and H. Schmid, *Helv. Chim. Acta*, **51**, 846 (1968); J. A. Berson, *Accounts Chem. Res.*, **1**, 152 (1968).
[3b] R. B. Woodward and R. Hoffmann, *J. Amer. Chem. Soc.*, **87**, 395 (1965).
[4] A. D. Ketley and J. L. McClanahan, *J. Org. Chem.*, **30**, 940 (1965); A. D. Ketley, J. L. McClanahan, and L. P. Fisher, *ibid.*, p. 1659.
[5] J. A. Berson and J. W. Patton, *J. Amer. Chem. Soc.*, **84**, 3406 (1962); J. A. Berson and E. S. Hand, *ibid.*, **86**, 1978 (1964).
[6] J. A. Berson, IUPAC Symposium, Karlsruhe, Sept. 1968.
[7] J. J. Drysdale, W. W. Gilbert, H. K. Sinclair, and W. H. Sharkey, *J. Amer. Chem. Soc.*, **80**, 245 (1958); S. J. Rhoads, *Molecular Rearrangements*, Vol. I, ed. P. de Mayo, Interscience Publishers, New York, 1963.
[8] W. von E. Doering, M. R. Willcott, and M. Jones, Jr., *J. Amer. Chem. Soc.*, **84**, 1224 (1962).
[9] W. von E. Doering and J. B. Lambert, *Tetrahedron*, **19**, 1989 (1963).
[10] F. B. Mallory and J. D. Roberts, *J. Amer. Chem. Soc.*, **83**, 393 (1961).

[11] J. F. Arens, *Angew. Chem.*, **70**, 631 (1958); E. Vogel, *ibid.*, **72**, 14, 15 (1960).
[12] E. F. Silversmith, Y. Kitahara, M. C. Caserio, and J. D. Roberts, *J. Amer. Chem. Soc.*, **80**, 5840 (1958).
[13] A. C. Cope, A. C. Haven, Jr., F. L. Ramp, and E. R. Trumbull, *J. Amer. Chem. Soc.*, **74**, 4867 (1952).
[14] A. C. Cope and D. J. Marshall, *J. Amer. Chem. Soc.*, **75**, 3208 (1953).
[15] T. S. Cantrell and H. Shechter, *J. Amer. Chem. Sc.*, **87**, 136 (1965).
[16] H. Höver, *Tetrahedron Letters*, **1962**, 255.
[17] R. Anet, *Tetrahedron Letters*, **1961**, 720; D. H. R. Barton, *Helv. Chim. Acta*, **42**, 2604 (1959).
[18] E. Vogel, M. Biskup, W. Pretzer, and W. A. Böll, *Angew. Chem. Intern. Edit., Engl.*, **3**, 642 (1964).
[19] F. Sondheimer and A. Shain, *J. Amer. Chem. Soc.*, **86**, 3168 (1964).
[20] E. Vogel, W. A. Böll, and H. Günther, *Tetrahedron Letters*, **1965**, 609; H. Günther, *ibid.*, p. 4085.
[21] E. Vogel and H. Günther, *Angew. Chem.*, **79**, 429 (1967).
[22] E. Vogel and H. D. Roth, *Angew. Chem., Intern. Edit. Engl.*, **3**, 228 (1964).
[23] K. Hafner and R. König, *Angew. Chem.*, **75**, 89 (1963); K. Hafner, *ibid.*, p. 1041; W. Lwowski, T. J. Maricich, and T. W. Mattingly, Jr., *J. Amer. Chem. Soc.*, **85**, 1200 (1965); R. J. Cotter and W. F. Beach, *J. Org. Chem.*, **29**, 751 (1964); A. S. Kende, P. T. Izzo, and J. E. Lancaster, *J. Amer. Chem. Soc.*, **87**, 5044 (1965); J. E. Baldwin and R. A. Smith, *ibid.*, p. 4819; F. D. Marsh and H. E. Simmons, *ibid.*, p. 3529.
[24] Review: G. Maier, *Chem. Ber.*, **98**, 2438 (1965); E. Ciganek, *J. Amer. Chem. Soc.*, **89**, 1454 (1967); D. Schönleber, *Angew. Chem.*, **81**, 83 (1969); M. Jones, Jr., *ibid.*, p. 83.
[25] E. Vogel, D. Wendisch, and W. R. Roth, *Angew. Chem., Intern. Edit. Engl.*, **3**, 443 (1964).
[26] E. Ciganek, *J. Amer. Chem. Soc.*, **87**, 1149 (1965).
[27] G. Maier, *Chem. Ber.*, **98**, 2446 (1965).

4

β-Eliminations

α-ELIMINATIONS, characterized by elimination of two substituents from the same carbon atom, have been discussed above on page 137.

A β-elimination is, in the general case, formation of a multiple bond by abstraction of one substituent from each of two adjacent atoms. Usually, one of the substituents is a hydrogen atom.

β-Eliminations may proceed by three mechanisms[1]:

1. If a carbonium ion is formed in the rate-determining step by separation of substituent X from a carbon atom, the elimination is a unimolecular reaction and is designated $E1$ elimination. The carbonium ion may be stabilized by loss of a proton from the β-position or, in analogy to S_N1 substitution, by bond formation with a substituent Y, which thus replaces X. The rate-determining step of both the $E1$ elimination and S_N1 substitution is therefore the formation of a carbonium ion. The rate depends on the tendency of X to separate from the carbon atom, on the stability of the developing carbonium ion, and on the polarity of the solvent:

$$E1: \quad -\underset{\underset{X}{|}}{\overset{H}{\underset{|}{C}}_\beta}-\overset{H}{\underset{|}{C}}_\alpha- \quad \underset{}{\overset{\text{Slow}}{\rightleftarrows}} \quad -\overset{H}{\underset{|}{C}}-\overset{H}{\underset{+}{\underset{|}{C}}}- \; + \; X^-$$

$$-\overset{H}{\underset{|}{C}}-\overset{}{\underset{+}{\underset{|}{C}}}- \quad \longrightarrow \quad \overset{}{\underset{}{}}C=C\overset{}{\underset{}{}} \; + \; H^+$$

If the $E1$ elimination can lead to the formation of several olefins, the most highly substituted olefin is predominantly formed in accordance with Saytzeff's rule. This indicates that the transition state **(448)** already has considerable olefin character and is therefore stabilized by the alkyl substituents:

$$H-\overset{R}{\underset{H}{\overset{|}{C}}}-\overset{X}{\underset{H}{\overset{|}{C}}}-\overset{H}{\underset{H}{\overset{|}{C}}}-H \longrightarrow H-\overset{R}{\underset{H}{\overset{|}{C}}}-\overset{H}{\underset{+}{\overset{|}{C}}}-\overset{H}{\underset{H}{\overset{|}{C}}}-H \longrightarrow$$

$$H\cdots\overset{R}{\underset{\overset{+}{H}}{\overset{|}{C}}}=\overset{H}{\underset{}{\overset{|}{C}}}-\overset{H}{\underset{H}{\overset{|}{C}}}-H \longrightarrow \overset{R}{\underset{H}{}}C=\overset{H}{\underset{}{\overset{|}{C}}}-\overset{H}{\underset{H}{\overset{|}{C}}}-H$$

(448)

Special cases are, however, known in which resonance and inductive effects are dominated by steric effects which favor the formation of the least substituted olefin[2].

2. The second elimination mechanism is bimolecular, designated $E2$. In this a molecule of base participates in the rate-determining step of the reaction. The base abstracts a proton from the β-position simultaneously with the separation of X.

$$E2: \quad B^- + \underset{X}{\underset{|}{-\overset{H}{\underset{|}{C}}_\beta - \overset{|}{\underset{|}{C}}_\alpha -}} \longrightarrow \underset{X}{\underset{|}{-\overset{\overset{B^-}{\vdots}}{\overset{H}{\overset{\vdots}{C}}} = \overset{|}{C} -}} \longrightarrow \overset{\diagdown}{\diagup}C = C\overset{\diagup}{\diagdown} + BH + X^-$$

3. According to the third mechanism a proton is abstracted by base in a bimolecular step, while the C–X bond is still intact. In the rate-determining step, the carbanion loses the substituent X. This mechanism is called $E1cB$ (elimination, unimolecular, conjugate base).

$$E1cB: \quad \underset{X}{\underset{|}{-\overset{H}{\underset{|}{C}}_\beta - \overset{|}{\underset{|}{C}}_\alpha -}} + B^- \rightleftarrows \underset{X}{\underset{|}{-\overset{|}{\bar{C}} - \overset{|}{\underset{|}{C}} -}} + BH$$

$$\underset{X}{\underset{|}{-\overset{|}{\bar{C}} - \overset{|}{\underset{|}{C}} -}} \longrightarrow \overset{\diagdown}{\diagup}C = C\overset{\diagup}{\diagdown} + X^-$$

The rate-determining step of a "carbanion" elimination mechanism could also be abstraction of a proton, followed by rapid loss of an anion from the carbanion, the latter having a real but very short lifetime. In this case the reaction would be of the second order, as in the synchronous $E2$ elimination. The carbanion elimination mechanism is still a subject of controversy. Recent studies[3] with 4-*tert*-butyl-2-(*p*-tolylsulfonyl) cyclohexanol conformers indicate that, if a carbanion is involved as intermediate, then at least in this case the proton abstraction must be the rate-determining step*.

Saytzeff and Hofmann rules

$E2$ eliminations frequently do not proceed by completely synchronous loss of H and X. The elimination may adopt the character of the $E1$ mechanism if the α-carbon is already rendered strongly positive by partial breaking of the C–X bond before β-attack by the base takes place; or it may resemble the carbanion mechanism in that separation of a proton from the β-carbon, caused by attacking base, begins before X separates from the molecule. The deviations from completely synchronous elimination are the reason why β-eliminations may follow either the

---

* For a recent review, see McLennan[3a].

Saytzeff or the Hofmann rule*. While the Saytzeff rule dominates in cases that are close to $E1$ in mechanism, the Hofmann elimination is favored if the mechanism resembles the carbanion mechanism. In the latter case the transition state of the elimination is on the side of the starting materials, in contrast to the transition state that yields Saytzeff products and already has considerable olefin character (cf. Hammond's rule, p. 237). In the Hofmann elimination, the base attacks the most acidic hydrogen, which in alkyl-substituted compounds is at the carbon with the smallest number of alkyl substituents.

The character of an $E2$ elimination is usually determined by electronic factors. Both α- and β-substituents influence the stability of the forming carbonium ion or carbanion by electronic effects. The stability of the C–X bond and the tendency of X to separate from the α-carbon may also be determined by electronic and steric effects. An α-aryl substituent increases the $E1$ character of the $E2$ elimination, as do α-alkyl substituents, since both stabilize the developing carbonium ion. A very readily separating substituent X, such as tosyloxy, has an effect in the same direction. Since halide ions separate more readily than 'onium groups, dehydrohalogenation follows Saytzeff's rule preferentially, whereas decomposition of 'onium bases follows the Hofmann rule preferentially.

A strong $-I$ effect (electron-attracting inductive effect) of X inhibits the formation of a carbonium ion. However, it facilitates accumulation of negative charge on the β-carbon, resulting in preferential Hofmann elimination. This has been

$$\begin{array}{c} {}^{\delta+}\text{H} \\ \| \\ -\underset{|}{\overset{|}{\text{C}}}_\beta -\underset{\downarrow}{\overset{|}{\text{C}}}_\alpha - \\ \text{X}_{\delta-} \end{array}$$

demonstrated in the reaction of 1-methylbutyl halides with sodium ethoxide in refluxing ethanol. Table 4 shows the results[4]: With decreasing inductive effect the

Table 4. Products formed on decomposition of 1-methylbutyl halides by refluxing ethanolic sodium ethoxide

Halide	1-Pentene	2-Pentene
F	82	18
Cl	35	65
Br	25	75
I	20	80

proportion of Saytzeff product (2-pentene) increases. An even more important factor than the increase of carbanion character with increasing inductive effect is the

---

* The Hofmann rule, formulated as a result of studies of ammonium hydroxides, states that decomposition of ammonium salts by a base yields predominantly the least substituted olefin.

ease of heterolysis of the C–X bond, which increases from F to I. The course of the reaction is influenced by the polarity of the solvent.

The influence of steric effects on the course of the $E2$ elimination in competition with electronic effects is still a matter of controversy. In certain cases, however, they probably predominate (see, *e.g.* reference 5).

In the transition state of the $E2$ elimination, the five participating atoms lie in a plane. The groups to be eliminated adopt the *trans* position, if they can (Barton's rule: bimolecular elimination from acyclic compounds proceeds readily only if the

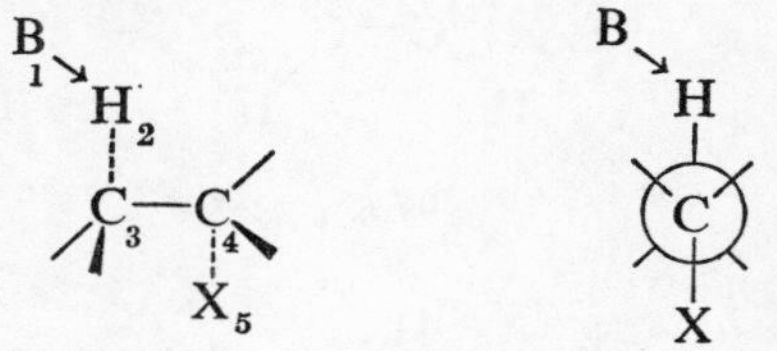

two substituents to be eliminated are in *trans* axial positions). If the attacking base is a very bulky group, or if the β-substituents or X are bulky, the reaction passes through the least sterically hindered transition state. Thus, despite the $E1$ character of an $E2$ elimination, one obtains Hofmann products if the *trans*-coplanar arrangement that would lead to Saytzeff products is sterically unfavorable.

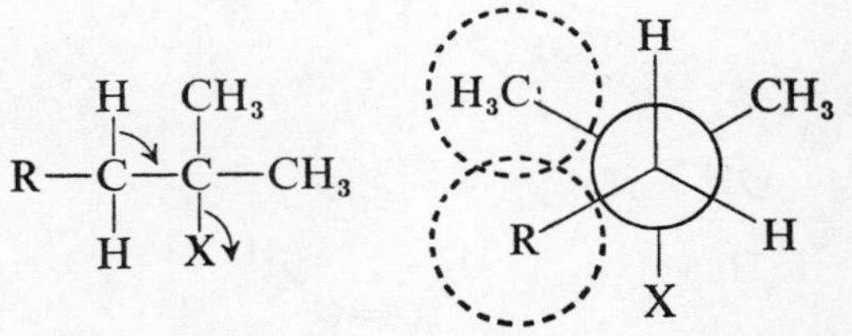

Steric interaction in Saytzeff elimination

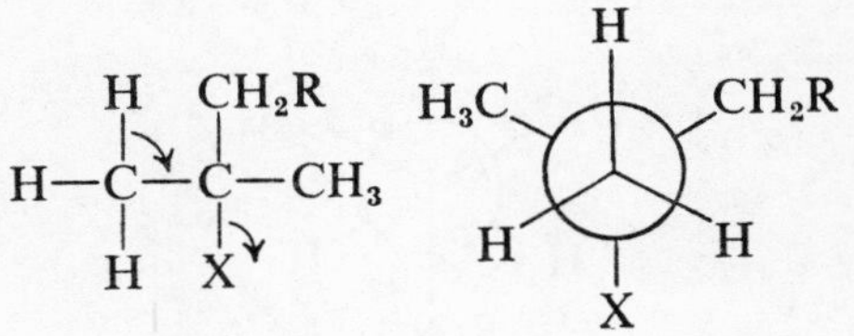

Sterically favorable Hofmann elimination

The influence of electronic and steric effects on the decomposition of ammonium bases is shown in example E-1.

E-1. Decomposition of tetraalkylammonium hydroxides (449) and (452)[6]

Abstraction of a proton from the β-carbon atom of the propyl group is inhibited by the electron-donating effect $(+I)$ of the terminal methyl group. An additional effect in the same direction is the steric hindrance between the $-CH_3$ and the NR_3^+ groups in conformation (451). Consequently, a β-hydrogen is abstracted from the ethyl group *via* conformation (450). As a result the least substituted olefin is formed

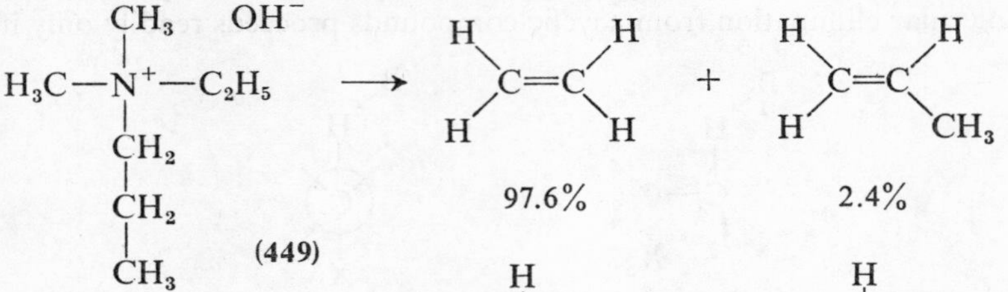

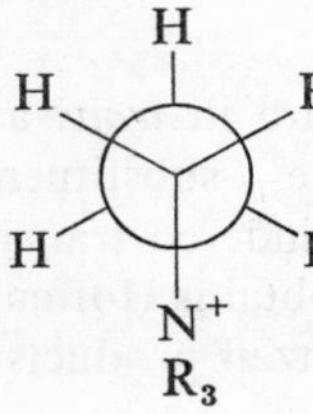

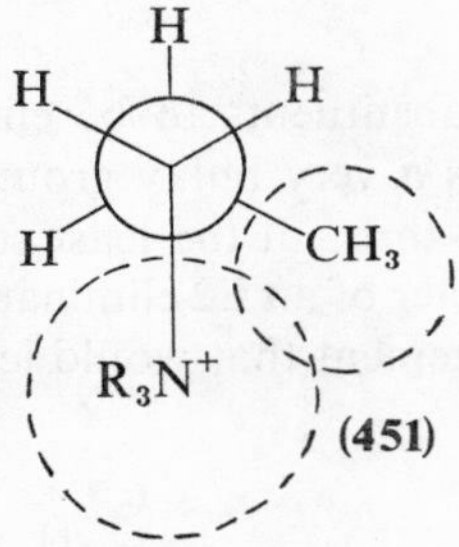

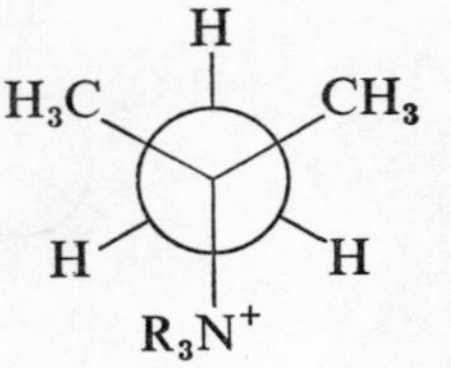

in accordance with Hofmann's rule. In (452), the two α-methyl groups of the *tert*-butyl group cause partial separation of the $(CH_3)_2NC_2H_5$ group with its bonding electron pair. The resulting positive charge at the central carbon atom of the *tert*-butyl group is stabilized by the methyl groups; consequently, the Saytzeff product is formed preferentially.

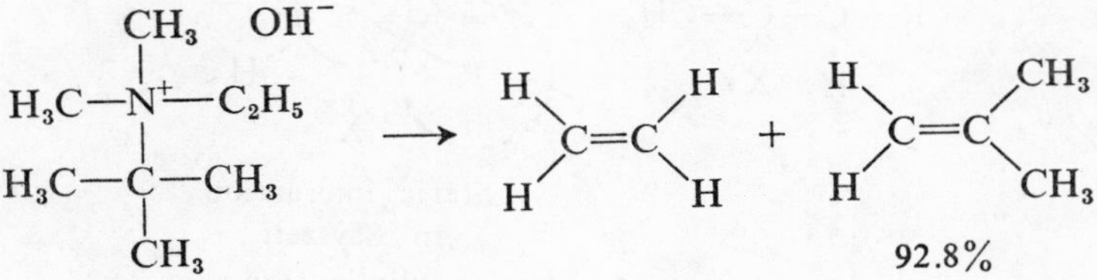

E-2. Reaction of *trans*-2-phenylcyclohexyl *p*-toluenesulfonate with potassium hydroxide[7]

Although $E2$ elimination is normally a *trans*-elimination, there are cases in which *cis*-elimination is favored. *cis*-Elimination can be important if the elimination proceeds by a largely stepwise mechanism, *i.e.*, if the mechanism is close to $E1$ or $E1cB$.

In the elimination of the trimethylamine group from compound (**453**) the initial attack of base at the β-proton is facilitated by the β-aryl group and the fact that $N(CH_3)_3$ is a poor leaving group. Abstraction of the proton takes place, therefore, somewhat before the separation of the $N(CH_3)_3$ group.

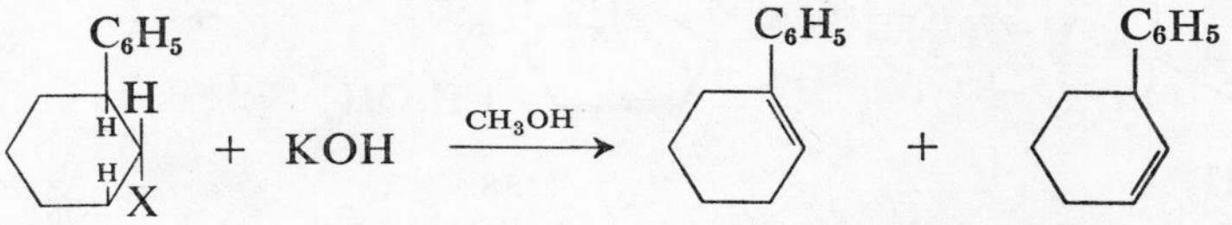

	(454)	(455)
X = $^+N(CH_3)_3$	64%	2%
X = OTs	20%	53%

Despite the *cis*-arrangement of H and $N(CH_3)_3$, product (**454**) predominates. In the case of the relatively readily separating –OTs group, the elimination approaches a truly synchronous $E2$ mechanism, which yields predominantly the *trans*-elimination product (**455**).

E-3, E-4, & E-5. Thermal *cis*-eliminations

Thermal *cis*-elimination is observed in numerous cases where compounds are in the position to form planar, cyclic five- or six-membered transition states that allow migration of hydrogen atoms. *cis*-Elimination takes place more readily according as coplanarity can be more easily achieved. It is likely, at least when hydrogen migrates to or from oxygen, that the bonds involved are cleaved and formed heterolytically; hydrogen consequently migrates as a proton.

E-3. *Intramolecular thermal* cis-*elimination from 3-ethyl-6-phenyl-5-hexen-3-ol* (**456**)[8]

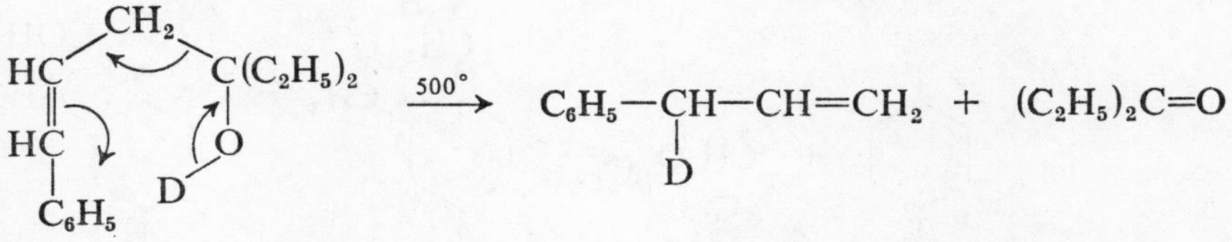

E-4. *Thermal rearrangement of (+)-3-hydroxymethyl-4-carene* **(457)**[9]

Thermal *cis*-eliminations have been observed in some isomerizations of α,β-unsaturated ketones[10] and other unsaturated hydrocarbons[11].

In conformation **(458)** of (+)-3-hydroxymethyl-4-carene **(457)**, which has an equatorial CH_2OH group, the 1 → 5 migration of hydrogen, resulting in product **(459)**, is sterically favored. Conformation **(460)**, however, yields (+)-3-carene **(461)** by loss of formaldehyde *via* a cyclic transition state.

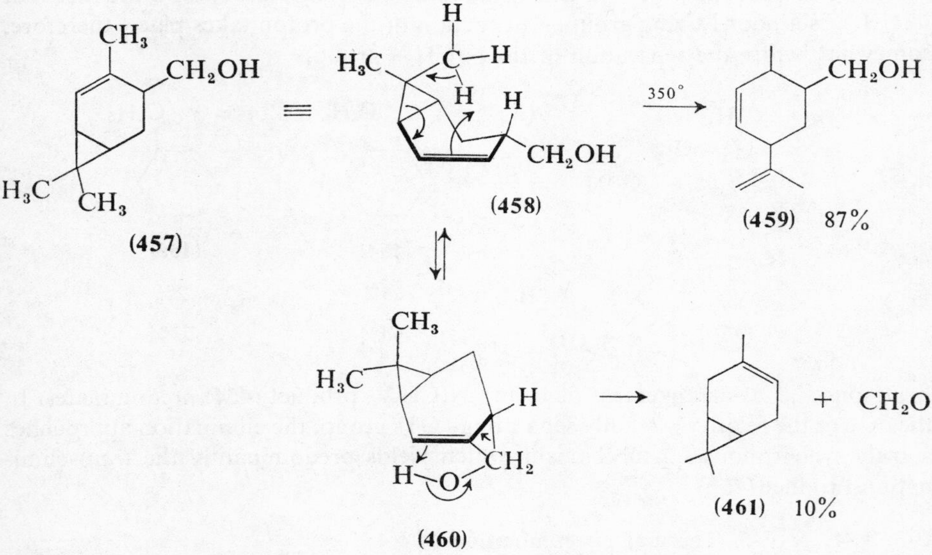

The latter reaction, which is a thermal retro-Prins reaction, has been observed also in other cases[9,12].

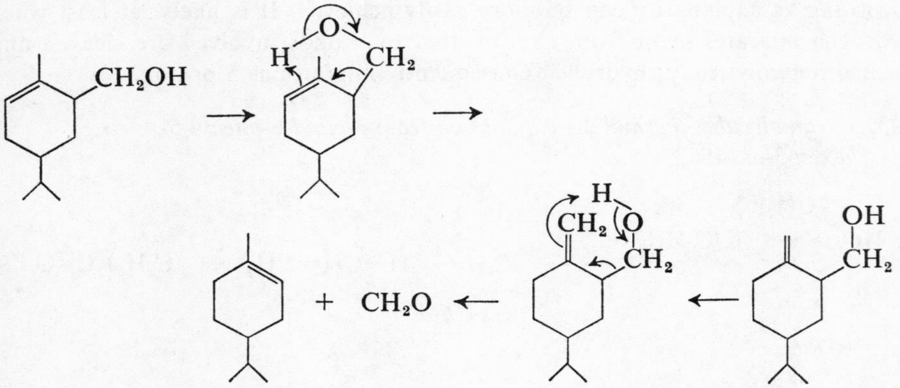

Conversely, the hydrocarbon (**461**) and CH_2O form the starting material (**457**) in acidic medium by the Prins reaction[13].

$$H_2C{=}O + H^+ \rightleftharpoons H_2\overset{+}{C}{-}OH \leftrightarrow H_2C{=}\overset{+}{O}H$$

(**461**) (**457**)

E-5. *Pyrolysis of* 1-*cyclopropylethyl acetate*[14]

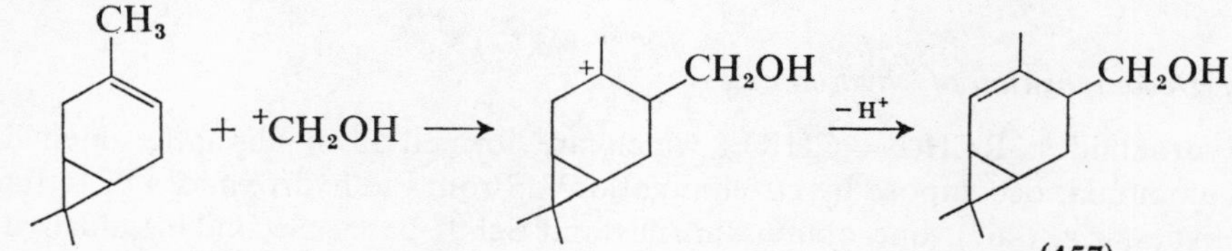

Pyrolysis of esters may also lead to *cis*-elimination. In general, the least substituted olefin is obtained, *i.e.* the Hofmann product. Ester (**462**) yields first vinylcyclopropane (**463**), which rapidly rearranges to the valence-isomeric cyclopentene.

(**462**) (**463**)

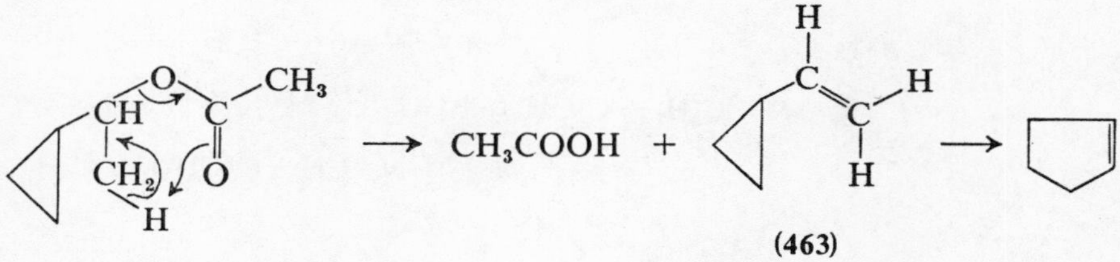

E-6. Cope degradation of amine oxides[6,15]

In the Cope degradation of amine oxides, olefins are formed *via* cyclic five-membered transition states. The Cope degradation and the decomposition of ammonium hydroxides resemble one another in selectivity of olefin formation, except when steric factors have special effects[16].

(**464**) (**465**) (**466**)

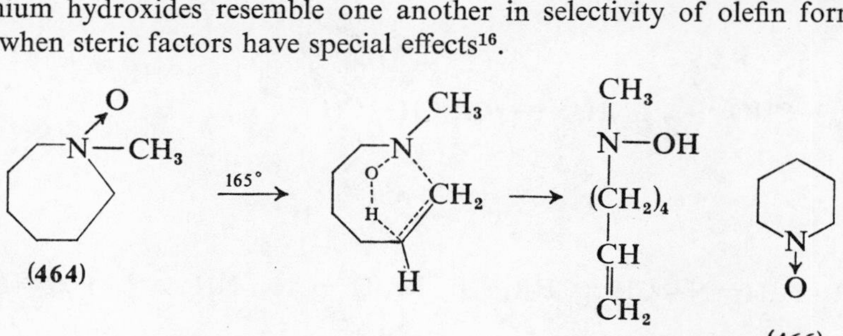

Decomposition of the cyclic amine oxide (**464**) gives the substituted hydroxylamine (**465**), Decomposition of the homologous eight-membered ring occurs at an even faster rate. Piperidine oxide (**466**) is, however, thermally stable, the ring size preventing formation of the necessary coplanar transition state.

E-7. *Decomposition of "hydroamides"*

Hydroamides, $RCH(N{=}CHR)_2$, which are formed from aliphatic aldehydes and ammonia, decompose by *cis*-elimination[17]. From the hydroamides of tertiary aldehydes, *e.g.* (**467**), one obtains nitriles and Schiff bases. Secondary aldehydes yield unsaturated Schiff bases[18].

$$3C_3H_7{-}C(CH_3)_2{-}CHO + 2NH_3 \longrightarrow$$

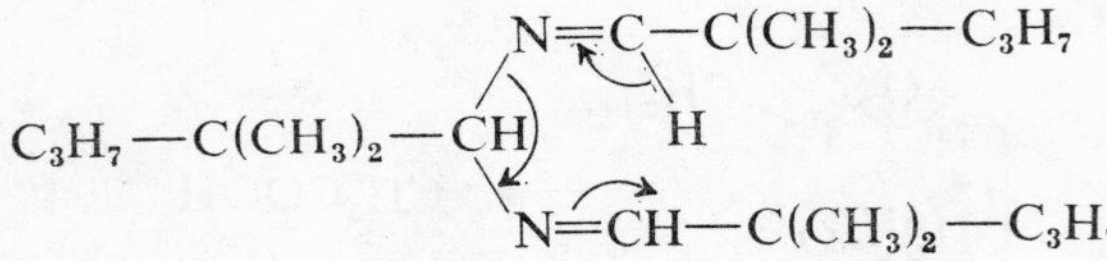

$$C_3H_7{-}C(CH_3)_2{-}CH\begin{matrix} N{=}C{-}C(CH_3)_2{-}C_3H_7 \\ H \\ N{=}CH{-}C(CH_3)_2{-}C_3H_7 \end{matrix}$$

(**467**) 62%

$\downarrow \Delta$

$C_3H_7{-}C(CH_3)_2{-}CN \quad + \quad C_3H_7{-}C(CH_3)_2{-}CH{=}N{-}CH_2{-}C(CH_3)_2{-}C_3H_7$

(**468**) 67% (**469**) 73%

$$3CHR_2{-}CHO + 2NH_3 \longrightarrow R_2CH{-}CH\begin{matrix} N{=}CH{-}CHR_2 \\ H \\ N{=}C{-}CR_2 \\ H \end{matrix}$$

$\downarrow$

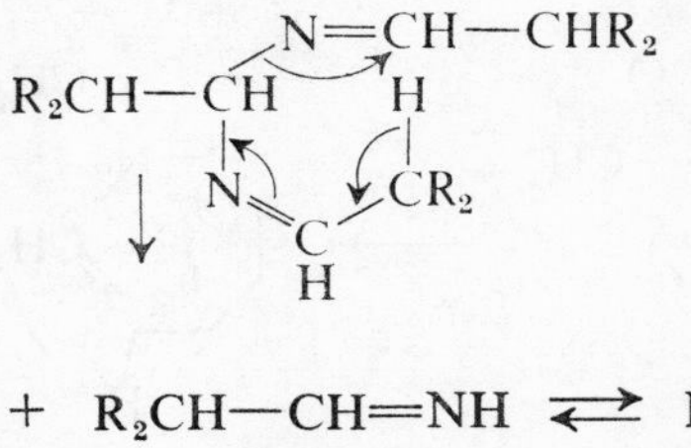

$CR_2{=}CH{-}N{=}CH{-}CHR_2 + R_2CH{-}CH{=}NH \rightleftarrows R_2C{=}CHNH_2$

$2CR_2{=}CH{-}NH_2 \longrightarrow CR_2{=}CH{-}N{=}CH{-}CHR_2 + NH_3$

Primary aliphatic aldehydes afford imines, which readily trimerize. Acetaldehyde and ammonia, for example, yield "aldehyde ammonia" (470).

$$H_3C-CHO + NH_3 \longrightarrow H_3C-CH(OH)-NH_2 \longrightarrow$$

$$H_3C-CH=NH \longrightarrow CH_3CH\underset{NH-CHCH_3}{\overset{NH-CHCH_3}{\diagup\diagdown}}NH$$

(470)

Formaldehyde and aqueous ammonia give hexamethylenetetramine. Aromatic aldehydes yield hydroamides, which are converted into arylimidazolines (471) by heat. A general review of reactions of aldehydes with ammonia has been given by Sprung[19].

$$3C_6H_5CHO + 2NH_3 \longrightarrow$$

$$C_6H_5-CH\underset{N=CH-C_6H_5}{\overset{N=CH-C_6H_5}{\diagup\diagdown}} \longrightarrow C_6H_5-C\underset{N-CH-C_6H_5}{\overset{NH-CH-C_6H_5}{\diagup\diagdown}}|CH-C_6H_5$$

(471)

E-8. *Decomposition of 2-acyltetrazoles*[20]

A further example of thermal *cis*-elimination is the decomposition of 2-thiobenzoyltetrazole (472) (see p. 212). This compound can be isolated under mild conditions. Warming or preparation above 60° leads to elimination of N_2. The thermal instability is a consequence of the formation of stable molecular nitrogen on decomposition and of the resonance-stabilized intermediate (474). Possibly (474) is preceded by another intermediate (473). The *C*-phenyl-*N*-thiobenzoylnitrile imine (474) cyclizes to 2,5-diphenyl-1,3,4-thiadiazole (475).

Decomposition of 2-acyltetrazoles is a first-order reaction, independent of an excess of acid chloride. The ring cleavage must therefore be the rate-determining step[21].

Unsubstituted tetrazole is a stable compound, which melts at 156° without decomposition, despite the four nitrogen atoms. This is good evidence for its aromatic character. The aromaticity[22] is also demonstrated by diazotization of the 5-amino derivative to yield a diazonium salt. Tetrazole exists in two tautomeric forms. Its acidity is comparable with that of carboxylic acids.

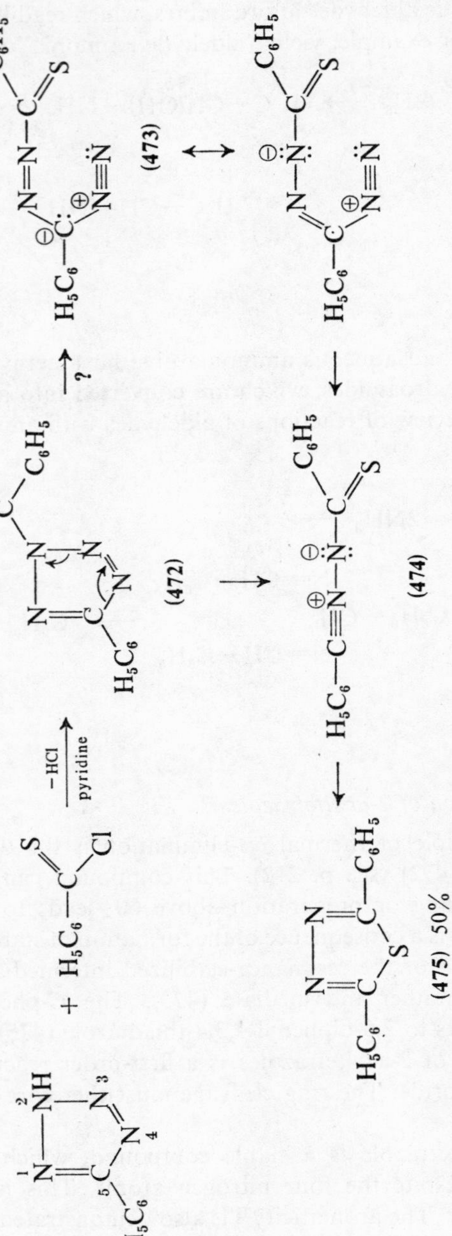

Scheme with structures (472), (473), (474), (475):

2-phenyl-5-phenyl-2H-tetrazole (numbered N1-NH, N2, N3, N4, C5-C6H5) + H5C6—C(=S)—Cl →(−HCl, pyridine)→ (472): 1,2,4-thiadiazole-type intermediate with C(=S)C6H5 on N and C6H5 on ring carbon, showing arrow pushing from N=N to C=N.

(472) → (474): H5C6—C⁺≡N—N̈:⁻—N̈:—C(=S)—C6H5 (open-chain diazo/nitrilium species)

(474) ↔ (473): resonance forms with H5C6—C⁻(=N⁺=N)—N=N—C(=S)—C6H5 and H5C6—C⁻—N⁺≡N ... N:⁻—N̈—C(=S)—C6H5

(474) → (475) 50%: 1,3,4-thiadiazole with C6H5 at C2 and C6H5 at C5 (ring: N=C(C6H5)—S—C(C6H5)=N).

E-9. Reaction of 1,1-dimethylpyrrolidinium bromide (476) with phenyllithium[23]

In some cases the decomposition of 'onium bases is initiated by abstraction of α-hydrogen. In the second step, the ylide-carbon abstracts a proton from the β-position of an adjacent alkyl group. This mechanism, which is designated α',β-elimination, has been observed in the decomposition of tetraalkylammonium[24] and trialkylsulfonium salts[25] by strong bases.

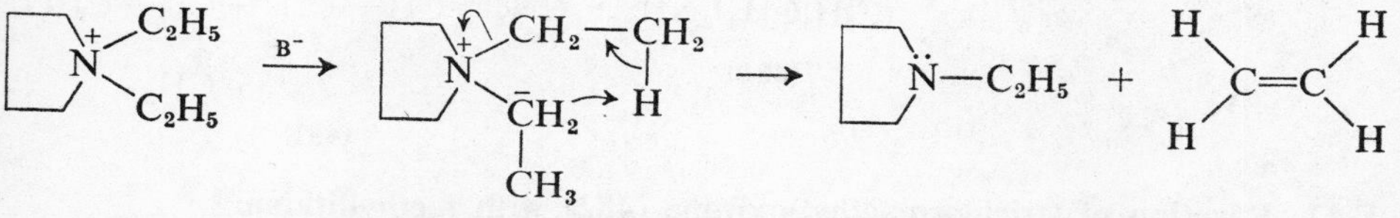

If 1,1-dimethylpyrrolidinium bromide (476) is treated with phenyllithium, the ylides (477) and (478) are formed. In (478), no proton exchange with a β-methylene group is possible for steric reasons[26]. Ylide (477) decomposes, however, by a fragmentation reaction according to the scheme a–b–c–d–X, which is discussed in detail on page 223. Re-ylidization converts ylide (478) *via* (477) also into the fragmentation products.

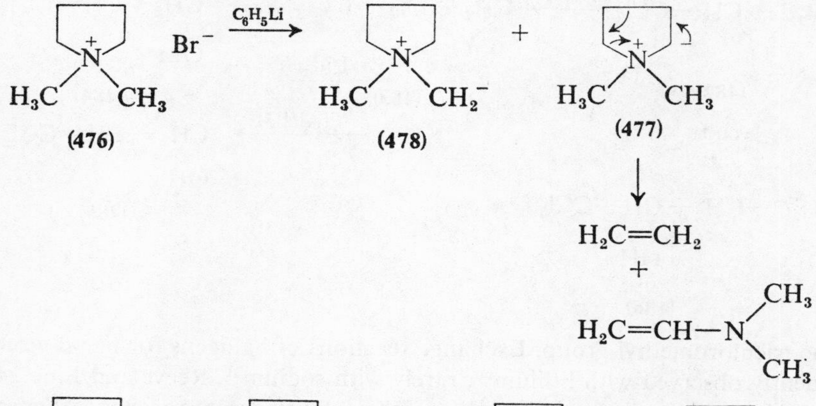

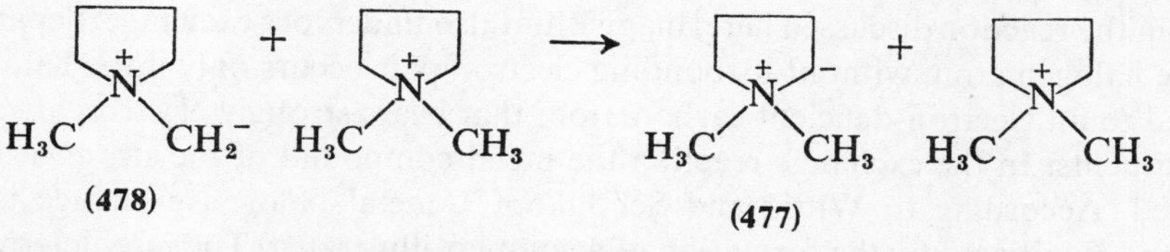

E-10. Decomposition of diallylammonium salts by base[27]

A special case of attack by base on the α-carbon of an ammonium salt is the decomposition of diallylammonium bases. For example, compound (479) yields 2-methyl-4-pentenal (481).

$(C_6H_5CH_2)_2\overset{+}{N}(CH_2-CH=CH_2)_2 \quad \xrightarrow{HO^-} \quad (C_6H_5CH_2)_2\overset{+}{N}\begin{smallmatrix}CH_2-CH=CH_2\\ \\CH=CH-CH_3\end{smallmatrix} \longrightarrow$

Br^-

HO^-

(479)

$(C_6H_5CH_2)_2NH \; + \; CH_2=CH-CH_2-\underset{CH_3}{\overset{|}{CH}}-CHO$

(480)

(481)

E-11. Reaction of (trichloromethyl)oxirane (482) with methyllithium[28]

An epoxide ring may be opened by an S_N1 or an S_N2 reaction. For steric reasons the latter occurs at the less substituted epoxide carbon atom[29]. The reaction of (482) with methyllithium should therefore lead to 1,1,1-trichloro-2-butanol (486), if primary attack is on the epoxide ring. The product, however, is 3,3-dichloroallyl alcohol (485). A plausible mechanism starts with exchange of chlorine for lithium

$H_2C-CH-CCl_3 \xrightarrow{CH_3Li} CH_2-CH-\bar{C}Cl_2 \longrightarrow CH_2-CH=CCl_2$
    $\underset{O}{\diagdown\diagup}$ $\underset{O}{\diagdown\diagup} \quad Li^+$ $\underset{OLi}{|}$

(482)   (483)   (484)

$\Big\downarrow CH_3Li$ $\xrightarrow{NH_4Cl/H_2O}$ $CH_2-CH=CCl_2$
                                                                $\underset{OH}{|}$

$H_3C-CH_2-\underset{OH}{\overset{|}{CH}}-CCl_3$ (485) 85%

(486)

at the trichloromethyl group. Exchange reactions of halogens for metal atoms are frequently observed with lithium[30], rarely with sodium[31]. Reeve and Fine[28] found that, in the reaction discussed here, magnesium also undergoes exchange. Separation of the halogen atom without its bonding electron pair occurs only if the halogen is bound to an electron-deficient carbon atom that bears strongly electron-attracting substituents. In the exchange reaction the metal compound of the stronger acid is formed. According to Wittig and Schöllkopf[32], metal–halogen exchange is very fast and is initiated by the formation of a complex illustrated. The rate-determining

$\begin{matrix} R\text{-----}Li \\ \vdots \quad \vdots \\ Br\text{-----}R' \end{matrix}$

step in the formation of the alcohol (485) may therefore be conversion of (483) into (484) in an E1cB reaction. Alternatives are the formation of the carbene (487) and the synchronous mechanism *via* (488). Secondary products originating from the carbene (487) have not been observed.

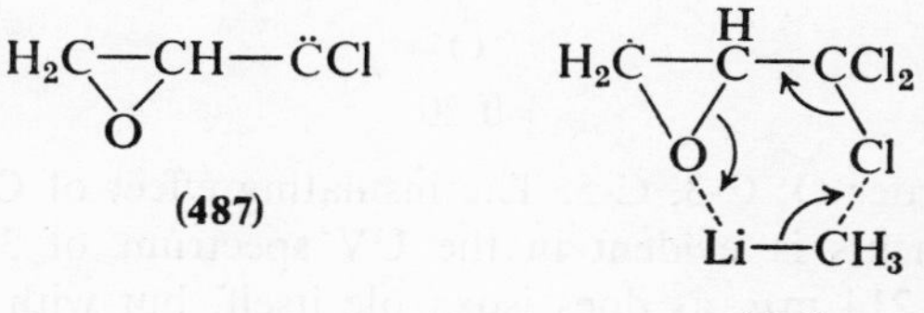

E-12. Rearrangement of 4-phenyl-1,2-diazabicyclo[3.2.0]hept-2-en-6-one (489)

Treatment of the bicyclic base (489) with dilute alkali leads to the diazepinone (490)[33]. The base presumably attacks the 4-hydrogen atom that is activated by the phenyl group. If the phenyl group is replaced by alkyl the rearrangement does not occur.

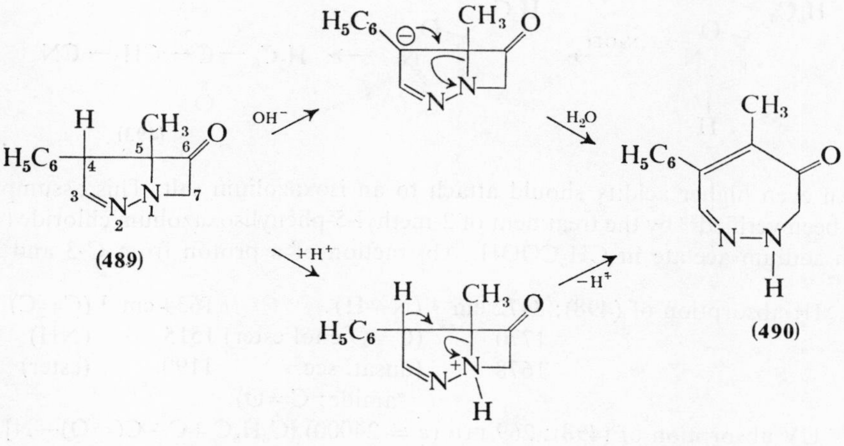

The rearrangement of (489) may also be initiated by acids. A plausible mechanism starts with the addition of a proton to N-1. In the next step, the 1,5-bond is cleaved, simultaneously with the loss of a proton from C-4. The interconversion of the bicyclo[3.2.0]heptene system and the cycloheptadiene system has been repeatedly described[34] (see also example V-2, p. 192).

E-13. Ring cleavage of 2-methyl-5-phenylisoxazolium chloride (494)[35]

MO calculations[36] of electron density for isoxazole have resulted in decreasing

$$+0.18 \quad C^3=C^4 \quad -0.09$$
$$-0.42 \quad N^2_1 \quad C^5 \quad +0.14$$
$$O$$
$$+0.20$$

densities in the order O, C-3, C-5. The insulating effect of C-3 against electron-attracting substituents is evident in the UV spectrum of 3,3'-bioxazole (491), which absorbs at 214 mμ, as does isoxazole itself, but with twice the intensity. 5,5'-Bioxazole (492), however, has a maximum at 260 mμ[37]. This electron dis-

(491)   (492)

tribution causes the acidity of the 3-hydrogen atom. Reaction with base, therefore, results in proton abstraction from C-3, which is followed by ring cleavage and formation of a cyanocarbonyl compound (493)[38].

$$H_5C_6\text{-isoxazole-H} \xrightarrow{NaOEt} H_5C_6\text{-isoxazole} \rightarrow H_5C_6-\underset{\underset{O}{\parallel}}{C}-CH_2-CN$$

(493)

An even higher acidity should attach to an isoxazolium salt. This assumption has been verified[39] by the treatment of 2-methyl-5-phenylisoxazolium chloride (494) with sodium acetate in CH_3COOH. Abstraction of a proton from C-3 and ring

IR absorption of (498): 3472 cm^{-1} (N—H)   1634 cm^{-1} (C=C)
                        1770   (C=O enol ester) 1515   (NH)
                        1678   (unsat. sec.   1190   (ester)
                                    amide; C=O)

UV absorption of (498): 269 mμ ($\varepsilon = 24000$) [$C_6H_5C=C-C(=O)-N$]

cleavage lead to an α-keto ketene imine (495). This can be observed directly in the IR spectrum if the salt (494) is treated with trimethylamine. Temporarily the absorption peak of cumulated double bonds appears at 2062 cm^{-1}. The imine (495) is converted into (496) on addition of acetic acid, and (496) finally rearranges to (498). The *cis*-arrangement of the acyloxy group and the amide group in (498) is demonstrated by the conversion of (498) into (499) in warm ethanol.

β-Eliminations

(494) → (495) → (496) ⇌ (497) → (498) → (498) → (499)

E-14. Reaction of phosphonium salts with base

Phosphonium salts react with base in one of three ways: (1) by α-hydrogen abstraction; (2) by β-hydrogen abstraction; or (3) by addition of the base to the positively charged phosphorus atom.

α-Hydrogen abstraction leads to ylides, in analogy to ammonium and sulfonium salts.

When β-hydrogen abstraction occurs, phosphonium hydroxides decompose, as do ammonium hydroxides, to olefins and phosphines. The formation of olefins is only observed, however, if the β-hydrogen atoms are sufficiently activated; for example, (2,2-diphenylethyl)triethylphosphonium hydroxide (**500**) decomposes to 1,1-diphenylethylene and triethylphosphine[40]. With less activation, decomposition

$$(C_6H_5)_2CH-CH_2-P^+(C_2H_5)_3 \; HO^- \longrightarrow (C_6H_5)_2C=CH_2 + P(C_2H_5)_3 + H_2O$$
$$(\mathbf{500})$$

to phosphine oxide and alkane, which is characteristic for phosphonium hydroxides, predominates.

The decomposition is of the third order, first-order in R_4P^+ and second order in OH^-. The following sequence of reactions fits the kinetics[41]:

$$R_4P^+ + HO^- \rightleftarrows R_4P-OH$$

$$R_4P-OH + HO^- \rightleftarrows R_4P-O^- + H_2O$$

$$R_4P-O^- \xrightarrow{\text{Slow}} R_3P^+-O^- + R^-$$

$$R^- + H_2O \longrightarrow RH + HO^-$$

E-14. *Reaction of (chloromethyl)triphenylphosphonium chloride* (**501**) *with base*[42]

This reaction starts with the formation of the quinquevalent phosphorus compound (**502**)*. The ease of decomposition of (**502**) increases with increasing stability of the forming carbanion[44]. $CH_3-S-CH_2^-$ separates more easily than $ClCH_2^-$ or $BrCH_2^-$. Formation of analogous products is not observed with ammonium salts, probably because nitrogen cannot form a quinquevalent intermediate. If cleavage of the phosphonium salt occurred by an S_N2 mechanism, *i.e.* by replacement of an alkyl group by a base molecule at the positively charged phosphorus atom, the reaction should also be observed for ammonium salts.

---

* For stable quinquevalent phosphorus compounds see reference 43.

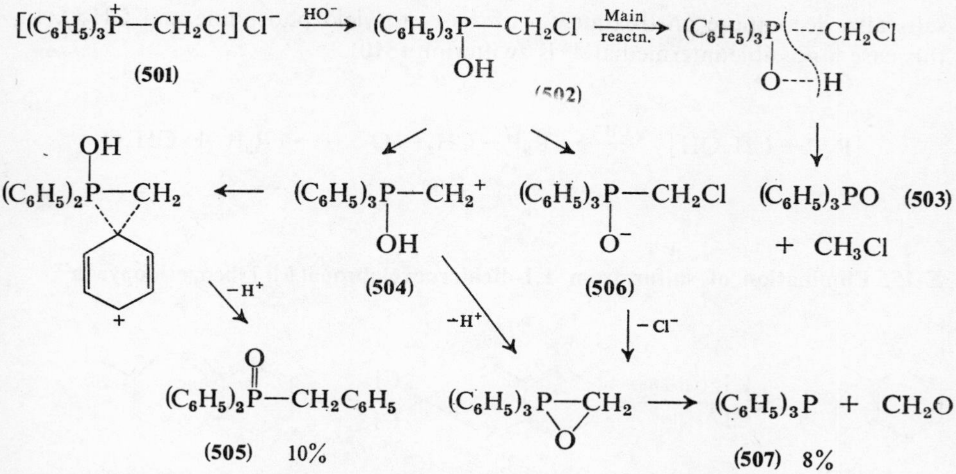

$$[(C_6H_5)_3\overset{+}{P}-CH_2Cl]Cl^- \xrightarrow{HO^-} (C_6H_5)_3P-CH_2Cl \xrightarrow[\text{reactn.}]{\text{Main}} (C_6H_5)_3P\overset{\displaystyle(-\!-\!-\!-CH_2Cl}{\underset{\displaystyle O-\!-\!-\!-|H}{}}$$
(501)   OH
         (502)

$$\underset{\text{(C}_6\text{H}_5\text{)}_2\text{P}-\text{CH}_2}{\overset{OH}{\diagup\!\diagdown}} \longleftarrow (C_6H_5)_3P-CH_2^+ \quad (C_6H_5)_3P-CH_2Cl \quad (C_6H_5)_3PO \quad (503)$$
                                    OH   O⁻   + CH₃Cl

      +   ↓ -H⁺   (504)   ↓ -H⁺   (506)   ↓ -Cl⁻

$$(C_6H_5)_2\overset{O}{\overset{\|}{P}}-CH_2C_6H_5 \qquad (C_6H_5)_3P\underset{O}{\overset{\diagup\!\diagdown}{-}}CH_2 \longrightarrow (C_6H_5)_3P + CH_2O$$
(505) 10%   (507) 8%

The formation of triphenylphosphine oxide (503) and methyl chloride is accompanied by rearrangements; these were first observed by H. Hellmann et al[42]. They lead to benzyldiphenylphosphine oxide (505) or triphenylphosphine (507) and formaldehyde. It is probable that (505) is formed by migration of a phenyl group with its bonding electron pair: (504) → (505). Since treatment of (chloromethyl)-triphenylphosphonium chloride with phenyllithium causes migration of a phenyl group[44], the formation of benzyldiphenylphosphine oxide (505) could alternatively have taken place through a carbene intermediate (509). The preceding ylide (508) may result from the decomposition of (502)[45] or from (chloromethyl)triphenyl-phosphonium chloride, as shown below. Triphenylphosphine (507) and formaldehyde are presumably formed from (504) or (506). (Hydroxymethyl)phosphonium

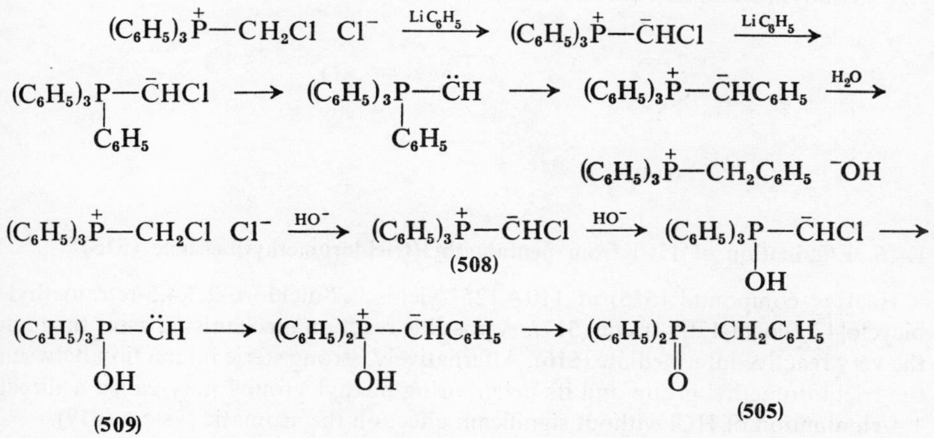

$$(C_6H_5)_3\overset{+}{P}-CH_2Cl \ Cl^- \xrightarrow{LiC_6H_5} (C_6H_5)_3\overset{+}{P}-\bar{C}HCl \xrightarrow{LiC_6H_5}$$

$$(C_6H_5)_3P-\bar{C}HCl \longrightarrow (C_6H_5)_3P-\ddot{C}H \longrightarrow (C_6H_5)_3\overset{+}{P}-\bar{C}HC_6H_5 \xrightarrow{H_2O}$$
$$\underset{C_6H_5}{|} \qquad\qquad \underset{C_6H_5}{|}$$
$$(C_6H_5)_3\overset{+}{P}-CH_2C_6H_5 \ ^-OH$$

$$(C_6H_5)_3\overset{+}{P}-CH_2Cl \ Cl^- \xrightarrow{HO^-} (C_6H_5)_3\overset{+}{P}-\bar{C}HCl \xrightarrow{HO^-} (C_6H_5)_3P-\bar{C}HCl \longrightarrow$$
                                    (508) |
                                                                           OH

$$(C_6H_5)_3P-\ddot{C}H \longrightarrow (C_6H_5)_2\overset{+}{P}-\bar{C}HC_6H_5 \longrightarrow (C_6H_5)_2\underset{O}{\overset{\|}{P}}-CH_2C_6H_5$$
$$\underset{OH}{|} \qquad\qquad \underset{OH}{|}$$
(509)   (505)

salts also decompose on treatment with base, to yield phosphine and CH_2O. In this case a possible intermediate[46] is zwitterion (510).

$$[R_3P-CH_2OH]^+ \xrightarrow{OH^-} R_3\overset{+}{P}-CH_2-O^- \longrightarrow R_3P + CH_2O$$
(510)

E-15. Elimination of sulfur from 1,1-dichlorocyclopropa[b][1]benzothiopyran[47]

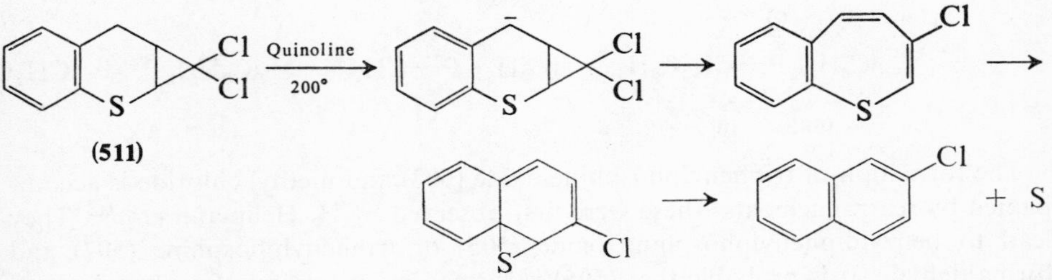

(511) → (512) → (513) + S

In this reaction, the base first attacks a benzylic hydrogen atom of (511). Loss of sulfur is comprehensible if the reaction passes through an intermediate (512). Ring contraction of heterocyclic compounds by elimination of sulfur is a common reaction, which in general proceeds through episulfides[48].

A well-known related elimination is the decomposition of certain cyclic sulfones, e.g. 2,5-dihydrothiophen dioxide (514).

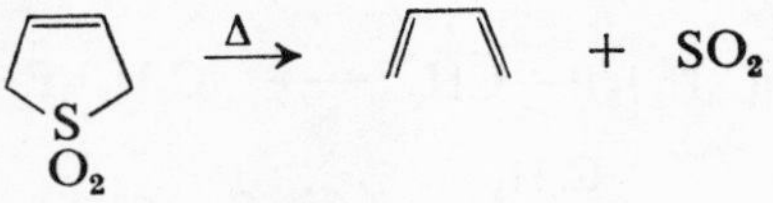

(514) $\xrightarrow{\Delta}$ + SO_2

E-16. Elimination of HCl from pentamethyl(trichloromethyl)benzene (515)[49]

Heating compound (515) at 110—125° yields 7,7-dichloro-2,3,4,5-tetramethylbicyclo[4.2.0]octa-1,3,5-triene (517). A possible reaction mechanism passes through the very reactive intermediate (516). Alternatively, strong steric interaction between the trichloromethyl group and its neighboring methyl groups may cause a direct 1,4-elimination of HCl without significant effect on the aromatic system (519).

REFERENCES

[1] J. F. Bunnett, *Angew. Chem., Intern. Edit. Engl.*, **1**, 225 (1962).
[2] E. S. Gould, *Mechanism and Structure in Organic Chemistry*, Henry Holt and Company, New York, 1959, p. 480.
[3] W. M. Jones, T. G. Squires, and M. Lynn, *J. Amer. Chem. Soc.*, **89**, 318 (1967).
[3a] D. J. McLennan, *Quart. Rev.*, **21**, 490 (1967).
[4] W. H. Saunders, Jr., S. R. Fahrenholtz, E. A. Caress, J. P. Cowe, and M. Schreiber, *J. Amer. Chem. Soc.*, **87**, 3401 (1965).
[5] H. C. Brown and I. Moritani, *J. Amer. Chem. Soc.*, **78**, 2203 (1956).
[6] A. C. Cope, N. A. LeBel, H. H. Lee, and W. R. Moore, *J. Amer. Chem. Soc.*, **79**, 4720 (1957).
[7] S. J. Christol and F. R. Stermitz, *J. Amer. Chem. Soc.*, **82**, 4692 (1960); A. C. Cope, G. A. Berchtold, and D. L. Ross, *ibid.*, **83**, 3859 (1961); R. T. Arnold and P. N. Richardson, *ibid.*, **76**, 3649 (1954); J. Weinstock and F. G. Bordwell, *ibid.*, **77**, 6706 (1955).
[8] R. T. Arnold and G. Smolinsky, *J. Org. Chem.*, **25**, 129 (1960).
[9] G. Ohloff, *Chem. Ber.*, **93**, 2673 (1960).
[10] G. Ohloff, J. Osiecki, and C. Djerassi, *Chem. Ber.*, **95**, 1400 (1962).
[11] G. Ohloff, J. Seibl, and E. sz. Kovats, *Ann. Chem.*, **675**, 83 (1964).
[12] R. L. Webb and J. P. Bain, *J. Amer. Chem. Soc.*, **75**, 4279 (1953).
[13] G. Ohloff, *Ann. Chem.*, **613**, 43 (1958).
[14] C. G. Overberger and A. E. Borchert, *J. Amer. Chem. Soc.*, **82**, 1007, 4896 (1960).
[15] A. C. Cope and N. A. LeBel, *J. Amer. Chem. Soc.*, **82** 4656 (1960).
[16] A. C. Cope, C. L. Bumgardner, and E. E. Schweizer, *J. Amer. Chem. Soc.*, **79**, 4729 (1957).
[17] R. H. Hasek, E. U. Elam, and J. C. Martin, *J. Org. Chem.*, **26**, 1822 (1961).
[18] A. Lipp, *Ann. Chem.* **211**, 345 (1882); *Chem. Ber.*, **14**, 1746 (1881).
[19] M. M. Sprung, *Chem. Rev.*, **26**, 297 (1940).
[20] R. Huisgen, H. J. Sturm, and M. Seidel, *Chem. Ber.*, **94**, 1555 (1961).
[21] R. Huisgen, *Angew. Chem.*, **72**, 359 (1960).
[22] J. D. Roberts and M. C. Caserio, *Basic Principles of Organic Chemistry*, W. A. Benjamin, Inc., New York, Amsterdam, 1965.
[23] F. Weygand and H. Daniel, *Chem. Ber.*, **94**, 1688 (1961).
[24] G. Wittig and R. Polster, *Ann. Chem.*, **599**, 13 (1956); **612**, 102 (1958).
[25] V. Franzen and C. Mertz, *Chem. Ber.*, **93**, 2819 (1960).
[26] G. Wittig and W. Tochtermann, *Chem. Ber.*, **94**, 1692 (1961).

[27] A. T. Babayan and M. H. Indjikyan, *Tetrahedron*, **20**, 1375 (1964).
[28] W. Reeve and L. W. Fine, *J. Amer. Chem. Soc.*, **86**, 880 (1964).
[29] S. Winstein and R. B. Henderson in *Heterocyclic Compounds*, ed. R. C. Elderfield, Vol. I, p. 48, John Wiley and Sons, Inc., New York, 1950.
[30] G. Wittig, *Z. Naturwiss.*, **30**, 702 (1942).
[31] M. Schlosser, *Angew. Chem.*, **76**, 134 (1964).
[32] G. Wittig and U. Schöllkopf, *Tetrahedron*, **3**, 91 (1958).
[33] J. A. Moore and R. W. Medeiros, *J. Amer. Chem. Soc.*, **81**, 6029 (1959); J. A. Moore and L. J. Pandra, *J. Org. Chem.*, **29**, 336 (1964).
[34] J. A. Moore, F. J. Marascia, R. W. Medeiros, and E. Wyss, *J. Amer. Chem. Soc.*, **84**, 3022 (1962); H. L. Dryden and B. E. Burgert, *ibid.*, **77**, 5633 (1955); W. G. Dauben, K. Koch, S. L. Smith, and O. L. Chapman, *ibid.*, **85**, 2616 (1963).
[35] R. B. Woodward and R. A. Olofson, *J. Amer. Chem. Soc.*, **83**, 1007 (1961).
[36] L. E. Orgel, T. L. Cottrell, W. Dick, and L. E. Sutton, *Trans. Faraday Soc.*, **47**, 113 (1951).
[37] S. Califano, G. Speroni, and C. P. Tafuri, Comm. IV, Meeting on Molecular Spectroscopy, Bologna, 1959.
[38] L. Claisen, *Chem. Ber.*, **36**, 3672 (1903); G. Speroni and P. Pino, Proc. XIth Intern. Congress Pure Appl. Chem., London, 1957.
[39] R. B. Woodward and R. A. Olofson, *Tetrahedron*, Memorial Vol. to H. Stephen, 1966, p. 415.
[40] G. W. Fenton and C. K. Ingold, *J. Chem. Soc.*, **1929**, 2342; **1933**, 531.
[41] W. E. McEwen, G. Axelrad, M. Zanger, and C. A. Vander Werf, *J. Amer. Chem. Soc.*, **87**, 3948 (1965).
[42] H. Hellmann and J. Bader, *Tetrahedron Letters*, 1961, 724; H. Hellmann, J. Bader, H. Birkner, and O. Schumacher, *Ann. Chem.*, **659**, 49 (1962).
[43] G. Wittig and E. Knochendoerfer, *Chem. Ber.*, **97**, 741 (1964).
[44] M. Schlosser, *Angew. Chem.*, **74**, 291 (1962).
[45] D. Seyferth, W. B. Hughes, and J. K. Heeren, *J. Amer. Chem. Soc.*, **87**, 2847, 3467 (1965).
[46] L. Horner, H. Hoffmann, H. G. Wippel, and G. Hassel, *Chem. Ber.*, **91**, 52 (1958).
[47] W. E. Parham and R. Koncos, *J. Amer. Chem. Soc.*, **83**, 4034 (1961); W. E. Parham and M. D. Bhavsar, *J. Org. Chem.*, **29**, 1575 (1964).
[48] J. D. Loudon, in *Organic Sulfur Compounds*, Vol. I, ed. N. Kharasch, Pergamon Press, London, 1961, p. 299.
[49] H. Hart and R. W. Fisch, *J. Amer. Chem. Soc.*, **83**, 4463 (1961); H. Hart, J. A. Hartlage, R. W. Fish, and R. R. Rafos, *J. Org. Chem.*, **31**, 2244 (1966).

5

Fragmentation

FRAGMENTATION is a decomposition reaction according to the following scheme[1]

$$a-b-c-d-X \rightarrow a-b + c=d + X$$

a, b, c, and d are atoms that are capable of forming double bonds, for example, C, N, or O. Fragmentations are usually heterolyses*, yielding the charged fragments

Table 5. Some Typical Fragmentation Patterns

a—b—	a—b	—c—d—	c=d	—X	X
HO—CR$_2$—	O=CR$_2$	—CR$_2$—CR$_2$—	R$_2$C=CR$_2$	—Cl	Cl$^-$
R$_2$N—CR$_2$—	R$_2\overset{+}{N}$=CR$_2$	—CR$_2$—NR—	R$_2$C=NR	—Br	Br$^-$
R—CO—	R—$\overset{+}{C}$=O	—CR=N—	RC≡N	—SO$_3$R	RSO$_3^-$
R$_2\bar{C}$—CR$_2$—	R$_2$C=CR$_2$	—CR$_2$—O—	R$_2$C=O	—OCOR	RCOO$^-$
		—N=N—	N$_2$	—OH$_2^+$	H$_2$O
				—NR$_3^+$	NR$_3$

a—b and X and the unsaturated fragment c=d. X is a group that readily separates with its bonding electrons. a is a group that stabilizes the positively charged atom b. If a—b$^+$ reacts with an anion, Y$^-$, the fragmentation represents an ethylogous S_N1 substitution. If a—b$^+$ is converted into a=b by loss of a proton, the reaction may be considered as an ethylogous β-elimination†.

$$a-b-Y \xrightarrow{+Y^-} a-b^+ \xrightarrow{-H^+} a=b$$

Detailed investigations of fragmentation mechanisms have been carried out with γ-amino halides. Solvolysis of γ-amino halides can, in principle, occur by any of

---

* Cracking processes in which olefins are formed by homolysis are not discussed in this Chapter.
† An ethylogous elimination probably occurs in the decomposition of polychloroisooctane on Al$_2$O$_3$–NiCl$_2$ at 200° (Höver[2]).

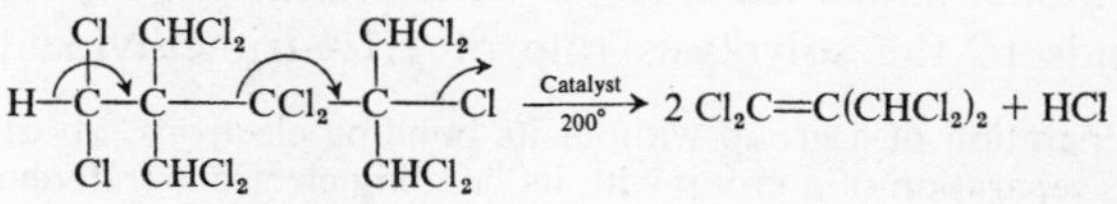

$$H-C(Cl)(CHCl_2)-CCl_2-C(CHCl_2)-Cl \xrightarrow[200°]{Catalyst} 2\ Cl_2C=C(CHCl_2)_2 + HCl$$

four mechanisms: substitution (S), 1,2-elimination (E), cyclization to an azetidinium ion (R), and fragmentation (F) to an olefin and a carbiminium cation[3]:

$$>\!\!N\!-\!C\!-\!C\!-\!C\!-\!X \xrightarrow{\substack{S \\ E \\ R \\ F}} \begin{array}{l} >\!\!N\!-\!C\!-\!C\!-\!C\!-\!Y \\ >\!\!N\!-\!C\!-\!C\!=\!C\!< \\ -\!\overset{+}{N}\!-\!C \\ \;\;\;|\;\;\;\;| \\ \;\;\;C\!-\!C \\ \overset{+}{N}\!=\!C\!< + >\!\!C\!=\!C\!< \end{array}$$

The investigations have shown that the fragmentation may become the main reaction if suitable starting materials are used.

For the fragmentation itself several mechanisms are conceivable:

1. Primary formation of a carbonium ion in the rate-determining step; decomposition of the carbonium ion in the second step.

$$>\!\!N\!-\!C\!-\!C\!-\!C\!-\!X \longrightarrow X^- + >\!\!N\!-\!C\!-\!C\!-\!C^+ \longrightarrow >\!\!N\!-\!C^+ + >\!\!C\!=\!C\!<$$

2. Primary separation of the "electrofugous" group*; decomposition of the anion

$$>\!\!N\!-\!C\!-\!C\!-\!C\!-\!X \longrightarrow >\!\!N\!-\!C^+ + {}^-C\!-\!C\!-\!X \longrightarrow >\!\!C\!=\!C\!< + X^-$$

in the second step.

3. Synchronous fragmentation.

$$>\!\!N\!-\!C\!-\!C\!-\!C\!-\!X \longrightarrow >\!\!\overset{+}{N}\!=\!C + >\!\!C\!=\!C\!< + X^-$$

4. Decomposition via an azetidinium ion.

$$>\!\!N\!-\!C\!-\!C\!-\!C\!-\!X \longrightarrow \begin{array}{c} |\;\;\;\;| \\ -\overset{+}{N}\!-\!C \\ |\;\;\;\;| \\ C\!-\!C \end{array} + X^- \longrightarrow \begin{array}{c} {}^+\!N \\ \|\! \\ C \end{array} + \begin{array}{c} C \\ \|\! \\ C \end{array}$$

In mechanisms 2, 3, and 4 the nitrogenous group must have an effect on the rate-determining step. The rate of fragmentation should therefore be different from the rate of solvolysis of a comparable compound in which R_2N- is replaced by R_2CH-. For the model halide (**520**) it has been found that the fragmentation rate largely corresponds to the solvolysis rate of 1,1,4-trimethylpentyl halide (**521**).

* This involves separation of a group without its bonding electrons. So-called "nucleofugous separation" involves separation of a group with its bonding electrons (reference 4).

$(CH_3)_2N-CH_2-CH_2-C(CH_3)_2Cl \xrightarrow[k_1]{\text{NaOH} \atop 80\% \text{EtOH}} (CH_3)_2N-CH_2-CH_2-\overset{+}{C}(CH_3)_2$
(520)

$\downarrow$

$(CH_3)_2N-CH_2^+ + CH_2=C(CH_3)_2$

$(CH_3)_2CH-CH_2-CH_2-C(CH_3)_2Cl \xrightarrow{k_2} (CH_3)_2CH-CH_2-CH_2-\overset{+}{C}(CH_3)_2$
(521) $k_1 \approx k_2$

It follows that γ-amino halides fragment by mechanism 1 if formation of a carbonium ion is favorable. The reaction is of the first order and corresponds to an $E1$ elimination.

The most favorable steric arrangement for the decomposition of the carbonium ion is where the $2p$-orbital of the carbonium carbon is in the same plane as the b–c bond, since the p-orbitals formed by cleavage of the b–c bond lie in the direction of the b–c axis.

Overlap of the developing p-orbital at carbon atom c with the p-orbital at carbon atom d causes a decrease of the transition-state energy and thus an increase of the rate of fragmentation (522).

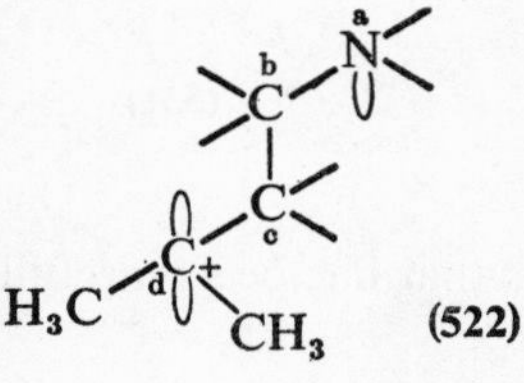

When primary formation of a carbonium ion is less favorable or inhibited, γ-amino halides may decompose by synchronous mechanism 3. The most favorable steric arrangement for this is an antiparallel arrangement of the orbital of the amino group which contains the free electron pair and the d–X bond to the b–c bond. Such conformations are realized in (523), (524), and (525).

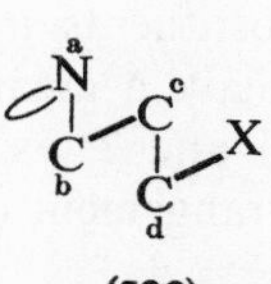

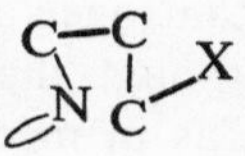

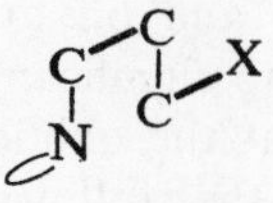

F-1. Fragmentation of 3-bromo-*N,N*-dimethyladamantan-1-amine (526)[5]

Structure (523) is present in the adamantane derivative (526). In 80% ethanol, this bromide decomposes quantitatively to compound (527) which is converted into (528) by hydrolysis. The effect of the amino group on the transition state is to

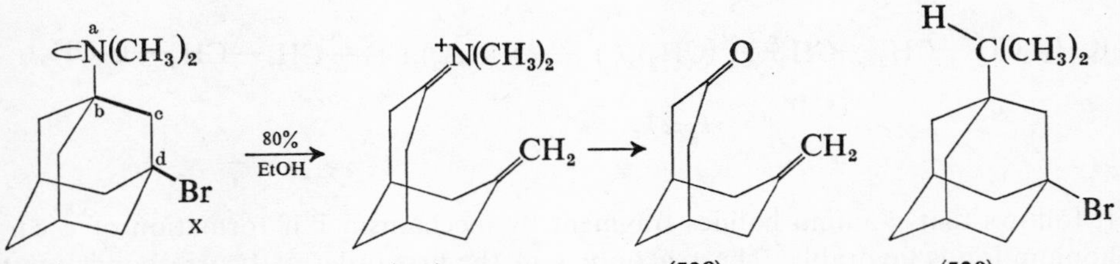

make the rate of synchronous fragmentation much higher than that of solvolysis of compound (529). The synchronous fragmentation often takes place even if formation of a carbonium ion is not particularly difficult, for example, the amine (530) decomposes by mechanism 3, yielding product (531)[6].

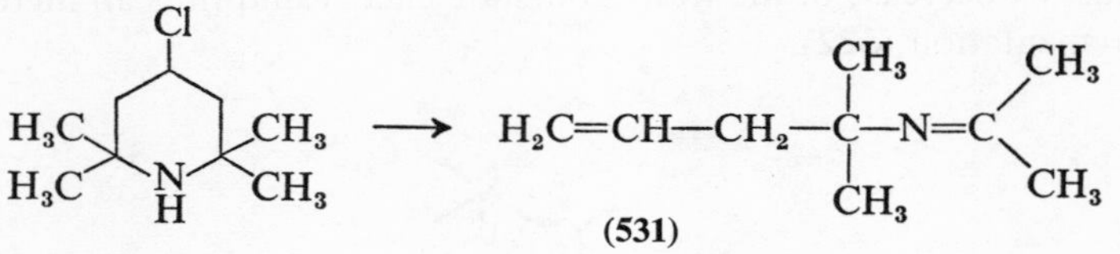

Another class of compounds that has been carefully investigated is the α-amino ketoximes.

F-2. Fragmentation of α-amino ketoximes[7]

Ketoximes may exist in a *syn-* (532) or an *anti*-configuration (533). Although theoretically (532) and (533) might undergo a normal Beckmann rearrangement, fragmentation is so fast that no rearrangement product is observed. Fragmentation of oximes is often called Beckmann fission or second-order Beckmann rearrangement. In analogy to example F-1, (532)* and (533) fragment by a synchronous mechanism. The rate of reaction depends on the types of substituent on the amino nitrogen atom and on the oxime nitrogen atom. Substituents in the aromatic nucleus have no significant effect, though they should have a great effect when a charge arises at the oxime carbon as in mechanisms 1 and 2, as well as in the fragmentation that follows a preceding Beckmann rearrangement (mechanism 5).

* The acid group at the oxime-nitrogen is different from that of (533).

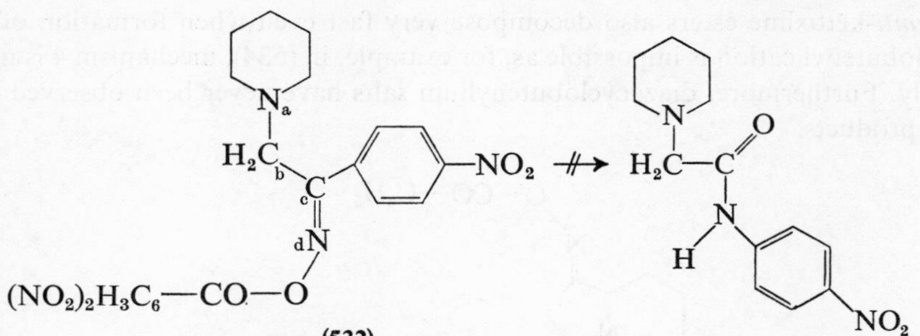

Fragmentation

(Scheme showing compound **(532)**: piperidine-N$_a$-CH$_2^b$-C^c(=N^d-O-CO-C$_6$H$_3$(NO$_2$)$_2$)-C$_6$H$_4$-NO$_2$ ⇏ piperidine-CH$_2$-CO-NH-C$_6$H$_4$-NO$_2$)

(Scheme showing compound **(533)**: piperidine-CH$_2$-C(=N-O-CO-C$_6$H$_5$)-C$_6$H$_4$-NO$_2$ ⇏ O$_2$N-C$_6$H$_4$-CO-NH-CH$_2$-piperidine)

Two parallel reactions both giving:

$$\text{piperidine}^+\text{=CH}_2 \;+\; O_2N\text{-}C_6H_4\text{-CN}$$

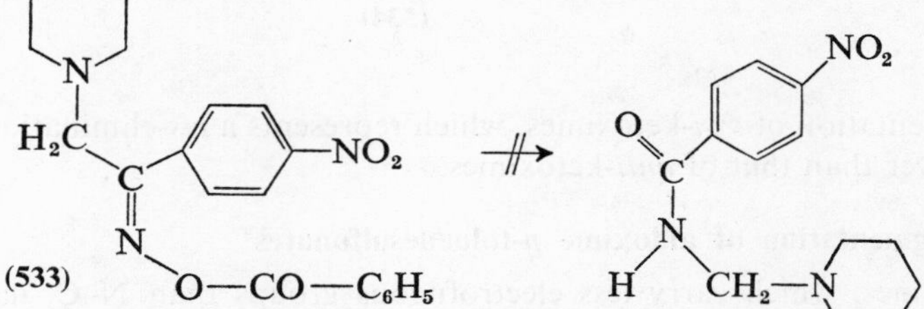

Left path: 80% ethanol, 80°, 99% (from piperidine-CH$_2$-C(=N-O-CO-C$_6$H$_5$)-C$_6$H$_4$-NO$_2$)

Right path: 84% (from piperidine-CH$_2$-C(=N-O-CO-C$_6$H$_3$(NO$_2$)$_2$)-C$_6$H$_4$-NO$_2$)

Mechanism:

piperidine-CH$_2$-C(=N-O-CO-C$_6$H$_5$)-C$_6$H$_4$-NO$_2$ $\xrightarrow{\text{Mech. 5}}$ piperidine-N-CH$_2$-N=C$^+$-C$_6$H$_4$-NO$_2$ →

$$\text{piperidine}^+\text{=CH}_2 \;+\; O_2N\text{-}C_6H_4\text{-CN}$$

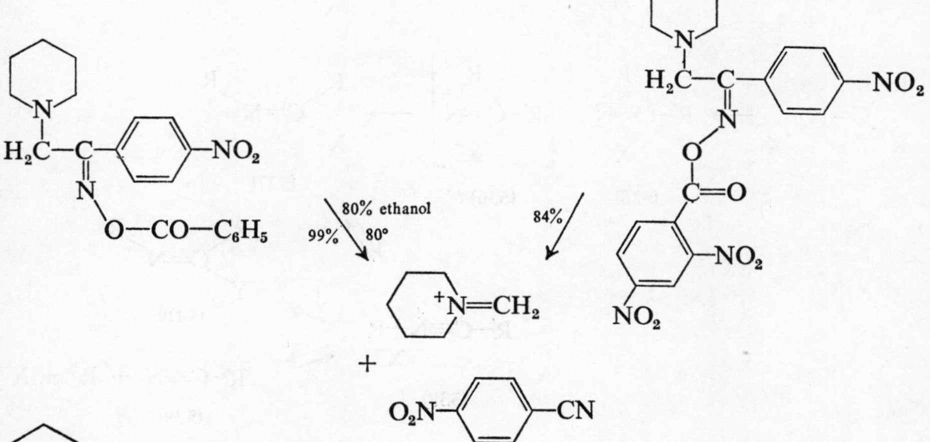

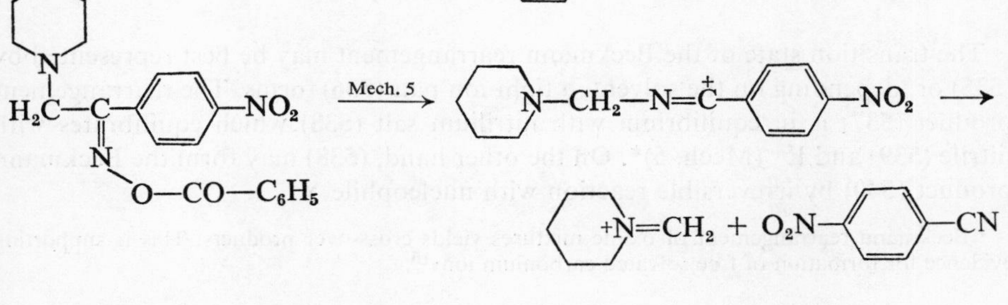

Since *anti*-ketoxime esters also decompose very fast even when formation of a diazacyclobutenyl cation is impossible as, for example, in (534), mechanism 4 is not very likely. Furthermore, diazacyclobutenylium salts have never been observed as reaction products.

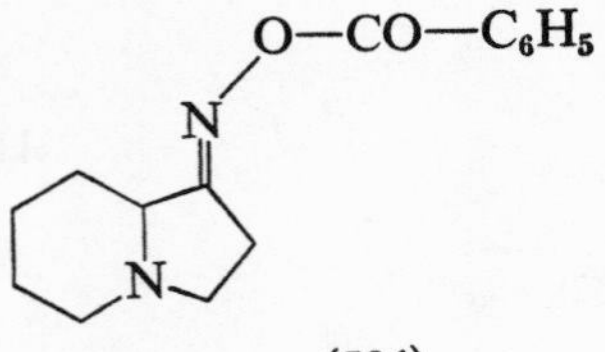

(534)

Fragmentation of *syn*-ketoximes, which represents a *cis*-elimination, is considerably slower than that of *anti*-ketoximes.

F-3. Fragmentation of aldoxime *p*-toluenesulfonates[8]

Ketoximes, which carry less electrofugous groups than N–C, namely, groups such as alkyl, cycloalkyl, or aryl, undergo preferentially the Beckmann rearrangement[9].

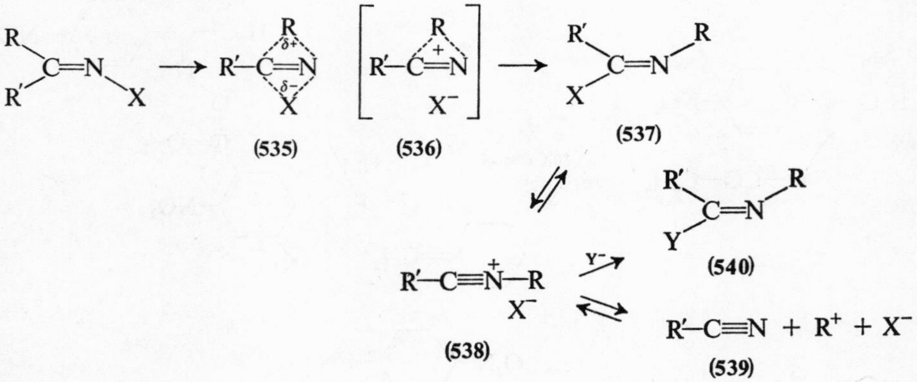

(535)   (536)   (537)

(538)   (539)   (540)

The transition state of the Beckmann rearrangement may be best represented by (535) or, depending on the solvent, a tight ion pair (536) forms. The rearrangement product (537) is in equilibrium with nitrilium salt (538) which equilibrates with nitrile (539) and R^+ (Mech. 5)*. On the other hand, (538) may form the Beckmann product (540) by irreversible reaction with nucleophile Y^-.

* Beckmann rearrangement of oxime mixtures yields cross-over products. This is supporting evidence for formation of free solvated carbonium ions[10].

Fragmentation

If the electrofugous group is a hydrogen atom readily expelled as a proton, Beckmann rearrangement products are not observed. Thus aldoxime (**541**) in which H is *anti* to X yields the nitrile (**542**), as expected. The *syn*-oxime (**543**) yields the

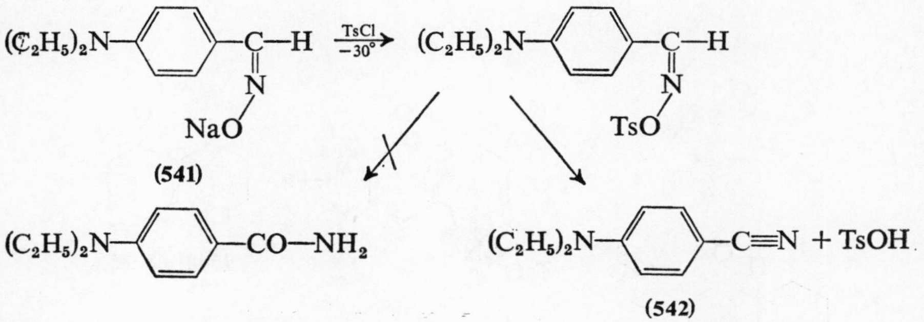

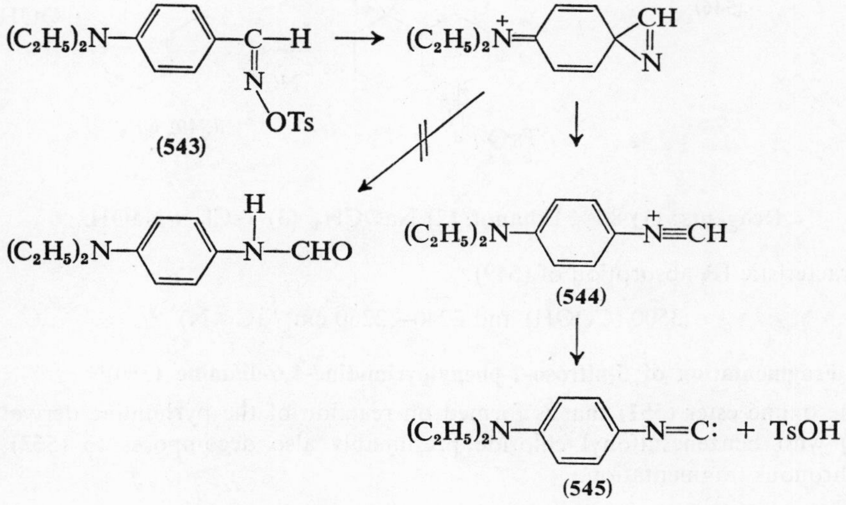

socyanide (**545**). The reaction presumably involves the same transition state as the Beckmann rearrangement. The nitrilium salt (**544**), however, does not lose R^+ from the nitrogen atom [(**538**) → (**539**)] or add a nucleophile [(**538**) → (**540**)] but loses H^+ from the carbon atom instead. Formamide derivatives, which could be formed by addition of OH^- to the C-atom of (**544**), are not observed.

F-4. Fragmentation of 5-(p-tosyloxyimino)-1-decalone (546)[11]

Fragmentation of this ester seems adequately explained by the Chart.

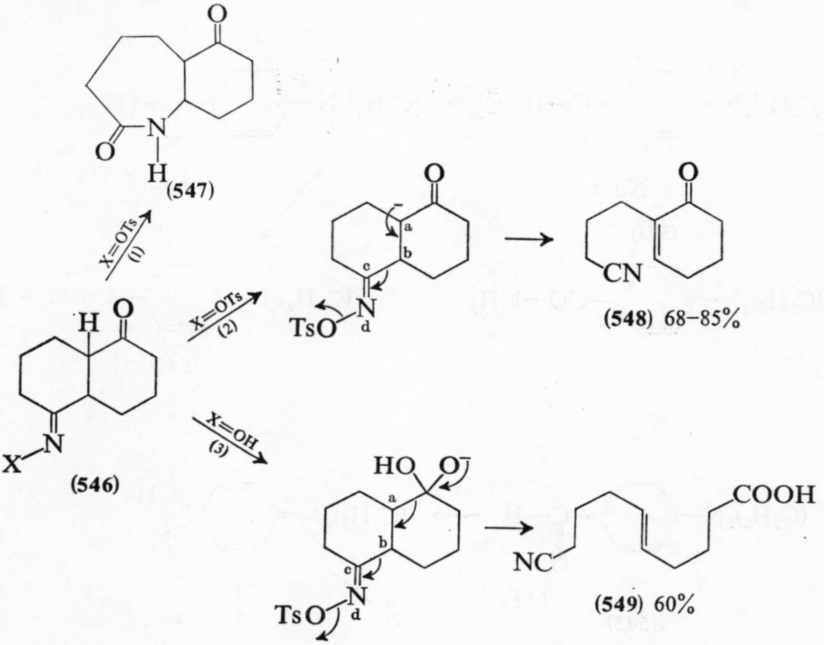

Reagents: (1) 80% Ethanol. (2) NaOCH$_3$. (3) TsCl, N-NaOH.

Characteristic IR absorption of (549):

3500 (COOH) and 2240—2260 cm^{-1} (C≡N)

F-5. Fragmentation of 5-nitroso-2-phenylpyrimidine-4,6-diamine (550)[12]

The oxime ester (551) that is formed on reaction of the pyrimidine derivative (550) with benzenesulfonyl chloride presumably also decomposes to (552) by synchronous fragmentation.

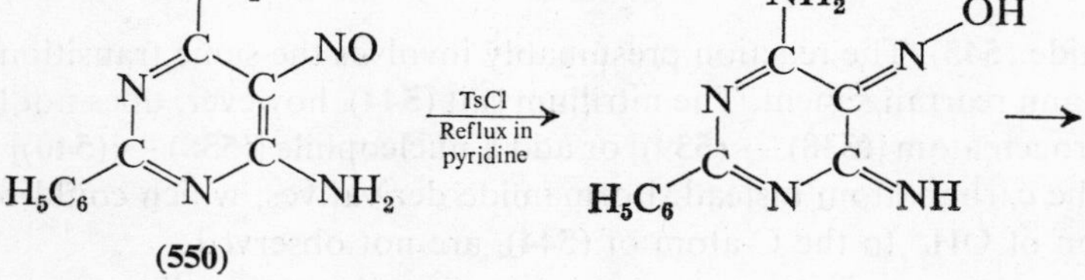

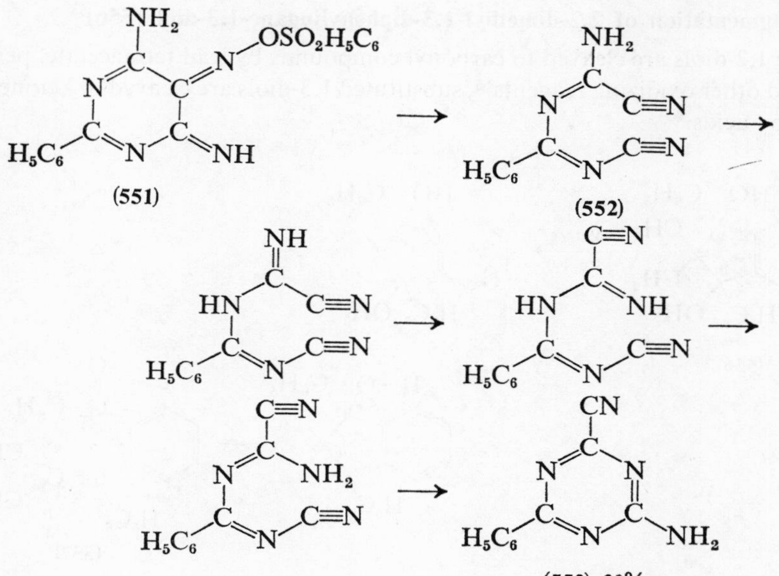

F-6. Fragmentation of 2-phenyl-1,3-dioxolane (554) on reaction with n-butyllithium[13]

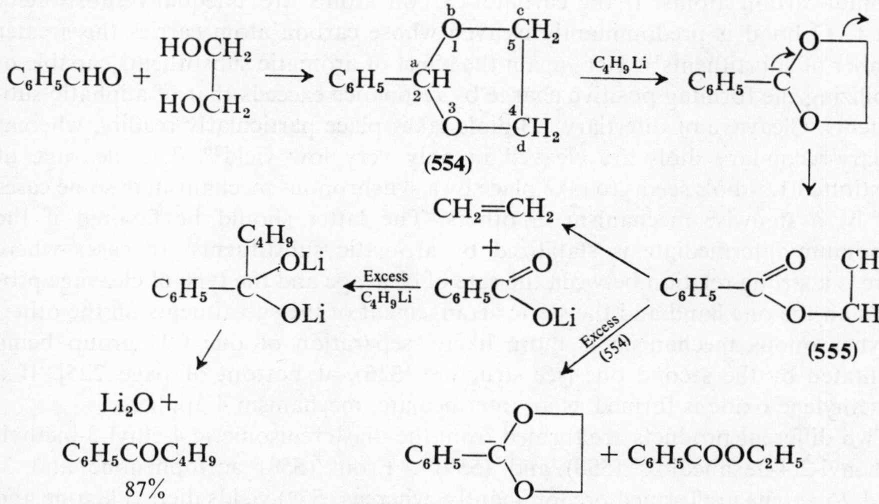

In treatment of the dioxolane (554) the base attacks at position 2, where the hydrogen atom is activated by a phenyl group and two oxygen atoms. It is believed that (555) is an intermediate, since ethyl benzoate is formed if a large excess of the dioxolane is used.

F-7. Fragmentation of 2,2-dimethyl-1,3-diphenylindane-1,3-diol (556)[14]

While 1,2-diols are cleaved to carbonyl compounds by lead tetraacetate, periodic acid, and other oxidizing reagents[15], substituted 1,3-diols are cleaved to ketones and olefins by acids.*

Cleavage increasingly predominates over the competing dehydration reactions as the number of substituents increases, in particular aryl substituents on the carbinol-carbon atoms. If the carbinol-carbon atoms are unequally substituted, that C–O bond is predominantly cleaved whose carbon atom carries the greater number of substituents[19]. Here again the effect of aromatic substituents capable of stabilizing the forming positive charge by resonance exceeds that of aliphatic substituents. Cleavage of ditertiary 1,3-diols takes place particularly readily, whereas tertiary-secondary diols are cleaved in only very low yield[20]. The cleavage of substituted 1,3-diols seems to take place by a synchronous mechanism in some cases and by a stepwise mechanism in others. The latter should be favored if the carbonium intermediate is stabilized by aromatic substituents. In cases where there is a strong relation between the rate of cleavage and the type of cleavage products on the one hand and the steric arrangement of the substituents on the other, a synchronous mechanism is more likely, separation of one OH group being facilitated by the second one [see structure (525), at bottom of page 225]. If a trimethylene oxide is formed as an intermediate, mechanism 4 applies.

Two different products are formed from the diastereoisomeric 4-ethyl-3-methyl-2-phenyl-2,4-hexanediols (558) and (559)[21]. From (558) acetophenone and 3-ethyl-2-pentene are formed predominantly, whereas (559) yields diethyl ketone and

---

* See reference 15, page 375. 1,3-Diols are also cleaved by base; in contrast to acid cleavage, ketones and alcohols are formed[16]. However, mono-p-toluenesulfonates of 1,3-diols yield ketones and olefins on cleavage by acid or base[17]. For a study of fragmentation of 3-aminocyclohexyl p-toluenesulfonates see reference 18.

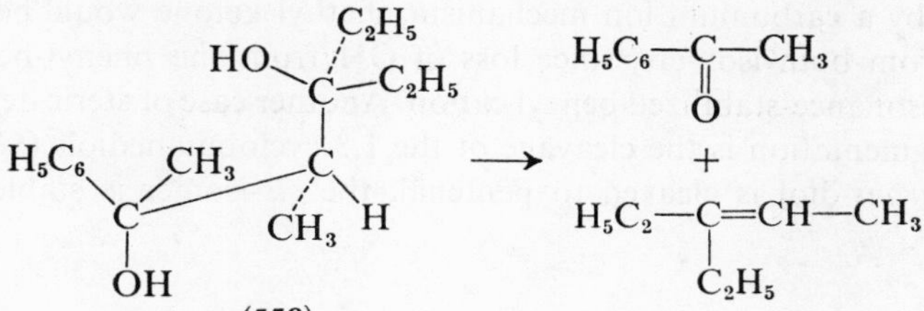

Fragmentation

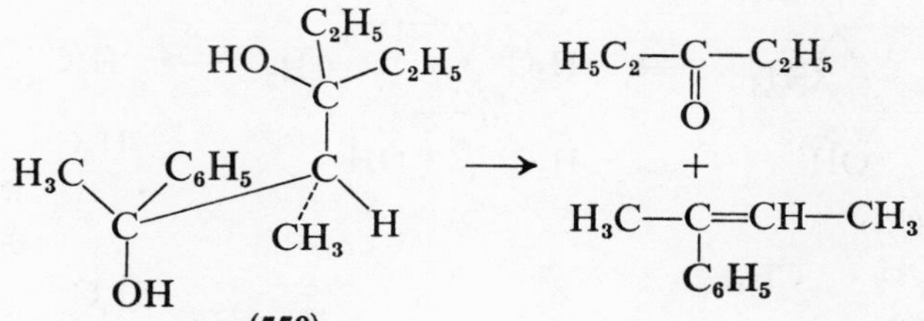

[Structure (558): pinacol with H₅C₆, CH₃, OH on left carbon; HO, C₂H₅, C₂H₅, CH₃, H on right carbon] → H₅C₆—C(=O)—CH₃ + H₅C₂—C(C₂H₅)=CH—CH₃

[Structure (559): pinacol with H₃C, C₆H₅, OH on left carbon; HO, C₂H₅, C₂H₅, CH₃, H on right carbon] → H₅C₂—C(=O)—C₂H₅ + H₃C—C(C₆H₅)=CH—CH₃

2-phenyl-2-butene. Consideration of the four possible transition states furnishes a rationalization of these results. The transition states (558a) and (559a) lead to diethyl ketone. However, (558a) is less stable because of steric interactions between the vicinal phenyl and methyl groups. The steric hindrance is avoided, if (558) decomposes through a transition state (558b) which leads to acetophenone and 3-ethyl-2-pentene. The situation is reversed in the case of isomer (559). If the reaction

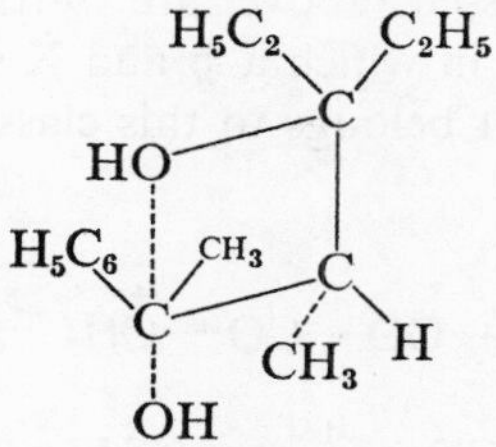

[Transition state structures (558a), (559a), (558b), (559b) shown]

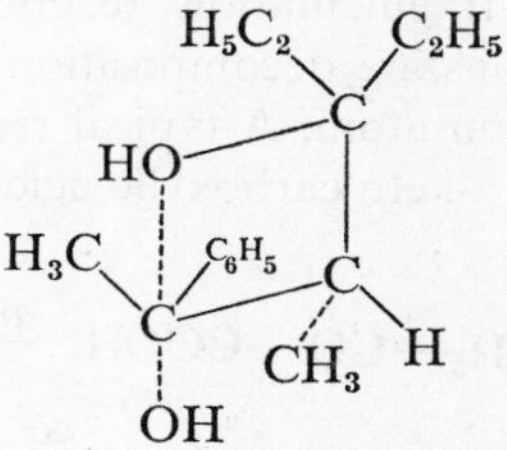

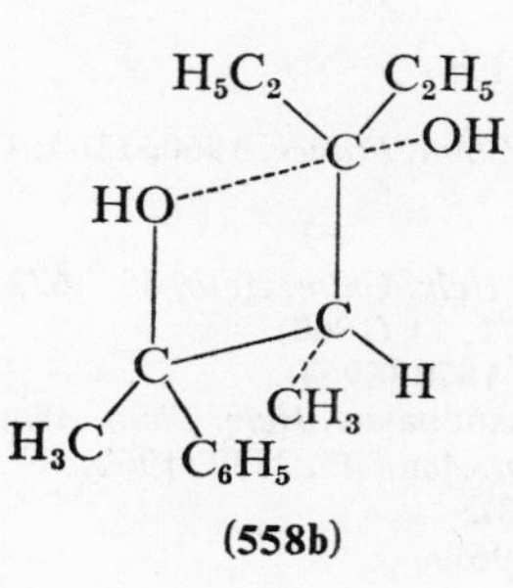

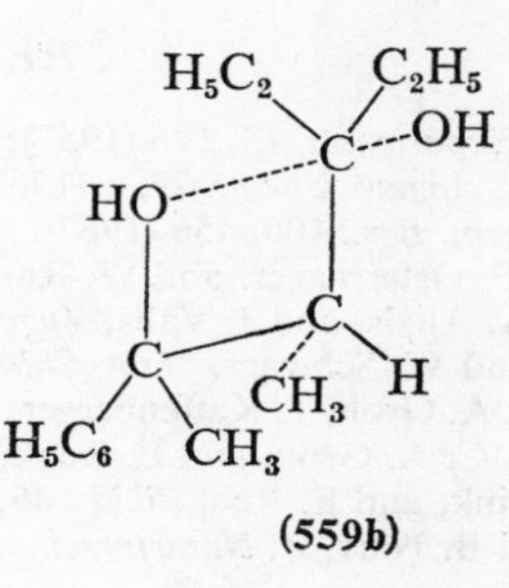

took place by a carbonium ion mechanism, diethyl ketone would be formed preferentially from both isomers, since loss of OH from the phenyl-bearing carbon leads to a resonance-stabilized benzyl cation. Another case of steric dependence of a 1,3-diol fragmentation is the cleavage of the 1,3-cyclobutanediols (560) and (561). While the *trans*-diol is cleaved to pentenal, the *cis*-isomer is stable under these conditions[22].

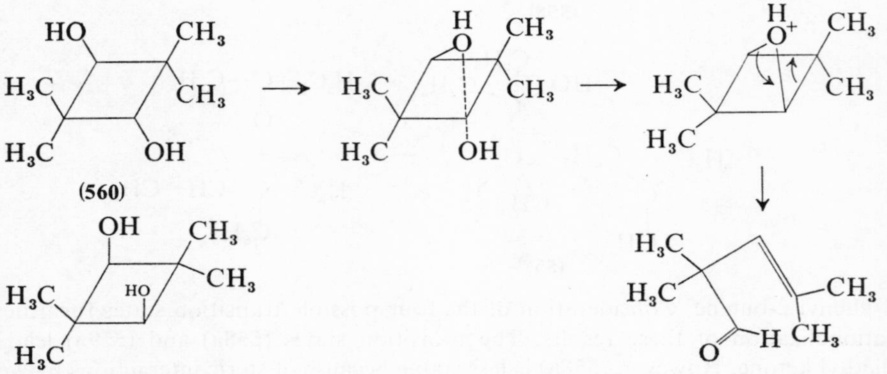

α-Fragmentations

While the fragmentation reactions discussed above are β-fragmentations, α-fragmentations are decomposition reactions in which a–b and X separate from the same carbon atom. A typical reaction that belongs to this class is the decarbonylation of α-keto carboxylic acids[23].

$$C_6H_5-CO-COOH \xrightarrow{H^+} C_6H_5-CO-CO-\overset{+}{O}H_2 \xrightarrow{-H_2O}$$

$$C_6H_5-CO-\overset{+}{CO} \xrightarrow{-CO} C_6H_5-\overset{+}{CO} \xrightarrow{H_2O} C_6H_5-COOH$$

REFERENCES

[1] C. A. Grob, *Experientia*, 13, 126 (1957); *Bull. Soc. Chim. France*, 1960, 1360; C. A. Grob and P. W. Schiess, *Angew. Chem.*, 79, 1 (1967).
[2] H. Höver, *Chem. Ber.*, 100, 456 (1967).
[3] C. A. Grob, F. Ostermayer, and W. Raudenbusch, *Helv. Chim. Acta*, 45, 1672 (1962).
[4] J. Mathieu, A. Allais, and J. Valls, *Angew. Chem.*, 72, 71 (1960).
[5] C. A. Grob and W. Schwarz, *Helv. Chim. Acta*, 47, 1870 (1964).
[6] R. D'Arcy, C. A. Grob, T. Kaffenberger, and V. Krasnobajew, *Helv. Chim. Acta*, 49, 185 (1966).
[7] H. P. Fischer, C. A. Grob, and E. Renk, *Helv. Chim. Acta*, 45, 2539 (1962); C. A. Grob, H. P. Fischer, H. Link, and E. Renk, *ibid.*, 46, 1190 (1963).
[8] E. Müller and B. Narr, *Z. Naturforsch.*, 16b, 845 (1961).

[9] C. A. Grob, H. P. Fischer, W. Raudenbusch, and J. Zergenyi, *Helv. Chim. Acta*, **47**, 1003 (1964).
[10] R. K. Hill, R. T. Conley, and O. T. Chortyk, *J. Amer. Chem. Soc.*, **87**, 5646 (1965).
[11] W. Eisele, C. A. Grob, and E. Renk, *Tetrahedron Letters*, **1963**, 75.
[12] E. C. Taylor and C. W. Jefford, *J. Amer. Chem. Soc.*, **84**, 3744 (1962).
[13] K. D. Berlin, B. S. Rathore, and M. Peterson, *J. Org. Chem.*, **30**, 226 (1965); P. S. Wharton, G. A. Hiegel, and S. Ramaswami, *ibid.*, **29**, 2441 (1964); L. J. Nehmsmann, *Diss. Abstr.*, **23**, 1929 (1962).
[14] F. V. Brutcher, Jr., and H. J. Cenci, *J. Org. Chem.*, **21**, 1543 (1956).
[15] H. H. Wasserman, *Steric Effects in Organic Chemistry*, ed. M. S. Newman, John Wiley and Sons, New York, 1956, p. 378.
[16] S. Searles, Jr., E. K. Ives, and S. Nukina, *J. Org. Chem.*, **24**, 1770 (1959).
[17] R. B. Clayton and H. B. Henbest, *Chem. Ind. (London)*, **1953**, 1315.
[18] U. Burckhardt, C. A. Grob, and H. R. Kiefer, *Helv. Chim. Acta*, **50**, 231 (1967).
[19] H. E. Zimmerman and J. English, Jr., *J. Amer. Chem. Soc.*, **76**, 2285 (1954).
[20] S. A. Ballard, R. T. Holm, and A. N. Williams, *J. Amer. Chem. Soc.*, **72**, 5734 (1950).
[21] H. E. Zimmerman and J. English, Jr., *J. Amer. Chem. Soc.*, **76**, 2291, 2294 (1954).
[22] R. H. Hasek, R. D. Clark, and J. H. Chandel, *J. Org. Chem.*, **26**, 3130 (1961).
[23] W. E. Elliott and D. Ll. Hammick, *J. Chem. Soc.*, **1951**, 3402; K. Bannholzer and H. Schmid, *Helv. Chim. Acta*, **39**, 548 (1956).

6

Aliphatic Substitution

I. S_N1 SUBSTITUTION

NUCLEOPHILIC ALIPHATIC substitutions mostly proceed by one of two mechanisms, the monomolecular S_N1 or the bimolecular S_N2 mechanism[1].

The rate-determining step of the S_N1 reaction is the heterolysis of the C–X bond*:

$$>\!\!C\!\!-\!\!X \longrightarrow >\!\!C^+ \quad X^-$$

X may be, for example, Cl, Br, I, OTs, OH, or $^+NR_3$. In the last case solvolysis leads to a carbonium ion and neutral NR_3.

The heterolysis is assisted by solvent molecules which associate with the leaving group and stabilize the developing ions by solvation. H^+ and Lewis acids act as catalysts by coordinating with the leaving group:

$$C\!-\!\underset{\underset{C}{|}}{\overset{\overset{C}{|}}{C}}\!-\!O\!-\!\overset{O}{\overset{\|}{C}}\!-\!R \xrightarrow{H^+} C\!-\!\underset{\underset{C}{|}}{\overset{\overset{C}{|}}{C}}\!\to\!O\!-\!\overset{OH}{\overset{|}{C^+}}\!-\!R \longrightarrow C\!-\!\underset{\underset{C}{|}}{\overset{\overset{C}{|}}{C^+}} + O\!=\!\overset{OH}{\overset{|}{C}}\!-\!R$$

One would expect the rate of the S_N1 substitution to increase with increasing stability of the carbonium ion and with increasing polarity of the solvent. In fact, the solvolysis of *tert*-butyl chloride is 3×10^4 times faster in H_2O than in ethanol.

If the reacting carbon atom is asymmetric, the S_N1 reaction should lead to a racemic product as it involves a planar carbonium intermediate. However, in many cases the racemic product is accompanied by inverted product. For example, heterolysis of α-methylbenzyl chloride in acetic acid leads to 15% of inverted acetate, although the reaction is unimolecular[2].

$$C_6H_5\!-\!\underset{H}{\overset{CH_3}{\overset{|}{C}}}\!-\!Cl \xrightarrow{CH_3COOH} CH_3\!-\!CO\!-\!O\!-\!\underset{H}{\overset{CH_3}{\overset{|}{C}}}\!-\!C_6H_5$$

15% inverted

*For a recent review, see Nenitzescu[1a].

By relating the rate of the S_N1 reaction to the stability of the carbonium ion to be formed, the implicit assumption is made that with increasing stability of the carbonium ion the energy of the transition state which leads to the carbonium ion also decreases, *i.e.* that the transition state already has some carbonium ion character. The transition state and the carbonium ion may indeed resemble one another rather closely[3]. If there is a large potential energy difference between the starting material and an unstable product such as a carbonium ion, but if the

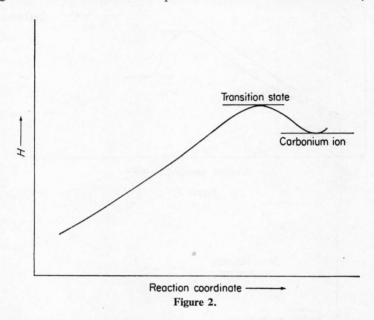

Figure 2.

reaction still proceeds at a measurable rate, it can be assumed that the potential energy of the transition state is not much higher than the energy of the intermediate which follows the transition state. As a result, the conversion of the transition state into the intermediate is accompanied by only little reorganization of the molecule. Consequently the transition state must resemble the unstable intermediate structurally and energetically (Figure 2).

In Figures 2—6, the reaction coordinate is plotted against the reaction enthalpy or potential energy. The enthalpy difference between the starting material and the transition state is the activation enthalpy of the reaction. According to absolute rate theory, the reaction rate depends also on the activation entropy. The relation between reaction rate, activation enthalpy, and activation entropy is given by the Eyring equation:

$$\ln k = \ln \frac{kt}{h} + \frac{\Delta S^{\ddagger}}{R} - \frac{\Delta H^{\ddagger}}{RT}.$$

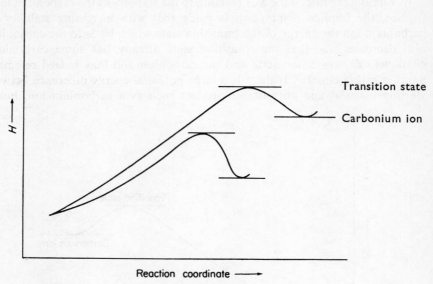

Figure 3.

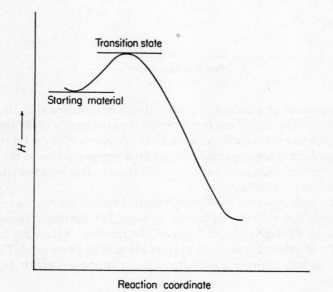

Figure 4.

The activation enthalpy is not exactly equal to the activation energy of the empirical Arrhenius equation E_A, but, as the difference is not significant, E_A is often used instead of the activation enthalpy.

$$\ln k = \ln PZ - (E_A/RT),$$

where P is a probability factor and Z is the number of collisions (per second).

In some energy diagrams the difference between starting materials and transition states is expressed by the free activation energy of the reaction. The free activation energy is related to activation enthalpy and entropy by the Gibbs–Helmholtz equation:

$$\Delta F^\ddagger = \Delta H^\ddagger - T \Delta S^\ddagger.$$

With increasing stability of the carbonium ion formed in the $S_N 1$ reaction, the structural and energetic differences between transition state and carbonium ion increase. As a result the transition state approaches the starting material as one passes along the reaction coordinate (Figure 3). In strongly exothermic (fast) reactions the other extreme exists, that the transition state closely resembles the starting material (Figure 4). If the transition state of an aliphatic substitution resembles the starting material, the reacting bond still possesses sp^3-character; as a consequence, the rearside of the reacting orbital is small and capable of only weak overlap with an approaching nucleophile. Furthermore, as a result of sp^3-hybridization the rearside is sterically shielded against attack of a nucleophile. Thus, if a relatively stable carbonium ion can be formed, the substitution is independent of the nucleophilicity of the medium and an $S_N 1$ mechanism applies.

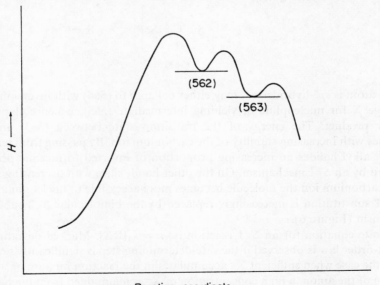

Figure 5.

The rate of an S_N1 reaction depends, not only on the stability of the carbonium ion and on the polarity of the solvent, but also on the nature of the leaving group. The tendency of X to separate from the carbon atom increases with decreasing basicity of X. The rate of the heterolysis thus increases in the following order:

$$-NR_2 < -OH < -OAc < -Cl < -Br < -I < -OSO_2C_6H_4CH_3$$

If the carbonium ion or ion pair is once formed, it is sp^2-hybridized and planar; the originally reacting sp^3-orbital has been converted into a p-orbital perpendicular to the plane of the molecule. While on the side where the leaving group is weak overlap still exists between the p-orbital and the leaving group, the rearside of the p-orbital is now in a position to react with an approaching nucleophile. Species (**562**) is converted into species (**563**) (see Figure 5). Intermediate (**563**), whose

$$-\overset{|}{\underset{|}{C}}-X \longrightarrow \overset{\delta+}{C}\cdots X^{\delta-} \xrightarrow{+S} S^{\delta-}\cdots \overset{\delta\delta+}{C}\cdots X^{\delta-} \xrightarrow{\text{Inv.}} {}^+S-\overset{|}{\underset{|}{C}}-$$

$$\quad\quad (562) \quad\quad\quad (563) \quad\quad\quad\quad\quad (564)$$

$$\downarrow {}^{-X^-}_{+S} \text{ Rac.}$$

$$[S\cdots \overset{|}{C}\cdots S] \longrightarrow {}^+S-\overset{|}{\underset{|}{C}}- \;+\; -\overset{|}{\underset{|}{C}}-S^+$$

$$(565)$$

carbon atom is sp^2-hybridrized, may either collapse to (**564**) with inversion or may exchange X for nucleophile S, yielding intermediate (**565**), which collapses to a racemic product. The energy of the transition state between (**563**) and (**565**) decreases with increasing stability of the carbonium ion. By passing from tertiary to primary alkyl halides an increasing proportion of inverted to racemic product is obtained by an S_N1 mechanism. On the other hand, since with decreasing stability of the carbonium ion the molecule becomes more accessible to nucleophilic attack, the S_N1 substitution is increasingly replaced by the bimolecular S_N2 substitution mechanism (Figure 6).

The rate equation for an S_N1 reaction is: $r = k_1[RX]$. Marked deviation from the first-order law is observed if the rate-determining step is significantly reversible. This is the case when sufficient X^- accumulates in the reaction mixture or when, on addition of the anion, a high concentration of X^- is maintained from the beginning of the reaction. Such a mass law effect is impossible in the irreversible S_N2 reaction.

In solvolyses in which the attacking nucleophile is a solvent molecule it is not possible to conclude from the observed reaction order whether the reaction follows an S_N1 or an S_N2 mechanism, because the large excess of S leads to first-order kinetics even if S is involved in the rate-determining step. Then the mass law effect and the degree of racemization are valuable tools for identification of the reaction mechanism.

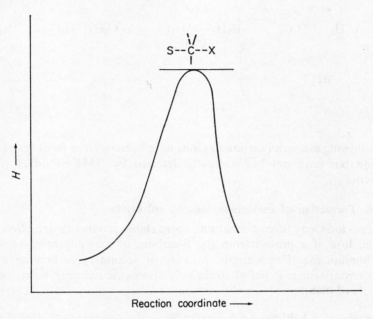

Figure 6.

II. CARBONIUM IONS

In the preceding discussion we have seen that solvolysis in ionizing solvents may lead to carbonium ions. Other methods of generating carbonium ions are the addition of cations to double bonds, the decomposition of diazo compounds by acids, and the reaction of amines with HNO_2.

Carbonium ions are highly reactive, sp^2-hybridized, planar intermediates, whose life-time depends on stabilization by substituents or by neighboring groups and on solvation. Excellent solvating media are strong acid systems such as SbF_5–SO_2, FSO_3H–SbF_5 and HF–SbF_5. These solvents were used by Olah to prepare long-lived, highly solvated carbonium ions and he was thus able to carry out reliable physical measurements on a variety of carbonium ions[4]. The method is of particular interest in the investigation of "non-classical" intermediates (see p. 265).

Aromatic systems such as substituted cyclopropenylium ions **(566)** or the tropylium ion **(567)** are exceptionally stable. The most stable carbonium ion yet known is tripropylcyclopropenylium which can even be recrystallized from water[5].

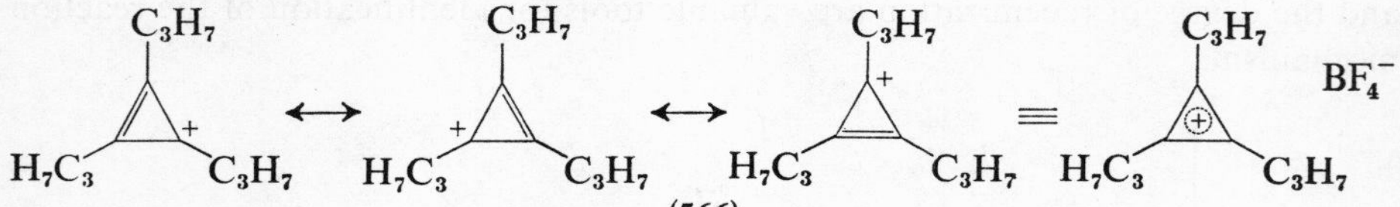

$$\underset{H_7C_3}{\overset{C_3H_7}{\triangle}} \underset{C_3H_7}{^+} \leftrightarrow \underset{H_7C_3}{\overset{C_3H_7}{\triangle}}{^+}\underset{C_3H_7}{} \leftrightarrow \underset{H_7C_3}{\overset{C_3H_7}{\triangle}} \underset{C_3H_7}{^+} \equiv \underset{H_7C_3}{\overset{C_3H_7}{\triangle}}\underset{C_3H_7}{} \quad BF_4^-$$

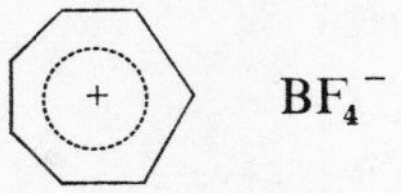

(566)

(567) with BF_4^-

In the following examples carbonium ions have been inferred from the course of the reaction but have not been actually detected by NMR or other physical measurements.

S-1 to S-6. Formation of carbonium ions by solvolysis

Carbonium ions may be converted into more stable products by reaction with a nucleophile, loss of a proton from the β-carbon, or rearrangement to a more stable carbonium ion. For example, primary or secondary carbonium ions rearrange to tertiary ones. Relief of strain may also cause rearrangement, as in the methanolysis of **(568)**.

S-1. Solvolysis of 5,6-dibromo-5,6-dihydrodibenzo[a,e]cyclooctene (568)[6]

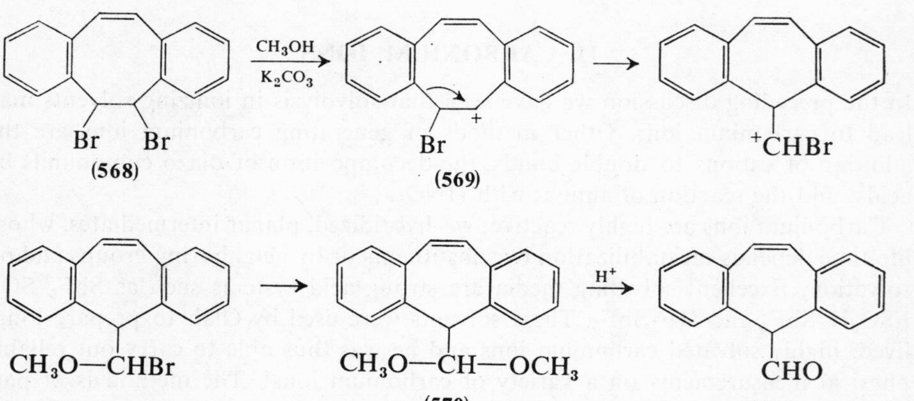

Aliphatic Substitution

Solvolysis of the dibromo-compound (**568**) is facilitated by the formation of a phenyl-conjugated secondary carbonium ion (**569**). The rearrangement involves ring contraction of the strained dibenzocyclooctatriene system to the less strained dibenzocycloheptatriene system. The free aldehyde (**571**) is converted into dibenzocycloheptatriene in a basic medium; a mechanism for this conversion has been proposed by the authors:

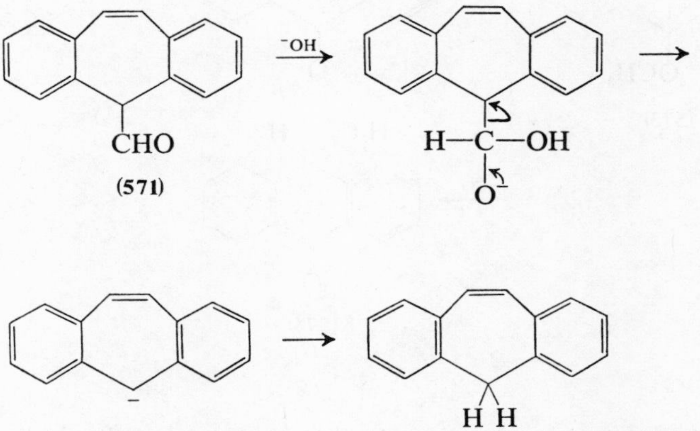

An analogous ring contraction takes place in the dehydrobromination of the monobromide (**572**):

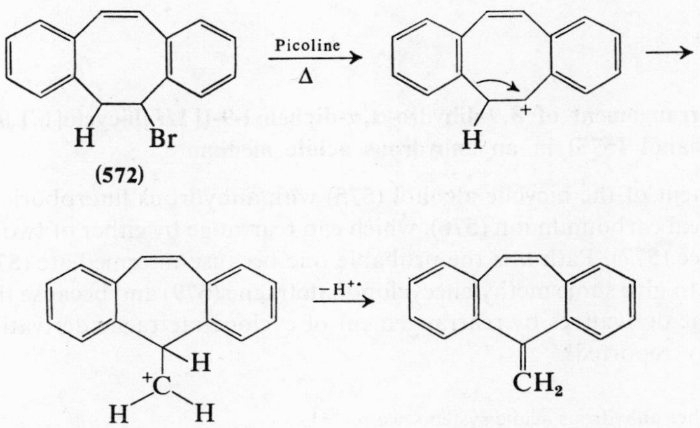

S-2. Rearrangement of 5,11-epoxy-10,11-dihydro-5-methoxy-10-phenyl[5H]-dibenzo[a,d]cyclohepten-10-ol (573)

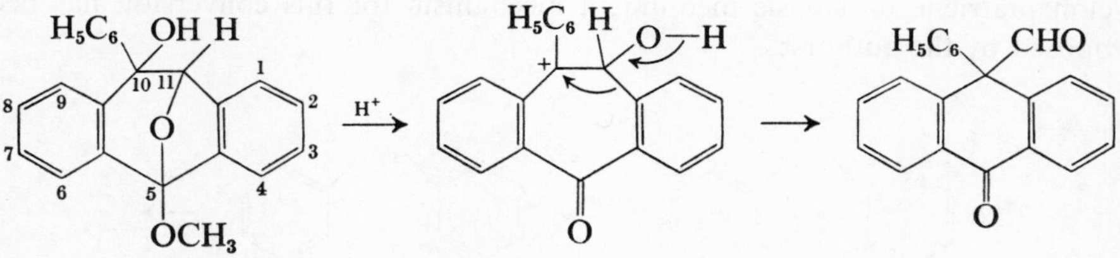

This rearrangement is analogous to that in example S-1, yielding 10-phenylanthrone[7].

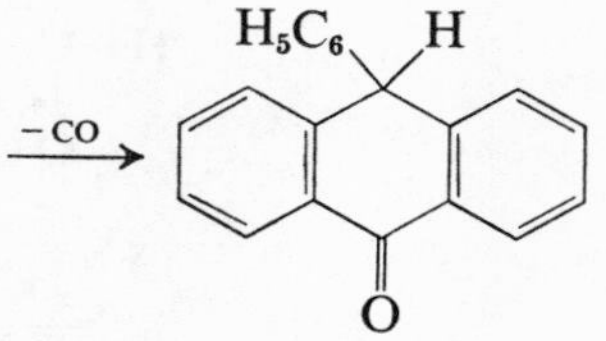

In (573) the most sensitive bonds towards acids are the acetal bonds at C-5. Hydrolysis leads to a product with a C-5 keto group and a C-10 OH group. Abstraction of the OH group from C-10 yields a more stable carbonium ion than abstraction of OH from C-11. Thus, the intermediate readily rearranges to (574), the driving force being formation of a strainless six-membered ring from a seven-membered ring.

S-3. Rearrangement of 8,9-dihydro-α,α-diphenyl-9-([1H]bicyclo[6.1.0]nonene)-methanol (575) in an anhydrous acidic medium[8]

Treatment of the bicyclic alcohol (575) with anhydrous fluoroboric acid* yields the classical carbonium ion (576), which can rearrange by either of two mechanisms to produce (577). Path *a* is the probable one because intermediate (578) would be expected to give some methylenecyclononatetraene (579) and because the formation of benzene derivatives by rearrangement of cyclooctatetraene derivatives has been repeatedly reported[9].

---
* For other anhydrous acidic systems see p. 241.

Aliphatic Substitution

S-4. S_N1 reactions of allyl halides

Allyl halides readily equilibrate in polar solvents with their isomers *via* a tight ion-pair as intermediate. Since the halide ion migrates from carbon 1 to carbon

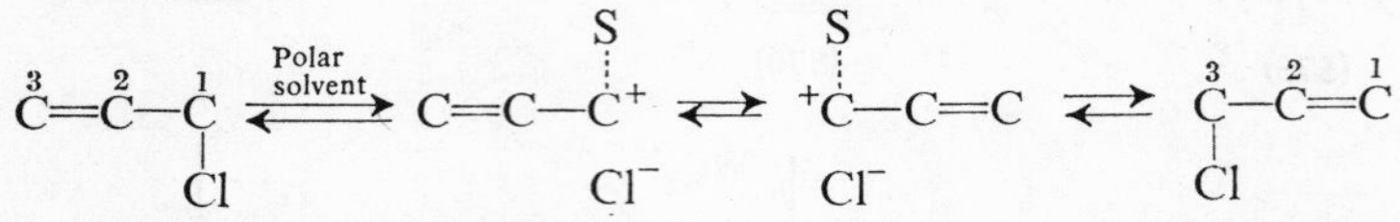

$$\overset{3}{C}=\overset{2}{C}-\underset{Cl}{\overset{1}{C}} \xrightleftharpoons[]{\text{Polar solvent}} C=C-\overset{S}{\overset{\vdots}{C^+}} \ \ Cl^- \rightleftharpoons \ \ ^+\overset{S}{\overset{\vdots}{C}}-C=C \ \ Cl^- \rightleftharpoons \underset{Cl}{\overset{3}{C}}-\overset{2}{C}=\overset{1}{C}$$

3 without equilibrating with the solvent, the rearrangement is called "internal return".

If other strong nucleophiles are available, the carbonium ion may react with them instead of with the halide ion. For example, in the solvolysis[10] of 3,4-dichloro-1,2,3,4-tetramethyl-1-cyclobutene (580) in tetrahydrofuran in the presence of H_2O_2, carbonium ion (581) reacts with H_2O_2, yielding the hydroperoxide (582).

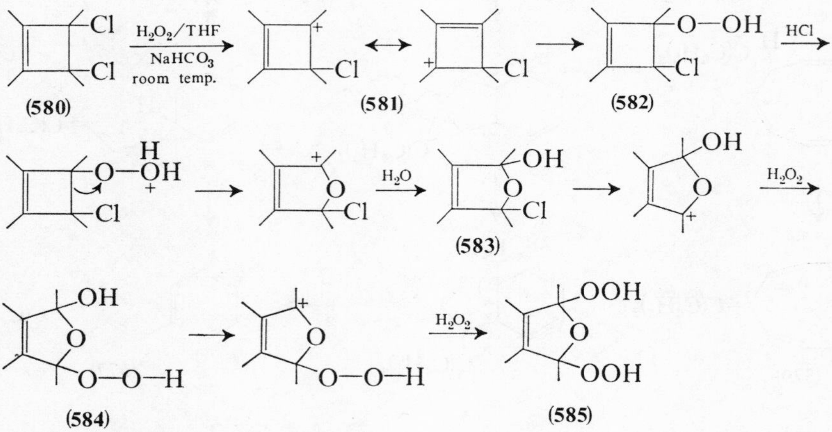

Bimolecular substitution, by either the S_N2 or the S_N2' mechanism, would also lead to (582). However, normal S_N2 substitution has never been observed in allylic systems, whereas S_N2' substitutions have been reported in only very few substitutions of cyclic compounds when the nucleophile attacks the ring at the side of the leaving group[11].

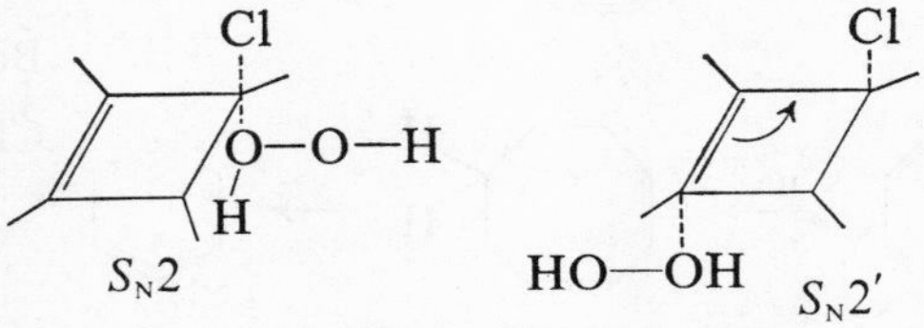

S_N1 and S_N2 substitutions of cyclopropyl and cyclobutyl compounds are slower than those of open-chain compounds because sp^2-hybridization of the carbonium carbon in the S_N1 displacement or sp^2-hybridization in the transition state of the S_N2 substitution causes considerable bond-angle expansion. The angle strain, which is already present in the ground state, is thus increased in the transition state. The formation of (**581**) by solvolysis of (**580**) has been demonstrated by several investigators[12]. In (**582**) heterolysis of the O–O bond takes place, followed by Criegee rearrangement (see p. 150). In general, this type of rearrangement takes place only if the hydroperoxide is esterified. The rate of rearrangement increases with increasing acidity of the acid used for esterification. In our example, hydrochloric acid, which is only partly neutralized by $NaHCO_3$, catalyses the rearrangement of (**582**). The driving force of the reaction is the release of ring strain. Displacement of Cl in (**583**) and of OH in (**584**) leads to the product (**585**).

In an acidic medium, H_2O_2 may react with C=C double bonds*; for example, it adds to the double bonds of 2-methyl-2-butene[17] and enol ethers[18]. However, the double bond of (**580**) is stable under the reaction conditions.

S-5. Formation and hydrolysis of acetals

The hydrolysis of acetals to hemiacetals, as well as the reverse reaction, are specifically acid-catalysed reactions†. The formation and hydrolysis of hemiacetals, which take place also in neutral media[19], are accelerated by general acid and base catalysis.

$$R-\underset{OR}{\overset{OR}{C}}-H + H^+ \underset{\text{Fast}}{\overset{\text{Fast}}{\rightleftharpoons}} R-\underset{OR}{\overset{\overset{+}{OR}H}{C}}-H \underset{+HOR \atop \text{Fast}}{\overset{-HOR \atop \text{Slow}}{\rightleftharpoons}} R-\underset{OR}{\overset{+}{C}}-H \underset{-H_2O \atop \text{Slow}}{\overset{+H_2O \atop \text{Fast}}{\rightleftharpoons}}$$

(**586**)   (**587**)

$$R-\underset{OR}{\overset{\overset{+}{OH_2}}{C}}-H \underset{+H^+ \atop \text{Fast}}{\overset{-H^+ \atop \text{Fast}}{\rightleftharpoons}} R-\underset{OR}{\overset{OH}{C}}-H \underset{\text{Fast}}{\overset{\text{Fast}}{\rightleftharpoons}} R-\underset{O}{\overset{H}{C}} + HOR$$

(**588**)

The slowest step in the hydrolysis of acetals is the unimolecular conversion of the conjugate acid (**586**) into the carbonium ion (**587**). Conversely, in formation of acetals, the transformation of the hemiacetal (**588**) into the ion (**587**) is the slowest

---

* In an acidic medium hydrogen peroxide reacts as ^+H_2OOH or OH^+, e.g. in the oxidation of sulfides or addition to the furan ring system[13]. In weakly acidic media it reacts with alcohols, alkyl sulfates, or tertiary halides in an S_N1 reaction, forming alkyl hydroperoxides[14]. In basic media[15] it affords the powerful nucleophile HOO^-. In the presence of salts such as $Fe^{2+}X^{2-}$ or on irradiation it provides OH radicals (Fenton's reagent)[16].

† A specifically acid-catalyzed reaction is catalyzed only by protons, whereas in general acid catalysis any type of acid acts catalytically.

step. The reactions (**586**) ⇌ (**588**) are S_N1 reactions. The rate of acetal hydrolysis depends on the concentration of the acetal and on the H⁺ concentration:

$$-dA/dt = k[\text{Acetal}][\text{H}^+]$$

Since the protons act as catalysts, their concentration remaining constant, a reaction of apparent first order is observed. A reaction, which depends on the concentration of two components, one of which does not change its concentration, is called a pseudo-first-order reaction.* Thus k in the equation above can be replaced by $k' = k[\text{H}^+]$. If the hydrolysis of the acetal is carried out at a different pH, the rate constant k' changes.

The same applies to other acid- or base-catalyzed reactions.

Thiols are, in general, more reactive than alcohols, and the products are more stable. For example, thiohemiacetals can be isolated in some instances, in contrast to hemiacetals.

Reaction of (n-butylthio)acetaldehyde dimethyl acetal with 1,2-ethanedithiol[20]

If the acetal (**589**) is treated with butane-1-thiol, the dithioacetal (**595**) is obtained[21]. However, attempts to prepare the cyclic dithioacetal (**592**) from the

$$C_4H_9S\text{—}CH_2\text{—}CH(OCH_3)_2 + 2\,C_4H_9SH \xrightarrow{H^+}$$
(**589**)

$$C_4H_9S\text{—}CH_2\text{—}CH(SC_4H_9)_2 + 2\,CH_3OH$$
(**595**)

dithiol (**590**) and the acetal (**589**) failed. Instead of (**592**), the dithian (**594**) was obtained in 82% yield. In the intermediate (**591**) the sulfur atom in the β-position to the methoxyl group probably participates in the separation of the latter group; the resulting episulfonium salt (**593**) rearranges to the product.

---

* See page 241.

Aliphatic Substitution

S-6. Rearrangements of 3,4-epoxy-2,2,4-trimethylpentyl isobutyrate in an acidic medium[22]

The products to be expected on acid-catalyzed ring-opening of the epoxide **(596)** are the esters **(597)** and **(598)**. The isomeric esters **(599)** and **(598)** are obtained together; they give the same alcohol **(598a)** on hydrolysis. It is probable that **(598)** is the source of **(599)** which in turn affords the furan derivative **(600)**.

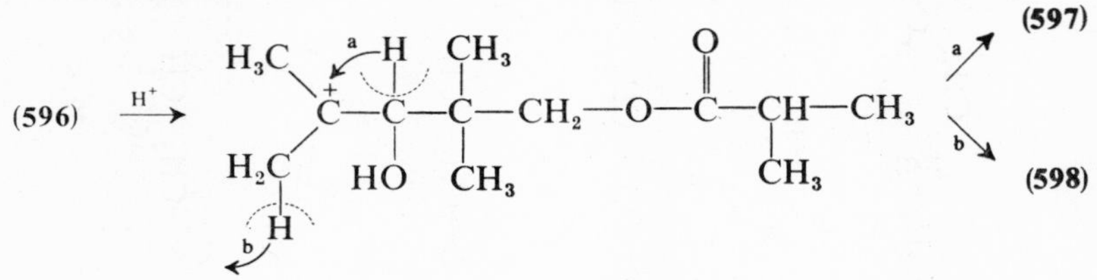

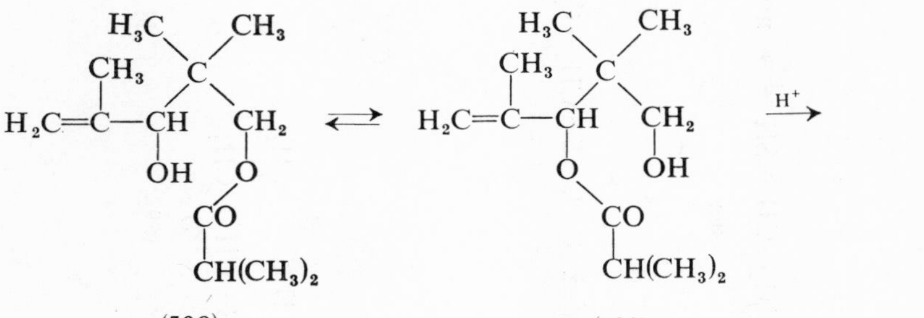

An equilibrium analogous to **(598)** ⇌ **(599)** that occurs under comparable conditions has been found between the γ-hydroxy esters **(601)** and **(602)**.

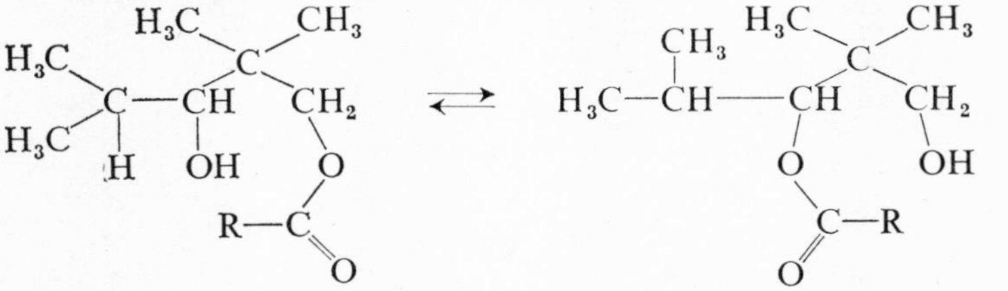

Aliphatic Substitution

251

(page consists of chemical structure diagrams labeled (596), (597), (597a), (598), (598a), (599), (600), (600a) with saponification reactions and an HCl/Ether reaction)

Unsaturated alcohols readily cyclize to tetrahydrofuran derivatives in a slightly warm acidic medium[23]. Concerted mechanisms have been proposed for formation of the products. The products (**597**), (**598**), and (**599**) are assumed to be formed by rearrangements of the chair conformation (**603**), whereas (**600**) could arise from the sterically favored boat conformation (**604**).

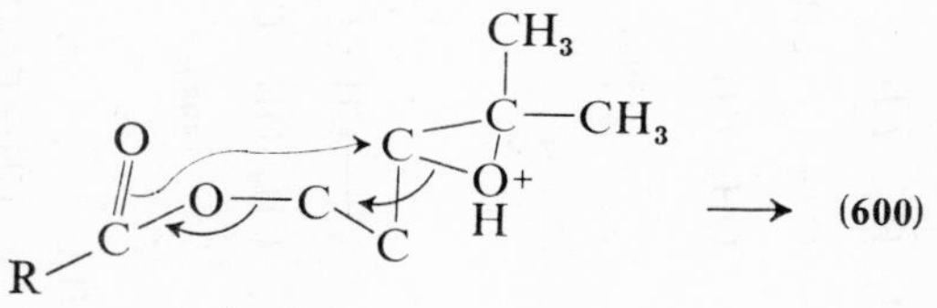

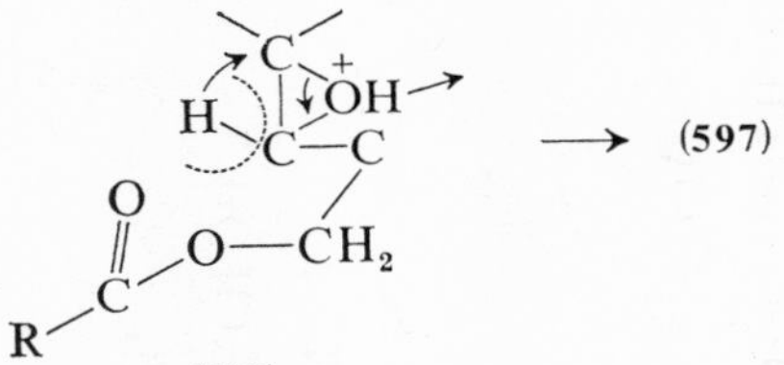

The δ-epoxy ester (**605**) yields a chlorohydrin (**606**) and an aldehyde (**607**) on treatment with HCl. If the ester is treated with BF_3 in dioxan, the tetrahydrofuran derivative (**608**) and a compound (**609**) are obtained[24] in addition to (**607**).

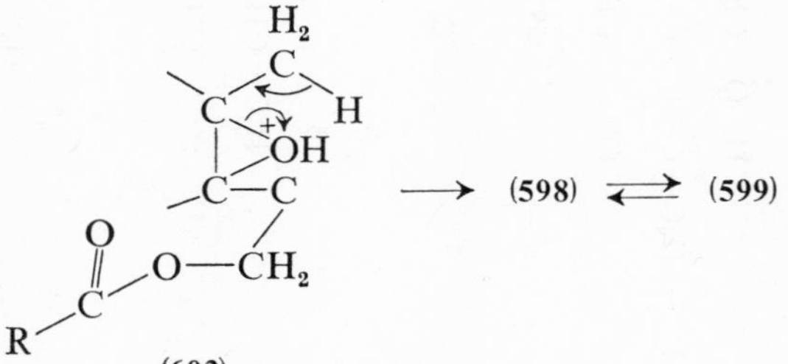

The furan derivative (**608**) is probably also formed by a synchronous mechanism through conformation (**610**). By cross-over experiments it has been demonstrated that these reactions are in fact intramolecular rearrangements.

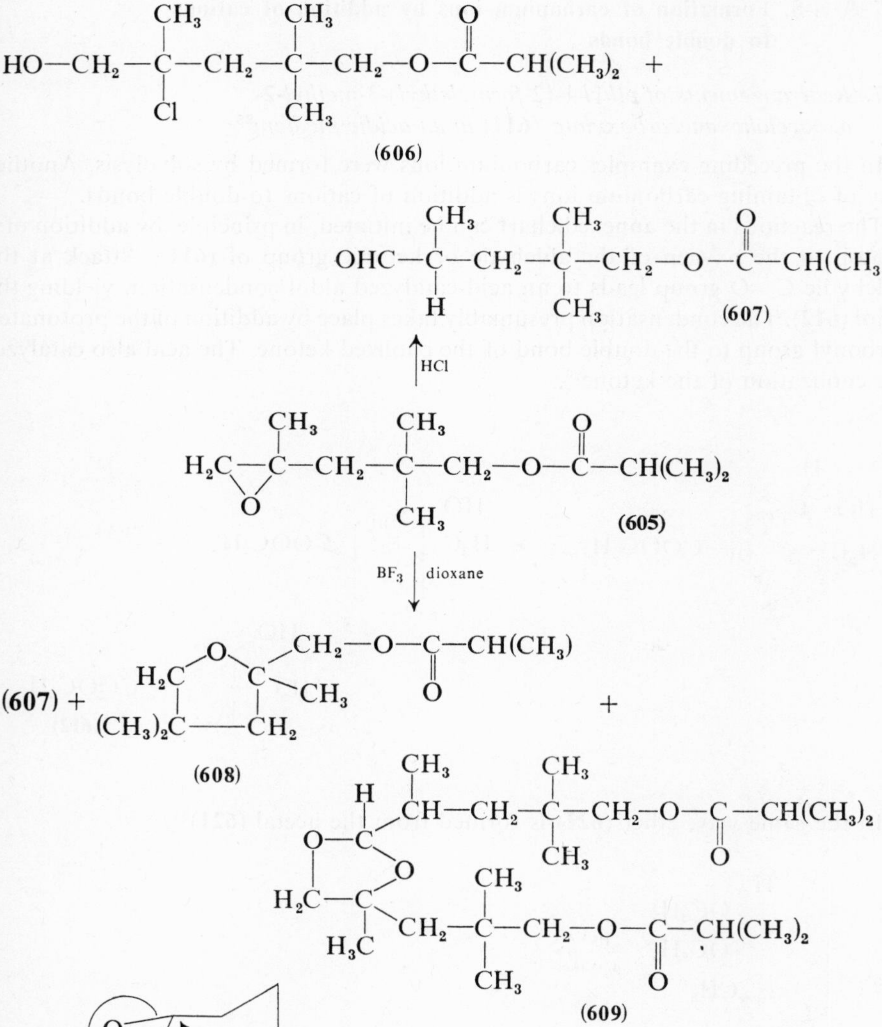

(606) (607) (605) (608) (609)

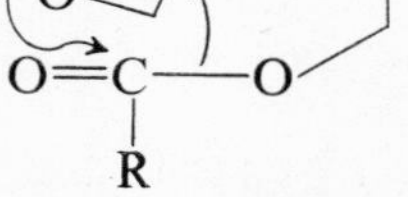

(610)

S-7 & S-8. Formation of carbonium ions by addition of cations to double bonds

S-7. *Rearrangements of ethyl 1-(2-formylethyl)-3-methyl-2-oxocyclohexanecarboxylate* **(611)** *in an acidic medium*[25]

In the preceding examples carbonium ions were formed by solvolysis. Another way of obtaining carbonium ions is addition of cations to double bonds.

The reactions in the annexed chart can be initiated, in principle, by addition of a proton to the oxygen of the aldehyde or ketonic group of **(611)**. Attack at the aldehydic C=O group leads to an acid-catalyzed aldol condensation, yielding the aldol **(612)**. The condensation presumably takes place by addition of the protonated carbonyl group to the double bond of the enolized ketone. The acid also catalyzes the enolization of the ketone[26].

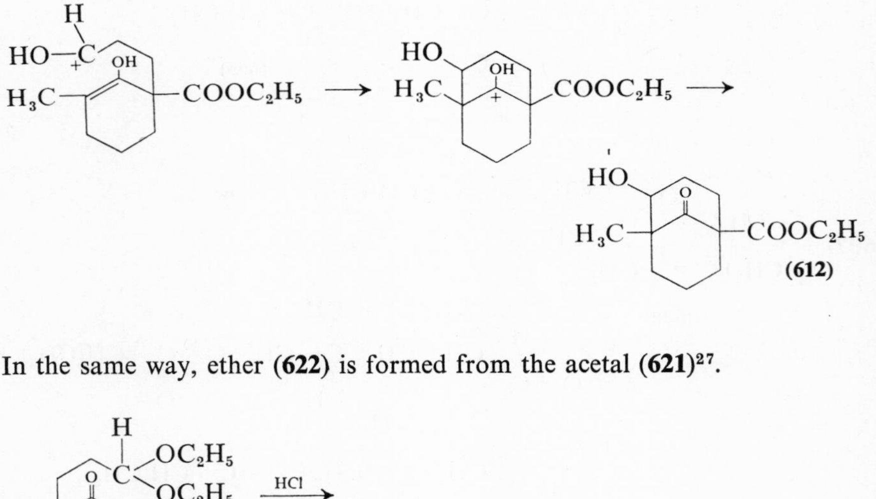

In the same way, ether **(622)** is formed from the acetal **(621)**[27].

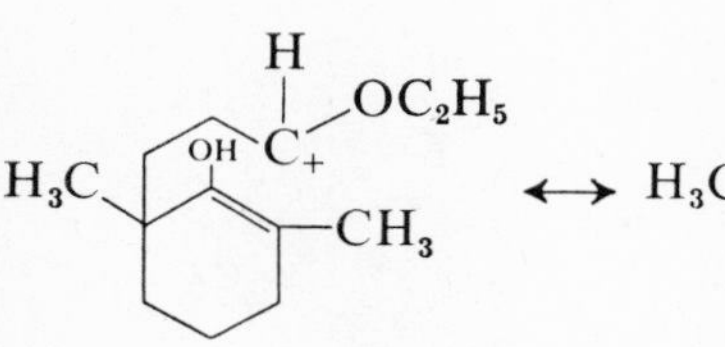

Aliphatic Substitution

Then the aldol **(612)** is converted by loss of water into **(614)** which can isomerize to **(615)**. The reaction furnishes a mixture of **(614)** and **(615)** in the ratio 2.8:1. If the methyl group is replaced by H, the ratio is 1.2:1[28].

The dehydration of **(612)** occurs by way of a carbonium ion **(613)** (*E*1 elimination) which either loses a proton to yield **(614)** and **(615)** or undergoes a Wagner–Meerwein rearrangement to compound **(617)**. Retroaldol condensation of **(617)** yields an intermediate **(618)**, which by aldol condensation forms the bicyclic compound **(619)**. In the final step, product **(620)** is formed from **(619)** by dehydration.

Characteristic absorption of **(620)**:
  IR, 1728 cm^{-1} (C=O of COOR); UV, 252 mμ (ε = 12000) (C=C—C=O)
  1667   (C=O, unsat.)

If the keto ester **(614)** is protonated at the keto group, it is converted by two-fold Wagner–Meerwein rearrangement into a product **(616)**. In an analogous reaction[29] tetralin derivative **(622b)** is obtained from compound **(622a)**.

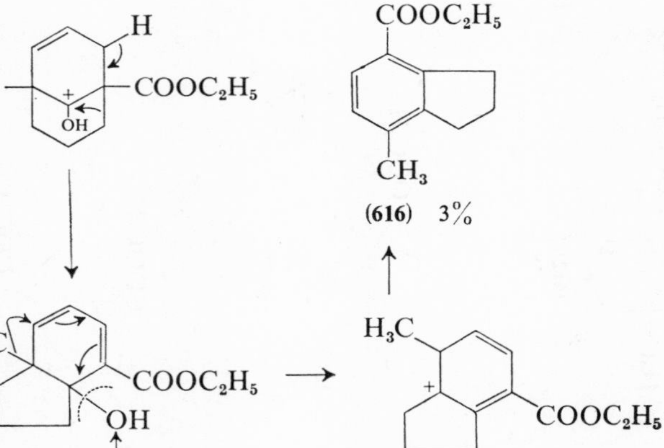

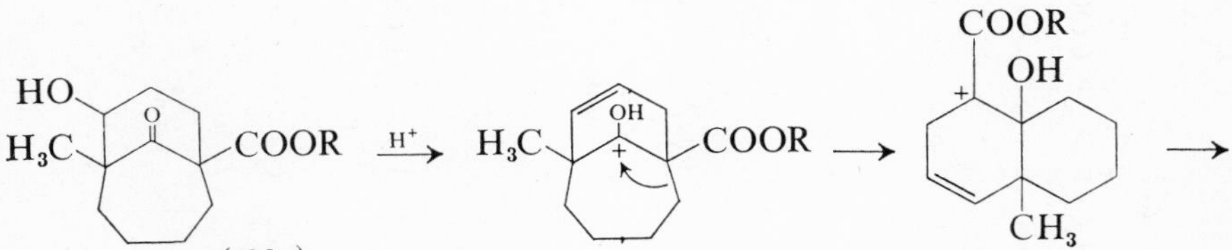

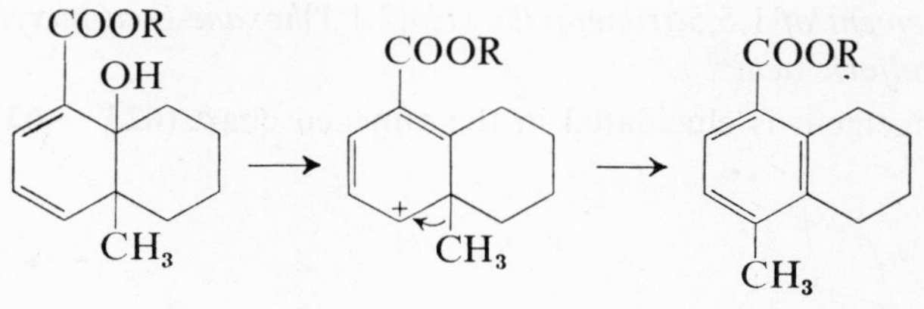

Aliphatic Substitution

[Structures showing conversion to **(622b)**]

The product **(616)** could in principle also be formed by protonation of the keto-carbonyl group of **(611)**.

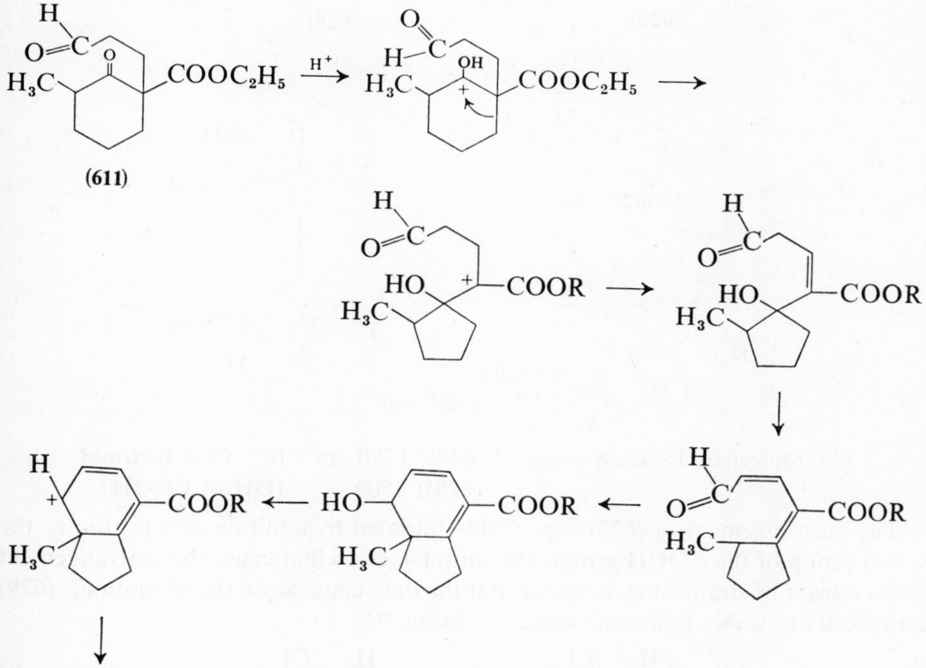

[Reaction scheme from **(611)** through intermediates to **(616)**]

The products **(620)** and **(616)** are obtained in the same ratio if the bicyclic compound **(612)** is first isolated and then treated with sulfuric acid.

S-8. *Rearrangement of* 1,5,5-*trimethylbicyclo*[2.1.1]*hexane*-6-*carboxylic acid* (**623**) *in sulfuric acid*[30]

This rearrangement is elucidated in the annexed chart (**623** → **627** and **628**).

(623) (624) (625) (626) (627) 50% (628) 20%

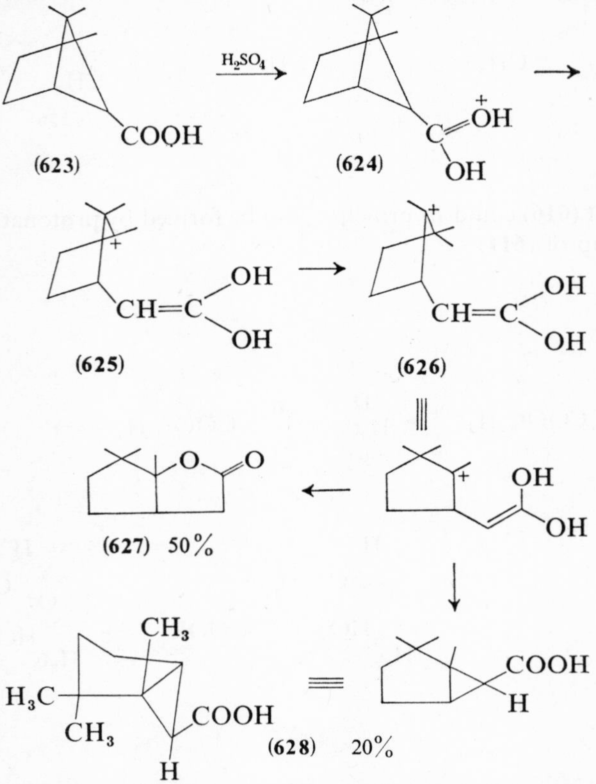

Characteristic IR absorption of (**627**): 1770 cm^{-1} (C=O of lactone)
(**628**): 3500 (OH of COOH)

The rearrangement of (**623**) is probably initiated by addition of a proton to the C=O group of the COOH group. One of the factors that cause the rearrangement is the release of strain; it is, however, not the only cause since the compounds (**629**) and (**630**) are stable under the same conditions[31].

(629) (630)

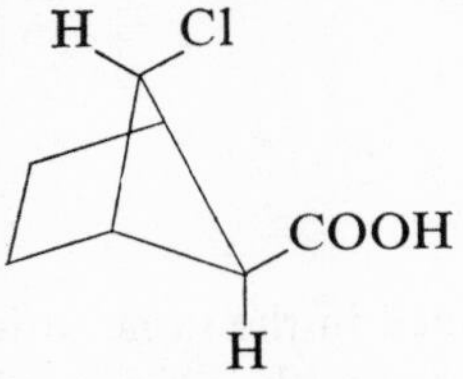

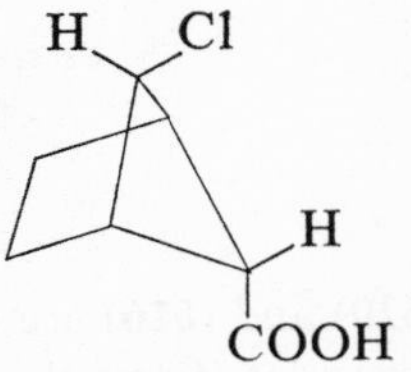

The decisive factor is probably conversion of the ion (**624**) into a tertiary carbonium ion (**625**), which then rearranges to (**626**), which cyclizes to product (**627**) and (**628**). The acid (**628**) probably exists in the sterically favorable *exo*-form.

Treatment of the starting material (**623**) with sulfuric acid could in principle lead to decarboxylation. The cationic intermediate (**631**) could rearrange to several cations, which by recarbonylation would yield carboxylic acids.

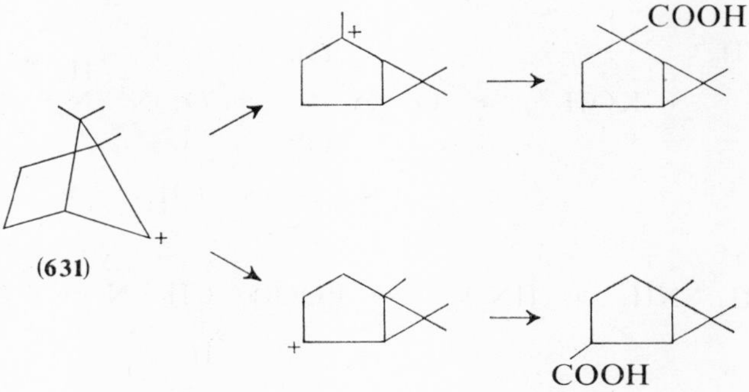

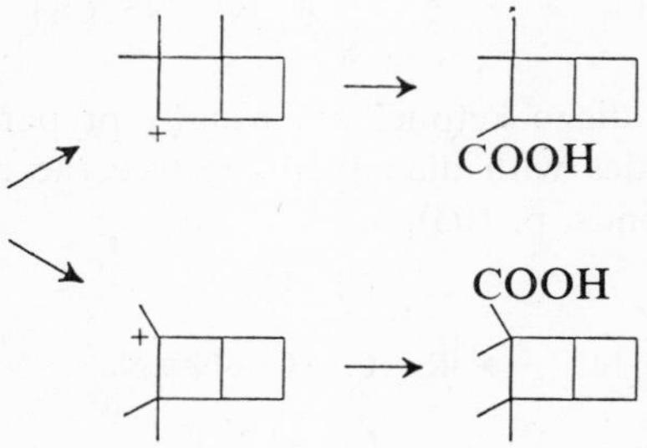

A de- and re-carbonylation has recently been described[32], namely:

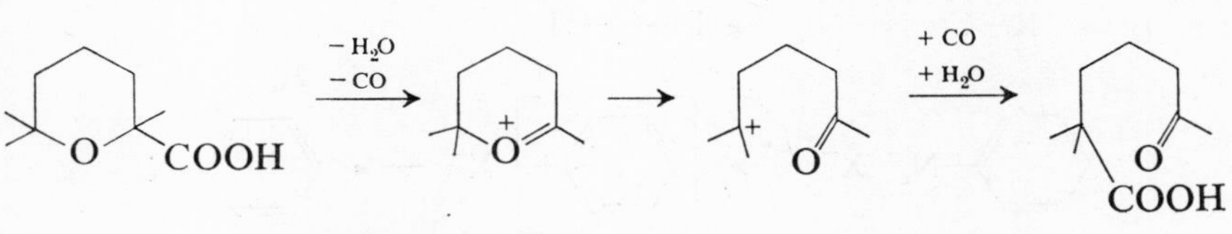

S-9 & S-10. Formation of carbonium ions from aliphatic diazo compounds

Simple aliphatic diazo compounds are rather unstable. They are prepared by reaction of *N*-alkyl-*N*-nitrosourethane, *N*-alkyl-*N*-nitrosourea, or similar substances with base or sometimes by reaction of primary amines with HNO_2. The latter method is only successful in some special cases of relatively stable diazo compounds, for example, diazoacetic esters. (However, the last method is very important in the preparation of aromatic diazonium salts.)

$$O=C(OR)(N(NO)CH_3) + KOH \rightarrow O=C(OR)(OK) + O=N-N(H)-CH_3 \rightarrow$$

$$HO-N=N-CH_3 \rightarrow \ddot{N}-\ddot{N}=CH_2$$

$$RO-\underset{O}{\underset{\|}{C}}-CH_2-NH_2 + HNO_2 \rightarrow ROOC-\underset{H}{\underset{|}{CH}}-N=N-OH \rightarrow$$

$$ROOC-CH=\ddot{N}-\ddot{N}$$

$$\text{Ph}-NH_2 + HNO_2 \rightarrow \text{Ph}-\ddot{N}=\ddot{N}-OH \rightarrow \text{Ph}-N_2^+ \; X^-$$

The relatively stable diazo ketones are usually prepared by the Arndt–Eistert reaction of acid chlorides with diazomethane (see the reaction of ketones with aliphatic diazo compounds, p. 103).

$$R-CO-Cl + CH_2N_2 \rightarrow R-\underset{O^-}{\underset{|}{C}}(Cl)-\underset{H}{\underset{|}{C}}(H)-\overset{+}{N}\equiv N: \rightarrow R-\underset{O}{\underset{\|}{C}}-CH=\ddot{N}-\ddot{N}$$

Thermal decomposition of aliphatic diazo compounds leads to carbenes. The aromatic diazonium salts, which decompose slowly already at room temperature, lose nitrogen rapidly when heated, forming an extremely reactive benzene carbonium ion, which immediately combines with an anion.

$$R-\underset{H}{\underset{|}{C}}=\ddot{N}-\ddot{N} \xrightarrow{\Delta} R-\ddot{C}-H$$

$$\text{Ph}-N_2^+ \; X^- \xrightarrow{\Delta} [\text{Ph}^+] \; X^- \rightarrow \text{Ph}-X$$

Aliphatic Substitution 261

In acidic media, aliphatic diazo compounds produce carbonium ions by addition of a proton to the α-carbon, accompanied by loss of nitrogen.

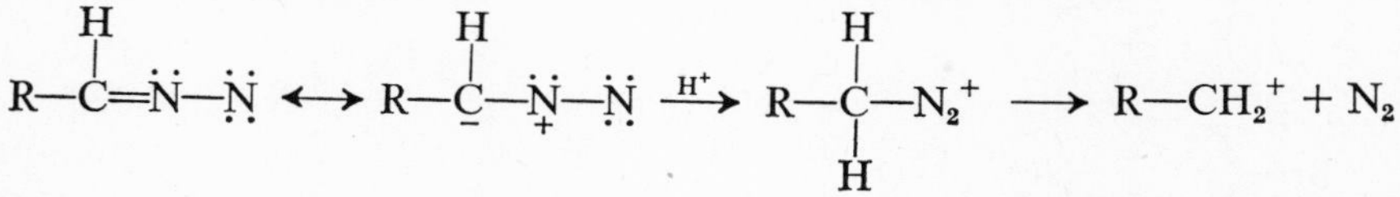

$$R-\overset{H}{C}=\overset{..}{N}-\overset{..}{N} \longleftrightarrow R-\overset{H}{\underset{-}{C}}-\overset{..}{\underset{+}{N}}-\overset{..}{N} \xrightarrow{H^+} R-\overset{H}{\underset{H}{C}}-N_2^+ \longrightarrow R-CH_2^+ + N_2$$

Carbonium ions are formed directly on reaction of primary aliphatic amines with HNO_2.

$$R-NH_2 + HNO_2 \longrightarrow R-N=N-OH \xrightarrow{H^+} R^+ + N_2 + H_2O$$

The methods of generating carbonium ions by treatment of primary aliphatic amines with HNO_2 or decomposition of diazo compounds by acid are very often used in investigations of carbonium ion reactions.

An example is the acid-catalyzed decomposition of the diazo ketone (**632**).

S-9. *Acid-catalyzed decomposition of* (o-*nitrobenzoyl*)*diazomethane* (**632**)[33]

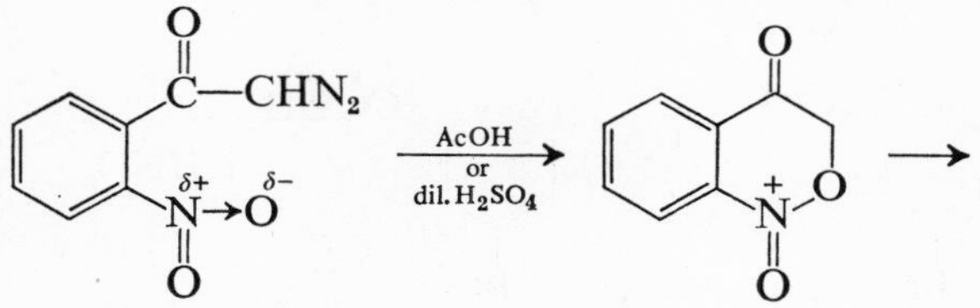

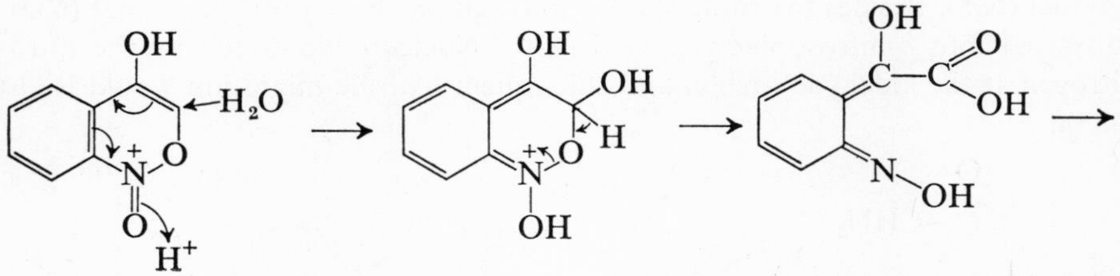

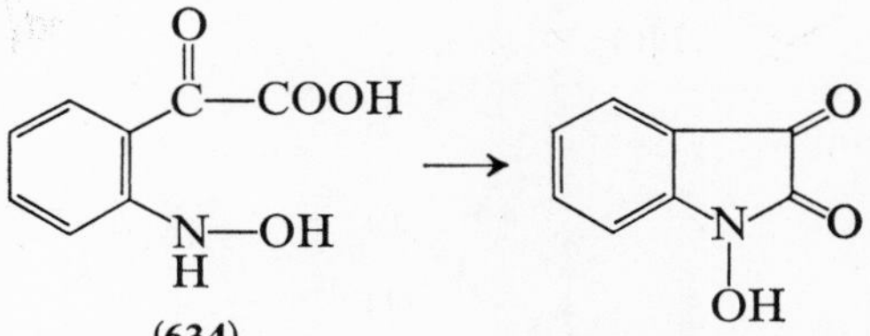

Intramolecular addition of the carbonium carbon to the *o*-nitro group yields intermediate (**633**). Numerous reactions between an *ortho*-substituent and a nitro group in benzene derivatives have been described[34]. A well-known example is the von Richter rearrangement:

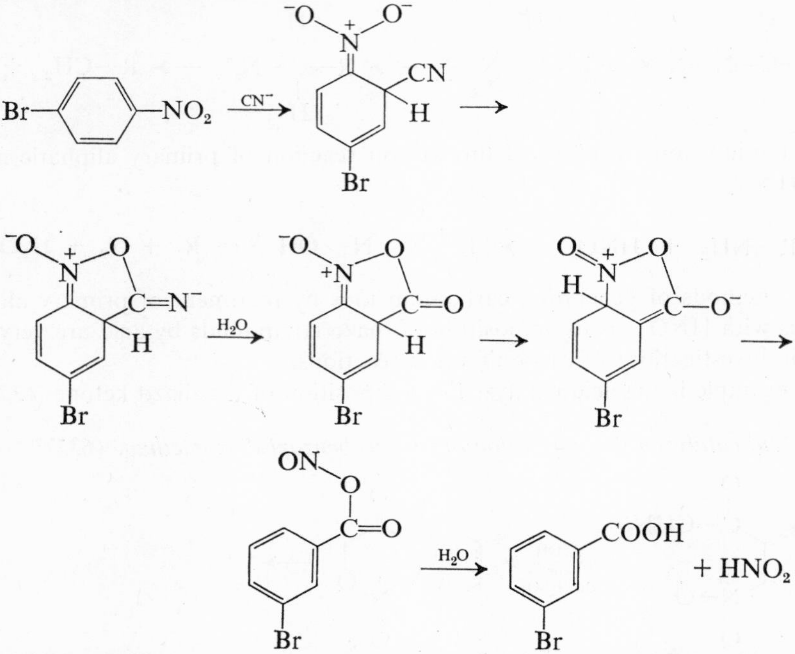

No agreement exists about the conversion of the intermediate (**633**) into the product (**635**). Besides the route shown above, it has been proposed[35] that (**633**) is converted into *o*-nitrosophenylglyoxal (**636**). Nucleophilic attack by the nitroso nitrogen at the aldehyde carbon and subsequent hydride migration would lead to (**635**).

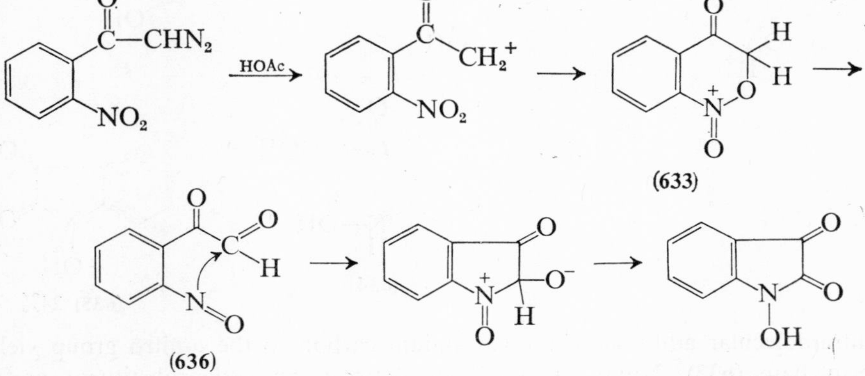

(633) (636) (635)

However, if compound (**636**) is generated by independent methods, the product (**635**) is not obtained. Further, the postulated reaction steps are not very likely. A reasonable alternative[33] is the addition of H_2O or CH_3COOH to the enol-form of (**633**). Ring cleavage leads to o-(hydroxyamino)phenylglyoxylic acid (**634**), whose spontaneous cyclization is known[36]. This mechanism also explains why the homologous compound (**637**) does not give the corresponding compound (**639**), since (**638**) can not rearrange analogously to (**633**).

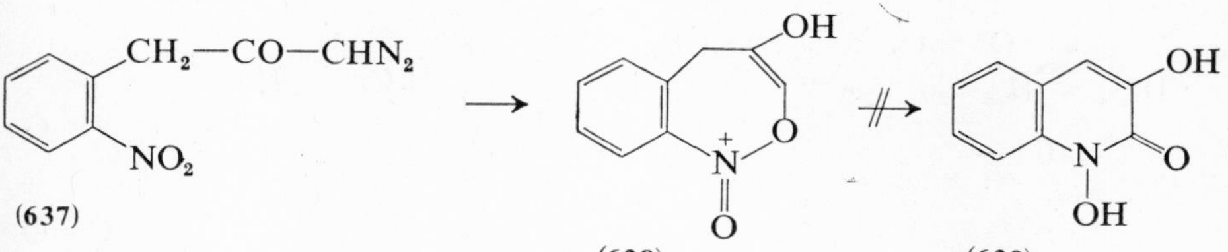

The possibility that Wolff rearrangement first takes place, yielding the ketene (**640**), was excluded by tracer studies. The reaction of labeled (**632**) showed that the hydroxyisatin (**635**) does not contain the labeled carbon in α-position, as would be expected from a Wolff rearrangement, but in position β to the phenyl ring.

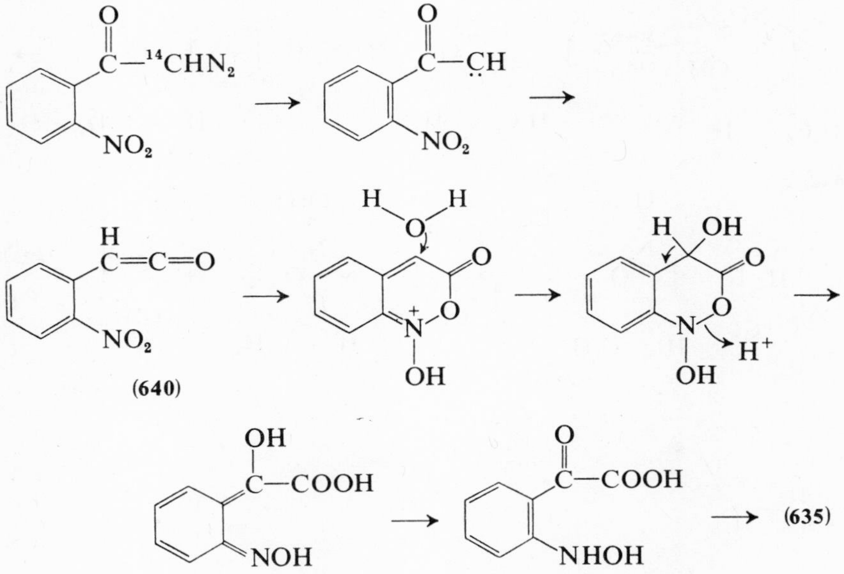

S-10. Reaction of o-nitrobenzhydrol (642) with thionyl chloride[37]

An intermediate similar to (633) is probably formed in the solvolysis of (641) and in the reaction of compound (642) with thionyl chloride.

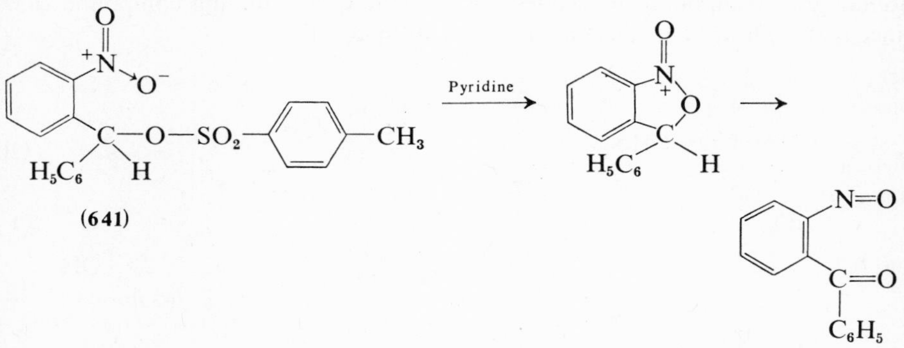

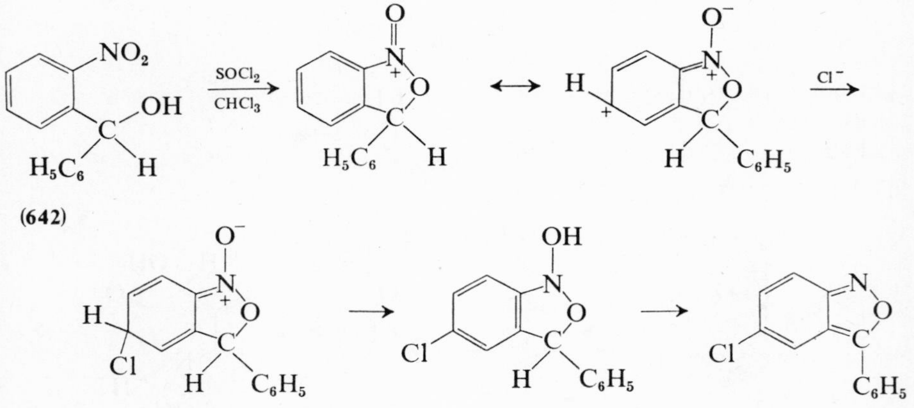

III. PARTICIPATION OF NEIGHBORING CARBON IN CARBONIUM ION REACTIONS

The cyclopropylcarbinyl system

An allylic carbonium ion is more stable than an alkyl carbonium ion. The transition state which leads to the allylic carbonium ion differs only little from the carbonium ion itself*; consequently the former should have less energy than the transition state that leads to the alkyl carbonium ion, and solvolysis of an allylic halide should be faster than that of an alkyl halide.

In the allylic carbonium ion the π-electrons are distributed over three $2p$-orbitals. A necessary precondition for electron delocalization is the planar arrangement of the $2p$-atomic orbitals (**644**).

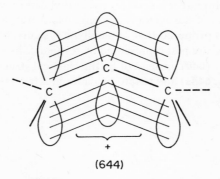

(644)

In halide (**645**), or in the 5-cholesten-3-yl system (**646**), the carbon atom that forms the positive center on solvolysis is separated from the double bond by a methylene group. As a homolog of allylic derivatives such a system is called a homoallylic system. As for allylic systems, so in the transition state which leads to

(645)   (646)

a homoallylic cation overlap of the developing p-orbital at the positive center with the p-orbitals of the double bond takes place. Since the pure π-electron delocalization rapidly decreases with increasing distance between the carbon atoms which are involved (**647**), better overlap is obtained if the $2p$-orbitals are arranged parallel to the 1,3-axis. This arrangement of overlapping orbitals greatly resembles the

---
* See Hammond's principle (page 237) and reference 38.

electron distribution of a σ-bond. Overlap is inversely proportional to the $C_1C_2C_3$ bond angle (**648**)[39]. The intermediate that arises from the transition state is a cyclopropylmethyl cation. The cyclopropyl ring strongly stabilizes the adjacent positive

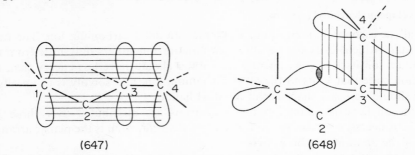

(647)   (648)

charge. The precise nature of the cyclopropylmethyl cation, which should account for the stabilizing effect of the cyclopropyl ring, has therefore evoked much recent interest. Winstein[39] has proposed that homoallylic systems, such as the 5-cholesten-3-yl and similar cations, are best represented by resonance hybrid (**648a** and **b**) or by the equivalent structure (**649**), instead of by the classical structure (**650**). Resonance between the structures (**648a**) and (**648b**) provides greater stability than attaches to (**650**).

A cation such as (**649**) exhibits typical features, which distinguish it from a classical carbonium ion. The electron-delocalization takes place between non-contiguous carbon atoms whose separation exceeds the length of a normal C–C bonds. Two delocalized electrons participate simultaneously in two σ-bonds or in one σ- and one π-bond. Cations of this type have been designated non-classical carbonium ions[40]. According to Bartlett[41], non-classical carbonium ions are intermediates with delocalized σ-electrons in the ground state.

Non-classical carbonium ions have been compared with the electron-deficient dimeric aluminum alkyls. The quantum-mechanical treatment of these "three-centre bonds" leads to one bonding and two antibonding orbitals. When the bonding orbital is filled with two electrons, the energy of the molecule is lower than that of a structure with normal bonds and an empty $2p$-orbital. Existence of a non-classical radical with one additional electron or of a non-classical anion with two additional electrons is not very likely since the additional electrons would have to be placed in an antibonding orbital[42].

In the reaction of cyclopropylmethylamine (**651**) with HNO_2, which should lead to the unsubstituted cyclopropylmethyl cation, 48% of cyclopropanemethanol (**652**), 47% of cyclobutanol (**653**), and 5% of 3-buten-1-ol (**654**) are obtained. The same product distribution is observed if cyclobutaneamine is treated with HNO_2. Labeling the α-carbon resulted in distribution of the label over the whole molecule[43].

▷—CH₂—NH₂ $\xrightarrow{HNO_2}$ ▷—CH₂OH + ☐—OH +

(**651**) (**652**) 48% (**653**) 47%

CH₂=CH—CH₂—CH₂OH ← ☐—NH₂

(**654**) 5%

These results have been rationalized by postulating a set of equilibrating non-classical ions (**655**). The energy barriers between the non-classical cations were assumed to be very low, but not exactly equal.

$$H_2C \cdots \overset{*}{C}H_2 \quad H_2C—CH_2 \quad H_2C \cdots CH_2$$
$$HC—CH_2 \quad HC=CH_2 \quad HC—CH_2$$
$$\qquad \qquad \qquad * \qquad \qquad \qquad *$$

(**655**)

Recent investigations indicate that the cyclopropylmethyl cation is better represented by structure (**656**) in which maximum overlapping of the carbonium orbital with the bent bonds of the cyclopropane ring is possible. In (**656**) the (linear) H–C–H axis of the CH_2^+ group bisects the plane of the cyclopropyl ring[44]. It is

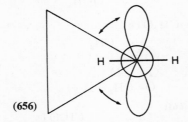

(**656**)

difficult to define quantitatively the degree of interaction between the methylium carbon and the cyclopropyl ring. However, there must be considerable interaction (delocalization) in order to explain why nucleophiles attack not only the methylium carbon but also to a considerable extent the carbon atoms of the three-membered ring. Reaction of 3-butenylamine with HNO_2 does not yield the same products as are obtained from cyclopropanemethylamine and cyclobutaneamine. Thus, the intermediate in this reaction must differ from the structures described above.

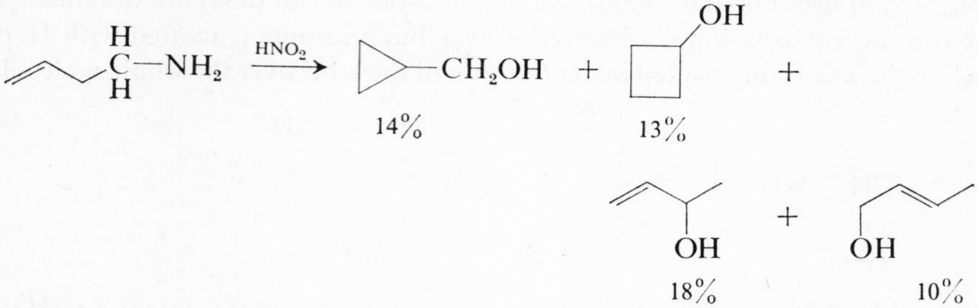

S-11. Solvolysis of 1,4-dihydro-3,5-dimethoxybenzyl *p*-toluenesulfonate[45]

If this ester (**657**) is solvolysed in pyridine, a cycloheptatriene derivative (**659**) is formed. It is very likely that cation (**658**) is formed as the first intermediate. Although one of the double bonds in (**657**) participates in the transition state of the solvolysis, (**658**) should be classical since the positive charge is strongly stabilized by the OCH_3

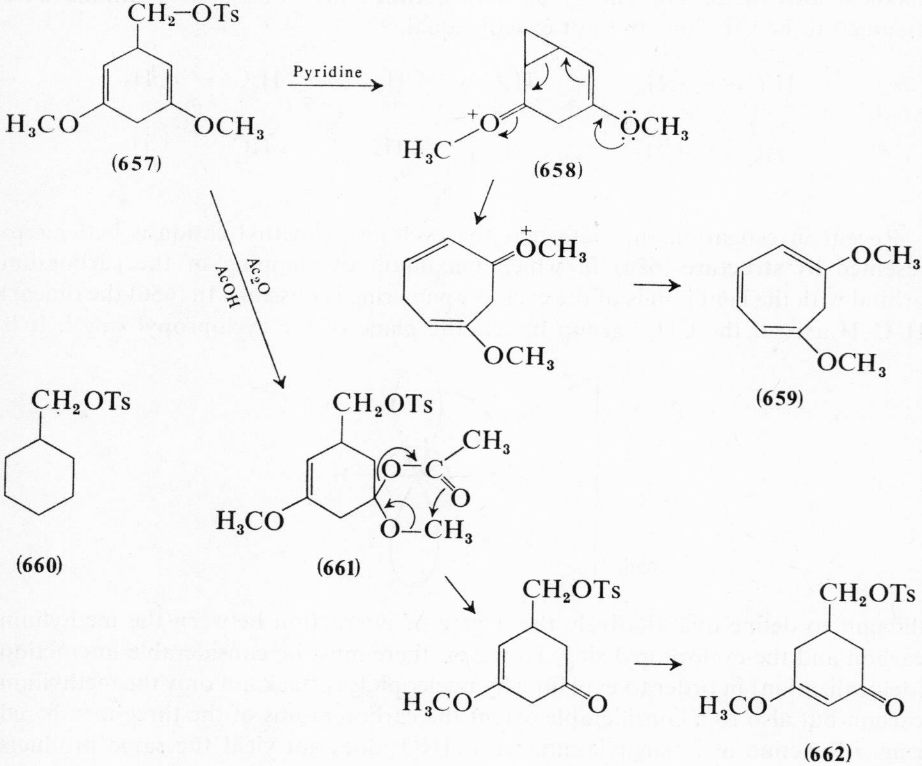

group. It is a basic principle of the theory of non-classical ions that the more stable the cationic center the less important non-classical participation should be.

The ester (**657**) is solvolyzed much more rapidly than cyclohexyl derivative (**660**). The rearrangement of (**657**) to (**659**) has been extended by Chapman to a general synthesis of tropolones.

The enol ether group is stable in pyridine. In acetic acid, however, the carboxylic acid adds in part to the enol double bond, yielding intermediate (**661**) which is converted into the product (**662**), probably by way of a cyclic transition state.

S-12. Rearrangement of 2-chloro-4,5-dihydro-1-benzothiepin-5-ol (663) in an acidic medium[46]

In analogy to the preceding example, the solvolysis of (**663**) is probably assisted by the 2,3-double bond. The intermediate (**664**), however, which arises from the transition state, is again a relatively stable carbonium ion in which participation by the cyclopropyl ring should be insignificant. The product is then the thiolactone (**665**), whose nature is proved by hydrolysis.

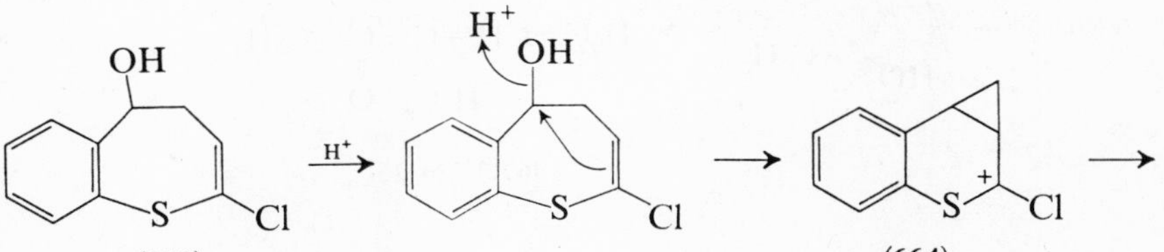

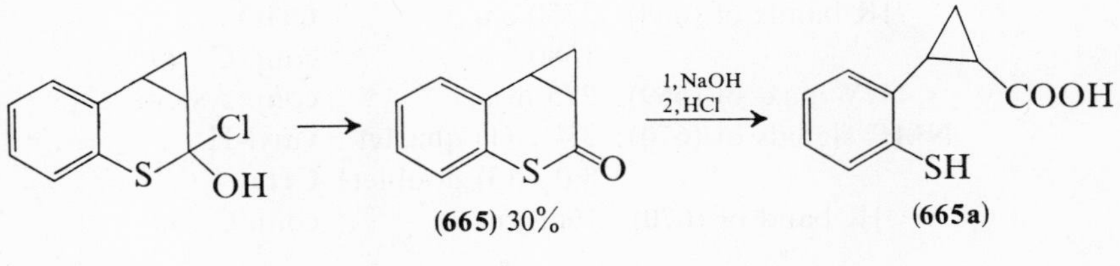

S-13. Solvolysis of (2,3-diphenyl-2-cyclopropenyl)methyl p-toluenesulfonate[47]

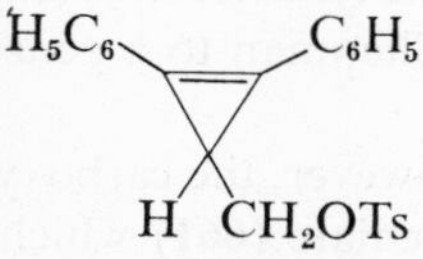

(666) H₅C₆, C₆H₅, H, CH₂OTs

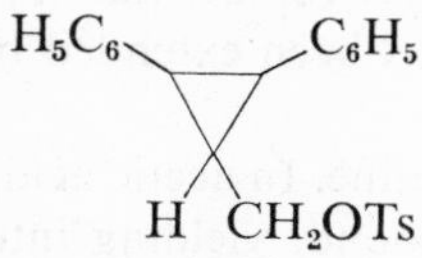

(667) H₅C₆, C₆H₅, H, CH₂OTs

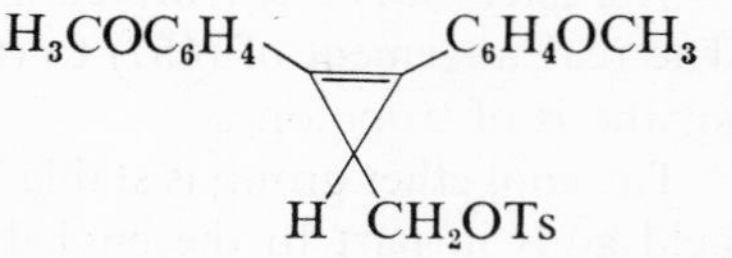

(668) H₃COC₆H₄, C₆H₄OCH₃, H, CH₂OTs

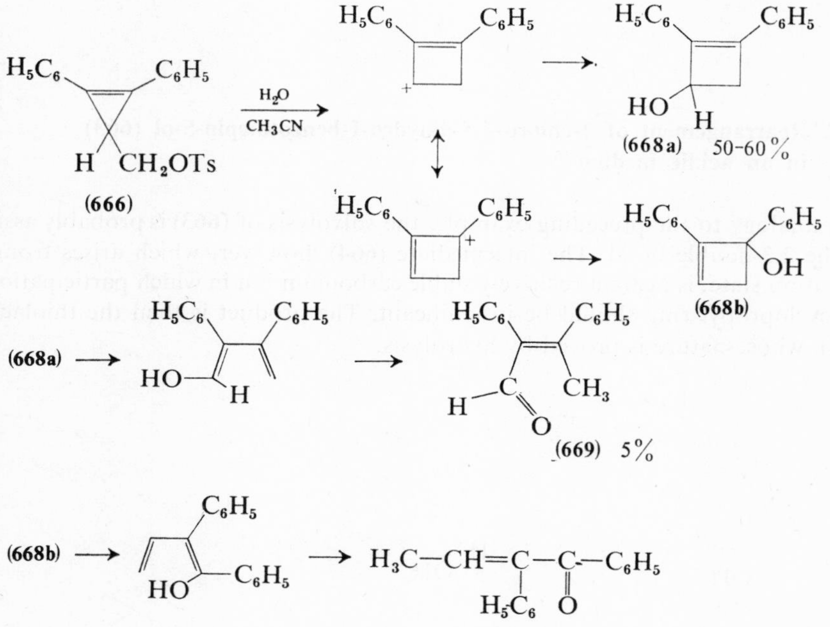

(666) →(H₂O / CH₃CN)→ [cyclobutenyl cation, H₅C₆, C₆H₅] → (668a) H₅C₆, C₆H₅, HO, H 50–60%

↕

[cyclobutenyl cation isomer, H₅C₆, C₆H₅] → (668b) H₅C₆, C₆H₅, OH

(668a) → H₅C₆, C₆H₅, HO, H (diene-ol) → H₅C₆, C₆H₅, CH₃, H–C(=O) (669) 5%

(668b) → C₆H₅, C₆H₅, HO (dienol) → H₃C—CH=C(C₆H₅)—C(=O)—C₆H₅ with H₅C₆

(670) 20–25%

NMR signals of (668a): 5.2 τ (1) quartet CHOH
6.5 (1) singlet OH
7.4 (2) multiplet CH₂
IR bands of (669): 2750 cm⁻¹ CHO
1680 conj. C=O
UV max. of (669): 275 mμ conj. system
NMR signals of (670): 3.4 τ (1) quartet vinyl-H
8.0 (3) doublet CH₃
IR band of (670): 1655 cm⁻¹ conj. C=O

Diphenylcyclopropenyl- (**666**), diphenylcyclopropyl- (**667**), and di-(*p*-methoxyphenyl)cyclopropenyl-methyl *p*-toluenesulfonate (**668**) are hydrolyzed at comparable rates. This shows that the solvolysis of the cyclopropenylmethyl derivatives must pass through a transition state similar to that formed on solvolysis of the diphenylcyclopropylmethyl ester. The double bond of the three-membered ring does not significantly affect the transition state. If the transition state collapses directly to the resonance-stabilized classical cyclobutenylium cation from which the products are formed, one must conclude that the cyclobutenylium cation and the cation formed from the cyclopropylmethyl derivative do not differ very much in energy. Alternatively, it may be assumed that similar cations are formed from the cyclopropyl and the cyclopropenyl derivative, no matter what their precise structure may be, the cyclopropenyl cation then rearranging to the more stable cyclobutenylium cation.

Since the true nature of the cyclopropylmethyl cation seems to be by no means clear at present some characteristic reactions of a particularly interesting cyclopropylmethyl carbonium system are presented in the following section. By studying the reactions (*a*)—(*g*) the reader may obtain an insight into the problems involved.

S-14 & S-15. The bicyclohepten-7-yl and the bicycloheptadien-7-yl system

Particularly favorable arrangements for homodelocalization are present in the bicyclohepten-7-yl and the bicycloheptadien-7-yl system. *anti*-7-Chlorobicycloheptene (**672**) is solvolysed 10^{11} and 7-chlorobicycloheptadiene (**673**) 10^{14} times faster than 7-chlorobicycloheptane (**675**)[48].

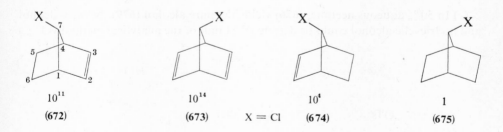

In the sequence illustrated open-chain secondary alkyl halides would be placed between *anti*- (**672**) and *syn*-7-chlorobicycloheptene (**674**).

The rates of solvolysis of (**672**) and (**673**) clearly show that in the transition state there must be strong participation by the π-electrons. The problem is again: what is the precise structure of the intermediates arising from the transition states? To differentiate between several conceivable intermediates the following reactions, among others, have been carried out:

(*a*) If *anti*-bicyclohepten-7-yl *p*-toluenesulfonate (**676**) is treated with 1.8M-$NaBH_4$ in 65% aqueous diglyme at room temperature, 70% of bicycloheptene (**677**), 15% of the tricyclic hydrocarbon (**678**), and 6—7% of *anti*-bicyclohepten-7-ol (**679**) are obtained[49] (the reaction with $NaBH_4$ is irreversible).

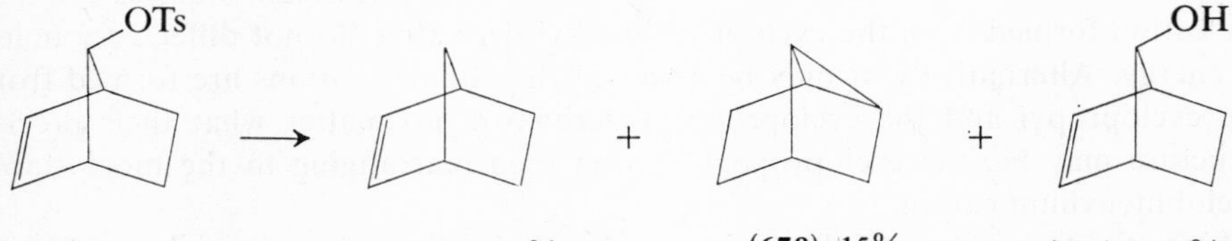

(676) **(677)** 70% **(678)** 15% **(679)** 6-7%

(*b*) In tetrahydrofuran or diglyme, in the absence of water, treatment of (**676**) with $LiAlH_4$ yields 60% of (**678**) and 34% of (**677**)[49].

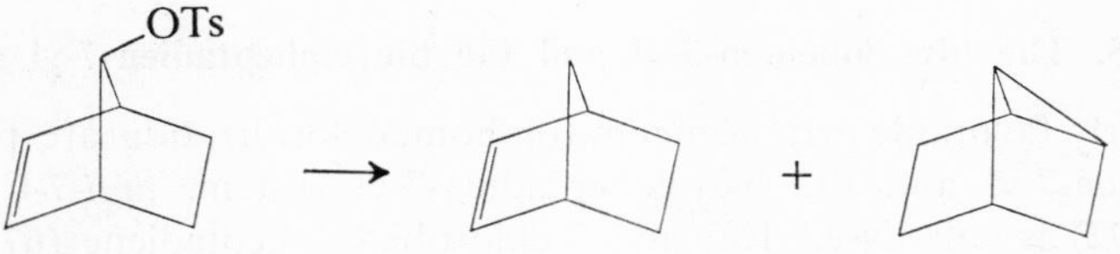

(676) **(677)** 34% **(678)** 60%

(*c*) In 50% aqueous acetone, (**676**) yields the pure alcohol (**679**). No *syn*-alcohol and no tricyclic alcohol could be detected[50]. (Limit of the analytical method 0.1%.)

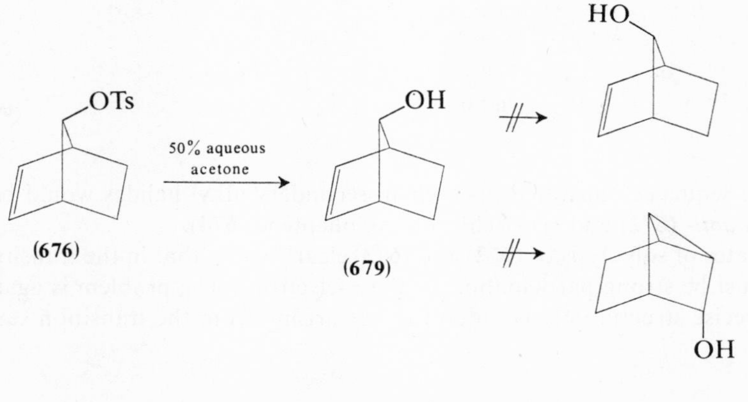

(676) **(679)**

(d) Neutral methanolysis of **(676)** at 25° yields, in a first-order reaction ($k = 1.31 \times 10^{-3}$ sec^{-1}), the ether **(680)** in 99.7% yield and 0.3% of an ether **(681)**. With

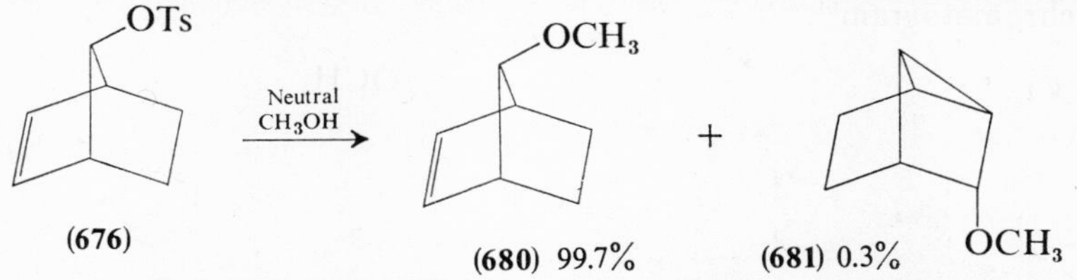

(676) →(Neutral CH$_3$OH) **(680)** 99.7% + **(681)** 0.3%

an increasing addition of NaOCH$_3$, increasing amounts of **(681)** are obtained [4M-NaOCH$_3$ results in 51.5% of **(681)** and 48% of **(680)**].

S-14

(e) Reaction of 7-chlorobicycloheptadiene **(673)** with 1.8M-NaBH$_4$ in 65% aqueous diglyme at room temperature[49] yields 12% of norbornadiene **(682)** and 83% of the tricyclic hydrocarbon **(683)**.

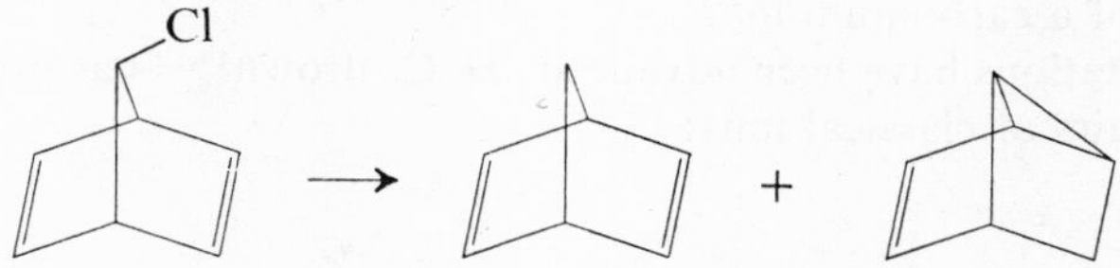

(673) → **(682)** 12% + **(683)** 83%

(f) In aqueous alkali the chloride **(673)** yields exclusively the alcohol **(684)**[49].

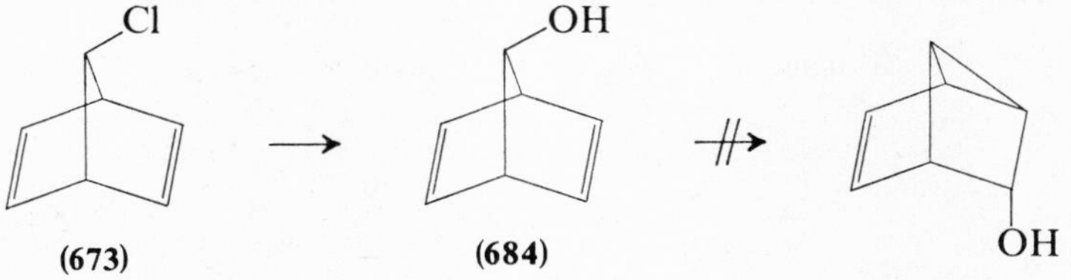

(673) → **(684)** ⇸ **(685)**

Absence of the tricyclic alcohol **(685)** from the product has been interpreted on the hypothesis that the kinetic product **(685)** rapidly rearranges to **(684)** [an ether **(686)** rearranges in acidic media to **(687)**].

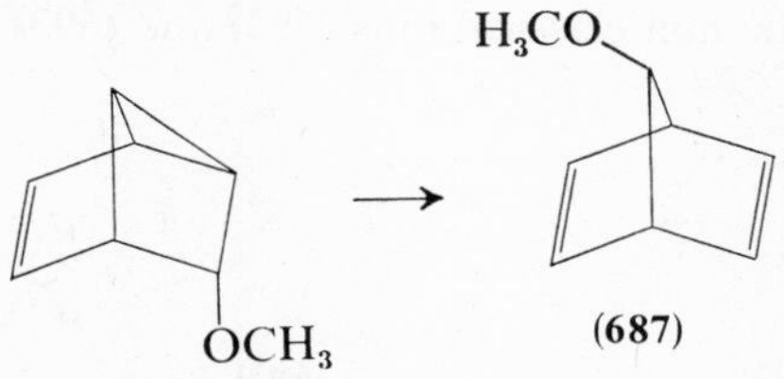

(686) → **(687)**

(g) Methanolysis of the chloride (673) in the presence of NaOCH$_3$ yields 50% of a mixture of (686), (687), and (688) with the relative peak areas of 54:24:22 on a paper chromatogram[51].

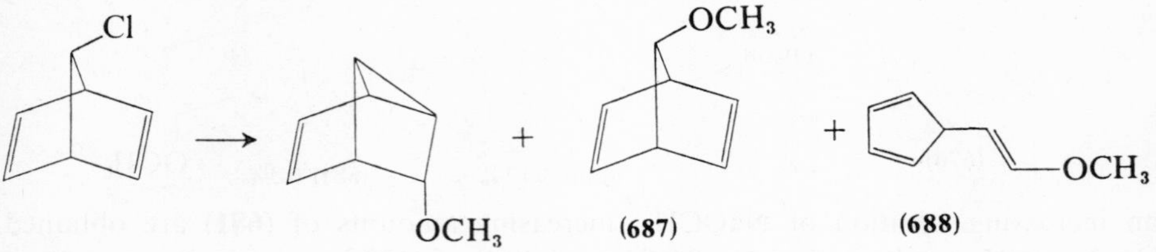

(673) → (686) + (687) + (688)

                54 : 24 : 22

Winstein et al.[50] showed that the first-order rate constant of the solvolysis of anti-7-chlorobicycloheptene (672) in 65% diglyme at 50° is 1.1×10^{-4} sec^{-1}, and that the reaction of (672) with 0.48M-NaBH$_4$ + 0.1M-NaOH has a rate constant of 1×10^{-4} sec^{-1}. These results show that the first step of both reactions must be the formation of a carbonium ion.

Two interpretations have been advanced: H. C. Brown[49,52] has proposed rapidly equilibrating pairs of classical ions:

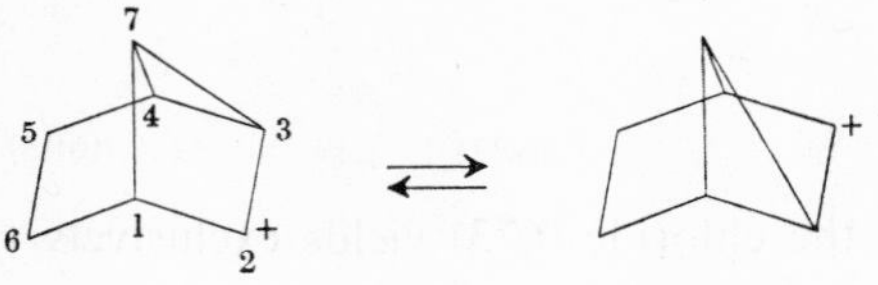

(689) ⇌ (690)

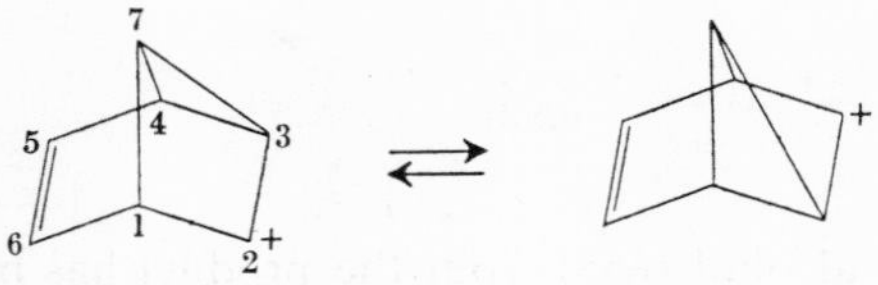

(691) ⇌ (691a)

Winstein[53] has suggested the non-classical ions (692) and (693):

(692)   (693)

In earlier papers[48,54] the non-classical structures (**694**), (**695**), and (**696**) were proposed, but these structures seem no longer to be taken seriously into consideration.

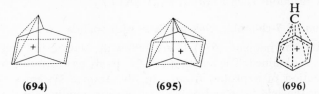

(**694**)   (**695**)   (**696**)

The products of reactions (*a*)—(*g*) show that attack of nucleophiles takes place at C-7 and at C-2 or C-3. Addition to C-2 and C-3 takes place exclusively at the *endo*-side, and addition to C-7 at the *anti*-side.

The non-classical ions (**692**) and (**693**) account qualitatively for the products (shielding of the *syn*- and the *exo*-side), whereas on the basis of the equilibrating classical ions (**690**) and (**691**) it is difficult to rationalize predominant attack at C-7 in the reactions (*a*) and (*d*) and exclusive attack at C-7 in the reactions (*c*) and (*f*). It is not reasonable to presume intermediate formation of the tricyclic alcohol in reaction (*c*) and of (**685**) in reaction (*f*), followed by rapid rearrangement to (**679**) and (**684**), since not even a trace of the tricyclic alcohol could be detected in reaction (*c*). A more quantitative interpretation of the results is, however, much more difficult.

As the reactions show, the product distribution changes with changing nucleophilicity of the attacking species. In the case of reaction (*d*), attempts were made[53] to explain the increasing formation of (**681**) with increasing proportions of NaOCH$_3$. Addition of CH$_3$O$^-$ to the carbonium ion is more exothermic than addition of the weaker nucleophile CH$_3$OH. The transition state on the addition of CH$_3$O$^-$ should therefore be on the side of the carbonium ion on the reaction coordinate, whereas the transition state of the CH$_3$OH addition should be determined by the products*. Since the tricyclic ether (**681**) has a considerably higher ground-state energy than the bicycloheptenyl ether, the latter should be predominantly formed in the reaction with CH$_3$OH. In the addition of CH$_3$O$^-$, which is determined by the carbonium ion, the product distribution is subject to a statistical factor of 2 in favor of addition to C-2 and C-3, and is also affected by the positive charge distribution which is believed to be higher at the "cyclopropylmethyl" carbon atoms 2 and 3.

The compounds (**679**) and (**681**) yield the same carbonium ion in SO$_2$–SbF$_5$–FSO$_3$H at $-35°$ to $-60°$. The NMR spectrum of the ion is in accord with structure (**692**), as well as with rapidly equilibrating cations. If classical equilibrating ions are involved, the energy barrier separating them must be very low since even at $-60°$ the hydrogen atoms at C-2 and C-3 are identical[55].

In conclusion it may be said that in the case of solvolyzing 7-substituted bicycloheptene and bicycloheptadiene derivatives the experimental results clearly show that

* See Hammond's principle (p. 237).

π-participation is involved in the transition state as in other homoallylic systems. The view that a non-classical carbonium ion arises from the transition state is supported by reactions such as (a), (c), (d), and (f).

S-15. Reaction of 7-chlorobicycloheptadiene with sodium cyanide[51,56]

If 7-chloronorbornadiene (**673**) is treated with NaCN in aqueous ethanol, product (**699**) is obtained. The reaction very probably passes through intermediate (**697**). Abstraction of a proton from a highly strained system such as (**697**) is plausible since the acidity of hydrocarbons increases rapidly with the angle strain. For example, the pK_a of the strained compound (**700**) is lower than that of tertiary butyl alcohol, i.e. (**700**) is the more acidic (see p. 329). The abstraction of the proton is additionally facilitated by formation of a cyclopentadienyl anion (**698**).

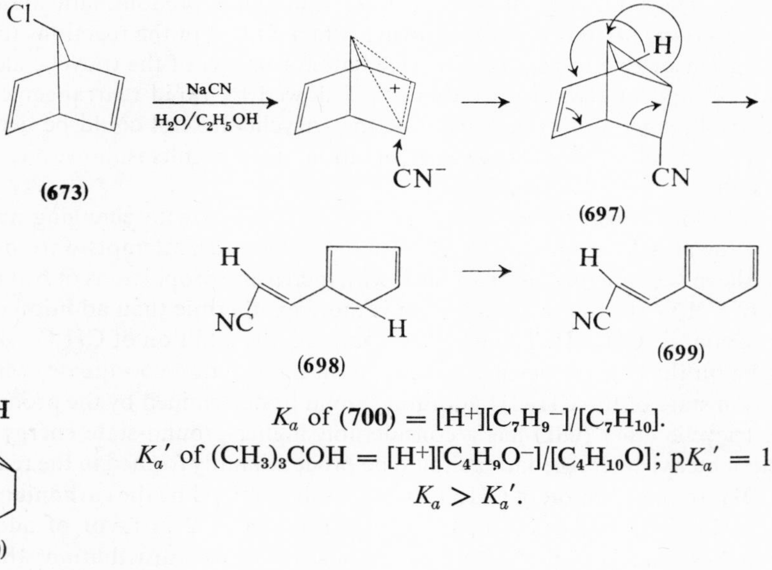

K_a of (**700**) = [H$^+$][C$_7$H$_9^-$]/[C$_7$H$_{10}$].
K_a' of (CH$_3$)$_3$COH = [H$^+$][C$_4$H$_9$O$^-$]/[C$_4$H$_{10}$O]; pK_a' = 1.9.
$K_a > K_a'$.

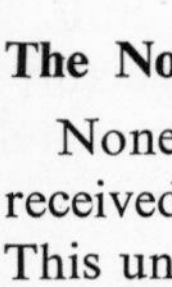

The Norbornyl system

None of the cations for which non-classical structures have been proposed has received more attention than the σ-bridged non-classical norbornyl cation (**702**). This unusual interest has resulted from three characteristic properties:

Additions to the norbornyl cation take place from the *exo*-side.

exo-Norbornyl derivatives are solvolyzed unusually rapidly and high *exo*/*endo* rate ratios are observed. These findings have been rationalized by postulating σ-electron participation in the transition state of the solvolysis of *exo*-norbornyl derivatives and a non-classical bridged intermediate arising from the transition state. According to this hypothesis the rapid solvolysis of *exo*-derivatives is caused by anchimeric assistance of the C_6–C_1 σ-electrons in the non-classical transition state. In addition it has been postulated that in *endo*-derivatives such σ-participation is impossible for stereoelectronic reasons. *endo*-Addition to (702) is prevented by the shielding effect of these electrons[57].

As an alternative, H. C. Brown has proposed a set of rapidly equilibrating classical cations (701). In a systematic exploration of the chemistry of norbornyl

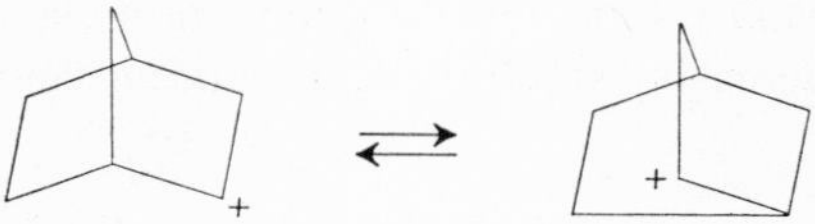

(701)

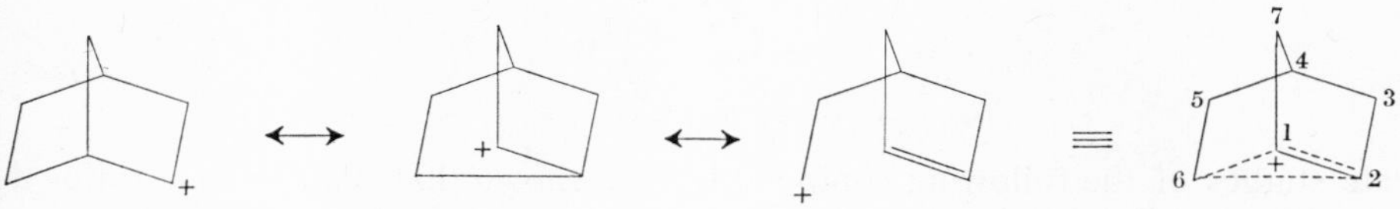

(702)

derivatives, he has urged[58] that the non-classical ion does not provide a satisfactory explanation of the experimental results. Some of the principal objections can be summarized as follows.

(1) The high *exo*-solvolysis rates may be due to relief of steric strain. For example, 1-methylcyclopentyl chloride (703) and 2-methylnorbornyl chloride (704) are solvolyzed at comparable rates[59].

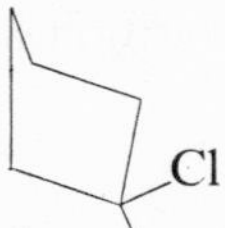

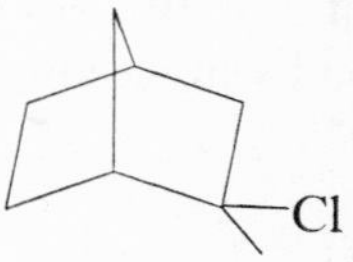

(703)   Rate   ratio   (704)
         66   :   355

Schleyer recently suggested that torsional effects exert an important influence on the reactivity of norbornyl derivatives. In the transition state for *exo*-solvolysis steric interactions between the leaving group X and the hydrogen atom at position 1 as well as non-bonded strain between R and the hydrogen atom at position 6 are relieved, whereas the opposite situation pertains to the *endo*-transition state[60].

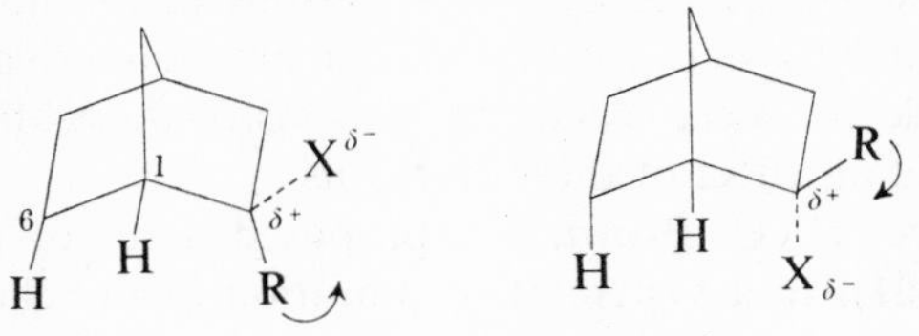

exo   *endo*

(2) If the high *exo/endo* rate ratios are due to non-classical delocalization they should decrease with increasing stability of the cationic centre. However, solvolysis

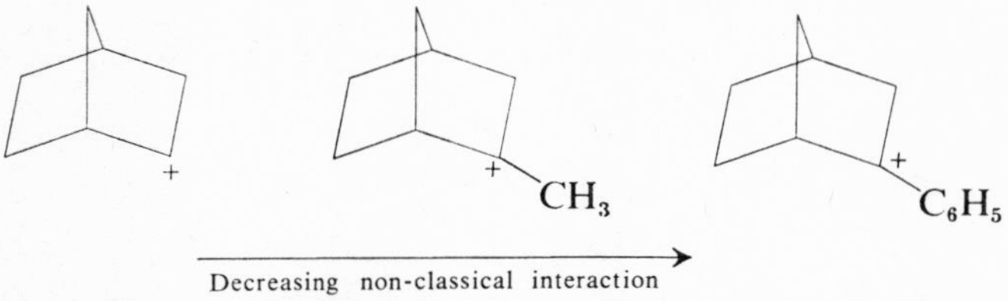

Decreasing non-classical interaction

rate studies of the following compounds have shown that the *exo/endo* ratios do not differ significantly[61] (see p. 269).

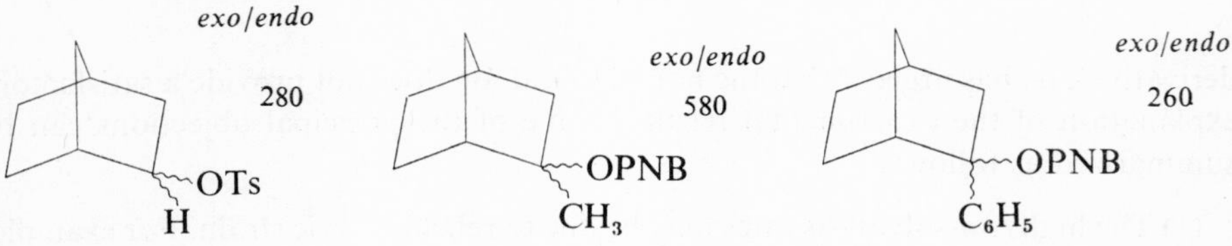

exo/endo 280   *exo/endo* 580   *exo/endo* 260

OTs = $p\text{-}CH_3C_6H_4SO_2\text{—}O\text{—}$;   OPNB = $p\text{-}NO_2\text{—}C_6H_4CO\text{—}O\text{—}$.

(3) A 1-phenyl substituent increases the rate by a factor of 3.9, a 2-phenyl substituent by one of 3×10^5. However, the two derivatives should lead to the same non-classical ion by proceeding through transition states that closely resemble the non-classical ion[62].

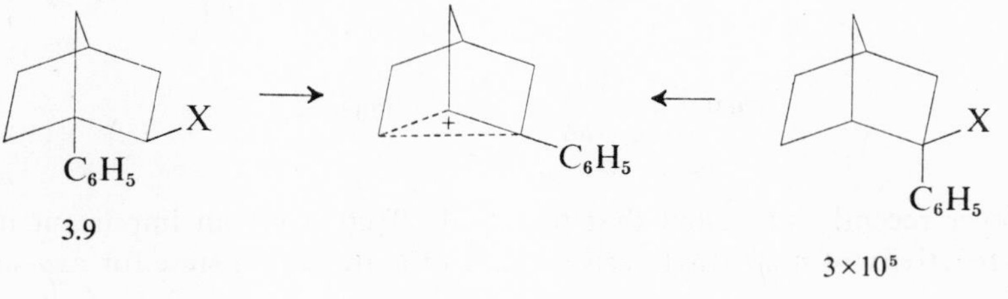

3.9 3×10^5

(4) The 2-*p*-methoxyphenyl-7,7-dimethylnorbornyl cation should undergo *endo*-addition since *exo*-approach is hindered by the 7-methyl groups, whereas *endo*-approach should be unrestricted because non-classical interaction between C-6 and C-2 should be insignificant in this case where the positive centre is strongly stabilized by the aryl group. Surprisingly, predominant *exo*-addition (87%) was found[63].

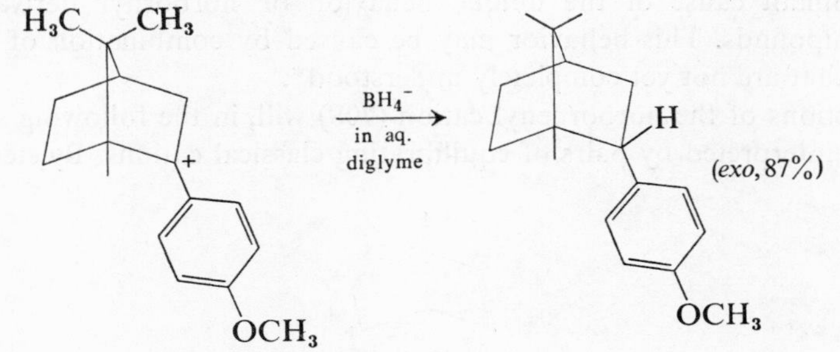

(5) On hydrochlorination of compound (705) under conditions that cause minimum exposure of the product to the further action of HCl, predominantly unscrambled product is obtained. A symmetrical non-classical ion should lead[64] to 50% of (706) and 50% of (707).

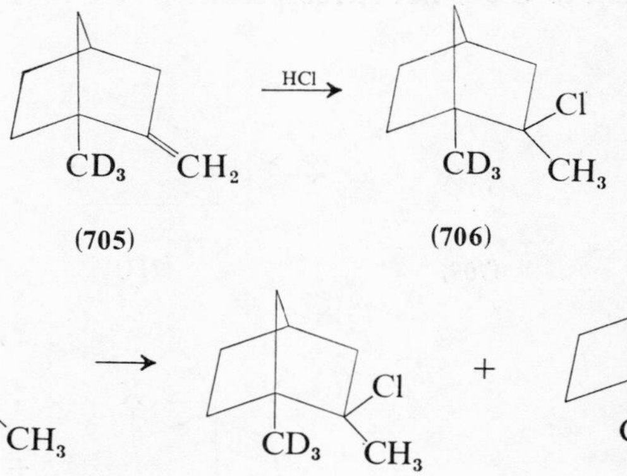

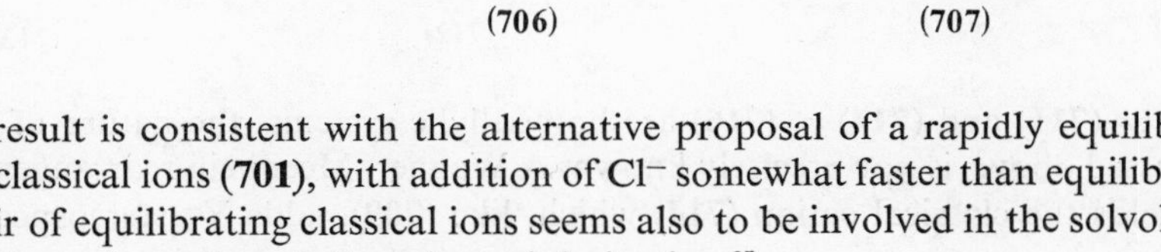

The result is consistent with the alternative proposal of a rapidly equilibrating pair of classical ions (701), with addition of Cl⁻ somewhat faster than equilibration.

A pair of equilibrating classical ions seems also to be involved in the solvolysis of 1,2-bis-(*p*-methoxyphenyl)-2-norbornyl derivatives[65].

Although predominantly *exo*-addition takes place in many carbonium ion reactions of norbornyl derivatives, *exo*-attack is also observed in non-carbonium ion reactions[66]. It therefore appears that *exo*-attack is a general characteristic of norbornyl derivatives whose reactions do not depend on the formation of a nonclassical ion. At present no convincing evidence is available that σ-participation either in the transition state or in the subsequent cationic intermediate is the sole or predominant cause of the unique behavior of norbornyl derivatives and related compounds. This behavior may be caused by combination of a number of factors that are not yet completely understood*.

The reactions of the norbornenyl cation (708) will, in the following section, be tentatively interpreted by pairs of equilibrating classical cations. By stereospecific

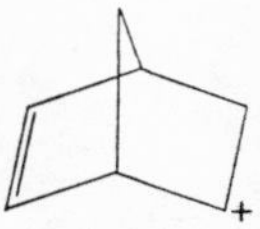

(708)

cis-addition of anion Y^- to C-2 of the classical cation (709), the product (710) is formed. *exo-cis*-Addition occurs only if the approaching Y^- is not sterically hindered by substituent X (steric hindrance is not observed if X is hydrogen). The products (713) and (712) are formed by addition of Y^- to C-5 of the classical cation (711). Addition to C-5 is not stereospecific.

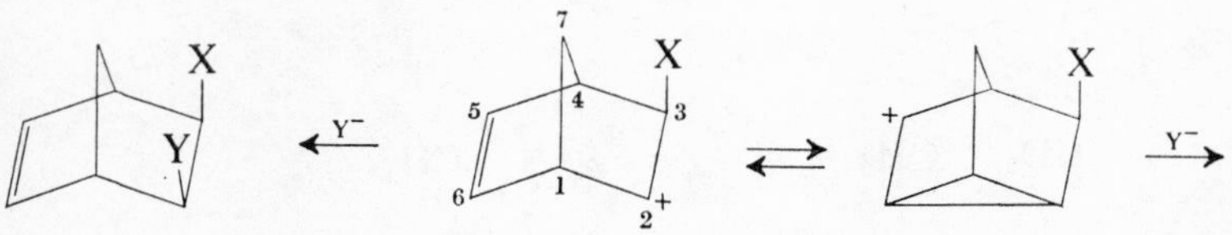

(710)   (709)   (711)

(712)   +   (713)

(709) ⇌ (711) and (714) ⇌ (716) are homoallylic systems, the nature of which, as discussed above, is not precisely known. A Wagner–Meerwein rearrangement of (709) yields the classical cation (714) which, like (709), adds Y^- stereospecifically

* Recent results have been reported by Olah[66a].

to yield (715). The classical cation (716), which is in equilibrium with (714), adds Y⁻ like (711) to yield the stereoisomeric products (717) and (717a). Alternatively,

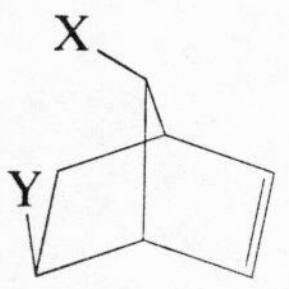

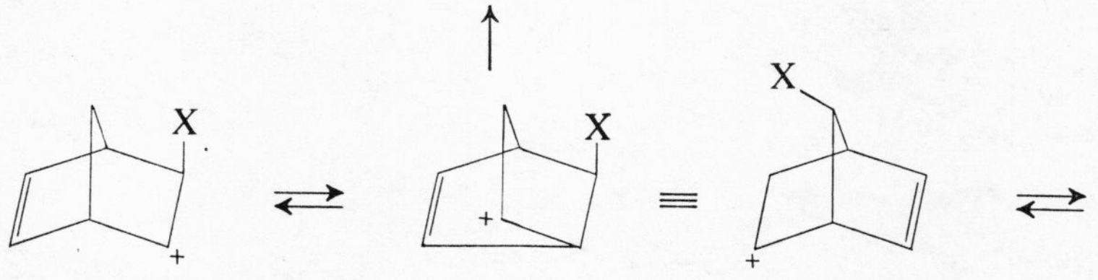

the product distribution may be qualitatively explained by addition to the single non-classical intermediate (718).

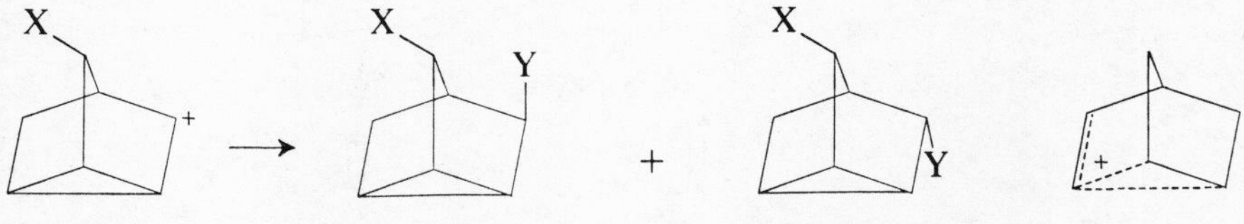

Quantitative investigations have been rationalized by a mixture of classical and non-classical ions; it was not possible to interpret them by the sole intervention of (718)[67].

S-16. Reaction of norbornadiene with *N*-chlorodiethylamine in H_2SO_4–CH_3COOH[68]

After reaction of norbornadiene with *N*-chlorodiethylamine in H_2SO_4–CH_3COOH five products have been isolated. The first step is probably *exo*-addition of Cl⁺. The resulting pair of ions (719) yields the rearranged product (720) and the nortricyclene derivative (721). *exo-cis*-Addition is prevented by the steric requirements of the Cl and $N(C_2H_5)_2$ groups. If the cations and the "counter-ion" equilibrate with the solvent, acetic acid adds to the cations, yielding the products (722), (723), and (724).

+ $(C_2H_5)_2NCl$ $\xrightarrow[CH_3COOH]{H_2SO_4}$

S-17. Reaction of 6,6-dimethylbicyclo[2.1.1]hexan-2-amine with HNO_2[69]

In the final example of this section the bicyclo[2.1.1]hexenyl system is presented though it has not been widely studied. The reaction of the optically active amine (**725**) with HNO_2 in acetic acid yields 14% of optically active acetate (**726**), 56% of optically active acetate (**727**), 21% of partially active acetate (**728**), and 9% of inactive tertiary alcohol (**729**).

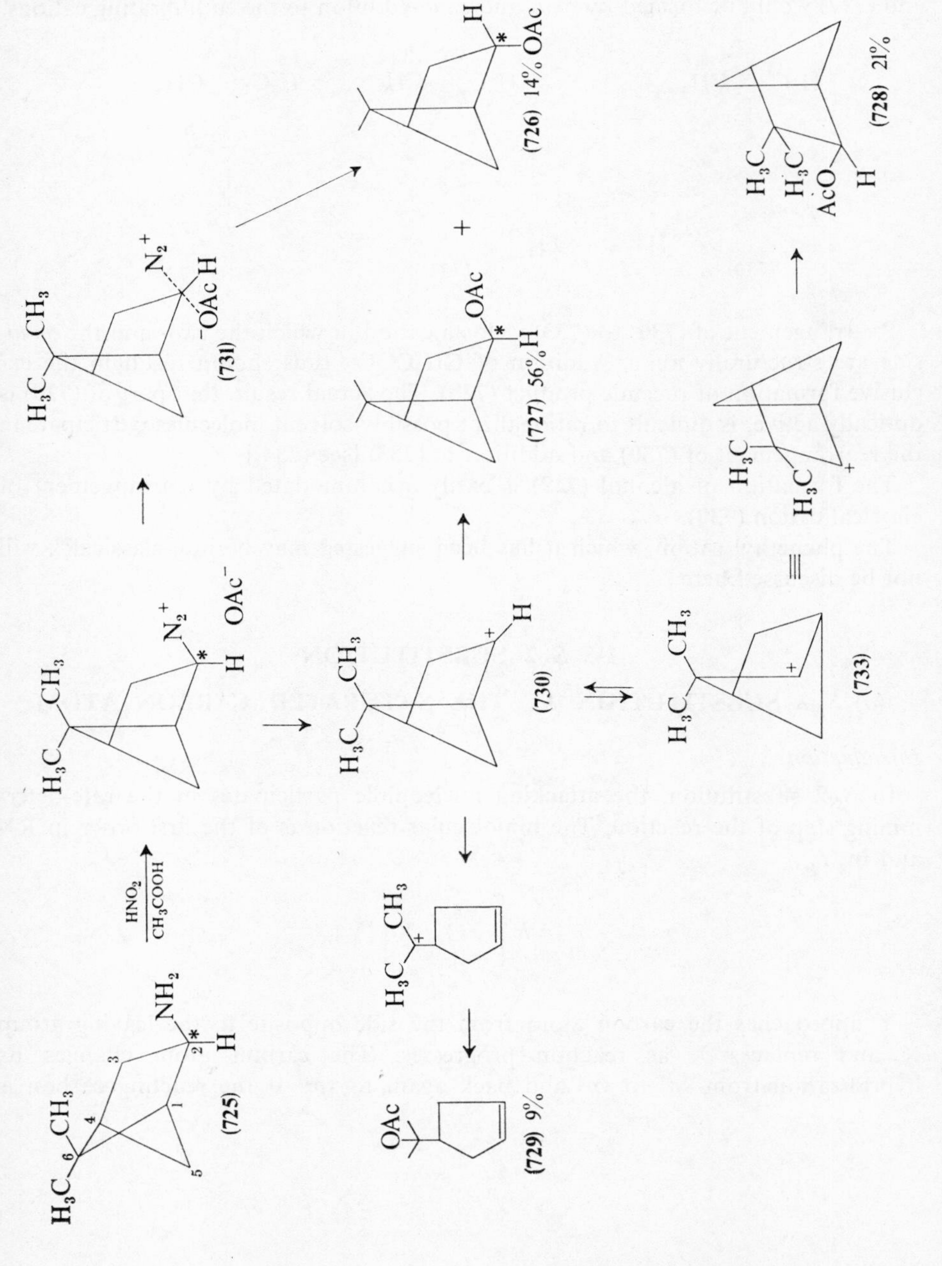

By analogy to the norbornyl system, *exo*-addition to the cation (**730**) should predominate, yielding optically active *exo*-acetate (**727**). (Classical cations are more likely to be produced by deamination than by solvolysis.) *endo*-Addition to (**730**) gives optically active acetate (**726**). The latter may be alternatively formed by S_N2 substitution in (**731**). Addition to (**730**) must take place much faster than Wagner–Meerwein rearrangement of (**730**) to (**732**), since otherwise racemic (**726**) and (**727**) would be formed by *exo*- and *endo*-addition to the equilibrating cations.

Rearrangement of (**730**) to (**733**) yields a cation in which the *exo*- and the *endo*-side are structurally alike. Addition of CH_3COO^- thus should result in the exclusive formation of racemic product (**728**). The actual result, that part of (**728**) is optically active, is difficult to rationalize; possibly solvent molecules participate in the rearrangement of (**730**) and addition to (**733**) [see (**734**)].

The formation of alcohol (**729**) is easily accommodated by rearrangement of classical cation (**730**).

The phenethyl cation, which it has been suggested may be non-classical[70], will not be discussed here.

IV. S_N2 SUBSTITUTION

(a) S_N2 SUBSTITUTION AT THE SATURATED CARBON ATOM

Introduction

In S_N2 substitution, the attacking nucleophile participates in the rate-determining step of the reaction. The bimolecular reaction is of the first-order in RX and in Y.

$$v = k \left[-\overset{|}{\underset{|}{C}} - X \right] [Y]$$

Y approaches the carbon atom from the side opposite to the leaving group X and replaces X as reaction progresses. The carbon atom changes its hybridization from sp^3 to sp^2 and back again to sp^3. If the reacting carbon is asymmetric, an inverted product is formed. The transition state (737) resembles

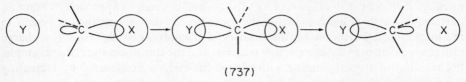

(737)

intermediate (563) of the S_N1 reaction. In contrast to intermediate (563), (737) is a transition state, in which exchange of X for Y does not take place.

In some cases the S_N2 reaction starts with the formation of an ion-pair, as in the S_N1 reaction, the ion-pair being attacked by the nucleophile in the rate-determining step[71].

Since the nucleophile participates in the rate-determining step of an S_N2 reaction, the rate depends on the nucleophilicity of the attacking Y.

Nucleophilicity

Nucleophilicity is the ability of a molecule or ion to form a bond with a carbon atom by simultaneously displacing a group X from the carbon.

Nucleophilicity and basicity often parallel one another. In supplying the electrons for the new bond, the nucleophile behaves like a base, forming a bond with a proton. Nucleophilicity, therefore, depends on the polarizability of the electrons of the attacking group in the direction of the Y—C—X axis.

Polarizability is a measure of the ease of moving electrons under the influence of an electric field. It can be determined by measuring the molecular refraction of a compound.

A quantitative relation between basicity, polarizability, and nucleophilicity has been advanced by Edwards[72]:

$$E_n = aP + bH,$$

where E_n = nucleophilicity, P = polarizability, H = basicity, and a and b are constants.

A nucleophilicity factor also appears in the Swain equation[73], which is an attempt to provide a quantitative correlation between the rate of nucleophilic substitution, nucleophilicity n, and electrophilicity e. s_n and s_e represent the sensitivity of the reaction to nucleophilic push and electrophilic pull:

$$\log (k/k_o) = ns_n + es_e.$$

Nucleophilicity is a kinetic quantity which is determined by the rate of a reaction. Basicity and acidity, however, are thermodynamic properties which are determined by the equilibrium constant of an acid–base equilibrium. The statement that A–H is more acidic than B–H means that in the equilibrium

$$A^- + B—H \rightleftarrows A—H + B^-$$

anion A^- and compound B–H are the thermodynamically more stable pair. Another factor affecting the nucleophilicity is solvation of the nucleophile. Since the nucleophile has to strip off part of its solvent shell in order to approach its reaction partner, its nucleophilicity is higher the more weakly the solvent shell is bound to the nucleophile. Furthermore, the polarity of the solvent is important if a charge difference exists between the starting material and the transition state. For example, the reaction of trimethylamine with methyl bromide is accelerated by increasing the polarity of the solvent, and the reverse reaction is retarded.

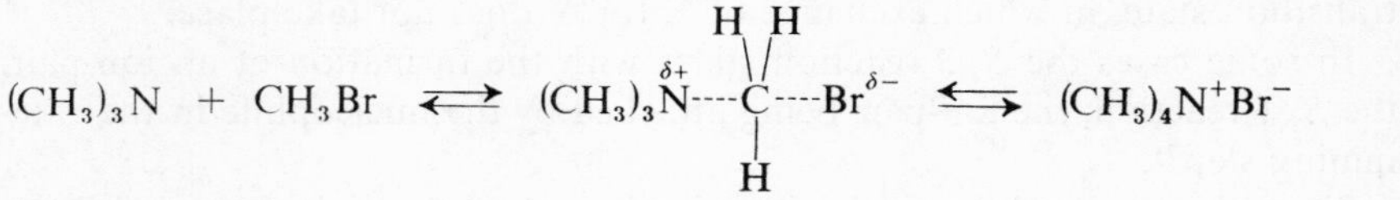

$$(CH_3)_3N + CH_3Br \rightleftarrows (CH_3)_3\overset{\delta+}{N}\text{---}\underset{H}{\overset{H\ \ H}{\overset{|}{C}}}\text{---}Br^{\delta-} \rightleftarrows (CH_3)_4N^+Br^-$$

S-18. Reaction of cyclic carbonates with potassium thiocyanate[74]

Five- and six-membered cyclic carbonates can be converted into episulfides and thietanes, respectively, by reaction with thiocyanates. The yields decrease with increasing substitution at positions 4 and 5 (or 4 and 6) of the ring. Whereas un-

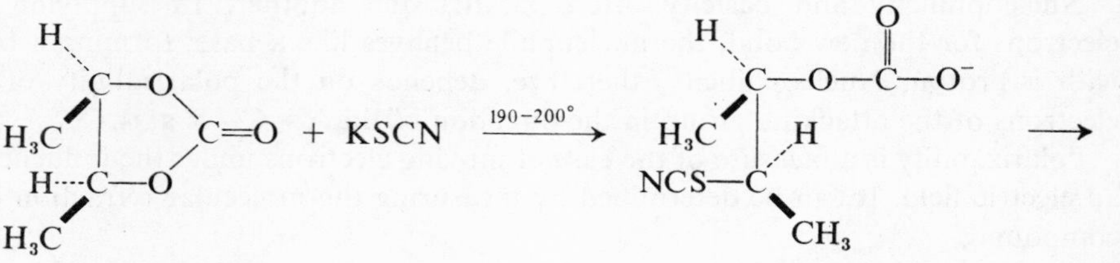

(738) + KSCN $\xrightarrow{190-200°}$ (739) $\rightarrow$

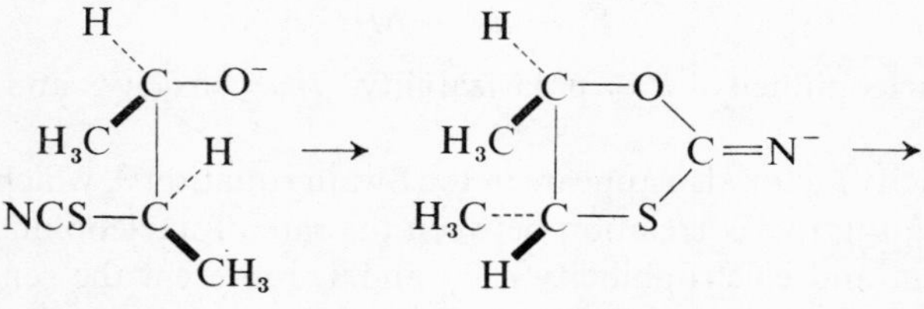

(740) $\rightarrow$ (741) $\rightarrow$

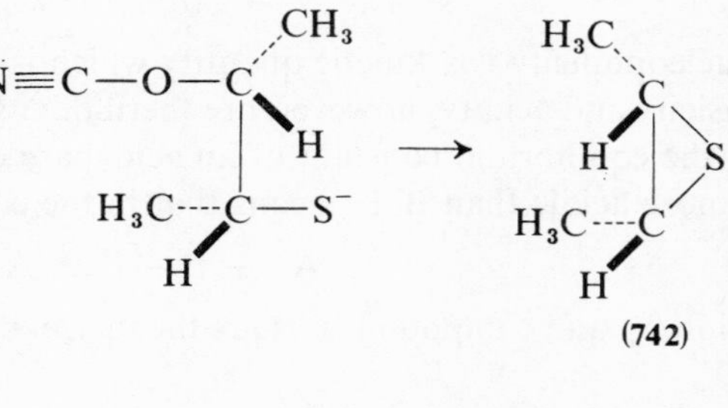

$\rightarrow$ (742)

substituted ethylene carbonate (1,3-dioxolan-2-one) reacts at 90–100°, the tetramethyl-substituted compound is unaffected even at 280° (the boiling point of the carbonate). The dependence of the reaction rate on the steric shielding of the carbon atoms 4 and 5 (or 4 and 6) indicates that the reactions are initiated by nucleophilic attack of SCN^-.

In the reaction of **(738)** with KSCN one obtains pure *cis*-dimethylthiirane **(742)** in 31% yield. The reaction is accompanied by a double Walden inversion. Anion **(739)**, which is formed by nucleophilic attack of the thiocyanate ion (first inversion) is converted into anion **(740)** by loss of CO_2, and this ion cyclizes to **(741)**. Ring cleavage of **(741)** and replacement of the ^-OCN group by the thiolate (second inversion) yields the product **(742)**.

An alternative is initial loss of CO_2, resulting in the formation of an epoxide or an oxetane (four-membered cyclic ether). Such a reaction is supported by the known ability of epoxides to form episulfides with thiocyanate and in contrast to cyclic carbonates also with thiourea[75]. Oxetanes, however, which would be formed by

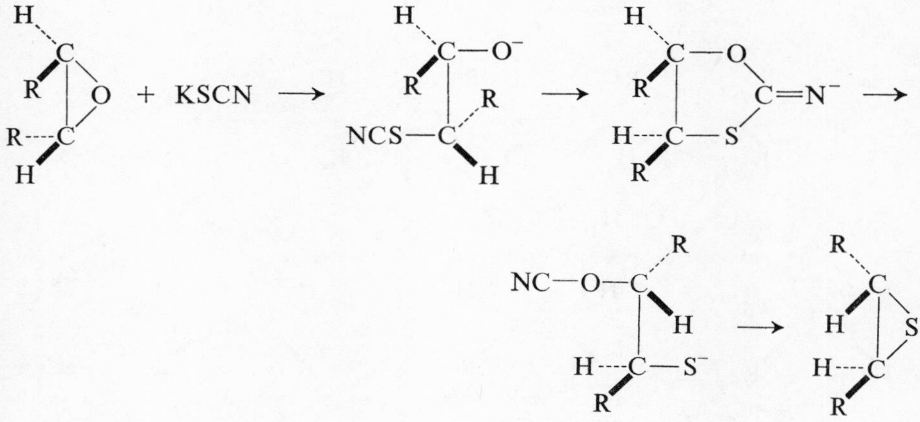

expulsion of CO_2 from the six-membered cyclic carbonates, do not react with thiocyanates under these conditions. Furthermore, evolution of CO_2 from the cyclic carbonates does not start below 240° and yields only 10% of *trans*-epoxide. Since two Walden inversions are involved, the formation of the episulfide from the *trans*-epoxide thus formed would lead to the *trans*-thiirane.

S-19. Reaction of *trans*-1,2-dimethyloxirane with potassium methyl xanthate[76]

Considerations here are as in the preceding example and will be evident from the annexed chart.

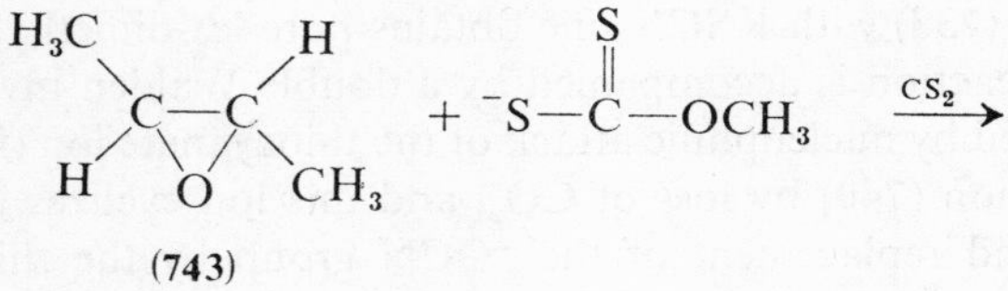

S-20. Reaction of epichlorohydrin with cyclopentadienylsodium[77]

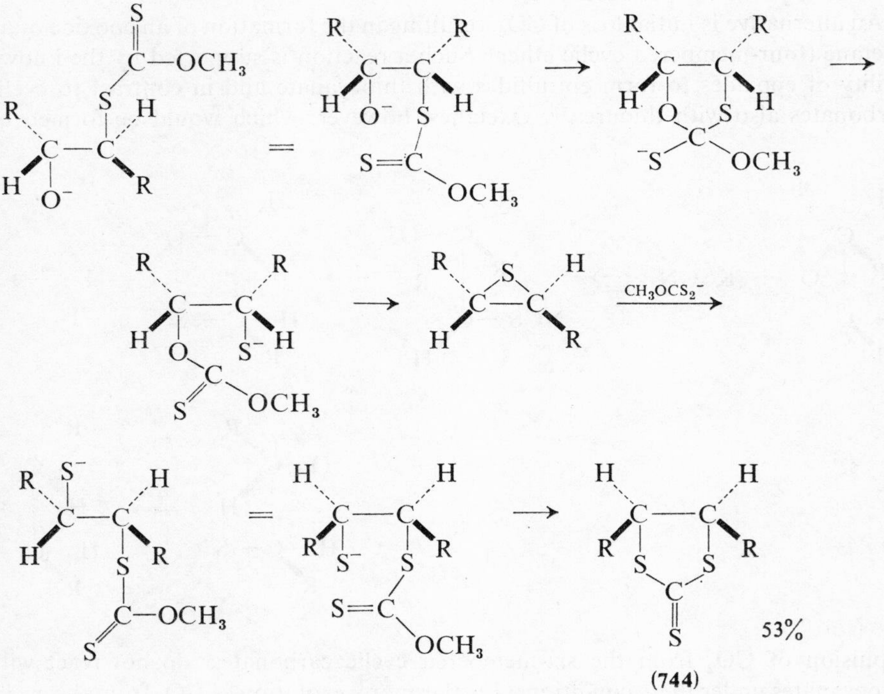

The reaction of epichlorohydrin (745) with cyclopentadienylsodium yields the alcohol (749). A probable reaction mechanism starts with replacement of the allylic halogen atom, resulting in the epoxide (746). By proton exchange with a second molecule of cyclopentadienylsodium, an anion (747) is obtained which, by cleavage of the epoxide ring, yields compound (**748**). A possible alternative is

(745) (746) (747) (748) (749)

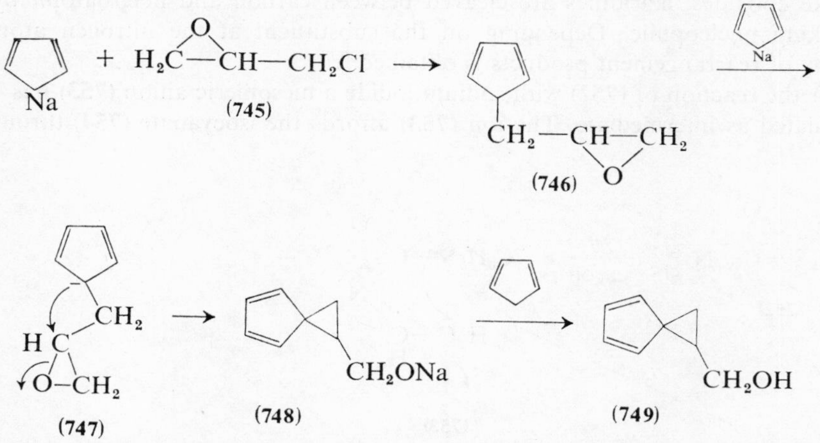

initial opening of the epoxide ring. Since in S_N2 reactions of unsymmetrically substituted epoxides the least substituted ring carbon atom is preferentially attacked by the nucleophile (see example **C**-1), one would obtain intermediate (**750**). This can easily form (**746**), which is converted into (**749**) as described above.

(750) **(746)**

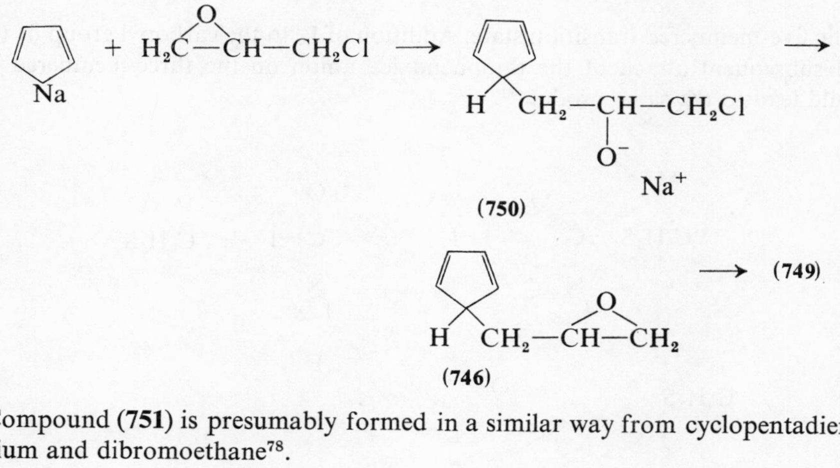

Compound (**751**) is presumably formed in a similar way from cyclopentadienyl-sodium and dibromoethane[78].

(751)

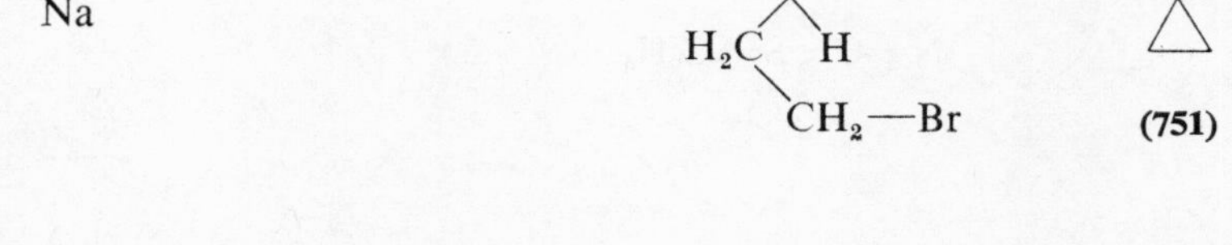

S-21. Reaction of 1-[(phenylthio)carbonyl]aziridine (752) with sodium iodide[79]

Like epoxides, aziridines are cleaved between carbon and heteroatom, by an attacking nucleophile. Depending on the substituent at the nitrogen atom, a variety of rearrangement products is obtained[80].

For the reaction of (752) with sodium iodide a mesomeric anion (753) has been postulated as intermediate. The ion (753) affords the isocyanate (754) through a

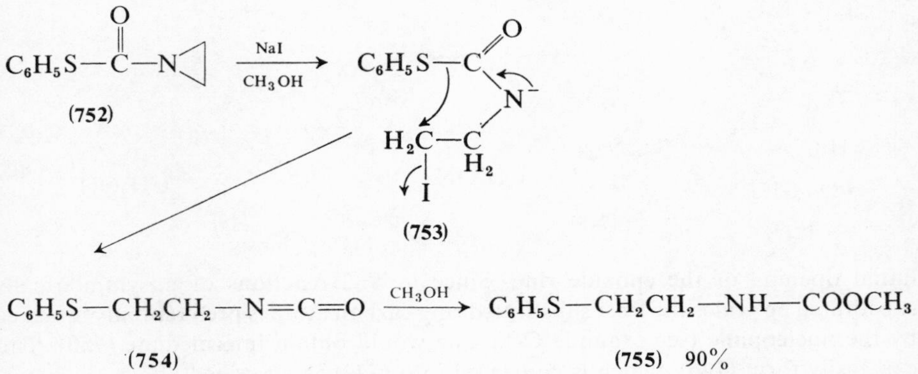

$$C_6H_5S-\overset{O}{\overset{\|}{C}}-N\triangleleft \quad \xrightarrow[CH_3OH]{NaI} \quad C_6H_5S-\overset{O}{\overset{\|}{C}}\begin{smallmatrix}\\ N^-\\ H_2C-C\\ |\quad H_2\\ I\end{smallmatrix}$$

(752)   (753)

$$C_6H_5S-CH_2CH_2-N=C=O \xrightarrow{CH_3OH} C_6H_5S-CH_2CH_2-NH-COOCH_3$$

(754)   (755)   90%

cyclic five-membered transition state. Addition of I$^-$ to the carbonyl group of (752) and subsequent attack of the thiophenoxide anion on the three-membered ring would lead to the same product[81].

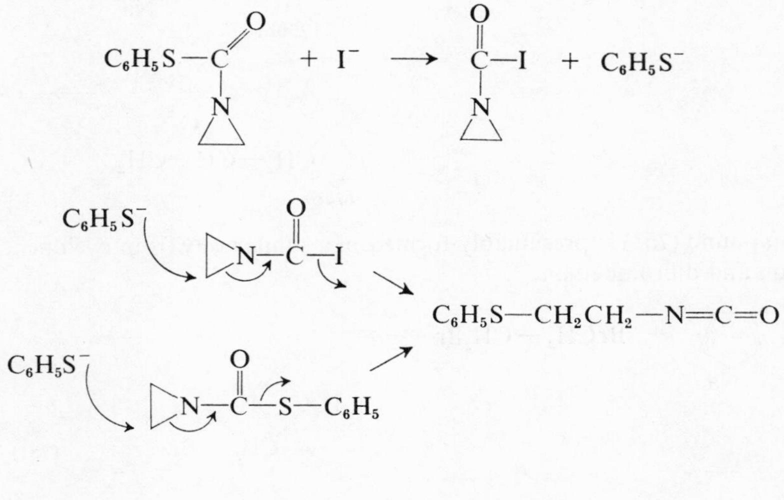

$$C_6H_5S-\overset{O}{\overset{\diagup}{C}}\underset{N\triangle}{} + I^- \longrightarrow \overset{O}{\overset{\|}{C}}-I \; \underset{N\triangle}{} + C_6H_5S^-$$

$$C_6H_5S^- \curvearrowright \triangleright N\!\!-\!\!\overset{O}{\overset{\|}{C}}\!\!-\!\!I$$

$$C_6H_5S^- \curvearrowright \triangleright N\!\!-\!\!\overset{O}{\overset{\|}{C}}\!\!-\!\!S\!-\!C_6H_5 \quad \longrightarrow \quad C_6H_5S-CH_2CH_2-N=C=O$$

S-22. Reaction of 1-(arylazo)aziridines (756) with sodium iodide

The explanation offered by the authors[80] is indicated in the chart.

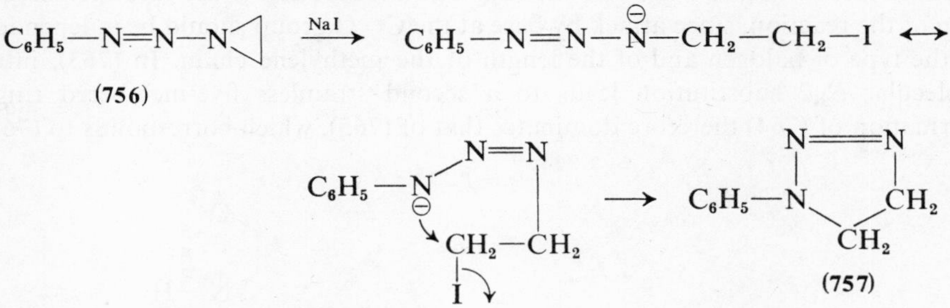

$C_6H_5-N=N-N\triangleleft$ $\xrightarrow{NaI}$ $C_6H_5-N=N-\overset{\ominus}{N}-CH_2-CH_2-I$ ⟷

(756)

$C_6H_5-\underset{\ominus}{N}\begin{matrix}N=N\\ |\\ CH_2-CH_2\\ |\\ I\end{matrix}$ ⟶ $C_6H_5-N\begin{matrix}N=\!=\!N\\ \diagdown\!\!\diagup\quad CH_2\\ CH_2\end{matrix}$

(757)

S-23. Reaction of 2(3H)-benzofuranone derivatives with base[82]

The benzofuranone derivative (758) and morpholine form a product (759) by an S_N1 reaction, since a relatively stable phenyl-conjugated tertiary carbonium ion is

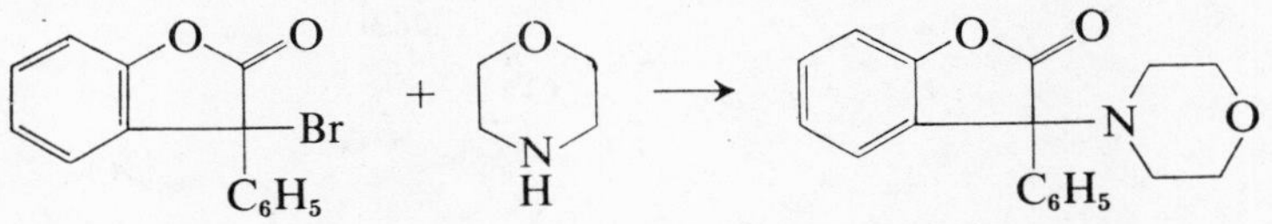

(758) + morpholine ⟶ (759) 76%

formed as intermediate. In (760), however, no such stabilization is possible; S_N2 substitution of the halogen is sterically inhibited; intramolecular S_N2 substitution in compounds of the neopentyl type does, however, occur[83]. Primary attack by morpholine at the carbonyl group leads to the cleavage of the furanone ring; subsequent S_N2 substitution yields the product (761). When $NaOCH_3$ is the

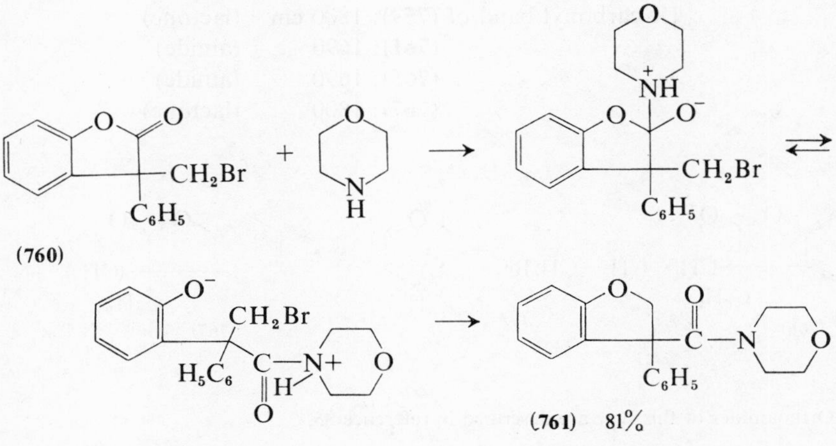

(760) + morpholine ⟶ intermediate ⇌

⟶ (761) 81%

attacking nucleophile, the bromide reacts faster than the chloride[84]. The rate depends, furthermore, on the size of ring in the product.

These results indicate that the internal S_N2 substitution is the rate-determining step of the reaction, since attack by base at the C═O group should be independent of the type of halogen and of the length of the methylene chain. In (**763**), intramolecular S_N2 substitution leads to a second strainless five-membered ring*. Formation of (**764**) therefore dominates that of (**765**), which corresponds to (**761**).

(762) (763)

(764) 92% (765)

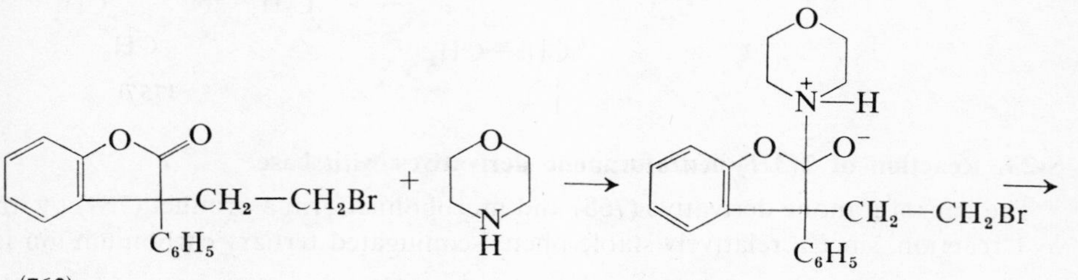

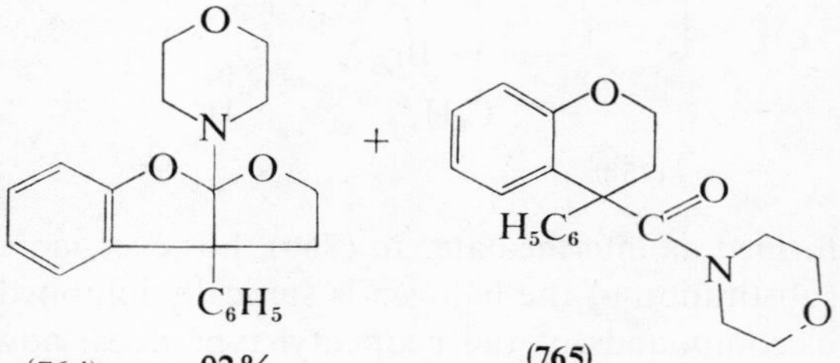

IR carbonyl band of	(**759**):	1800 cm^{-1}	(lactone)
	(**761**):	1690	(amide)
	(**765**):	1690	(amide)
	(**767**):	1800	(lactone)

(766) (767) 85%

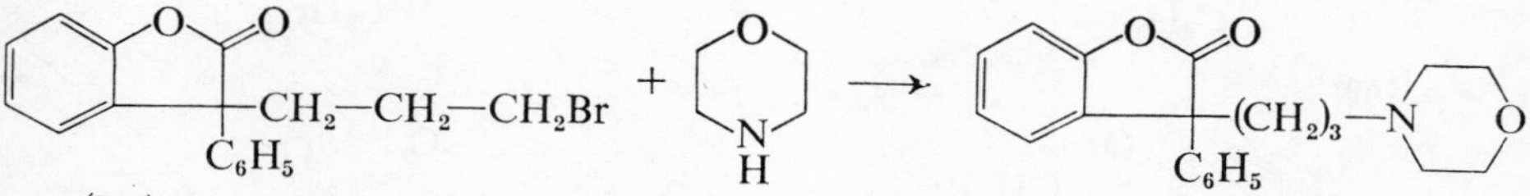

* Orthoamides of this type are described in reference 85.

An alternative to the formation of **(764)** through **(763)** is formation through **(768)**[86]. A stable salt **(770)** is formed if 2-chloroethyl *p*-nitrobenzoate **(769)** is treated with $SbCl_5$; however, no such salt has been observed on reaction of **(762)** with $SbCl_5$. This mechanism is, therefore, not very likely.

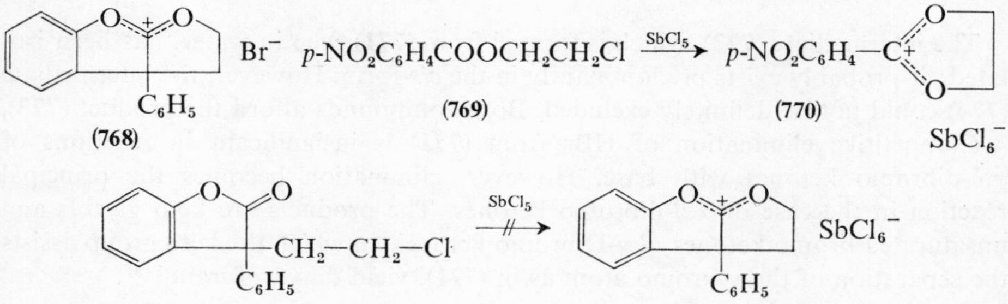

[structures: **(768)** with Br⁻, **(769)** *p*-NO₂C₆H₄COOCH₂CH₂Cl $\xrightarrow{SbCl_5}$ **(770)** *p*-NO₂C₆H₄—C with $SbCl_6^-$]

[structure $\xrightarrow{SbCl_5}\not\to$ structure with $SbCl_6^-$]

The derivative **(766)** undergoes normal S_N2 substitution, giving the product **(767)** in 85% yield. Rearrangement in analogy to the formation of **(761)** and **(765)** would result in an unfavored seven-membered ring.

S-24. Solvolysis of γ,δ-dibromo ketones[87]

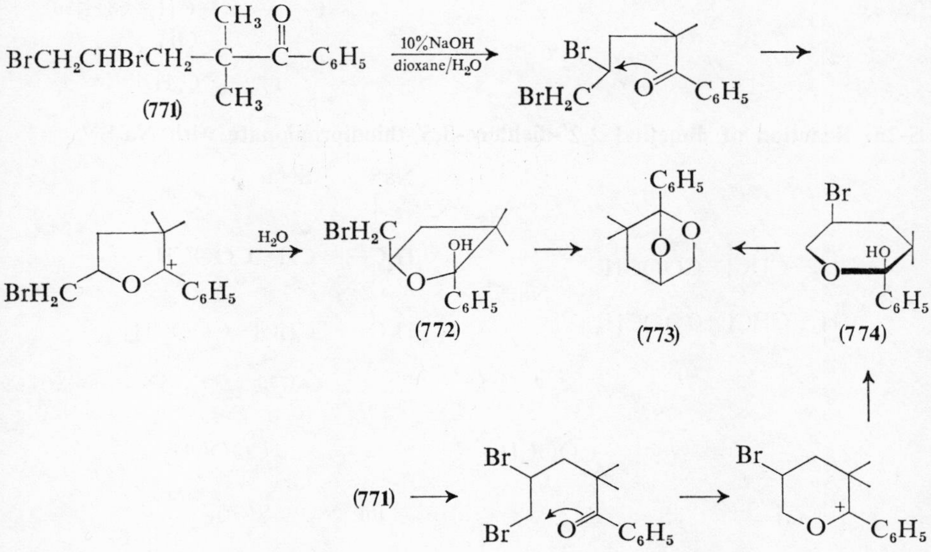

$$BrCH_2CHBrCH_2-\underset{\underset{CH_3}{|}}{\overset{\overset{CH_3}{|}\overset{O}{\|}}{C}}-C-C_6H_5 \xrightarrow[dioxane/H_2O]{10\% NaOH}$$

(771)

[cyclic intermediates leading through **(772)**, **(773)**, **(774)**]

$$(771) \longrightarrow \text{[Br,Br intermediate]} \longrightarrow \text{[cyclic oxocarbenium]}$$

Loss of both the γ- and δ-bromine atoms in **(771)** is probably assisted by the C=O group. The positive charge at the carbonyl-carbon atom is stabilized by the phenyl group. Possibly the reaction is initiated by attack of OH⁻ on the carbonyl group.

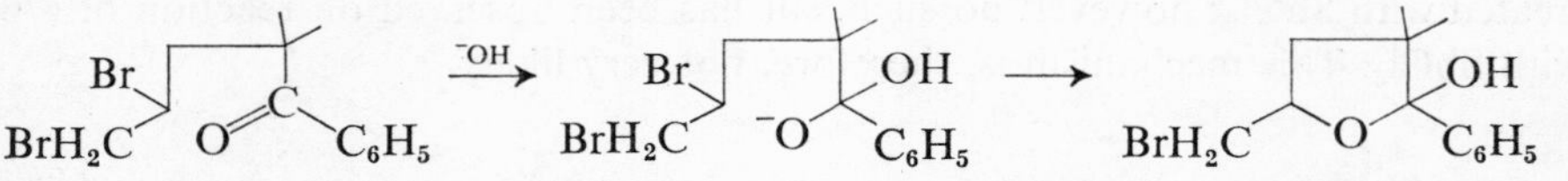

$$\underset{BrH_2C}{Br}\underset{O}{\diagdown}\underset{C_6H_5}{C} \xrightarrow{^-OH} \underset{BrH_2C}{Br}\underset{O}{\diagdown}\underset{C_6H_5}{\diagup}\overset{OH}{\diagup} \longrightarrow \underset{BrH_2C}{\diagdown}\underset{O}{\diagdown}\underset{C_6H_5}{\diagup}\overset{OH}{\diagup}$$

The intermediate (772), which is formed from (771) even in water, has been isolated; it probably exists predominantly in the *cis*-form. However, the intermediate (774) could not be definitely excluded. Both compounds afford the product (773).

Competitive elimination of HBr from (771) is insignificant in reactions of γ,δ-dibromo ketones with base. However, elimination becomes the principal reaction in the case of α,β-dibromo ketones. The products are keto glycols and unsaturated bromo ketones. β,γ-Dibromo ketones, in which the keto group assists the separation of the γ-bromo atom as in (771), yield dihydrofuranols[88].

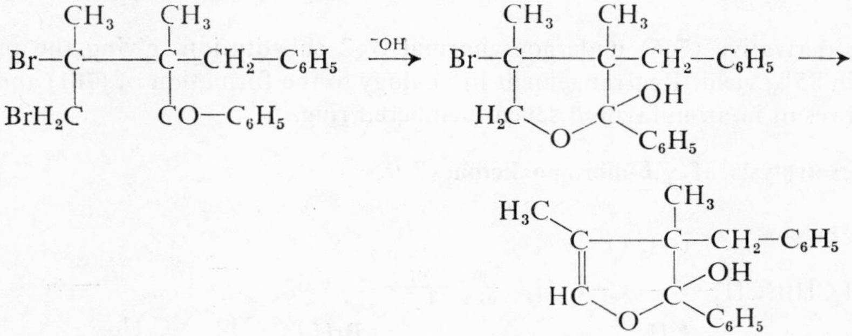

S-25. Reaction of dimethyl 2,2'-dichloro-3,3'-thiodipropionate with Na₂S[89]

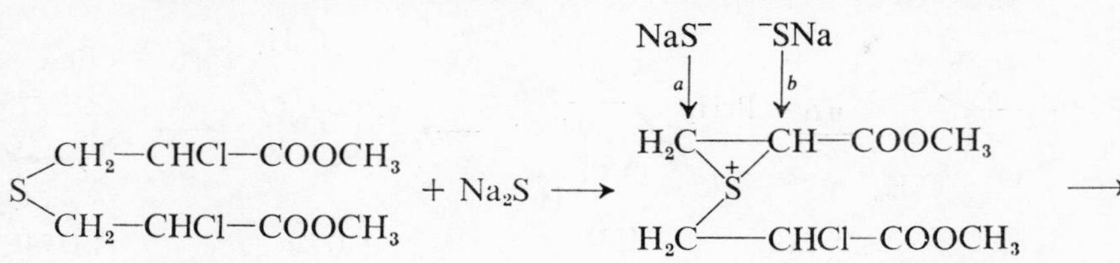

$$S\underset{CH_2-CHCl-COOCH_3}{\overset{CH_2-CHCl-COOCH_3}{\diagup}} + Na_2S \longrightarrow$$

(774)

(775)

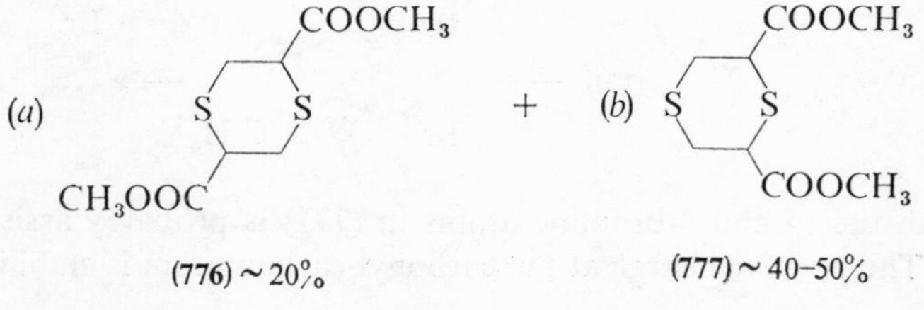

(776) ~ 20%   (777) ~ 40–50%

In the reaction of β,β'-dichloro sulfide (774) with Na_2S two isomeric dicarboxylic esters (776) and (777) are obtained. The reaction probably involves an intermediate (775). A similar intermediate is formed in the solvolysis of (778)[90].

$$R-S-CH_2CH_2Cl \rightleftarrows R-\overset{+}{S}\underset{CH_2}{\overset{CH_2}{\diagup}}\overset{Cl^-}{\underset{H_2O}{\xrightarrow{X^-}}} \begin{array}{l} R-S-CH_2CH_2X \\ \\ R-S-CH_2CH_2OH \end{array}$$

(778)

S-26. Arbusow–Michaelis reaction

The reaction of phosphites with alkyl halides yields phosphonic esters (780). The

$$(RO)_3P + R'-X \longrightarrow \left[(RO)_3P^+-R'\right]X^- \longrightarrow$$

(779)

$$X^-\cdots R\cdots O-{}^+PR'(OR)_2 \longrightarrow (RO)_2P\overset{O}{\underset{R'}{\diagdown}} + RX$$

(780)

reaction has been named after its discoverers the Arbusow–Michaelis reaction[91]. In the first step of the reaction a phosphonium salt (779) is formed by S_N2 displacement of the halide anion, and it can be isolated in some cases. In the second step the halide ion attacks the α-carbon atom of an alkoxy group. If the α-carbon is optically active, this replacement is accompanied by inversion of configuration. In some cases the second step of the Arbusow–Michaelis reaction may be S_N1; for example, tri-*tert*-butyl phosphite and methyl iodide yield di-*tert*-butyl methylphosphonate and isobutene[92].

In the reaction of triethyl phosphite with ethyl bromofluoroacetate the fluorinated phosphonic ester (781) is obtained since the weaker base Br^- is preferentially displaced[93].

$$(C_4H_9O)_3P + CH_3I \longrightarrow [(C_4H_9O)_3P^+CH_3]\,I^- \longrightarrow$$
$$(C_4H_9O)_2P(CH_3)\rightarrow O + CH_2\!\!=\!\!C(CH_3)_2 + HI$$

$$(C_2H_5O)_3P + CHFBr-COOC_2H_5 \longrightarrow$$
$$(C_2H_5O)_2P(O)-CHF-COOC_2H_5 + C_2H_5Br$$

(781)　　　　83%

S-27. Perkow reaction

Alkyl or aryl halides and aliphatic or aromatic acid halides are suitable starting materials for the Arbusow–Michaelis reaction. α-Halo ketones and some α-halo esters afford enol esters as side-products along with phosphonates as principal products[94]. Perkow[95] *et al.* discovered that enol esters such as **(782)** are obtained exclusively if α-halo aldehydes are used. Several mechanisms have been proposed

$$(C_2H_5O)_3P + Cl_2HC-CHO \longrightarrow (C_2H_5O)_2P(O)-O-CH=CHCl + C_2H_5Cl$$

(782)

for the Perkow reaction. Theoretically, attack of the phosphite may occur at several positions:

(1) *Attack at α-carbon*

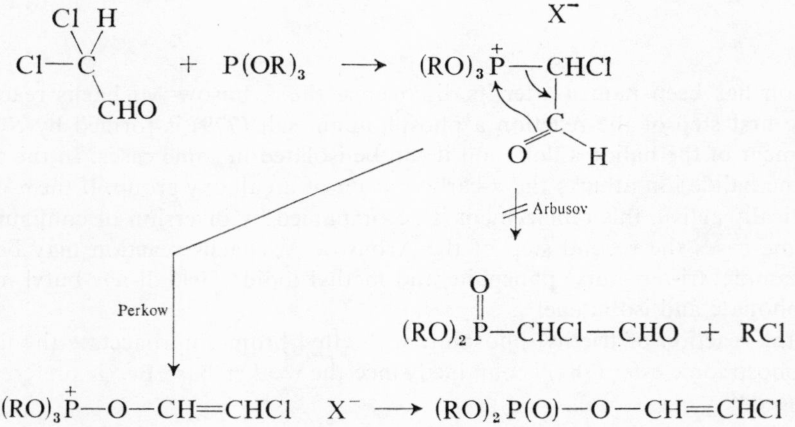

$$\underset{\text{CHO}}{\overset{\text{Cl}\ \ \text{H}}{Cl-C}} + P(OR)_3 \longrightarrow (RO)_3\overset{+}{P}-CHCl-\underset{H}{\overset{O}{C}} \quad X^-$$

$$\downarrow \text{Arbusov}$$

$$(RO)_2\overset{O}{\underset{\|}{P}}-CHCl-CHO + RCl$$

Perkow ↓

$$(RO)_3\overset{+}{P}-O-CH=CHCl \ \ X^- \longrightarrow (RO)_2P(O)-O-CH=CHCl$$

This mechanism is questionable because trialkyl-(β-oxoalkyl)phosphonium salts **(783)** do not rearrange to enol esters.

$$X^- \ R_3P^+-CHR'-COR'' \ \ \not\longrightarrow \ R_3P^+-O-CR''=CHR' \ \ X^-$$

(783)

(2) *Attack at the α-halogen atom*

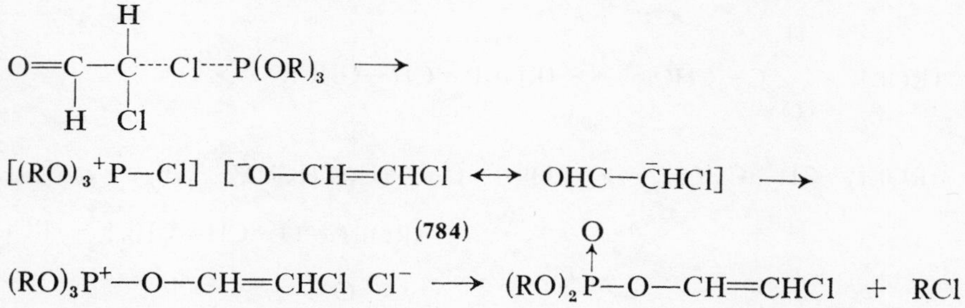

$$O=C-\underset{\underset{Cl}{|}}{\overset{\overset{H}{|}}{C}}\cdots Cl \cdots P(OR)_3 \longrightarrow$$
(with H on the left carbon)

$$[(RO)_3{}^+P-Cl]\ [^-O-CH=CHCl \longleftrightarrow OHC-\bar{C}HCl] \longrightarrow$$
$$\qquad\qquad\qquad\qquad (784)$$

$$(RO)_3P^+-O-CH=CHCl\ Cl^- \longrightarrow (RO)_2\overset{\overset{O}{\uparrow}}{P}-O-CH=CHCl + RCl$$

Ions such as (**784**) react readily with alcohols:

$$[(RO)_3P^+-Cl][OHC-\bar{C}HCl] + CH_3OH \longrightarrow$$
$$\quad (784)$$
$$\qquad\qquad\qquad\qquad (RO)_2P(O)-OCH_3 + OHC-CH_2Cl + RCl$$

Since, on the other hand, high yields of enol esters are obtained in C_2H_5OH, this mechanism is not very likely either.

Loss of positive halogen is facilitated by electron-attracting halogen atoms at the same carbon atom. This mechanism may apply to reactions of α-trihalogenated carbonyl compounds with phosphites.

(3) *Attack at the carbonyl-oxygen*

(*a*) nucleophilic attack

$$(RO)_3P \rightarrow OHC-CHCl_2 \longrightarrow (RO)_3P^+-O-CH=CHCl\ X^- \longrightarrow$$
$$\qquad\qquad\qquad\qquad\qquad\qquad (RO)_2P-O-CH=CHCl + RCl$$

(*b*) electrophilic attack

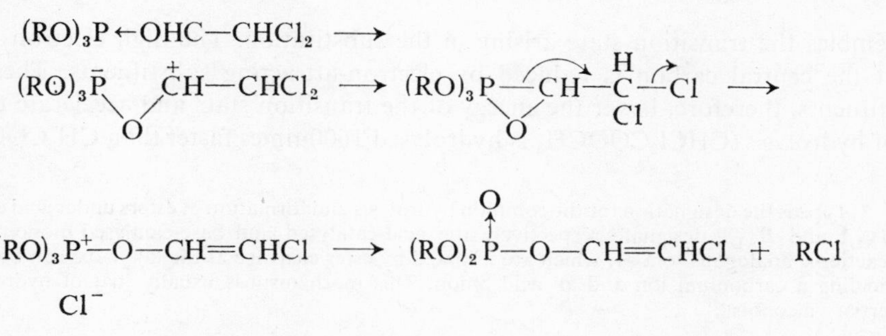

$$(RO)_3P \leftarrow OHC-CHCl_2 \longrightarrow$$

$$(RO)_3\bar{P}\diagdown\overset{+}{C}H-CHCl_2 \longrightarrow (RO)_3P\diagdown CH-\overset{\overset{H}{}}{\underset{\underset{Cl}{}}{C}}-Cl \longrightarrow$$
$$\qquad O \qquad\qquad\qquad\qquad O$$

$$(RO)_3P^+-O-CH=CHCl \longrightarrow (RO)_2\overset{\overset{O}{\uparrow}}{P}-O-CH=CHCl + RCl$$
$$Cl^-$$

(4) *Attack at the carbonyl-carbon*

$$(RO)_3P + \underset{O}{\overset{H}{C}}-CHCl_2 \longrightarrow (RO)_3\overset{+}{P}-\underset{O^-}{CH}-CHCl_2 \longrightarrow$$

$$(RO)_3P-\underset{O}{CH}-CHCl_2 \longrightarrow (RO)_3\overset{+}{P}-O-CH=CHCl\ X^- \longrightarrow$$

$$(RO)_2\underset{\underset{O}{\downarrow}}{P}-O-CH=CHCl + RCl$$

The results available so far do not permit a final decision between these mechanisms.

(b) S_N2 SUBSTITUTION AT CARBONYL-CARBON

The base-catalysed hydrolysis of esters, designated as $B_{AC}2$ reaction by Sir Christopher Ingold[96], usually proceeds by S_N2 substitution at the carbonyl-carbon atom (B denotes basic, AC indicates cleavage between acyl-carbon and oxygen, 2 indicates the bimolecularity of the reaction)*. In contrast to S_N2 substitution at a saturated carbon atom, the carbon is quadrivalent in the transition state. An anion (785) is an intermediate with a definite lifetime, and closely re-

$$R-\overset{\overset{O}{\|}}{C}-OR' \xrightarrow{\ ^-OH\ } R-\underset{\underset{OH}{|}}{\overset{\overset{O^-}{|}}{C}}-OR' \longrightarrow R-C\overset{\diagup O}{\diagdown OH} + {}^-OR'$$

(785)

sembles the transition state arising in the substitution. The high electron density at the central carbon is reduced by electron-attracting substituents. These substituents, therefore, lower the energy of the transition state and accelerate the rate of hydrolysis ($CHCl_2COOCH_3$ is hydrolysed 1600 times faster than CH_3COOCH_3).

---

* $A_{AC}2$ is the designation for the common hydrolysis and formation of esters under acid catalysis. $A_{AL}1$ and $B_{AL}1$ designate respectively the acid-catalysed and base-catalysed monomolecular reactions, analogous to S_N1, which are initiated by ester cleavage at the alkyl–oxygen bond, thus yielding a carbonium ion and an acid anion. This mechanism is usually that of hydrolysis of tertiary alcohols.

S-28. Thermolysis of piperidinomethyl thioacetate[97]

In the thermolysis of (**786**) in refluxing benzene the nitrogen atom reacts with the carbonyl-carbon atom.

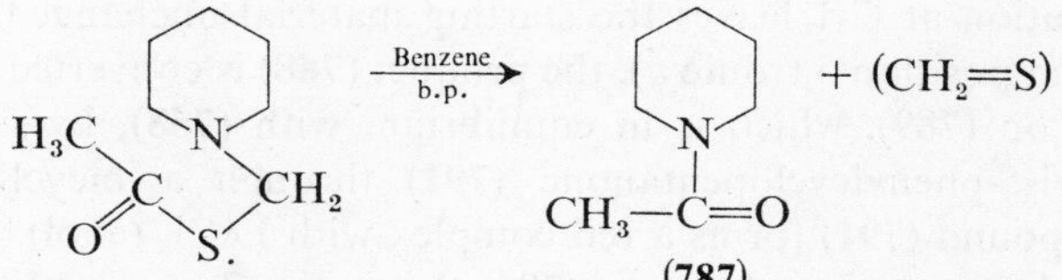

In solvents such as hexane and acetone, the reaction is of the first order and consequently intramolecular. However, in $CHCl_3$ or CH_3NO_2 it is of the second order in (**786**) and proceeds by an intermolecular mechanism[98].

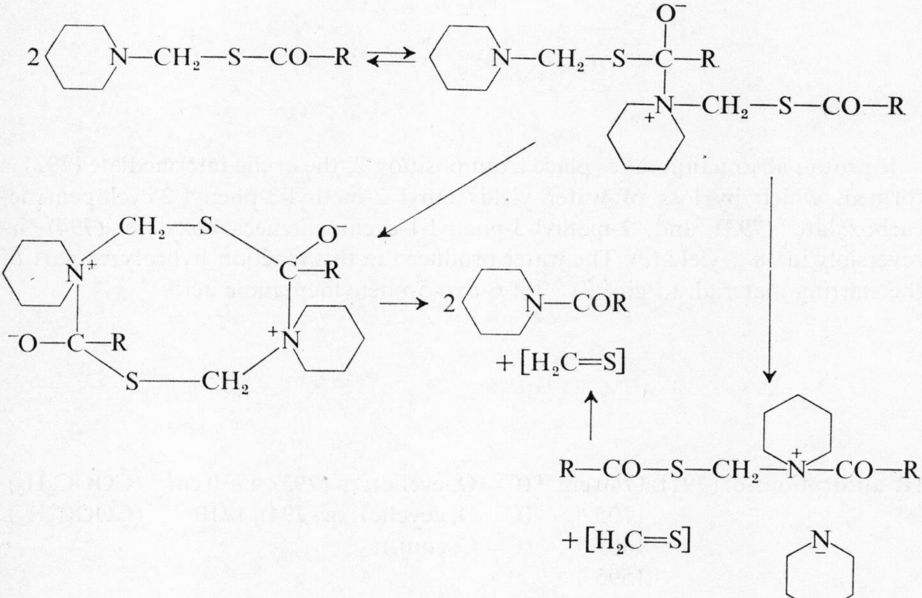

S-29. Base-catalyzed cyclizations of ethyl 6-oxo-5-phenylheptanoate (787)[99] (see facing page)

Attack by the base can occur at four positions: at C-1, C-2, C-5, or C-7. The acidity of the hydrogen atoms decreases in the following sequence: $C_5 > C_7 > C_2$. Direct substitution at C-1 leaves the starting material unchanged. If a proton is abstracted from position 5 (route *a*), the product (**788**) is converted into a diketone (**789**). Attack on (**789**), which is in equilibrium with (**788**), by sodium ethoxide yields 2-acetyl-5-phenylcyclopentanone (**791**) through a bicyclic intermediate (**790**)[100]. Compound (**791**) forms a red complex with $FeCl_3$ (enol).

A more likely mechanism leads to (**791**) through a 7-membered ring (route *b*). It was shown that 1,3-cycloheptanedione rearranges under these conditions to 2-acetylcyclopentanone.

$$\text{cycloheptane-1,3-dione} \xrightarrow[C_2H_5OH]{NaOC_2H_5} \text{2-acetylcyclopentanone}$$

If proton abstraction takes place from position 2, the cyclic intermediate (**792**) is formed, which by loss of water yields ethyl 2-methyl-3-phenyl-2-cyclopentenecarboxylate (**793**) and 2-methyl-3-phenyl-1-cyclopentenecarboxylate (**794**) irreversibly in 38% yield (*c*). The water produced in this reaction hydrolyzes part of the starting material, to give 31% of 6-oxo-5-phenylheptanoic acid.

IR absorption: of (**791**), 1740 cm^{-1} (C=O, cyclic); of (**793**), 1730 cm^{-1} (COOC$_2$H$_5$).
            1705   (C=O, acyclic); of (**794**), 1710   (COOC$_2$H$_5$).
            1605   (C=C, conj.);
            1595
            1500
            760
            700

NMR absorption of (**791**), 2.9–2.76 τ (5) (phenyl)
                       7.8 τ (3) (H$_3$C—CO—)

Aliphatic Substitution

S-30. Reaction of 4-oxocyclohexyl benzoate (794) with $KOC(CH_3)_3$[101]

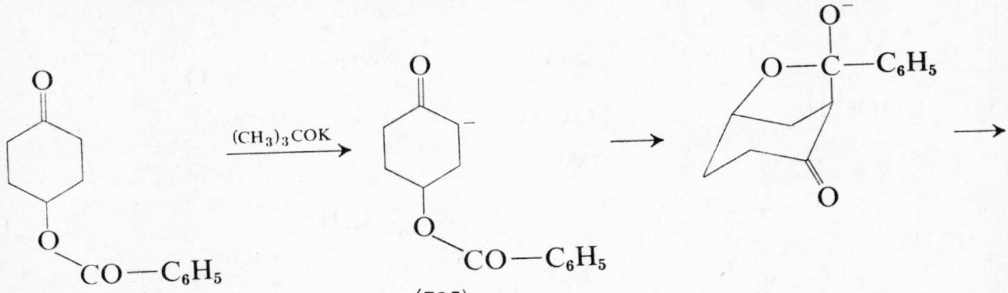

Characteristic absorption of (**798**):
        1720 cm^{-1} (COOH);   244 mμ ($\varepsilon = 16700$);
        1685 cm^{-1} (C=O, conj.);
        -0.98 τ (1) (COOH);   8.42 τ (3) (H_a)
        2.1   (2) (o-H of Ph); 9.12   (2) (H_c)
        7.52   (3) (H_b)

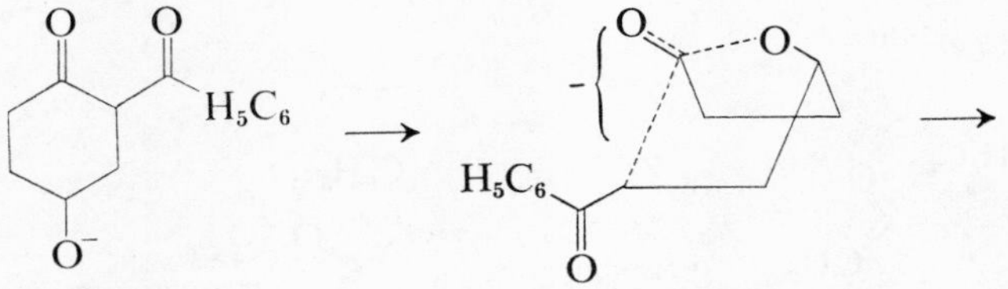

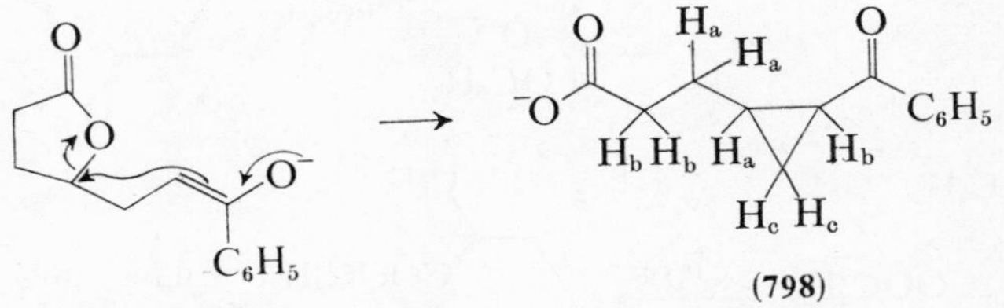

The base abstracts a proton from the α-position to the keto group. Reesterification, which would yield *tert*-butyl benzoate, is unlikely because of the size of the *tert*-butyl group.

The mechanism illustrated is supported by the isolation of intermediates; the enolate anion (**795**) attacks the ester-carbonyl group intramolecularly (cf. the Baker–Venkataraman rearrangement[102]).

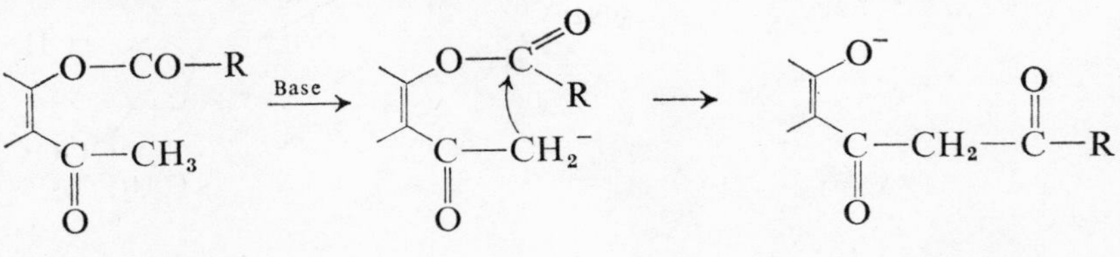

Aliphatic Substitution

A similar intramolecular substitution has been described by Newman and Yu[103].

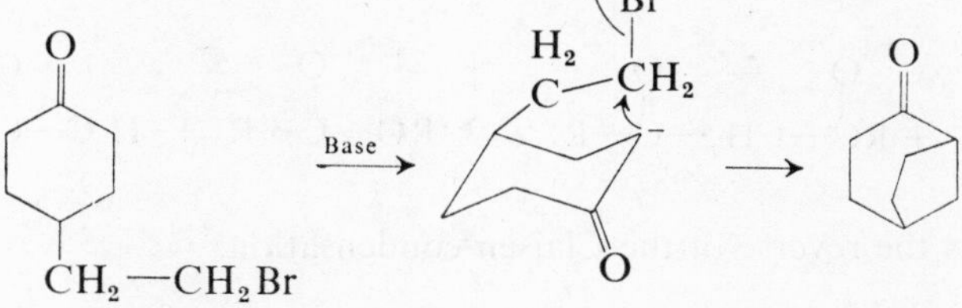

The intermediate **(796)** above is converted intramolecularly into a lactone **(797)**, which was isolated[104]. A closely related 1,4-substitution is the formation of 1,4-

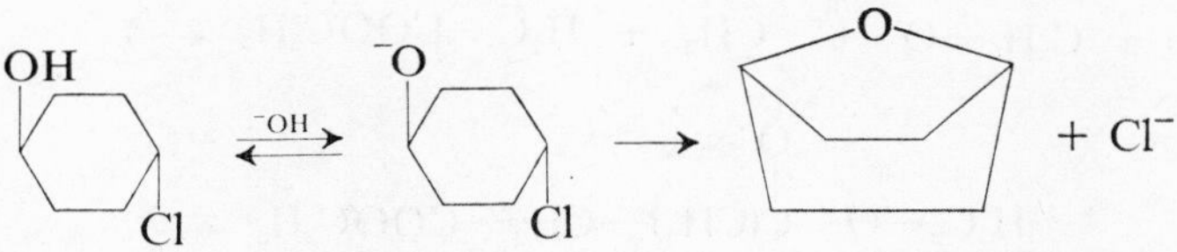

epoxycyclohexane[105]. The predominant attack of the alkoxide-oxygen on the ring carbonyl group instead of the C=O group in the side chain has also been observed in other reactions[106]. Lactone **(797)** rearranges in a known manner to the cyclopropane derivative **(798)**. An analogous reaction is the formation of 2-vinylcyclopropane-1,1-dicarboxylic ester **(800)** from (4-bromo-2-butenyl)malonic ester **(799)**[107].

$$BrCH_2CH=CHCH_2CH(COOR)_2 \xrightarrow{Base} BrCH_2CH=CHCH_2-\bar{C}(COOR)_2 \longleftrightarrow$$
$$(799)$$

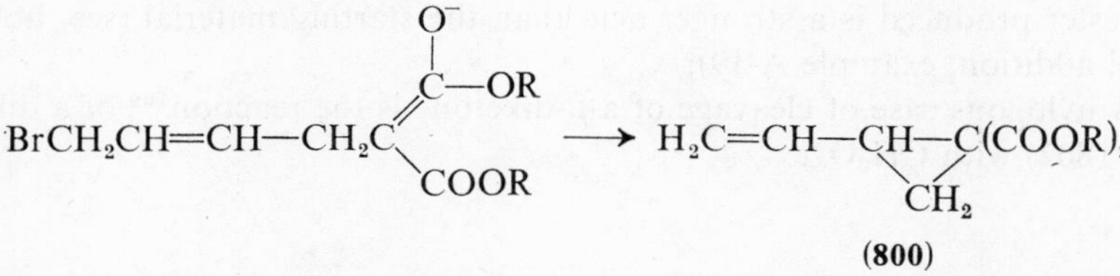

$$BrCH_2CH=CH-CH_2\overset{O^-}{\underset{COOR}{C-OR}} \longrightarrow H_2C=CH-\underset{CH_2}{\underset{|}{CH}}-C(COOR)_2$$
$$(800)$$

A mechanism involving an anion **(801)** is unlikely, since the strongly basic ⁻OH is a bad leaving group[108].

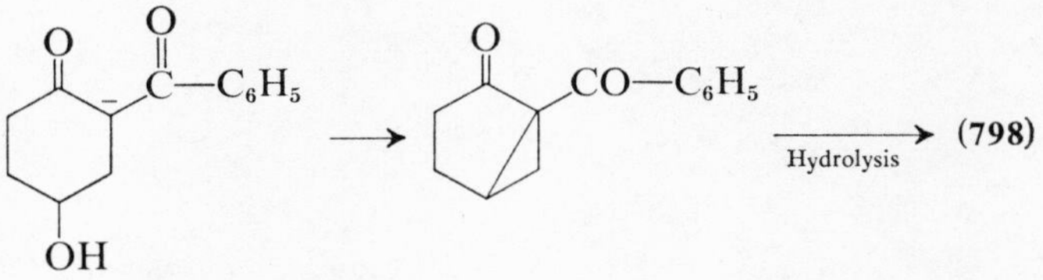

$$\xrightarrow{\text{Hydrolysis}} (798)$$

(801)

S-31. Cleavage of β-diketones

β-Diketones are cleaved by strong bases into ester and enolate:

$$RO^- + R\overset{O}{\underset{\|}{C}}-CH_2-\overset{O}{\underset{\|}{C}}-R \rightleftarrows RO-\overset{O}{\underset{\|}{C}}-R + H_2\bar{C}-\overset{O}{\underset{\|}{C}}-R$$

The reaction is the reverse of the Claisen condensation.

$$H_3C-COOC_2H_5 + {}^-OC_2H_5 \rightleftarrows H_2\bar{C}-COOC_2H_5 + C_2H_5OH$$

$$C_2H_5-O-\overset{O}{\underset{\|}{C}}-CH_3 + H_2\bar{C}-COOC_2H_5 \rightleftarrows$$

$$H_5C_2-O-\overset{O^-}{\underset{|}{C}(CH_3)}-CH_2-COOC_2H_5 \rightleftarrows$$

$$C_2H_5O^- + H_3C-\overset{O}{\underset{\|}{C}}-CH_2-COOC_2H_5 \rightleftarrows$$

$$H_3C-\overset{O}{\underset{\|}{C}}-\bar{C}H-COOC_2H_5 + C_2H_5OH$$

An excess of base shifts the Claisen reaction to the side of the products, since the β-keto ester produced is a stronger acid than the starting material (see, however, Michael addition, example A-19).

The vinylogous case of cleavage of a β-diketone is the reaction[109] of a diketone such as (802) with CH_3O^-.

S-31. *Reaction of 8a-methyl-1,2,3,4,6,7,8,8a-octahydro-1,6-naphthalenedione* **(802)** *with sodium methoxide*

In the second step of this reaction anion **(803)** forms the product **(805)** by attack of the negative carbon on the ester group. The second product **(804)** can be obtained in 10% yield.

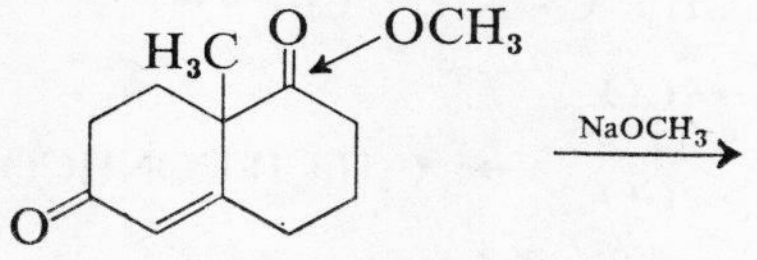

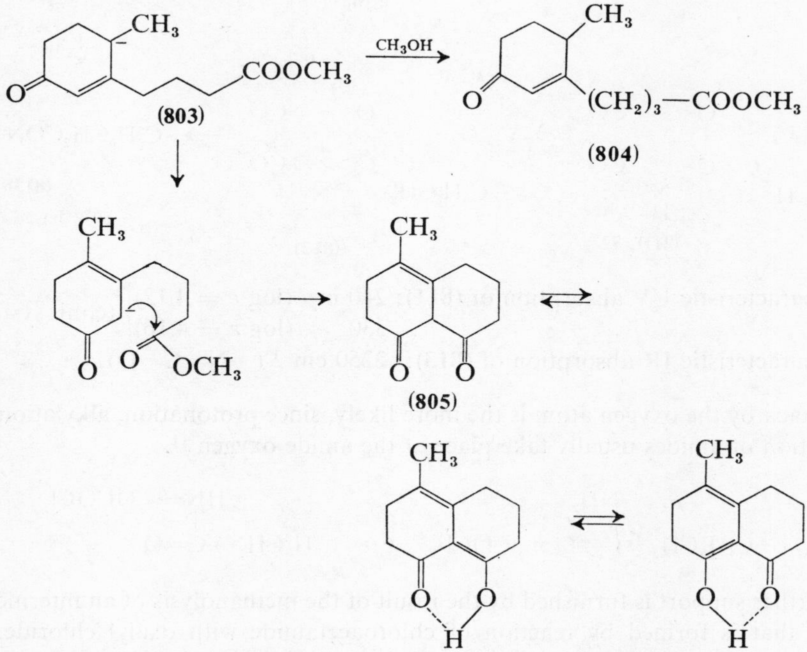

If **(802)** is treated with OH⁻ instead of CH_3O^-, the carboxylic acid **(807)** that corresponds to product **(804)** is formed exclusively. Cyclization of **(806)** (negative charge at the COO⁻ group) is much less favored than that of **(803)**.

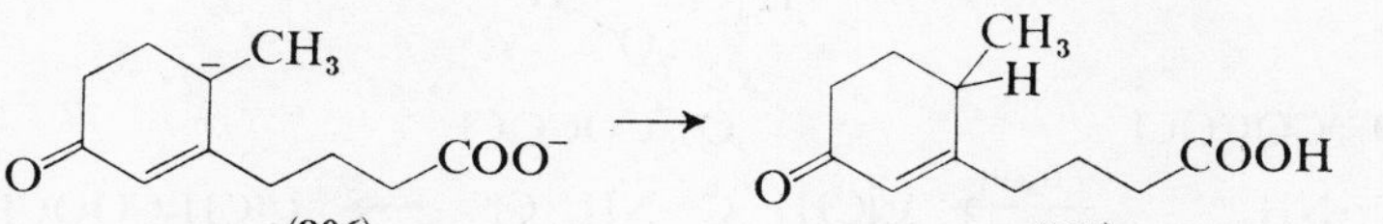

S-32. Reaction of oxalyl chloride with amides[110]

Nucleophilic attack on oxalyl chloride can, in principle, take place by the oxygen or by the nitrogen atom of an amide.

$$C_6H_5CH_2CONH_2 + ClCOCOCl \rightleftarrows C_6H_5CH_2-\underset{\underset{O-COCOCl}{|}}{C}=NH_2^+ \;\; Cl^- \underset{+HCl}{\overset{-HCl}{\rightleftarrows}}$$

$$C_6H_5CH_2-\underset{\underset{O-COCOCl}{|}}{C}=NH \longrightarrow \text{(810)} \longrightarrow C_6H_5CH_2CONHCOCOCl \;\; \text{(814)}$$

(810): 5-membered ring with C$_6$H$_5$CH$_2$ group, O—CO, CO, N$^+$H, Cl$^-$

(811) 72%: 5-membered ring with H$_5$C$_6$/H on exocyclic C=C, O—CO, CO, NH

(812): 5-membered ring with C$_6$H$_5$CH$_2$, O—CO, CO, N

$$\text{(812)} \xrightarrow{-CO, -HCl} C_6H_5CH_2CON=C=O \quad \text{(813)}$$
50% based on (811)

(811) → (812) via Δ Heat, −HCl

Characteristic UV absorption of (811): 240 mμ (log ε = 4.12)
330 (log ε = 4.16) } (conj. system)

Characteristic IR absorption of (813): 2250 cm⁻¹ (—N=C=O)

Attack by the oxygen atom is the more likely, since protonation, alkylation, and acylation of amides usually take place at the amide oxygen[111].

$$C_6H_5CH_2-\underset{\underset{NH_2}{|}}{C}=O + (COCl)_2 \not\longrightarrow C_6H_5CH_2-\underset{\underset{HN-COCOCl}{|}}{C}=O$$

Further support is furnished by the result of the methanolysis of an intermediate (808) that is formed by reaction of chloroacetamide with oxalyl chloride. The methanolysis products are NH$_4$Cl and methyl chloroacetate, which are the products expected from nucleophilic attack by the amide-oxygen atom on oxalyl

$$ClCH_2-\underset{\underset{Cl^-}{+}}{\overset{O-COCOCl}{\underset{|}{C}-NH_2}} \xrightarrow{CH_3OH} ClCH_2-\underset{\underset{OCH_3}{|}}{\overset{O-COCOCl}{\underset{|}{C}-\overset{+}{N}H_3}} Cl^- \longrightarrow ClCH_2COOCH_3 + NH_4Cl$$

(808) $\quad\quad\quad\quad\quad\quad\quad\quad\quad\quad\quad\quad\quad\quad\quad\quad\quad\quad\quad\quad\quad$ [+ H$_3$COCOCOCl]

chloride. The reaction path *via* enol (**809**) with enol formation as rate-determining step can be excluded, because the reaction rate is reduced by a phenyl group on the nitrogen atom whereas enol formation should be independent of the substituent on nitrogen.

$$C_6H_5CH=\underset{\underset{(809)}{}}{\overset{OH}{C}}-NH_2$$

By loss of one of the relatively acidic benzylic hydrogens from intermediate (**810**), the product (**811**) is formed and this is converted into product (**813**), presumably through an intermediate (**812**). If (**810**) did not possess acidic hydrogen atoms, (**812**) would be formed directly from (**810**) by loss of a proton from the nitrogen.

The intermediate (**814**) cannot be isolated if amides without a substituent on the nitrogen are used as starting materials. From *N*-monosubstituted amides, however, compounds of type (**814**) have been obtained. It is likely, therefore, that (**814**) is also formed as an intermediate from unsubstituted amides. The reaction of *N*-unsubstituted amides with oxalyl chloride is a useful method for the preparation of acyl isocyanates.

S-33. Preparation of cyanic esters[112]

$$C_2H_5O-\underset{\underset{S}{\|}}{C}-Cl + NaN_3 \longrightarrow C_2H_5O-C\underset{(815)}{\overset{\begin{array}{c}S\\ \| \\ N \quad N \\ \diagdown N \diagup \end{array}}{}} \xrightarrow{20°} \underset{(816)\ 60\%}{C_2H_5O-C\equiv N} + N_2 + S$$

Nucleophilic attack of N_3^- on *O*-ethyl chloro(thioformate), followed by cyclization, yields 5-ethoxy-1,2,3,4-thiatriazole (**815**). At room temperature this changes into ethyl cyanate (**816**). Numerous aryl and alkyl cyanates have been prepared by this method.

In contrast to isocyanic esters, cyanic esters were unknown for a long time*, probably because of the strong tendency of the latter to trimerize to cyanuric esters or to rearrange to isocyanic esters.

$$3\,RO-C\equiv N \longrightarrow \begin{array}{c} OR \\ | \\ C \\ N\diagup \ \diagdown N \\ | \qquad | \\ C \qquad C \\ RO\diagdown N \diagup OR \end{array}$$

* The first aryl cyanate was prepared by Stroh and Gerber.[113] Secondary reactions of cyanic esters are prevented by the presence of bulky substituents, as in 2,6-di-*tert*-butylphenyl cyanate[114].

Recently some aryl cyanates have been prepared by reaction of phenols with ClCN or BrCN[115]. Aliphatic esters are obtained by this method only if strongly

$$RONa + HalCN \rightarrow ROCN \quad (R = C_6H_5, CH_3C_6H_4, NO_2C_6H_4, \text{ or } Hal_3CCH_2)$$

acidic alcohols such as CCl_3CH_2OH or CBr_3CH_2OH are used.

V. S_Ni SUBSTITUTION

Treatment of alcohols with thionyl chloride yields alkyl halides by way of alkyl chlorosulfites (**817**).

$$ROH + SOCl_2 \longrightarrow RO\text{—}SOCl \longrightarrow R^+ \overset{O}{\underset{Cl}{\cdots}} S{=}O \longrightarrow RCl + SO_2$$
(**817**)

If alcohols with an asymmetric α-carbon atom are used, the reaction may proceed with retention of configuration. It is likely that a tight ion-pair is first formed by a S_N1 reaction, and that the anion adds to the asymmetric carbon by "internal return" with simultaneous loss of SO_2. Recombination occurs faster than a nucleophile from the solution could react with the asymmetric centre[116]. The internal nucleophilic substitution is referred to as S_Ni reaction. Since a carbonium ion is formed as an intermediate, the S_Ni reaction may be accompanied by rearrangements. For example, reaction of labeled butanol (**818**) with $SOCl_2$ in HCOOH yields 23% of rearranged butyl chloride (**819**) as well as the butyl chloride (**820**) with the substituent in the original position[117].

$$H_3{}^{14}C\text{—}\underset{OH}{CH}\text{—}CH_2\text{—}CH_3 \xrightarrow[HCOOH]{SOCl_2} H_3{}^{14}C\text{—}\overset{H}{\underset{\underset{\underset{O}{\parallel}}{\underset{S}{\overset{}{|}}}}{CH}}{\cdots}\underset{Cl}{CH}\text{—}CH_3$$

(**818**)

$$H_3{}^{14}C\text{—}\underset{Cl}{CH}\text{—}CH_2\text{—}CH_3 + H_3{}^{14}C\text{—}CH_2\text{—}CHCl\text{—}CH_3$$
(**820**)   (**819**) 23%

If an optically active alcohol reacts with $SOCl_2$ in pyridine or other base, inversion of configuration is observed. The high concentration of Cl^- ions evidently causes an S_N2 substitution.

$$ROH + SOCl_2 \xrightarrow{Pyridine} RO\text{—}SOCl + \bigcirc^+N\text{—}H \;\; Cl^-$$

$$RO\text{—}SOCl + Cl^- \longrightarrow Cl\cdots R\cdots O\text{—}SOCl \longrightarrow RCl + SO_2 + Cl^-$$

Decomposition of chloroformates (**821**) and thiocarbonates (**822**) also occurs by an $S_N i$ mechanism.

$$\text{RO}-\underset{\underset{\text{O}}{\|}}{\text{C}}-\text{Cl} \xrightarrow{\Delta} \text{RCl} + \text{CO}_2 \qquad \text{RS}-\underset{\underset{\text{O}}{\|}}{\text{C}}-\text{OR} \xrightarrow{\Delta} \text{R}_2\text{S} + \text{CO}_2$$

(**821**)   (**822**)

S-34. Pyrolysis of thiocarbonates[118]

(a) $\text{RO}-\text{CO}-\text{SR}' \underset{k_{-1}}{\overset{k_1}{\rightleftarrows}} [\text{R}^+ \ ^-\text{O}-\text{CO}-\text{SR}'] \xrightarrow{k_2}$
   (**823**)   (**824**)

$$\text{CO}_2 + [\text{R}^+ \ ^-\text{SR}'] \longrightarrow \text{R}-\text{S}-\text{R}'$$
                                              (**825**)

(b) $\text{RO}-\text{CO}-\text{SR}' \longrightarrow \underset{\underset{\underset{R'}{|}}{S}}{R}\diagdown\text{C}=\text{O} \longrightarrow \text{R}-\text{S}-\text{R}' + \text{CO}_2$

$$R = p\text{-ClC}_6\text{H}_4-\overset{*}{\text{C}}\text{HC}_6\text{H}_5; \quad R' = -\text{CH}_3$$

The dependence of the decomposition rate of thiocarbonates on the substituent R indicates that a carbonium ion is formed in the rate-determining step. It has been shown, however, that the rate depends, not only on R, but also on R'.

Table 6. Relative rates of decomposition of O- S-alkyl or S-aryl thiocarbonates, $(C_6H_5)_2\text{CHO–CO–SR}'$

R'	Rel. rate
CH_3CH_2-	0.6
CH_3-	1.0
$\text{C}_6\text{H}_5\text{CH}_2-$	2.2
$(\text{C}_6\text{H}_5)_2\text{CH}-$	4.3
$\text{ClCH}_2\text{CH}_2-$	5.7
$\text{CH}_3\text{OOCCH}_2-$	10.0
C_6H_5-	23

The influence of R' is presumably due to an inductive electron-attracting effect $(-I)$, which facilitates the separation of R^+. However, the value for phenyl deviates strikingly from that expected for an inductive effect (Figure 7). If the decomposition rate is plotted for constant R against the Taft σ^*-values of R'^*, one obtains

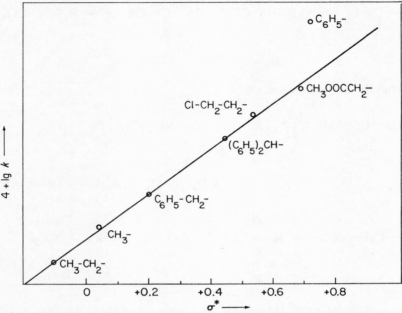

Figure 7. Correlation of the Taft σ^* values of groups R' with the rates of decomposition of thiocarbonates, RO·CO·SR'.

an approximately straight line. Since the phenyl group is the only one of the groups in question that can stabilize negative charge on the sulfur by resonance, it follows that in the rate-determining transition state the carbonyl–sulfur bond (and the carbonyl–oxygen bond) must be effectively broken.

To clarify the question whether the reaction proceeds by a two-step mechanism (*a*) or a synchronous mechanism (*b*) the thermal decomposition of optically active *p*-chlorobenzhydryl *S*-methyl thiocarbonate (**823**) was studied. In benzonitrile at 166° the *p*-chlorobenzhydryl methyl sulfide obtained was almost completely racemized. On the other hand, in the decomposition of chloroformates 15—99% of retention has been observed[119].

The result shows that the decomposition must proceed by a two-step mechanism, whose first step is reversible and in which k_2 is rate-determining ($k_2 < k_1$). Return of the ion pair (**824**) to (**823**) leads to racemization of R. If the decomposition of chloroformates takes place by the same mechanism, then it must be one in which $k_2 > k_{-1}$, as otherwise the decomposition products should be much more highly racemized.

* The σ-value of the Hammett equation can be divided into an inductive (σ^*) and a resonance component (σ_{res}). The normal Hammett σ values are the sum of these two: $\sigma = \sigma^* + \sigma_{\text{res}}$.

VI. S_N2 SUBSTITUTION AT HETERO ATOMS

(a) S_N2 substitution at oxygen[120]

S-35. *Reaction of peroxides with phosphines*

The reaction of hydrogen peroxide with a tertiary phosphine is of the first order in both hydrogen peroxide and in phosphine. Like other S_N2 substitutions the reaction is characterized by a strongly negative activation entropy. It is assumed that the nucleophile displaces the OR^- (R = H, alkyl, or aroyl) group by attacking a peroxidic oxygen atom in the direction of the O–O axis. Other nucleophiles presumably react in a similar way.

$$R\text{—}O\text{—}O\text{—}H + R'_3P \longrightarrow \underset{H}{\overset{R}{O}}\text{----}O\text{----}PR'_3 \longrightarrow ROH + O{\leftarrow}PR'_3$$

The rate of reaction generally depends only insignificantly on the polarity of the solvent; but increases greatly on change from a hydroxyl-free to a hydroxylic solvent. The solvent presumably participates in the transition state by hydrogen bonding.

Aroyl peroxides react with tertiary phosphines by a similar mechanism, yielding acid anhydrides. (Dialkyl peroxides react by a different mechanism.) In the reaction of benzoyl peroxide with the optically active phosphine (**826**)* in acetonitrile, 88% of racemic phosphine oxide was obtained[121] (49% racemization in light petroleum; 78% racemization in CH_3OH).

Horner and Jurgeleit[122] have proposed a mechanism in which initially a tight ion-pair (**827**) is formed, which either collapses rapidly to a product retaining the optical activity or yields racemic product *via* a trigonal-bipyramidal intermediate (**828**). Since the life-time of the tight ion-pair, and hence the formation of (**828**), depend on the polarity of the solvent, use of highly polar solvents leads to an increase in the proportion of racemic product.

---

* In contrast to amines, which cannot be separated into enantiomers, chiral phosphines are relatively stable optically; for instance, the phosphine (**826**) is racemized only after several hours' boiling in toluene. Phosphine oxides and sulfonium salts are also optically stable.

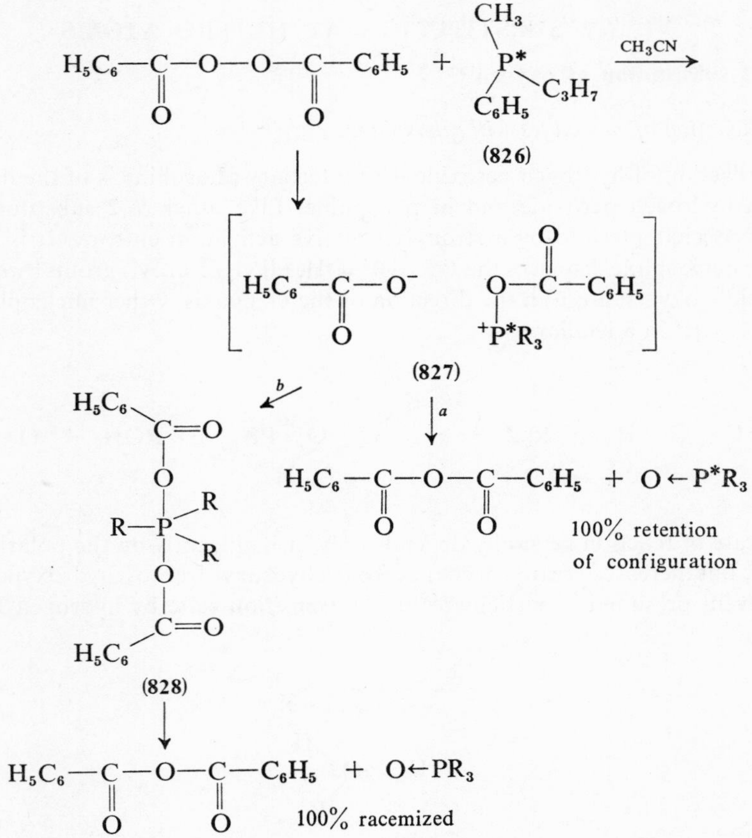

$$H_5C_6-\underset{\underset{O}{\|}}{C}-O-O-\underset{\underset{O}{\|}}{C}-C_6H_5 + \underset{C_6H_5}{\overset{CH_3}{P^*}}C_3H_7 \xrightarrow{CH_3CN}$$

(826)

$$\downarrow$$

$$\left[H_5C_6-\underset{\underset{O}{\|}}{C}-O^- \quad \underset{{}^+P^*R_3}{\overset{\overset{O}{\|}}{O-C-C_6H_5}} \right]$$

(827)

$\overset{b}{\swarrow} \quad \downarrow a$

$H_5C_6-\underset{\underset{O}{\|}}{C}-O-\underset{\underset{O}{\|}}{C}-C_6H_5 \;+\; O{\leftarrow}P^*R_3$

100% retention of configuration

$$\begin{array}{c} H_5C_6\diagdown \\ C=O \\ | \\ O \\ R\diagdown \;\;|\;\; \diagup R \\ R-P \\ \diagup \;\;|\;\; \diagdown R \\ O \\ | \\ C=O \\ H_5C_6\diagup \end{array}$$

(828)

$\downarrow$

$H_5C_6-\underset{\underset{O}{\|}}{C}-O-\underset{\underset{O}{\|}}{C}-C_6H_5 \;+\; O{\leftarrow}PR_3$

100% racemized

When benzoyl peroxide is labelled on the carbonyl-oxygen atom the labelled oxygen appears in the anhydride exclusively in accordance with the above mechanism[123]. The reaction is insensitive to the addition of oxygen; and polymerization of added styrene does not occur; a radical mechanism can therefore be excluded.

(b) S_N2 **substitution at bivalent sulfur**

S-36. *Reaction of triphenylphosphine with 1,3-dimethyl-2-butenyl ethyl disulfide*[124]

In the reaction of triphenylphosphine with S_8, a sequence of nucleophilic substitutions converts the sulfur ring into triphenylphosphine sulfide[125].

$$(C_6H_5)_3P + \begin{matrix} S-S \\ | \quad | \\ S-S_5 \end{matrix} \longrightarrow (C_6H_5)_3\overset{+}{P}-S-S \overset{P(C_6H_5)_3}{\swarrow} \longrightarrow$$
$$ {}^-S-S_5$$

$$(C_6H_5)_3P{=}S + \begin{matrix} S-\overset{+}{P}(C_6H_5)_3 \\ | \\ {}^-S-S_5 \end{matrix} \quad \text{etc.}$$

Reaction of alkyl, aryl, and aralkyl tetrasulfides with triphenylphosphine yields the corresponding disulfide. The reaction shows that the S_N2 substitution at a

$$RS-S-S-SR \xrightarrow{(C_6H_5)_3P} RS-\underset{{}^+P(C_6H_5)_3}{S} {}^-S-SR \xrightarrow{(C_6H_5)_3P} RS-\underset{}{\overset{P(C_6H_5)_3}{\underset{|}{S}}}-SR \longrightarrow$$
$$ + (C_6H_5)_3P{=}S$$

$$RS-\underset{}{\overset{{}^+P(C_6H_5)_3}{\underset{|}{S}}} {}^-SR \longrightarrow RS-SR + (C_6H_5)_3P{=}S$$

sulfur atom with simultaneous displacement of a disulfide or sulfide anion is favored over the corresponding substitution at the α-carbon atom of a di- or poly-sulfide. Triphenylphosphine thus preferentially attacks a sulfur atom of 1,3-dimethyl-2-butenyl ethyl disulfide (**829**).

$$\begin{matrix}(CH_3)_2C{=}CH-CHCH_3 \\ | \\ C_2H_5-S-S\end{matrix} \xrightarrow[(C_6H_5)_3P]{80°} \begin{matrix} & \text{CH} & \\ (CH_3)_2C\overset{\diagup}{}\overset{\diagdown}{}CHCH_3 \\ | | \\ S\cdots\cdots S\cdots P(C_6H_5)_3 \\ {}^{\delta-}{}^{\delta+} \\ | \\ C_2H_5 \end{matrix} \longrightarrow$$

(**829**) \hspace{4cm} (**830**)

$$(CH_3)_2\underset{\underset{SC_2H_5}{|}}{C}-CH{=}CHCH_3$$

(**832**) 80%

Since, in contrast to the reaction of triphenylphosphine with S_8, the reaction rate is not markedly accelerated by increasing polarity of the solvent, the reaction cannot proceed by an S_N2–S_N2' sequence.

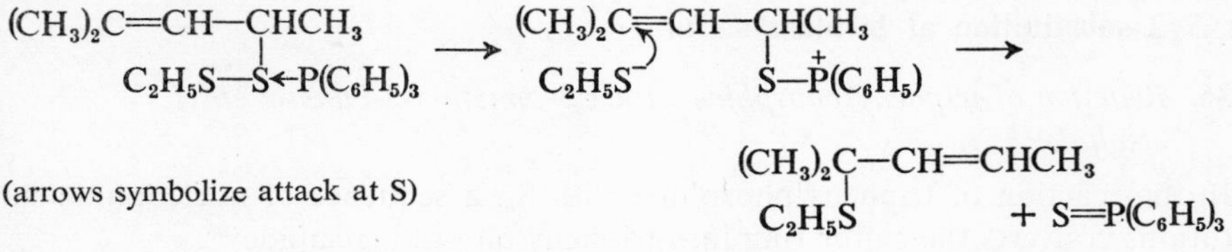

$$(CH_3)_2C=CH-CHCH_3 \atop C_2H_5S-S\leftarrow P(C_6H_5)_3 \longrightarrow {(CH_3)_2C=CH-CHCH_3 \atop C_2H_5S^{\frown} \ \ S-\overset{+}{P}(C_6H_5)} \longrightarrow$$

$$(CH_3)_2C-CH=CHCH_3 \atop C_2H_5S \qquad + S=P(C_6H_5)_3$$

(arrows symbolize attack at S)

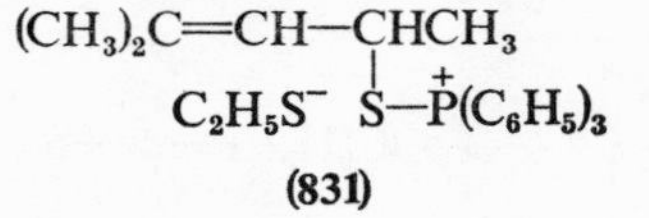

$$(CH_3)_2C=CH-CHCH_3 \atop C_2H_5S^- \ \ S-\overset{+}{P}(C_6H_5)_3$$
(831)

It is likely, however, that the reaction proceeds by a $S_N i'$ mechanism* *via* a polar transition state of only slight charge separation **(830)** or *via* a tight ion-pair **(831)**. Compound **(833)** is the product of an $S_N i$ substitution *via* a cyclic three-membered transition state; it is obtained only in 1.5% yield.

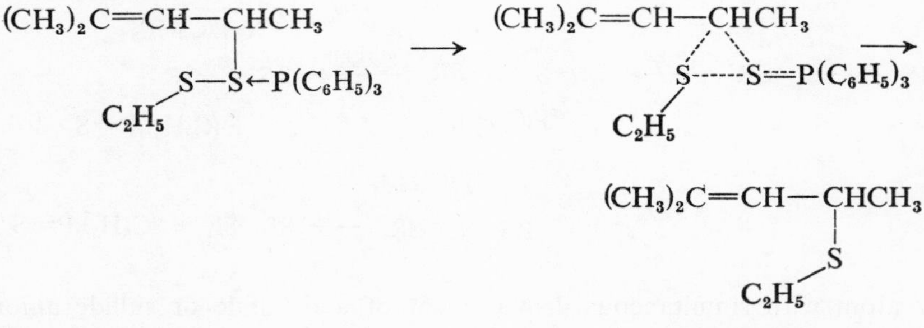

$$(CH_3)_2C=CH-CHCH_3 \atop {S-S\leftarrow P(C_6H_5)_3 \atop C_2H_5}} \longrightarrow {(CH_3)_2C=CH-CHCH_3 \atop {S\cdots\cdots S=P(C_6H_5)_3 \atop C_2H_5}} \longrightarrow$$

$$(CH_3)_2C=CH-CHCH_3 \atop {S \atop C_2H_5}$$
(833) 1.5%

$$*S_N 2: \ Y^- + \ \rangle C-X \longrightarrow Y\cdots C\cdots X \longrightarrow Y-C\langle + X^-$$

$$S_N 2': \ Y^- + \ \rangle C=C-C-X \longrightarrow Y\cdots C=C=C\cdots X \longrightarrow Y-C-C=C\langle + X^-$$

$$S_N i: \ \rangle C-O-SO-X \longrightarrow \rangle C\overset{O}{\underset{X}{\diagdown}}SO \longrightarrow \rangle C-X + SO_2$$

$$S_N i': \ \overset{3}{\rangle}C=\overset{2}{C}-\overset{1}{C}-O-SOX \longrightarrow \overset{-C=C\diagdown O}{\underset{-C\diagdown X}{\diagup}}SO \longrightarrow -\overset{3}{\underset{X}{C}}-\overset{2}{C}=\overset{1}{C}\langle + SO_2$$

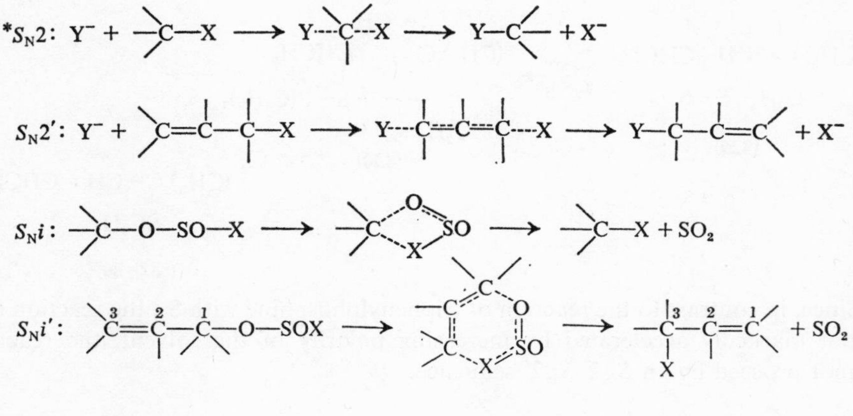

(c) S_N2 substitution at chlorine

S-37. *Reaction of trialkylphosphines with α-halo acetamides*[126]

α-Halo aldehydes, α-halo ketones, and α-halo esters form enol esters on reaction with phosphites (see the Perkow reaction, p. 296). α-Haloacetamides, however, yield enamines or imines on reaction with trialkylphosphines in even higher yields.

$$(C_4H_9)_3P + CCl_3-CO-NHC_6H_5 \longrightarrow$$
$$(834)$$

$$[R_3\overset{+}{P}Cl][\overset{-}{C}Cl_2-CO-NHC_6H_5] \longleftrightarrow CCl_2=\underset{\underset{}{|}}{\overset{\overset{O^-}{|}}{C}}-NHC_6H_5 \longrightarrow$$

$$CCl_2=C\underset{ClPR_3}{\overset{NHC_6H_5}{\diagup}} \longleftrightarrow \overset{-}{C}Cl_2-\underset{Cl\diagdown\diagup PR_3}{\overset{NHC_6H_5}{\overset{+}{\diagup}}}_{O} \longrightarrow$$

$$(835)$$

$$\underset{Cl}{\overset{Cl}{\diagdown}}\overset{-}{C}-\underset{Cl}{\overset{NHC_6H_5}{\underset{|}{C}}}-O\overset{+}{P}R_3 \longrightarrow CCl_2=C\underset{NHC_6H_5}{\overset{Cl}{\diagup}} + OPR_3$$

$$\downarrow$$

$$HCCl_2-CCl=NC_6H_5$$

$$(836) \quad 33\%$$

Since the reaction of trichloroacetamide (834) is strongly inhibited if a chlorine atom is replaced by a fluorine atom (presumably because fluorine is less capable than chlorine of stabilizing an adjacent negative charge*, it is concluded that the nucleophilic attack by the trialkylphosphine takes place at a chlorine atom in the first step of the reaction. In contrast to normal enol esters, the intermediate (835) forms the imine (836) by rearrangement and loss of $R_3P=O$. A similar mechanism

---

* In contrast to electronegativity which increases from I to F, the $-R$ effect (electron-attracting resonance effect) which can be expressed by participation of the resonance structure (*a*) in the resonance hybrid, increases in the reverse sequence.

$$RF \longleftrightarrow {}^+R=F^-$$
$$(a)$$

has been formulated for the reaction of amides with PCl_5[127].

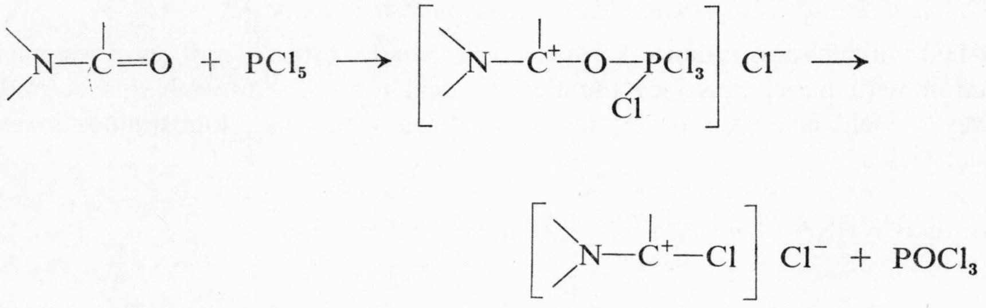

$$\ce{>N-C(=O)- + PCl5 -> [>N-C^+(-O-PCl3)(Cl)] Cl^- ->}$$

$$\ce{[>N-C^+-Cl] Cl^- + POCl3}$$

The rearrangement of (835) is facilitated by the olefinic chlorine atoms and by the amide group. The chlorine atom presumably migrates by a concerted mechanism, since Cl^- is not replaced by Br^- in the presence of bromide ions. Nucleophilic attack on chlorine has also been postulated for the reaction of phosphines and phosphites with α-halo ketones[128] (see Perkow reaction, p. 296).

S-38. Preparation of perchlorofulvalene (838)[129]

A nice example of nucleophilic attack on chlorine is the reaction of compound (837) with trialkyl phosphites.

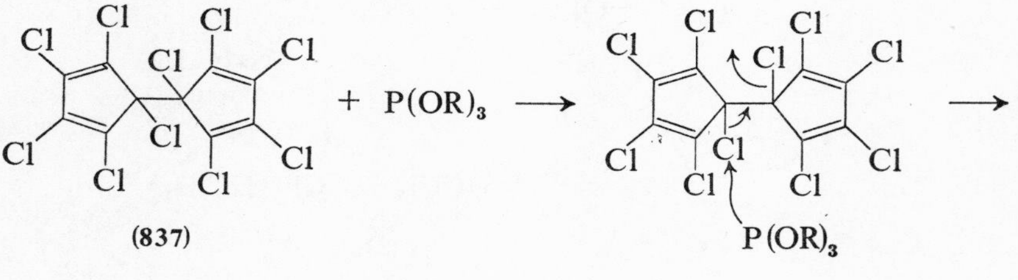

(837) + P(OR)₃ ⟶ [intermediate with P(OR)₃] ⟶

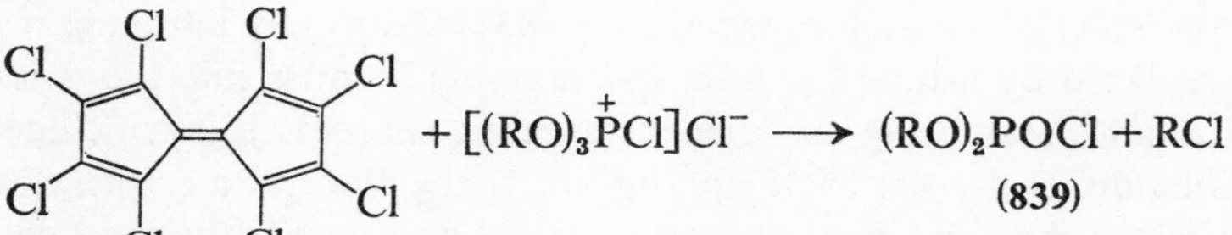

(838) + $[(RO)_3\overset{+}{P}Cl]Cl^-$ ⟶ $(RO)_2POCl + RCl$
(839)

[R = CH(CH₃)₂]

Perchlorofulvalene (838) forms dark blue crystals, which decompose at 200°. The cyclopentadiene rings are not coplanar, as would be expected, but are at a 41° angle to each other, a compromise between conjugation and steric effects of the chlorine atoms[130].

VII. S_E1 SUBSTITUTION

Nucleophilic aliphatic substitution is one of the most thoroughly investigated reactions in organic chemistry. Many fewer experimental results are available for electrophilic displacement reactions of aliphatic compounds.

Evidence has accumulated, however, that in analogy to nucleophilic substitutions two limiting mechanisms apply to electrophilic displacements: the S_E1 and the S_E2 reaction. Important progress in the understanding of these substitutions has been made in particular by the investigations of Cram and his group[131].

(1) In the S_E1 reaction a carbanion or intimate ion-pair is formed in the rate-determining unimolecular step by separation of a cation from a carbon atom. If the substitution takes place at an asymmetric carbon atom *via* a free solvated carbanion the resulting product should be a racemate, since the carbanion intermediate either approaches planarity as a consequence of resonance stabilization by its substituents or has a tetrahedral structure. The tetrahedral non-resonance-stabilized carbanion rapidly equilibrates with its mirror image (840), as do amines. In both cases attack of an electrophile is equally probable from either side.

$$RM_1 \xrightarrow{\text{Slow}} R^- M_1^+ \xrightarrow{M_2^+} RM_2 + M_1^+$$

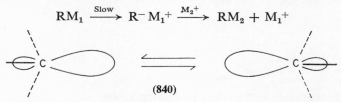

(840)

When the carbanion is little stabilized by resonance, or not at all so stabilized, the substituents are located at the corners of a trigonal pyramid whose apex is the negative carbon atom. Carbonium ions, on the other hand, are always planar. The bonds of the carbanion intermediate may have essentially *p*-character, while the free electron pair is located in the 2*s*-orbital. The carbanion thus has bond angles of approximately 90°. In an alternative structure the carbon atom is sp^3-hybridized with the free electron pair occupying one sp^3-orbital. sp^3-Hybridization of a carbanion is energetically less favorable than of a saturated carbon atom, because the necessary promotion and hybridization energies are not compensated by formation of only three sp^3 σ-bonds, whereas in normal saturated hydrocarbons four sp^3-bonds are formed. There is lack of general agreement as to which of the two structures apply to carbanions[131,132]. In certain cases steric effects favor sp^3-hybridization[133]. As a consequence of their pyramidal structure bridgehead carbanions are relatively stable intermediates in contrast to bridgehead carbonium ions[134].

(2) In the rate-determining step of the S_E2 substitution a positive group M_1 is displaced by an attacking cation M_2. In contrast to S_N2 substitution, in which two electron pairs are involved in the transition state, only one electron pair participates

in S_E2 substitution. In the S_N2 case the nucleophile consequently attacks the carbon atom from the side opposite to the bonding electron pair of the leaving group, whereas in the S_E2 substitution the electrophile attacks from the side of highest electron density, that is, from the same side from which the leaving group is displaced. Thus in the S_E2 substitution at an asymmetric carbon atom retention (*b*) is observed instead of inversion (*a*).

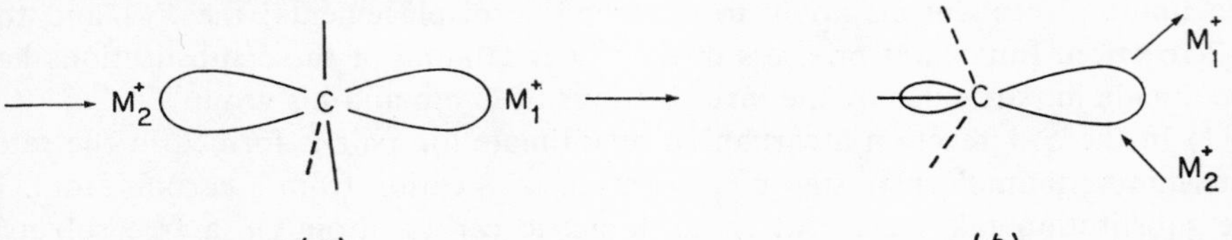

(*a*)   (*b*)

Good evidence for mechanism (*b*) is, for example, the ready acetolysis of compound (**841**), the rate of whose solvolysis is intermediate between that of C_4H_9HgCl and neophylmercury chloride. Since attack from the rear-side is impossible, the mercury atom must be displaced from the front[135].

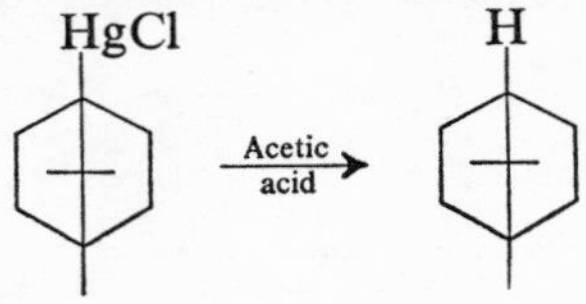

(**841**)

S-39. S_E1 Substitution with carbon as leaving group

Although base-catalyzed decomposition of the optically active alcohol (**842**) takes place by an S_E1 mechanism, retention or inversion is observed in the product depending on the reaction conditions[136].

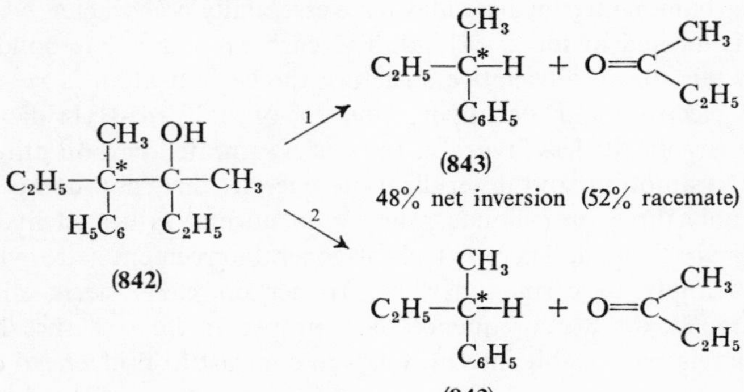

$$C_2H_5-\overset{CH_3}{\underset{C_6H_5}{\overset{|}{C^*}}}-H + O=\overset{CH_3}{\underset{C_2H_5}{C}}$$

(**843**)

48% net inversion (52% racemate)

$$C_2H_5-\overset{CH_3\; OH}{\underset{H_5C_6\;\; C_2H_5}{\overset{|\;*\;\;\;|}{C-C}}}-CH_3$$

(**842**)

$$C_2H_5-\overset{CH_3}{\underset{C_6H_5}{\overset{|}{C^*}}}-H + O=\overset{CH_3}{\underset{C_2H_5}{C}}$$

(**843**)

96% net retention (4% racemate)

Reagents: 1, $KOCH_2CH_2OH$ in $HOCH_2CH_2OH$
2, t-C_4H_9OK, little t-BuOH, in dioxane

Aliphatic Substitution

The reaction resembles S_N1 substitution where the inverted product is formed along with racemate. It has been proposed that, in analogy to the S_N1 reaction, a tight ion-pair is formed and coordinated to a proton-donating solvent molecule, as (844). This species rearranges to an ion-pair (845) faster than it equilibrates with the solvent if a solvent of low polarity is used. The rearrangement consists of the rotation of the metal ion with its ligands at the front face of the carbanion. In solvents of low dielectric constant such as dioxane ($\varepsilon = 2.29$) (with K *tert*-butoxide as base and a little *t*-butyl alcohol as proton donor) a high degree of retention of configuration is observed on conversion of (845) into (846). The retention depends, furthermore, on the ability of the metal ion to coordinate with proton donors.

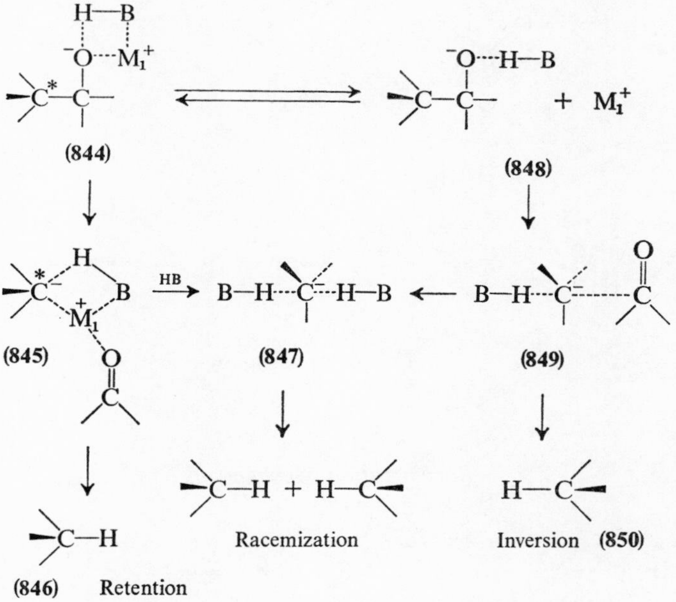

In a competing reaction some dissociation of (845) occurs, followed by addition of a solvent molecule to the back-face of the carbanion. The resulting symmetric solvated carbanion (847) yields a racemic product. A bulky cation such as a tetraalkylammonium group strongly facilitates the formation of (847) from (845).

Predominant racemization in decompositions of alcohols such as (842) is observed if a poorly proton-donating but highly polar solvent is used, *e.g.* dimethyl sulfoxide ($\varepsilon = 45$) (+ a small amount of proton-donating alcohol). Dimethyl sulfoxide strongly accelerates the dissociation of the tight ion-pair (845) and gives the carbanion a long enough life-time to form (847).

In highly polar solvents that are also good proton donors, for example in diethylene glycol or ethylene glycol ($\varepsilon = 46.6$), considerable inversion is observed. Under these conditions, **(844)** dissociates to **(848)**. Whilst in **(849)** the front side is still shielded by the carbonyl compound, a proton-donating solvent molecule attacks the carbanion at the back face, yielding inverted product **(850)**. In a highly polar solvent that is only a weak proton donor, **(849)** may have sufficient life-time to be converted into **(847)**.

Retention Mechanisms[131]

S_N1 (S_Ni): →C*—O—S(=O)—Cl → asymmetric ion-pair (C+ ··· O—S=O, Cl) → →C*—Cl + SO_2

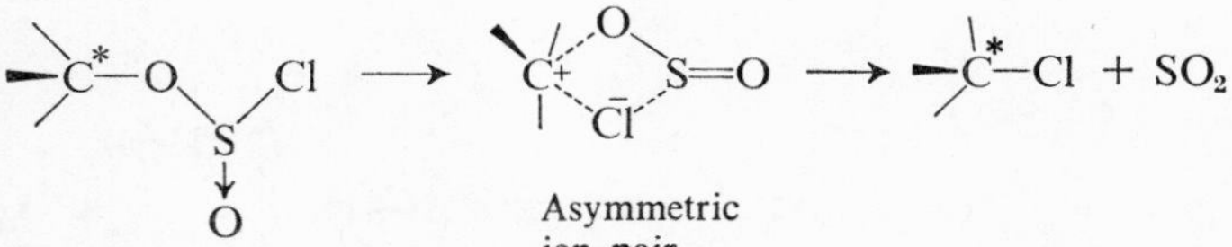

Asymmetric ion–pair

S_E1: →C—C(—O⁻···M⁺, H—B)— → asymmetric ion-pair (C⁻···H—B, M⁺) → →C*—H

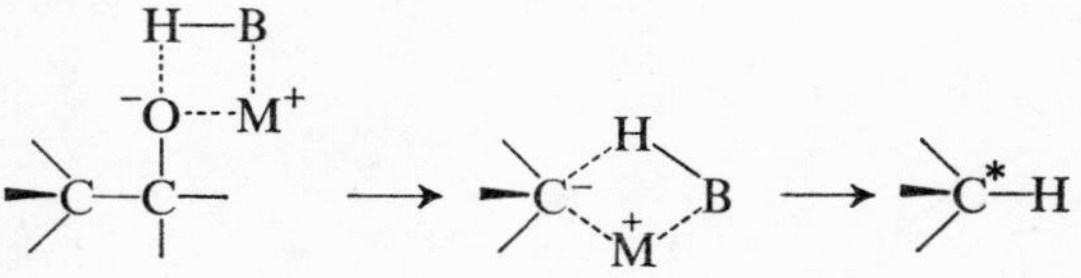

Asymmetric ion–pair

Racemization Mechanisms

S_N1 →C*—X $\xrightarrow{S}$ S···C⁺···S → →C—S + S—C←

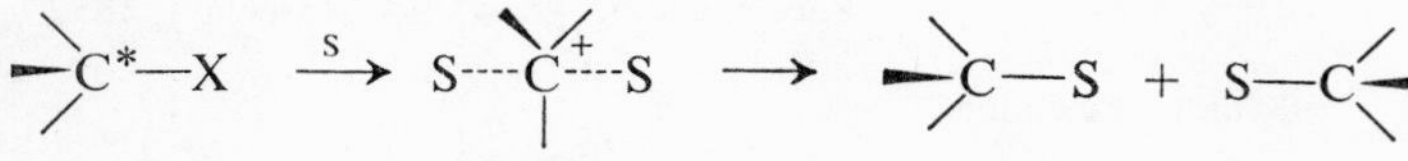

Symmetrically solvated carbonium ion

S_E1 →C*—C(—O⁻)— $\xrightarrow{DMSO}$ B—H···C⁻···H—B $\xrightarrow{HB}$ →C—H H—C←

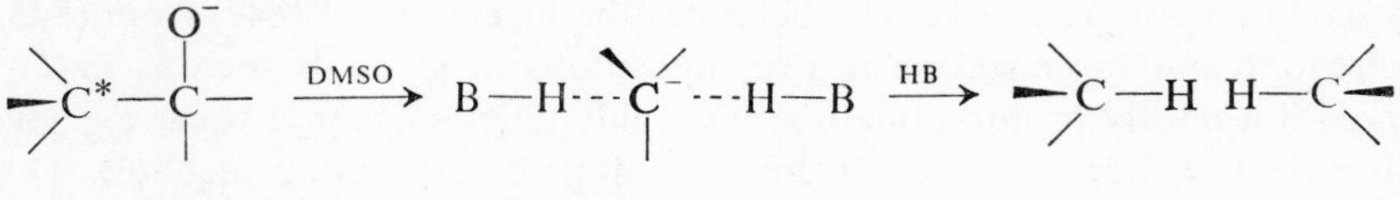

Symmetrically solvated carbanion

Inversion Mechanisms

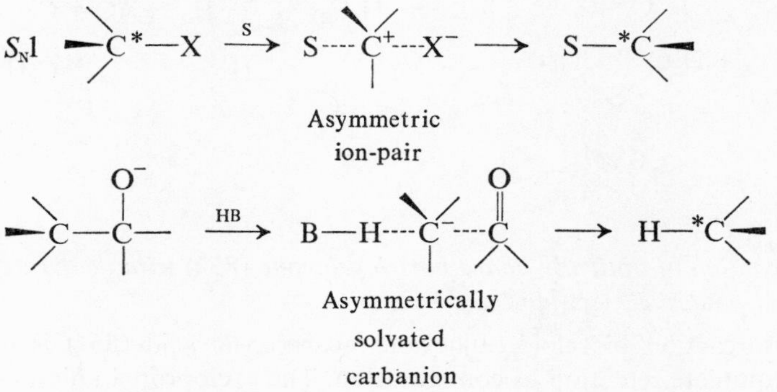

$S_N 1$ mechanism: R–C*–X → S----C⁺----X⁻ → S—*C–R (Asymmetric ion-pair)

$S_E 1$ mechanism: R–C–C(O⁻)– →(HB) B—H----C----C(=O) → H—*C–R (Asymmetrically solvated carbanion)

Similar mechanisms apply to $S_E 1$ substitution with hydrogen, deuterium, nitrogen, or oxygen as leaving group.

Ketones are cleaved in dimethyl sulfoxide by a special mechanism that leads to product with retention of configuration instead of racemization.

If optically active 2-methyl-1,2-diphenyl-1-butanone **(851)** (0.16M) is treated with potassium *tert*-butoxide (0.51M) and *tert*-butyl alcohol (0.42M) in DMSO, isobutylbenzene is obtained in 74% yield with 40% net retention of configuration[137].

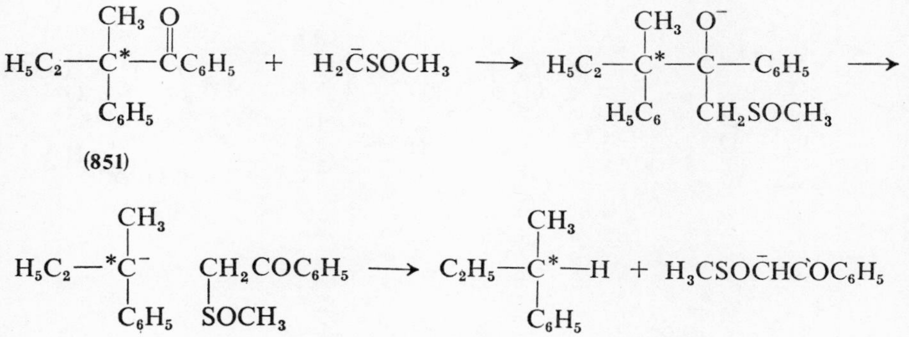

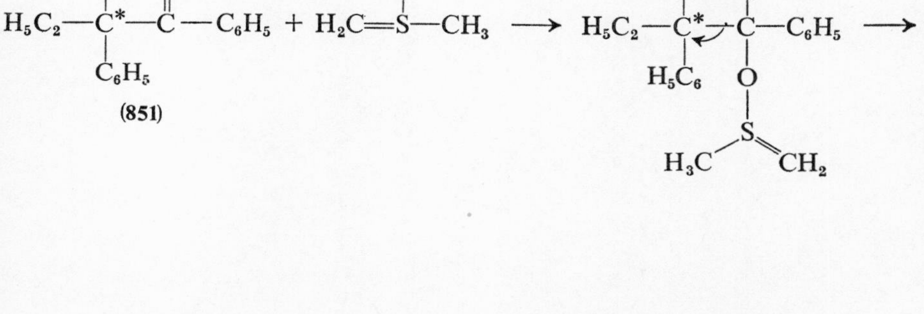

$$H_5C_2-\underset{C_6H_5}{\underset{|}{C^*}}(CH_3)-\overset{O}{\overset{\|}{C}}C_6H_5 + H_2\bar{C}SOCH_3 \longrightarrow H_5C_2-\underset{H_5C_6}{\underset{|}{C^*}}(CH_3)-\underset{CH_2SOCH_3}{\underset{|}{C}}(O^-)-C_6H_5 \longrightarrow$$

(851)

$$H_5C_2-\underset{C_6H_5}{\underset{|}{{}^*C^-}}(CH_3) \quad \underset{SOCH_3}{\underset{|}{CH_2COC_6H_5}} \longrightarrow C_2H_5-\underset{C_6H_5}{\underset{|}{C^*}}(CH_3)-H + H_3CSO\bar{C}HCOC_6H_5$$

$$H_5C_2-\underset{C_6H_5}{\underset{|}{C^*}}(CH_3)-\overset{O}{\overset{\|}{C}}-C_6H_5 + H_2C=\overset{O^-}{\underset{|}{S}}-CH_3 \longrightarrow H_5C_2-\underset{H_5C_6}{\underset{|}{C^*}}(CH_3)-\underset{O-S(CH_3)=CH_2}{\underset{|}{C}}(O^-)-C_6H_5 \longrightarrow$$

(851)

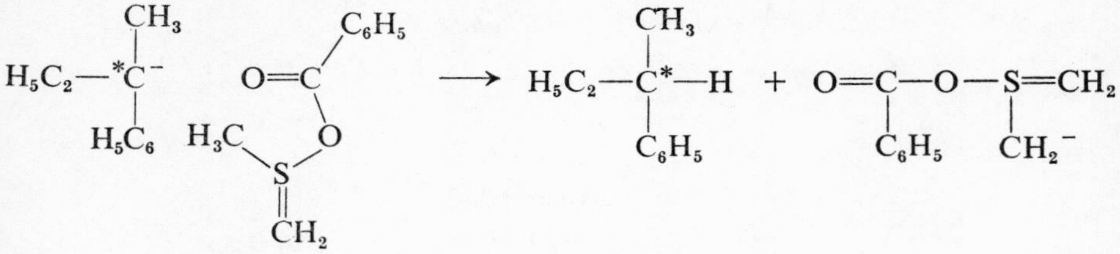

$$H_5C_2-\overset{*}{\underset{H_5C_6}{C}}-\overset{CH_3}{\underset{H_3C}{|}} \quad \underset{\underset{CH_2}{\overset{\|}{S}}}{\overset{C_6H_5}{\underset{O}{\overset{|}{C}}}}=C \longrightarrow H_5C_2-\overset{CH_3}{\underset{C_6H_5}{\overset{|}{C^*}}}-H + O=\overset{CH_3}{\underset{C_6H_5}{\overset{|}{C}}}-O-\underset{CH_2^-}{\overset{\|}{S}}=CH_2$$

S-40. *Reaction of optically active nortricyclanone* **(852)** *with deuterated DMSO and potassium* tert-*butoxide*[138]

In this reaction bicyclo[3.1.0]hexane-3-carboxylic acid **(854)** is obtained with almost complete retention of configuration. The cyclopropyl anion as in **(853)** has a strong tendency to preserve its configuration.

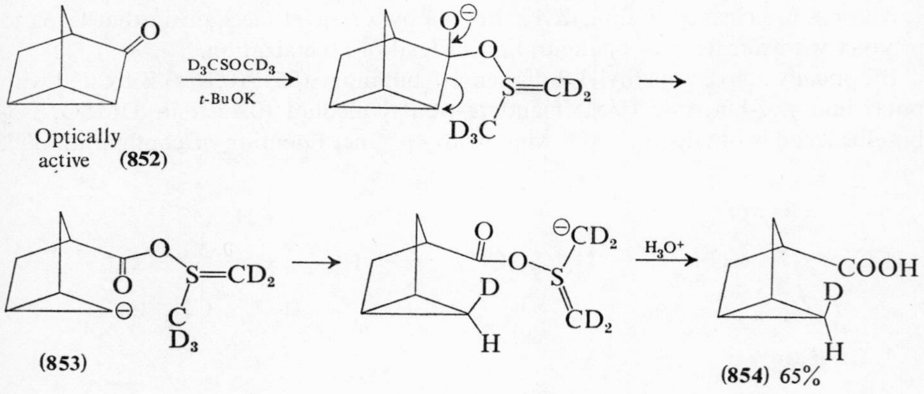

Optically active **(852)** → (D₃CSOCD₃, *t*-BuOK) → intermediate →

(853) → intermediate → $\xrightarrow{H_3O^+}$ **(854)** 65%

Oxygen as leaving group

S-41. *Reaction of benzyl α-ethyl-α-methylbenzyl ether* **(855)** *with potassium N-methylanilide*[139]

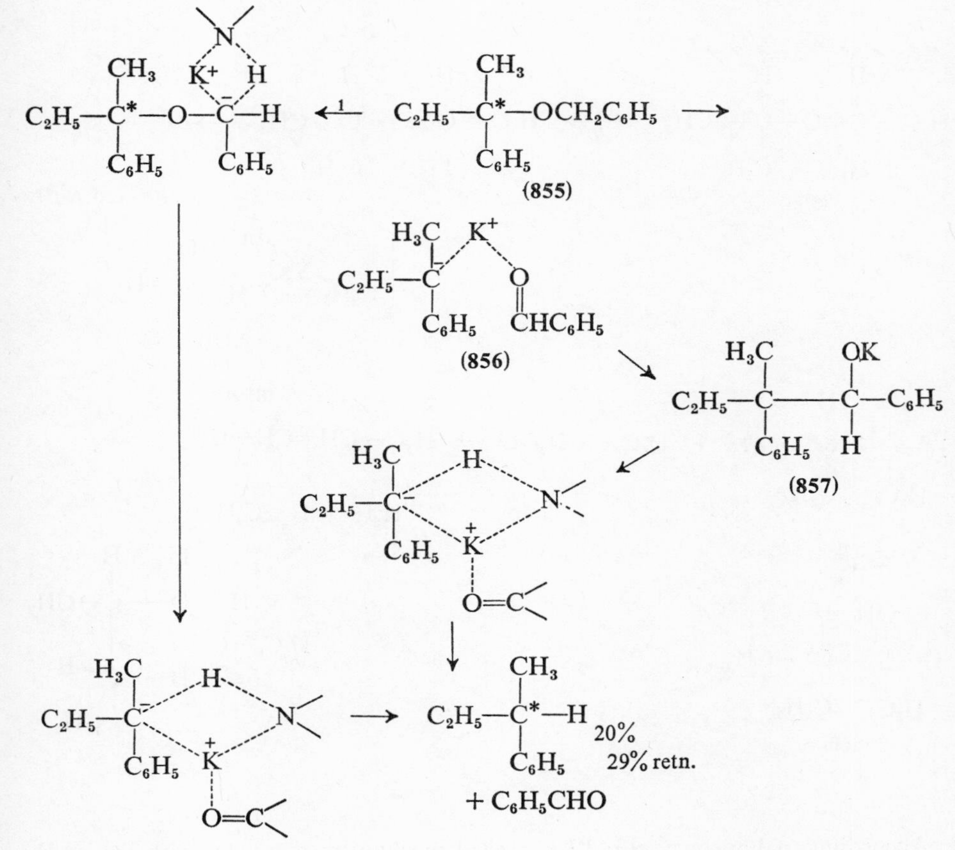

Reagent: 1, C_6H_5NMeK in C_6H_5NHMe at 180°.

The ether (**855**) decomposes on treatment with potassium *N*-methylanilide in *N*-methylaniline by a mechanism analogous to that in the decomposition of alcohol (**842**). Isobutylbenzene is obtained with 29% retention of configuration along with racemic product. It is probable that some of the ether (**855**) first dissociates to the tight ion-pair (**856**) which rearranges to alcohol (**857**) (Wittig rearrangement). In the absence of proton-donors Wittig rearrangement takes place exclusively.

S-42. Wittig rearrangement of benzyl isobutyl ether (858)[140]

If the ether (858) is treated with butyllithium in benzene, rearrangement takes place with partial retention of configuration (see pages 319 and 323). Butenes and benzyl alcohol are obtained as by-products.

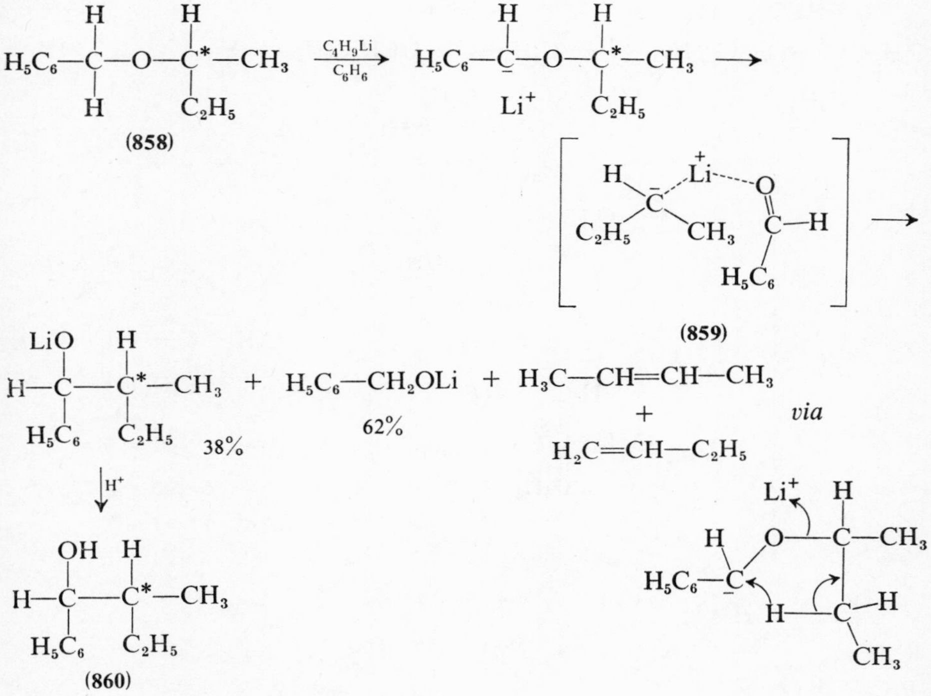

$$H_5C_6-\underset{H}{\overset{H}{C}}-O-\underset{C_2H_5}{\overset{H}{C^*}}-CH_3 \xrightarrow[C_6H_6]{C_4H_9Li} H_5C_6-\underset{Li^+}{\overset{H}{C}}-O-\underset{C_2H_5}{\overset{H}{C^*}}-CH_3 \longrightarrow$$

(858)

$$\left[\begin{array}{c} H \\ \underset{C_2H_5}{\overset{-}{C}}\underset{CH_3}{\overset{Li^+\cdots O}{\diagdown}}\overset{}{\underset{H_5C_6}{C}}-H \end{array} \right] \longrightarrow$$

(859)

$$\underset{H_5C_6\ \ C_2H_5}{\overset{LiO\ \ \ H}{H-\underset{|}{C}-\overset{|}{C^*}-CH_3}} + H_5C_6-CH_2OLi + H_3C-CH=CH-CH_3 \quad via$$

38%   62%   + H_2C=CH-C_2H_5

↓ H⁺

$$\underset{H_5C_6\ \ C_2H_5}{\overset{OH\ \ H}{H-\underset{|}{C}-\overset{|}{C^*}-CH_3}}$$

(860)

[structure showing cyclic transition state with Li⁺, O, H₅C₆, CH₃, H arrangement]

According to Lansbury et al.[141] a radical mechanism competes with the carbanion mechanism, at least in the case of secondary and tertiary alkyl benzyl ethers.

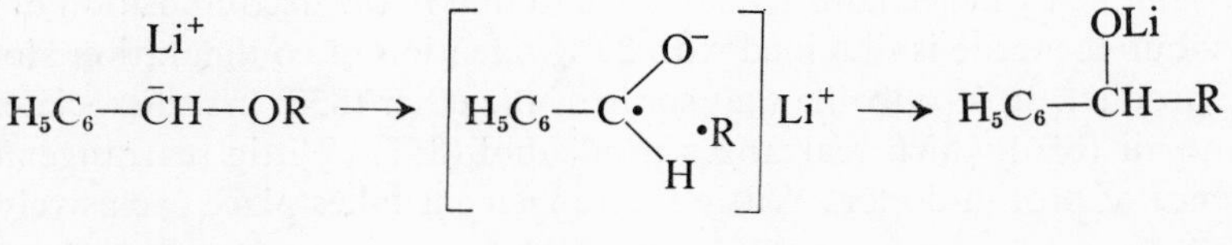

$$\underset{}{\overset{Li^+}{H_5C_6-\bar{C}H-OR}} \longrightarrow \left[H_5C_6-\underset{H}{\overset{O^-}{\underset{|}{C\cdot}}}\ \ \cdot R \right] Li^+ \longrightarrow H_5C_6-\underset{}{\overset{OLi}{\underset{|}{CH}}}-R$$

Reactions closely related to the Wittig rearrangement are the following three (a)—(c).

(a) Stevens rearrangement[142]

$$H_5C_6-\underset{O}{\overset{CH_3}{\underset{|}{C}}}-CH_2-\overset{+}{\underset{CH_3}{\underset{|}{N}}}-\underset{CH_3}{\overset{H}{\underset{|}{C}}}-C_6H_5 \xrightarrow{Base} H_5C_6-\underset{O}{\overset{CH_3}{\underset{|}{C}}}-\overset{-}{\underset{H}{\underset{|}{C}}}-\overset{+}{\underset{CH_3}{\underset{|}{N}}}-\underset{CH_3}{\overset{H}{\underset{|}{C}}}-C_6H_5 \longrightarrow$$

$$\left[\begin{array}{c} H-\underset{CH_3}{\overset{C_6H_5}{C}}\\ H_5C_6-\underset{O}{\overset{|}{C}}-\underset{H}{\overset{|}{C}}=\overset{+}{\underset{CH_3}{N}}\diagdown CH_3 \end{array}\right] \longrightarrow H_5C_6-\underset{O}{\overset{|}{C}}-CH-N\diagup^{CH_3}_{CH_3}$$
$$ \underset{H_3C\ \ C_6H_5}{CH}$$

(b) Isomerization of sulfonium ylides[143].

$$R\bar{C}H-\overset{+}{S}R_2 \longrightarrow [RCH=\overset{+}{S}R\ \ \bar{R}] \longrightarrow R_2CH-SR$$

S-43 (c) Meisenheimer rearrangement of amine oxides[144]

$$H_5C_6CH_2-\underset{CH_3}{\overset{CH_3}{\underset{|}{N}}}\rightarrow O \xrightarrow{80-165°} \left[\begin{array}{c} H_5C_6CH_2^-\\ H_3C\diagdown\overset{+}{N}=O\\ H_3C\diagup \end{array}\right] \longrightarrow H_3C\diagdown N-O-CH_2C_6H_5\\ H_3C\diagup$$

(861) \hspace{4cm} (862) 61%

Amine oxides that have no β-hydrogen atom rearrange above 100° to trisubstituted hydroxylamines. If β-hydrogen atoms are present, β-elimination becomes a competing reaction (see example E-6). Cross-over experiments (heating mixtures of amine oxides) did not yield mixed trisubstituted hydroxylamines, showing that the rearrangement takes place *via* a tight ion-pair without equilibration with the solvent. In certain cases, as in the Wittig rearrangement, the reaction may also take place *via* radical pairs[145].

VIII S_E2 AND S_Ei SUBSTITUTION

Optically active organometallic compounds are suitable models for investigation of electrophilic substitution mechanisms. The first optically active organometallic compound to be reported was 1-methylheptyllithium[146].

Optically active cyclopropyl-lithium and -magnesium compounds have been prepared by Walborsky and his group[147].

During metallation and carbonation of compound (863) 100% retention of configuration is observed. With the exception of mercury compounds, however, relatively

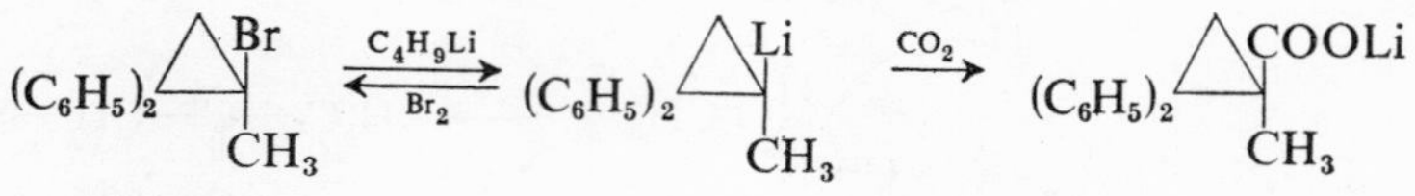

$$(C_6H_5)_2\overset{Br}{\underset{CH_3}{\triangle}} \underset{Br_2}{\overset{C_4H_9Li}{\rightleftarrows}} (C_6H_5)_2\overset{Li}{\underset{CH_3}{\triangle}} \overset{CO_2}{\longrightarrow} (C_6H_5)_2\overset{COOLi}{\underset{CH_3}{\triangle}}$$

(863)

few optically active organometallic compounds are known[131]. In many cases, S_E2 reactions have therefore been studied for mercury compounds.

Bimolecular electrophilic substitution at asymmetric carbon atoms leads to products with complete retention of configuration. If the attack of the electrophile on the front-side is accompanied by simultaneous pull of a nucleophile at the leaving group, the reaction mechanism is designated as an S_Ei substitution.

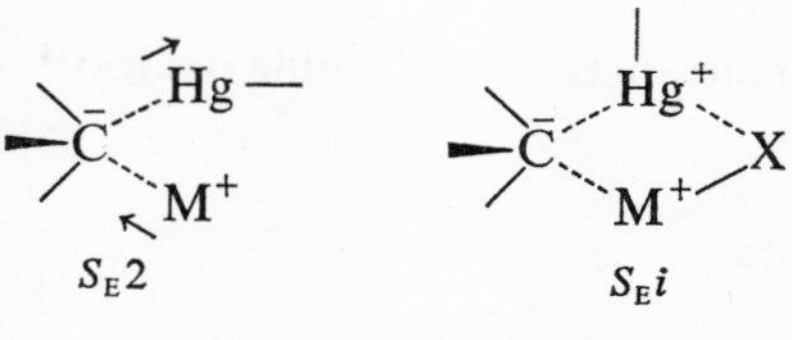

$S_E2 \qquad S_Ei$

It is often difficult to distinguish between an S_E2 and an S_Ei mechanism.

S-44. Reaction of *trans*-4-methylcyclohexylmercuric chloride with bromine

Bromination of *trans*-4-methylcyclohexylmercuric chloride (864) takes place by a bimolecular S_E mechanism. The product, *trans*-4-methylcyclohexyl bromide (865), is formed with retention of configuration[148].

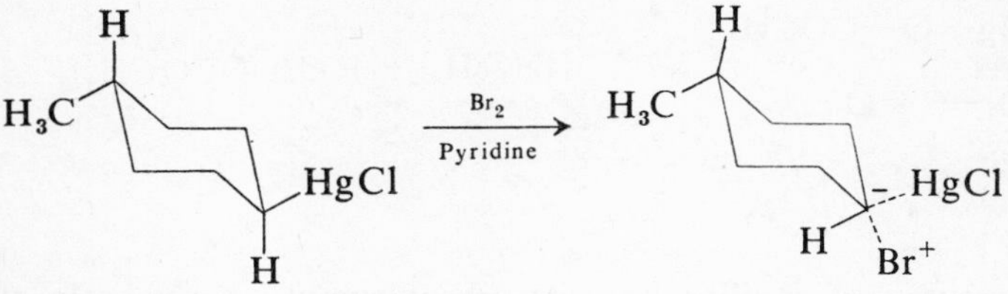

Another typical bimolecular S_E reaction is the symmetrization of alkylmercury salts, resulting in the equilibrium concentration of the corresponding dialkylmercury compound (special case of the Schlenk equilibrium). The equilibrium which normally lies completely on the side of the alkylmercury salt may be shifted to the right by complex-formation or by reduction of the components on the right-hand side.

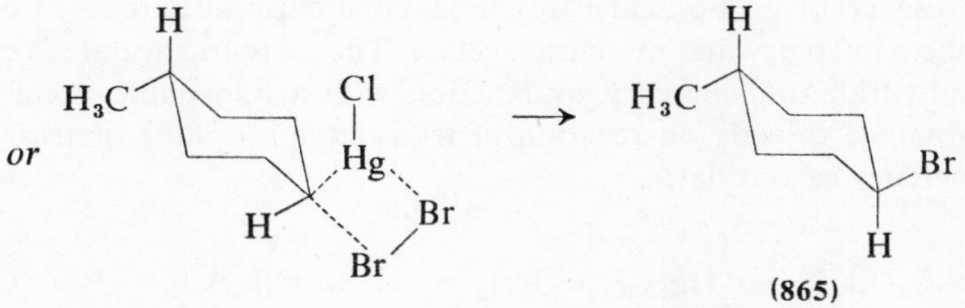

$$2RHgX \rightleftarrows HgR_2 + HgX_2$$

S-45. Reaction of diphenylmercury with tri-*n*-butylphosphine and acetic acid[149]

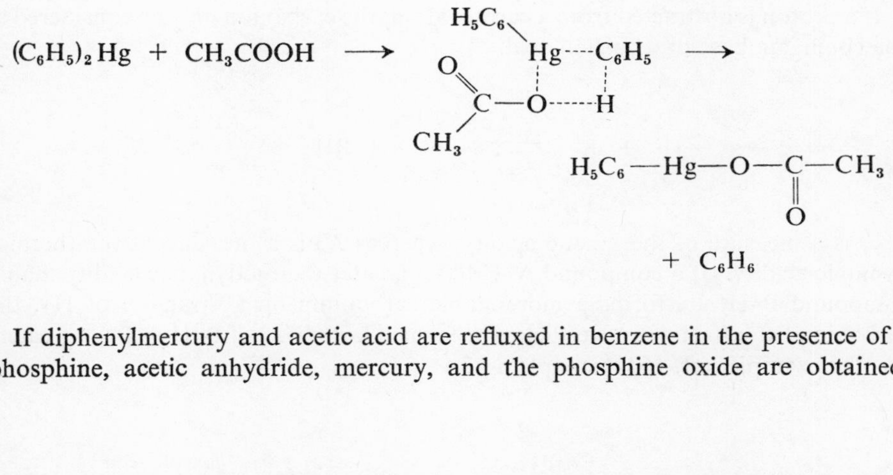

If diphenylmercury and acetic acid are refluxed in benzene in the presence of a phosphine, acetic anhydride, mercury, and the phosphine oxide are obtained.

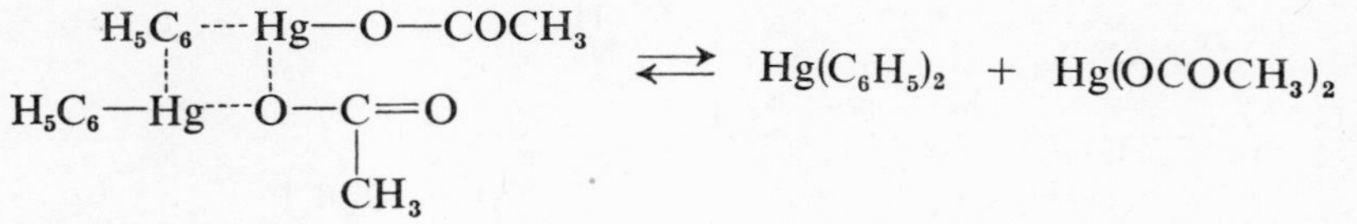

$$\begin{array}{c} H_5C_6\cdots Hg-O-COCH_3 \\ H_5C_6-Hg\cdots O-C=O \\ | \\ CH_3 \end{array} \rightleftarrows Hg(C_6H_5)_2 + Hg(OCOCH_3)_2$$

Diphenylmercury presumably equilibrates with phenylmercuric acetate in the presence of acetic acid ($S_E i$). Phenylmercuric acetate is in equilibrium with diphenylmercury and mercuric acetate. The mercuric acetate is converted into acetic anhydride and mercury by reaction with a phosphine. Acid anhydrides may be obtained directly on reaction of tri-n-butylphosphine or triphenylphosphine with mercury carboxylates.

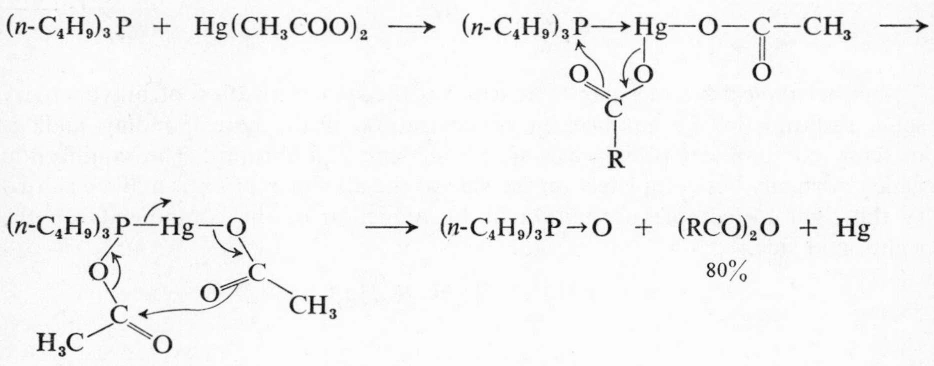

$$(n\text{-}C_4H_9)_3P + Hg(CH_3COO)_2 \longrightarrow (n\text{-}C_4H_9)_3P\rightarrow Hg-O-\underset{\underset{R}{|}}{\overset{\overset{O}{\|}}{C}}-CH_3 \longrightarrow$$

$$(n\text{-}C_4H_9)_3P\rightarrow O + (RCO)_2O + Hg \quad 80\%$$

IX. CARBANIONS

If a proton is abstracted from a carbon atom, the carbanion may be considered as the conjugate base of a carbon acid.

$$-\overset{|}{\underset{|}{C}}-H + B^- \underset{k_{-1}}{\overset{k_1}{\rightleftarrows}} -\overset{|}{\underset{|}{C}}{}^- + BH \qquad \frac{k_1}{k_{-1}} = K_a$$

k_1 is a measure of the kinetic acidity, whereas K_a is a measure of the thermodynamic acidity. (If a compound A–H has a greater thermodynamic acidity than a compound B–H, *i.e.* forms a more stable carbonium on abstraction of H⁺, the carbanion from B–H will nevertheless be formed faster if k_1 of B–H $\rightarrow$ B⁻ is greater; see, for example, ref. 162, example S-47.)

k_1 and K_a of carbon acids depend on several effects, some of which will here be briefly discussed[131].

(1) s-*Orbital effects*

The acidity of the C–H bond of aliphatic hydrocarbons increases with increasing s-character of the C–H σ-bond (see Table 7). The acidity changes in this

Table 7. Acidity of aliphatic hydrocarbons

	Bond hybridization	s-Character (%)	Estimated pK_a
H—C≡C—H	sp	50	25
$H_2C=CH_2$	sp^2	33	36.5
$H_3C—CH_3$	sp^3	25	42

direction because the spherosymmetrically arranged s-electrons are more strongly attracted by the positively charged nucleus than are the p-electrons. It follows that the greater the s-character the easier the proton abstraction and the more stable the carbanion. In cyclic compounds the s-character of a C_1–H bond increases with decreasing CC_1C bond angle. Bonds angles in these compounds can therefore be used as a measure of the C_1H acidity in highly strained systems such as cyclopropane and bicyclobutane[150]; the p-character of the cyclic C–C bonds increases with the bending of these bonds. At the same time the s-character of the C–H bonds must increase. The HCH or CCH bond angle spreads and in cyclopropane amounts to approximately 120° as in sp^2-hybridized σ-bonds (see S-15 and p. 276).

(2) *Electronic effects*

Carbanions are strongly resonance-stabilized by nitro, carbonyl, cyano, and other electron-attracting groups. Resonance effects are often difficult to distinguish from inductive effects. A purely inductive stabilizing effect is observed in nitrogen ylides. Quaternary ammonium nitrogen cannot stabilize the neighboring negative charge by expansion of its octet shell, forming an ylene structure. Stabilization

$$\underset{\text{ylide}}{R-\overset{\overset{R}{|}}{\underset{\underset{R}{|}}{N^+}}-\bar{C}HR} \quad \leftarrow\!\!\!/\!\!\!\rightarrow \quad \underset{\text{ylene}}{R-\overset{\overset{R}{|}}{\underset{\underset{R}{|}}{N}}=CHR}$$

probably takes place by electron-attraction through the N–C bond and electrostatic interaction through space.

As a consequence of inductive effects, and in contrast to carbonium ions, primary carbanions are more stable than secondary ones and the latter more stable than tertiary ones. The acidity of saturated hydrocarbons consequently decreases from primary to tertiary C–H bonds (see discussion of k_1 and K_a and ref. 6a of Chapter 8).

(3) d-*Orbital stabilization*

If second-row elements such as P or S are attached to a carbanionic carbon atom, the negative charge is stabilized by interaction between the orbital containing the electron pair and an empty *d*-orbital of the heteroatom. A good example of this effect is furnished by the relative rates and activation parameters for the deuteroxide-catalyzed H–D exchange between NR_4^+, PR_4^+, or SR_3^+ and D_2O[151] (see Table 8).

Table 8. H–D exchange between quaternary salts and D_2O

	Relative rates at 62°	$\Delta H^{\ddagger}$ (kcal mole^{-1})	$\Delta S^{\ddagger}$ (e.u.)	C–X bond distance (Å)
$(CH_3)_4N^+$	1	32.2 ± 0.6	−15 ± 2	1.47
$(CH_3)_4P^+$	2.4 + 10^6	25.6 ± 0.2	+4 ± 1	1.87
$(CH_3)_3S^+$	2.0 + 10^7	22.4 ± 0.5	−1 ± 2	1.81

From the striking rate difference between carbanion formation from NR_4^+ and PR_4^+ it is evident that *d*-orbital interaction strongly stabilizes the transition state for carbanion formation by PR_4^+ and SR_3^+. Correspondingly the activation enthalpy

$$(CH_3)_3\overset{+}{P}-CH_3 \xrightarrow{\text{Base}} (CH_3)_3\overset{+}{P}-\overset{-}{CH_2} \leftrightarrow (CH_3)_3P=CH_2$$
$$\text{ylide} \qquad \text{ylene}$$

for proton abstraction from PR_4^+ is 6.6 kcal lower than from NR_4^+.

According to Table 7 the basicity of the carbanions increases from acetylene to alkane. Therefore, when the alkali-metal compound of a saturated hydrocarbon reacts with an olefin, the saturated hydrocarbon and the organometallic compound of the olefin are obtained, corresponding to an acid–base equilibrium*. Since the

$$\overset{\diagdown}{\underset{\diagup}{C}}=\overset{H}{\underset{\diagdown}{C}} + M-C_2H_5 \rightleftarrows \overset{\diagdown}{\underset{\diagup}{C}}=\overset{M}{\underset{\diagdown}{C}} + C_2H_6$$

acidity of a saturated hydrocarbon is much lower than that of an olefin, the equilibrium lies completely on the side of the metal–olefin compound. Exchange reactions between an alkylpotassium and saturated hydrocarbons have been observed by Finnegan[152]. A detailed study has been carried out by Höver, Mergard, and Korte[153]. On reaction of alkylmercury with potassium and saturated hydro-

---

* The insolubility of many organometallic compounds prevents achievement of true equilibrium.

carbons only the terminal monocarboxylic acids of the saturated hydrocarbons could be isolated after carboxylation.

In exchange reactions of this type, the nature of the metal plays an important part. Exchange takes place only if the metal–carbon bond has sufficient ionic character. According to Pauling[154] the ionic character increases in the sequence:

$$\text{Li } 45\%; \quad \text{Na } 47\%; \quad \text{K, Rb } 51\%; \quad \text{Cs } 55\%$$

(the ionic character of the Hg–C bond is comparatively low, 9%)

More recent investigations indicate that these values are too high. According to Ebel[155], the sp^3-bond of C with Li has an ionic character of 27%, and C–Na one of 29%.

Lithium alkyls are well known to add to double bonds: there is little exchange with vinyl hydrogen. Sodium alkyls, however, yield mainly exchange products. Exchange reactions between saturated hydrocarbons and metal alkyls are possible only with potassium or cesium alkyls.

K_a scales of carbon acids have been suggested by McEwen[156] and later by Streitwieser[157], Applequist[158], and Dessy[159], and their collaborators. An amalgamated scale, the MSAD scale (McEwen–Streitwieser–Applequist–Dessy) has been introduced by Cram[131].

Table 9. The MSAD scale of K_a for carbon acids

Compound	pK_a	Compound	pK_a
Fluordene	11	Ethylene	36.5
Cyclopentadiene	15	Benzene	37
9-Phenylfluorene	18.5	Cumene (α-posn.)	37
Indene	18.5	Triptycene (α-posn.)	38
Phenylacetylene	18.5	Cyclopropane	39
Fluorene	22.9	Methane	40
Acetylene	25	Ethane	42
1,3,3-Triphenylpropene	26.5	Cyclobutane	43
Triphenylmethane	32.5	Neopentane	44
Toluene (α-posn.)	35	Propane (posn. 2)	44
Propene (α-posn.)	35.5	Cyclopentane	44
Cycloheptatriene	36	Cyclohexane	45

S-46. Reaction of bicyclo[2.2.1]heptadiene with pentylsodium[160]

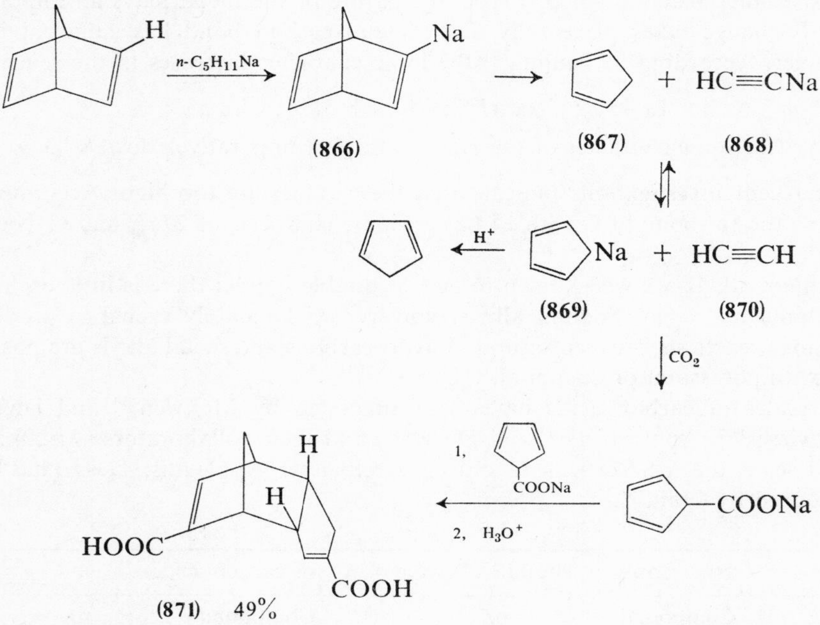

[structures (866), (867), (868), (869), (870), (871)]

In the reaction of bicycloheptadiene with pentylsodium, the base presumably attacks a vinyl-hydrogen atom. According to Bredt's rule, the tertiary allylic hydrogen is relatively unreactive[134]. The sodium compound (866) decomposes by a retro-Diels–Alder reaction to cyclopentadiene and acetylene. Both compounds have been detected by gas-liquid chromatography. Carboxylation yields sodium cyclopentadienecarboxylate, which is converted into the product (871). Since propiolic acid has not been found as a product, the equilibrium between (867), (868), (869), and (870) lies largely on the side of (869) and (870).

S-47. Reaction of 6-phenyl-1-hexene (872) with organoalkali compounds[161]

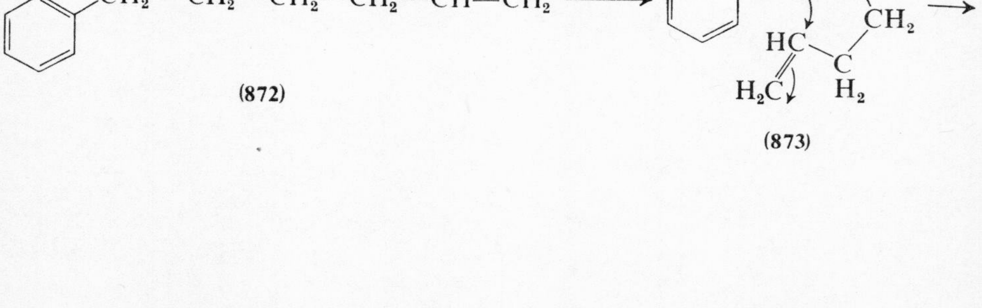

[structures (872), (873)]

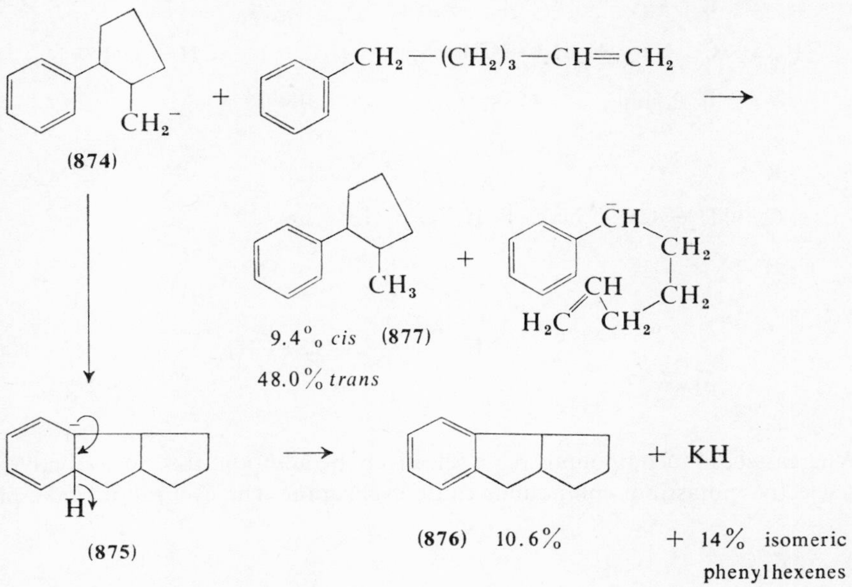

Aliphatic Substitution

[Scheme showing structures (874), (875), (876), (877) with 9.4% cis (877), 48.0% trans, (876) 10.6% + KH + 14% isomeric phenylhexenes]

The anionic chain reaction is started by catalytic amounts of strong base capable of abstracting a proton from the hydrocarbon. The benzylic-hydrogen atom is obviously the most acidic one.

Investigations by Benkeser et al.[162] indicate that α-metallations are preceded by kinetically controlled metallations of the aromatic nucleus in the *para-* and *meta-* positions. The metallated compounds rearrange to the α-metal derivatives.

The sterically favorable position of the double bond in **(873)** leads to intramolecular addition of the carbanion to the double bond, yielding a five-membered ring **(874)**. The energetically unfavorable conversion of a benzylic carbanion into a primary aliphatic carbanion is compensated in total energy balance by formation of another benzylic carbanion in the subsequent step of the chain reaction (chain reactions usually proceed by a radical mechanism, see p. 388). The actual reaction that takes place is consequently **(872)** → **(877)**. Additions of α-metallated aromatic compounds to olefins such as ethylene or propene have been carried out with good yields. The annexed mechanism has been postulated[163].

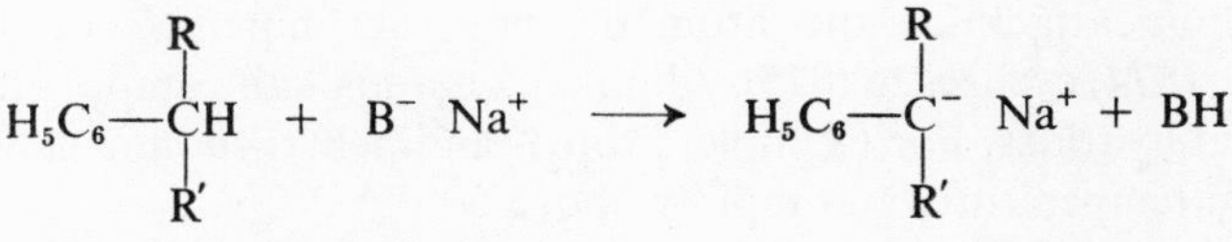

$$H_5C_6-\underset{R'}{\overset{R}{C}}H + B^- Na^+ \longrightarrow H_5C_6-\underset{R'}{\overset{R}{C}}^- Na^+ + BH$$

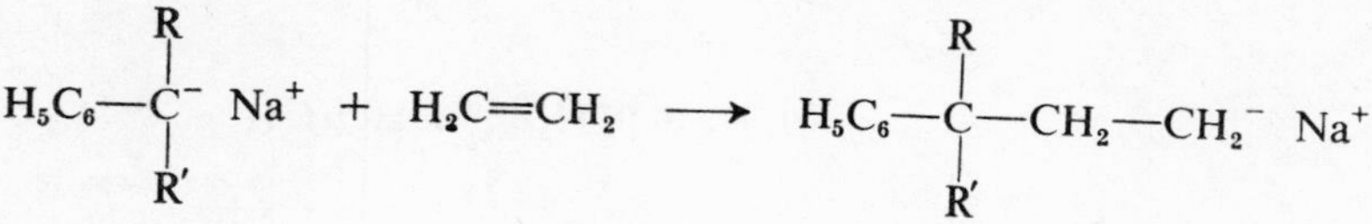

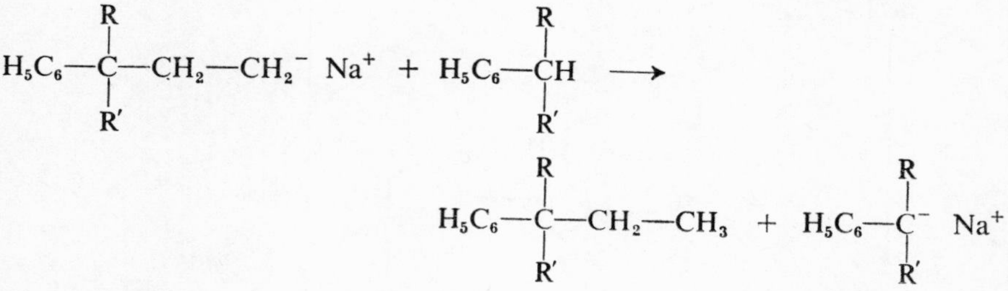

$$\underset{R'}{\overset{R}{H_5C_6-\underset{|}{\overset{|}{C}}^-}} Na^+ + H_2C=CH_2 \longrightarrow \underset{R'}{\overset{R}{H_5C_6-\underset{|}{\overset{|}{C}}-CH_2-CH_2^-}} Na^+$$

$$\underset{R'}{\overset{R}{H_5C_6-\underset{|}{\overset{|}{C}}-CH_2-CH_2^-}} Na^+ + \underset{R'}{\overset{R}{H_5C_6-\overset{|}{C}H}} \longrightarrow$$

$$\underset{R'}{\overset{R}{H_5C_6-\underset{|}{\overset{|}{C}}-CH_2-CH_3}} + \underset{R'}{\overset{R}{H_5C_6-\underset{|}{\overset{|}{C}}^-}} Na^+$$

Whereas sodium compounds react selectively by addition, the more reactive but less selective potassium compounds cause cyclization. The cyclizations take place

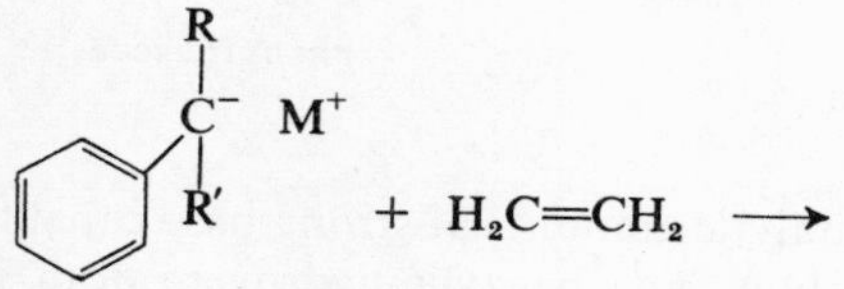

[Scheme showing cyclization reaction with $M = Na$ or K, producing open-chain product with RH (M = Na) and cyclic indane product (M = K)]

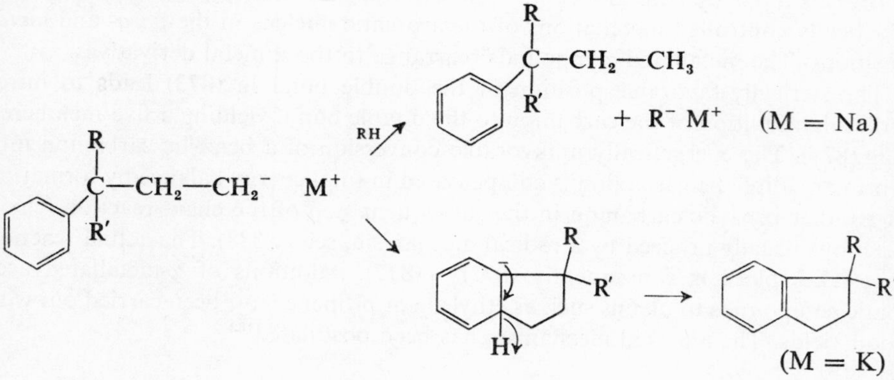

by nucleophilic attack at the aromatic ring, accompanied by loss of KH. Analogously, **(874)** cyclizes to **(875)**. Aliphatic organoalkali compounds also readily split off metal hydrides. For example, propyl- and pentyl-sodium slowly lose NaH even at room temperature, and rapidly above 50° [164].

As well as the products (876) and (877), isomeric phenylhexenes are obtained which presumably are formed *via* allylic carbanions.

S-48. Reaction of vinyl ethers with phenyllithium[165]

$$C_6H_5O\text{—}\overset{\alpha}{C}H\text{=}\overset{\beta}{C}H_2 + C_6H_5Li \longrightarrow HC\equiv CH + C_6H_5OLi + C_6H_6$$

Vinyl ethers are usually cleaved by organolithium compounds between the vinyl-carbon and the oxygen atom. The reaction is initiated by attack of $C_6H_5^-$ on the β-vinyl hydrogen.

S-49. Reaction of vinyl sulfides by organolithium compounds[166]

The behavior of vinyl sulfides towards organolithium compounds differs from that of vinyl ethers.

$$C_6H_5S\text{—}CH\text{=}CH_2 + RLi \longrightarrow C_6H_5S\text{—}CHLi\text{—}CH_2R$$
$$\quad\quad\quad (879) \quad\quad\quad\quad\quad\quad\quad\quad\quad\quad (880)\ 55\%$$

$$C_2H_5S\text{—}CH\text{=}CH_2 + RLi \longrightarrow C_2H_5S\text{—}CHLi\text{—}CH_2R$$
$$\quad\quad\quad\quad\quad\quad\quad\quad\quad\quad\quad\quad\quad\quad 5\%$$

Phenyl vinyl sulfide adds butyllithium in good yields. Addition also occurs, though in low yield, to *n*-alkyl vinyl sulfides. By addition of the primary carbanion $C_4H_9^-$, a secondary carbanion is formed adjacent to the sulfur atom. Since the acidity of a methyl C–H bond is greater than that of a methylene C–H bond or a tertiary C–H bond, the sulfur atom must have a dominating influence on the reaction by 3*d*-orbital interaction (see p. 330).

S-50. Unusual reaction of a vinyl sulfide with *n*-butyllithium[166]

$$H_3C\text{—}(CH_2)_8\text{—}\underset{\underset{CH_3}{|}}{\overset{\overset{CH_3}{|}}{C}}\text{—}S\text{—}CH\text{=}CH_2 + n\text{-}C_4H_9Li \longrightarrow$$

(881)

The reaction of 1,1-dimethyldecyl vinyl sulfide (881) with *n*-butyllithium is an exception to the additions discussed above. Instead of addition, cleavage occurs between alkyl-carbon and sulfur. This may be due to the presence of two relatively acidic CH_3 groups, to the formation of a six-membered cyclic transition state, as well as to the formation of a β-methyl-substituted α-olefin.

$$\text{R-C(CH}_3\text{)(S-CH=CH}_2\text{)(CH}_2\text{-CH(Li)-CH}_2\text{-C}_3\text{H}_7\text{)} \longrightarrow H_3C-(CH_2)_8-\underset{\underset{\text{(882) 64\%}}{|}}{C}=CH_2 +$$

$$LiS-CH=CH_2 + C_4H_{10}$$
(883)

S-51. Reaction of 1-bromoethyl ethyl sulfone (884) with base (Ramberg–Bäcklund reaction)[167]

$$H_3C-\underset{H}{\overset{H}{C}}-SO_2-\underset{H}{\overset{CH_3}{C}}-Br \underset{\text{Fast}}{\overset{\text{NaOH 50°} \atop 2:3 \text{ dioxane/H}_2\text{O}}{\rightleftharpoons}} H_3C-\underset{H}{\overset{\bar{C}}{C}}-SO_2-\underset{H}{\overset{CH_3}{C}}-Br + H_2O$$
(884)   (885)

$$H_3C-\underset{H}{\overset{\bar{C}}{C}}-SO_2-\underset{H}{\overset{CH_3}{C}}-Br \xrightarrow{\text{Slow}} \underset{\underset{O_2}{S}}{\overset{H_3C \quad\quad CH_3}{\underset{H \quad\quad H}{C-C}}} + \underset{\underset{O_2}{S}}{\overset{H \quad\quad CH_3}{\underset{H_3C \quad\quad H}{C-C}}}$$
(885)   cis 79%   (886)   trans 21%

$$\underset{\underset{SO_2}{}}{\overset{H_3C \quad\quad CH_3}{\underset{H \quad\quad H}{C-C}}} + 2OH^- \xrightarrow{\text{Fast}} \overset{H_3C}{\underset{H}{>}}C=C\overset{CH_3}{\underset{H}{<}} + SO_3^{2-} + H_2O$$

The attacking base may either abstract a hydrogen atom from the position alpha to the sulfone group or replace the halogen atom. Although the SO_2 group is in many cases comparable to other electron-attracting groups such as C=O or C=N, for example, in its directing effect on aromatic substitution and in the acidity of the α-hydrogen atoms of alkyl sulfones, the α-halogen is somewhat inert to nucleophilic replacement. To explain this behavior, reference has been made to steric hindrance by comparing the O=S=O group with the neopentyl system[168]; this is supported by the ready replacement of halogen in vinylogous chloroalkyl sulfones.

If the base were to attack the α-hydrogen atom bound to the same carbon atom as the halogen, a carbene would be formed. Attempts to trap a carbene intermediate were, however, unsuccessful. The reaction is probably initiated by abstraction of an α-hydrogen from the methylene group. The carbanion (**885**) is converted by intramolecular substitution into the cyclic sulfones (**886**). Similar mechanisms have been found to apply to the Favorsky reaction[169] (see p. 79), the Neber rearrangement (see example S-52), and related reactions[170].

This mechanism is supported by the fact that cyclic sulfones, independently prepared, also decompose to olefins. *cis*-2,3-Dimethylthiirane dioxide yields pure *cis*-butene. The reaction is of the second order—first order in OH⁻ and first order in sulfone, *i.e.* in accord with the postulated mechanism. The intramolecular nucleophilic replacement is the rate-determining step of the reaction. The predominant formation of *cis*-episulfone is surprising, since the *trans*-isomer is thermodynamically more stable. Bordwell and Neureiter[171] have attempted to rationalize this result: it corresponds to the recent finding[172] that the base-catalyzed, kinetically controlled isomerization of olefins yields preferentially *cis*-olefins. It is assumed that the *cis*-olefins are formed owing to the greater stability of the anionic *cis*-transition state (**887**), which results from the more favorable dipole–dipole and van der Waals interactions between the substituents. Similar effects are assumed

(**887**)

to be effective in the transition state for ring closure of the sulfone. The cyclic sulfones can be cleaved purely thermally. That base participates in the reaction discussed above is, however, supported by isolation of unsaturated sulfonic acids and vinyl halides in the cleavage of chlorinated cyclic sulfones[173].

S-52. Neber rearrangement[174]

In the Neber rearrangement α-amino ketones are obtained on treatment of alkyl- and aryl-substituted oxime esters with a strong base. In example **(888)**, the reaction is already initiated by pyridine. A competing reaction in polar solvents is the Beckmann rearrangement (see pages 147 and 228).

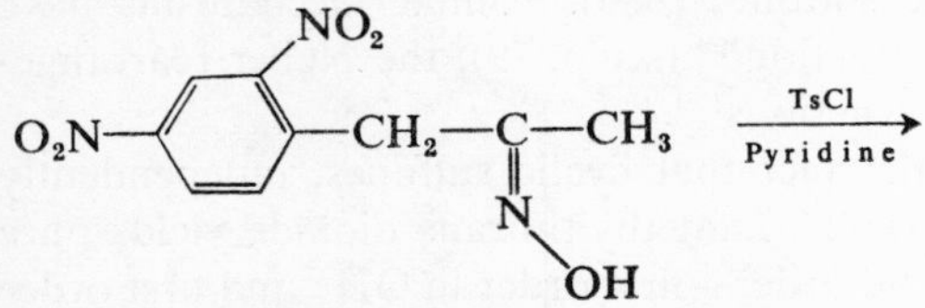

O_2N—C$_6$H$_3$(NO$_2$)—CH$_2$—C(=N—OH)—CH$_3$ $\xrightarrow{\text{TsCl, Pyridine}}$

(888)

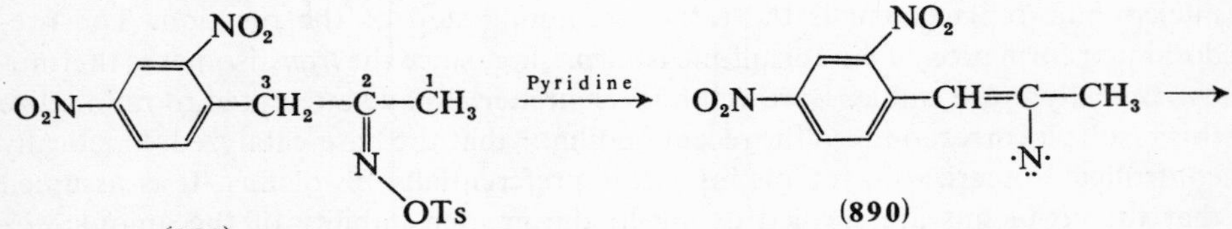

O_2N—C$_6$H$_3$(NO$_2$)—$\overset{3}{\text{CH}_2}$—$\overset{2}{\text{C}}$(=N—OTs)—$\overset{1}{\text{CH}_3}$ $\xrightarrow{\text{Pyridine}}$ O_2N—C$_6$H$_3$(NO$_2$)—CH=C(—$\ddot{\text{N}}\cdot$)—CH$_3$ ⟶

(889)   **(890)**

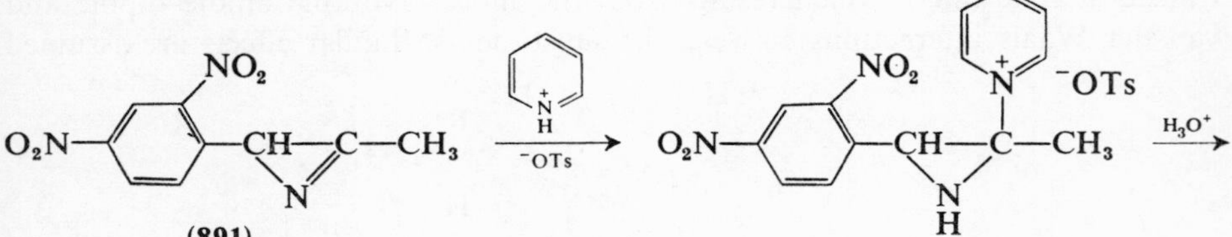

O_2N—C$_6$H$_3$(NO$_2$)—CH—C(—CH$_3$) (aziridine with N) $\xrightarrow[-\text{OTs}]{\text{C}_5\text{H}_5\text{NH}^+}$ O_2N—C$_6$H$_3$(NO$_2$)—CH—C(Py$^+$)(—CH$_3$) (with NH in ring), $^-$OTs $\xrightarrow{H_3O^+}$

(891)   **(892)**

O_2N—C$_6$H$_3$(NO$_2$)—CH(NH$_2$)—C(=O)—CH$_3$ + H$^+$

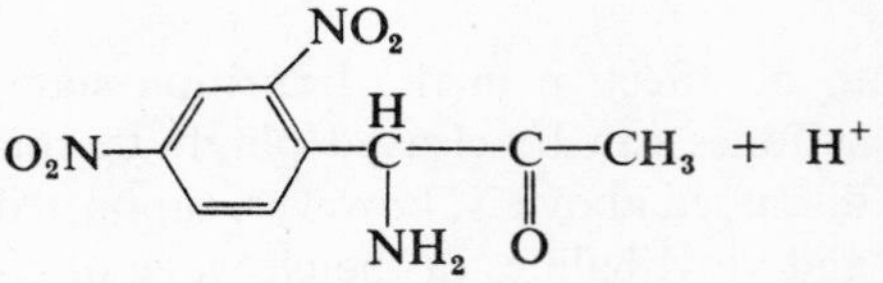

(893)

Initial attack of the base on C-2 would lead to a nitrene intermediate **(894)**. As an

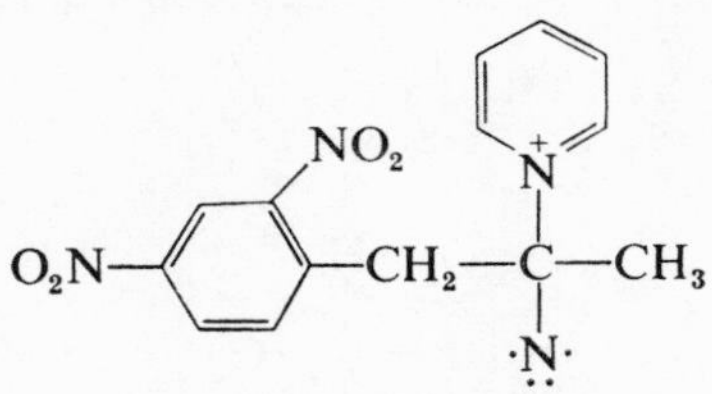

O_2N—C$_6$H$_3$(NO$_2$)—CH$_2$—C(Py$^+$)(—$\ddot{\text{N}}\cdot$)—CH$_3$

(894)

electrophilic agent the nitrene–nitrogen should be inserted into the C–H bond of highest electron density. House and Berkowitz[175] have, however, shown that the opposite takes place: 1-*p*-methoxyphenyl-3-*p*-nitrophenylacetone (**895**) yields exclusively (**896**).

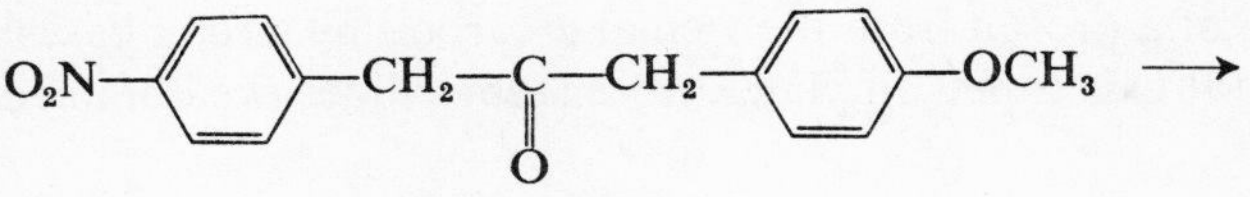

$O_2N-\langle\text{C}_6H_4\rangle-CH_2-\underset{\underset{O}{\|}}{C}-CH_2-\langle\text{C}_6H_4\rangle-OCH_3 \longrightarrow$

(**895**)

$O_2N-\langle\text{C}_6H_4\rangle-\underset{NH_2}{CH}-\underset{\underset{O}{\|}}{C}-CH_2-\langle\text{C}_6H_4\rangle-OCH_3$

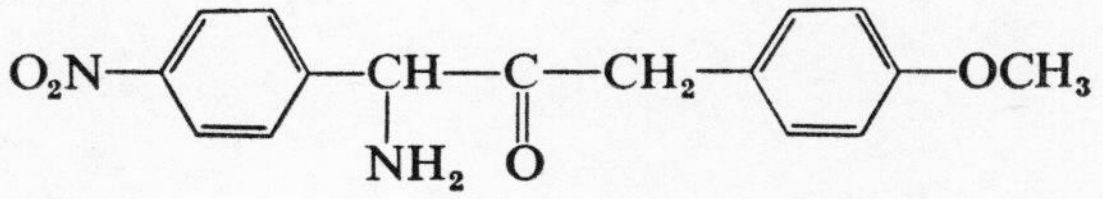

(**896**)

Alkoxyaziridines such as (**897**) and an intermediate (**892**) that could in principle arise by attack on C-2 have been isolated; they are, however, very likely formed by addition of alcohol or pyridinium hydrochloride to the azirine intermediate (**891**), which can also be isolated in the Neber rearrangement[176].

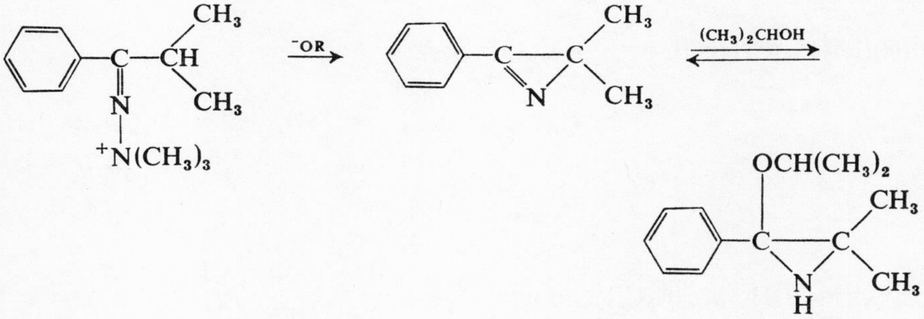

(**897**)

The Neber rearrangement very probably begins with abstraction of a proton from C-3. If two methylene groups of differing acidity are available, the more acidic H atoms participate in the reaction, independently of the position of the tosylate group. The azirine intermediate may be formed either by synchronous loss of the tosylate group or by a two-step mechanism through a vinylnitrene (**890**). The latter mechanism is supported by the fact that no particular configuration of the tosylate group is necessary. Furthermore, the decomposition of α-azidostyrene (**898**) yields 3-phenyl-2*H*-azirine (**899**)[177].

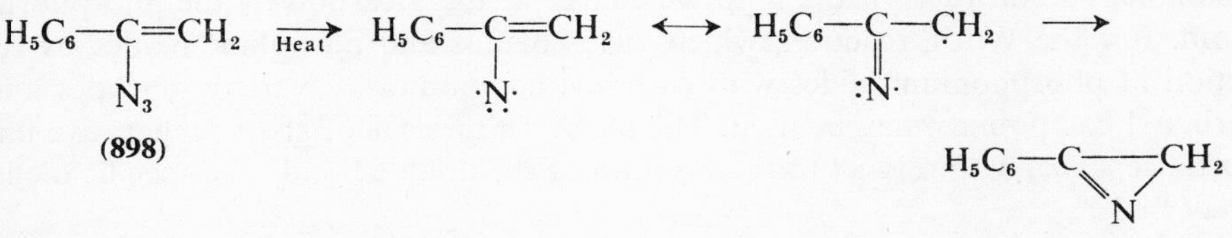

$H_5C_6-\underset{N_3}{\underset{|}{C}}=CH_2 \xrightarrow{\text{Heat}} H_5C_6-\underset{\cdot\ddot{N}\cdot}{\underset{|}{C}}=CH_2 \longleftrightarrow H_5C_6-\underset{:\ddot{N}\cdot}{\overset{\|}{C}}-\dot{C}H_2 \longrightarrow$

(**898**)

$H_5C_6-\underset{N}{C\diagdown\diagup}CH_2$

(**899**)

S-53 & S-54. Ylides of nitrogen and phosphorus

The electron-attracting inductive effect of the ammonium group in (**900**) facilitates the abstraction of a proton from the vicinal α-carbon by strong bases[178]. The resulting zwitterion (**901**) is called an ylide. More stable ylides are obtained by the

$$[(CH_3)_3\overset{+}{N}\text{—}CH_3]Br^- + C_6H_5Li \longrightarrow (CH_3)_3\overset{+}{N}\text{—}\bar{C}H_2$$
$$\quad\quad\quad (900) \quad\quad\quad\quad\quad\quad\quad\quad\quad\quad (901)$$

reaction of phosphonium salts with phenyllithium.

S-53. *Wittig reaction* (i)

Synthesis of olefin (**903**) from methyltriphenylphosphonium bromide, phenyllithium, and diphenyl ketone is an example of the Wittig reaction[179].

$$[(H_5C_6)_3\overset{+}{P}\text{—}CH_3] \; Br^- + H_5C_6Li \longrightarrow$$
$$(902)$$
$$(H_5C_6)_3\overset{+}{P}\text{—}\bar{C}H_2 \longleftrightarrow (H_5C_6)_3P\text{=}CH_2$$
$$\quad\quad\text{ylide} \quad\quad\quad\quad\quad\quad \text{ylene}$$

$$(H_5C_6)_3\overset{+}{P}\text{—}\bar{C}H_2 + (H_5C_6)_2C\text{=}O \longrightarrow \begin{array}{c}(H_5C_6)_3P\text{—}CH_2\\|\quad\quad\quad|\\O\text{—}C(C_6H_5)_2\end{array} \longrightarrow$$

$$(H_5C_6)_3P\rightarrow O \; + \; H_2C\text{=}C\overset{\displaystyle C_6H_5}{\underset{\displaystyle C_6H_5}{}}$$

$$(903) \quad 84\%$$

The greater stability of phosphorus ylides than of nitrogen ylides is caused by resonance stabilization of the negative charge at the α-carbon by the phosphorus atom. For the Wittig reaction, which yields olefins and phosphine oxides by reaction of phosphonium ylides with carbonyl compounds, a variety of ylides and carbonyl compounds may be used. The olefin synthesis is of great preparative importance since, with respect to the position of the double bond, one specific olefin is formed[180].

Aliphatic Substitution

Phosphorus ylides have been divided into three classes: (1) Stable ylides, which can be isolated in substance. (2) Moderately stable ylides; a moderately stable ylide is, for example, benzylidenetriphenylphosphorane. (3) Reactive ylides, mainly alkylidenetriphenylphosphoranes.

Stereochemistry of the Wittig reaction[181]

If unsymmetrically substituted ylides react with unsymmetrically substituted carbonyl compounds, *cis-trans*-isomeric olefins are obtained. Although many olefin preparations *via* the Wittig reaction do not proceed stereoselectively, reactions with stable ylides yield predominantly *trans*-olefins, whereas reactive ylides free from lithium salts yield predominantly *cis*-ylides (see Table 10).

Table 10.

$C_6H_5CHO + (C_6H_3)_5P=CH-CH_3 \rightarrow C_6H_5-CH=CH-CH_3$

Reaction conditions	Yield (%)	*cis/trans* Ratio
Salt-free	88%	96:4
LiI adduct	80%	83:17
Ethanol as solvent	12%	10:90
Intramolecular epimerization	69%	3:97

(1) *Stable ylides*

General agreement seems to exist as to the course of the reaction of stabilized ylides with aldehydes[182] (ketones are much less reactive).

S-54. (ii) *Reaction of benzaldehyde with (methoxycarbonylmethylene)-triphenylphosphorane.*

Stabilized ylides always yield predominantly *trans*-olefins. In the reaction of ylide (**904**) with benzaldehyde a *trans/cis* ratio of 84:16 is obtained. To explain this it is assumed that k_3 and k_6 are rate-determining and that the betaines (**905**) and (**906**) are rapidly interconverted. Clearly in the transition state between the *threo*-betaine and the *trans*-olefin there must be less steric hindrance than in the other case. Further, better conjugation is possible between the substituents R and R′ and the incipient double bond since coplanarity is sterically less hindered. Thus the more stable *trans*-olefin is formed predominantly.

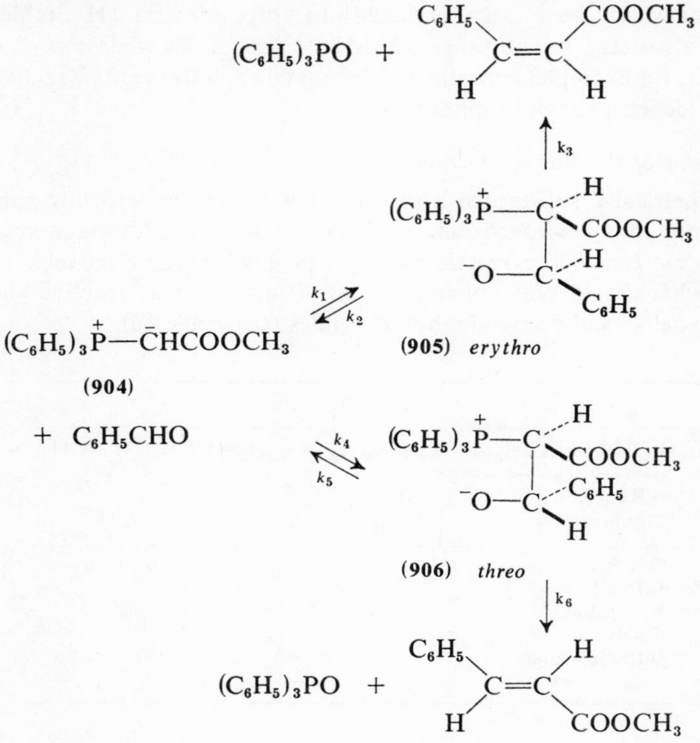

$$(C_6H_5)_3PO + \underset{H}{\overset{C_6H_5}{>}}C=C\underset{H}{\overset{COOCH_3}{<}}$$

$$\uparrow k_3$$

$$(C_6H_5)_3\overset{+}{P}-\underset{\overset{|}{-O}-C\cdots C_6H_5}{C\cdots COOCH_3}^{\cdots H}$$

(905) *erythro*

$$(C_6H_5)_3\overset{+}{P}-\bar{C}HCOOCH_3 \quad \underset{k_2}{\overset{k_1}{\rightleftarrows}}$$

(904)

$$+ \; C_6H_5CHO \qquad \underset{k_5}{\overset{k_4}{\rightleftarrows}} \quad (C_6H_5)_3\overset{+}{P}-\underset{\overset{|}{-O}-C\cdots C_6H_5}{C\cdots COOCH_3}^{\cdots H}$$

(906) *threo*

$$\downarrow k_6$$

$$(C_6H_5)_3PO + \underset{H}{\overset{C_6H_5}{>}}C=C\underset{COOCH_3}{\overset{H}{<}}$$

2. *Reactive and moderately reactive ylides*

A much more complicated situation is encountered in the reaction of moderately reactive or reactive ylides with carbonyl compounds. The reaction of a reactive ylide, for example, of ethylidenetriphenylphosphorane, when free from lithium salts[183], with benzaldehyde (if the formation of an ylide is accompanied by the formation of lithium salts such as LiBr or LiI, these salts form adducts with the ylide), leads to an 88% yield of β-methylstyrene with a *cis-trans* ratio of 96:4. In this case, interconversion between the betaines is much slower than for the stable ylides, whereas transformation of the betaines into the olefins is relatively fast. The *cis-trans* ratio is therefore determined by the irreversible, kinetically controlled betaine formation. It is not yet known why formation of the *erythro*-betaine (907) is so obviously favored over that of the *threo*-form (908)[181].

$$(C_6H_5)_3\overset{+}{P}-\overset{-}{C}HR + C_6H_5CHO$$

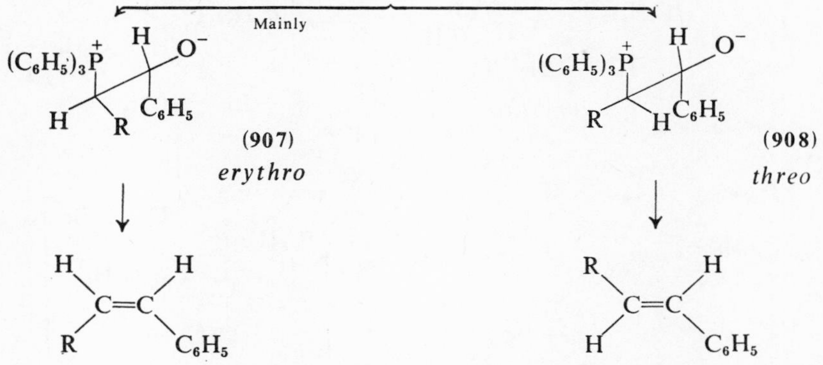

erythro (907) — threo (908)

All factors that facilitate interconversion of the betaines lead to a higher proportion of *trans*-olefin, since in the equilibrium between *erythro*- and *threo*-betaine[184], the latter is thermodynamically favored and the activation energy between the *threo*-betaine and the *trans*-olefin is lower.

Such factors are:
(1) Stabilization of the negative charge at the α-carbon by suitable substituents. The result is a more stable ylide, but a less stable *erythro*-betaine (because of steric hindrance). Consequently the *erythro*-betaine is more easily cleaved to the starting materials and thus epimerizes to the *threo*-betaine. Thus moderately reactive ylides yield less *cis*-isomer than do reactive ylides, and stable ylides yield predominantly *trans*-olefins.
(2) Addition of lithium halides[181,185]. Adducts of ylides and betaines with lithium salts are more stable than free ylides and free betaines. Since only the energies of the transition states between ylide and betaines are lowered, whereas the transition states between betaines and olefins are unaffected, interconversion between the betaines *via* the starting materials takes place more rapidly.
(3) Use of an alcohol as solvent stabilizes the ylides and betaines by hydrogen bonding.
(4) Addition of a second equivalent of phenyllithium to the betaines causes rapid interconversion of the latter [intramolecular epimerization *via* betaine ylide (909)].

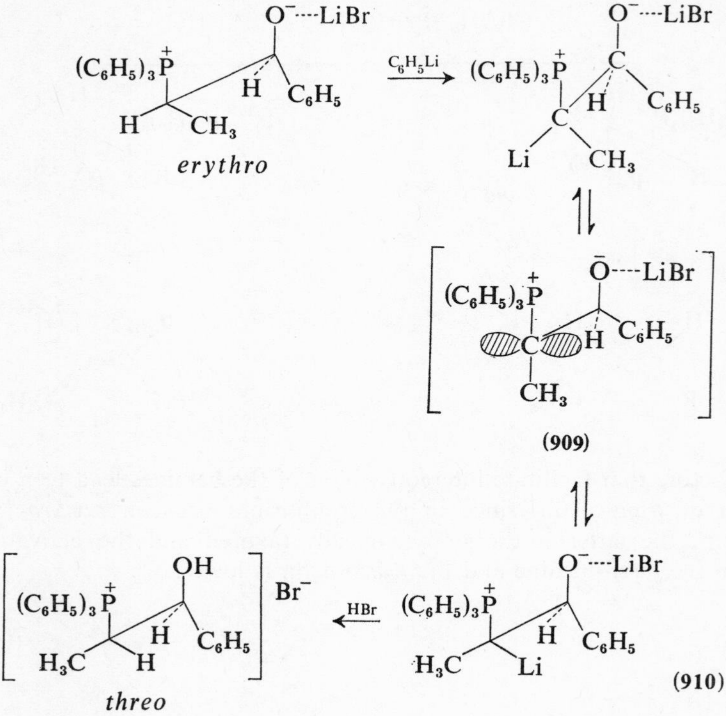

[Scheme showing erythro to threo isomerization via lithiation, with structures (909) and (910)]

S-55. Reaction of sulfur ylides with electrophiles[186, 187]

The reactivity of ylides decreases with increasing resonance stabilization of the negative charge on the ylide-carbon atom by substituents. For example the ylides (912)[188], (913)[189], (914)[190], and (915)[191] no longer react with carbonyl compounds.

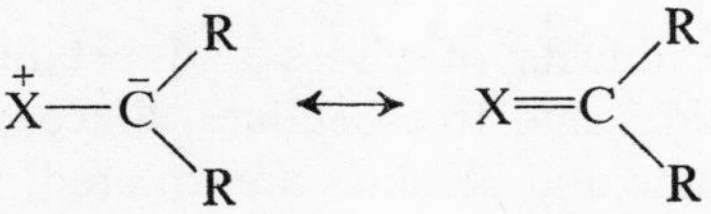

[Structures (911), (912), (913), (914), (915)]

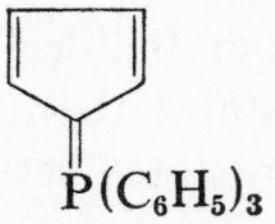

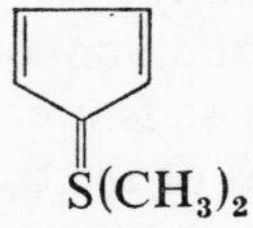

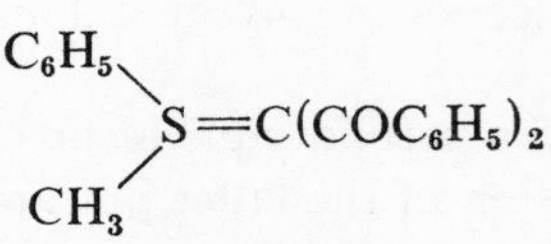

Further, the stability of the ylides depends on the nature of the hetero atom and of its substituents. With increasing ability of the hetero atom to stabilize the negative charge of the carbon, *i.e.* with increasing importance of resonance structure (911), the reactivity towards electrophiles decreases. pK values of the following 'onium salts show decreasing acidity from left to right[192]:

$$R_2S^+\!-\!CH\!\!<\quad > \quad (C_6H_5)_3P^+\!-\!CH\!\!<\quad >$$

$$(C_6H_5)_3As^+\!-\!CH\!\!<\quad > \quad R_3P^+\!-\!CH\!\!<\quad (R = alkyl)$$

In this series, the sulfur ylide is the weakest base, *i.e.* it possesses the strongest resonance interaction (2p–3d resonance) between ylide-carbon and hetero atom. This interaction is affected by the substituents on the hetero atom in such a way that electron-attracting substituents stabilize the ylide, whereas electron-donating substituents increase its reactivity. The trialkylphosphorus ylide is thus the most reactive ylide in the above sequence.

S-55. *Reaction of dimethylsulfonium fluorenylide with* p-*nitrobenzaldehyde*[193]

The ylide (916) does not react with benzaldehyde, but it reacts with the stronger electrophile *p*-nitrobenzaldehyde. Whereas phosphonium ylides yield olefins with carbonyl compounds *via* a four-membered cyclic transition state, the intermediate (917) is converted into an epoxide by loss of dimethyl sulfide*. The different

behavior may be caused by the different bond energies of the P–O and the S–O bonds and by the easy separation of R_2S from the ylide-carbon. For example, compounds (916a) are readily hydrolysed by water while phosphonium salts are unaffected.

* Epoxides are also formed from dimethylsulfoxonium methylide and ketones[194]. Sulfenes formed by abstraction of hydrogen halide from sulfonic acid halides also belong to this class of compound[195] (see p. 159):

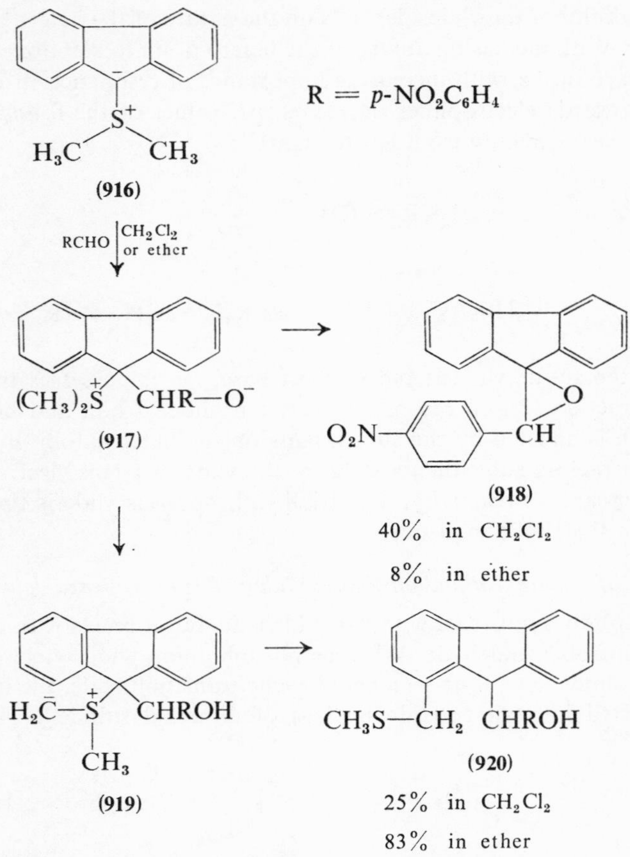

R = p-NO$_2$C$_6$H$_4$

Characteristic IR absorption of (918): 850 cm^{-1} (epoxide)
Characteristic IR absorption of (920): 3450 (OH)

In polar solvents, an ylide (919), apparently formed by proton exchange, rapidly rearranges (Sommelet–Hauser rearrangement) to the product (920). Sulfur ylides are probably capable of decomposing to carbenes and sulfides[196]. For example, the benzyl diphenyl sulfonium salt (921) with butyllithium in tetrahydrofuran at −40° yields stilbene and diphenyl sulfide. An alternative mechanism leading to the same products is nucleophilic attack by the ylide on unchanged sulfonium salt.

$(C_6H_5)_2S^+CH_2C_6H_5$ X^- + LiR $\longrightarrow$ $(C_6H_5)_2S=CHC_6H_5$ + RH + LiX

(921)

$\downarrow$ via C_6H_5CH:

$(C_6H_5)_2S$ + $C_6H_5CH=CHC_6H_5$

$(C_6H_5)_2\overset{+}{S}—\overset{-}{C}HC_6H_5$ $(C_6H_5)_2\overset{+}{S}—CHC_6H_5$
                                                              $|$ $\xrightarrow{LiR}$
X^- $(C_6H_5)_2\overset{+}{S}—CH_2C_6H_5$ X^- $CH_2C_6H_5$

$(C_6H_5)_2S$ + $C_6H_5CH=CHC_6H_5$
+ LiX + RH

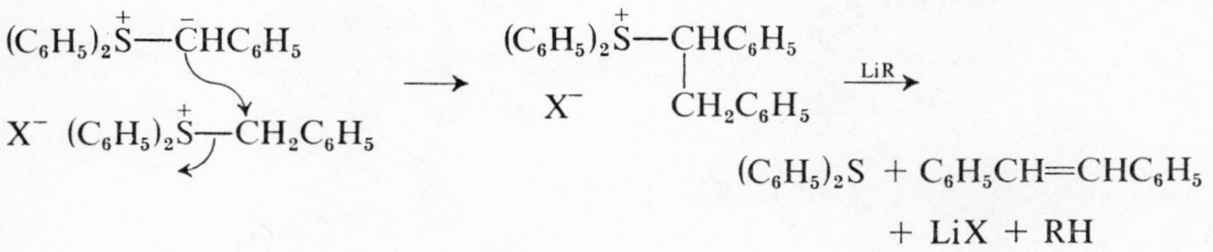

A decomposition that definitely yields carbenes is the photolysis[197] of compound (922) in presence of cyclohexene.

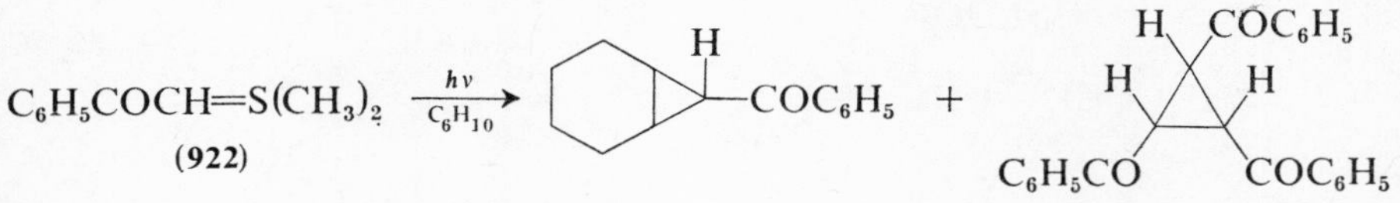

$C_6H_5COCH=S(CH_3)_2$ $\xrightarrow[C_6H_{10}]{h\nu}$ [bicyclic structure]—COC_6H_5 + [cyclopropane with H, COC$_6$H$_5$, H, H, C$_6$H$_5$CO, COC$_6$H$_5$]

(922)

S-56. Reactions of ambident ions

β-Chlorovinyl ketones can be regarded as vinylogous acid chlorides[198]. Like these they are acylation agents. The reactivity of the carbonyl-carbon atom towards nucleophilic reagents is transmitted through the double bond to the β-carbon atom:

$$H_3C—\overset{O}{\underset{\|}{C}}—Cl \longleftrightarrow H_3C—\overset{O^-}{\underset{|}{\overset{+}{C}}}—Cl$$

$$H_3C—\overset{O}{\underset{\|}{C}}—CH=CHCl \longleftrightarrow H_3C—\overset{O^-}{\underset{|}{C}}=CH—{}^+CHCl$$

Vinylogs differ from the parent substance by having one or several additional CH=CH groups and have similar chemical properties. For example, crotonaldehyde is capable of condensation at the terminal CH_3 group, as is acetaldehyde. Pentenedione monoenol (923), which is a vinylog of formic acid, has an acidity comparable with that of carboxylic acids.

$$O=CH\!-\!CH=CH—CH=CH\!-\!OH$$

(923)

S-56. *Reaction of 2-chlorovinyl methyl ketone* **(924)** *with ethyl acetoacetate* **(925)**

2-Chlorovinyl methyl ketone **(924)** reacts like an acid chloride with ethyl acetoacetate[199]. The condensation is preceded by conversion of the ester into its enolate.

$$CH_3COCH_2COOC_2H_5 + CH_3COCH=CHCl \xrightarrow[C_6H_6]{K_2CO_3}$$
(925) (924)

CH$_3$COCH=CH—CH(COCH$_3$)COOC$_2$H$_5$
(926)

⇅

CH$_3$COCH=CHC(COOC$_2$H$_5$)=C(CH$_3$)OH

→

(927) [3-methyl-substituted benzene ring with COOC$_2$H$_5$, H$_3$C, OH substituents]

→

(928) [cyclohexene with C$_2$H$_5$OOC, H$_3$C, OH, COCH$_3$, H, H, Cl, CH$_3$CO, H substituents]

↓

(929) [benzene ring with C$_2$H$_5$OOC, CH$_3$, COCH$_3$, COCH$_3$ substituents]

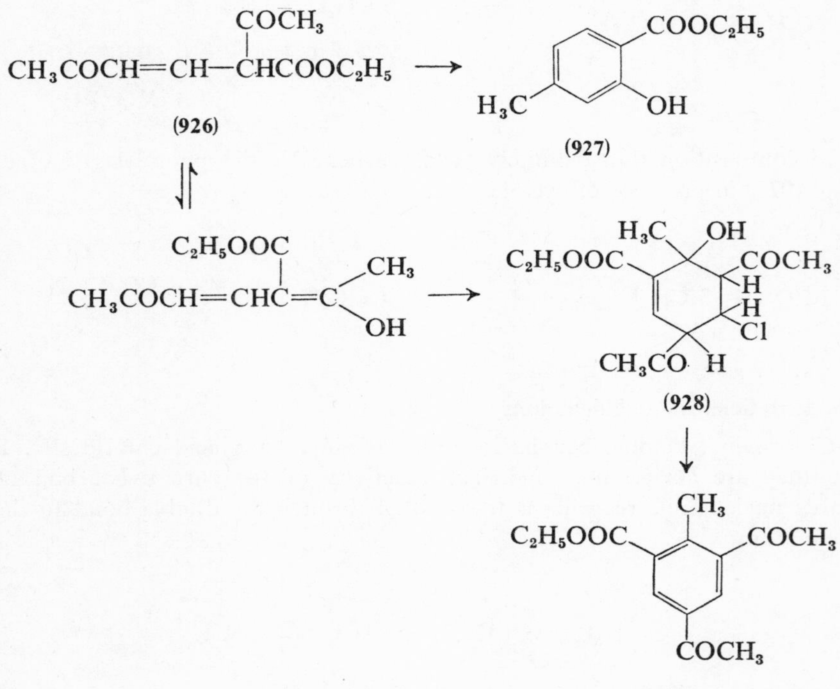

Acylation and alkylation of enolates may occur at the enolate-carbon or at the oxygen atom. Systems such as the enolate system, that possess two nucleophilic or, in other cases, two electrophilic centres, are called ambident ions[200].

Preferential reaction at the nucleophilic carbon is observed when the enolate reacts with alkyl or acyl halides by an S_N2 mechanism, since, as a consequence of its higher polarizability, the enolate carbon is a stronger nucleophile than the enolate oxygen (see p. 285). Further, the *C*-alkylated or *C*-acylated product is thermodynamically the more stable one. The transition state of alkylation or acylation at the enolate carbon probably lies towards the side of the products, that is, the thermodynamic stability of the products affects the transition state (see p. 237).

Aliphatic Substitution 349

On the other hand, reactions of the enolate ion with protons, carbonium ions, or highly polarized bonds take place preferentially at the oxygen. The transition state of these highly exothermic reactions is towards the side of the starting materials on the reaction co-ordinate. The course of the reaction is determined by electrostatic attraction, which is stronger at the oxygen anion because of its higher electron density. For example, acetyl chloride in pyridine reacts with an enolate anion at the oxygen, since the carbonyl-carbon of the 1-acetylpyridinium cation bears a relatively high positive charge compared with acetyl chloride. Likewise, in the reaction of diazomethane with an enol the methyl ether is obtained.

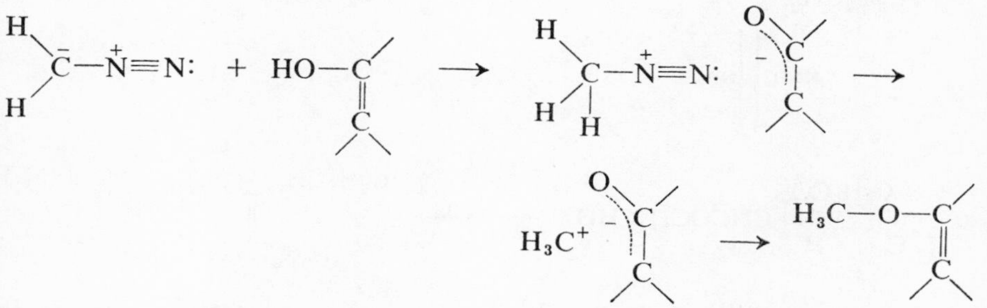

If an enolate is acidified, the enol is formed first and subsequently rearranges to the ketone. Solvent and the counter-ion of the enolate have a marked effect on the course of the reactions discussed above by shielding the location of highest electron density.

Intermediate (**926**) of our example either cyclizes to the aromatic compound (**927**) or forms compound (**928**) with another molecule of (**924**) by a Diels–Alder reaction. The product (**928**) is finally transformed into (**929**) by loss of H_2O and HCl.

A similar Diels–Alder reaction has been observed by Braude *et al.*[201] in the base-catalyzed dimerization of mesityl oxide.

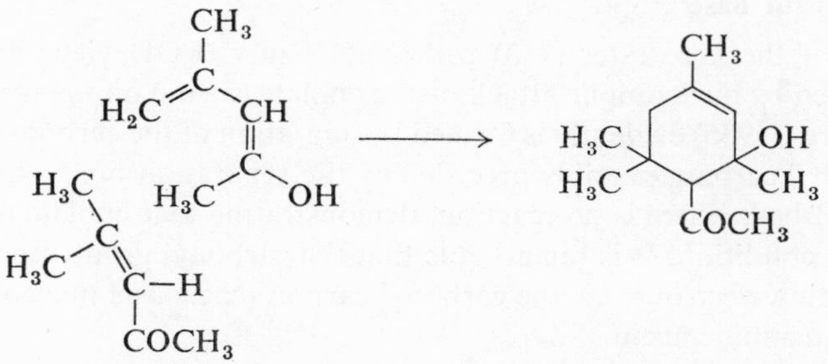

S-57. Substitution of β-keto carbonyl compounds at the γ-carbon atom[202]

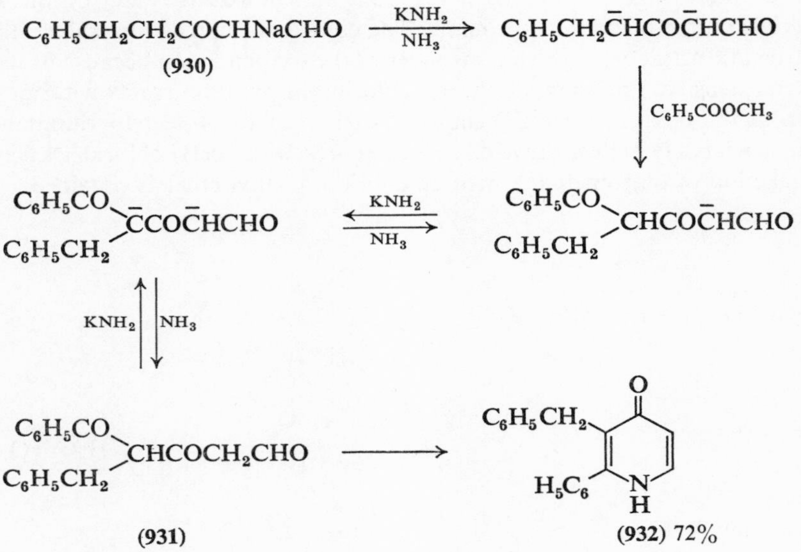

$$C_6H_5CH_2CH_2COCHNaCHO \xrightarrow[NH_3]{KNH_2} C_6H_5CH_2\bar{C}HCO\bar{C}HCHO$$
(930)

$$\downarrow C_6H_5COOCH_3$$

$$\begin{array}{c}C_6H_5CO\\C_6H_5CH_2\end{array}\!\!\!\!>\bar{C}CO\bar{C}HCHO \xleftarrow[NH_3]{KNH_2} \begin{array}{c}C_6H_5CO\\C_6H_5CH_2\end{array}\!\!\!\!>\bar{C}HCO\bar{C}HCHO$$

$$KNH_2 \updownarrow NH_3$$

$$\begin{array}{c}C_6H_5CO\\C_6H_5CH_2\end{array}\!\!\!\!>CHCOCH_2CHO \longrightarrow$$
(931)

(932) 72% — pyridinone with $C_6H_5CH_2$ and C_6H_5 substituents

Substitution at the γ-carbon of β-keto carbonyl compounds can be accomplished by first treating the enolate with KNH_2 in liquid ammonia and subjecting the resulting dianion to acylation or alkylation. For example, the enolate of 3-oxo-5-phenylpentanal (930) is benzoylated at the 4-position by ethyl benzoate. The condensation product (931) yields compound (932) by cyclization with NH_3.

S-58. Reaction of (1-methyl-2-oxocyclohexyl)methyl p-toluenesulfonate (933) with base[203]

Reaction of the keto ester (933) with NaOH in CH_3OH yields two products. (934) is formed by nucleophilic attack by the enolate carbon on the tosyloxymethyl-carbon, whereas (935) evidently is formed by migration of the carbonyl group to the tosyloxymethyl carbon, possibly preceded by the separation of the tosylate anion. In absence of base, there is no reaction, demonstrating that enolate formation is a necessary precondition. It is remarkable that the carbonyl group seems to migrate with its bonding electrons, *i.e.* the carbonyl carbon reacts as a nucleophilic instead of as an electrophilic agent.

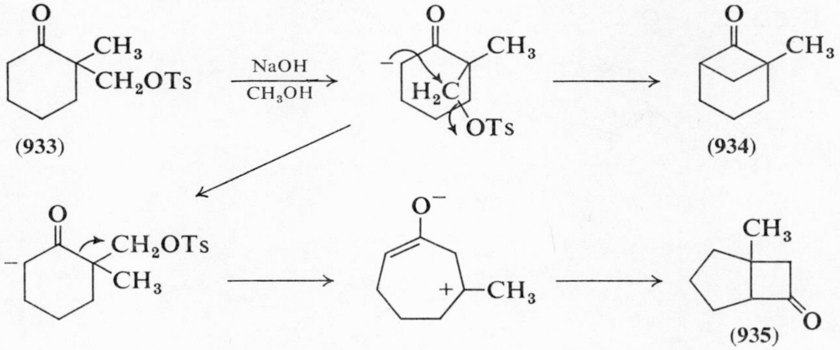

S-59. Reaction of methyl (o-{N-benzoyl-N-[o-(methoxycarbonyl)phenyl]-amino}phenyl)acetate (936) with sodium methoxide[204]

Attack by base takes place at the methylene group of (936), these hydrogen atoms being activated by a phenyl ring and an ester group (attack at the ester group or at the amide group has not been observed.) The expected Dieckmann condensation, which would lead to (937), does not occur, though it takes place readily[205] in the case of compound (941). Instead of (937), the oxindole enol (940) is obtained in 72% yield. The authors assume a mechanism that involves the amide anion (939) as intermediate. The preferred attack by the negative carbon on the amide-carbonyl carbon in (938) is presumably caused by the ready formation of a five-membered ring, whereas the Dieckmann condensation would lead to a seven-membered ring.

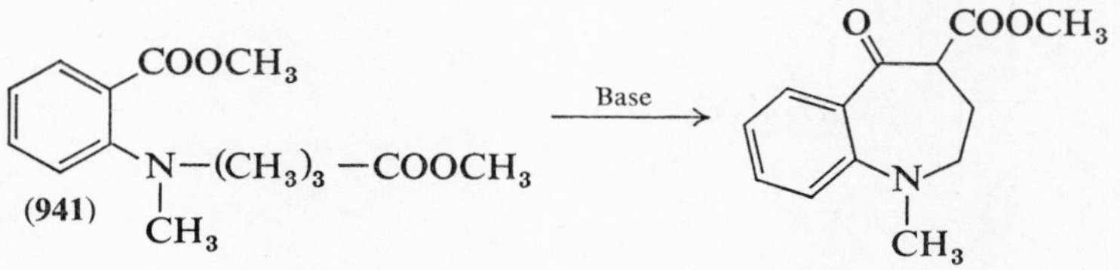

The formation of (939) is facilitated by the resonance stabilization of the negative charge by two phenyl rings.

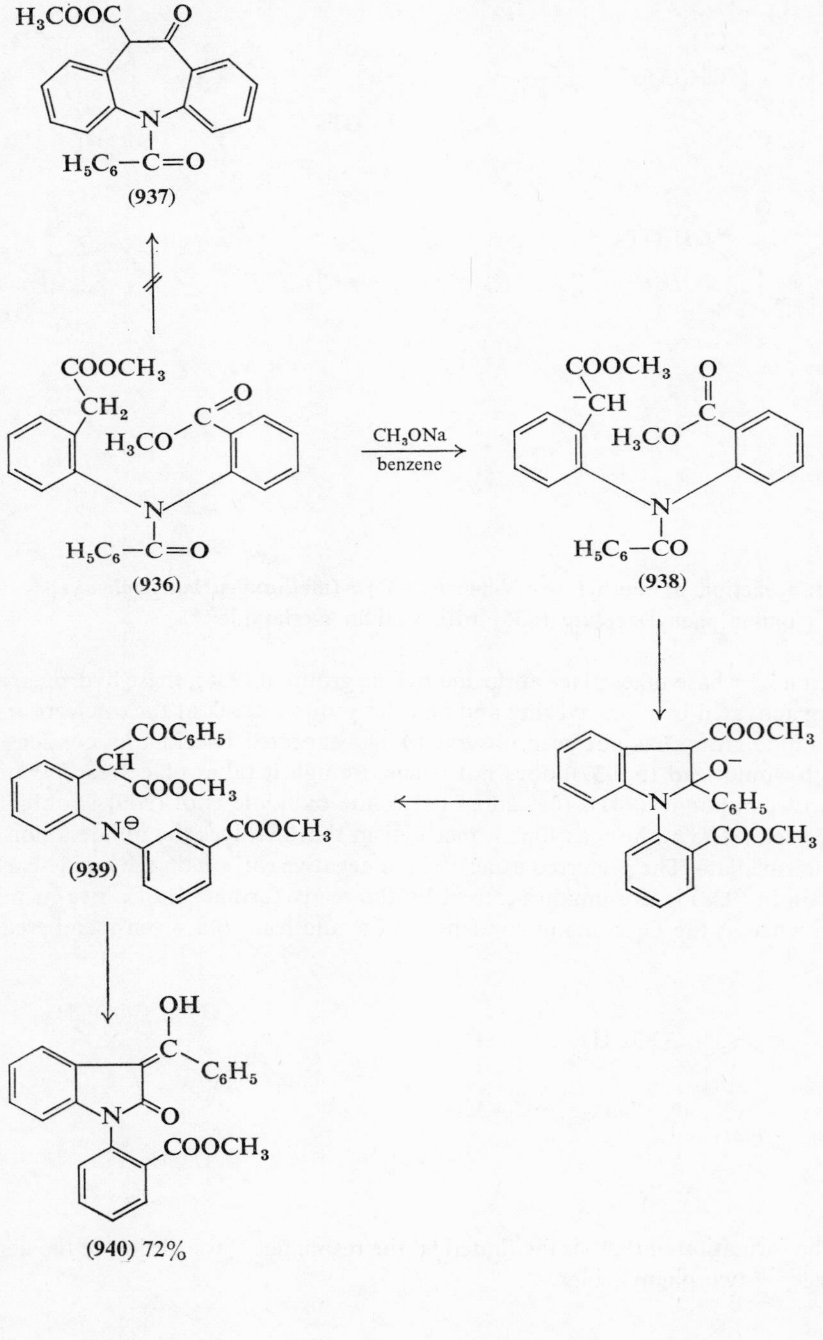

Problems in Organic Reaction Mechanisms

(937): tricyclic compound with H₃COOC and C=O groups on central 7-membered ring fused to two benzene rings, N substituted with H₅C₆—C=O

↑̸

(936): two ortho-substituted phenyl rings linked by N—C(=O)—C₆H₅; one ring bears CH₂—COOCH₃, the other bears C(=O)—OCH₃

→ CH₃ONa / benzene →

(938): similar to 936 but with CH⁻ (anion) bearing COOCH₃

↓

(right structure): indoline-type bicyclic with COOCH₃, O⁻, C₆H₅, and N-aryl bearing COOCH₃

←

(939): ortho-disubstituted aniline anion N⊖; one aryl ring bears CH(COC₆H₅)(COOCH₃), the other aryl bears COOCH₃

↓

(940) 72%: 3-(α-hydroxybenzylidene)-2-oxoindoline with N-aryl bearing COOCH₃; exocyclic C(OH)—C₆H₅

Absorption of (**940**): IR, 1735 cm^{-1} (COOR); UV, 268 mμ (ε 12600)
    1650 ($>$N—C=O)   320 (ε 11300) } conj.
    1630   (C=C)
NMR, -1.7τ (1) enolic H
    1.67—3.5 (13) aromatic H
    6.37 (3) OCH_3

A similar mechanism has been proposed for the abnormal Chapman rearrangement illustrated[206].*

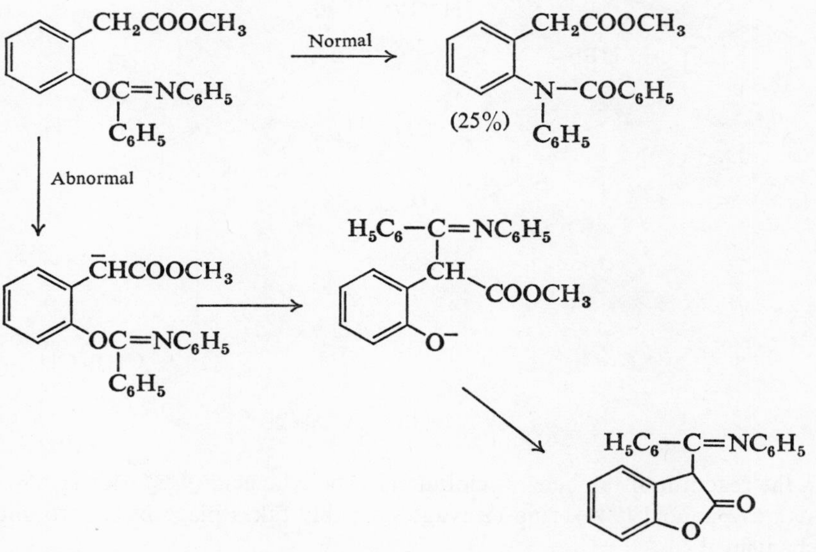

[Scheme: ortho-substituted benzene with CH$_2$COOCH$_3$ and OC(=NC$_6$H$_5$)C$_6$H$_5$ groups undergoes Normal rearrangement (25%) to give N(COC$_6$H$_5$)(C$_6$H$_5$) with CH$_2$COOCH$_3$; Abnormal pathway gives ĊHCOOCH$_3$ intermediate → H$_5$C$_6$—C=NC$_6$H$_5$ with CH(COOCH$_3$) and O$^-$ → benzofuran-2(3H)-one with H$_5$C$_6$—C=NC$_6$H$_5$ substituent]

---

* The Chapman rearrangement is a useful method[207] for preparation of unsymmetrically substituted diarylamines from imidic esters:

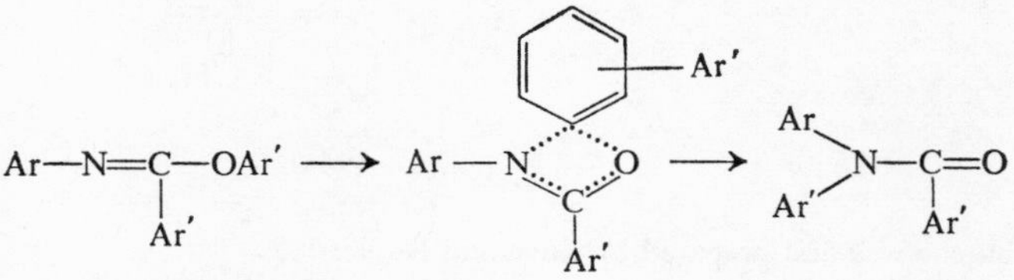

Ar—N=C(Ar')—OAr' → Ar—N⋯O cyclic intermediate with Ar' and Ar' → (Ar)(Ar')N—C(=O)Ar'

S-60. Reactions of bicyclo[4.2.0]octa-1,3,5-triene (942) and its derivatives

The four-membered ring of benzocyclobutene and its derivatives is unstable in the presence of strong acids and other electrophilic reagents*. Ring cleavage always occurs between a carbon atom of the aromatic nucleus and an α-carbon atom[209]; for example, benzocyclobutene yields phenethyl bromide with HBr in refluxing CH_3COOH. Possibly the dealkylation proceeds by a multicentre mechanism in which the developing positive charge on the α-carbon is stabilized by the bromide anion. Styrene and α-methylbenzyl bromide were not observed[210].

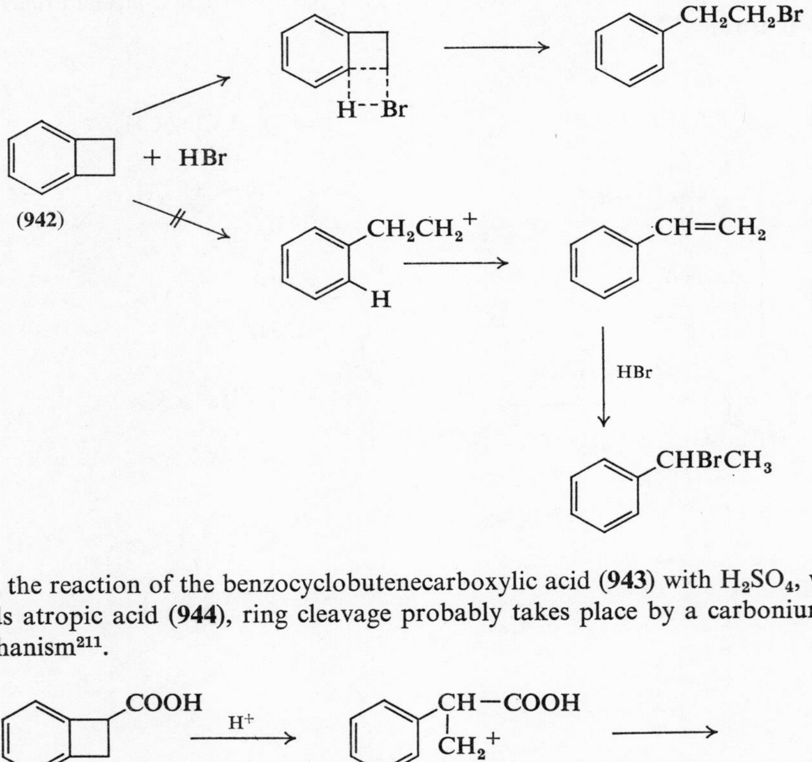

In the reaction of the benzocyclobutenecarboxylic acid (943) with H_2SO_4, which yields atropic acid (944), ring cleavage probably takes place by a carbonium ion mechanism[211].

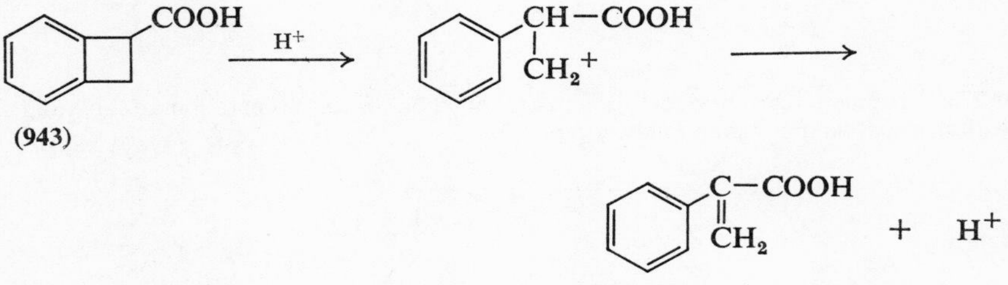

* Benzocyclobutene was first prepared by Cava and Napier[208].

S-60. Reaction of bicyclo[4.2.0]octa-1,3,5-trien-7-ol and -7-one with base[212]

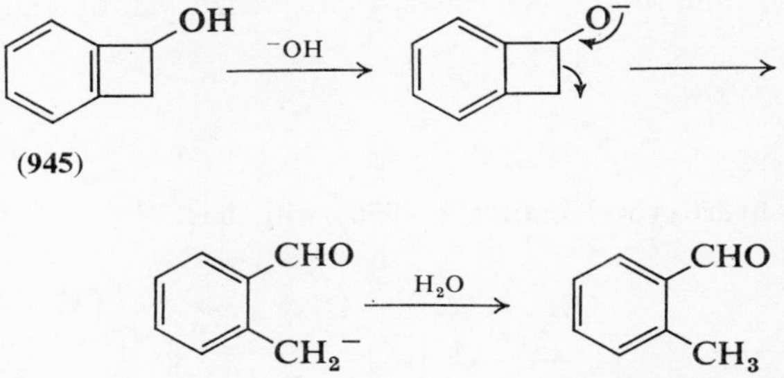

(945)

(947) 72%

Bicyclo[4.2.0]octa-1,3,5-trien-7-ol (benzocyclobutenol) **(945)**, the derived ketone **(946)**, and the 7,8-dione[213] are attacked not only by electrophiles but also by nucleophiles. The latter cleave the four-membered ring between the two aliphatic carbon atoms.

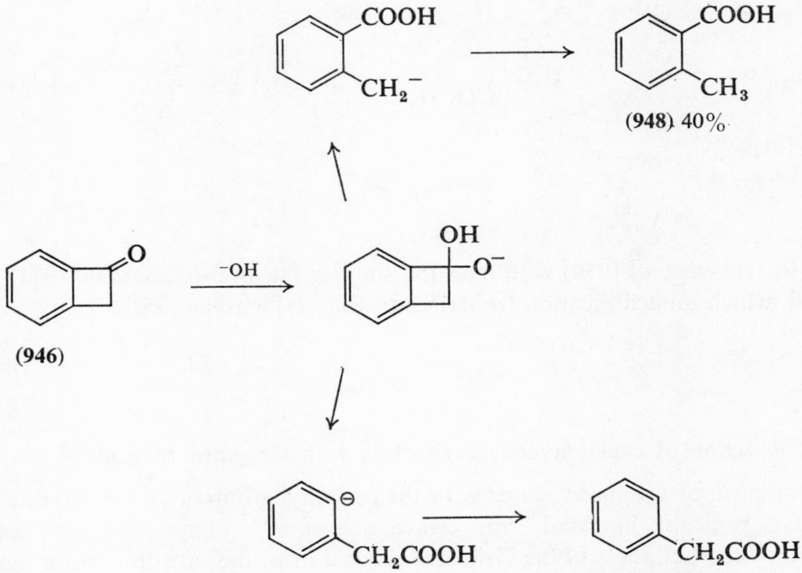

(946)

(948) 40%

(949) 40%

Benzocyclobutene is attacked by radicals without cleavage of the four-membered ring. For example, reaction with N-bromosuccinimide yields benzocyclobutenyl bromide. The bromide can be converted into the carboxylic acid by magnesium and CO_2. Reaction of the bromide with NH_3 yields benzocyclobutylamine.

S-61. Reaction of 3-hydroxycyclobutanone (950) with base[214]

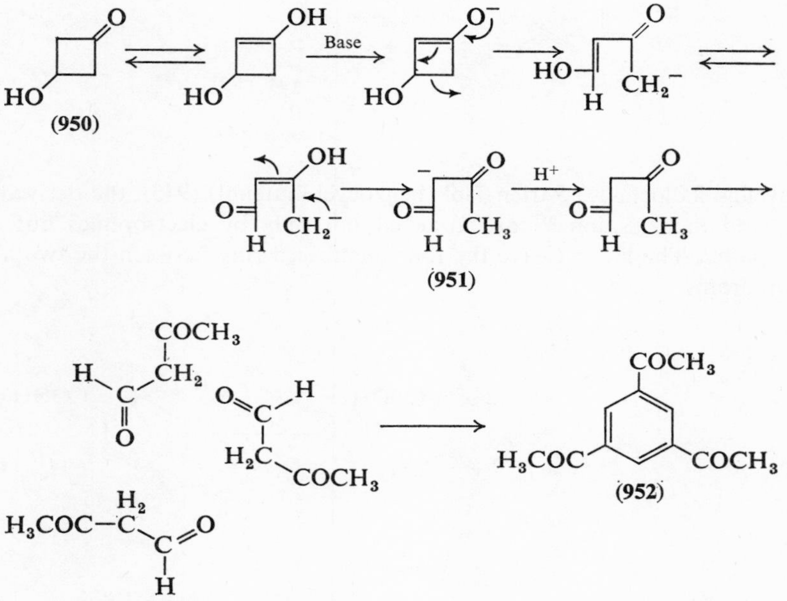

In the cleavage of (950) with base the sodium salt of 3-oxobutanal (951) is obtained, which on acidification trimerizes to triacetylbenzene (952).

S-62. Reaction of cyclobutanone derivatives with Grignard reagents[215]

If addition of Grignard reagents to the carbonyl group of cyclobutanone derivatives is sterically hindered, ring cleavage occurs[216]. Steric hindrance may be caused by the bulkiness of the Grignard reagent or of the substituents on the four-membered ring.

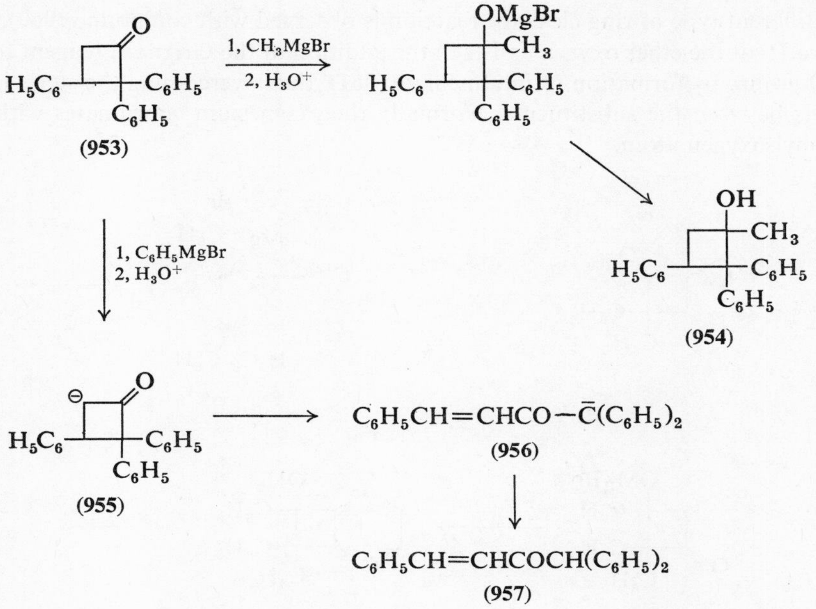

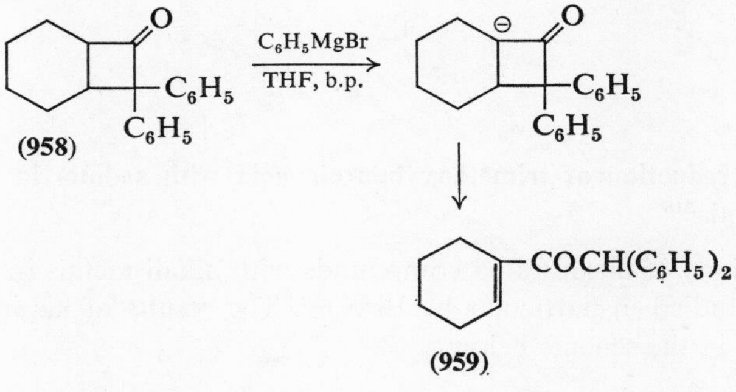

For example, reaction of 2,2,3-triphenylcyclobutanone (953) with methylmagnesium bromide yields the normal alcohol (954). Phenylmagnesium bromide, however, yields compound (957). The Grignard reagent abstracts a proton from the position α to the carbonyl group; the resulting carbanion (955) rearranges to a more stable carbanion (956). Analogously, (958) yields (959)[217].

A different type of ring cleavage reaction is observed with compound (960). It is believed that the ether oxygen assists in the addition of the Grignard reagent to the C=O group, by formation of intermediate (961), thus overcoming the steric interactions between the substituents. Normally the magnesium coordinates with the carbonyl-oxygen atom.

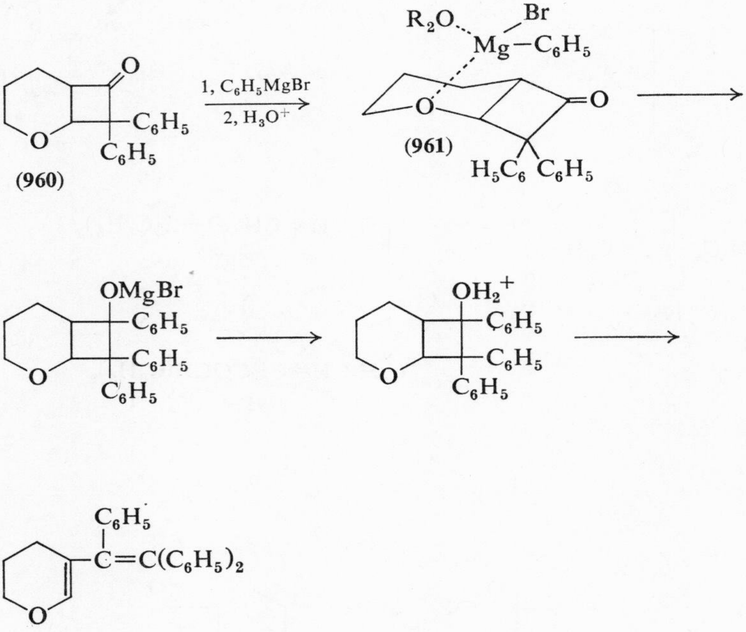

S-63. Birch reduction of trimethoxybenzoic acid with sodium in liquid ammonia[218]

The reactions of unsaturated compounds with alkali metals in liquid ammonia have been studied in particular by Birch[219]. The results of his investigations are summarized in the scheme below.

It is assumed that the alkali metal is present in liquid NH_3 in the form of solvated metal cations and solvated electrons. The electrons are of high enough energy to initiate the reaction by addition to a double bond. The very low acidity of NH_3 permits the formation of anions and anion radicals. Since NH_3 is only a very weak acid, it is usual to add compounds to the reaction mixture to serve as proton donors for conversion of the anions into the saturated compounds. Such proton donors are, for example, alcohols or ammonium salts. In the absence of these

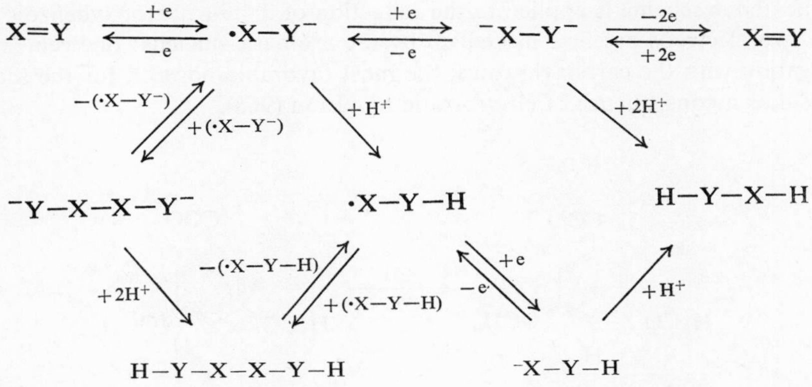

$$X=Y \underset{-e}{\overset{+e}{\rightleftarrows}} \cdot X-Y^- \underset{-e}{\overset{+e}{\rightleftarrows}} {}^-X-Y^- \underset{+2e}{\overset{-2e}{\rightleftarrows}} X=Y$$

$$\begin{array}{ccc} \Bigg\downarrow\!\!\!\!{\overset{-(\cdot X-Y^-)}{\underset{+(\cdot X-Y^-)}{}}} & \Bigg\downarrow +H^+ & \Bigg\downarrow +2H^+ \\ {}^-Y-X-X-Y^- & \cdot X-Y-H & H-Y-X-H \end{array}$$

$$\begin{array}{ccc} \Bigg\downarrow +2H^+ & \Bigg\downarrow\!\!\!\!{\overset{-(\cdot X-Y-H)}{\underset{+(\cdot X-Y-H)}{}}} \quad \underset{-e}{\overset{+e}{\rightleftarrows}} & \Bigg\uparrow +H^+ \\ H-Y-X-X-Y-H & & {}^-X-Y-H \end{array}$$

substances, more and more NH_2^- accumulates in the solution, so that base-catalysed rearrangements then occur.

If the negative charge arising is resonance-stabilized, a second reaction may take place, along with the reduction of multiple bonds, namely, reductive cleavage of saturated bonds[220].

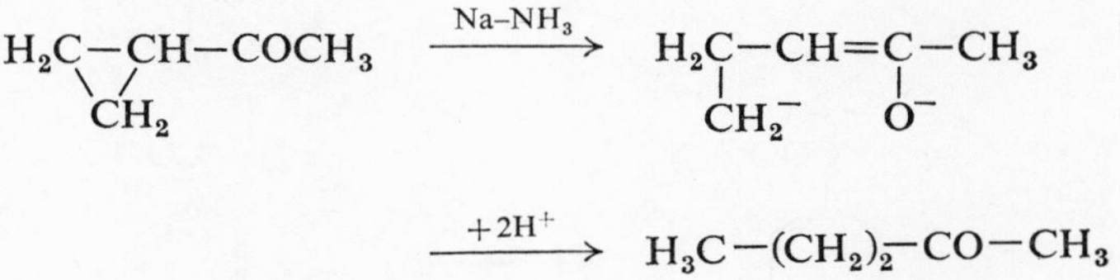

$$\underset{CH_2}{H_2C-CH-COCH_3} \xrightarrow{Na-NH_3} \underset{CH_2^- \quad O^-}{H_2C-CH=C-CH_3}$$

$$\xrightarrow{+2H^+} H_3C-(CH_2)_2-CO-CH_3$$

If the above scheme is applied to the reduction of 3,4,5-trimethoxybenzoic acid (**962**), two electrons are first picked up by the aromatic nucleus; one remains in conjugation with the carboxyl group; the most favorable position for the second is at C-4 as a consequence of electrostatic repulsion (**963**).

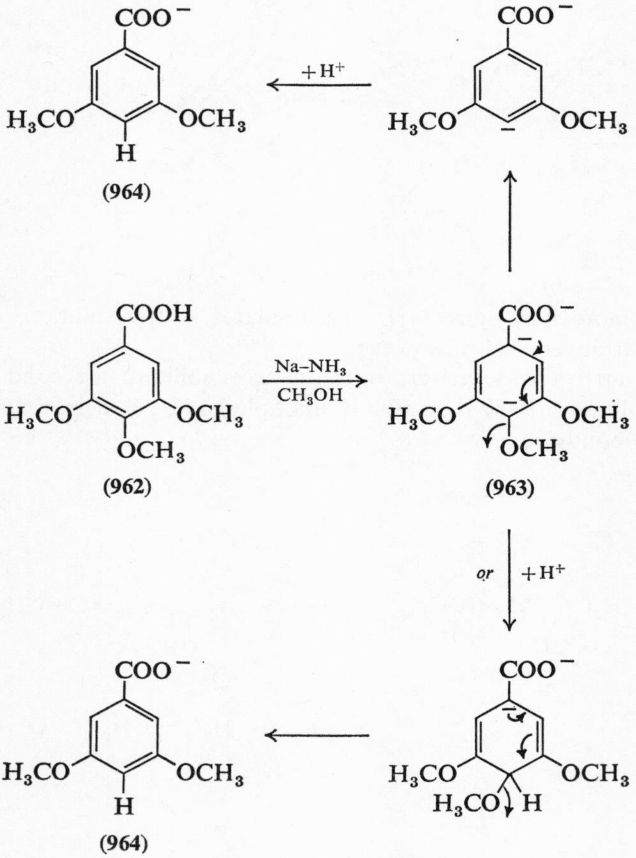

Loss of a methoxyl group and subsequent addition of a proton, or first addition of a proton to the most basic carbon atom (C-4), followed by separation of the OCH₃ group, yield 3,5-dimethoxybenzoic acid (**964**) which is subsequently reduced to the cyclohexadiene derivative (**965**). Reduction of the enol ether double bond has not been observed.

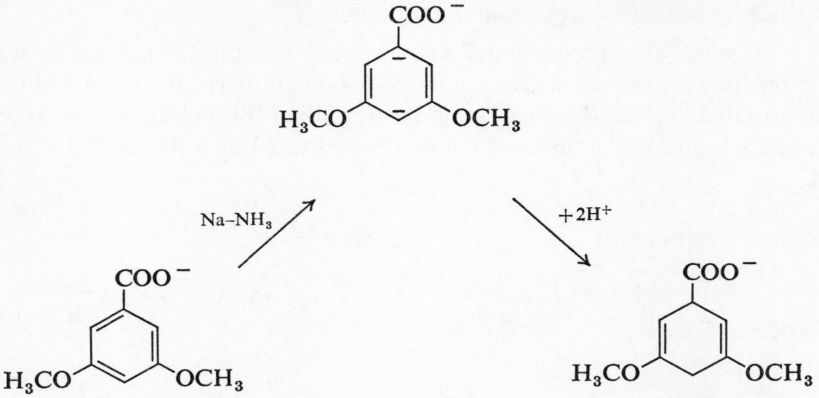

S-64. Birch reduction of 10-propargyl-*trans*-decalin-2,7-dione (966)[221]

If the diketone (966) is treated with lithium in liquid ammonia and ether, containing some $(NH_4)_2SO_4$, compound (969) is obtained. The radical anion (967) reacts with the adjacent carbonyl group, yielding radical anion (968) which picks up another electron and then a proton from $(NH_4)_2SO_4$.

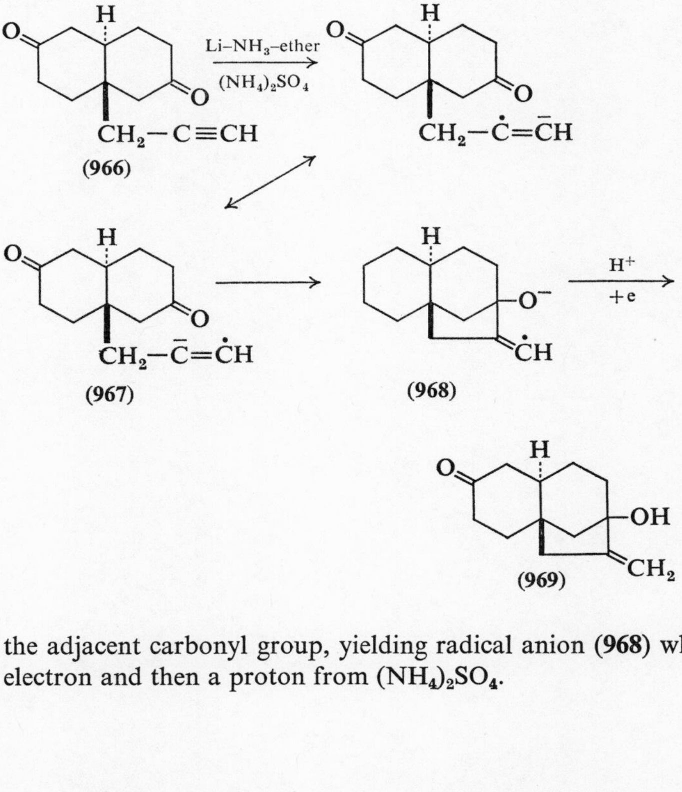

S-65. Birch reduction of α,β-unsaturated ketones[222]

Birch reduction of α,β-unsaturated ketones leads to saturated ketones by way of an intermediate containing a nucleophilic β-carbon atom. (In the usual addition to α,β-unsaturated ketones the β-carbon is electrophilic.) It is not known whether this intermediate is a radical anion or a dianion[223] [indicated by a star at the oxygen of (971)].

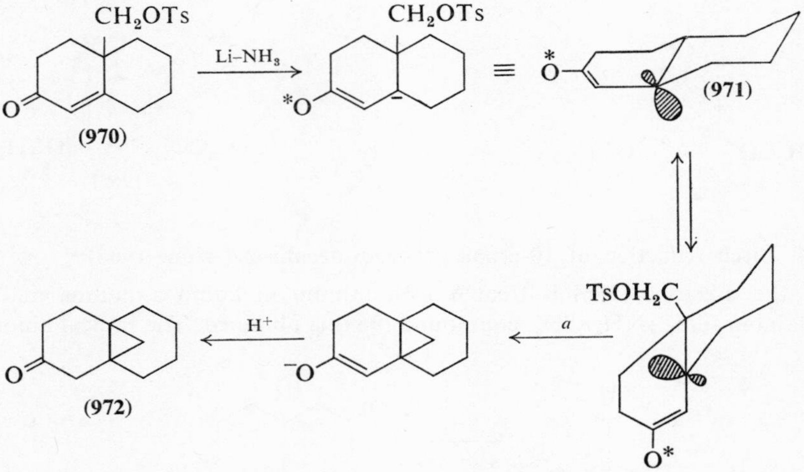

Aliphatic Substitution

The nucleophilic β-carbon can either react with an electrophilic center in the near proximity (*a*) or abstract a proton from a proton donor (*b*). The enolate obtained by path (*b*) reacts with alkylating agents at the α-carbon atom. The exclusive or predominant α-alkylation (at the α-carbon of the original α,β-double bond) takes place only if the alkylation rate is much faster than equilibration of the theoretically possible enolate ions. This equilibration is particularly fast if Na or K is used as the alkali-metal. With lithium the equilibration rate is much reduced.

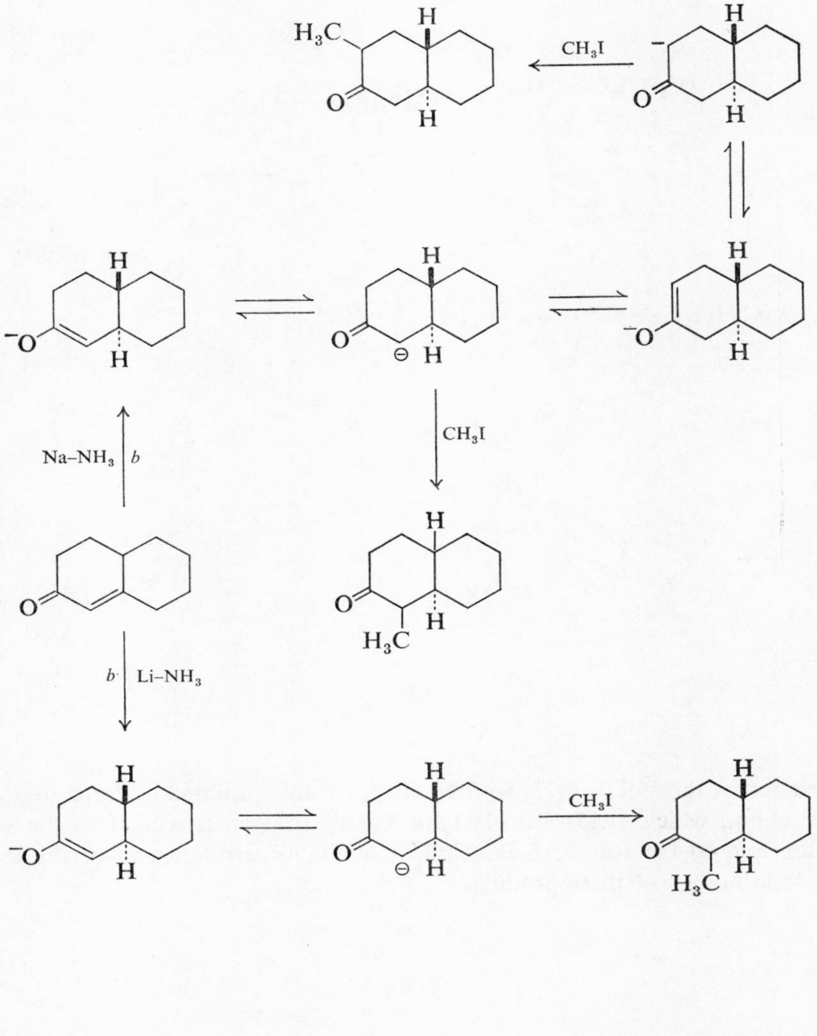

S-66. Reaction of 2-chloro-2,2-diphenylacetanilide with sodium hydride[224]

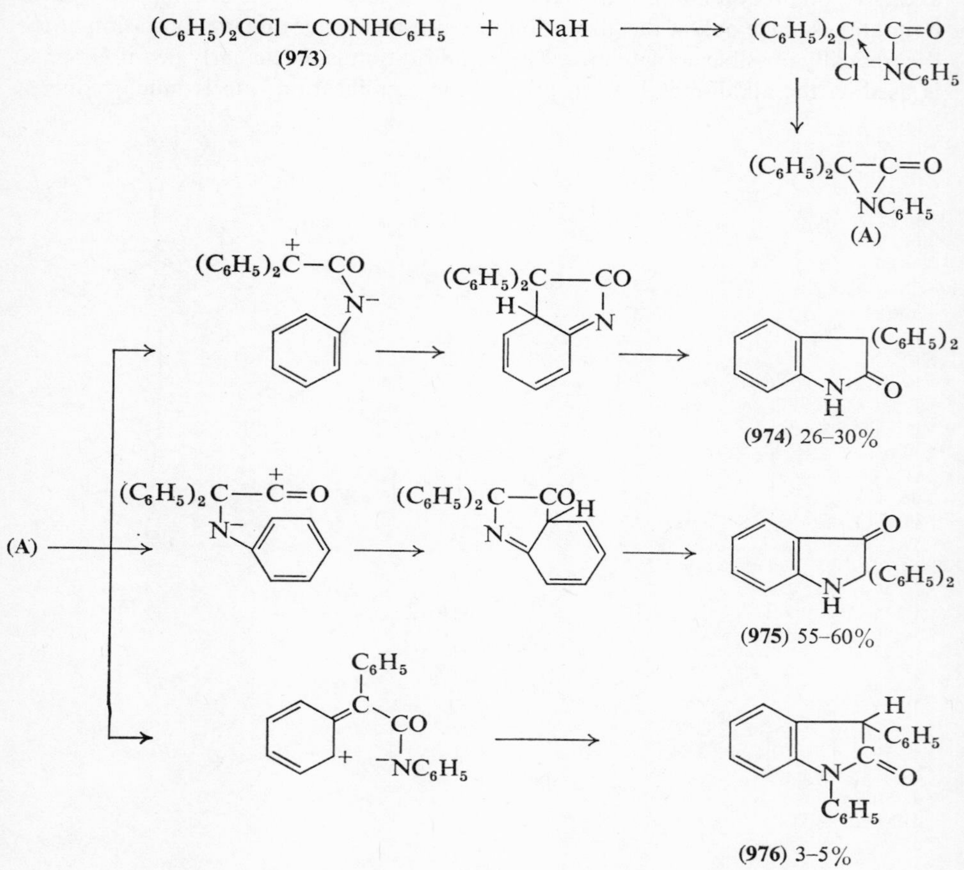

$(C_6H_5)_2CCl-CONHC_6H_5$ + NaH ⟶ $(C_6H_5)_2\underset{Cl}{C}-\underset{{}^-NC_6H_5}{C}=O$

⟶ $(C_6H_5)_2C\overset{\diagdown}{-}\underset{NC_6H_5}{C}=O$
(A)

(A) ⟶ $(C_6H_5)_2\overset{+}{C}-CO$ with N^-–phenyl ⟶ $(C_6H_5)_2C\text{---}CO$ cyclohexadienyl-N= ⟶ (974) 26–30% (3,3-diphenyloxindole)

(A) ⟶ $(C_6H_5)_2C-\overset{+}{C}=O$ with N^-–phenyl ⟶ $(C_6H_5)_2C-CO$ cyclohexadienyl ⟶ (975) 55–60% (2,2-diphenylindolin-3-one type)

⟶ $\underset{{}^-NC_6H_5}{\overset{C_6H_5}{C}}=\text{...}CO$ ⟶ (976) 3–5%

Reaction of the anilide (973) with NaH is probably initiated by formation of an amide anion, which yields an α-lactam. The cyclization is similar to the Neber reaction and to the Ramberg–Bäcklund reaction. Rearrangement of the lactam leads to a mixture of three products.

REFERENCES

[1] A. Streitwieser, Jr., *Chem. Rev.*, **56**, 571 (1965).
[1a] C. D. Nenitzescu, in *Carbonium Ions*, ed. G. A. Olah and P. von R. Schleyer, Interscience Publ., Inc., New York, 1968, p. 59.
[2] J. Steigman and L. P. Hammett, *J. Amer. Chem. Soc.*, **59**, 2536 (1937).
[3] G. S. Hammond, *J. Amer. Chem. Soc.*, **77**, 334 (1955).
[4] G. A. Olah, *Chem. Eng. News*, **45**, 77 (1967).
[5] R. Breslow, H. Höver, and H. Chang, *J. Amer. Chem. Soc.*, **84**, 3168 (1962).
[6] M. P. Cava, R. Pohlke, B. W. Erickson, J. C. Rose, and G. Fraenkel, *Tetrahedron*, **18**, 1005 (1962).
[7] G. L. Buchanan and D. B. Jhaveri, *J. Org. Chem.*, **26**, 4295 (1961).
[8] K. Bangert and V. Boekelheide, *J. Amer. Chem. Soc.*, **86**, 1159 (1964).
[9] W. Reppe, O. Schlichting, K. Klager, and T. Toepel, *Ann. Chem.* **560**, 1 (1948).
[10] R. Criegee and D. Seebach, *Chem. Ber.*, **96**, 2704 (1963).
[11] G. Stork and W. N. White, *J. Amer. Chem. Soc.*, **78**, 4609 (1956); J. R. Wolfe, Jr., and W. G. Young, *Chem. Rev.*, **56**, 769 (1956).
[12] T. J. Katz, J. R. Hall, and W. C. Neikam, *J. Amer. Chem. Soc.*, **84**, 3199 (1962); T. J. Katz and E. H. Gold, *ibid.*, **86**, 1600 (1964).
[13] L. S. Levitt, *J. Org. Chem.*, **20**, 1297 (1955); D. Seebach, *Chem. Ber.*, **96**, 2712 (1963).
[14] R. Criegee, *Fortschr. chem. Forsch.*, **1**, 531 (1949—50).
[15] K. B. Wiberg, *J. Amer. Chem. Soc.*, **75**, 3961 (1953).
[16] J. H. Merz and W. A. Waters, *J. Chem. Soc.*, **1949**, 15.
[17] N. A. Milas, R. L. Peeler, Jr., and O. L. Mageli, *J. Amer. Chem. Soc.*, **76**, 2322 (1954).
[18] A. G. Davies, R. V. Foster, and A. M. While, *J. Chem. Soc.*, **1953**, 1541.
[19] G. W. Meadows and B. De B. Darwent, *Can. J. Chem.*, **30**, 502 (1952); *Trans. Faraday Soc.*, **48**, 1015 (1952).
[20] H. J. Schneider and J. J. Bagnell, *J. Org. Chem.*, **26**, 3009 (1961).
[21] W. E. Parham, H. Wynberg, and F. L. Ramp, *J. Amer. Chem. Soc.*, **75**, 2065 (1953).
[22] M. V. McConnell and W. H. Moore, *J. Org. Chem.*, **28**, 822 (1963).
[23] O. Wallach, *Ann. Chem.* **275**, 169 (1893); B. K. Wasson, C. H. Gleason, J. Levi, J. M. Parker, L. M. Thompson, and C. H. Yates, *Can. J. Chem.*, **39**, 923 (1961).
[24] W. V. McConnell and W. H. Moore, *J. Org. Chem.*, **30**, 3480 (1965).
[25] R. D. H. Murray, W. Parker, R. A. Raphael, and D. B. Jhaveri, *Tetrahedron*, **18**, 55 (1962).
[26] J. E. Leffler, *Reactive Intermediates of Organic Chemistry*, Interscience Publishers, New York, 1956.
[27] J. Martin, W. Parker, and R. A. Raphael, *J. Chem. Soc.*, **1964**, 289.
[28] E. W. Colvin and W. Parker, *J. Chem. Soc.*, **1965**, 5764.
[29] G. L. Buchanan, A. McKillop, and R. A. Raphael, *J. Chem. Soc.*, **1965**, 833.
[30] J. Meinwald, A. Lewis, and P. G. Gassman, *J. Amer. Chem. Soc.*, **84**, 977 (1962).
[31] J. Meinwald, *Rec. Chem. Progress* 22 39 (1961).
[32] J. Meinwald, H. C. Hwang, A. P. Wolf, and D. Christman, *J. Amer. Chem. Soc.*, **82**, 483 (1960).
[33] E. C. Taylor and D. R. Eckroth, *Tetrahedron*, **20**, 2059 (1964).
[34] J. D. Loudon and G. Tennant, *Quart. Rev.*, **18**, 389 (1964).
[35] J. A. Moore and D. H. Ahlstrom, *J. Org. Chem.* **26**, 5254 (1961).
[36] E. Bamberger and A. Lindberg, *Chem. Ber.*, **43**, 122 (1910); G. Heller, *ibid.*, **44**, 2418 (1911).
[37] W. B. Dickinson, *J. Amer. Chem. Soc.*, **86**, 3580 (1964).
[38] M. G. Evans and M. Polanyi, *Trans. Faraday Soc.*, **34**, 11 (1938).
[39] M. Simonetta and S. Winstein, *J. Amer. Chem. Soc.*, **76**, 18 (1954); A. Ehret and S. Winstein, *ibid.*, **88**, 2048 (1966).
[40] H. C. Brown, *Chem. Eng. News*, **45**, 87 (1967); G. D. Sargent, *Quart. Rev.*, **20**, 301 (1966).
[41] P. D. Bartlett, *Non-classical Ions*, W. A. Benjamin Inc., New York, 1965.
[42] W. N. Lipscomb, *J. Chem. Phys.*, **22**, 985, 989 (1954); **61**, 23 (1957); H. E. Zimmerman and A. Zweig, *J. Amer. Chem. Soc.*, **83**, 1196 (1961); J. Warkentin and E. Sandford, *ibid.*, **90**, 1667 (1968); H. C. Longuet-Higgins, *Quart. Rev.*, **11**, 121 (1957).

[43] J. D. Roberts and R. H. Mazur, *J. Amer. Chem. Soc.*, **73**, 2509 (1951); R. H. Mazur, W. N. White, D. A. Semenow, C. C. Lee, M. S. Silver, and J. D. Roberts, *ibid.*, **81**, 4390 (1959).
[44] P. von R. Schleyer and G. W. Van Dine, *J. Amer. Chem. Soc.*, **88**, 2321 (1966); H. C. Brown and J. D. Cleveland, *ibid.*, p. 2051 (1966).
[45] O. L. Chapman and P. Fitton, *J. Amer. Chem. Soc.*, **85**, 41 (1963).
[46] V. J. Traynelis and J. R. Livingston, Jr., *J. Org. Chem.*, **29**, 1092 (1964).
[47] R. Breslow, J. Lockhart, and A. Small, *J. Amer. Chem. Soc.*, **84**, 2793 (1962).
[48] S. Winstein and C. Ordronneau, *J. Amer. Chem. Soc.*, **82**, 2084 (1960).
[49] H. C. Brown and H. M. Bell, *J. Amer. Chem. Soc.*, **85**, 2324 (1963).
[50] S. Winstein, A. H. Lewin, and K. C. Pande, *J. Amer. Chem. Soc.*, **85**, 2324 (1963).
[51] B. Tanida, T. Tsuji, and T. Jrie, *J. Amer. Chem. Soc.*, **88**, 864 (1966).
[52] H. C. Brown, "Nonclassical Intermediates" in the Organic Reaction Mechanism Conference, Brookhaven, New York, N.Y., Sept. 5—8, 1962.
[53] A. Diaz, M. Brookhart, and S. Winstein, *J. Amer. Chem. Soc.*, **88**, 3133, 3135 (1966).
[54] P. R. Story and M. Saunders, *J. Amer. Chem. Soc.*, **84**, 4876 (1962); P. R. Story, L. C. Snyder, D. C. Douglass, E. W. Anderson, and R. L. Kornegay, *ibid.*, **85**, 3630 (1963); P. R. Story and M. Saunders, *ibid.*, **82**, 6199 (1960).
[55] H. G. Richey, Jr., and R. K. Lustgarten, *J. Amer. Chem. Soc.*, **88**, 3136 (1966).
[56] H. Tanida and Y. Hala, *J. Org. Chem.*, **30**, 977 (1965).
[57] S. Winstein and D. S. Trifan, *J. Amer. Chem. Soc.*, **74**, 1147, 1154 (1952).
[58] H. C. Brown, *Chemistry in Britain*, **2**, 199 (1966).
[59] H. C. Brown and F. J. Chloupek, *J. Amer. Chem. Soc.*, **85**, 2322 (1963).
[60] P. von R. Schleyer, *J. Amer. Chem. Soc.*, **89**, 699, 701 (1967).
[61] H. C. Brown, F. J. Chloupek, and M.-H. Rei, *J. Amer. Chem. Soc.*, **86**, 1248 (1964).
[62] H. C. Brown, F. J. Chloupek, and M.-H. Rei, *J. Amer. Chem. Soc.*, **86**, 1246 (1964); H. C. Brown and M.-H. Rei, *ibid.*, p. 5008.
[63] H. M. Bell and H. C. Brown, *J. Amer. Chem. Soc.*, **86**, 5007 (1964).
[64] H. C. Brown and K.-T. Liu, *J. Amer. Chem. Soc.*, **89**, 466 (1967).
[65] P. von R. Schleyer, D. C. Kleinfelter, and H. G. Richey, *J. Amer. Chem. Soc.*, **85**, 429 (1963).
[66] H. C. Brown, *J. Amer. Chem. Soc.*, **89**, 6381 (1967).
[66a] G. A. Olah, *J. Amer. Chem. Soc.*, **90**, 3882 (1968).
[67] S. J. Cristol, T. C. Morrill, and R. A. Sanchez, *J. Org. Chem.* **31**, 2719 (1966), and subsequent papers.
[68] R. S. Neale and E. B. Whipple, *J. Amer. Chem. Soc.*, **86**, 3130 (1964).
[69] J. Meinwald, P. G. Gassman, and J. J. Hurst, *J. Amer. Chem. Soc.*, **84**, 3722 (1962); J. Meinwald and P. G. Gassman, *ibid.*, **85**, 57 (1963).
[70] H. C. Brown, K. J. Morgan, and F. J. Chloupek, *J. Amer. Chem. Soc.*, **87**, 2137 (1965).
[71] R. A. Sneen and J. W. Larsen, *J. Amer. Chem. Soc.*, **88**, 2593 (1966); R. A. Sneen, J. V. Carter, and P. S. Kay, *ibid.*, p. 2595.
[72] J. O. Edwards, *J. Amer. Chem. Soc.*, **78**, 1819 (1956).
[73] E. S. Gould, *Mechanism and Structure in Organic Chemistry*, Henry Holt and Company, New York, 1959, p. 299.
[74] S. Searles, Jr., H. R. Hays, and E. F. Lutz, *J. Org. Chem.* **27**, 2828, 2832 (1962).
[75] M. G. Ettlinger, *J. Amer. Chem. Soc.*, **72**, 4792 (1950); E. E. van Tamelen, *ibid.*, **73**, 3444 (1951); C. C. Price and P. F. Kirk, *ibid.*, **75**, 2396 (1953); F. G. Bordwell and H. M. Anderson, *ibid.*, p. 4959.
[76] C. G. Overberger and A. Drucker, *J. Org. Chem.*, **29**, 360 (1964); S. M. Iqbal and L. N. Owen, *J. Chem. Soc.*, **1960**, 1030; J. R. Lowell, Jr., and G. K. Helmkamp, *J. Amer. Chem. Soc.*, **88**, 768 (1966); G. K. Helmkamp, D. J. Pettitt, J. R. Lowell, Jr., W. R. Mabey, and R. G. Wolcott, *ibid.*, p. 1030.
[77] K. Bangert and V. Boeckelheide, *Tetrahedron Letters.* **1963**, 1119.
[78] G. Chirurdoglu and B. Tursch, *Bull. Soc. Chim. Belges*, **66**, 600 (1957).
[79] H. W. Heine, *J. Amer. Chem. Soc.*, **85**, 2743 (1963).
[80] H. W. Heine and D. A. Tomalia, *J. Amer. Chem. Soc.*, **84**, 993 (1962).
[81] H. W. Heine, *Angew. Chem.*, **74**, 772 (1962).
[82] H. E. Zaugg, F. E. Chadde, and R. J. Michaels, *J. Amer. Chem. Soc.*, **84**, 4567 (1962).
[83] G. S. Hammond in *Steric Effects in Organic Chemistry*, ed. M. S. Newman, J. Wiley and Sons, New York, 1956, p. 464.

Aliphatic Substitution 367

[84] H. E. Zaugg, R. W. DeNet, and R. J. Michaels, *J. Org. Chem.*, **26**, 4821 (1961).
[85] H. Meerwein, W. Florian, H. Schön, and G. Stopp, *Ann. Chem.*, **641**, 1 (1961); T. Taguchi and Y. Kawazoe, *J. Org. Chem.*, **26**, 2699 (1961).
[86] H. E. Zaugg and R. J. Michaels, *Tetrahedron*, **18**, 899 (1962).
[87] J. R. Holum, D. Jorenby, and P. Mattison, *J. Org. Chem.*, **29**, 769 (1964).
[88] L. J. Smith and J. R. Holum, *J. Amer. Chem. Soc.*, **78**, 3417 (1956).
[89] K. D. Gundermann and C. Burba, *Chem. Ber.*, **94**, 2157 (1962).
[90] P. D. Bartlett and C. G. Swain, *J. Amer. Chem. Soc.*, **71**, 1406 (1962).
[91] A. Michaelis and R. Krehne, *Chem. Ber.*, **31**, 1048 (1898); A. Arbusow, *J. Russ. Phys. Chem. Ges.*, **38**, 687 (1906).
[92] G. M. Kosolapoff, *J. Amer. Chem. Soc.*, **74**, 4953 (1952).
[93] H. Machleidt and R. Wessendorf, *Ann. Chem.*, **674**, 1 (1964).
[94] F. W. Lichtenthaler, *Chem. Rev.*, **61**, 607 (1961); R. F. Hudson, *Structure and Mechanism in Organic Phosphorus Chemistry*, Academic Press, New York-London, 1965, p. 153.
[95] W. Perkow, E. W. Krockow, and K. Knoevennagel, *Chem. Ber.*, **88**, 662 (1955).
[96] Sir Christopher Ingold, *Structure and Mechanism in Organic Chemistry*, Cornell University Press, New York, 2nd edn., 1958.
[97] E. E. Smissman and J. R. J. Sorenson, *J. Org. Chem.*, **30**, 300 (1965).
[98] S. Searles, Jr., and S. Nukina, *J. Amer. Chem. Soc.*, **87**, 5656 (1965).
[99] I. Maclean and R. P. A. Sneeden, *Tetrahedron*, **19**, 1307 (1963).
[100] A. R. Pinder and R. Robinson, *J. Chem. Soc.*, **1955**, 3341; N. H. Bromham and R. Pinder, *ibid.*, **1959**, 2688.
[101] P. Yates and C. D. Anderson, *J. Amer. Chem. Soc.*, **80**, 1264 (1958); **85**, 2937 (1963).
[102] J. E. Gowan and T. S. Wheeler, *J. Chem. Soc.*, **1950**, 1925.
[103] M. S. Newman and Y. T. Yu, *J. Amer. Chem. Soc.*, **74**, 507 (1952).
[104] R. L. Clarke and W. T. Hunter, *J. Amer. Chem. Soc.*, **80**, 5304 (1958).
[105] H. W. Heine, *J. Amer. Chem. Soc.*, **79**, 6268 (1957).
[106] C. R. Hauser, F. W. Swamer, and B. I. Ringler, *J. Amer. Chem. Soc.*, **70**, 4023 (1948).
[107] R. P. Linstead, R. W. Kierstead, and B. C. L. Weedon, *J. Chem. Soc.*, **1952**, 3616, 3610; **1953**, 1799, 1803; E. Vogel, *Fortschr. Chem. Forsch.*, **3**, 430 (1955).
[108] A. Streitwieser, Jr., *Chem. Rev.*, **56**, 571 (1956).
[109] M. S. Newman and A. B. Mekler, *J. Amer. Chem. Soc.*, **82**, 4039 (1960).
[110] A. J. Speziale and L. R. Smith, *J. Org. Chem.*, **27**, 4361 (1962); **28**, 1805 (1963); A. J. Speziale, L. R. Smith, and J. E. Fedder, *ibid.*, **30**, 4306 (1965).
[111] W. D. Kumler, *J. Amer. Chem. Soc.*, **83**, 4983 (1961); A. R. Katritzky and R. A. Y. Jones, *Chem. Ind. (London)*, **1961**, 722; H. K. Hall, *J. Amer. Chem. Soc.*, **78**, 2717 (1956).
[112] D. Martin, *Chem. Ber.*, **97**, 2689 (1964); *Tetrahedron Letters*, **1964**, 2829; D. Martin and W. Mucke, *Chem. Ber.*, **98**, 2059 (1965); K. A. Jensen and A. Holm, *Acta Chem. Scand.*, **18**, 826, 2417 (1964).
[113] R. Stroh and H. Gerber, *Angew. Chem.*, **72**, 1000 (1960).
[114] J. C. Kauer and W. W. Henderson, *J. Amer. Chem. Soc.*, **86**, 4732 (1964).
[115] E. Grigat and R. Pütter, *Chem. Ber.*, **97**, 3012 (1964).
[116] D. J. Cram. *J. Amer. Chem. Soc.*, **75**, 332 (1953).
[117] C. C. Lee, J. W. Clayton, D. G. Lee, and A. J. Finlayson, *Tetrahedron*, **18**, 1395 (1962).
[118] J. L. Kice, R. A. Bartsch, M. A. Dankleff, and S. L. Schwartz, *J. Amer. Chem. Soc.*, **87**, 1734 (1965); J. L. Kice and M. A. Dankleff, *Tetrahedron Letters*, **1966**, 1783.
[119] K. B. Wiberg and T. M. Shryne, *J. Amer. Chem. Soc.*, **77**, 2774 (1955); K. L. Olivier and W. G. Young, *ibid.*, **81**, 5811 (1959).
[120] J. O. Edwards, *Peroxide Reaction Mechanisms*, Interscience Publishers Inc., New York, 1962, p. 67.
[121] L. Horner and H. Winkler, *Tetrahedron Letters*, **1964**, 3275.
[122] L. Horner and W. Jurgeleit, *Ann. Chem.*, **591**, 138 (1955).
[123] M. A. Greenbaum, D. B. Denney, and A. K. Hoffmann, *J. Amer. Chem. Soc.*, **78**, 2563 (1956).
[124] C. G. Moore and B. R. Trego, *Tetrahedron*, **18**, 205 (1962); **19**, 1251 (1963).
[125] P. D. Bartlett and G. Meguerian, *J. Amer. Chem. Soc.*, **78**, 3710 (1967).
[126] A. J. Speziale and L. R. Smith, *J. Amer. Chem. Soc.*, **84**, 1868 (1962); A. J. Speziale and R. C. Freeman, *ibid.*, **82**, 903 (1960).

[127] H. Eilingsfeld, M. Seefelder, and H. Weidinger, *Angew. Chem.*, **72**, 836 (1960).
[128] R. D. Partos and A. J. Speziale, *J. Amer. Chem. Soc.*, **87**, 5068 (1965).
[129] V. Mark, *Tetrahedron Letters*, **1961**, 333; *Org. Synth.*, **46**, 93 (1966).
[130] P. J. Wheatley, *J. Chem. Soc.*, **1961**, 4936.
[131] D. J. Cram, *Fundamentals of Carbanion Chemistry*, Academic Press, New York, 1965; G. Köbrich, *Angew. Chem.*, **74**, 453 (1962).
[132] H. A. Staab, *Einführung in die theoretische organische Chemie*, Verlag Chemie, Weinheim/ Bergstrasse, 1959.
[133] H. C. Brown, H. Bartholomay, Jr., and M. D. Taylor, *J. Amer. Chem. Soc.*, **66**, 435 (1944).
[134] U. Schöllkopf, *Angew. Chem.*, **72**, 153 (1960); P. T. Lansbury and J. D. Sidler, *Chem. Commun.*, **1965**, 373; *Tetrahedron Letters*, **1965**, 691; R. C. Fort, Jr., and P. von R. Schleyer, *Adv. Alicyclic Chem.*, **1**, 284 (1966).
[135] S. Winstein and T. G. Traylor, *J. Amer. Chem. Soc.*, **78**, 2597 (1956).
[136] D. J. Cram, A. Langemann, J. Allinger, and K. R. Kopecky, *J. Amer. Chem. Soc.*, **81**, 5740 (1959); D. J. Cram, A. Langemann, and F. Hauck, *ibid.*, p. 5750; D. J. Cram, A. Langemann, W. Lwowski, and K. R. Kopecky, *ibid.*, p. 5760; D. J. Cram, K. R. Kopecky, F. Hauck, and A. Langemann, *ibid.*, p. 5754; D. J. Cram, F. Hauck, K. R. Kopecky, and W. D. Nielsen, *ibid.* p. 5767; D. J. Cram, J. L. Mateos, F. Hauck, A. Langemann, K. R. Kopecky, W. D. Nielsen, and J. Allinger, *ibid.*, p. 5774.
[137] D. J. Cram and H. P. Fischer, see ref. 131, p. 148.
[138] P. G. Gassmann and F. V. Zalar, *Tetrahedron Letters*, 1964, 3251.
[139] D. J. Cram, C. A. Kingsbury, and A. Langemann, *J. Amer. Chem. Soc.*, **81**, 5758 (1959).
[140] U. Schöllkopf and W. Fabian, *Ann. Chem.*, **642**, 1 (1961); U. Schöllkopf and D. Walter, *Angew. Chem.*, **73**, 545 (1961); U. Schöllkopf, *Angew. Chem. Intern. Ed. Engl.*, **1**, 126 (1962).
[141] P. T. Lansbury, V. A. Pattison, J. D. Sidler, and J. B. Bieber, *J. Amer. Chem. Soc.*, **88**, 78 (1966).
[142] T. Thomson and T. S. Stevens, *J. Chem. Soc.*, **1932**, 1932; E. F. Jenny and J. Deney, *Angew. Chem. Intern. Edit., Engl.*, **1**, 155 (1962).
[143] K. Krollpfeiffer and H. Hartmann, *Chem. Ber.*, **83**, 90 (1950).
[144] A. C. Cope and P. H. Towle, *J. Amer. Chem. Soc.*, **71**, 3423 (1949); A. C. Cope, T. T. Foster, and P. H. Towle, *ibid.*, p. 3929; A. H. Wragg, T. S. Stevens, and D. M. Ostle, *J. Chem. Soc.*, **1958**, 4057; U. Schöllkopf, *Angew. Chem. Intern. Edit., Engl.*, **2**, 161 (1963).
[145] U. Schöllkopf, M. Patsch, and H. Schäfer, *Tetrahedron Letters*, **1964**, 2515.
[146] R. L. Letsinger, *J. Amer. Chem. Soc.*, **72**, 4842 (1950).
[147] H. M. Walborsky, F. J. Impastato, and A. E. Young, *J. Amer. Chem. Soc.*, **86**, 3283 (1964); H. M. Walborsky and A. E. Young, *ibid.*, p. 3288.
[148] F. R. Jensen and L. H. Gale, *J. Amer. Chem. Soc.*, **81**, 1261 (1959).
[149] T. Mukaiyama, H. Nambu, and I. Kuwajiama, *J. Org. Chem.*, **28**, 912 (1963); T. Mukaiyama, I. Kuwajiama, and Z. Suzuki, *ibid.*, p. 2024; R. E. Dessy, G. F. Reynolds, and J.-Y. Kim, *J. Amer. Chem. Soc.*, **81**, 2683 (1959).
[150] D. Seebach, *Angew. Chem. Intern. Edit., Engl.*, **4**, 121 (1965).
[151] W. von E. Doering and A. K. Hoffmann, *J. Amer. Chem. Soc.*, **77**, 521 (1955).
[152] R. A. Finnegan, *Tetrahedron Letters*, **1963**, 429, 851.
[153] H. Höver, H. Mergard, and F. Korte, unpublished results.
[154] L. Pauling, *The Nature of the Chemical Bond*, Cornell University Press, New York, 1960.
[155] H. F. Ebel, *Tetrahedron*, **21**, 699 (1965).
[156] W. K. McEwen, *J. Amer. Chem. Soc.*, **58**, 1124 (1936); A. A. Morton, *Chem. Rev.*, **35**, 1 (1944).
[157] A. Streitwieser, Jr., J. I. Brauman, J. H. Hammons, and A. H. Dudjaalmaka, *J. Amer. Chem. Soc.*, **87**, 384 (1965).
[158] D. E. Applequist and D. F. O'Breen, *J. Amer. Chem. Soc.*, **85**, 743 (1963).
[159] R. M. Salinger and R. E. Dessy, *Tetrahedron Letters*, **1963**, 729.
[160] R. A. Finnegan and R. S. McNees, *J. Org. Chem.* **29**, 3234 (1964).
[161] E. Lewicki, H. Pines, and N. C. Sih, *Chem. Ind. (London)*, **1964**, 154; H. Pines, N. C. Sih, and E. Lewicki, *J. Org. Chem.*, **30**, 1457 (1965); N. C. Sih and H. Pines, *ibid.*, p. 1462.
[162] R. E. Benkeser, A. E. Trevillyan, and J. Hooz, *J. Amer. Chem. Soc.*, **84**, 258 (1962); R. E. Benkeser, J. Hooz, T. V. Liston, and A. E. Trevillyan, *ibid.*, **85**, 3984 (1963).

[163] H. Pines and L. A. Schaap, *Advan. Catalysis*, **12**, 116 (1960).
[164] M. Schlosser, *Angew. Chem.*, **76**, 127 (1964).
[165] G. Wittig and G. Harborth, *Ber. Deut. Chem. Ges.*, **77**, 306 (1944).
[166] W. E. Parham, M. A. Kalnius, and D. R. Theissen, *J. Org. Chem.*, **27**, 2698 (1962); W. E. Parham and R. F. Motter, *J. Amer. Chem. Soc.*, **81**, 2146 (1959); P. D. Bartlett, S. Friedman, and M. Stiles, *ibid.*, **75**, 1771 (1953); F. G. Bordwell and G. D. Cooper, *ibid.*, **73**, 5187 (1951).
[167] N. P. Neureiter and F. G. Bordwell, *J. Amer. Chem. Soc.*, **85**, 1209 (1963); L. A. Paquette and L. S. Wittenbrook, *ibid.*, **90**, 6783 (1968).
[168] F. G. Bordwell and G. D. Cooper, *J. Amer. Chem. Soc.*, **73**, 5184 (1951).
[169] R. B. Loftfield, *J. Amer. Chem. Soc.*, **72**, 632 (1950); T. S. Stevens and E. Farkas, *ibid.*, **74**, 5352 (1952); H. O. House and W. F. Gilmore, *ibid.*, **83**, 3980 (1961).
[170] E. S. Gould, *Mechanism and Structure in Organic Chemistry*, Henry Holt and Company, New York, 1959, p. 562; W. Kirmse and L. Horner, *Chem. Ber.*, **89**, 1674 (1956).
[171] N. P. Neureiter, *J. Amer. Chem. Soc.*, **88**, 558 (1966).
[172] A. Schriesheim and C. A. Rowe, Jr., *Tetrahedron Letters*, **1962**, 405.
[173] L. A. Paquette, *J. Amer. Chem. Soc.*, **86**, 4085, 4089 (1964).
[174] D. J. Cram and M. S. Halch, *J. Amer. Chem. Soc.*, **75**, 33, 38 (1953); C. O'Brien, *Chem. Rev.*, **64**, 81 (1964); see also Cram, ref. 131, p. 249.
[175] H. O. House and W. F. Berkowitz, *J. Org. Chem.*, **28**, 307 (1963).
[176] R. F. Parcell, *Chem. Ind. (London)*, **1963**, 1396.
[177] G. Smolinsky, *J. Amer. Chem. Soc.*, **83**, 4483 (1961); *J. Org. Chem.*, **27**, 3557 (1962).
[178] G. Wittig, *Angew. Chem.*, **68**, 505 (1956); **63**, 15 (1951).
[179] G. Wittig and G. Geissler, *Ann. Chem.*, **580**, 44 (1953); *Angew. Chem.*, **66**, 10 (1954).
[180] Reviews: U. Schöllkopf, *Angew. Chem.*, **71**, 260 (1959); S. Tripett, *Advan. Org. Chem.*, **1**, 83 (1960); L. A. Yanovskaya, *Russ. Chem. Rev.*, **30**, 347 (1961); S. Tripett, *Quart. Rev.*, **17**, 406 (1963); G. Wittig, *Pure Appl. Chem.*, **9**, 245 (1964); H. J. Bestmann, *Angew. Chem.*, **77**, 609, 651, 850 (1965); H. O. House, *Modern Synthetic Reactions*, W. A. Benjamin, New York, 1965, p. 245. A. Maerker, *Organic Reactions*, Vol. 14, p. 270, J. Wiley and Sons, New York, 1965; A. W. Johnson, *Ylid Chemistry*, Academic Press, New York, 1966, 177; L. D. Bergelson and M. M. Shemyakin, *Neuere Methoden der präparativen organischen Chemie*, Vol. 5, p. 135, Verlag Chemie, 1967.
[181] M. Schlosser and K. F. Christmann, *Ann. Chem.*, **708**, 1 (1967).
[182] A. J. Speziale and D. E. Bissing, *J. Amer. Chem. Soc.*, **85**, 3878 (1963)
[183] M. G. Wittig, H. Eggers, and P. Duttner, *Ann. Chem.*, **619**, 10 (1958); C. Ruchardt, S. Eichler, and P. Panse, *Angew. Chem.*, **75**, 858 (1963).
[184] M. Schlosser and K. F. Christmann, *Angew. Chem., Intern. Ed. Engl.* **5**, 126 (1966).
[185] L. D. Bergelson and M. M. Shemyakin, *Angew. Chem. Intern. Ed. Engl.*, **3**, 250 (1964).
[186] A. W. Johnson, *Ylid Chemistry*, Academic Press, New York, 1966.
[187] A. W. Johnson and R. B. LaCount, *J. Amer. Chem. Soc.*, **83**, 417 (1961).
[188] F. Ramirez and S. Levy, *J. Amer. Chem. Soc.*, **79**, 69 (1959).
[189] H. Behringer and F. Scheidl, *Tetrahedron Letters*, **1965**, 1757.
[190] W. J. Middleton, E. L. Buhle, J. G. NcNally, Jr., and M. Zanger, *J. Org. Chem.*, **30**, 2384 (1965).
[191] A. W. Johnson and R. T. Amel, *Tetrahedron Letters*, **1966**, 819.
[192] A. W. Johnson and R. B. LaCount, *Tetrahedron Letters*, **9**, 130 (1960).
[193] A. W. Johnson and R. B. LaCount, *Chem. Ind. (London)*, **1958**, 1440.
[194] E. J. Corey and M. Chaykovsky, *J. Amer. Chem. Soc.*, **84**, 867 (1962); E. J. Corey and W. Oppolzer, *ibid.*, **86**, 1899 (1964); H. König and H. Metzger, *Tetrahedron Letters*, **1964**, 3003.
[195] J. F. King and T. Durst, *J. Amer. Chem. Soc.*, **86**, 287 (1964); W. E. Truce, R. W. Campbell, and J. R. Norell, *ibid.*, p. 288.
[196] A. W. Johnson, V. J. Hruby, and J. L. Williams, *J. Amer. Chem. Soc.*, **86**, 918 (1964).
[197] B. M. Trost, *J. Amer. Chem. Soc.*, **88**, 1587 (1966).
[198] A. E. Pohland and W. R. Benson, *Chem. Rev.*, **66**, 161 (1966).
[199] N. K. Kochetkov, L. J. Kudryashov, and B. P. Gottich, *Tetrahedron*, **12**, 63 (1961).
[200] R. Gompper, *Angew. Chem. Intern. Ed. Engl.*, **3**, 560 (1964); S. Hünig, *ibid.*, p. 548.
[201] E. Braude, B. Jofton, L. Lowe, and E. Waight, *J. Chem. Soc.*, **1956**, 4054.

[202] T. M. Harris, S. Boatman, and C. R. Hauser, *J. Amer. Chem. Soc.*, **87**, 3186 (1965).
[203] K. B. Wiberg and G. W. Klein, *Tetrahedron Letters*, **1963**, 1043; E. Wenkel and D. P. Strike, *J. Org. Chem.*, **27**, 1883 (1962).
[204] J. W. Schulenberg and S. Archer, *J. Amer. Chem. Soc.*, **83**, 3091 (1961).
[205] B. D. Astill and V. Boeckelheide, *J. Amer. Chem. Soc.*, **77**, 4079 (1955).
[206] J. W. Schulenberg and S. Archer, *J. Amer. Chem. Soc.*, **82**, 2035 (1960).
[207] J. W. Schulenberg and S. Archer, *Organic Reactions*, **14**, 1 (1965); F. Möller, in *Methoden der organischen Chemie*, Vol. XI/1, 4th ed., Georg Thieme Verlag, Stuttgart, pp. 910–913.
[208] M. P. Cava and D. R. Napier, *J. Amer. Chem. Soc.*, **80**, 2255 (1958).
[209] L. Horner, H. G. Schmelzer, and B. Thompson, *Chem. Ber.*, **93**, 1774 (1960).
[210] J. B. F. Lloyd and P. A. Ongley, *Tetrahedron*, **20**, 2185 (1964); **21**, 245, 2281 (1965).
[211] L. Horner, W. Kirmse, and K. Muth, *Chem. Ber.*, **91**, 430 (1958).
[212] M. P. Cava, D. R. Napier, and R. J. Pohl, *J. Amer. Chem. Soc.*, **85**, 2076 (1963).
[213] M. P. Cava and K. Muth, *J. Amer. Chem. Soc.*, **82**, 652 (1960).
[214] E. Vogel and K. Hasse, *Ann. Chem.*, **615**, 22 (1958).
[215] R. D. Kimbrough, Jr., *J. Org. Chem.*, **28**, 3577 (1963); H. Staudinger and E. Suter, *Ber.*, **53**, 1029 (1920).
[216] H. Staudinger and A. Rheiner, *Helv. Chim. Acta*, **7**, 8 (1924).
[217] R. D. Kimbrough, Jr., and R. D. Hancock, Jr., *Chem. Ind.* (*London*), **1965**, 1180; **1920**, 1092.
[218] M. E. Kuehne and B. F. Lambert, *J. Amer. Chem. Soc.*, **81**, 4278 (1959).
[219] A. J. Birch, *Quart. Rev.*, **4**, 69 (1950); H. Smith, *Organic Reactions in Liquid Ammonia*, in *Chemie in nichtwässrigen, ionisierenden Lösungsmitteln*, Vol. I/2, ed., G. Jander, H. Spandau, and C. C. Addison, Friedrich Vieweg und Sohn, Brunswick, 1963, p. 248.
[220] R. van Volkenburgh, K. W. Greenlee, J. M. Derfer, and C. E. Boord, *J. Amer. Chem. Soc.*, **71**, 3595 (1949).
[221] G. Stork, S. Malhotra, H. Thompson, and M. Uchibayashi, *J. Amer. Chem. Soc.*, **87**, 1148 (1965).
[222] G. Stork and S. D. Darling, *J. Amer. Chem. Soc.*, **86**, 1761 (1964).
[223] G. Stork, P. Rosen, N. Goldman, R. V. Coombs, and J. Tsuji, *J. Amer. Chem. Soc.*, **87**, 275 (1965).
[224] J. C. Sheehan and J. W. Frankenfeld, *J. Amer. Chem. Soc.*, **83**, 4792 (1961).

7
Aromatic Substitutions

I. ELECTROPHILIC AROMATIC SUBSTITUTION[1]

(a) Reaction intermediates

It was explained in the preceding chapter that nucleophilic substitution is the predominant substitution mechanism at saturated aliphatic carbon. Substitution at aromatic carbon, however, takes place preferentially by an electrophilic displacement mechanism.

In the initial step of many electrophilic substitutions of aromatic compounds a π-complex is formed. For example, solubility studies of HCl in benzene have shown that HCl and benzene form a 1:1 adduct. In the adduct only weak interaction exists between the π-system of the aromatic nucleus and the attacking electrophile[2]. π-Complexes are formed by aromatic systems with hydrogen halides, halogens, or Ag^+; they are colorless in solution and are electrically non-conducting.

The π-complexes are converted, usually in the rate-determining step of the substitution, into a second intermediate, the σ-complex or Wheland intermediate. Solutions of σ-complexes are colored and electrically conducting. 1:1:1 Complexes of aromatic compounds, alkylating agents, and BF_3, which are presumably σ-complexes, have been isolated[3] at $-20°$ to $-80°$ *; their nature is, however, still a matter of controversy[5].

In the reaction of (trifluoromethyl)benzene with nitrylium fluoroborate $[NO_2][BF_4]$ at $-50°$ a yellow crystalline compound has been observed, which is presumed to be

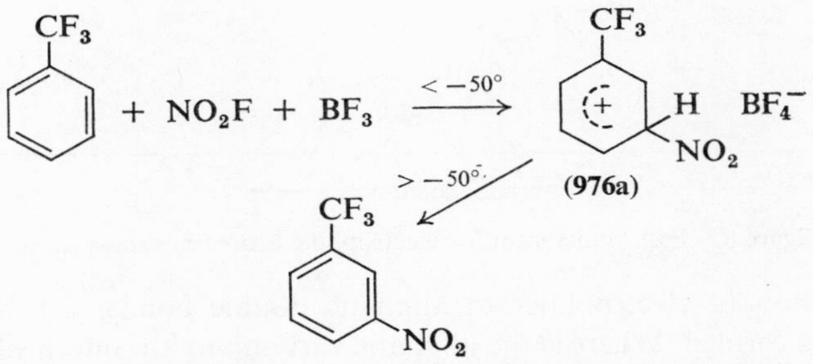

---

* Physical methods of investigating H^+ complexes of aromatic compounds have been reviewed by Perkampus and Baumgarten[4].

the σ-complex (**976a**). Above −50° (**976a**) is converted into *m*-nitro(trifluoromethyl)benzene.

Under the usual conditions of the electrophilic aromatic substitution, the σ complex affords the substitution product spontaneously by loss of a proton. The aromatic substitution accordingly can be described by equation (*a*).

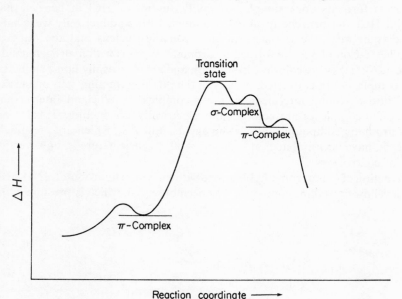

The energy diagram of the substitution is presented in Figure 8.

Figure 8. Energy diagram for electrophilic aromatic substitution.

As in addition of electrophiles to aliphatic double bonds, a carbonium ion intermediate is formed. Whereas the aliphatic carbonium ion intermediate may be stabilized by loss of a proton, by addition of a nucleophile, or by rearrangement, the aromatic cation forms the substitution product exclusively by loss of a proton. This differing behavior is due to the gain in energy on formation of the aromatic system.

(b) Substituent effects and the Hammett equation

If a substituent is introduced into an aromatic compound that already bears a substituent, the rate and position of the second substitution are affected characteristically by the first substituent. A quantitative relation between the effect of the substituent and the rate of the reaction has been advanced by Hammett[6]. He originally measured the dependence of the ionization constant of benzoic acid on the type of substituent at the aromatic nucleus and found the relation:

$$\log K - \log K_0 = \sigma\rho$$

where K is the ionization constant of the substituted benzoic acid and K_0 that of the unsubstituted benzoic acid. ρ was arbitrarily set equal to 1. The ratio $\log K/K_0$ then affords the σ-values, which represent the ability of the substituents to attract or repel electrons by their inductive or resonance effects. A positive σ-value indicates that the substituent has a stronger electron-attracting power than hydrogen, whereas for a substituent with a negative σ-value it is weaker than for hydrogen. The σ-values have been successfully applied to many aromatic substitution reactions. Since the ρ-value of the ionization of benzoic acid is by definition equal to 1, other reactions have their characteristic ρ-values, which are different from 1. The ρ-value is a measure of the sensitivity of a reaction to the influence of substituents: with increasing effect of the substituents, the ρ-values become more negative.

The Hammett σ-values are valid only for *meta-* and *para-*substitution. They are invalid for reactions in which there is steric interaction between the first substituent and the reaction centre of the second substituent, as in *ortho-*substitution.

The equilibrium constants used in the Hammett equation are proportional to the free energies:

$$\ln K = -\Delta F/RT$$

It follows from the Gibbs–Helmholtz equation that

$$\ln K = -(\Delta H - T\Delta S)/RT$$

The electronic effects of substituents influence the reaction enthalpy, ΔH, whereas steric interactions are reflected in the entropy term ΔS.

Instead of the differences in enthalpies and entropies between starting material and products, the activation enthalpies and activation entropies, *i.e.* the enthalpy and entropy differences between ground state and transition state can be used in the Hammett equation. The equilibrium constants are thus replaced by the rate constants:

$$\log k - \log k_0 = \sigma\rho$$

Since the Hammett σ-values have been determined for *meta-* and *para-*substituents and are consequently entropy- and temperature-independent, they can be used for kinetic investigations at different temperatures, but they cannot be used for substitutions involving steric interaction.

Substituents exert an inductive effect (I-effect), which is transmitted through σ- and π-bonds, and a resonance effect, which is transmitted through conjugated π-bonds (R-effect). The Hammett σ-values include both effects. They have been separated into an inductive and a resonance term by Taft[7] (see p. 310).

The inductive effects of the halogens, as well as of substituents that are bound to the aromatic nucleus by an oxygen or nitrogen atom, are negative ($-I$ effect), *i.e.* they attract electrons from the aromatic nucleus more powerfully than hydrogen. (Note that the directions of I- and R-effects and those of the σ-values are opposed to each other.)

The negative σ values for OCH_3 and $N(CH_3)_2$ in Table 11 show that the $-I$ effects of OCH_3 and $N(CH_3)_2$ are overcompensated by their resonance effects in *para*-substitutions. The $-I$ effect of Cl, however, is not compensated by its $+R$ effect, and for NO_2 the two effects act in the same direction. Negative σ-values indicate that a substituent has a weaker electron-attracting power than the H

Table 11. Hammett σ- and Hammett–Brown σ^+-values for *meta*- and *para*-substitution

	σ_m	σ_m^+	σ_p	σ_p^+
OCH_3	0.115	0.047	−0.268	−0.778
$N(CH_3)_2$			−0.83	−1.8
CH_3	−0.069	−0.066	−0.170	−0.311
C_2H_5	−0.07	−0.064	−0.15	−0.295
$CH(CH_3)_2$			−0.151	−0.28
Cl	0.373	0.399	0.227	0.114
NO_2	0.71	0.674	0.778	0.79

atom and that the electrophilic substitution occurs faster than for the unsubstituted aromatic nucleus.

(c) Transition state of electrophilic aromatic substitution

Consideration of the transition states (in fact, of the σ-complexes, which differ only little from the transition states) of electrophilic substitution of anisole shows that resonance structures can be written for *ortho*- and *para*-substitution in which the first substituent participates in the electron delocalization and stabilization of the positive charge. This indicates that the methoxyl group and other substituents with free electron pairs decrease the energy of the transition state, *i.e.* facilitate the second substitution. No such structure can be written in the case of *meta*-substitution. The weakly deactivating effect of the methoxyl group in *meta*-substitution, which is revealed in the positive σ-value of 0.115, is probably mainly due to the inductive effect of the substituent.

Analogous considerations can be applied to other substituents.

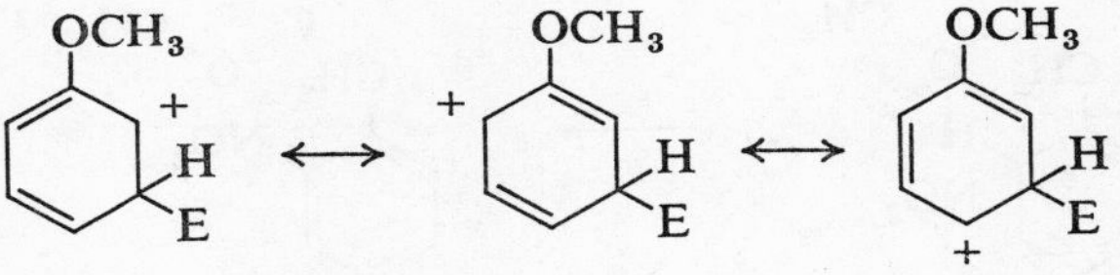

(d) o/p-Ratio

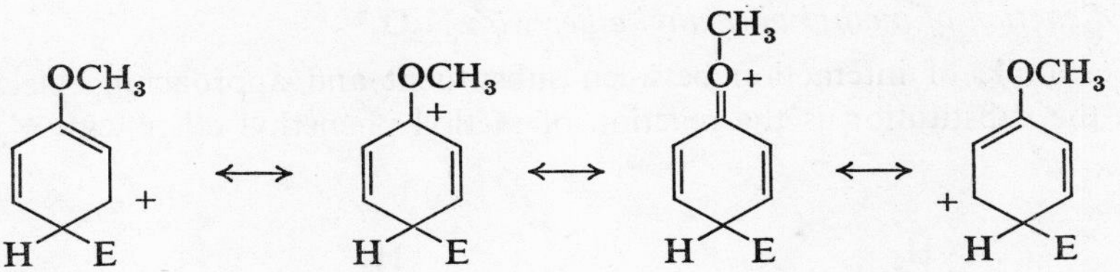

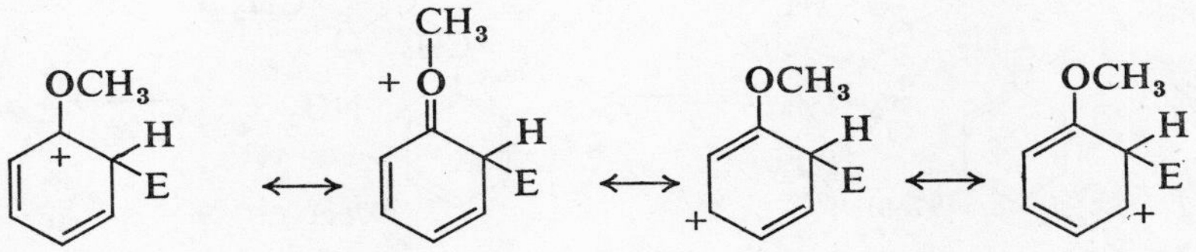

One might expect that the rate of *ortho*-substitution should be equal to the rate of *para*-substitution. Since two *ortho*-positions are available, but only one *para*-position, the ratio of *ortho*- to *para*-substitution should be 2:1. In many substitutions this ratio is, however, not obtained. Whereas the *p/m* ratio is essentially determined by electronic effects, the *o/p* ratio is affected by additional factors, in particular by steric interactions, interactions between substituent and attacking electrophile before the actual electrophilic substitution takes place, and by inductive factors, which have a stronger effect on *ortho*- than on *para*-substitution.

AS-1. Reaction of methyl phenethyl ether with N_2O_5[8]

An example of interaction between substituent and approaching electrophile before the substitution is the reaction of methyl phenethyl ether with N_2O_5 (see **976b**).

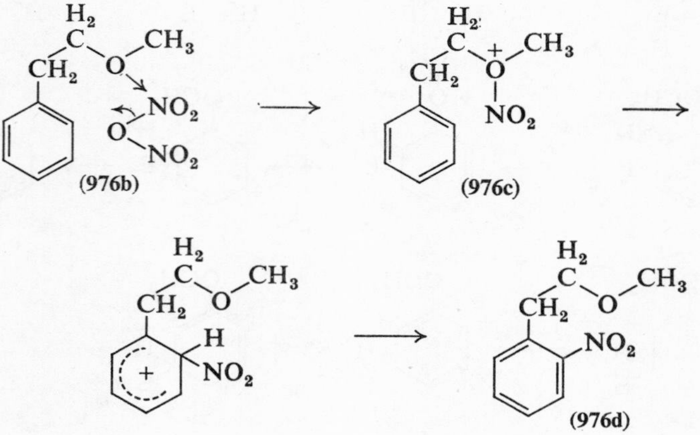

	ortho	meta	para
HNO_3/H_2SO_4 at 0°C	28.9%	8.7%	62.4%
Acetyl nitrate or N_2O_5 at 0°C	66.0%	4.2%	29.8%

Nitration of toluene with HNO_3 or acetyl nitrate leads to the same o/p ratio. However, nitration of the ether (**976b**) with acetyl nitrate or N_2O_5 (which is formed from acetyl nitrate) leads to a much higher o/p ratio than is obtained with HNO_3. It is believed that N_2O_5 partially dissociates into $NO_2^+ NO_3^-$; NO_2^+ then nitrates (**976b**) in the usual manner in *ortho*- and *para*-positions. Additional *ortho*-substitution product is obtained by nucleophilic attack of the ether-oxygen on N_2O_5, which leads first to an oxonium salt (**976c**). From (**976c**) the *ortho*-substitution product is formed through a six-membered cyclic transition state. This interpretation is supported by nitration of methyl 3-phenylpropyl ether with HNO_3 and N_2O_5; the two reagents yield the same o/p ratio, whereas benzyl methyl ether, which can be *ortho*-nitrated *via* a five-membered cyclic transition state, yields an even higher o/p ratio with N_2O_5 than does (**976b**).

(e) Modifications of the Hammett equation

The Hammett σ-values lead to equilibrium and rate constants with a mean deviation of $\pm 15\%$. Values that better reflect the relation between substituent and reaction rate have been calculated by McGary, Okamoto, and Brown[9]. The solvolysis of α,α-dimethylbenzyl chloride (977) was chosen as a reference reaction, the transition state resembling that of electrophilic aromatic substitution. (It is assumed, that the transition states differ only slightly from the cationic intermediates below.)

The ρ value of the reaction was set equal to -4.54, thus permitting the use of the original Hammett ρ values in the modified equation:

$$\log k - \log k_0 = \sigma^+ \rho$$

The σ^+-values correlate the substituent effects with the reaction rates more satisfactorily than do the original Hammett values (see table 11, p. 374).

AS-2. Partial rate factors

In aromatic substitution the ratio k/k_0 is often called the partial rate factor f. The symbol f_p denotes the ratio of the rate of substitution in *para*-position of a substituted benzene derivative to the rate of the initial substitution of benzene. The factor f can be calculated from σ^+ and ρ, or from the amounts of *meta*- and *para*-substitution products if the ratio of the reactivities towards a particular substitution (independently of the position of substitution) of unsubstituted benzene to that of the substituted benzene is known.

For example, chlorination of toluene with elemental chlorine in acetic acid at 25° yields 59.78% of *o*-product, 0.48% of *m*-product, and 39.74% of *p*-product[10]. The ratio of the reactivities of benzene to toluene is 1:344. This leads to a reactivity of toluene in *para*-position of 344(39.74/100) = 136.7. Since benzene has six positions for initial substitution compared to only one *para*-position in toluene, the reactivity of the *para*-position is six times as high as that of a single benzene carbon. f_p is therefore 6 × 136.7 = 820.

Since benzene has three times as many reaction centres as toluene has *meta*-positions the reactivity of the *meta*-position, f_m, is 344 × 3 × 0.48/100 = 4.95.

In some reactions even the modified σ-values are unsatisfactory. For example, the σ^+-value for *para*-fluorine is -0.073. Fluorine should therefore accelerate *para*-substitution. Some substitutions, however, *e.g.* the nitration of fluorobenzene, are retarded by a fluoro-substituent. The cause of this deviation is not considered either in the Hammett or Hammett–Brown equation. It is well known that the fluorine atom exerts an electron-attracting inductive and an electron-donating resonance effect. The transition state for electrophilic *para*-substitution of fluorobenzene may in one extreme case resemble the starting material, and in the other extreme the σ-complex (**978**). In the first case, the inductive effect, which hinders the approach of the electrophile, obviously dominates. In the latter case, however, the resonance effect of the fluorine atom is more important. It is probable that with

etc.

(**978**)

increasing electrophilicity of the attacking reagent (bond formation already advanced in the transition state), the transition state approaches (**978**). This means that the resonance effect of the fluorine atom increases with increasing electrophilicity of the attacking species. Consequently, substitution of fluorobenzene by relatively poor electrophiles is retarded relative to benzene by the inductive effect of fluorine, whereas reactions with strong electrophiles are accelerated. A parameter r, which takes these interactions into account, has been introduced by Yukawa and Tsuno[11].

$$\log k - \log k_0 = \rho[\sigma + r(\sigma^+ - \sigma)]$$

σ and ρ have the original Hammett values, σ^+ has the modified Hammett value. The equation gives the best correlation between substituent effects and reaction rates. Since ρ is a measure of the sensitivity of an aromatic substitution reaction to substituent effects, and since resonance effects of substitutents are particularly effective where bond formation is advanced in the transition state, strongly negative ρ values indicate substitution by strong electrophiles (see Table 12).

Table 12. ρ Values for some reactions

Reaction	ρ
Ethylation with C_2H_5Br and $GaBr_3$ at 25°	-2.4
Nitration with HNO_3 in CH_3NO_2 at 25°	-6.0
Bromination with $HOBr$–$HClO_4$ in 50% dioxane–H_2O at 25°	-6.2
Friedel–Crafts acetylation with CH_3COCl–$AlCl_3$ in $C_2H_4Cl_2$ at 25°	-9.1
Chlorination with Cl_2 in acetic acid at 25°	-10.0
Bromination with bromine in CH_3COOH–H_2O at 25°	-12.1

Another interpretation of the invalidity of the σ^+-values in certain substitutions may be mentioned: in most substitutions conversion of the π-complex into the σ-complex is rate-determining. In some reactions, however, formation of the π-complex may be rate-determining. Since in a π-complex only a weak interaction exists between the electrophile and the π-electron system of the aromatic compound, the σ^+-effect which is based on definite positions in the aromatic ring cannot reflect the rate of reaction.

An example is nitration with nitrylium salts, in which the π-complex formation is assumed to be rate-determining[12].

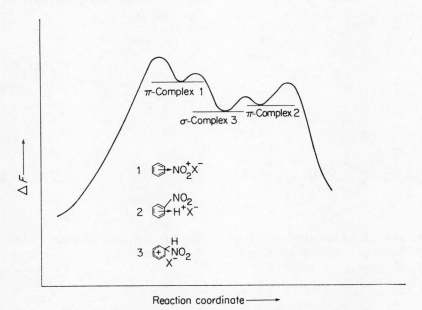

Figure 9. Energy diagram for electrophilic aromatic substitution.

AS-3. Electrophilic substitution by OH^+

The hydroxyl group can be introduced into aromatic compounds by electrophilic substitution with H_2O_2 in acidic medium[13] or by peracids. Usually, however, the reaction does not stop at monosubstitution, since the newly formed phenols are much more reactive than the starting materials. The electrophilic character of the attacking species follows from the dependence of the substitution rate on the first substituent and from attack by the OH group at *ortho*- and *para*-positions relative to the initial substituent. An example is the reaction of 1,3,5-trimethoxybenzene with perbenzoic acid at 0°.

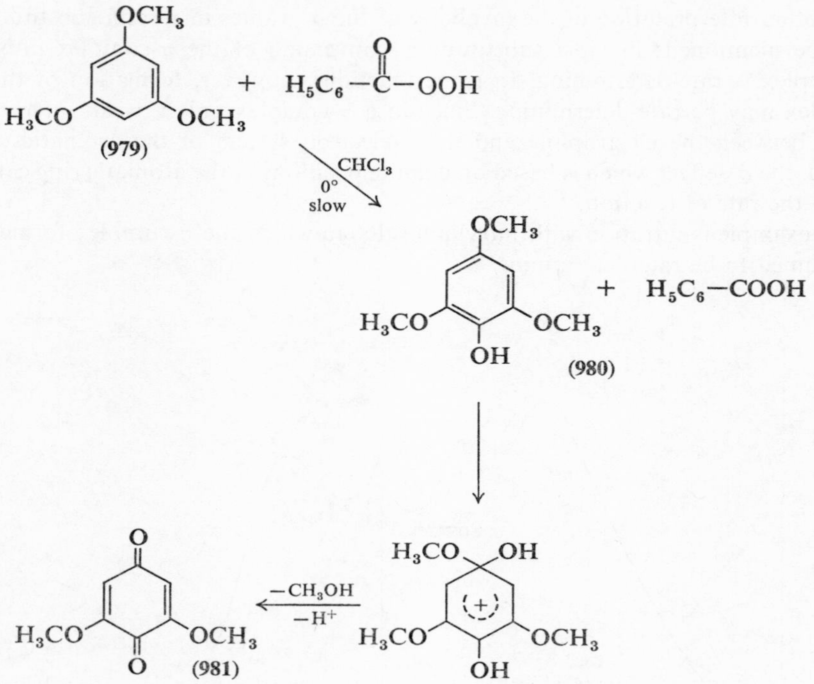

[Scheme showing reaction of 1,3,5-trimethoxybenzene (979) with perbenzoic acid ($H_5C_6-C(=O)-OOH$) in $CHCl_3$ at 0°, slow, giving 2,4,6-trimethoxyphenol (980) + H_5C_6-COOH, which via a cyclohexadienone cation intermediate ($-CH_3OH$, $-H^+$) gives 2,6-dimethoxy-p-benzoquinone (981).]

In the bimolecular rate-determining step of the reaction, 2,4,6-trimethoxyphenol (980) is formed; it is then converted in a fast reaction into 2,6-dimethoxy-p-benzoquinone (981).

AS-4. Reaction of 2,4,6-trimethylphenoxide (982) with chloramine[14]

Reaction of alkoxides with chloramine leads to alkoxylamines[15]. Accordingly, the salt (982) might be expected to yield 2,4,6-trimethylphenoxylamine (983). Since aroxylamines are very unstable compounds[16], (983) could not be isolated under the reaction conditions.

On the other hand, it is well known that phenoxides are rapidly substituted by electrophilic reagents in solvents of high polarity, for example, in phenol or water[17]. Unsubstituted sodium phenoxide and chloramine yield o- and p-aminophenol[18]. It is likely, therefore, that compound (982) would undergo predominantly electrophilic amination. Whereas 4-amino-2,4,6-trimethyl-2,5-cyclohexadien-1-one probably polymerizes rapidly, 2-amino-2,4-6-trimethyl-3,5-cyclohexadien-1-one (984) can rearrange to a more stable compound (986).

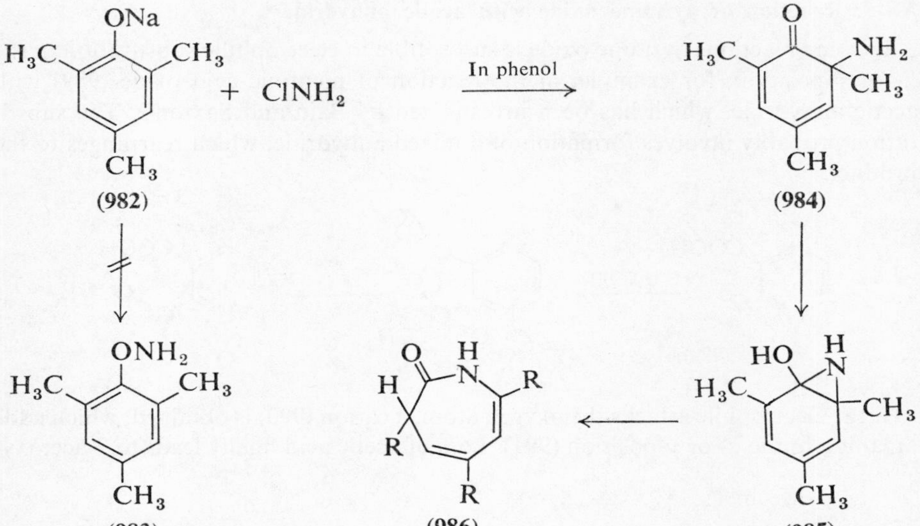

Aromatic Substitutions

[Structures (982), (984), (983), (986), (985) with reaction scheme:]

(982) H₃C-C₆H₂(CH₃)(CH₃)-ONa + ClNH₂ —In phenol→ (984)

(982) ⇏ (983)

(985) → (986) → (984)

The mechanism proposed is supported by the analogous rearrangement of (987). However, the tertiary amine (988) does not rearrange because formation of the ethylenimine intermediate [analogous to (985)] is impossible.

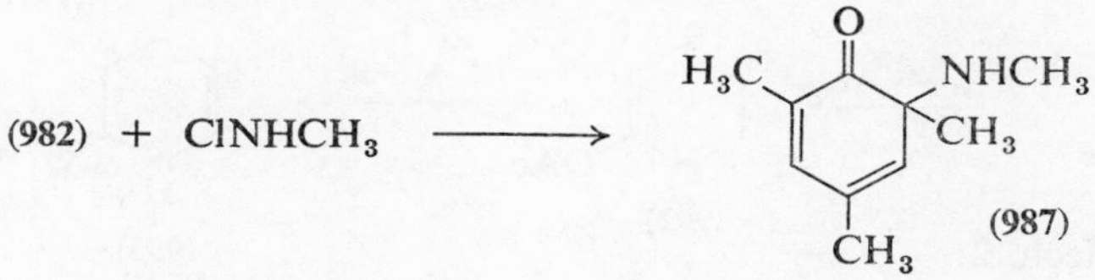

(982) + ClNHCH₃ ⟶ (987)

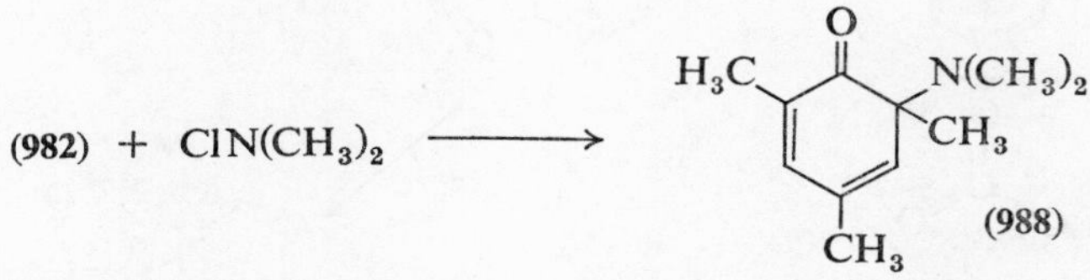

(982) + ClN(CH₃)₂ ⟶ (988)

The reverse of the azepinone formation has recently been described[19], namely:

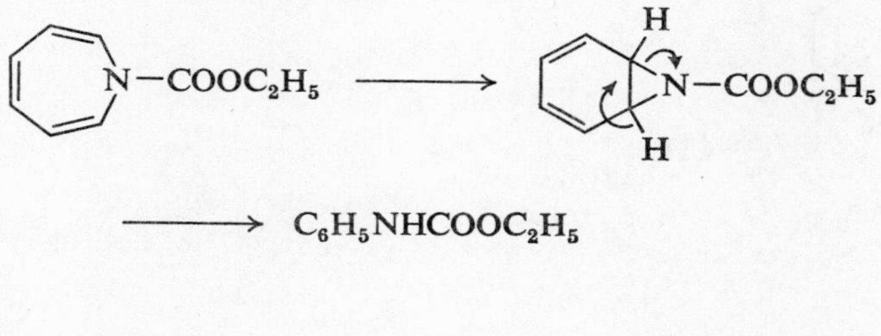

[azepine]N—COOC₂H₅ ⟶ [bicyclic intermediate]N—COOC₂H₅ ⟶ $C_6H_5NHCOOC_2H_5$

AS-5. Reaction of pyridine oxide with acetic anhydride[20]

In some reactions pyridine oxide is susceptible to electrophilic substitution in the α- or γ-position, for example, in the reaction of nicotinic acid oxide (989) with acetic anhydride, which has been investigated by Bain and Saxton[21]. The substitution probably involves formation of a mixed anhydride, which rearranges to the product.

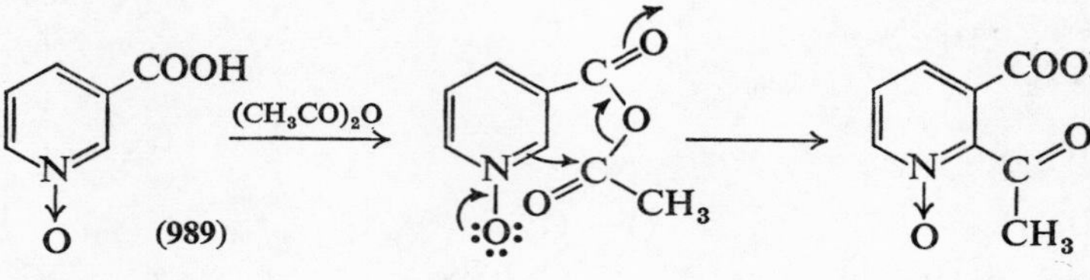

If the electrophile attacks the oxygen atom, a cation (990) is obtained, which adds an anion in the α- or γ-position (991). Loss of acetic acid finally leads to 2-acetoxy-

Mech. 1:

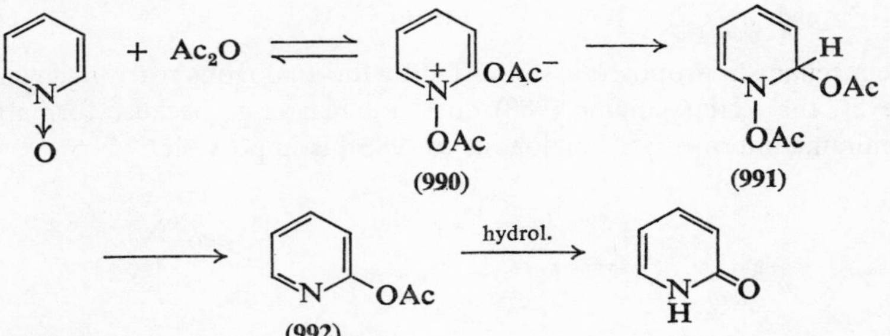

Mech. 2:

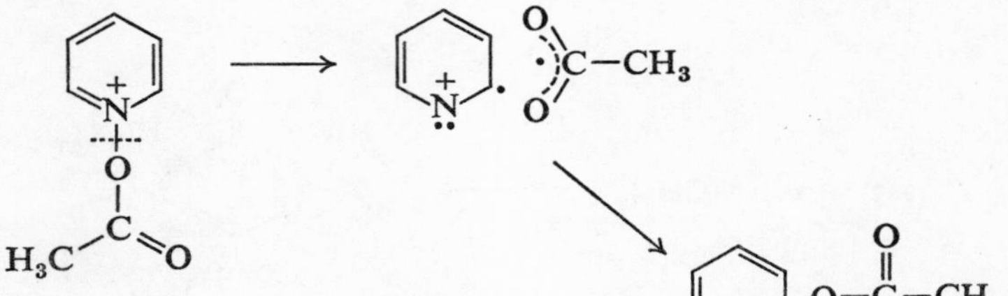

Mech. 3:

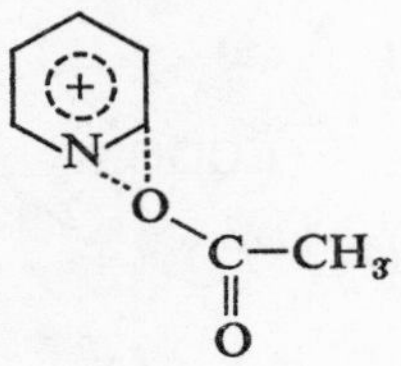

pyridine (**992**), which is converted into 2-pyridone (**993**) on hydrolysis.

The formation of 2-acetoxypyridine can, in principle, occur by two other mechanisms, either by homolysis of the N–O bond of 1-acetoxypyridinium acetate (**990**), followed by recombination of the radicals in the solvent cage (mech. 2)[22], or by rearrangement of an intimate ion-pair (mech. 3)[23].

Oae et al.,[24] using ^{18}O-labelled acetic anhydride, have shown elegantly that mechanism 1 applies: each oxygen in the anhydride contained 0.89 atom % of ^{18}O (natural abundance 0.2 atom %). Mechanism 1 must yield a pyridone with 0.72 atom % of ^{18}O, because mixing of the acetoxy group (which contains the natural oxygen of the pyridine oxide) with the ^{18}O-enriched acetate ion takes place:

$$\frac{3 \times 0.89 + 1 \times 0.20}{4} = 0.72$$

Mech. 2 would lead to 0.55 atom %:

$$\frac{1 \times 0.89 + 1 \times 0.20}{2} = 0.55$$

Mech. 3 must lead to a pyridone with a natural ^{18}O distribution of 0.20 atom %. Since 0.72 atom % of ^{18}O were found, the reaction must follow mechanism 1.

II. NUCLEOPHILIC AROMATIC SUBSTITUTION[25]

Nucleophilic aromatic substitution may proceed by one of four mechanisms:

1. The most important mechanism is the bimolecular reaction of a nucleophile with an aromatic compound. In the rate-determining step an addition product [e.g. (**994**)] is formed, which corresponds to the σ-complex of the electrophilic aromatic substitution and is called a Meisenheimer complex[26].

The reaction takes place under mild conditions only if the negative charge which develops on the aromatic ring during the formation of (**994**) is stabilized by electron-attracting substituents. Evidence for the formation of (**994**) is the isolation of an addition product (**995**) in the reaction of 2,4,6-trinitroanisole with potassium ethoxide[27]. In this reaction, the formation of a π-complex which precedes compound (995) has been demonstrated. The addition of a nucleophile to an aromatic compound, resulting in the formation of a π-complex or a charge-transfer complex, presumably takes place also in many other nucleophilic substitutions.

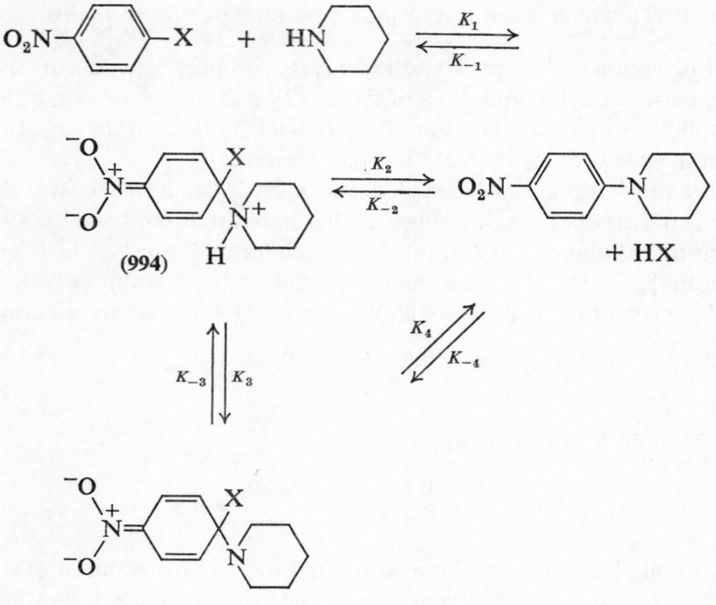

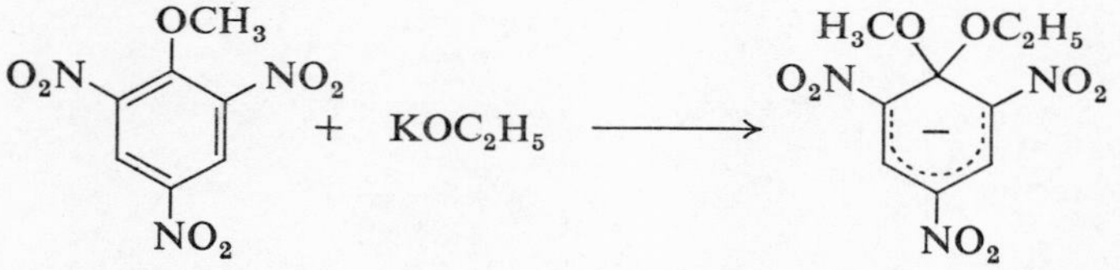

The rate of nucleophilic substitution depends on the charge density at the reacting atom of the aromatic ring, on the stability of the leaving group, and the nucleophilicity of the attacking nucleophile.

Table 13 shows[28] that in the reaction of *p*-nitrobenzene derivatives with piperidine in DMSO, which has been thoroughly investigated, fluorine is replaced particularly readily; this is a consequence of its strong electron-attracting inductive effect. In contrast to fluorine, hydrogen is rarely replaced by nucleophiles.

Table 13. Relative rates of replacement of X in p-X-C_6H_4-NO_2 by nucleophiles, specifically piperidine

X	Relative rate
F	412
Cl	1.0
NO_2	8.7
Tosyl	5.5
OC_6H_5	0.030

In aliphatic nucleophilic substitutions the C–F bond is less reactive than the other C–halogen bonds. The high reactivity of fluorine in nucleophilic aromatic substitutions indicates that the rate-determining step is not S_N2 displacement, but formation of addition product (**994**). Cleavage of the C–F bond occurs in the subsequent step.

Variation of the nucleophile in the reaction of 1-chloro-2,4-dinitrobenzene in 60% dioxane leads to the following sequence, in which the reaction rate decreases from left to right[29]. Since the conversion of the starting material into the first

$$C_6H_5S^- \gg \text{piperidine} > C_6H_5O^- > H_2N\text{—}NH_2 > OH^- > C_6H_5NH_2$$

intermediate is accompanied by an increase in polarity, the reaction is promoted by polar solvents.

(2) *S_N2 mechanism in aromatic substitution*

Nucleophilic aromatic substitutions by an S_N2 mechanism are unusual.

In the case of the conversion of 1- and 2-bromonaphthalene by piperidine at 230° into 1-(1-naphthyl)- and 1-(2-naphthyl)-piperidine it has been assumed that an S_N2 substitution competes with the "cine" mechanism (3)[30].

(3) *Cine substitution*

In reactions of aromatic halogen compounds with strong bases, substitution products are formed in which at least part of the attacking group occupies a position different from that which had been occupied by the leaving group. Such substitutions are called "cine" substitutions. In general, an aryne (see p. 177) is formed as intermediate by loss of hydrogen halide[31]. It is characteristic of these substitutions that it is always hydrogen that is replaced by the second substituent, and that this substituent enters at a position adjacent to the first substituent.

(4) *Aromatic S_N1 substitution*

The only known example of S_N1 substitution in aromatic chemistry is the thermal decomposition of arenediazonium salts.

AS-6. Smiles rearrangement[32]

An example of mechanism 1 is the Smiles rearrangement. Treatment of 2-hydroxy-5-methylphenyl 2′-nitrophenyl sulfone (**996**) with NaOH in 50% aqueous dioxane yields an intermediate (**997**), which is converted by intramolecular nucleophilic substitution into the ether (**998**).

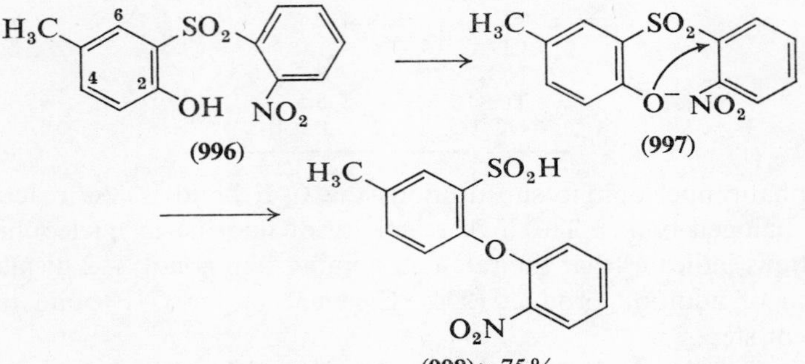

The rearrangement is enormously accelerated by substituents in position 6, independently of the direction of their inductive or resonance effects[33]. Substituents in position 4 which, like position 6, is *meta* to the OH group but *para* to the SO_2 group, do not accelerate the reaction (Table 14). An explanation is furnished by consideration of the conformations (**999**) and (**1000**). The former is the more

Table 14. Effect of substituents on the rate of the Smiles rearrangement of compound (**996**) at 0°

Substituent	k (min^{-1})
6-Cl	9.2×10^{-1}
4-Cl	1.8×10^{-6}
6-CH_3	>3
4-CH_3	3.6×10^{-5}

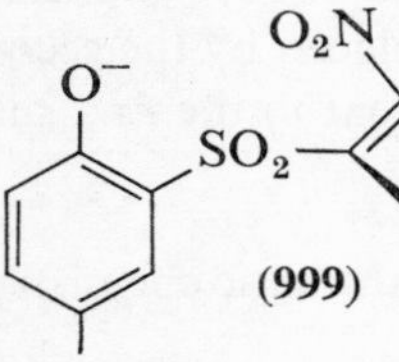

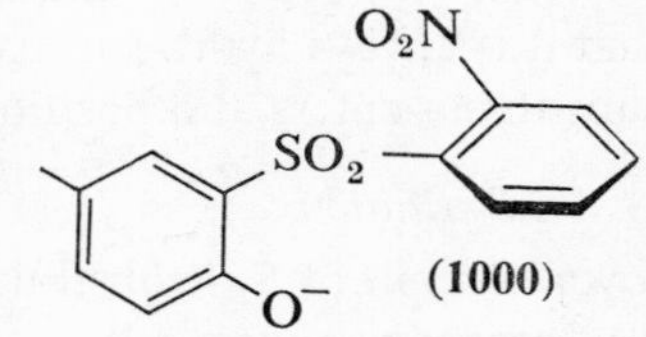

stable if C-6 does not bear a substituent. If, however, C-6 carries a bulky substituent, the proportion of (**1000**) increases. Since (**1000**) is the conformation that rearranges to the product, the rate of rearrangement also increases.

REFERENCES

[1] R. O. C. Norman and R. Taylor, *Electrophilic Substitution in Benzenoid Compounds*, Elsevier Publ. Co., Amsterdam-London-New York, 1965.
[2] H. C. Brown and J. J. Melchiore, *J. Amer. Chem. Soc.*, **87**, 5269 (1965).
[3] G. A. Olah and S. J. Kuhn, *J. Amer. Chem. Soc.*, **80**, 6541 (1958).
[4] H. H. Perkampus and E. Baumgarten, *Angew. Chem.*, **76**, 955 (1964).
[5] R. Nakane, A. Natsubori, and O. Kurihara, *J. Amer. Chem. Soc.*, **87**, 3597 (1965); R. Nakane and A. Natsubori, *ibid.*, **88**, 3011 (1966).
[6] L. P. Hammett, *Physical Organic Chemistry*, McGraw-Hill Book Company, Inc., New York, 1940, p. 184.
[7] R. W. Taft, Jr., in *Steric Effects in Organic Chemistry*, ed. M. S. Newman, John Wiley and Sons, New York, 1956, p. 556.
[8] R. O. C. Norman and G. K. Radda, *J. Chem. Soc.*, **1961**, 3030.
[9] C. W. McGary, Y. Okamoto, and H. C. Brown, *J. Amer. Chem. Soc.*, **77**, 3037 (1955).
[10] H. C. Brown and L. M. Stock, *J. Amer. Chem. Soc.*, **79**, 5157 (1957).
[11] Yukawa and Tsuno, *Bull. Chem. Soc. Japan*, **32**, 971 (1959).
[12] G. A. Olah and S. J. Kuhn, *J. Amer. Chem. Soc.*, **83**, 4564 (1961); G. A. Olah, S. J. Kuhn, and S. H. Flood, *ibid.*, p. 4571, 4581; G. A. Olah and S. J. Kuhn, *ibid.*, **84**, 3684 (1962); G. A. Olah, S. J. Kuhn, S. H. Flood, and J. C. Evans, *ibid.*, p. 3687.
[13] S. L. Fries, A. H. Soloway, B. K. Morse, and W. C. Ingersoll, *J. Amer. Chem. Soc.*, **74**, 1305 (1952).
[14] L. A. Paquette, *J. Amer. Chem. Soc.*, **85**, 3288 (1963); **84**, 4987 (1962).
[15] W. Theilacker and E. Wegner, *Angew. Chem.*, **72**, 130 (1960).
[16] C. L. Bumgardner and R. L. Lilly, *Chem. Ind. (London)*, **1962**, 559.
[17] J. S. Nicholson and D. A. Peak, *Chem. Ind. (London)*, **1962**, 1244.
[18] N. Kornblum, P. J. Berrigan, and W. J. LeNoble, *J. Amer. Chem. Soc.*, **82**, 1257 (1960); **85**, 1141 (1963); N. Kornblum, R. Seltzer, and P. Haverfield, *ibid.*, p. 1148.
[19] K. Hafner and C. Koenig, *Angew. Chem.*, **75**, 89 (1963); W. Lwowski, T. J. Maricich, and T. W. Mattingly, Jr., *J. Amer. Chem. Soc.*, **85**, 1200 (1963).
[20] S. Oae and S. Kozuka, *Tetrahedron*, **21**, 1971 (1965); J. H. Markgraf, H. B. Brown, S. C. Mohr, and R. G. Peterson, *J. Amer. Chem. Soc.*, **85**, 958 (1963).
[21] B. M. Bain and J. E. Saxton, *J. Chem. Soc.*, **1961**, 5216.
[22] V. Boekelheide and W. J. Linen, *J. Amer. Chem. Soc.*, **76**, 1286 (1954).
[23] S. Oae and S. Kozuka, *Tetrahedron*, **20**, 2691 (1964).
[24] S. Oae, T. Kitao, and Y. Kitaoka, *J. Amer. Chem. Soc.* **84**, 3359, 3362 (1962); see also T. Cohen and J. H. Fager, *J. Amer. Chem. Soc.*, **87**, 5701, 5710 (1965).
[25] J. F. Bunnett, *Quart. Rev.*, **12**, 1 (1958); J. F. Bunnett and R. E. Zahler, *Chem. Rev.*, **49**, 273 (1951); H. Sauer and R. Huisgen, *Angew. Chem.*, **72**, 294 (1960).
[26] J. F. Bunnett, E. W. Garbisch, Jr., and K. M. Pruit, *J. Amer. Chem. Soc.*, **79**, 385 (1957).
[27] R. Foster and L. Hammick, Jr., *J. Chem. Soc.*, **1954**, 2153; J. B. Ainscough and E. F. Caldin, *ibid.*, **1956**, 2528.
[28] H. Suhr, *Chem. Ber.*, **97**, 3268 (1964).
[29] J. F. Bunnett and G. T. Davis, *J. Amer. Chem. Soc.*, **76**, 3011 (1954).
[30] J. F. Bunnett and T. K. Brotherton, *J. Amer. Chem. Soc.*, **78**, 155, 6265 (1956).
[31] H. Heany, *Chem. Rev.*, **62**, 81 (1962); M. Panar and J. D. Roberts, *J. Amer. Chem. Soc.*, **82**, 3629 (1966).
[32] C. S. McClement and S. Smiles, *J. Chem. Soc.*, **1937**, 1016.
[33] T. Okamoto and J. F. Bunnett, *J. Amer. Chem. Soc.*, **78**, 5357 (1956); J. F. Bunnett and T. Okamoto, *ibid.*, p. 5363.

8

Radical Reactions

RADICALS are substances that possess one or more unpaired electrons. Molecules in excited singlet states, in which two electrons are in different orbitals but have opposite spin, may also behave like diradicals (see Chapter 9, p. 406).

Both stable and unstable radicals are known; their stability depends on the substituents at the site of the unpaired electron.

The triphenylmethyl radical, which is stable in solution, has long been known—since M. Gomberg's investigations in 1900. Its stability results from the delocal-

$$H_5C_6-\underset{\underset{H_5C_6}{|}}{\overset{\overset{H_5C_6}{|}}{C}}-\underset{\underset{C_6H_5}{|}}{\overset{\overset{C_6H_5}{|}}{C}}-C_6H_5 \quad \rightleftharpoons \quad 2 H_5C_6-\underset{\underset{C_6H_5}{|}}{\overset{\overset{C_6H_5}{|}}{C}}\cdot$$

ization of the unpaired electron over three phenyl rings, and from steric hindrance to recombination that yields hexaphenylethane.

Resonance-stabilized radicals are assumed to be planar. Depending on the substituents, a radical may however possess a pyramidal structure, in analogy to carbanions. In most cases it is not known with certainty to what degree radicals possess one structure or the other.

The stability of radicals depends further on inductive effects, as with carbonium ions. Tertiary are more stable than secondary radicals, and the latter more stable than primary radicals.

Radical reactions may be subdivided, like polar reactions, into substitutions, eliminations, additions, and rearrangements. In contrast to polar reactions, many of these reactions have a chain-mechanism, which means that during the reaction new radicals are generated that propagate the reaction until it comes to a standstill by consumption of the starting components or by termination reactions. In order to start a reaction, radicals have to be generated. Methods for generating radicals are homolytic decomposition of thermally unstable radical initiators, irradiation of suitable initiators, or reactions of hydrogen peroxide with certain inorganic salts such as $FeCl_2$. Important radical initiators are organic peroxides, for example, dibenzoyl peroxide; and also azo compounds.

Presence of inhibitors prevents radical reactions. They usually convert a radical that is capable of carrying on a chain reaction into another but less reactive radical; inhibitors are, for example, I_2 and O_2.

Radical reactions in which oxygen is not completely excluded pass through an incubation period. As soon as the oxygen is completely consumed, the rate of reaction increases rapidly. On the other hand, in the absence of other initiators, oxygen may serve as an initiator as a consequence of its diradical character.

I. RADICAL SUBSTITUTION

R-1. Selective chlorination of *n*-heptane[1]

Radical substitution usually proceeds by a chain mechanism. The atom most frequently replaced is hydrogen bound to carbon. In contrast to ionic substitution, no direct displacement* by an attacking radical takes place as in an S_N2 substitution. The reaction consists instead of two bimolecular steps. In order to start the replacement of hydrogen on a saturated hydrocarbon by chlorine, chlorine molecules are cleaved to atoms by irradiation. The chlorine atoms propagate the chain reaction by abstraction of a hydrogen atom from the hydrocarbon.

$$Cl_2 \xrightarrow{h\nu} 2\,Cl\cdot$$

$$Cl\cdot + RH \longrightarrow R\cdot + HCl$$

The carbon radical attacks a chlorine molecule in the second step, thus forming RCl and again a chlorine atom, which propagates the chain.

$$R\cdot + Cl_2 \rightarrow RCl + Cl\cdot$$

If a branched saturated hydrocarbon is treated, under irradiation, with molecular chlorine, the relative rates[3] of chlorination of primary, secondary, and tertiary hydrogen are 1:3.3:4. Under these conditions all secondary hydrogen atoms of an unbranched hydrocarbon, for example, of *n*-heptane, are replaced at the same rate, independently of their position in the chain. Chlorination with *N*-chlorosuccinimide leads to the same result, since the actual chlorination agent is elemental chlorine which is continuously supplemented from *N*-chlorosuccinimide[4]:

$$(CH_2CO)_2NCl \longrightarrow (CH_2CO)_2N\cdot + Cl\cdot$$

$$Cl\cdot + RH \longrightarrow R\cdot + HCl$$

$$HCl + (CH_2CO)_2NCl \longrightarrow (CH_2CO)_2NH + Cl_2$$

$$R\cdot + Cl_2 \longrightarrow RCl + Cl\cdot$$

* Reaction of iodine with *sec*-butyl iodide seems to be an exception[2].

A different result is obtained, however, if a more selective, less reactive chlorinating agent is chosen. In chlorination with *tert*-butyl hypochlorite[5] the hydrocarbon is attacked, not by a chlorine atom alone, but also by the more stable *tert*-butoxyl radical:

$$t\text{-}C_4H_9O\cdot + HR \longrightarrow t\text{-}C_4H_9OH + R\cdot$$

$$R\cdot + t\text{-}C_4H_9OCl \longrightarrow RCl + t\text{-}C_4H_9O\cdot$$

tert-Butyl hypochlorite and *n*-heptane, with irradiation at 0°, yield the following distribution of monochlorides: 1-chloro 6.4%, 2-chloro 45.3%, 3-chloro 33.8%, 4-chloro 14.1%.

Similar results are obtained with *N*-bromosuccinimide.

Clearly the secondary hydrogen at C-2 is preferentially attacked and the rate of replacement decreases from outer CH_2 groups to the inner ones. On the basis of our present knowledge of inductive effects in saturated hydrocarbons, this result was to be expected. The 2-methylene C–H bonds are affected by the electron-donating inductive effect of the *n*-pentyl group and by the somewhat stronger effect of the terminal methyl group. The two unequal effects lead to the highest electron density at C-2 and a slowly decreasing electron density towards C-4. Radicals preferentially attack the side of highest electron density as long as they are not too reactive; this can be seen, for example, in the effect of peroxides on the direction of addition of HBr to terminal double bonds; HBr adds under these conditions by a radical mechanism and since the terminal C-atom is the side of highest electron density, Br· adds to C-1 in opposition to Markownikoff's rule. Thus radicals attack predominantly the C–H bonds at position 2. The results of Fell and Kung[1] are in contrast to the opinion recently advocated by Korte and Höver[6] that in a normal saturated hydrocarbon the highest electron density exists at the terminal C-atoms. Korte and Höver used the analogy between an unbranched saturated hydrocarbon and a metal rod: since the latter has a slightly increased charge density at the surface as a consequence of electrostatic repulsion, it was postulated that a hydrocarbon "rod" may similarly possess increased charge density at the chain ends. There is, however, no sound experimental evidence for this assumption (but see ref. 6a).

Chlorination of *n*-heptane with chlorine in benzene[7] at 20°, with irradiation, shows that radicals can be generated that possess nucleophilic properties and consequently attack at the side of lowest electron density: the proportions of monochloroheptanes obtained are: 1-chloro 11.8%; 2-chloro 34.5%; 3-chloro 36.5%; 4-chloro 17.2% (the number of H atoms at C-4 is only half as great as at C-3 or C-2).

(1001)

Since in this case C-2 is attacked to a smaller extent than C-3, it is assumed that the attacking species is, not a chlorine atom, but a radical complex of a chlorine atom with benzene. In (1001) the chlorine atom possesses an unusually high electron density, following from the electron-donor properties of benzene. Analogous results are obtained in chlorination with CH_3SCl^8 and with PCl_5 in the presence of peroxides[9] (in the latter reaction the proportions are: 1-chloro 2.5%; 2-chloro 36.4%; 3-chloro 41.5%; 4-chloro 19.3%).

The investigations show that radicals may be surprisingly selective[10] and that reactive intermediates can be generated that represent transitions between radicals and ionic species.

II. RADICAL ADDITION

R-2. Addition of CCl_4 to 1-heptyne[11]

A variety of compounds can be added to C=C double bonds by a radical chain mechanism, for example, halogens, HBr, polyhalomethanes, aldehydes, alcohols, amines, and thiols. HCl and HI, however, do not add by a chain mechanism. The reason is that one of the reaction steps X· + C=C → X–C–C· or X–C–C· + HX → X–C–C–H + X· has a positive reaction enthalpy, and a chain reaction is observed only if both steps are exothermic (see Table 15).

Table 15. Heats of reaction (kcal mole^{-1} at 25°) of some addition reactions

A–B	B· + C=C ΔH	B–C–C· + A–B → B–C–C–A + B· ΔH
H–Cl	−26	+5
H–Br	−5	−11
H–I	+7	−27
Cl–Cl	−26	−19
Br–Br	−5	−17
I–I	+7	−13
H–CCl$_3$	−14	−8
Cl–CCl$_3$	−14	−8

Reactions competing with the 1:1 addition are radical polymerization, telomerization, rearrangement, or abstraction of allylic hydrogen atoms. Working in high dilution retards polymerization in favour of 1:1 addition.

An example of a 1:1 radical addition is the reaction of CCl_4 with 1-heptyne initiated by benzoyl peroxide at 77°. Besides the normal addition product (**1002**), product (**1007**) is obtained in 20% yield. The latter is probably formed by intramolecular hydrogen abstraction through a cyclic six-membered transition state (**1005**). The hydrogen abstraction is a result of the high reactivity of the vinyl radical (**1004**); the corresponding secondary alkyl radical, which is formed by addition of ·CCl$_3$ to 1-heptene is not sufficiently reactive. Radical (**1006**) is converted into (**1007**) by elimination of chlorine which evidently takes place much faster than abstraction of a chlorine atom from CCl$_4$. The minor product (**1003**) is formed by addition of Cl$_2$ to 1-heptyne.

$$H_3C-(CH_2)_4-C\equiv CH + CCl_4 \xrightarrow{(C_6H_5COO)_2}$$

$$H_3C-(CH_2)_4-CCl=CH-CCl_3 + H_3C-(CH_2)_4-CCl=CHCl$$
40% (**1002**)   6% (**1003**)

+ [cyclopentane with CH$_3$ and C=C(Cl)(Cl) substituents]

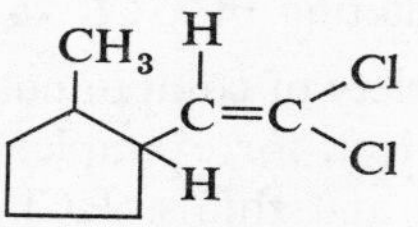

(**1007**) 20%

via $H_3C-CH_2-(CH_2)_3-\overset{\bullet}{C}=CH-CCl_3 \longrightarrow$ (**1002**)
(**1004**)

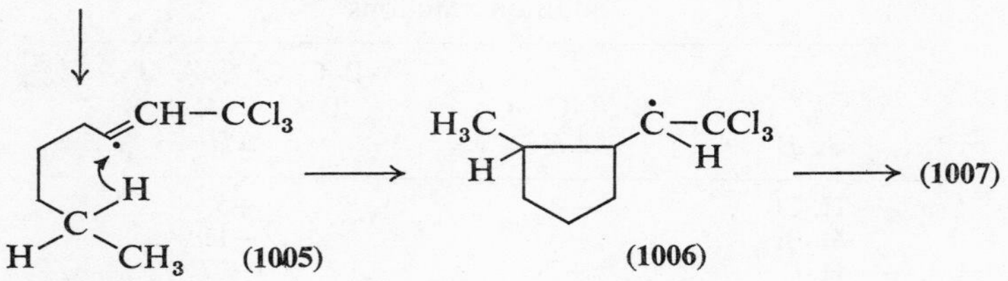

↓

[structure **1005**] → [structure **1006**] ⟶ (**1007**)

R-3. Addition of oxygen to 1,3-dienes

Oxygen possesses two unpaired electrons in its ground state (triplet state) and is therefore strongly paramagnetic (for a discussion of the oxygen molecule see reference 12). As a consequence of its diradical character, oxygen can add to conjugated dienes; however, it has not yet been possible to prove the existence of addition products of oxygen to isolated double bonds, *i.e.* the so-called moloxides (**1008**). For the reaction with oxygen, the conjugated double bond has

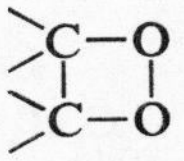

$$\begin{array}{c} \diagup C-O \\ \diagup C-O \end{array}$$
(**1008**)

to be excited to the first triplet state, in which the molecule possesses diradical character. This is usually achieved by photoexcitation of sensitizers, which transfer their triplet energy to the diene by collision (see Chapter 9).

A well-known example is the addition of oxygen to α-terpinene (**1009**), yielding ascaridole (**1010**). Other examples are addition to pyrrole, to 1,3-diphenylisobenzofuran (**1011**), and to 1,2,4,6-tetraphenylthiabenzene (**1012**).

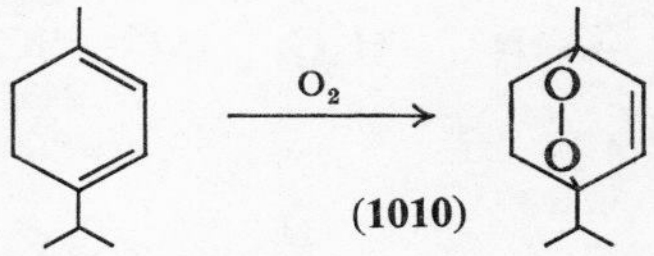

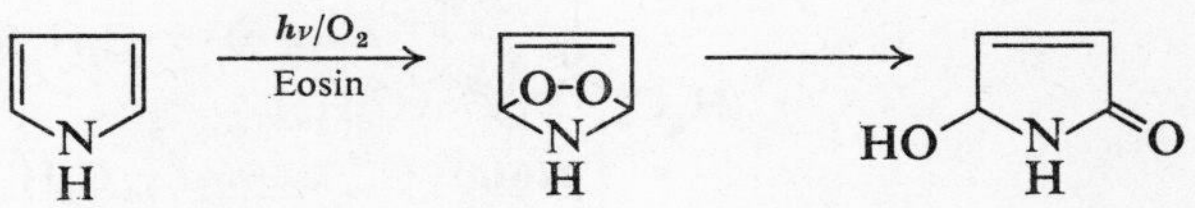

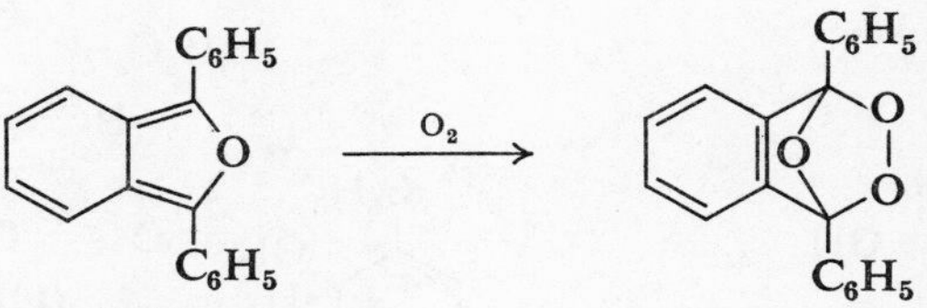

R-3. Addition of oxygen to 1,2,4,6-tetraphenylthiabenzene[13]

The stable peroxide (**1013**), formed in this reaction and isolated as a white residue on evaporation of the solvent (ether), yields in an acidic medium the zwitterion (**1014**), which with maleic anhydride affords the adduct (**1015**).

(1012) (1013) (1014) (1015)

or via

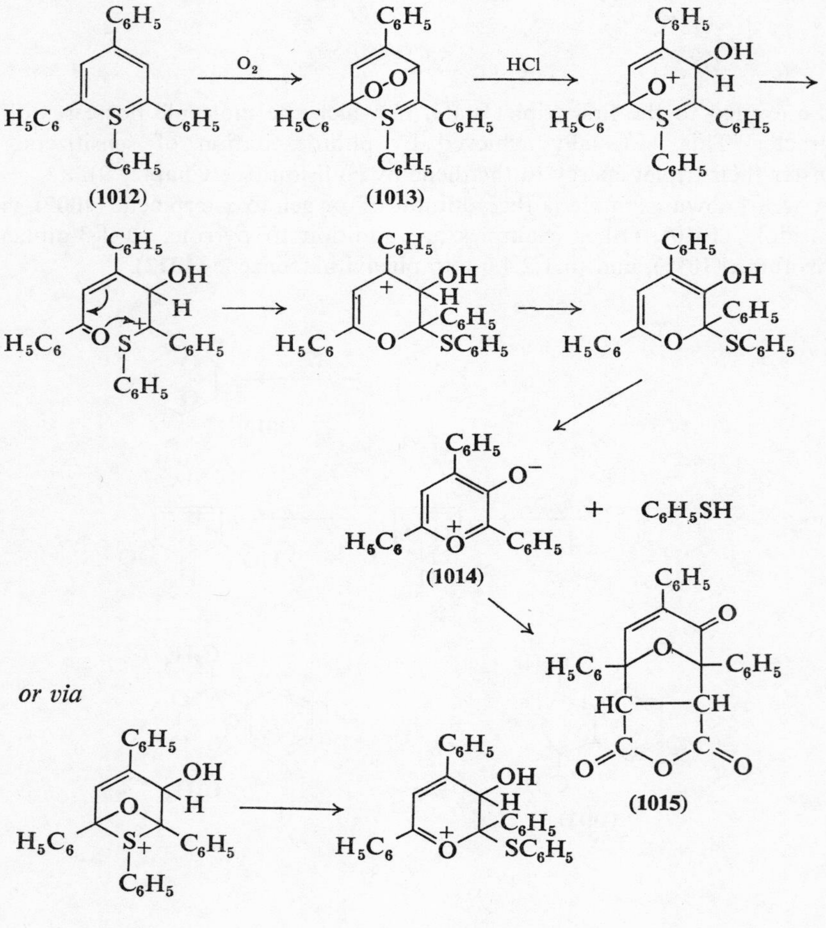

III. AUTOXIDATION

Even under mild conditions, molecular oxygen oxidizes many compounds, including certain saturated and unsaturated hydrocarbons, ethers, and aldehydes. Such oxidations, which are referred to as autoxidations, are started by radical initiators. In the case of saturated hydrocarbons, the rate of autoxidation increases from primary to tertiary C–H bonds. One of the most important industrial autoxidation processes is the manufacture of phenol and acetone from isopropylbenzene (cumene).

R-4. Autoxidation of 2,4-dimethylpentane[14] and 2,4-pentanediol[15]

2,4-Dimethylpentane (**1016**) yields the 2,4-bis(hydroperoxide) (**1017**). If the

$$C-\underset{H}{\overset{C}{C}}-C-\underset{H}{\overset{C}{C}}-C \xrightarrow{R\cdot} C-\underset{\cdot}{\overset{C}{C}}-C-\underset{H}{\overset{C}{C}}-C \xrightarrow{O_2} C-\underset{O-O\cdot}{\overset{C}{C}}-C-\underset{H}{\overset{C}{C}}-C \longrightarrow$$
(**1016**)

$$C-\underset{O-OH}{\overset{C}{C}}-C-\underset{\cdot}{\overset{C}{C}}-C \xrightarrow{O_2} C-\underset{OOH}{\overset{C}{C}}-C-\underset{OO\cdot}{\overset{C}{C}}-C \xrightarrow{RH} C-\underset{OOH}{\overset{C}{C}}-C-\underset{OOH}{\overset{C}{C}}-C$$
(**1017**)

$$C-\underset{H}{\overset{OH}{C}}-C-\underset{H}{\overset{OH}{C}}-C \xrightarrow{R\cdot} C-\underset{\cdot}{\overset{OH}{C}}-C-\underset{H}{\overset{OH}{C}}-C \xrightarrow{O_2} C-\underset{OO\cdot}{\overset{OH}{C}}-C-\underset{}{\overset{OH}{C}}-C \nrightarrow$$
(**1018**)   (**1019**)

$$C-\underset{OOH}{\overset{OH}{C}}-C-\underset{OOH}{\overset{OH}{C}}-C \longrightarrow C-\overset{O}{\underset{\parallel}{C}}-C-\overset{O}{\underset{\parallel}{C}}-C + 2H_2O_2$$

methyl groups are replaced by hydroxyl groups, as in (**1018**), reaction with oxygen occurs at only one of the CHOH positions. 2,4-Pentanediol yields 4-hydroxy-2-pentanone (**1021**). Hydroperoxides are found in the reaction mixture only to a small extent; thus the reaction does not follow the same path as the oxidation of 2,4-dimethylpentane. It has been proposed that the peroxy-radical which is formed as an intermediate is stabilized by hydroxyl-hydrogen as symbolized in (**1019**). As a result, the radical concentration increases, causing formation of the dimer (**1020**) which decompose by loss of O_2 instead of H_2O_2 and thus gives the hydroxy ketone (**1021**).

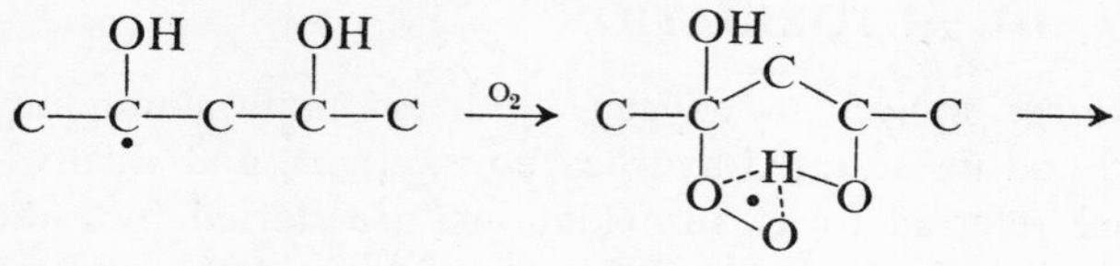

[Structures (1019), (1020), (1021) — radical oxidation/tetroxide decomposition scheme shown as chemical diagrams]

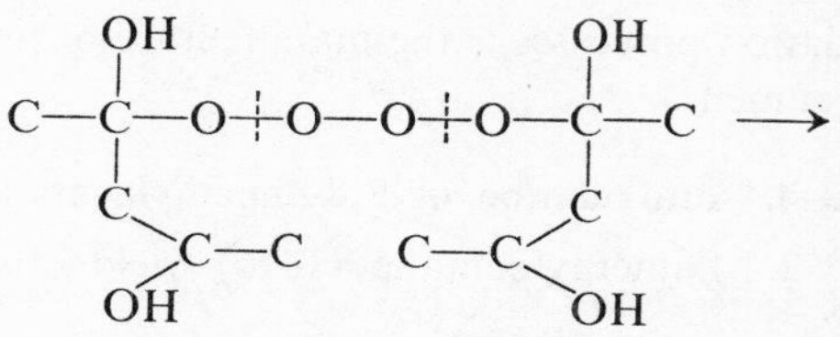

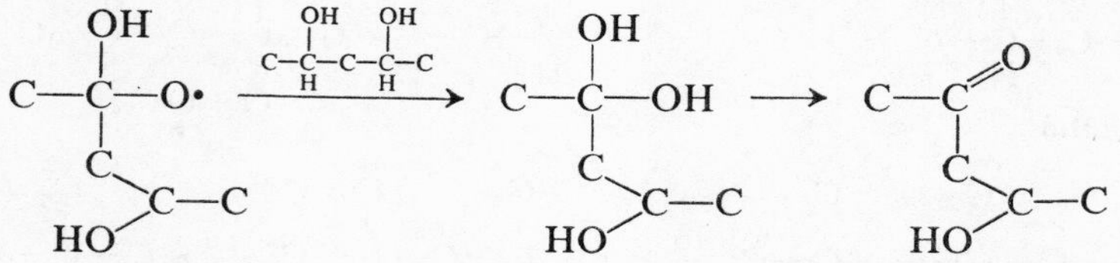

IV. DECOMPOSITION OF HYDROPEROXIDES

Tertiary hydroperoxides decompose above 100°, and in the presence of various radical initiators below 100°, yielding alcohols and oxygen. Such radical-forming reagents are, for example, nitriles in combination with a base, peroxide, or metal ion. The decomposition presumably takes place by the following reaction sequence:

$(CH_3)_3C-O-O-H \longrightarrow (CH_3)_3C-O\cdot + \cdot OH$ ... (1)

$(CH_3)_3C-O\cdot + (CH_3)_3C-O-O-H \longrightarrow$
$(CH_3)_3C-OH + (CH_3)_3C-O-O\cdot$ ... (2)

$2(CH_3)_3C-O-O\cdot \longrightarrow (CH_3)_3C-O\vdash O-O\vdash O-C(CH_3)_3 \longrightarrow$
$2(CH_3)_3C-O\cdot + O_2$ ... (3)

For the peroxide-catalyzed decomposition of cumyl hydroperoxide it has been demonstrated[16] that the oxygen is in fact formed according to reaction (3). The reaction represented by (3) is one of the rare cases in which the combination of two radicals does not necessarily lead to chain termination. However, combination of two *tert*-alkoxy radicals in the solvent cage, in which decomposition of the tetraoxide has taken place, may be a significant termination reaction: definitive evidence for cage recombination has been obtained[17] in the decomposition of di-*tert*-butyl peroxyoxalate in the liquid phase:

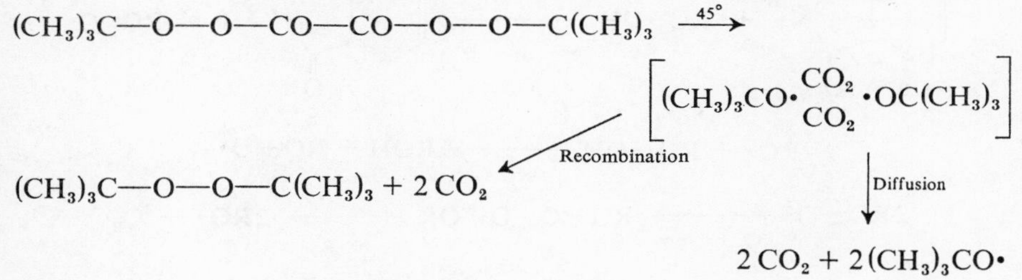

$$(CH_3)_3C-O-O-CO-CO-O-O-C(CH_3)_3 \xrightarrow{45°}$$

$$\left[(CH_3)_3CO\cdot \begin{matrix} CO_2 \\ CO_2 \end{matrix} \cdot OC(CH_3)_3 \right]$$

$$(CH_3)_3C-O-O-C(CH_3)_3 + 2\ CO_2 \xleftarrow{\text{Recombination}}$$

$$\downarrow \text{Diffusion}$$

$$2\ CO_2 + 2(CH_3)_3CO\cdot$$

R-5. Reaction of *tert*-butyl hydroperoxide with KOH and *o*-phthalonitrile[18]

tert-Butyl hydroperoxide decomposes at room temperature in the presence of a nitrile and a base, the initial step being attack by the peroxide anion on the nitrile[19].*

$$R = (CH_3)_3C-$$

$$RO-OH + KOH \rightleftarrows RO-O^-K^+ + H_2O$$

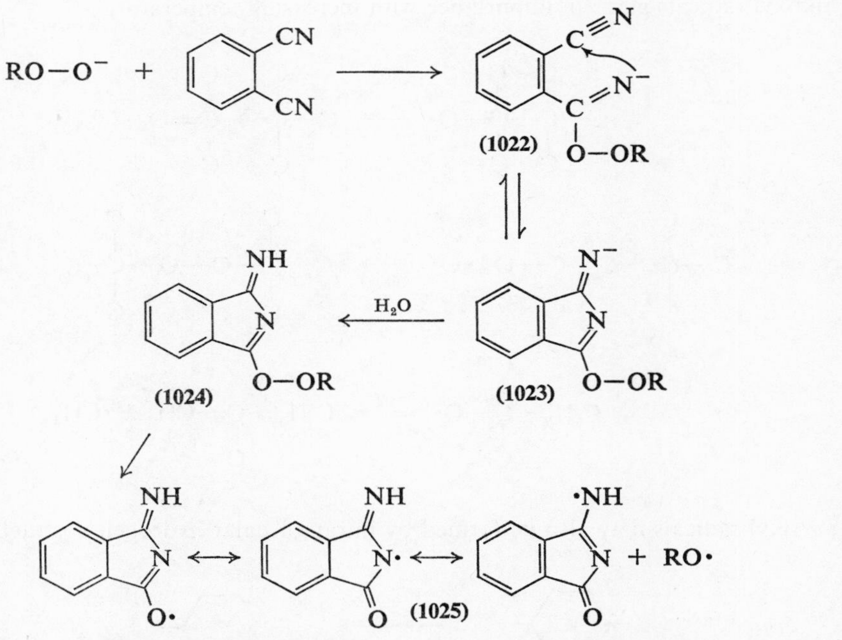

[Reaction scheme showing RO-O⁻ + o-phthalonitrile → (1022) → (1023) ⇌ (1024) via H₂O, leading to (1025) + RO·]

* In an alkaline medium nitriles and hydrogen peroxide yield amides[20]. It has not been possible to isolate an adduct of H_2O_2 to a nitrile.

$$\text{(1025)} \quad \underset{\text{NH}}{\underset{\text{O}}{\text{C}_6\text{H}_4}}\!\text{N}\cdot \; + \; \text{RO-OH} \longrightarrow \underset{\text{NH}}{\underset{\text{O}}{\text{C}_6\text{H}_4}}\!\text{N-H} \; + \; \text{RO-O}\cdot$$

$$\text{RO}\cdot \; + \; \text{RO-OH} \longrightarrow \text{ROH} \; + \; \text{RO-O}\cdot$$

$$2\,\text{RO-O}\cdot \longrightarrow \text{RO-O-O-OR} \longrightarrow 2\,\text{RO}\cdot \; + \; \text{O}_2$$

Cyclization of the initial anion **(1022)** leads to anion **(1023)**, which readily decomposes *via* **(1024)** to the resonance-stabilized radical **(1025)** and a *tert*-butoxyl radical. Both radicals abstract hydrogen from *tert*-butyl hydroperoxide. The peroxy radicals decompose by the mechanism shown, giving *tert*-butyl alcohol and oxygen. As expected, the reaction is inhibited by styrene.

The evolution of oxygen depends on the structure of the hydroperoxide. A reaction competing with the sequence (1)—(3) above is observed with some *tert*-alkyl hydroperoxides. β-Cleavage of the alkoxyl radical can give an alkyl radical and a ketone. Alkyl radicals abstract hydrogen less readily from hydroperoxides, recombining preferably with peroxy radicals to give peroxides. The β-cleavage of alkoxyl radicals gains in importance with increasing temperature.

$$\text{C}-\underset{\underset{\text{C}}{|}}{\overset{\overset{\text{C}}{|}}{\text{C}}}-\underset{\underset{\text{C}}{|}}{\overset{\overset{\text{C}}{|}}{\text{C}}}-\text{O}\cdot \longrightarrow \text{C}-\underset{\underset{\text{C}}{|}}{\overset{\overset{\text{C}}{|}}{\text{C}}}\cdot \; + \; \underset{\underset{\text{C}}{|}}{\overset{\overset{\text{C}}{|}}{\text{C}}}=\text{O}$$

$$\text{C}-\underset{\underset{\text{C}}{|}}{\overset{\overset{\text{C}}{|}}{\text{C}}}\cdot \; + \; \text{C}-\underset{\underset{\text{C}}{|}}{\overset{\overset{\text{C}}{|}}{\text{C}}}-\text{O}-\text{O}\cdot \longrightarrow \text{C}-\underset{\underset{\text{C}}{|}}{\overset{\overset{\text{C}}{|}}{\text{C}}}-\text{O}-\text{O}-\underset{\underset{\text{C}}{|}}{\overset{\overset{\text{C}}{|}}{\text{C}}}-\text{C}$$

$$\text{C}_6\text{H}_5-\underset{\underset{\text{CH}_3}{|}}{\overset{\overset{\text{CH}_3}{|}}{\text{C}}}-\text{O}\cdot \; \xrightarrow{120°} \; \text{C}_6\text{H}_5-\underset{\underset{\text{O}}{\|}}{\text{C}}-\text{CH}_3 \; + \; \cdot\text{CH}_3$$

Alkyl radicals may also be formed by intramolecular hydrogen abstraction.

At low temperatures the dimerization of alkoxyl radicals may become the predominant reaction.

$$2C_6H_5-\underset{\underset{CH_3}{|}}{\overset{\overset{CH_3}{|}}{C}}-O\cdot \longrightarrow C_6H_5-\underset{\underset{CH_3}{|}}{\overset{\overset{CH_3}{|}}{C}}-O-O-\underset{\underset{CH_3}{|}}{\overset{\overset{CH_3}{|}}{C}}-C_6H_5$$

When the hydroperoxide has an α-hydrogen atom, as have secondary and primary hydroperoxides, the corresponding peroxy radical may disproportionate to ketone or aldehyde and alcohol[21]. Then the aldehyde may react with hydroperoxide, yielding an α-hydroxy hydroperoxide, which decomposes presumably through a cyclic transition state with evolution of hydrogen[22].

$$2R-\underset{\underset{H}{|}}{\overset{\overset{R}{|}}{C}}-O-O\cdot \longrightarrow \underset{R}{\overset{R}{>}}C=O + R-\underset{\underset{H}{|}}{\overset{\overset{R}{|}}{C}}-OH + O_2$$

$$RCHO + RCH_2-O-OH \longrightarrow R-\underset{\underset{H}{|}}{\overset{\overset{H}{|}}{C}}\overset{O-O}{\underset{H}{\diagdown}}\underset{OH}{\overset{R}{C}}$$

$$\longrightarrow RCHO + RCOOH + H_2$$

V. RADICAL MECHANISMS IN LEAD TETRAACETATE OXIDATIONS

R-6. Oxidation of primary alcohols by lead tetraacetate[23]

The primary intermediate formed in the numerous reactions of lead tetraacetate with compounds possessing reactive hydrogen has the structure X–Pb(OAc)$_3$ (X = alkyl, aryl, alkoxyl, etc.)[24]. In some cases this intermediate could be isolated[24,25]. The next step, which is the actual oxidation step, is heterolysis of the X–Pb bond, during which the Pb(OAc)$_3$ takes over the bonding electrons. At higher temperature, however, the X–Pb bond may undergo homolysis.

The ease of decomposition increases with increasing stability of the cation X$^+$ or radical X· formed.

The reaction of primary alcohols with lead tetraacetate in refluxing benzene, *i.e.* in a temperature range where thermolysis of the RO–Pb(OAc)$_3$ bond occurs, probably involves a radical mechanism[26]. Supporting evidence is that ether formation also takes place at very low temperatures under irradiation[27], and that the reversible fragmentation of alcohols that have a structure fixed in a polycyclic system is not compatible with an ionic mechanism (see below).

Homolysis of (**1027a**), followed by a hydrogen shift from C-4 to the oxygen atom*, explains the loss of optical activity at C-4†. Recombination of the carbon

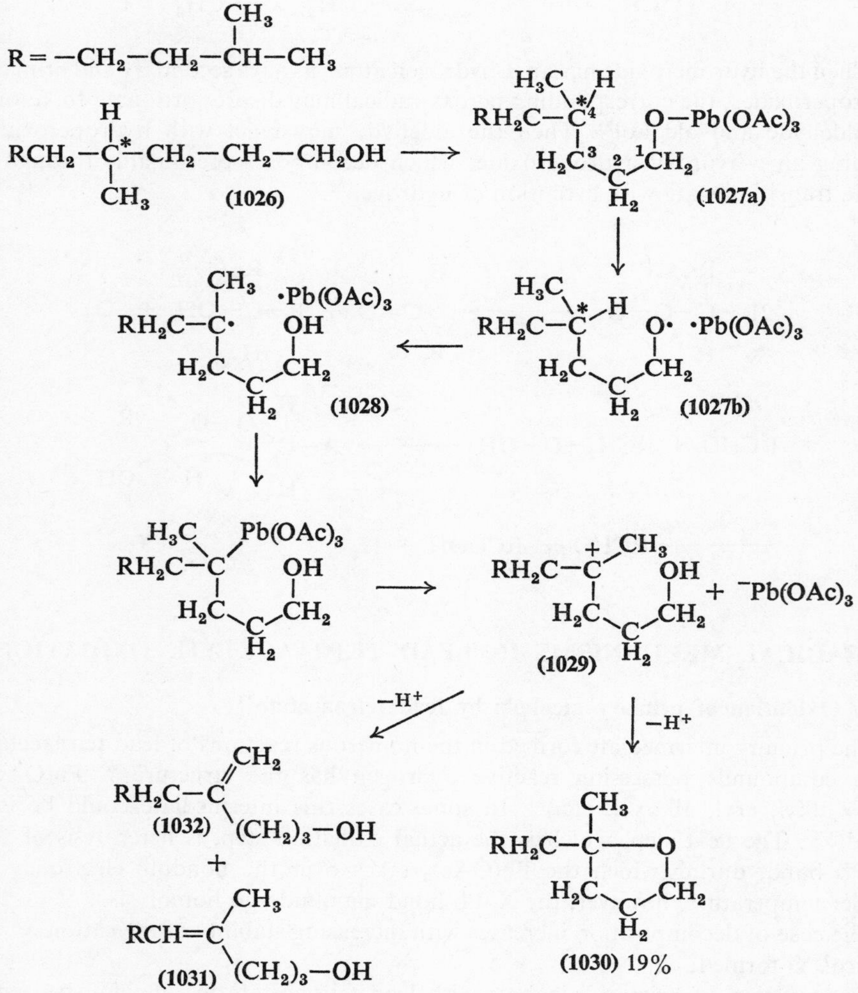

$$R = -CH_2-CH_2-\underset{\underset{CH_3}{|}}{CH}-CH_3$$

* For reactions of alcohols that do not possess a δ-hydrogen atom see reference 28.
† In polar solvents such as pyridine heterolysis of the CH_2O-Pb bond may occur[29].

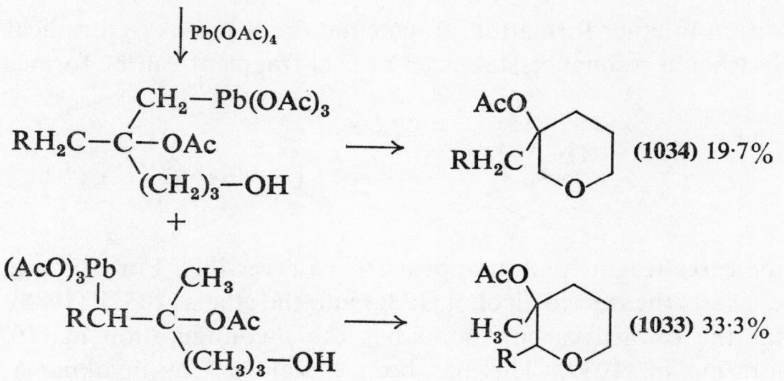

$$\Big\downarrow Pb(OAc)_4$$

$$\begin{array}{c} CH_2-Pb(OAc)_3 \\ RH_2C-C-OAc \\ (CH_2)_3-OH \end{array} \longrightarrow \text{AcO-[tetrahydropyran with } RH_2C\text{]} \quad (1034)\ 19.7\%$$

$$+$$

$$\begin{array}{c} (AcO)_3Pb \\ | \\ RCH-C-OAc \\ (CH_3)_3-OH \end{array} \begin{array}{c} CH_3 \\ \end{array} \longrightarrow \text{AcO-[tetrahydropyran with } H_3C, R\text{]} \quad (1033)\ 33.3\%$$

radical in (1028) with ·Pb(OAc)$_3$ and subsequent heterolysis of the C–Pb bond leads to a carbonium ion (1029), which yields the ether (1030) or the olefins (1031) and (1032). Addition of lead tetraacetate to the double bonds of these olefins leads to the cyclic ethers (1033) and (1034).

Other oxidations by lead tetraacetate

In principle, formation of an ether (1030) from (1027b) can be envisaged as occurring by way of an intermediate (1035); however, loss of optical activity cannot be explained by such a mechanism.

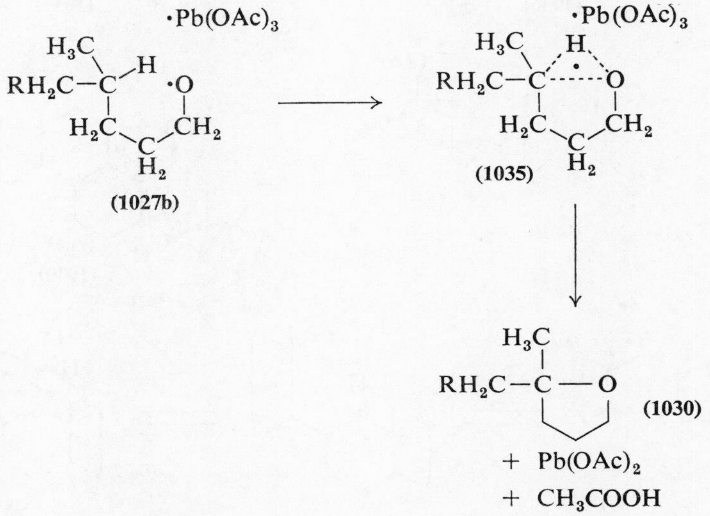

(1027b) ·Pb(OAc)$_3$ → (1035) ·Pb(OAc)$_3$

↓

$$\begin{array}{c} H_3C \\ | \\ RH_2C-C-O \\ \underline{} \end{array} \quad (1030)$$

$$+\ Pb(OAc)_2$$
$$+\ CH_3COOH$$

In addition to ether formation, fragmentation of the oxygen radical may occur, especially when a resonance-stabilized radical fragment can be formed:

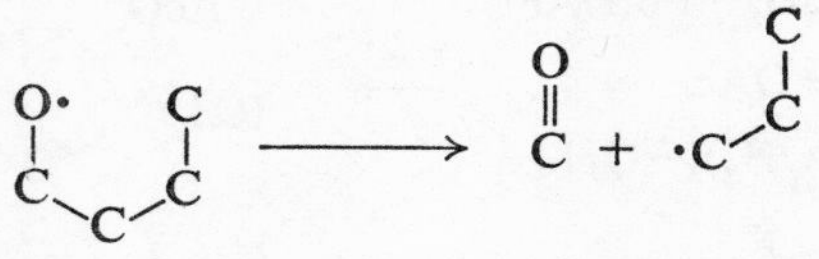

In some cases fragmentation appears to be reversible. For example, lead tetraacetate converts the steroid alcohol **(1036)** into the ethers **(1037)**, **(1038)**, and **(1039)**. Note that the oxygen–carbon bond has the β-configuration in **(1038)** and the α-configuration in **(1039)**. This has been explained[30] by invoking a free-radical intermediate which can yield either **(1040)** or **(1041)**. If the primary intermediate were to cleave by an ionic mechanism it would be necessary to postulate formation of **(1043a)** and **(1043b)** from **(1044)** and this does not seem a likely reaction (addition of a secondary carbonium ion to the electrophilic carbonyl-carbon atom).

The isomerization **(1040)** $\rightleftarrows$ **(1041)**, though referring to secondary alcohols, is cited as evidence for the radical mechanism of oxidation also of primary alcohols by lead tetraacetate.

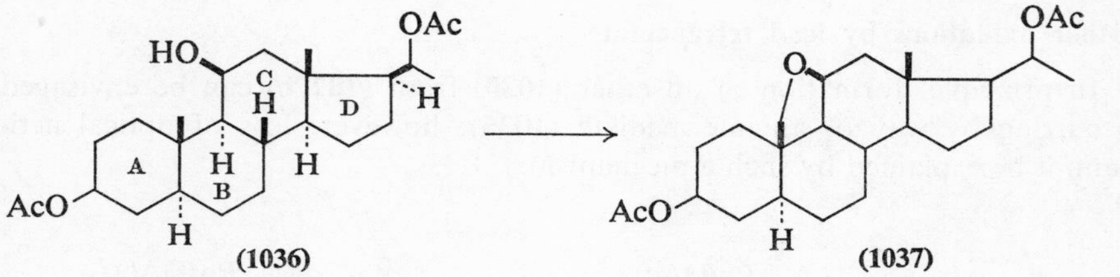

(1036) **(1037)**

\+

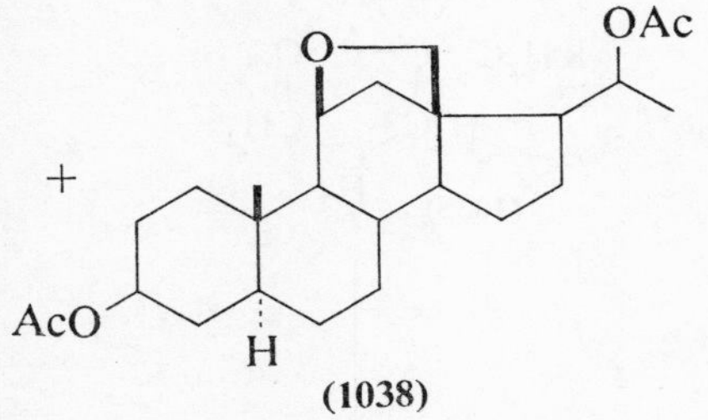

(1038)

\+

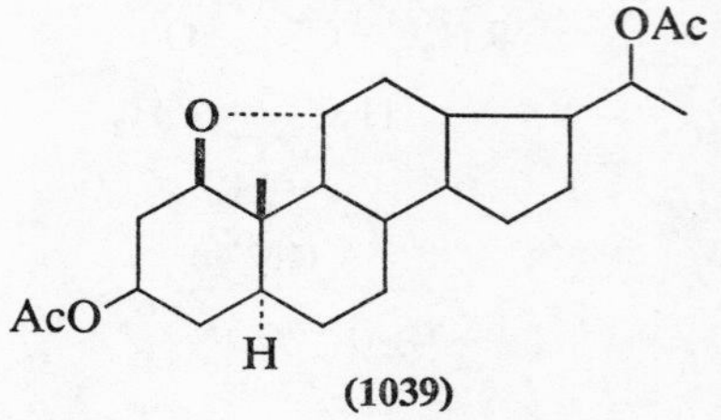

(1039)

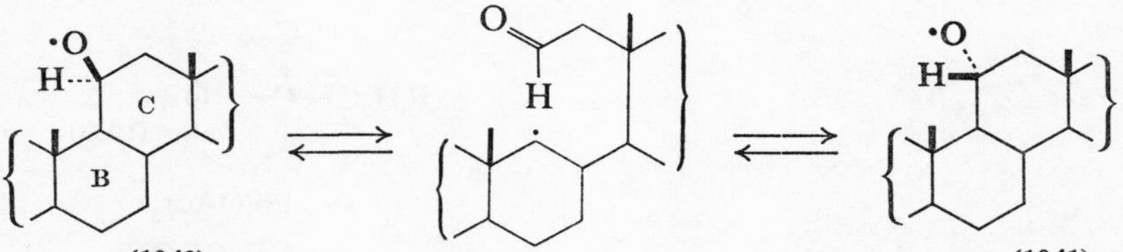

(1040) **(1041)**

(1042) (1043a) (1044)

(1043b)

REFERENCES

[1] B. Fell and L.-H. Kung, *Chem. Ber.*, **98**, 2871 (1965).
[2] R. A. Herrmann and R. M. Noyes, *J. Amer. Chem. Soc.*, **78**, 5764 (1956); S. Levine and R. M. Noyes, *ibid.*, **80**, 2401 (1958).
[3] F. Asinger and B. Fell, *Erdoel Kohle*, **17**, 74 (1964).
[4] C. Chiltz, P. Goldfinger, G. Huybrechts, G. Martins, and G. Verbeke, *Chem. Rev.*, **63**, 355 (1963).
[5] Ch. Walling and B. B. Jacknow, *J. Amer. Chem. Soc.*, **82**, 6108, 6113 (1960).
[6] F. Korte and H. Höver, *Tetrahedron*, **21**, 1287 (1965).
[6a] J. I. Brauman and L. K. Blair, *J. Amer. Chem. Soc.*, **90**, 6561 (1968).
[7] G. A. Russel, *J. Amer. Chem. Soc.*, **80**, 4987 (1958); Ch. Walling and M. F. Mayahi, *ibid.*, **81**, 1485 (1959).
[8] E. S. Huyser, *J. Amer. Chem. Soc.*, **82**, 5246 (1960); E. S. Huyser and B. Giddings, *J. Org. Chem.*, **27**, 3391 (1962); E. S. Huyser, H. Schinke, and R. L. Burham, *ibid.*, **28**, 2141 (1963); H. Kloosterziel, *Rec. Trav. chim.*, **82**, 497 (1963).
[9] D. P. Wyman, J. Y. C. Wang, and W. R. Freeman, *J. Org. Chem.*, **28**, 3173 (1963).
[10] W. Thaler, *J. Amer. Chem. Soc.*, **85**, 2607 (1963).
[11] El-Ahmadi, I. Heiba, and R. M. Dessau, *J. Amer. Chem. Soc.*, **88**, 1589 (1966).
[12] E. Müller, *Fortschr. Chem. Forsch.*, **1**, 401 (1949–50); L. Pauling, *The Nature of the Chemical Bond*, Cornell University Press, New York, 1960.
[13] G. Suld and C. C. Price, *J. Amer. Chem. Soc.*, **84**, 2094 (1962).
[14] F. F. Rust, *J. Amer. Chem. Soc.*, **79**, 4000 (1957).
[15] F. F. Rust and E. A. Youngman, *J. Org. Chem.*, **27**, 3778 (1962); E. A. Youngman, F. F. Rust, G. M. Coppinger, and H. E. de la Mare, *ibid.*, **28**, 144 (1963).
[16] T. G. Traylor and P. D. Bartlett, *Tetrahedron Letters*, **24**, 30 (1960); P. D. Bartlett and T. G. Traylor, *J. Amer. Chem. Soc.*, **85**, 2407 (1963).
[17] R. Hiatt and T. G. Traylor, *J. Amer. Chem. Soc.*, **87**, 3766 (1965).
[18] D. B. Denney and J. D. Rosen, *Tetrahedron*, **20**, 271 (1964).
[19] G. B. Payne, P. H. Demming, and P. H. Williams, *J. Org. Chem.*, **26**, 659 (1961).
[20] K. B. Wiberg, *J. Amer. Chem. Soc.*, **75**, 3961 (1953).

[21] C. A. Russell, *J. Amer. Chem. Soc.*, **77,** 4583 (1955).
[22] L. J. Durham and H. S. Mosher, *J. Amer. Chem. Soc.*, **84,** 2811 (1962).
[23] D. Hauser, K. Schaffner, and O. Yeger, *Helv. Chim. Acta*, **47,** 1883 (1964).
[24] R. Criegee, *Angew. Chem.*, **70,** 173 (1958).
[25] R. Preuss, *Angew. Chem.*, **71,** 747 (1959).
[26] K. Heusler and J. Kalvoda, *Angew. Chem. Intern. Ed. Engl.*, **3,** 525 (1964); V. M. Micovic, R. I. Mamuzic, D. Jeremic, and M. Lj. Mihailovic, *Tetrahedron*, **20,** 2279 (1964); M. Lj. Mihailovic, Z. Maksimovic, D. Jeremic, Z. Cekovic, A. Milovanovic, and Lj. Lorenc, *ibid.*, **21,** 1395, 2799 (1965).
[27] V. Franzen and R. Edens, *Angew. Chem.*, **73,** 579 (1961).
[28] M. L. Mihailovic, Z. Cekovic, and D. Jeremic, *Tetrahedron*, **21,** 2813 (1965).
[29] M. Lj. Mihailovic, J. Bosujah, Z. Maksimovic, Z. Cekovic, and Lj. Lorenc, *Tetrahedron*, **21,** 955 (1966).
[30] K. Heusler, J. Kalvoda, G. Anner and W. Wettstein, *Helv. Chim. Acta*, **46,** 352 (1963).

9
Photochemistry

I. THEORETICAL BACKGROUND

PHOTOCHEMICAL processes essentially involve an excited electronic state of a molecule resulting from its interaction with light.

Light (electromagnetic radiation) is quantized in photons of the energy $h\nu$ (h = Planck's constant, ν = frequency of light). Since the relation between the frequency and the wavelength (λ) of light is given by $\nu = c/\lambda$, where c = the velocity of light, the energy of a photon is directly proportional to its frequency and inversely proportional to its wavelength.

The process of absorption (as well as emission) of light follows the equation (Bohr's condition):

$$\Delta e = E_f - E_i = h\nu,$$

where Δe = the energy difference per molecule,
E_i = the initial energy level of molecule, and
E_f = the final energy level of molecule.

The energy difference per molecule (Δe) has to be multiplied by the number of molecules per mole (N = Avogadro's number) to give the energy change per mole (ΔE):

$$\Delta E = N\Delta e = Nh\nu = Nhc/\lambda.$$

This energy change (ΔE) (in kcal/mole) depends on the wavelength of absorbed (or emitted) light and if the wavelength is given in mμ (10^{-7} cm) is termed an einstein:

$$\Delta E \text{ (kcal/mole)} = \frac{2.86 \times 10^4}{\lambda \text{ (in m}\mu\text{)}} = 1 \text{ einstein (at wavelength } \lambda\text{)}$$

Absorption

Absorption of electromagnetic radiation can lead to changes in rotational (Δe_r), vibrational (Δe_v), and electronic (Δe_{el}) energy levels of a molecule:

$$\Delta e = \Delta e_r + \Delta e_v + \Delta e_{el}$$

The amount of energy required to effect an electronic transition is considerably greater than that required to effect a vibrational transition, which, in turn, is greater than that required to cause rotational transition. Consequently, absorption of energy that effects an electronic transition also results in changes of vibrational and rotational energy states.

At room temperature sufficient thermal energy is available to induce rotational transitions of molecules.

Vibrational transitions require the additional energy that can be provided by light of wavelengths corresponding to the infrared region of the electromagnetic spectrum, *i.e.* between 25 and 2.5 μ. (The value of 1 einstein in this range varies from 1.144 kcal/mole at 25 μ to 11.44 kcal/mole at 2.5 μ.) At room temperature a molecule is in thermodynamic equilibrium with its environment, *i.e.* in the state of lowest vibrational energy.

Electronic transitions require the still greater energy provided by light of the visible (800—400 mμ), ultraviolet (400—200 mμ), or far- or vacuum-ultraviolet (below 200 mμ) region. (The value of 1 einstein at 800 mμ is 35.75 kcal/mole, at 400 mμ 71.5 kcal/mole, and at 200 mμ 143.0 kcal/mole.)

A molecule in the ground state can absorb a photon to allow electronic transition if the energy of the photon equals the energy difference (Δe) between two energy levels of the molecule that meet the following requirements. The lower energy level (E_i) is that of the ground state; the ground state of ordinary organic compounds is generally represented by an arrangement of electrons such that orbitals of lowest energy are occupied ("Aufbau" principle) by pairs of electrons with antiparallel spin (Pauli principle). The high energy level (E_f) is that of the excited state. The transition to the excited state is achieved if an electron from one of the energetically upper orbitals of the ground state (*i.e.* valence electrons, which are responsible for chemical reactions) is promoted to a previously unoccupied orbital of higher energy.

The resulting excited state is represented by one additional occupied orbital; two of the orbitals of this state are occupied each by a single electron. The spins of these two single electrons can be either antiparallel with respect to each other (singlet state) or parallel (triplet state). By this definition a ground state that consists of pairs of electrons with antiparallel spin is also a singlet state, each pair occupying one orbital, whereas in the singlet excited state one of these pairs is distributed over two orbitals. Singlet and triplet excited states can therefore be regarded as "biradical" chemical species.

In principle, transitions from the ground state (*i.e.* the lowest electronic and vibrational energy level) to diverse vibrational levels of excited electronic states are possible; certain transitions, however, take place preferentially, according to the Franck–Condon principle (for simplicity a diatomic molecule is considered). This principle states that during an electronic transition (10^{-15} sec required for excitation of an electron) nuclear and vibrational motion (10^{-10} to 10^{-12} sec is required) can be neglected. Consequently the most probable electronic transitions are those that display the highest probability of residence of nuclei in ground and excited states without change in geometry (interatomic distance). This condition is, in the zero vibrational state, near the equilibrium distance of the atomic nuclei; in higher vibrational states, however, it is close to the turning points of the vibrating atoms (Figure 10).

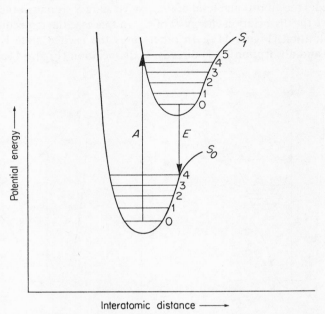

Figure 10. Potential curves of ground state (S_0) and first excited singlet state (S_1); the numbers indicate vibrational sublevels. A = absorption; E = emission.

Because transitions from singlet to triplet state are "forbidden" by spin conservation rules, the excited state obtained from a ground state singlet by absorption of light is also a singlet state. This singlet excited state can be the first or any higher excited singlet state; of course, any vibrational sublevel of these electronic states can be attained in accordance with the Franck–Condon principle.

A molecule in an excited state loses its energy by various mechanisms, which will be briefly discussed in the following sections.

Deactivation

By absorption of radiation energy molecules can be brought to states which can lose their excess of energy in several ways. From the primary excited state the molecule can cascade down to a lower excited state, or even return to the ground state, by transformation of its excess of energy into thermal or radiant energy, usually both. Alternatively, the absorbed energy can be converted into chemical energy, and one or more new chemical ground-state species will be formed; in most cases this process is accompanied by one or both of the other relaxation modes. Deactivation may occur by several processes:

(*a*) Dissociation. Absorption of energy can lead directly to photodecomposition; when this occurs the vibrational level of the excited state S_n is one of the continuous set lying above the dissociation energy D of S_n. In this case the molecule dissociates at its first vibration (Figure 11a). In other cases the excited state is completely unstable (S_r) and absorption leads directly to dissociation (Figure 11b).

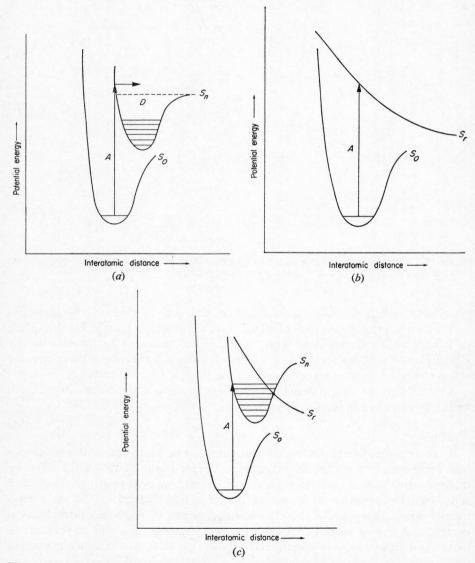

Figure 11. Potential curves of ground states (S_0) and excited singlet states (S_n) leading to dissociation. S_r = repulsive state; D = dissociation energy of S_n.

Alternatively an electronically excited molecule in a state of high vibrational energy may lose vibrational energy and cross over to another potential energy curve that leads to dissociation; this case is called predissociation (Figure 11c).

All these dissociation processes lead to new chemical species which, particularly in condensed media, are not always isolated because of their rapid recombination.

(b) *Internal conversion.* In a fluid system a process called internal conversion seems to occur in all excited molecules. It consists in radiationless conversion of electronic energy of an excited electronic state into vibrational energy of the next lower electronic excited state (*e.g.* $S_2 \rightarrow S_1$); no spin conversion is involved. This is followed by rapid (10^{-12} sec) relaxation of the high vibrational level to the lowest one of the same electronic state and transfer of energy to the environment (solvent) (Figure 12).

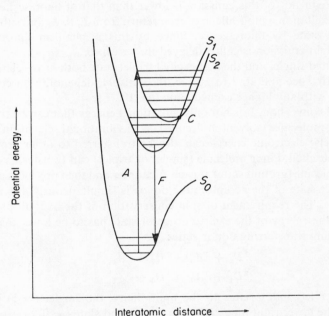

Figure 12. Potential curves of ground state (S_0) and first and second excited states (S_1 and S_2); C = point of internal conversion; F = fluorescence.

These processes can be repeated until the lowest vibrational level of S_1 is reached, which may relax further to the ground state with high vibrational energy but often does not relax further in the same manner, mainly because too large an amount of electronic energy would have to be transferred into vibrational energy.

(c) *Fluorescence.* Emission of light accompanying the transition from S_1 to S_0 is called fluorescence. Emission starts from the zero vibrational level of S_1 leading to higher vibrational levels of the ground state. The emission wavelength is therefore longer than that of absorption, except the 0–0 band which nearly coincides (under the Franck–Condon principle) (Figure 12).

(*d*) **Intersystem crossing.** Although transitions between singlet and triplet states are "forbidden" by spin conservation rules these transitions do occur (particularly when higher excited states are involved). Numerous compounds are known in which radiationless transitions from S_1 to T_1 take place. The triplet state is energetically lower than its corresponding singlet; intersystem crossing (like predissociation and internal conversion) is an isoenergetic process, leading to a vibrationally excited triplet state that falls to its lowest vibrational level by vibrational relaxation.

In the same manner the triplet T_1 can be deactivated again by intersystem crossing to the singlet ground state S_0. This radiationless process competes with other deactivation modes.

(*e*) **Phosphorescence.** The radiative transition from T_1 to S_0 is only observed in rigid media, and not in fluid solution where radiationless deactivation modes are favored. The frequency of this emission is lower than that of fluorescence.

By thermal excitation a molecule may even return from T_1 to S_1 and subsequently to the ground state by fluorescence. Since by this mechanism fluorescence is retarded, the light emission is called delayed fluorescence.

The first excited singlet and the first excited triplet state both have relatively long lives (10^{-9} to 10^{-6} sec and 10^{-4} to 10 sec, respectively, depending on conditions) in comparison with any higher excited state (*ca.* 10^{-12} sec).

(*f*) **External conversion.** It is not only vibrational energy that can be transferred to surrounding molecules (solvent molecules), as in vibrational relaxation of higher vibrational levels; electronic energy can also be transferred to other ground-state molecules. An excited donor molecule (singlet or triplet) can transfer its energy to an acceptor molecule (ground state) at each encounter and then produces an excited singlet or triplet state of the acceptor; the donor is simultaneously deactivated to its ground state. The requirement of spin conservation of the over-all process has to be met and the energy of the excited acceptor state has to be lower than that of the corresponding transferring donor state:

$$^1D_1 + {}^1A_0 \to {}^1D_0 + {}^1A_1$$
$$^3D_1 + {}^1A_0 \to {}^1D_0 + {}^3A_1$$

Here, the upper left index refers to the multiplicity, the singlet ($= 1$) or triplet ($= 3$) state*; the lower right index refers to the ground state ($= 0$) or excited state ($= 1, 2, \ldots, n$).

This bimolecular process is more effective in fluid than in rigid media. Since long life of the excited donor species endows it with a greater probability of encountering an acceptor molecule, triplet–triplet transfers are preferred to singlet

---

* The multiplicity indicates the number of possible orientations of the total spin momentum with respect to the total angular momentum, resulting in the total momentum of the molecule.

The total spin momentum of the singlet state is zero; the total momentum is therefore equal to the total angular momentum; one observes a single absorption system. For the triplet state, in which the total spin momentum is 1, three orientations with respect to the total angular momentum are possible, namely, in the same direction, in opposite direction, and perpendicular to the total angular momentum; the three resulting total momenta correspondingly cause three absorption systems.

transfers. In organic photochemistry, wide use is made of sensitization, mostly of triplet states which are often available only with some difficulty by means of intersystem crossing. Excitation of the acceptor singlet must, therefore, be avoided and all light energy has to be used to excite the energetically lower sensitizer (donor) singlet. For an effective sensitizer a high probability of intersystem crossing to its triplet is desirable. Transfer of its triplet energy to the acceptor excites the acceptor triplet directly (Figure 13).

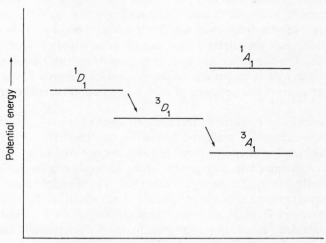

Figure 13. Energy transfer in external conversion.

(g) *Deactivation by chemical reaction.* In the first instant after absorption of light the geometry of the excited molecule is the same as in the ground state, according to the Franck–Condon principle. The ground-state geometry, however, is usually not the most stable one for the excited state, so that the molecule tends quickly to reach the new equilibrium geometry of the excited state. Therefore, bond distances in the excited state generally differ from those in the ground state.

The transitions which are most important for photochemical reactions are $\pi-\pi^*$ transitions in which a π-electron of a double bond is excited to the next higher unoccupied antibonding orbital, and $n-\pi^*$ transitions of electrons in non-bonding orbitals (for example, the free electron pairs of oxygen in carbonyl groups) to antibonding π-orbitals. After these transitions the electron distribution of the excited state is different from that of the ground state. The case of a ketone may be illustrated as in Figure 14[1].

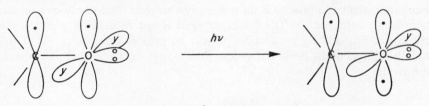

$n-\pi^*$ Transition
Figure 14. Excitation of a carbonyl group.

Both π–π^* and n–π^* excited states can exist as triplet or singlet state. In general, the lowest excited singlet and triplet state of a molecule will undergo photochemical reactions.

These facts lead one to expect that the geometry of the molecule, the distribution of valence electrons in the excited states (singlet and triplet state can be regarded as a "biradical"), and the chemistry of the excited state will differ from those of the ground state. Since the life-time of T_1 is greater than that of S_1, one would also expect photochemical reactions to take place preferentially from the triplet state.

In many cases it can be determined whether a reaction takes place in the singlet or the triplet state, since the triplet energy can often be supplied directly with the aid of sensitizers. On the other hand, energy-quenching substances are known which quench triplet molecules but not singlet species (for example, 1,3-pentadiene). Reactions taking place in the presence of such a compound therefore occur in the singlet state.

Thermal activation of a molecule increases its over-all energy, and transitions of low activation energy are preferred (ground-state chemistry). Photochemical activation allows crossing of energy barriers which are not commonly accessible by thermal energy: particular bonds or groups of the molecule can be activated when light of appropriate energy (frequency) is absorbed. Thermal activation leads to a vibrationally excited ground state ("hot" ground state) which can be also produced by internal conversion from S_1 or intersystem crossing from T_1. Photochemical reactions in condensed phases, however, start from an electronically excited state (S_1 or T_1) that is in its lowest vibrational level owing to rapid vibrational relaxation. Despite the selectivity of photochemical activation more than one probable reaction path may lead to several products. Furthermore, the primary photochemical conversion may result in a reactive ground-state species that may react further in so-called "dark" reactions (thermal processes). Also the primary photoproduct may undergo chemical conversion in a secondary photochemical process. Finally the photochemical reaction path may be determined by the nature of solvent: its ability to affect the photoreaction (quenching, internal conversion, intersystem crossing) or to serve as a reactant has to be taken into consideration.

Much information about the fate of the energy of an electronically excited molecule is obtained from the quantum yield (Φ) which is defined as follows:

$$\Phi = \frac{\text{Number of molecules reacting}}{\text{Number of quanta absorbed}} \quad \text{or} \quad \frac{\text{Number of moles reacting}}{\text{einsteins taken up}}$$

The quantum yield of a process is unity if every absorbed photon causes a molecule to afford its photoproduct. The quantum yield is not necessarily unity: if chain reactions are initiated by irradiation, Φ can be greater than unity; Φ of many organic photoreactions, however, is less than unity, owing to competing deactivation modes.

P-1. Photoisomerization of 4,4-diphenyl-2,5-cyclohexadienone (1045)[1a]

The diene (**1045**) is excited by irradiation to its lowest excited singlet $S_1(n-\pi^*)$ (**1046**). It should be emphasized that the three (of six possible) resonance structures drawn for (**1046**) are not resonance contributors of the ground-state molecule (**1045**) but those of a new species, the $n-\pi^*$ singlet molecule (**1046**). (Circles represent sp hybrid electrons, y's p_y electrons and dots the π system electrons.)

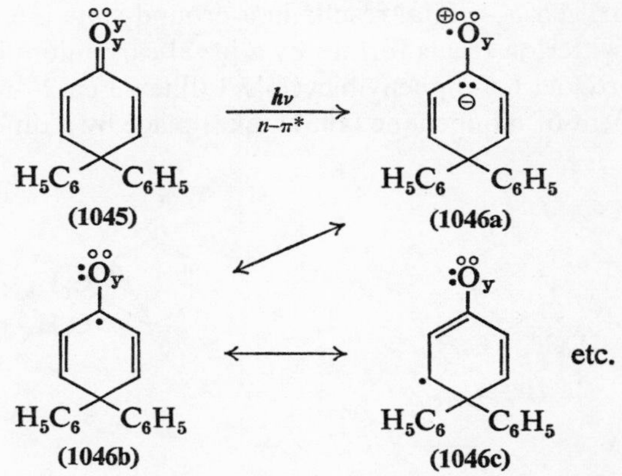

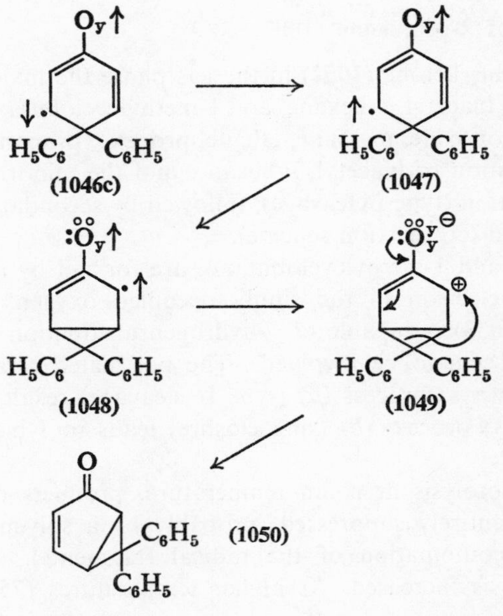

Intersystem crossing from $S_1(n-\pi^*)$ leads to the chemically active state $T_1(n-\pi^*)$, as is assumed because acetophenone (a triplet excitation donor) sensitizes the reaction. (Arrows illustrate spin momentum, parallel or antiparallel.) Structure (**1046c**) is chosen to illustrate intersystem crossing (accompanied by spin flip) to one of six possible resonance structures of the new species $T_1(n-\pi^*)$ (**1047**).

The next step is the formation of a new σ-bond between C-3 and C-5, leading to triplet species (**1048**). Intersystem crossing of the excited state (**1048**) by electron demotion to the oxygen p_y-orbital results in a ground-state intermediate (**1049**), a zwitterion. This zwitterion reacts further by a usual carbonium ion rearrangement to give the final product 6,6-diphenylbicyclo[3.1.0]hex-3-en-2-one (**1050**).

The rearrangement of monoenone (**1051**) takes place by a different mechanism[2].

(**1051**)

P-2. Photoreactions of 2-pentanone (1052)[3]

On irradiation of this ketone (**1052**) in the gas phase the major photoproducts are ethylene, acetone, biacetyl, *n*-hexane, and 1-methylcyclobutanol; much smaller amounts of carbon monoxide, methane, ethane, propene, propane, and butane are also obtained. Formation of biacetyl, *n*-hexane, and the minor products may be explained by dissociation (type I cleavage), followed by secondary reactions of the radicals thus produced (cf. reaction scheme).

Ethylene, acetone, and 1-methylcyclobutanol are formed by an intramolecular rearrangement. The electron in the singly occupied oxygen p_y-orbital in the excited $n-\pi^*$ state (**1053**) is capable of γ-hydrogen abstraction (both triplet and singlet excited state seem to be involved). The postulated biradical (**1054**) can react further in two ways: process (*a*) (type II cleavage) results in ethylene and acetone (as observed); process (*b*) (ring closure) leads to 1-methylcyclobutanol (**1055**).

In liquid-phase photolysis at room temperature, products caused by type I cleavage are almost entirely suppressed (probably by a solvent-cage effect that causes immediate recombination of the radical fragments), and the yield of 1-methylcyclobutanol is increased. At higher temperatures (75°) type I photoproducts are again obtained (biacetyl, propane). Cyclobutanol formation seems to be a stereospecific process, presumably involving a synchronous mechanism[4].

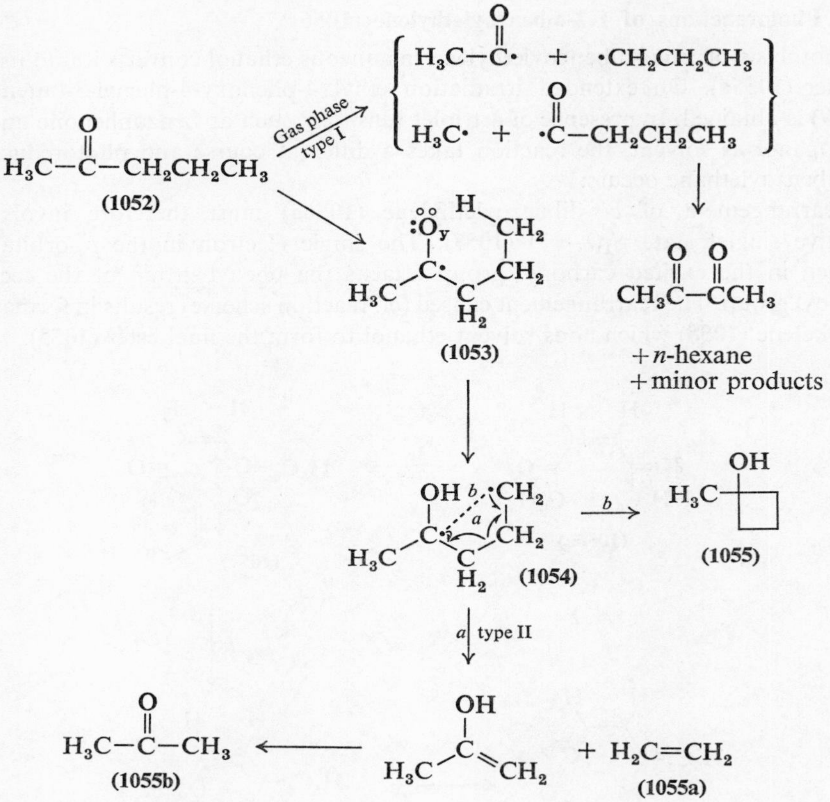

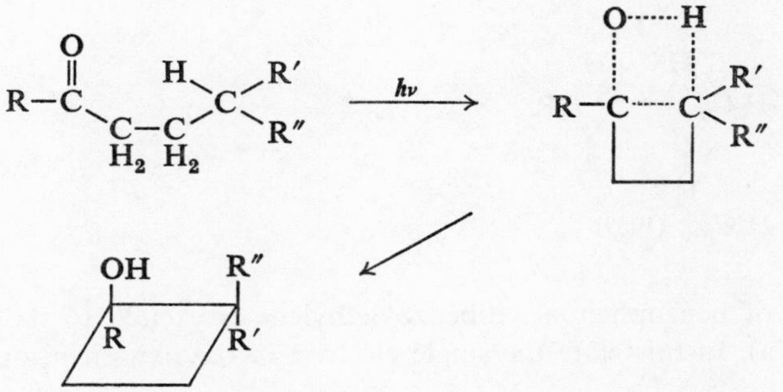

P-3. Photoreactions of 1,2-dibenzoylethylene (1056)[5]

Photolysis of *trans*-dibenzoylethylene in aqueous ethanol converts it into its *cis*-isomer (1056a). On extended irradiation ethyl 4-phenoxy-4-phenyl-3-butenoate (1059) is obtained. In presence of a triplet sensitizer such as benzophenone and in 2-propanol as solvent, the reaction takes a different course and photoreduction to dibenzoylethane occurs.

Rearrangement of *cis*-dibenzoylethylene (1056a) must therefore involve a reactive singlet state $S_1(n-\pi^*)$ (1057). The single electron in the p_y-orbital of oxygen in the excited carbonyl group attacks the phenyl group of the second benzoyl group. The rearrangement caused (cf. reaction scheme) results in formation of a ketene (1058) which adds solvent ethanol to form the final ester (1059).

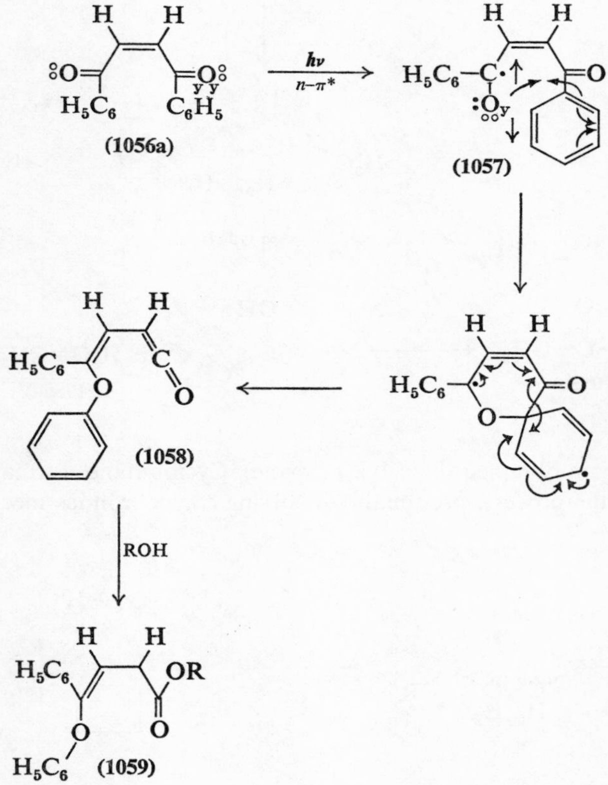

In presence of benzophenone, dibenzoylethylene is excited to its triplet state $T_1(n-\pi^*)$ (1057a). In this state the single electron in the oxygen p_y-orbital causes hydrogen abstraction from the solvent (2-propanol), leading to the reduced product (**1060**).

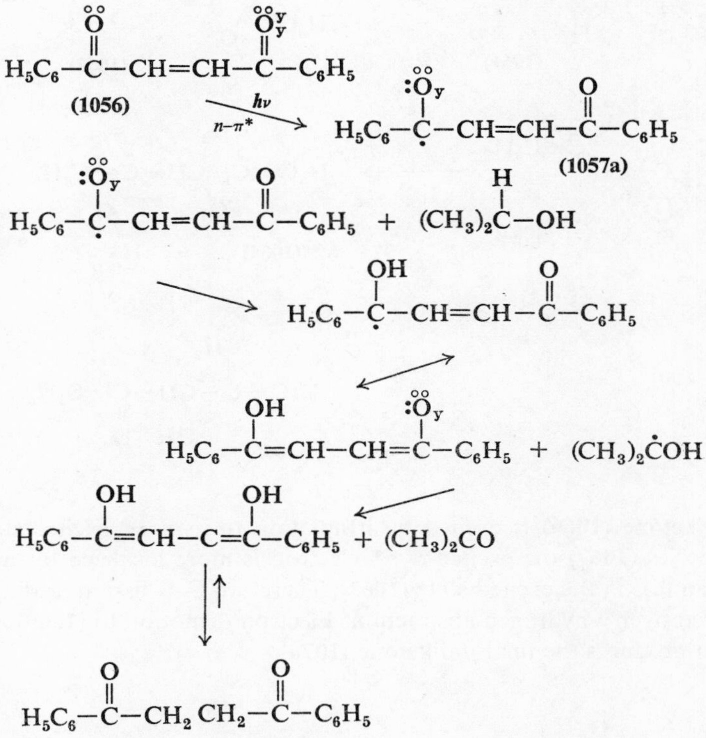

$$H_5C_6-\overset{\overset{\circ\circ}{O}}{\underset{\|}{C}}-CH=CH-\overset{\overset{\overset{\circ\circ}{O}_y^y}{\|}}{C}-C_6H_5 \xrightarrow[n-\pi^*]{h\nu} H_5C_6-\overset{:\overset{\circ\circ}{O}_y}{\underset{|}{\underset{\cdot}{C}}}-CH=CH-\overset{O}{\underset{\|}{C}}-C_6H_5$$

(**1056**)   (**1057a**)

$$H_5C_6-\overset{:\overset{\circ\circ}{O}_y}{\underset{|}{\underset{\cdot}{C}}}-CH=CH-\overset{O}{\underset{\|}{C}}-C_6H_5 + (CH_3)_2\overset{H}{\underset{|}{C}}-OH$$

$$\longrightarrow H_5C_6-\overset{OH}{\underset{|}{\underset{\cdot}{C}}}-CH=CH-\overset{O}{\underset{\|}{C}}-C_6H_5$$

$$H_5C_6-\overset{OH}{\underset{|}{C}}=CH-CH=\overset{:\overset{\circ\circ}{O}_y}{\underset{|}{C}}-C_6H_5 + (CH_3)_2\overset{\cdot}{C}OH$$

$$H_5C_6-\overset{OH}{\underset{|}{C}}=CH-\overset{OH}{\underset{|}{C}}=\overset{}{C}-C_6H_5 + (CH_3)_2CO$$

$$\updownarrow$$

$$H_5C_6-\overset{O}{\underset{\|}{C}}-CH_2\ CH_2-\overset{O}{\underset{\|}{C}}-C_6H_5$$

(**1060**)

P-4. Photorearrangement of epoxy ketones[6]

Irradiation of 2-benzoyl-3,3-dimethyloxirane (**1061**) leads to $n-\pi^*$ excitation. The single electron in the oxygen p_y-orbital (which is not promoted in $n-\pi^*$ excitation) of the excited state (**1062**) is capable of γ-hydrogen abstraction to give (**1063**). The epoxide ring is opened, affording a diradical (**1064**) which after intramolecular disproportion leads to the product (**1065**). Proximity of the benzoyl and the methyl groups seems to be required in this mechanism.

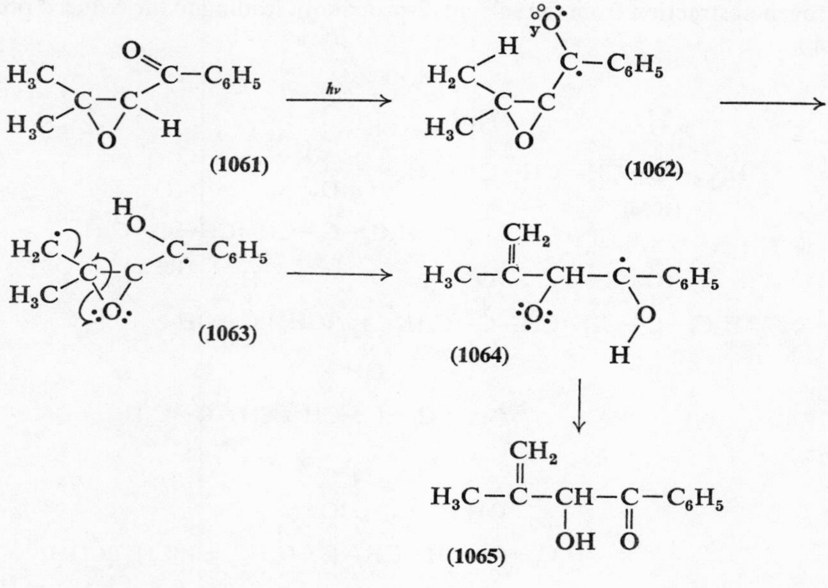

418   Problems in Organic Reaction Mechanisms

$$\underset{(1061)}{\underset{H_3C}{\overset{H_3C}{>}}\!\!\underset{O}{\overset{O}{\underset{|}{C}}}\!\!\underset{H}{\overset{C_6H_5}{\underset{|}{C}}}} \xrightarrow{h\nu} \underset{(1062)}{\underset{H_3C}{\overset{H_2C}{>}}\!\!\underset{O}{\overset{H}{\underset{|}{C}}}\!\!\underset{}{\overset{\cdot\overset{\circ\circ}{\overset{y}{O}}\cdot}{\underset{|}{C}}}\!\!-C_6H_5} \longrightarrow$$

$$\underset{(1063)}{\underset{H_3C}{\overset{H_2\dot C}{>}}\!\!\underset{\overset{\cdot}{\underset{\cdot O\cdot}{}}}{\overset{H}{\underset{|}{C}\!\!-\!\!C}}\!\!\underset{}{\overset{O}{\underset{}{\overset{H}{C}}}\!\!-\!\!C_6H_5}} \longrightarrow \underset{(1064)}{H_3C\!-\!\overset{\overset{CH_2}{\|}}{C}\!-\!CH\!-\!\overset{\cdot}{\underset{\underset{H}{\overset{|}{O}}}{C}}\!-\!C_6H_5}$$

$$\downarrow$$

$$\underset{(1065)}{H_3C\!-\!\overset{\overset{CH_2}{\|}}{C}\!-\!\underset{\underset{OH}{|}}{CH}\!-\!\overset{\overset{}{\|}}{\underset{O}{C}}\!-\!C_6H_5}$$

Epoxy ketone (**1066**) is excited by irradiation to its $n\text{--}\pi^*$ excited state (**1067**); in this species (**1067**) the excited $n\text{--}\pi^*$ electron is more localized in the carbonyl group than it is in the excited state (**1062**). Therefore C–O fission leading to (**1068**) is preferred over γ-hydrogen abstraction. Electron demotion to (**1069**) and methyl migration produces the final β-diketone (**1070**).

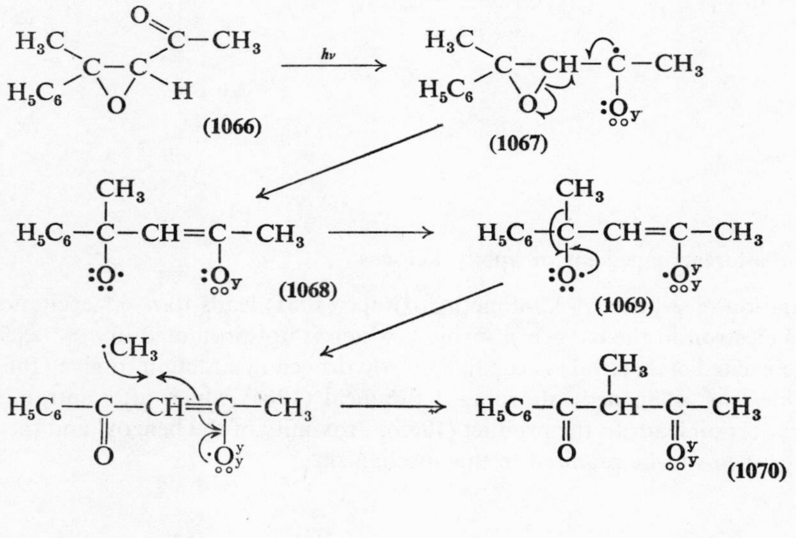

P-5. Photoreactions of stilbene[7]

Irradiation of olefins causes *cis–trans*-isomerization. Depending on the intensity of absorption of the different isomers, a shift of the thermal equilibrium between the isomers takes place. In many cases, the *trans*-isomer shows stronger absorption than the *cis*-isomer, whereupon irradiation increases the concentration of the *cis*-isomer.

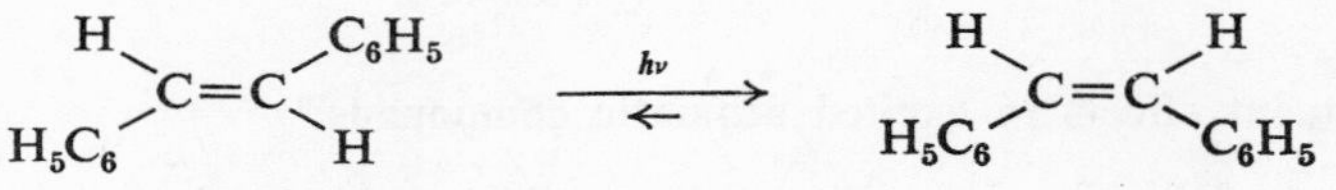

The stilbene system has been thoroughly investigated. The quantum yield of the *trans–cis*-conversion depends on the temperature, but this is not the case for the reverse *cis–trans*-conversion. It was concluded that there is a relatively high activation energy barrier between the S_1 state and the T_1 state of the *trans*-isomer, (which can rearrange to the *cis*-isomer by free rotation), whereas the *cis*-S_1 state is separated from the T_1 state only by a relatively low energy "saddle". For the *trans*-isomer high temperatures favor crossing of the S_1–T_1 transition state.

If very dilute solutions of *trans*-stilbene are irradiated, isomerization first results. In the presence of oxygen, the *cis*-isomer yields phenanthrene in 50% yield *via* dihydrophenanthrene (**1072**)[8]. Since no phenanthrene is obtained on irradiation in the presence of a sensitizer, it is probably the S_1 state that leads to phenanthrene.

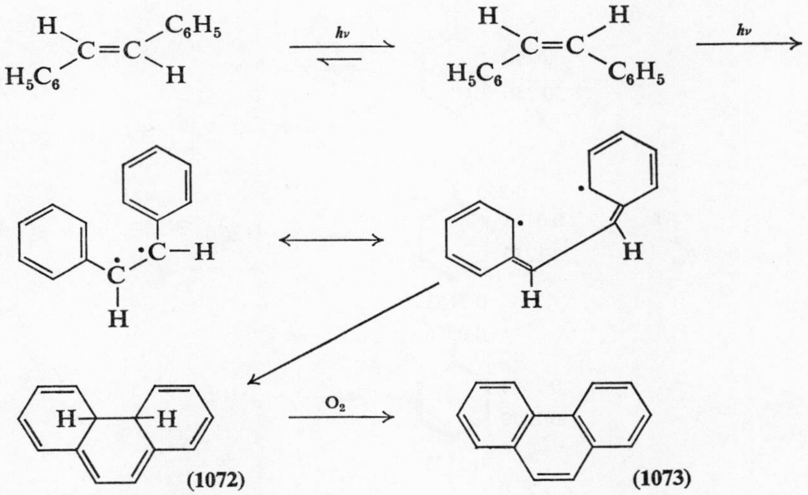

Irradiation of a saturated solution of stilbene in benzene yields tetraphenyl-cyclobutane[9] **(1074)**.

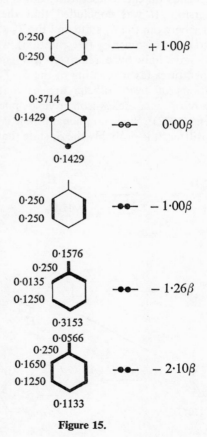

P-6. Substituent effects in excited aromatic compounds[10]

Calculation of the energies of five of the seven benzyl orbitals and of the electron densities at the aromatic C-atoms by the HMO method leads to the result summarized in Figure 15. For the ground state of the benzyl anion, an electron density

Figure 15.

in the *para*-position of $2 \times 0.1133 + 2 \times 0.3153 + 2 \times 0.1429 = 1.1430$ is obtained. The total electron density in the *meta*-position is: $2 \times 0.1250 + 2 \times 0.1250 + 2 \times 0.2500 = 1.000$.

In the excited state, one electron of the orbital 0.00β is located in the next higher orbital with the energy $+1.00\beta$. This leads to the following electron densities in the excited state:

para: $2 \times 0.1133 + 2 \times 0.3153 + 1 \times 0.1429 = 1.000$

meta: $2 \times 0.1250 + 2 \times 0.1250 + 2 \times 0.2500 + 1 \times 0.2500 = 1.2500$

At the *para*-position the excited molecule thus has a lower electron density than in the ground state, but a higher one in the *meta*-position. It can be assumed with some reservation that the $-CH_2^-$ group causes an electron distribution that is characteristic for electron-donating substituents. It was therefore expected that the OCH_3 group in compound (**1076**) would increase the electron density at the *meta*-position of the excited state, which facilitates separation of substituents such as Cl or CH_3COO from the side chain as anion. In contrast, as a consequence of the lower electron density at the *para*-position of compound (**1075**), a radical cleavage is to be expected, this being the normal reaction in photoreactions. These considerations were verified by experiments. Photolysis of (**1075**) leads to products that are typical for a radical reaction, whereas the *m*-isomer (**1076**) gives the alcohol (**1077**) predominantly, this being expected for an ionic reaction.

$H_3CO-\langle\ \rangle-CH_2OAc \xrightarrow[\pi \to \pi^*]{h\nu} H_3CO-\langle\ \rangle-CH_2\cdot\ +\ \cdot OAc$

(**1075**)

$H_3CO-\langle\ \rangle-CH_2OAc \xrightarrow[\pi \to \pi^*]{h\nu} H_3CO-\langle\ \rangle-CH_2^+\ +\ OAc^-$

(**1076**)

$\xrightarrow{H_2O} H_3CO-\langle\ \rangle-CH_2OH$

(**1077**)

The increase in electron density at the *meta*-position has been rationalized by resonance structure (**1078**).

(**1076**) $\xrightarrow{h\nu}$ [structure with CH₂OAc and OCH₃] ⟷ [structure **1078**] ⟶

[structure with CH₂ and ⁺OCH₃] ⟷ [structure with ⁺CH₂ and OCH₃]

Resonance effects between *meta*-substituents in the excited state follow also from the acidities of excited monosubstituted phenols[11]. Analogous results have been obtained in calculations by de Bie and Havinga for phenols[12].

REFERENCES

[1] H. E. Zimmerman, *Science*, **153**, 837 (1966).
[1a] H. E. Zimmerman and D. I. Schuster, *J. Amer. Chem. Soc.*, **84**, 4527 (1962); H. E. Zimmerman and J. S. Swenton, *ibid.*, **86**, 1436 (1964); **89**, 906 (1967).
[2] H. E. Zimmerman, R. G. Lewis and J. J. McCullough, A. Padwa, S. Staley, and M. Semmelhack, *J. Amer. Chem. Soc.*, **88**, 159 (1966).
[3] C. H. Bamford and R. G. W. Norrish, *J. Chem. Soc.*, **1938**, 1544; N. C. Yang and D. D. H. Yang, *J. Amer. Chem. Soc.*, **80**, 2913 (1958); P. Ausloos and R. E. Rebberl, *ibid.*, **83**, 4897 (1961); G. R. McMillan, J. G. Calvert, and J. N. Pitts, Jr., *ibid.*, **86**, 3602 (1964); P. S. Wagner and G. S. Hammond, *ibid.*, **87**, 4009 (1965); T. J. Dougherty, *ibid.*, p. 4011; C. R. Mason, V. Boekelheide and W. A. Noyes Jr., A. Weissberger, *Techniques of Organic Chemistry*, Interscience Publishers, Inc., New York, 1956, Vol. II, p. 257.
[4] N. C. Yang, A. Mordachowitz, and D. D. H. Yang, *J. Amer. Chem. Soc.*, **85**, 1017 (1963).
[5] G. W. Griffin and E. J. O'Connell, *J. Amer. Chem. Soc.*, **84**, 4148 (1962); H. E. Zimmerman, H. G. C. Dürr, R. G. Lewis, and S. Bram, *ibid.*, **84**, 4149 (1962); H. E. Zimmerman, H. G. C. Dürr, R. S. Givens, and R. G. Lewis, *ibid.*, **89**, 1863 (1967).
[6] H. E. Zimmerman, B. R. Cowley, C. Y. Tseng, and J. W. Wilson, *J. Amer. Chem. Soc.*, **86**, 947 (1964).
[7] D. Schulte-Frohlinde, H. Blums, and H. Gusten, *Ann. Chem.* **612**, 138 (1958); *J. Phys. Chem.*, **66**, 2486 (1962); H. Dyck and D. S. McClure, *J. Chem. Phys.*, **36**, 2326 (1962).
[8] P. Hugelshofer, J. Kalvoda, and K. Schaffner, *Helv. Chim. Acta*, **43**, 1322 (1960); W. M. Moore, D. D. Morgan, and F. R. Stermitz, *J. Amer. Chem. Soc.*, **85**, 829 (1963).
[9] M. Paiter and U. Müller, *Monatsh. Chem.*, **79**, 615 (1948).
[10] H. E. Zimmerman and V. R. Sandel, *J. Amer. Chem. Soc.*, **85**, 915, 922 (1963).
[11] E. L. Wehry and L. B. Rogers, *J. Amer. Chem. Soc.*, **87**, 4234 (1965).
[12] D. A. de Bie and E. Havinga, *Tetrahedron*, **21**, 2363 (1965).

10

Miscellaneous Reactions

M-1. Acid-catalyzed rearrangement of *N*-allylaniline[1]

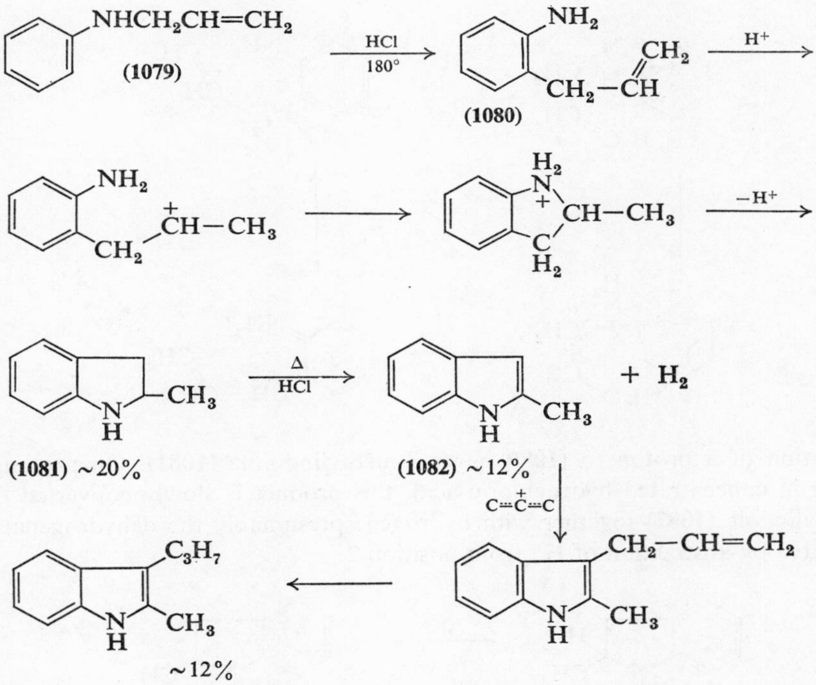

If *N*-allylaniline (**1079**) is heated with ZnCl$_2$ in xylene, *o*-allylaniline (**1080**) is obtained in good yield[2]. It is likely, therefore, that the reaction of (**1079**) with concentrated hydrochloric acid at 180°, which leads to 2-methylindoline, also proceeds through the intermediate (**1080**). From the reaction of *N*-2-butenylaniline with HCl, which leads to 2,3-dimethylindoline, it follows that the acid-catalyzed rearrangement is related to the thermal Claisen rearrangement.

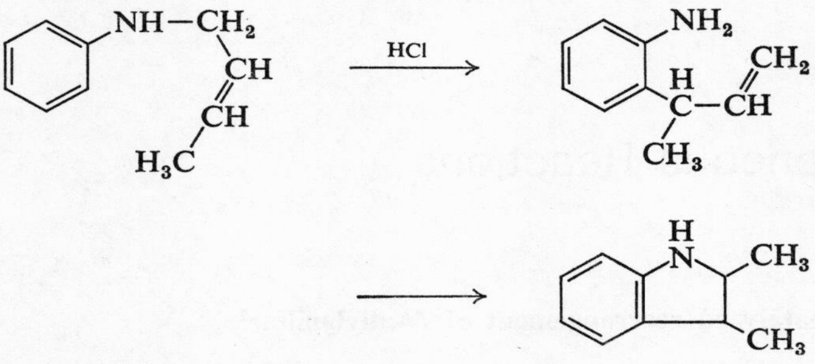

A likely mechanism for the acid-catalyzed rearrangement is:

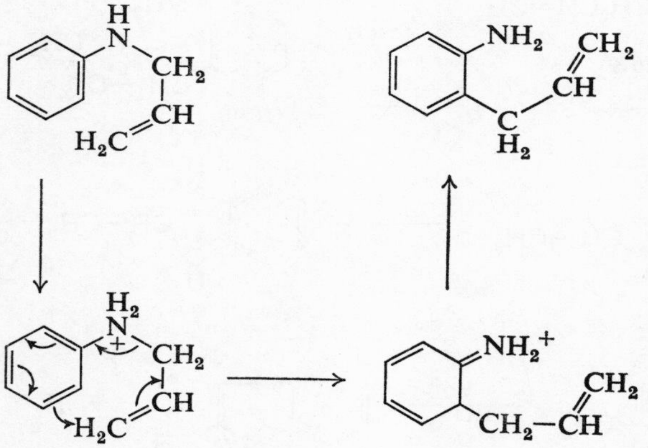

Addition of a proton to (**1080**) gives 2-methylindoline (**1081**). On prolonged heating in concentrated hydrochloric acid, this product is slowly converted into 2-methylindole (**1082**) together with hydrogen; presumably the dehydrogenation is initiated by abstraction of H⁻ from position 2.

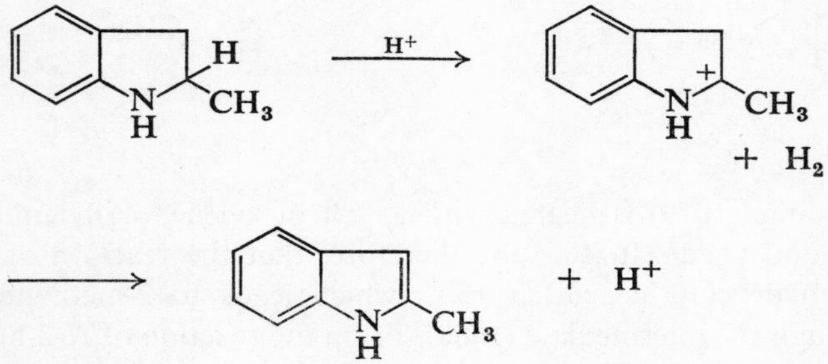

2-Methylindole (**1082**) may alternatively be formed by the following mechanism:

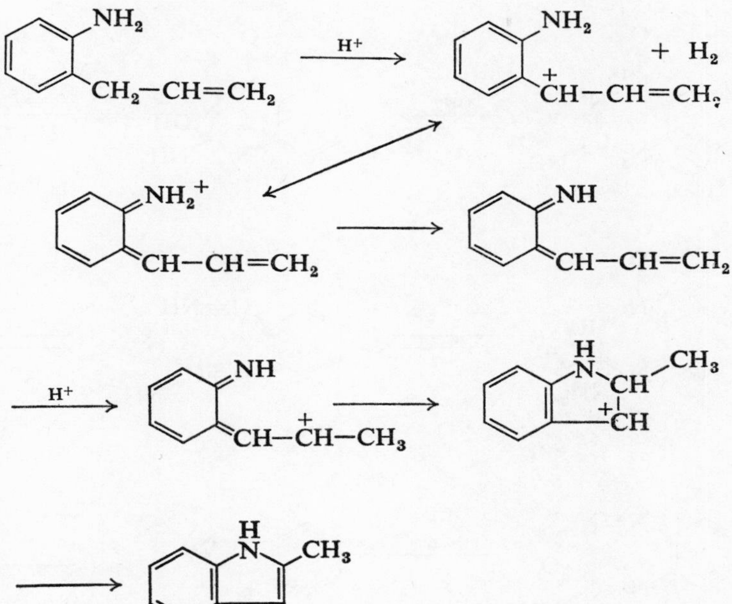

[Reaction scheme showing stepwise conversion of 2-allylaniline to 2-methylindole via protonation, cyclization, and tautomerization steps]

This mechanism is supported by the finding that 2,3-dimethylindole is formed from N-2-butenylaniline, although 2,3-dimethylindoline is stable on prolonged heating with HCl at 240°.

In a side reaction, N-allylaniline (**1079**) may be cleaved by concentrated hydrochloric acid to aniline and an allylic cation. The latter substitutes 2-methylindole (**1082**) in the β-position; the resulting 3-allyl derivative is hydrogenated to 2-methyl-3-propylindole. The hydrogen is furnished by the dehydrogenation of (**1081**).

A related thermal rearrangement of an N-allyl system has been described by Makisumi[3], namely:

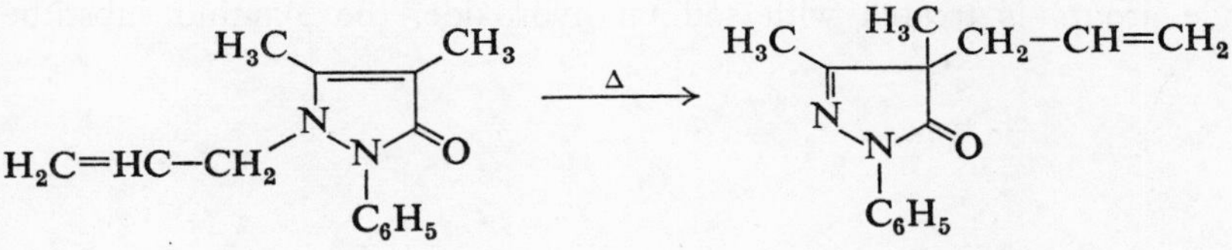

[Scheme: 1-allyl-2-phenyl pyrazolone rearranging thermally (Δ) to the C-allyl isomer]

M-2. Base-catalyzed rearrangement of 3-amino-1,2,4-benzotriazine 1-oxide (1083)[4]

(1083) **(1084)** **(1085)** 16% **(1086)** 37%

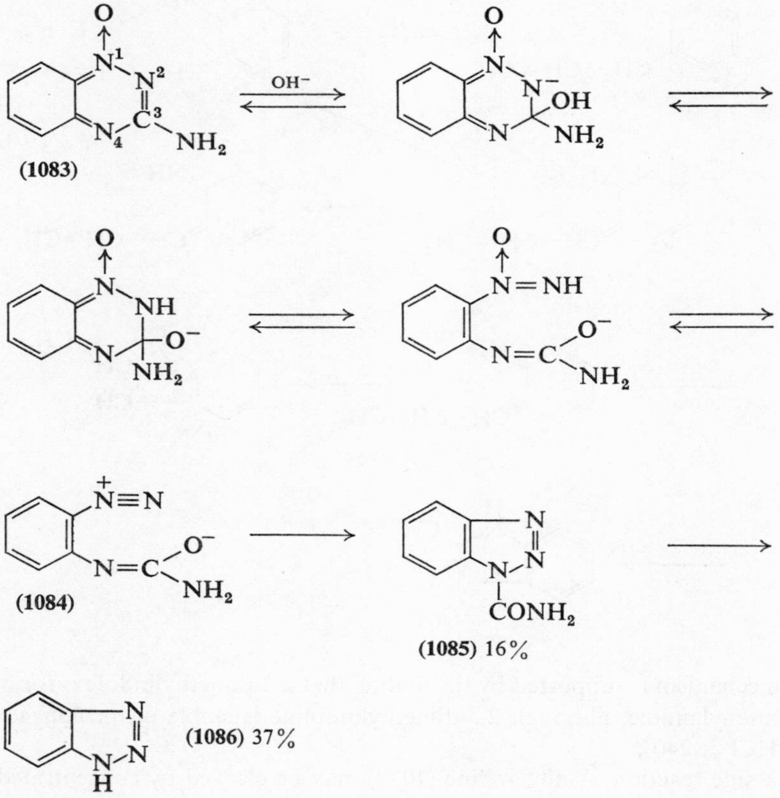

Refluxing the oxide (**1083**) with 10% NaOH solution yields benzotriazole-1-carboxamide (**1085**) in 16% yield and benzotriazole (**1086**) in 37% yield. Since the amide is readily converted into the parent base (**1086**) by treatment with base, it is evident that the latter is formed by way of the former. Attack of the base presumably takes place at the electrophilic C_3. If compound (**1087**), which has no *N*-oxide group, is treated with sodium hydroxide, the 3-amino substituent is

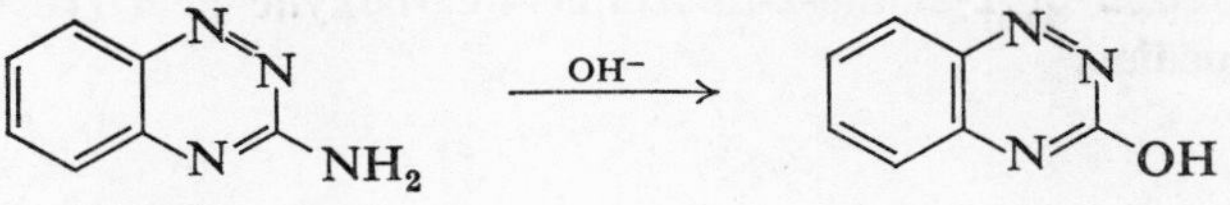

(1087)

replaced by hydroxyl; rearrangement products have not been observed. Other 1,2,4-triazines are also substituted at position 3; for example, 3-chloro-5,6-diphenyl-1,2,4-triazine (1088) yields 3-amino-5,6-diphenyl-1,2,4-triazine on treatment with ammonia[5].

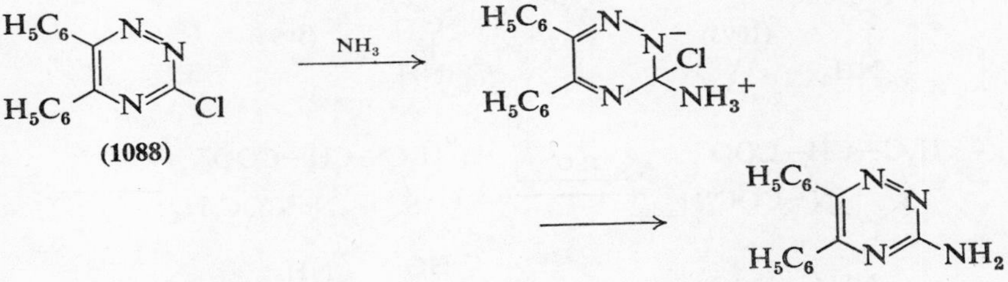

(1088)

For the triazole formation, the N-oxide group is evidently decisive. It is assumed that diazonium compound (1084) is formed through several intermediates; it cyclizes to compound (1085). Supporting evidence is the deep red color observed during the reaction, also the formation of triazole from o-diamines by way of diazonium salts[6].

An analogous rearrangement has been observed in the reaction of the 3-amino 1-oxide (1089) with sodium hydroxide[7].

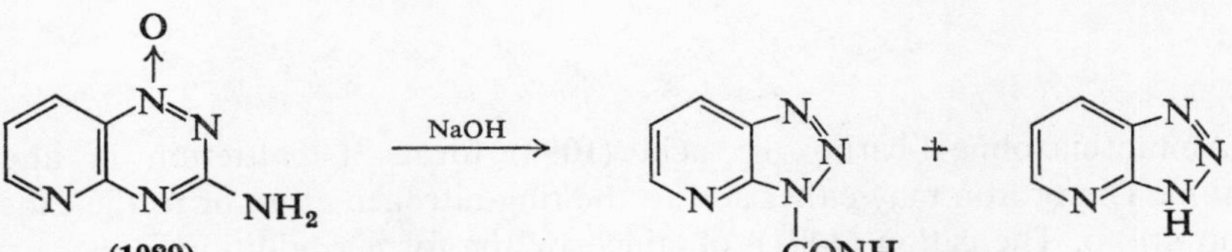

(1089)

M-3. Benzoylation of 2-amino-2-thiazoline-4-carboxylic acid (1090) in a basic medium[8]

$$\begin{array}{c} H_2C-CH-COOH \\ | \quad\quad | \\ S \quad\quad N \\ \diagdown C \diagup \\ | \\ NH_2 \end{array} \quad (1090) \quad \underset{}{\overset{H_2O}{\rightleftharpoons}} \quad \begin{array}{c} H_2C-CH-COO^- \\ | \quad\quad | \\ S \quad\quad ^+NH \\ \diagdown C \diagup \\ | \\ NH_2 \end{array} \quad (1091) \quad \underset{H^+}{\overset{OH^-}{\rightleftharpoons}}$$

$$\begin{array}{c} H_2C-COO^- \\ | \quad\quad | \\ S \quad\quad N \\ \diagdown C \diagup \\ | \\ NH_2 \end{array} \quad (1093) \quad \overset{C_6H_5COCl}{\longrightarrow} \quad \begin{array}{c} H_2C-CH-COO^- \\ | \quad\quad | \\ S \quad\quad ^+N-COC_6H_5 \\ \diagdown C \diagup \\ | \\ NH_2 \end{array} \quad (1094) \quad \overset{-H^+}{\longrightarrow}$$

$$\begin{array}{c} H_2C-CH-COO^- \\ | \quad\quad | \\ S \quad\quad N-COC_6H_5 \\ \diagdown C \diagup \\ \| \\ NH \end{array} \quad \underset{}{\overset{H_2O}{\rightleftharpoons}} \quad \begin{array}{c} H_2C-CH-COO^- \\ | \quad\quad | \\ S \quad\quad N-COC_6H_5 \\ \diagdown C \diagup \\ HO \quad NH_2 \end{array} \longrightarrow$$

$$\begin{array}{c} H_2C-CH-COO^- \\ | \quad\quad | \\ HS \quad N \\ H_5C_6OC \quad CONH_2 \end{array} \quad (1095) \quad \rightleftharpoons \quad \begin{array}{c} H_2C-CH-COO^- \\ | \quad\quad | \\ S \quad N-CONH_2 \\ \diagdown C \diagup \\ HO \quad C_6H_5 \end{array} \longrightarrow$$

$$\begin{array}{c} H_2C-CH-COO^- \\ | \quad\quad | \\ H_5C_6CO-S \quad NHCONH_2 \end{array} \quad (1096)$$

2-Aminothiazoline-4-carboxylic acid (**1090**) forms a zwitterion in aqueous solution. The proton may either add to the ring-nitrogen atom or to the exocyclic amino group. The cation (**1091**), obtained by the former addition is more stable than cation (**1092**); this follows from a consideration of their resonance structures: (**1091**) can be described by two resonance structures of approximately equal energy, whereas compound (**1092**) cannot (an electron decet at the *exo*-nitrogen is impossible). It follows that the ring-nitrogen must be more basic than the *exo*-nitrogen.

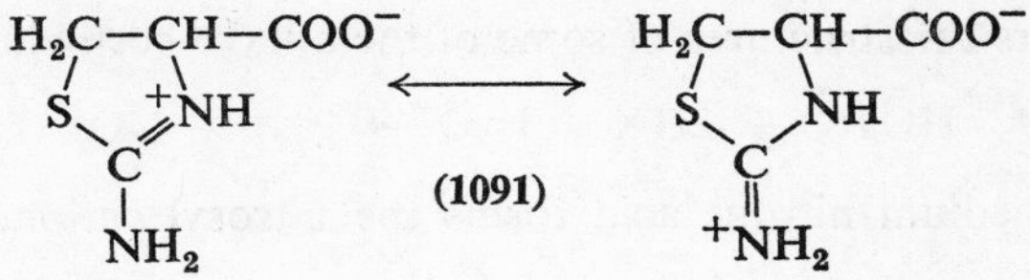

$$\text{(1091)}$$

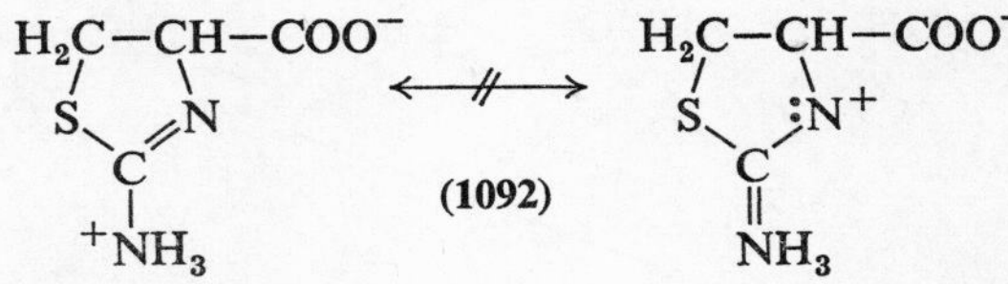

$$\text{(1092)}$$

Protonation at the ring-nitrogen atom is also proved by the NMR spectrum which shows two N–H peaks at 1.30 and 1.89 with the relative areas 1 and 2. If (1093) is treated with benzoyl chloride in a basic medium, reaction takes place preferentially at the more nucleophilic ring-nitrogen. The resulting compound (1094) can be isolated. Treatment with base converts it into the end product (1096).

The conversion of the cyclic compound (1094) into the open-chain amide (1096) via (1095) resembles the acyl-lactone rearrangement, by which α-acyl lactones rearrange in an acidic or a basic medium to cyclic hemiacetals[9].

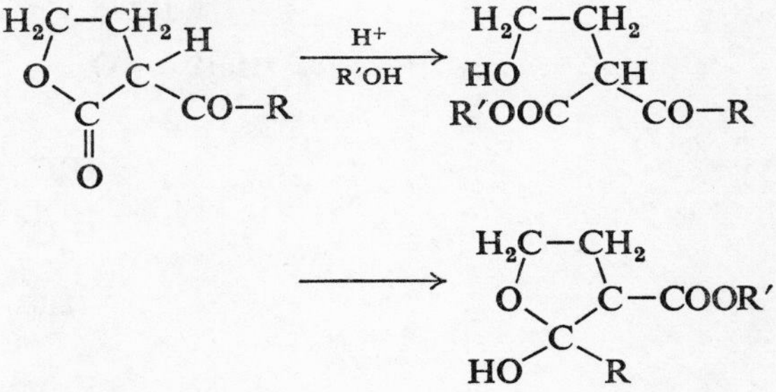

M-4 & M5. Rearrangement of "pernitroso" compounds[10]

Ketones and aldehydes form oximes on reaction with hydroxylamine. Treatment of the ketoximes with nitrous acid regenerates the ketone. With aldoximes, however, the reaction does not take place so readily[11].

The first step of the reaction of ketoximes with HNO_2 is presumably nitrosation of the oxime which leads to a pernitroso compound of the formula $R_2C{=}N_2O_2$. The constitution of the pernitroso compounds has been obscure for a long time, but in recent years the structures of some of them have been elucidated[12].

$$HONO + 2HX \rightarrow H_3O^+ + 2X^- + NO^+$$

In an acidic medium nitrous acid forms the nitrosyl cation, which yields an

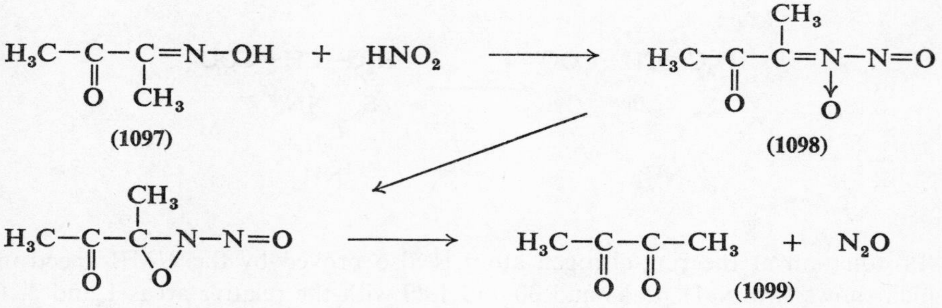

$$\underset{(1097)}{H_3C-\underset{\underset{O}{\|}}{C}-\underset{\underset{CH_3}{|}}{C}=N-OH} + HNO_2 \longrightarrow \underset{(1098)}{H_3C-\underset{\underset{O}{\|}}{C}-\underset{\underset{\downarrow}{\overset{CH_3}{|}}}{C}=N-N=O}$$

$$\underset{(1101\text{ analog})}{H_3C-\underset{\underset{O}{\|}}{C}-\underset{\overset{CH_3}{\underset{O}{\diagdown\!\diagup}}}{C}-N-N=O} \longrightarrow \underset{(1099)}{H_3C-\underset{\underset{O}{\|}}{C}-\underset{\underset{O}{\|}}{C}-CH_3} + N_2O$$

unstable nitroso nitrone (**1098**) (pernitroso compound) by reaction at the oxime-nitrogen atom. That such a compound is formed was proved in the case of mesityl oxide oxime (**1100**); here, the primarily formed nitroso nitrone (**1101**) cyclizes to the stable 3*H*-pyrazole dioxide (**1102**)[13].

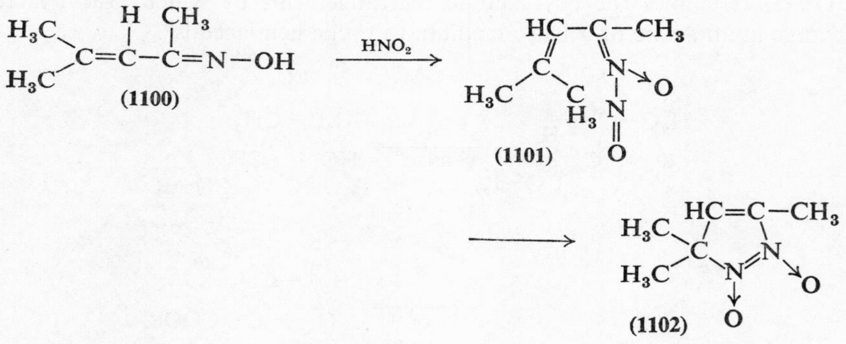

$$\underset{(1100)}{\underset{H_3C}{\overset{H_3C}{\diagdown}}C=\underset{\overset{|}{H}}{C}-\underset{\overset{|}{CH_3}}{C}=N-OH} \xrightarrow{HNO_2} \underset{(1101)}{\text{cyclic intermediate}}$$

$$\longrightarrow \underset{(1102)}{\text{3H-pyrazole dioxide}}$$

$$\underset{R_3C-\underset{\|}{\overset{\overset{N=O}{|}}{N\rightarrow O}}}{\overset{}{C}-R} \longrightarrow \underset{R}{\overset{R_3C}{\diagdown}}C=\overset{+}{N}\underset{\diagdown}{\overset{\diagup O}{\underset{O^-}{N}}}$$

$$\longrightarrow \underset{R}{\overset{R_3C}{\diagdown}}C=N-NO_2 \quad (1103)$$

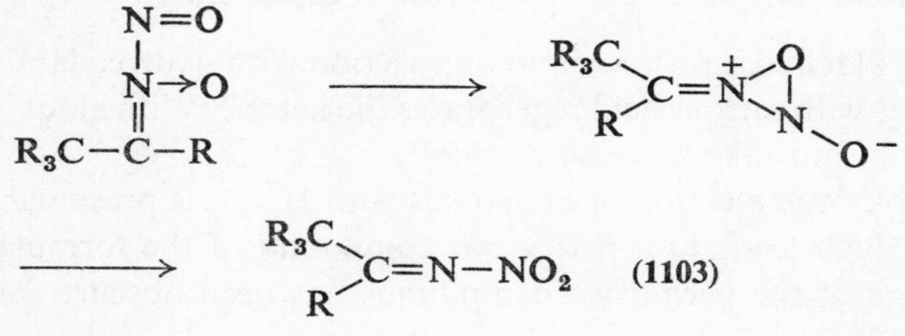

If the original carbonyl-carbon atom carries a tertiary alkyl group having a strong electron-donating effect, the nitroso nitrone rearranges to a nitroimine (**1103**). Nitroimines can be isolated in several cases. They decompose readily to the ketone and N_2O. However, if the substituent is a group with a weak electron-donating effect, the nitroso nitrone rearranges directly to ketone and N_2O. It has been shown with biacetyl monoxime as model that the decomposition takes place through a three-membered ring compound.

$$\begin{array}{c}
\text{>C=N-N}^{18}\text{O} \xrightarrow{c} \text{>C=N} \quad \text{or} \quad \text{>C=N} \\
\qquad\downarrow \qquad\qquad\qquad\qquad |\qquad\qquad\qquad\qquad | \\
\qquad\text{O} \qquad\qquad\qquad\qquad ^{18}\text{O=N}\to\text{O} \quad\qquad \text{O=N}\to^{18}\text{O}
\end{array}$$

The rearrangement could, in principle, occur by path a or b, or through a nitrimine, which could decompose via a four-membered cyclic transition state (path c). Labeling of the nitrosyl-oxygen would lead to completely labeled N_2O by path a. Path b would lead to unlabeled N_2O, and path c would result in 50% of labeled N_2O. Since 89% of labeled N_2O has been found, the pernitroso compound decomposes essentially by path a.

M-6. Base-catalyzed acyloin rearrangement

The acyloin rearrangement takes place when secondary or tertiary α-ketols are heated or treated with acid or base[14].

In the case of secondary ketols a hydrogen atom and in the case of tertiary ketols an alkyl group migrates.

$$\text{RCHOH-COR} \longrightarrow \text{RCO-CHR} \\
\qquad\qquad\qquad\qquad\qquad\qquad |\\
\qquad\qquad\qquad\qquad\qquad\text{OH}$$

$$\begin{array}{c}
\text{R}\\
\text{>C-COR'}\\
\text{R} \;|\\
\quad\text{OH}
\end{array} \longrightarrow \text{RCO-C} \begin{array}{c} \text{R}\\ \text{R'} \\ \end{array}$$

The rearrangement has been thoroughly investigated with steroids containing an α-ketol group. The base-catalyzed rearrangement of the α-ketol (**1104**)[15] is initiated

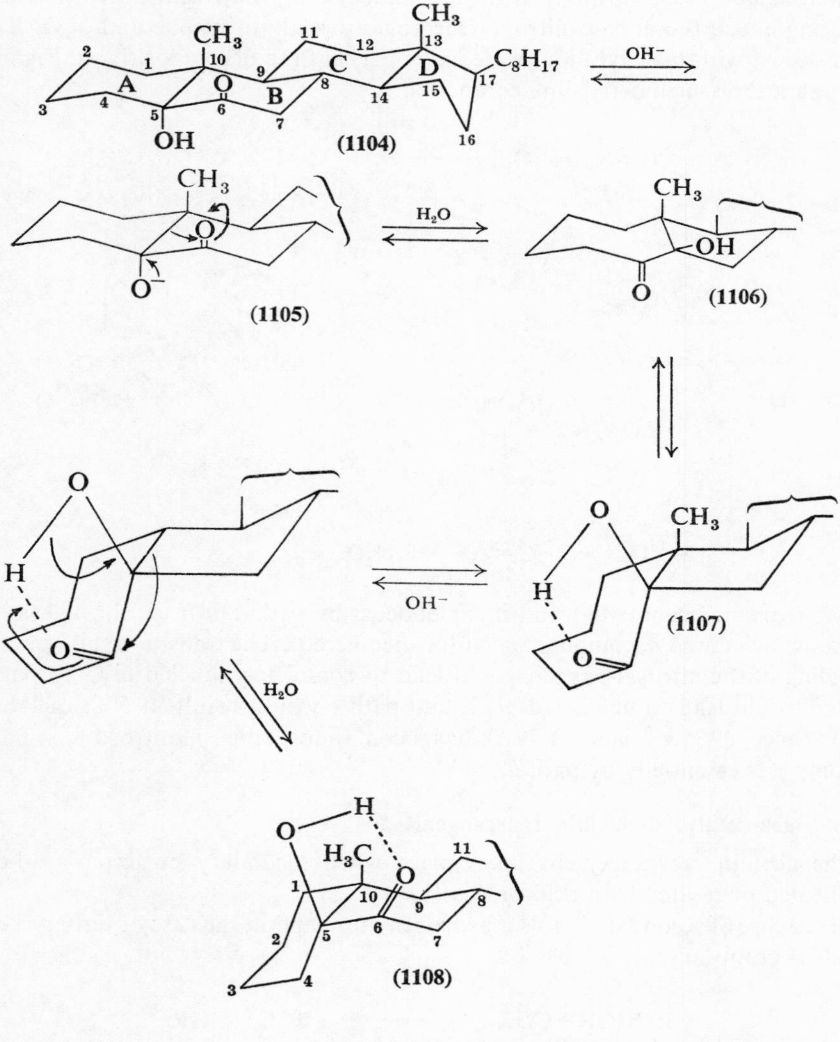

by the formation of the anion (**1105**), which rearranges to the perhydroazulene (**1106**). The latter equilibrates with conformation (**1107**), which is stabilized by hydrogen bonding; and, in the presence of base, (**1107**) rearranges to (**1108**) which is more stable than (**1104**) since hydrogen bonding is possible in the former case.

The compound (**1107**) was synthesized independently. In refluxing CH$_3$OH containing 2% of NaOH it rearranges to (**1108**) within an hour. Therefore it could not be isolated on isomerization of (**1104**).

The acyloin rearrangement is related to the benzilic acid rearrangement, in which an α-diketone is converted into an α-hydroxy carboxylic acid on treatment with base (see also p. 79).

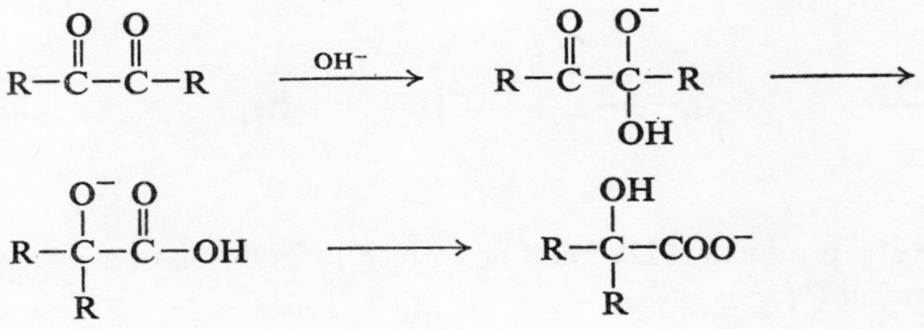

$$R-\underset{O}{\underset{\|}{C}}-\underset{O}{\underset{\|}{C}}-R \xrightarrow{OH^-} R-\underset{O}{\underset{\|}{C}}-\underset{OH}{\underset{|}{\underset{O^-}{\underset{|}{C}}}}-R \longrightarrow$$

$$R-\underset{R}{\underset{|}{\underset{O^-}{\underset{|}{C}}}}-\underset{}{\underset{\|}{\underset{O}{C}}}-OH \longrightarrow R-\underset{R}{\underset{|}{\underset{OH}{\underset{|}{C}}}}-COO^-$$

M-7. Pummerer rearrangement

The rearrangement of β-oxo sulfoxides in aqueous acid involves formal migration of the sulfoxide-oxygen to the α-carbon; it is often called, after its discoverer, the Pummerer rearrangement[16]. It occurs under very mild conditions in a neutral, acidic, or basic medium depending on the structure of the β-oxo sulfoxide. For example, compound (**1108**) rearranges even on recrystallization from absolute ethanol[17].

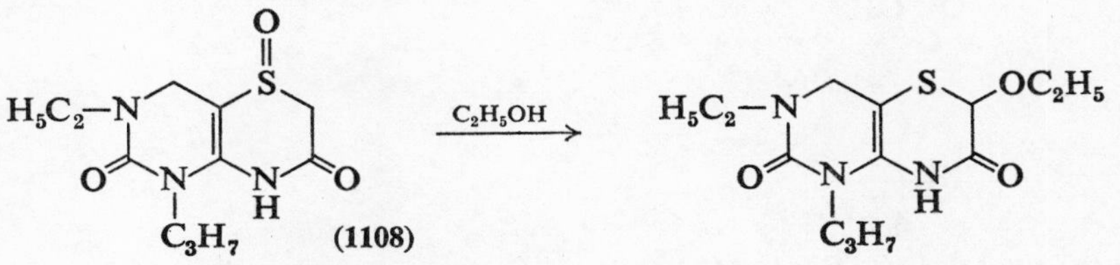

The rearrangement of methylsulfinylacetophenone (**1109**) which results in the formation of the methyl hemithioacetal (**1112**) of phenylglyoxal takes place at room temperature in aqueous hydrochloric acid[18]. Presumably the initial step is protonation of the sulfoxide-oxygen[19].

Migration of OH in (**1111a**) would lead directly to the product (**1112**) (mechanism *a*).

$$H_5C_6-\underset{O}{\overset{\|}{C}}-CH=\underset{\cdot\cdot}{S}-CH_3 \quad \xrightarrow{} \quad H_5C_6-\underset{O}{\overset{\|}{C}}-\underset{H}{\overset{H}{\underset{|}{C}}}-\overset{/+\backslash}{\underset{\cdot\cdot}{S}}-CH_3$$
(**1111a**)

$$\xrightarrow{\text{Mech. } a} \quad H_5C_6-\underset{O}{\overset{\|}{C}}-\underset{|}{\overset{OH}{\underset{|}{CH}}}-SCH_3 \quad (\mathbf{1112})$$

Alternatively, product (**1112**) could be formed by addition of a solvent molecule as in mechanism (*b*).

$$H_5C_6-\underset{O}{\overset{\|}{C}}-CH_2-\underset{\downarrow}{\overset{\cdot\cdot}{S^+}}-CH_3 \xrightarrow[\text{HCl}]{\text{Aq.}} H_5C_6-\underset{O}{\overset{\|}{C}}-CH_2-\underset{\cdot\cdot}{\overset{OH}{\underset{|}{S^+}}}-CH_3$$

(**1109**)   O⁻   −H⁺   (**1110**)

$$H_5C_6-\underset{O}{\overset{\|}{C}}-\bar{C}H-\underset{\cdot\cdot}{S^+}-CH_3 \xrightarrow{H_2O} H_5C_6-\underset{O}{\overset{\|}{C}}-CH=\underset{\cdot\cdot}{\overset{OH}{S}}-CH_3$$

(**1111**)   Mech. *b*   (**1111a**)   O—H   H

$$H_5C_6-\underset{O}{\overset{\|}{C}}-\underset{OH}{\overset{|}{CH}}-\underset{\cdot\cdot}{S}-CH_3 \quad (\mathbf{1112})$$

For the model compound (**1113**) it has recently been shown[20] that the reaction very probably takes place by addition of a solvent molecule. 2-Methylsulfinyl-1,3-indanedione (**1113**) forms a product (**1114**) in the presence of HCl. In ethanol, however, a product (**1115**) is obtained. The analogous compound (**1116**) could not be converted into (**1114**) by HCl and so is not an intermediate, as would be expected if mechanism (*a*) applied. (**1114**) is thus formed by direct attack of Cl⁻ on the C=S double bond.

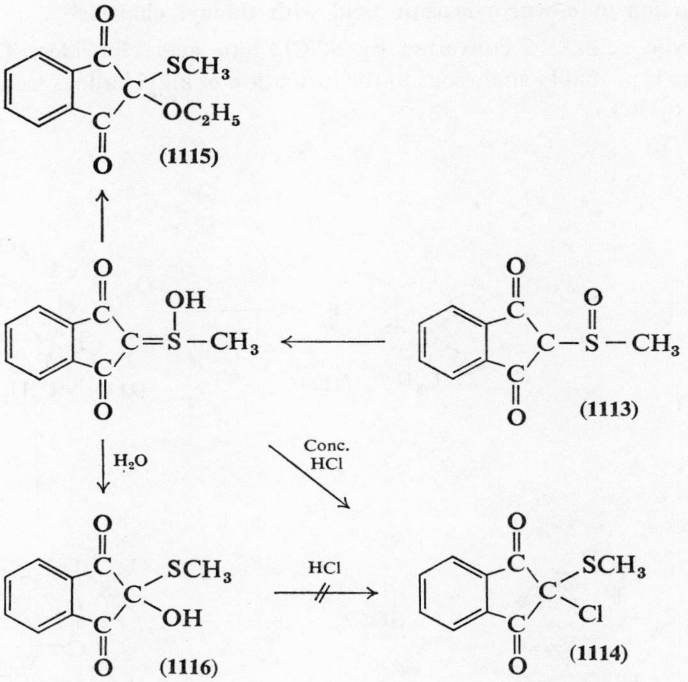

[Structures 1115, 1113, 1116, 1114 shown in scheme]

A mechanism proposed by Kenney, Walsh, and Davenport[21] involves the replacement of OH in (**1117**) by OAc. Since the OH group in (**1116**) has been shown not to be replaceable by other nucleophiles, this mechanism is not very likely (for related reactions see reference 22).

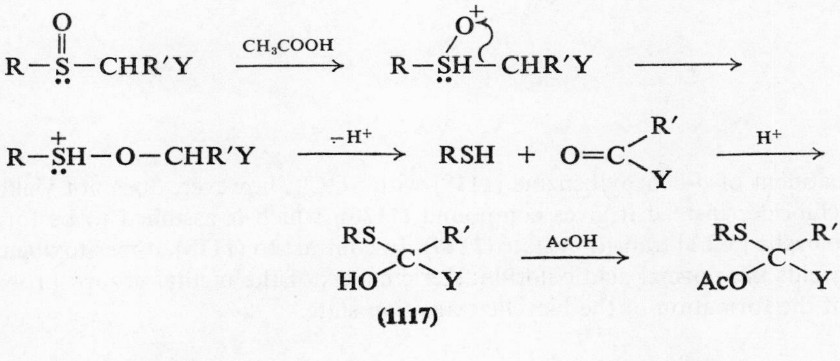

$$R-\overset{O}{\underset{..}{\overset{\|}{S}}}-CHR'Y \xrightarrow{CH_3COOH} R-\overset{\overset{+}{O}}{\underset{..}{S}H}-CHR'Y \longrightarrow$$

$$R-\overset{+}{\underset{..}{S}}H-O-CHR'Y \xrightarrow{-H^+} RSH + O=C\overset{R'}{\underset{Y}{}} \xrightarrow{H^+}$$

$$\underset{HO}{\overset{RS}{}}C\overset{R'}{\underset{Y}{}} \xrightarrow{AcOH} \underset{AcO}{\overset{RS}{}}C\overset{R'}{\underset{Y}{}}$$

(**1117**)

M-8. Reaction of *o*-benzoylbenzoic acid with thionyl chloride[23]

Carboxylic acids are converted by $SOCl_2$ into acid chlorides. The reaction mechanism is probably analogous to the formation of alkyl halides from alcohols[24]. (See also p. 308.)

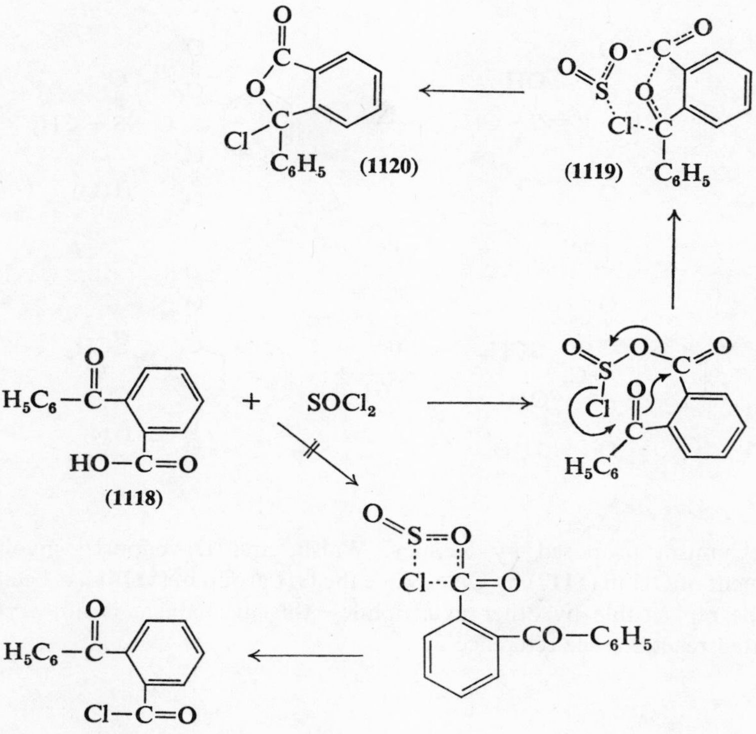

Treatment of *o*-benzoylbenzoic (**1118**) with $SOCl_2$, however, does not yield the acid chloride; instead it gives compound (**1120**), which is assumed to be formed by a bicyclic [3.2.1] transition state (**1119**). In contrast to (**1118**), *o*-mesitoylbenzoic acid yields the normal acid chloride: steric effects of the methyl groups probably inhibit the formation of the bicyclic transition state.

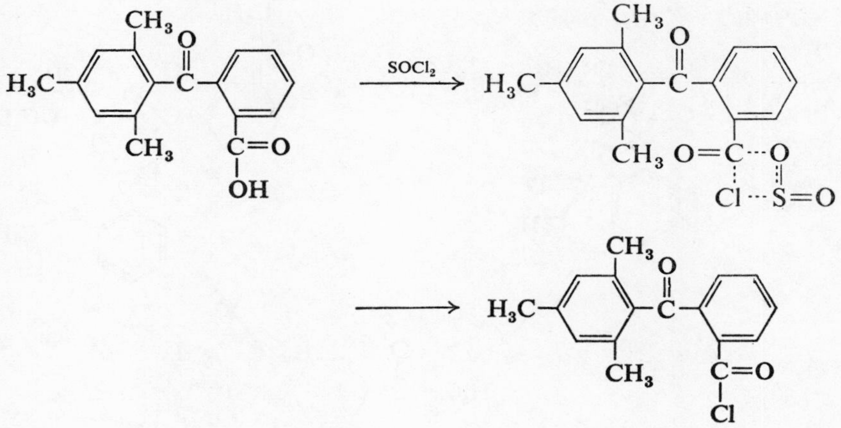

When *o*-benzoylbenzoic acid is treated with ethyl propiolate, the 2:1 adduct (**1121**) rearranges, probably by the mechanism described above, this evidently being preferred to a six- or four-membered cyclic transition state, both of which would lead to an anhydride (**1122**).

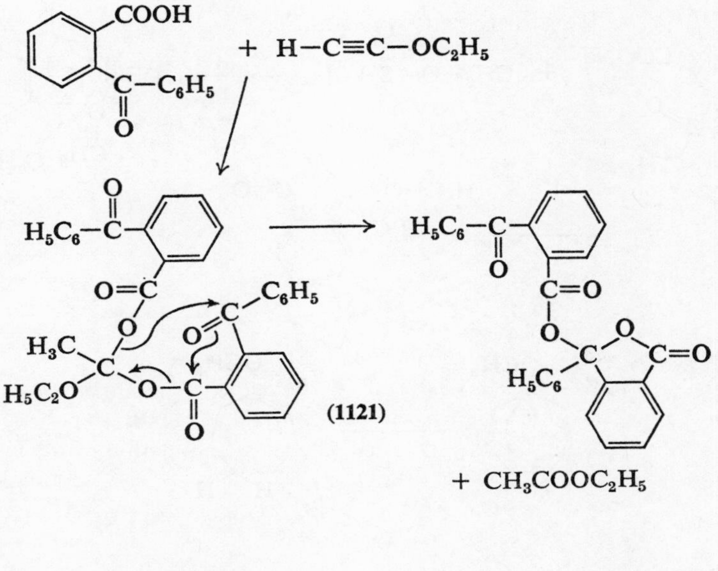

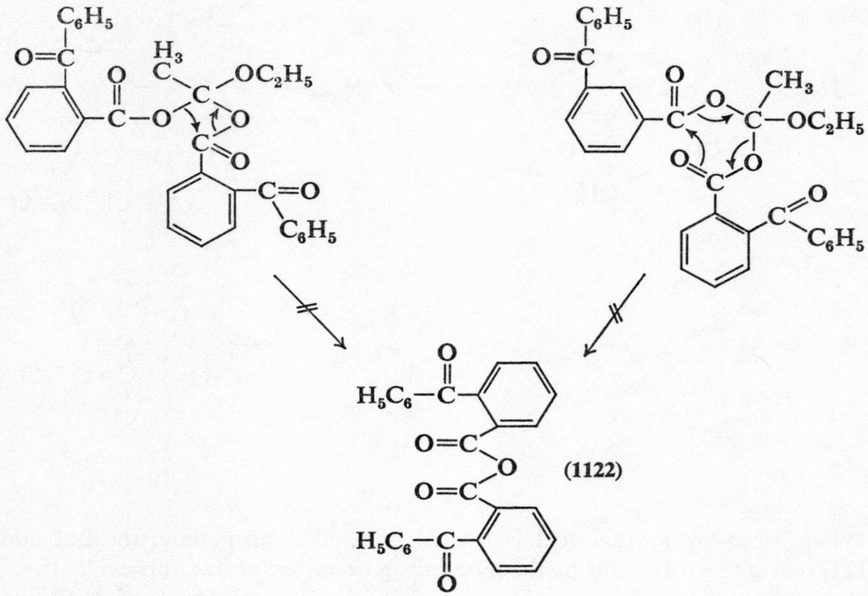

Further examples of this mechanism are the reaction of sodium o-benzoylbenzoate with methyl chlorosulfite and the transannular hydride shift of 2,4,6-cycloheptatrienyl methyl ether[25]

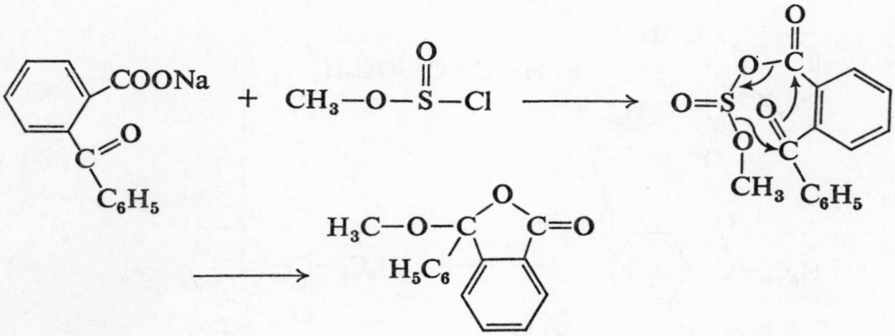

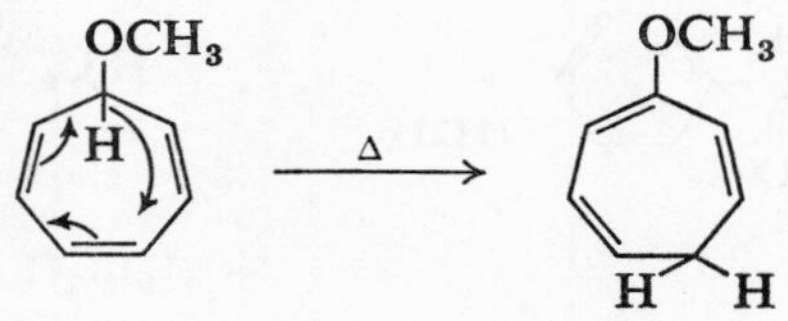

M-9 Rearrangement of o-glycylsalicylamide[26]

Rearrangement of this amide proceeds by a path similar to that discussed in

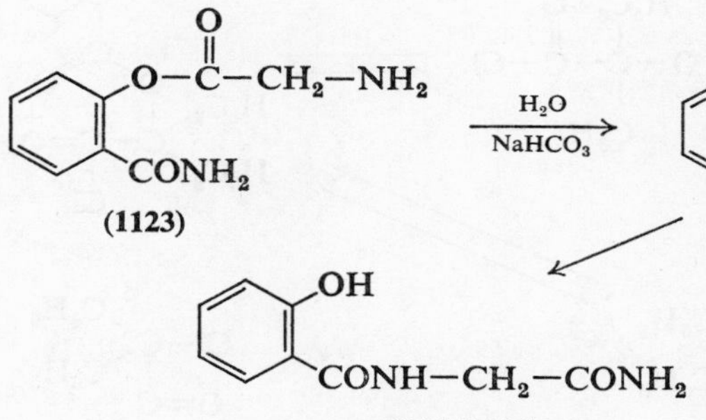

[Structures: (1123) o-glycylsalicylamide → intermediate → (1124) salicyloylglycinamide, 48%, via H₂O/NaHCO₃]

the preceding section. A [3.1.1] transition state has been proposed[27] for the rearrangement of compound (1125).

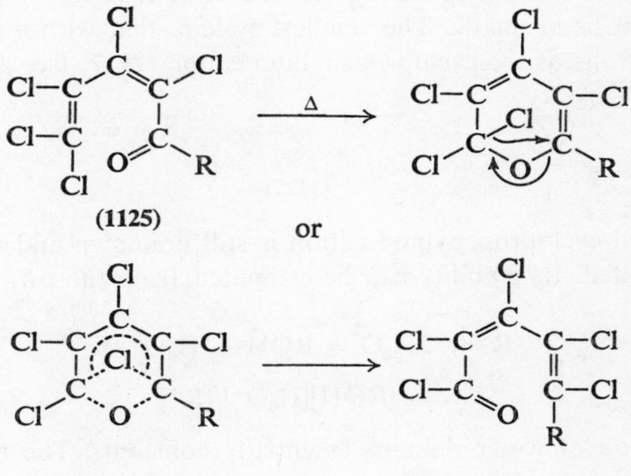

[Structures showing rearrangement of (1125) with Δ, and alternative "or" pathway]

The rearrangement of **(1126)** probably takes place through a [3.1.0] transition state[28].

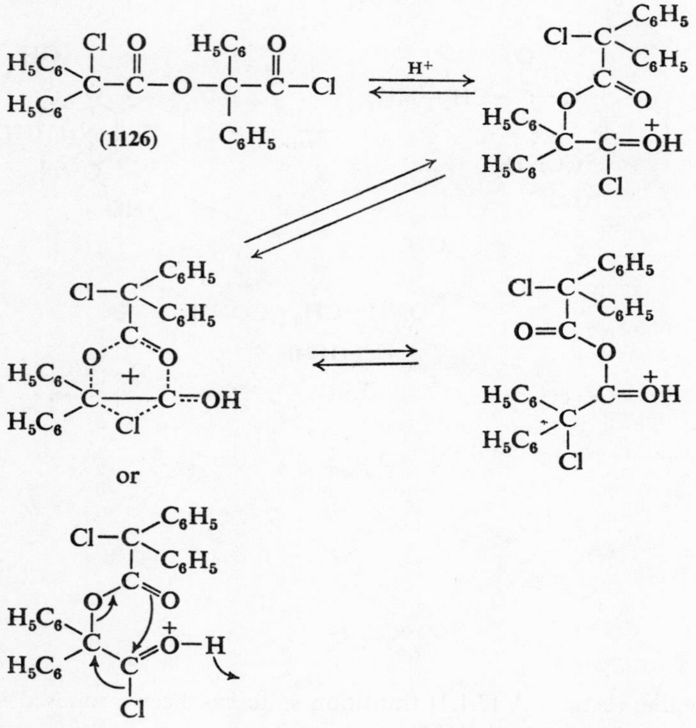

M-10. Cyclopropenylium cations

Cyclic compounds containing conjugated double bonds and $(4n + 2)$ π-electrons are considered to be aromatic. The smallest system, that with $n = 0$ and consequently 2 π-electrons, is the cyclopropenylium cation **(1127)** (see also p. 242):

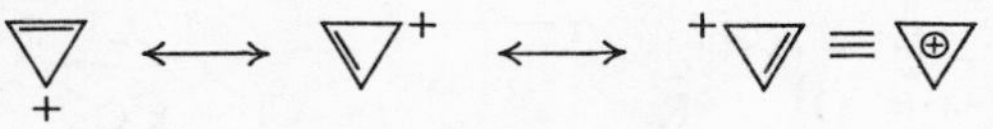

The unsubstituted cyclopropenylium cation is still unknown and may be too unstable to be isolated. Its stability can be estimated from the pK_{R^+} of some substituted cations[29].

$$R^+ + 2H_2O = ROH + H_3O^+ \qquad (a)$$

$$K_{R^+} = [ROH][H_3O^+]/[R^+]$$

(The concentration of water remains essentially constant.) The more stable the cation, the more the equilibrium lies to the left-hand side in equation (*a*). Thus K_{R^+} decreases with increasing stability of R^+, whereas pK_{R^+} increases.

Table 16. Equilibrium constants for some cyclopropenylium ions [equation (*a*)]

Ion	pK_{R^+}	Medium
Dipropylcyclopropenylium (perchlorate)[30]	2.7	50% Aq. acetonitrile
Tripropylcyclopropenylium (perchlorate)	7.2	50% Aq. acetonitrile
Diphenylcyclopropenylium (bromide)	−0.67	23% Aq. ethanol
Diphenylpropylcyclopropenylium (fluoroborate)	3.8	23% Aq. ethanol

Table 16 shows that pK_{R^+} of the cation changes by 4.5 units per propyl group. Extrapolation leads to a value of −6.3 for the unsubstituted cation. The cation can thus be compared with the triphenylmethyl cation, which has $pK_{R^+} = -6.6$ and is stable only in very strong acids.

The Table shows also that the tripropyl cation is strikingly stable, even in water (see p. 242).

The cyclopropenylium cations and the cyclopropenes undergo very interesting reactions. An example is the reaction of the triphenylcyclopropenylium cation with phenyldiazomethane[31].

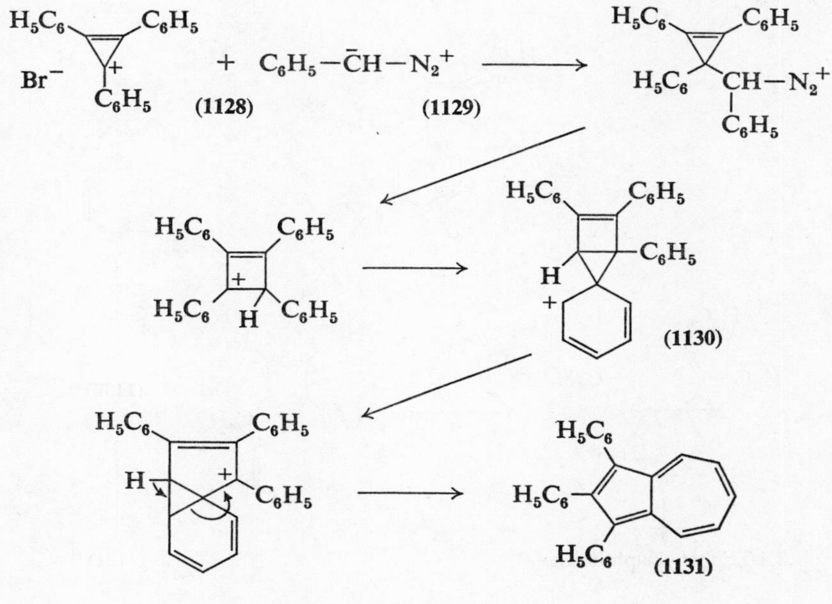

With an excess of phenyldiazomethane pentaphenylcyclopentadiene is obtained.

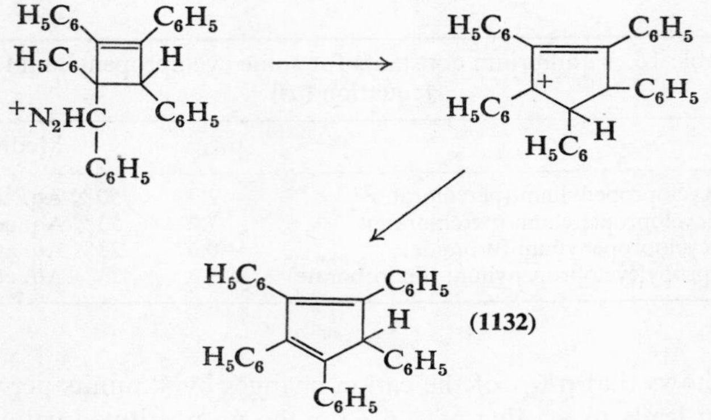

An analogous reaction takes place when compound (1133) is treated with acetyl chloride[32].

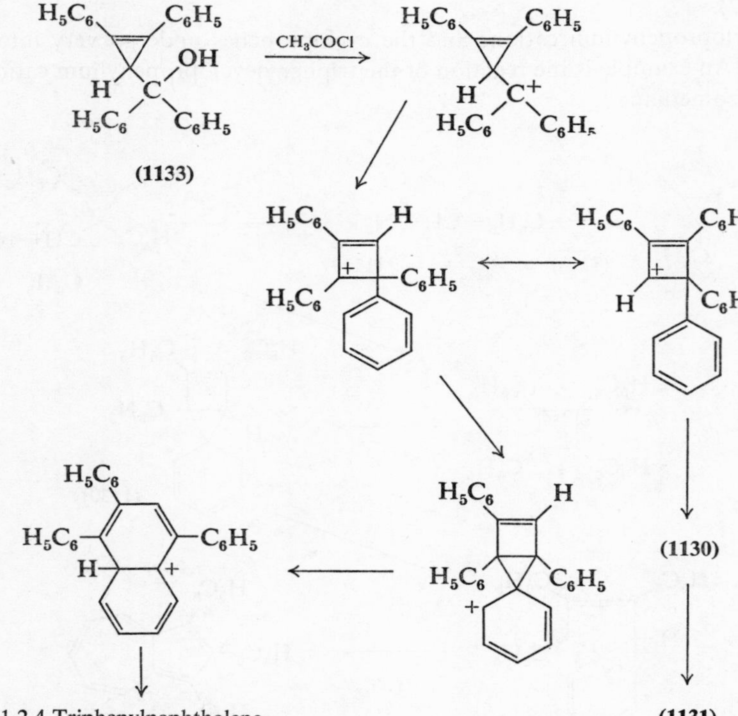

1,2,4-Triphenylnaphthalene

REFERENCES

[1] J. E. Hyre and A. R. Bader, *J. Amer. Chem. Soc.*, **80**, 437 (1958); A. R. Bader, R. J. Bridgwater, and P. R. Freeman, *ibid.*, **83**, 3319 (1961).
[2] C. D. Hurd and W. W. Jenkins, *J. Org. Chem.*, **22**, 1418 (1957).
[3] Y. Makisumi, *Tetrahedron Letters*, **1966**, 6413.
[4] J. A. Carbon, *J. Org. Chem.*, **27**, 185 (1962).
[5] J. P. Horwitz, in *Heterocyclic Compounds*, Vol. 7, ed. R. C. Elderfield, J. Wiley and Sons, New York, London, 1961, p. 763.
[6] F. R. Benson and W. L. Savell, *Chem. Rev.*, **46**, 1 (1950).
[7] J. A. Carbon and S. H. Tabata, *J. Org. Chem.*, **27**, 2504 (1962).
[8] O. Gawron, J. Fernando, J. Keil, and T. J. Weismann, *J. Org. Chem.*, **27**, 3117 (1962).
[9] F. Korte and K. H. Büchel, *Neuere Methoden der präparativen organischen Chemie*, Verlag Chemie, Weinheim, 1961, Vol. 3, p. 157.
[10] Th. Wieland and D. Grimm, *Chem. Ber.*, **96**, 275 (1963).
[11] J. H. Boyer and H. Alul, *J. Amer. Chem. Soc.*, **81**, 4237 (1959); L. Horner, L. Hockenberger, and W. Kirmse, *Chem. Ber.*, **94**, 290 (1961).
[12] J. P. Freeman, *J. Org. Chem.*, **26**, 4190 (1961); H. A. Stansbury, Jr., *J. Amer. Chem. Soc.*, **81**, 4885 (1959).
[13] J. P. Freeman, *J. Org. Chem.*, **27**, 1309 (1962).
[14] N. L. Wendler, in *Molecular Rearrangements*, ed. P. de Mayo, Vol. 2, p. 1114, Interscience Publ., New York, 1964; J. Elphimoff-Felkin, *Bull. Soc. Chim. France*, **1956**, 1845.
[15] Y. Mazur and M. Nussin, *Tetrahedron Letters*, **1961**, 817.
[16] R. Pummerer, *Chem. Ber.*, **42**, 2282 (1909); **43**, 1401 (1910).
[17] E. F. Schroeder and R. M. Dodson, *J. Amer. Chem. Soc.*, **84**, 1904 (1962).
[18] H. D. Becker, G. J. Mikol, and G. A. Russell, *J. Amer. Chem. Soc.*, **85**, 3410 (1963).
[19] R. G. Laughlin, *J. Org. Chem.*, **25**, 864 (1960).
[20] H. D. Becker, *J. Org. Chem.*, **29**, 1358 (1964).
[21] W. J. Kenney, J. A. Walsh, and D. A. Davenport, *J. Amer. Chem. Soc.*, **83**, 4019 (1961).
[22] H. Böhme, H. Fischer, and R. Frank, *Ann. Chem.* **563**, 54 (1949); F. G. Bordwell and B. M. Pitt, *J. Amer. Chem. Soc.*, **77**, 572 (1955).
[23] M. S. Newman and D. Courduvelis, *J. Amer. Chem. Soc.*, **86**, 2942 (1964); **88**, 781 (1966); M. S. Newman, N. Gill, and B. Darré, *J. Org. Chem.*, **31**, 2713 (1966).
[24] E. S. Gould, *Mechanism and Structure in Organic Chemistry*, Henry Holt, New York, 1959, pp. 267, 295.
[25] E. Weth and A. S. Dreiding, *Proc. Chem. Soc.*, **1964**, 59.
[26] M. Brenner and J. P. Zimmermann, *Helv. Chim. Acta*, **40**, 1933 (1957); **41**, 467 (1958); M. Brenner and J. Wehrmüller, *ibid.*, **40**, 2374 (1957); M. Brenner, H. Dahn, R. Menasse, and J. Rosenthaler, *ibid.*, **42**, 2249 (1959).
[27] A. Roedig, *Angew. Chem.*, **76**, 276 (1964).
[28] H. H. Wasserman and P. S. Wharton, *J. Amer. Chem. Soc.*, **82**, 3457 (1960).
[29] R. Breslow, H. Höver, and H. W. Chang, *J. Amer. Chem. Soc.*, **84**, 3168 (1962).
[30] R. Breslow and H. Höver, *J. Amer. Chem. Soc.*, **82**, 2644 (1960).
[31] R. Breslow and M. Mitchell, *Molecular Rearrangements*, Vol. 1, p. 276, ed. P. de Mayo, Interscience, New York, 1963.
[32] R. Breslow and M. Battiste, *J. Amer. Chem. Soc.*, **82**, 3626 (1960).

Index

Absorption of light, 405
Abstraction of γ-hydrogen, 50, 51
Acetaldehyde, 2-(2,2,3-trimethyl-3-cyclopentenyl)-, 18, 167
Acetaldehyde dimethyl acetal, (n-butylthio)-, 30, 248
Acetals, dithio-, 30, 248
   formation and hydrolysis of, 247
   rate of hydrolysis of, 247
Acetamide, α-halogeno-, 37
   trichloro-, 37
Acetanilide, 2-chloro-2,2-diphenyl-, 44
Acetic acid ester, trichloro-, 14, 138, 139
Acetic anhydride, reaction with pyridine oxide, 46
Acetoacetate, ethyl, 42
Acetone, hexafluoro-, 17, 50, 161
Acetonitrile, 11, 118, 119
Acetophenone, 232, 233
   methylsulfinyl-, 433
2-Acetoxypyridine, 382
1-Acetoxypyridinium acetate, 383
Acetyl chloride, 442
2-Acetylcyclopentanone, 300
α-Acetylfuran, 72
Acetyl nitrate, 376
2-Acetyl-5-phenylcyclopentanone, 300
Acetylpyridinium cation, 349
Acetylene, *tert*-butylfluoro-, 20, 182
   dichloro-, 182
   diphenyl-, 8, 94
   ethoxy-, 17, 161
Acetylenedicarboxylate, dimethyl, 7, 89, 91
Acetylenes, cyclodimerization of, 17, 162
   electrophilic *vs.* nucleophilic addition to, 7, 88
   phenyl- and vinyl- in Mannich synthesis, 162
   trimerization of, 182

Acid catalysis, in additions to the C=O bond, 95
   general, 95
   specific, 95
Acidic media, strongly, non-aqueous, 117
Acidity, of acetylenic hydrogen, 88
   of hydrocarbons, effect of bond angles on, 329
   of hydrogen bound to carbon, 276, 328
   kinetic *vs.* thermodynamic, of olefins, 328
   of saturated hydrocarbons, 328
   of strained hydrocarbons, 276, 329
Acidity scale of hydrocarbons, 331
Acids, strong, 242, 244
Acrylic ester, 161
Acrylonitrile, 4, 70
   cyclodimerization of, 153
Activation, photochemical, 405
Activation enthalpy, 237
Activation entropy, 106, 165, 170, 237
Acylation, agents, 347
   of the C=C bond, 3, 63
   of enamines, 70, 75
   of enolates, 348
   of olefins, 3, 63
   of saturated hydrocarbons, 64
Acyl isocyanates, 307
α-Acyl ketone, 98
Acyl lactones, rearrangement of, 429
Acyloin rearrangement, 431
Adamantan-1-amine, 3-bromo-N,N-dimethyl-, 27, 226
Additions, 3, 59
   to the C=C bond, 59
   of aluminum alkyls, 6, 86
   of aluminum hydrides, 61, 86
   electrophilic, 3, 59
      of acetylenes to enamines, 4, 74
      of acrylonitrile to enamines, 4, 70
      of acyl halides, 3, 63

Additions (*Continued*)
  of bromine, 3, 4, 7, 59, 61, 67, 88
  of chlorine, 59, 61, 67
  of deuterobromide, 3, 60
  of halogens to butadiene, 4, 67
  of hydrogen halides, 59
  of hydrogen iodide to enols, 5, 75
  of ozone, 4, 64
  homolytic, 388, 391
  nucleophilic, 5, 78
    of the hydride anion, 6, 81
    to immonium salts, 70
    of the hydroxyl anion, 5, 78
    of LiAlH$_4$, 6, 81
    of α-metallated aromatic compounds, 333
    of phosphines, 99
    of sodium hydrazide, 6, 84, 85
  to the C≡C bond, 7, 88
    alcohols, 88
    aluminum alkyls, 94
    aluminum aryls, 8, 94
    amines, 88
    bromine, 7, 88
    hydrogen halides, 88
    malononitrile, 7, 89
    phenyllithium, 94
    poly-ynes, 89
    ylides, 7, 91
  to the C≡N bond, 11, 117
    tert-alcohols, 11, 117
    carbonium ions, 11, 117, 119
    nitrilium salts, 11, 120
    phenethyl chloride to benzonitrile, 11, 119
  to the C=O bond, 95, 110
    electrophilic, 10, 110
      of phosphorus pentachloride, 10, 110
    nucleophilic, 95
      of diazoethane, 9, 103
      of hydrogen iodide, 5, 75
      of hydrogen peroxide, 8, 95
      of lithium aluminum hydride, 6, 8, 81, 98
      of phospholin 1-oxide, 9, 105
      of phosphorus tris(dimethylamide), 8, 99
      of trimethyl phosphite, 9, 101
  to the C=S bond, 12, 121
  1,1-cyclo-, 13, 130

  of OH$^+$, 130
  of Br$^+$, 61
  of carbenoids, 13, 14, 15, 16, 134, 135, 136, 137, 138, 140
  1,2-cyclo-, 16, 153
  1,3-cyclo-, 17, 165
  1,4-cyclo-, 18, 170
  dipolar, 17, 165
  to furans, electrophilic *vs.* nucleophilic, 72
  heterolytic, 59
  in 1,4-position, 81
  in 1,8-position, 81
  stereospecific, 59
Alcohols, primary, oxidation with Pb(OAc)$_4$, 399
Alcohols, *tert*-, 11, 117
Aldehyde ammonia, 210
Aldehydes, reaction with hydroperoxides, 399
  reaction with sulfur ylides, 344
  α-halo-, reaction with phosphites, 296
Alder rule, 174
Aldol condensation, acid-catalyzed, 254
Aldoxime *p*-toluenesulfonates, 27, 228
Aldoximes, 429
Alkoxylamines, 380
Alkylating agents, 120
Alkylation, of carbon *vs.* nitrogen, 74
  of carbon *vs.* oxygen, 348
  of enamines, with alkyl halides, 5, 70, 74
    with electrophilic olefins, 70
  of enolates, 69
  of ketones, 69
Alkyl chlorosulfites, 308
Alkyl halides, 5, 74, 119
tert-Alkyl hydroperoxides, 151
Alkylidenetriphenylphosphoranes, 341
N-Alkyl-*N*-nitrosourea, 260
1-Alkyn-3-ol, 163
Allenes, cyclodimerization of, 17, 162
  formation from alkynols and SOCl$_2$, 163
Allyl alcohol, 3,3-dichloro-, 214
N-Allylaniline, rearrangement of, 423
Allyl halides, S_N1 reactions of, 29, 246
Allylic cation, 425
tert-Allylic hydrogen, 332
Aluminum alkyls, addition to C≡C bond, 8, 94
  dimeric, 266
  reaction with C=C bond, 6, 86

Aluminum aryls, 94
Aluminum hydrides, 6, 86
Aluminum isopropoxide, 11, 115
Aluminum, triphenyl-, 8, 94
Aluminole, 5-phenyl dibenz-, 95
Aluminoles, 94
Ambident ions, 348
Amides, acylation of, 306
  alkylation of, 306
  formation from H_2O_2 and nitriles, 397
  N-monosubstituted, 307
  protonation of, 306
  reaction with oxalyl chloride, 306
Amine oxides, Cope degradation of, 24, 209
  rearrangement of, 325
Amines, secondary, in Mannich syntheses, 162
  secondary, preparation of enamines with, 70
  reaction with HNO_2, 241
3-Amino-1,2,4-benzotriazine 1-oxide, 426
3-Amino-5,6-diphenyl-1,2,4-triazine, 427
γ-Amino halides, 223
α-Amino ketones, 338
α-Amino ketoximes, 27, 226
o- and p-Aminophenol, 380
2-Amino-2-thiazoline-4-carboxylic acid, 428
4-Amino-2,4,6-trimethyl-1,2-cyclohexadien-1-one, 380
Ammonia, acidity of, 358
  addition to pyrylium salts, 12, 124
  reaction with aliphatic aldehydes, 24, 210
  reaction with aromatic aldehydes, 211
  liquid, ozonization in, 4, 67
Ammonium bases, 24, 204, 206
  decomposition of tetraalkyl-, 24, 206
Ammonium salts, diallyl, 25, 213
  tetraalkyl, 24, 206, 213
Anchimeric assistance, 277
Aniline, N-allyl-, 423
  o-allyl-, 423
  N-2-butenyl-, 423
Anion, benzyl, 420
  radical, 358, 361, 362
Anisole, 2,4,6-trinitro-, 383
Annulene, 1,6-epoxy[10]-, 22, 198
Antimony pentachloride, 293
  pentafluoride–SO_2, 241
Anthracene, 9,10-dimethyl-, 175
Anthrone, 10-phenyl-, 244

Arbusow–Michaelis reaction, 295
Arenediazonium salts, 260
Arndt–Eistert reaction, 260
Aromaticity, of 1,6-methylenecyclodecapentaene, 199
  of tetrazole, 211
Aroxylamines, 380
Arrhenius equation, 239
Aryl cyanates, 307
Arylimidazolines, 211
Arynes, 19, 177
Ascaridole, 393
Asymmetric carbon atom, S_E1 substitution at, 317
  S_E2 substitution at, 318, 326
  S_N1 substitution at, 236
  S_N2 substitution at, 285
Asymmetry preservation in insertion reactions, 135, 146
Atomic distances in isonitriles, 143
Atropic acid, 354
Attack, nucleophilic, of ammonia at oxygen, 67
  at double bonds, 59, 78
  at triple bonds, 59, 88
Aufbau principle, 406
Autoxidation, 395
  of aldehydes, 395
  of ethers, 395
  of saturated hydrocarbons, 395
  of unsaturated hydrocarbons, 395
Autoxidation rates of primary, *sec*, and *tert* C–H bonds, 395
Avogadro's number, 405
Azepine, 199
Azepinone derivative, 381
Azetidinium ion, 224
Azide, lithium, 18, 169
  o-(2-methylbutyl)phenyl-, 15, 146
  phenyl-, 169
Azides, 146, 169
Azines, 9, 99, 103
Aziridine, 1-[(phenylthio)carbonyl]-, 290
  3-phenyl-2H-, 339
Aziridines, ring opening of, 290
  alkoxy-, 339
  1-(arylazo)-, 291
Azirine intermediate, 339
Azo compounds, 388
Azulene, perhydro-, 432

Baeyer–Villiger rearrangement, 97, 151
  in open-chain ketones, 97
Baker–Venkataraman rearrangement, 302
Barton reaction, 152; rule, 205
Basicity of carbanions, 285, 328, 331
Beckmann fission, 226
Beckmann rearrangement, 147, 226, 228
  second order, 226
Benzaldehyde, 8, 99
  reaction with N-phenylhydroxylamine, 17, 167
  reaction with phosphorus tris(dimethylamide), 8, 99
  reaction with ylides, 341
  1,2-dihydro-, 133
  p-nitro-, 345
Benzamidocyclohexene, 3, 62
Benzene, addition of maleic anhydride, 179
  adduct with HCl, 371
  o-bis(trifluoromethyl)-, 179
  Diels–Alder addition of, 179
  m-nitro(trifluoromethyl)-, 372
  pentamethyl(trichloromethyl)-, 26, 220
  1,2,4,5-tetrakis(trifluoromethyl)-, 179
  triacetyl-, 356
  trifluoromethyl-, 371
  1,3,5-trimethoxy-, 379
  1,2,4- and 1,3,5-tri-*tert*-butyl-, 185
Benzene derivatives, p-nitro-, 384
Benzenediazonium chloride, 18, 169
Benzenesulfonyl chloride, 150
Benzenonium-2-carboxylate, 143
Benzhydrol, 116
Benzhydryl p-toluenesulfonate, o-nitro-, 264
Benzilic acid rearrangement, 79
Benzocyclobutene, 354
Benzocyclobutenecarboxylic acid, 354
Benzocyclobutenyl bromide, 356
Benzocyclobutylamine, 356
2(3H)-Benzofuranone derivatives, 291
Benzoic acid, o-benzoyl-, 436
  3,5-dimethoxy-, 360
  o-mesityl-, 436
  trimethoxy-, 360
Benzoin, condensation, 146
  benzylidenedeoxy-, 112
Benzonitrile, 11, 18, 119, 169
Benzophenone, 13, 116, 135
p-Benzoquinone, 115
  2,6-dimethoxy-, 380
Benzothiazolium halides, 144

Benzothiepin-5-ol, 2-chloro-4,5-dihydro-, 31, 269
1,2,4-Benzotriazine 1-oxide, 3-amino-, 426
Benzotriazole-1-carboxamide, 426
1-Benzoxepin, 199
Benzoylation, 429
o-Benzoylbenzoic acid, 436
Benzoylcyclooctatetraene, 197
2-Benzoyl-3,3-dimethyloxirane, 417
α-Benzoylfurans, 73
Benzoyl peroxide, 392
Benzotriazole, 426
Benzvalene derivatives, 182, 185
Benzyl anion, 420
Benzyl chloride, α,α-dimethyl-, 377
  α-methyl-, 236
Benzyldiphenylphosphine oxide, 219
Benzyl diphenyl sulfonium salt, 346
Benzyl α-ethyl-α-methylbenzyl ether, 323
Benzylic hydrogen, reaction with strong base, 333
Benzylideneacetone, 97
Benzylidenemalononitrile, 101
Benzylidenetriphenylphosphorane, 341
Benzyl orbitals, 420
Benzyl p-toluenesulfonate, 1,4-dihydro-3,5-dimethoxy-, 31, 268
Benzyne, 19, 143, 177
Betaine formation in the Wittig reaction, 341, 342, 343
erythro-Betaine, 341, 342, 343
threo-Betaine, 341, 342, 343
Biacetyl monoxime, 430, 431
Bibenzyl, 86
Bicyclobutane, acidity of, 329
Bicyclo[1.1.0]butane derivatives, 140
endo-cis-Bicyclo[5.3.0]decan-2-ol, 141
Bicyclo[2.2.1]heptadiene, 332
  dimerization of, 182
  epoxidation of, 13, 131
  7-chloro-, 271, 273, 274, 276
Bicycloheptadien-7-yl system, 271
Bicycloheptane, 7-chloro-, 271
Bicycloheptene, *anti*-7-chloro-, 271
Bicyclo[3.2.0]heptene, tetrafluoro-, 193
Bicyclo[3.2.0]heptene system, 215
anti-Bicyclohepten-7-ol, 272
Bicyclo[3.2.0]hept-2-en-6-one, 4-phenyl-1,2-diaza-, 25, 215
Bicyclo[3.2.0]hept-2-en-6-yl acetate, 21, 192

Bicyclohepten-7-yl system, 271
anti-Bicyclohepten-7-yl *p*-toluenesulfonate, 272
Bicyclo[2.1.1]hexan-2-amine, 6,6-dimethyl-, 283
Bicyclo[3.1.0]hexane-3-carboxylic acid, 322
Bicyclo[2.1.1]hexane-6-carboxylic acid, 1,5,5-trimethyl-, 31, 258
Bicyclo[3.1.0]hex-3-en-2-one, 6,6-diphenyl-, 414
Bicyclo[2.1.1]hexenyl system, 283
Bicyclo[4.2.0]octa-2,4-diene, 7,8-dibromo-, 195
Bicyclo[4.2.0]octa-1,3,5-triene-7,8-dione, 355
Bicyclo[4.2.0]octa-1,3,5-triene, 354
  7,7-dichloro-2,3,4,5-tetramethyl-, 220
Bicyclo[4.2.0]octa-1,3,5-trien-7-ol, 355
Bicyclo[4.2.0]octa-1,3,5-trien-7-one, 355
Bi-(2-cyclopropenyl), hexaphenyl-, 20, 184
Bicyclo[4.4.1]undecapentaene, 199
Bi-(2-imidazolidinylidene), 1,1′,3,3′-tetraphenyl-, 144
3,3′-Bioxazole, 216
5,5′-Bioxazole, 216
Biphenyl, 2,2′-diacetyl-, 10, 109
Biradical species, 388, 406
Birch reduction, 88
Bis(trifluoromethyl)fumaronitrile, 157
Bis(trifluoromethyl)maleonitrile, 157
Bonds, fluctuating, 190
Bohr's condition, 405
Borohydride, sodium, 272, 273
Bredt's rule, 332
Bridgehead carbanions, 317
Bromination, of 2,3-dimethylbutadiene, 69
  of tetraphenylbutadiene, 4, 68
Bromine, addition to the C=C bond, 61, 88
  in S_E2 substitution, 326
(4-Bromo-2-butenyl)malonic ester, 303
4-Bromo-7-*tert*-butylindanone oxime, 147
trans-γ-Bromodypnone, 5, 78
1-Bromoethyl ethyl sulfone, 336
Bromoform, reaction with strong base, 137
Bromonium cation, 61, 67
2-Bromopropene, 163
Bullvalene, 191
1,3-Butadiene, 155
  derivatives, 195
  2,3-dimethyl-, 69
  1,1,2,3,4-pentaethyl-, 94
  perfluoro-2,3-dimethyl-, 140
  1,2,3,4-tetraphenyl-, 4, 68
Butane, 2-methyl-, 109
Butane-1-thiol, 248
Butanol (labelled), reaction with $SOCl_2$, 308
2-Butanol, 1,1,1-trichloro-, 214
1-Butanone, 2-methyl-1,2-diphenyl-, 321
2-Butanone, 3-methyl-, 109
  4-phenyl-, 111
1-Butene, 2-chloro-4-phenyl-, 111
2-Butene, *cis*-, 14, 136
  2,3-dimethyl-, 12, 121
  2-methyl-, 247
  2-phenyl-, 233
  trans-, 14, 136
3-Butenoate, 4-phenoxy-4-phenyl-, 416
3-Buten-1-ol, 267
3-Butenylamine, 267
N-2-Butenylaniline, 423
tert-Butoxy radical, 396
tert-Butyl chloride, 236
tert-Butylfluoroacetylene, 20, 182
Butyl halides, 1-methyl-, 204
tert-Butyl hydroperoxides, 396
tert-Butyl hypochlorite, 390
sec-Butyl iodide, 389
Butyllithium, 28, 231
n-(Butylthio)acetaldehyde dimethyl acetal, 30, 248
4-*tert*-Butyl-2-(*p*-tolylsulfonyl)cyclohexanol, 203
tert-Butyl vinyl sulfide, 157
1-Butyne, 1-fluoro-3,3-dimethyl-, 20, 182
2-Butyne, hexafluoro-, 15, 19, 140, 179

Cannizzaro reaction, 117
Carbanions, allylic, 335
  bridgehead, 317
  formation in *E1cB* mechanism, 203
  from halomethanes, 137
  stability of primary *vs*. secondary, 329
  stabilization by electronic and orbital effects, 329, 330
  structure of, 317
  trichloromethyl, 137
Carbenes, 13, 14, 15, 129, 134, 136, 137, 140, 141, 144, 263
  dichloro-, 137, 139, 140
  difluoro-, 140
  dihalo-, 140
  dimers, 144

Carbenes (*Continued*)
  electrophilic character of, 103
  intermediates, 219
  intramolecular addition of, 14, 136
  keto, 165
  monohalo-, 139
  nucleophilic, 15, 144
  vinyl, 165
  from ylides, 346, 347
Carbenoids, definition of, 129, 148
Carbiminium cation, 224
Carbonates, cyclic, 286
Carbonium ions (see also cations), 241
  acylium, 63
  alkyl, 265
  allylic, 265
  benzene, 260
  classical, 60, 241
  cyclopropylmethyl, 266
  formation of, 242
    by addition of cations to double bonds, 254
    from aliphatic amines and HNO_2, 261
    from alkyl halides, 119
    during Clemmensen reduction, 109
    from diazo compounds, 260
    in $E1$ elimination, 202
    by solvolysis, 242
  long-lived, 241, 242
  non-classical, 266
  primary, secondary, and tertiary, stability of, 242, 329
  properties of, 241
  rearrangement of, 414
  solvated, 3
  tertiary, 291
Carbon monoxide, 414
Carbon tetrachloride, 391
(+)-3-Carene, 208
4-Carene, (+)-3-hydroxymethyl-, 24, 208
Catalysis, acid, 78
Catalyst, Ziegler, 86
  $ZnCl_2$ as, 67
Cations (see also carbonium ions), acetyl-pyridinium, 349
  allylic, 246, 425
  cyclobutenylium, 271
  cyclopropenylium, 242, 440
  diazacyclobutenylium, 228
  equilibrating classical, 274, 277
  hydroxyl, 379

2-*p*-methoxyphenyl-7,7-dimethyl-norbornyl, 279
  nitrosyl, 430
  phenethyl, 284
  triphenylcyclopropenylium, 441
  triphenylmethyl, 441
  tripropylcyclopropenylium, 242, 441
Chain reaction, carbanionic, 333
Chain reactions, radical, 388
Chain termination, 388
Chapman rearrangement, abnormal, 353
Chloral, 96, 105
Chloramine, 46, 380
Chloranil, 100
Chlorination, with CH_3SCl, 391
  with irradiation, 389
  with PCl_5, 391
  of prim. or sec. hydrogen, 389
  radical, of *n*-heptane, 48, 389
  of tetraphenylbutadiene, 4, 68
  of toluene, 377
Chlorine, radical cleavage of, 389
Chloroacetamide, 306
Chloroalkyl sulfones, 337
p-Chlorobenzhydryl methyl sulfide, 310
7-Chlorobicyclo[2.2.1]heptadiene, 271, 273, 276
7-Chlorobicycloheptane, 271
anti-7-Chlorobicycloheptene, 271
N-Chlorodiethylamine, 32, 281
2-Chloro-4,5-dihydrobenzothiepin-5-ol, 31, 269
2-Chloro-2,2-diphenylacetanilide, 364
3-Chloro-2,5-diphenylfuran, 113
1-Chloro-2,3-diphenylindene, 112
3-Chloro-5,6-diphenyl-1,2,4-triazine, 427
2-Chloroethyl *p*-nitrobenzoate, 293
Chloroform and strong base, 137
Chloroformates, 309
Chlorohydrin, 252
5-Chloroisophthalic acid, 175
(Chloromethyl)triphenylphosphonium chloride, 218
Chloronium ions, 67
Chloro olefins, 111
2-Chloro-4-phenyl-1-butene, 111
2-Chloro-5-phenyl-1-pentene, 111
Chloroprene, 154
Chlorotrifluoroethylene, 195
β-Chlorovinyl ketones, 347
5-Cholesten-3-yl system, 266

Cine-substitution, 385
cisoid-Conformation of dienes, 170
cis-Rule, 170
Claisen condensation, 304
Claisen rearrangement, 424
Cleavage, of alkoxyl radicals, 398
　of cyclooctatrienes, 195
　radical, 421
　of vinyl sulfides, 335
Clemmensen reduction, 10, 107
Collision of methylene molecules, 134
π-Complexes, in aromatic substitution, 371
　in epoxidation, 131
　of chlorine with benzene, 390
　of halogens and dienes, 67
　of halogens and olefins, 61
　of hydrogen halides and olefins, 3, 59
　of nitriles, 119
　of ozone and olefins, 64
　of the triple bond, 88
Conformation, quasi-boat, 190
　quasi-chair, 190
Conjugate acid, 247
　of carbonyl compounds, 107
Conversion, *cis–trans*, 417
　external, 410
　internal, 409
　radiationless, 409
Cope degradation, 24, 209
Cope rearrangement, 133, 177, 189, 191
Copper powder, 136
Criegee mechanism of ozonization, 64
Criegee rearrangement, 97, 151, 247
Cross-over experiments, 228, 252
Crotonaldehyde, 347
Crotonate, methyl, 18, 174
Cumyl hydroperoxide, 396
Cyanate, 2,6-di-*tert*-butylphenyl, 307
Cyanic esters, 307
Cyanide, sodium, 276
Cyclization of a diazonium compound, 427
Cycloadditions, classification of, 129
　rules governing, 153, 158, 159
　1,1-cycloaddition, 129, 130
　1,2-cycloaddition, 129, 153
　　of 1,1-dichloro-2,2-difluoroethylene, 16, 153
　　Diels–Alder addition *vs.*, 154
　　of dienes, 155
　　diradical *vs.* ionic, 153, 155
　　heterocyclic, 159

　　four-membered rings by, 159, 161
　　of benzene, 179
　　of bis(trifluoromethyl)maleonitrile, 157
　　of bis(trifluoromethyl)fumaronitrile, 157
　　of butadiene, 155
　　of *tert*-butyl vinyl sulfide, 157
　　of chloroprene, 154
　　of 2,4-hexadiene, 155
　　of propyl propenyl ether, 157
　1,3-Cycloaddition, 165
　　of diazoacetic ester, 18, 169
　　of diphenyldiazomethane, 17, 166
　　of diphenylnitrimine, 166
　　of nitrones, 17, 18, 167
Cyclobutadiene, 183
Cyclobutadienediol, 195
　phenyl-, 195
Cyclobutane, 164
　cis-1,3-dibromo-1,3-dimethyl-*cis*- and -*trans*-divinyl-, 190
　monovinyl-, 190, 192
　octafluoro-, 153
　tetraphenyl-, 420
Cyclobutane amine, 267, 268
1,3-Cyclobutanediol, 232
1,3-Cyclobutanedione, 2,2,4,4-tetramethyl-, 101
Cyclobutanes, by 1,2-cycloaddition, 153 158, 159
　from enamines, 4, 73
Cyclobutanol, 267, 268
　1-methyl-, 414
Cyclobutanone, 3-hydroxy-, 356
Cyclobutanone derivatives, 356
　2,2,3-triphenyl-, 357
Cyclobutene, dichlorophenyl-, 195
　3,4-dichloro-1,2,3,4-tetramethyl-, 246
3-Cyclobutene-1,2-dione, 3-phenyl-, 22, 195
Cyclobutenylium cation, 271
Cyclobutyl compounds, S_N1 and S_N2 substitution of, 246
Cyclodecapentaene, 199
　1,6-epimino-, 199
　1,6-methylene-, 199
2-Cyclodecen-1-ol, 141
cis-Cyclodecene oxide, 141
　deuterated, 142
Cyclodimerization, of acetylenes, 17, 161, 162
　of allenes, 17, 162

Cycloheptadiene, 271
  7-chloro-, 271, 273, 276
  derivative, 74
Cycloheptadienyl acetate, 193
1,3-Cycloheptanedione, 300
Cycloheptatriene, 82, 200
  derivative, 268
Cycloheptatrienyl cyanide, 1-(trifluoromethyl)-, 200
2,4,6-Cycloheptatrienyl methyl ether, 438
Cyclohexadiene, epoxy-, 199
  derivative, 360
1,2-Cyclohexadien-1-one, 4-amino-2,4,6-trimethyl-, 380
2,5-Cyclohexadienone, 3,4-dimethyl-4-(trichloromethyl)-, 10, 113
2,5-Cyclohexadienone, 4,4-diphenyl-, 413
  4-methyl-4-(trichloromethyl)-, 10, 112
Cyclohexane, epoxy-, 303
Cyclohexanecarboxylate, 30, 254
  1-(2-formylethyl)-3-methyl-2-oxo-, 254
1,2-Cyclohexanedione, 76
1,2-Cyclohexanedione 2,3-monoenol, 3,5,5-trimethyl-, 5, 75
Cyclohexanol, 4-*tert*-butyl-2-(*p*-tolylsulfonyl)-, 203
Cyclohexanone, 115
Cyclohexene, 13, 135, 136
Cyclohexyl benzoate, 4-oxo-, 302
Cyclohexyl bromide, *trans*-4-methyl-, 326
Cyclohexyl hydroperoxide, 1,3,3-trimethyl-, 150
Cyclohexyl *p*-toluenesulfonate, *trans*-2-phenyl-, 23, 207
Cyclononatetraene, methylene-, 244
cis,*cis*-1,5-Cyclooctadiene, 190
Cyclooctatetraene, benzoyl-, 22, 197
  derivatives, 197, 244
Cyclooctatriene derivatives, 22, 195
Cyclopentadiene, 18, 172, 174, 193
  hexachloro-, 175
  pentaphenyl-, 442
Cyclopentadienone, 90
  tetrachloro-, 90
  tetraphenyl-, 90
Cyclopentadienylsodium, 288
Cyclopentanone, 300
  acetyl-, 300
  2-acetyl-5-phenyl-, 300
3-Cyclopentenecarbaldehyde, 133

2-Cyclopentenecarboxylate, 2-methyl-3-phenyl-, 300
3-(1-Cyclopentenyl)-2-methyl-2-propanol, 11, 118
3-(3-Cyclopentenyl)-2-methyl-2-propanol, 119
Cyclopentyl chloride, 1-methyl-, 277
Cyclopropa[*b*][1]benzothiopyran, 1,1-dichloro-, 26, 220
Cyclopropane, derivatives, 79, 99
  cis-dimethyl-, 135, 136
  trans-dimethyl-, 135, 136
  monovinyl-, 189
Cyclopropane-1,1-dicarboxylic acid, 2-vinyl-, 303
Cyclopropanemethanol, 267
Cyclopropanemethylamine, 267, 268
Cyclopropanones in Favorsky rearrangement, 70
Cyclopropenes, 441
Cyclopropenylium cations, 440
  triphenyl-, 441
  tripropyl-, 242
Cyclopropenyl *p*-toluenesulfonate, di-(*p*-methoxyphenyl)-, 271
Cyclopropyl anion, 322
Cyclopropylcarbinyl system, 265
Cyclopropyl derivatives, by addition of carbanions to olefins, 14, 137
  S_N1 and S_N2 substitution at, 246
1-Cyclopropylethyl acetate, 24, 209
Cyclopropyllithium, 326
Cyclopropylmagnesium, 326
Cyclopropylmethyl cation, 266
Cyclopropyl ring, effect on adjacent positive charge, 266

Deactivation, by chemical reaction, 411
  radiationless, 409
Deactivation processes, 407
Decalin-2,7-dione, 4a-propargyl-*trans*-, 361
cis,*cis*-1-Decalol, 141
cis,*trans*-1-Decalol, 141
1-Decalone, 28, 230
  5-(*p*-tosyloxyimino)-, 28, 230
Decarbonylation, 259
  of α-ketocarboxylic acids, 234
Decarboxylation, 259
Decomposition, thermal, of diazirine, 134
  of diazomethane, 134
Dehydration, 256

Index

Dehydrobenzene, 19, 143, 177
Dehydrobromination, 243
Delocalization of sigma-electrons, 266
Demotion, electron, 414, 418
Densities, electron, in aromatic molecules, 420
Deuteroacetic acid, 195
Deuterium bromide, 3, 60
Dewar benzene, hexamethyl-, 184
Dewar benzene derivatives, 182, 184, 185
2,2'-Diacetylbiphenyl, 10, 109
Diallylic systems, 21, 190
Dianion by Birch reduction, 362
Diarylamines, unsymmetrically substituted, 353, 354
1,2-Diazabicyclo[3.2.0]hept-2-en-6-one, 4-phenyl-, 25, 215
Diazacyclobutenyl cation, 228
Diazepine derivatives, 200
Diazepinone derivatives, 215
Diazoacetic ester, 18, 169, 260
Diazoalkanes, 165, 260
Diazo compounds, aliphatic, 165, 260
  nucleophilic character of, 103
  reaction with acids, 261
  thermal decomposition of, 260
Diazoethane, 9, 103
Diazo ketones, 260, 261, 263
Diazoles, 166
Diazomethane, 105, 134, 135, 260
  diphenyl-, 17, 166
  o-(nitrobenzoyl)-, 31, 261
  phenyl-, 441, 442
Diazonium fluoroborate, 121
Diazonium salts, aromatic, 260
3-Diazo-1-(1,2,3-triphenylcyclopropenyl)-2-propanone, 136
Dibenzocycloheptatriene system, 243
[5H]-Dibenzo[a,d]cyclohepten-10-ol, 5,11-epoxy-10,11-dihydro-5-methoxy-10-phenyl-, 29, 244
Dibenzocyclooctatriene system, 243
Dibenzo[a,e]cyclooctene, 5,6-dibromo-5,6-dihydro-, 29, 242
Dibenzoylethane, 79
1,2-Dibenzoylethylene, 10, 113
7,8-Dibromobicyclo[4.2.0]octa-2,4-diene, 198
5,6-Dibromo-5,6-dihydro-dibenzo[a,e]cyclooctene, 29, 242

cis-1,3-Dibromo-1,3-dimethylcyclobutane, 164
Dibromo ketones, α,β, 294
  β,γ, 294
  γ,δ, 293, 294
2,2-Dibromopropane, 163
Di-$tert$-butyl peroxyoxalate, 397
2,6-Di-$tert$-butylphenyl cyanate, 307
Dichloroacetylene, 182
3,3-Dichloroallyl alcohol, 214
Dichlorocarbene, 137, 139, 140
1,1-Dichlorocyclopropa[b][1]benzothiopyran, 26, 220
1,1-Dichloro-2,2-difluoroethylene, 153
Dichlorophenylcyclobutene, 195
3,5-Dichloro-2-pyrone, 175
7,7-Dichloro-2,3,4,5-tetramethylbicyclo[4.2.0]octa-1,3,5-triene, 220
3,4-Dichloro-1,2,3,4-tetramethylcyclobutene, 246
2,2'-Dichloro-3,3'-thiodipropionate, 294
α-Dicyclopentadien-1-ol, 19, 176
1,1-Dicyclopropylethylene, 21, 192
Dieckmann condensation, 351
Diels–Alder reaction, 170
  of benzene, 179
  of benzyne, 177
  of cyclopentadiene, 18, 174
  of 1,3-dienes, 153, 155
  of dienophiles, 90, 123, 170
  of hexafluoro-2-butyne, 19, 179
  of maleic anhydride, 19, 175
  of O_2, 392
Diethylamine, N-chloro-, 281
Diethylene glycol, 320
Diethyl ketone, 232
Difluorocarbene, 15, 140
Diglyme, 272, 273
1,2-Dihydrobenzaldehyde, 131
1,4-Dihydro-3,5-dimethoxybenzyl p-toluenesulfonate, 31, 268
Dihydrofuranols, 294
Dihydrofurans, 79
2,3-Dihydro-5-methylfurans, 139
9,10-Dihydronaphthalene, 179
Dihydrooxazine, 118
Dihydrophenanthrene, 419
Dihydropyran derivatives, 72
2,5-Dihydrothiophen dioxide, 220
Diimine, 86
α-Diketones, 75, 433

β-Diketones, 304, 418
Dimerization, during Clemmensen reduction, 108
  of fluorinated olefins, 141
  of ketene, 159
  of methylacetylene, 164
  of α-olefins, 86, 87
  of propene, 87
3,5-Dimethoxybenzoic acid, 360
Di-(p-methoxyphenyl)cyclopropenylium p-toluenesulfonate, 271
Dimethyl acetylenedicarboxylate, 7, 89, 91
p-(Dimethylamino)phenylpentazole, 170
9,10-Dimethylanthracene, 175
α,α-Dimethylbenzyl chloride, 377
6,6-Dimethylbicyclo[2.1.1]hexan-2-amine, 283
2,3-Dimethyl-1,3-butadiene, 69
2,3-Dimethyl-2-butene, 12, 121
1,3-Dimethyl-2-butenyl ethyl disulfide, 313
Dimethylcyclopropane, 135, 136
Dimethyl 2,2'-dichloro-3,3'-thiodipropionate, 294
2,2-Dimethyl-1,3-diphenylindane-1,3-diol, 232
Dimethylfulvene, 82
2,3-Dimethylindole, 425
Dimethylindoline, 424
trans-1,2-Dimethyloxirane, 288
2,4-Dimethylpentane, 395
3,3-Dimethyl-2,4-pentanedione, 8, 97
9,10-Dimethylphenanthrene, 109
1,1-Dimethylpyrrolidinium bromide, 25, 213
Dimethyl sulfide, 345
Dimethylsulfonium fluorenylide, 345
Dimethyl sulfoxide, 322
cis-Dimethylthiirane, 287
cis-2,3-Dimethylthiirane dioxide, 337
3,4-Dimethyl-4-(trichloromethyl)-2,5-cyclohexadienone, 10, 113
Dimorpholinoethylene, 159
Dinitrogen oxide
Dinitrogen pentoxide, 376
1,3-Dioxolane, 2-phenyl-, 28, 231
1,3-Dioxolan-2-one, 4,4,5,5-tetramethyl-, 287
1,2- and 1,3-Diols, 232
Diphenylacetylene, 8, 94
6,6-Diphenylbicyclo[3.1.0]hex-3-en-2-one, 414

4,4-Diphenyl-2,5-cyclohexadienone, 413
Diphenylcyclopropenylmethyl p-toluenesulfonate, 32, 270
Diphenylcyclopropylmethyl p-toluenesulfonate, 271
Diphenyldiazomethane, 17, 166
Diphenylethylene, 218
(2,2-Diphenylethyl)triethylphosphonium hydroxide, 218
1,3-Diphenylisobenzofuran, 393
Diphenyl ketone, 340
Diphenylmercury, 327, 328
Diphenylmethane, 116
Diphenylnitrimine, 166
1,3-Diphenyl-3-phospholin 1-oxide, 9, 105
Diphenyl sulfide, 346, 347
2,5-Diphenyl-1,3,4-thiadiazole, 211
1,3-Dipolar additions, 17, 165
Dipole–dipole interaction, 337
Dipole moment, of isonitriles, 143
  of thiobenzophenone vs. benzophenone, 122
Diradical character, of cyclobutadiene, 183
  of methylene, 135
Diradical intermediate in 1,2-additions, 153, 155
Diradicals, 417
Dissociation, 408
Disulfide, reaction with phosphines, 313
Dithians, 248
Dithioacetals, 248
Divinylcyclobutane, 190
Dypnone, trans-γ-bromo-, 5, 78

Edward's equation, 285
Einstein, 405
Electrofugous, 229
Electromagnetic radiation, 405
Electron densities, of isoxazole, 216
  in saturated hydrocarbons, 390
Electronegativity of the halogens, 315
Electronic effects in elimination reactions, 204
Electrophiles, reaction with enamines, 74
  reaction with the C=C bond, 3, 60
  reaction with the C=O bond, 8, 95
Electrophilicity, 107
Elimination, of aluminum hydride, 86
  ethylogous, 223
  of sulfur, 26, 220
  thermal, 23, 207

α-Elimination, 129, 137, 202
α′,β-Elimination, 213
β-Elimination, 23, 202
E1-Elimination, 202
E1cB-Elimination, 203
E2-Elimination, 203
cis- and trans-Elimination, 205, 207
1,4-Elimination, 220
Emission of light, 405
Enamines, 4, 69
  addition to acetylenes, 73, 74
  addition to methyl propiolate, 74
  alkylation of, 69, 74
  formation of cyclobutanes from, 73
  formation from α-haloacetamides and trialkylphosphines, 315
  preparation of, 70
  reaction with sulfenes, 16, 159
  stability of, 105
endo-Adduct in Diels–Alder addition, 172, 173
endo-Rule, 172
Ene-addition, 121
Enediamine, 159
Energy levels, electronic, 405
  rotational, 405
  vibrational, 405
Enolates, alkylation of, 363
  reaction with acyl and alkyl halides, 348
  reaction with carbonium ions and protons, 349
Enol esters, 296, 297, 315
Enol ethers, 269
Enolization, 254
  of cyclic vs. open-chain ketones, 76
Enols, reaction with FeCl$_3$, 300
Enthalpy, 237
Epichlorohydrin, 288
1,6-Epiminocyclodecapentaene, 199
Episulfides, 220
Episulfonium salt, 248
Epoxidations, 13, 130
Epoxides, acid-catalyzed ring-opening of, 250
  formation from phosphorus triamides and carbonyl compounds, 99
  formation from ozone zwitterion and olefins, 130
  reaction with thiocyanates, 287
  ring opening of, 214
  S_N2 reaction of, 214

1,6-Epoxy[10]annulene, 22, 198
Epoxycyclohexadiene, 199
Epoxycyclohexane, 303
5,11-Epoxy-10,11-dihydro-5-methoxy-10-phenyl[5H]dibenzo[a,d]cyclohepten-10-ol, 29, 244
Epoxy esters, 250
Epoxy ketones, photorearrangement of, 417, 418
3,4-Epoxy-2,2,4-trimethylpentyl isobutyrate, 30, 250
Esters, acid-catalyzed hydrolysis, 298
  base-catalyzed hydrolysis, 298
  monomolecular hydrolysis, 298
  α-halo-, 296
Ethane, 414
  dibromo-, 289
  diphenyl-, 85, 86
  hexaphenyl-, 388
1,2-Ethanedithiol, 30, 248
Ethanethiol, 2,2,2-trifluoro-, 122
Ether, benzyl isobutyl, 324
  benzyl α-ethyl-α-methylbenzyl, 323
  methyl phenethyl, 376
  methyl 3-phenylpropyl, 376
Ethoxyacetylene, 17, 161
p-Ethoxyphenylpentazole, 170
5-Ethoxy-1,2,3,4-thiatriazole, 307
Ethynylation, 162
Ethyl acetate, 1-cyclopropyl-, 24, 209
Ethyl acetoacetate, 348
Ethyl bromofluoroacetate, 295
O-Ethyl chlorothioformate, 307
Ethyl cyanate, 307
Ethylene, 86
  chlorotrifluoro-, 195
  1,2-dibenzoyl-, 10, 113
  1,1-dicyclopropyl-, 21, 192
  1,1-fluoro-2,2-dichloro-, 16, 153
  1,1-dimorpholino-, 159
  1,1-diphenyl-, 218
  tetracyano-, 131, 159
  tetrafluoro-, 153
  tetramethyl-, 12, 121
Ethylene glycol, 320
Ethylenimine, 381
Ethyl 1-(2-formylethyl)-3-methyl-2-oxocyclohexanecarboxylate, 30, 254
3-Ethyl-3-hexene, 94
1-Ethylidene-2-phenacylidenehydrazine, 103

Ethylidenetriphenylphosphorane, 342
4-Ethyl-3-methyl-2-phenyl-2,4-hexanediol, 232
Ethyl 6-oxo-5-phenylheptanoate, 300
3-Ethyl-2-pentene, 232
3-Ethyl-6-phenyl-5-hexen-3-ol, 23, 207
Ethyl propiolate, 347
Ethyl trichloroacetate, 139
Exchange of halogen for metal atoms, 214
Excitation, thermal, 410
Excited electronic states, 405
   bond distances in, 411
   electron distribution in, 411
   life-time of, 410
   singlet and triplet, 406, 410
exo-Addition to the norbornyl cation, 276
exo-Adduct in Diels–Alder addition, 172, 173
Eyring equation, 237

Favorsky reaction, 79
Fenton's reagent, 247
Ferric chloride, 300
Fission of the C–O bond, 418
Fluorenylide, dimethylsulfonium, 345
Fluorescence, 409
Fluorine, Hammett–Brown sigma-value of, 378
Fluoroboric acid, 244
1,1-Fluoro-2,2-dichloroethylene, 16, 153
1-Fluoro-3,3-dimethyl-1-butyne, 20, 182
Fluoro ketones, 161
Fluorosulfonic acid, 241
Five-membered rings, 129, 165
Formaldehyde, 162, 208, 219
   in Mannich synthesis, 162
   in Prins reaction, 208
   reaction with NH_3, 211
   cis- and trans-2-vinylcyclopropyl-, 133
Formamide derivatives, 229
Formic acid, vinylog of, 347
Formimide derivatives, 143
1-(2-Formylethyl)-3-methyl-2-oxocyclohexanecarboxylate, 30, 254
Four-membered rings, 129, 153
α-Fragmentation, 234
β-Fragmentation, 213, 223
   of aldoxime p-toluenesulfonate, 27, 228
   of α-amino ketoximes, 27, 226
   of 3-bromo-N,N-dimethyladamantan-1-amine, 27, 226
   of 2,2-dimethyl-1,3-diphenylindane-1,3-diol, 28, 232
   of 5-nitroso-2-phenylpyrimidine-4,6-diamine, 28, 230
   of 2-phenyl-1,3-dioxolane, 28, 231
   of 5-(p-toluenesulfonyloxy)-1-decalone, 28, 230
Franck–Condon principle, 406
Frequency of light, 405
Friedel–Crafts catalysis, 63
Fulminic acid, 130, 142
Fulvalene, perchloro-, 316
Fulvene, dimethyl-, 82
Fumaronitrile, bis(trifluoromethyl)-, 157
Furans, α-acetyl-, 72
   α-benzoyl-, 73
   3-chloro-2,5-diphenyl-, 113
   dihydro-, 79
   2,3-dihydro-5-methyl-, 139
   tetrahydro derivatives, 250
Furanols, dihydro-, 294
Furfuraldehyde, 144
Furoin, 144

General acid catalysis, 247
Geometry of phosphoranes, 92
Gibbs–Helmholtz equation, 239
Glycol, diethylene, 320
o-Glycylsalicylamide, 439
Glyoxal, o-nitrosophenyl-, 262
Glyoxylic acid, o-(hydroxyamino)phenyl-, 263
Grignard reagents, reaction with cyclobutanone derivatives, 356, 357
Ground state, of alkyl nitrenes, 146
   of benzyl anion, 420
Ground state geometry vs. excited state geometry, 411

α-Halo acetamides, 315
Halo carbenes, 137, 140
α-Halo ketones, 296, 315, 396
Hammett equation, 105
   modification of, 376
   sigma-values in, 373, 374
Hammett–Brown equation, 378
Hammond's principle, 237, 265
Hemiacetals, 247
Hemithioacetals, 248
Heptadien-2-one, 4-methyl-, 124
n-Heptane, chlorination of, 389, 390

2,6-Heptanedione, 4-methyl-, 124
Heptanoic acid, 6-oxo-5-phenyl-, 300
1-Heptyne, 391, 392
Heterolysis of the O–O bond, 246
Hexachlorocyclopentadiene, 175
2,4-Hexadiene, 155
Hexafluoroacetone, 17, 161
Hexafluoro-2-butyne, 19, 140, 179
Hexafluorothioacetone, 12, 121
Hexamethyl Dewar benzene, 184
Hexamethylenetetramine, 211
2,4-Hexanediol, 4-ethyl-3-methyl-2-phenyl-, 232
Hexaphenylbi-(2-cyclopropenyl)-, 20, 184
3-Hexene, 3-ethyl-, 94
1-Hexene, 6-phenyl-, 332
5-Hexen-3-ol, 3-ethyl-6-phenyl-, 23, 207
HMO method, 420
Hofmann rule, 203
Homoallylic system, 265
Homodelocalization, 271
Hückel rule, 199
Hund's rule, 134
Hunsdiecker reaction, 152
Hybridization, of carbanions, 317
 of phosphoranes, 92
Hydration of chloral vs. acetone, 96
Hydrazide, sodium, 83
Hydrazine, 84, 85
 1-ethylidene-2-phenacylidene-, 103
Hydride ion, addition to pyrylium salts, 12, 123
 in Clemmensen reduction, 107
 formation of, 81
Hydrides, metal, 334
Hydroamides, 24, 210
Hydrogen, nascent, 107
Hydrogen bonding in sodium hydrazide reactions, 84
Hydrogen bromide, addition of, 390
Hydrogen fluoride, 241
Hydrogen iodide, 5, 75
Hydrogen peroxide, 8, 95, 246
 addition to furans, 247
 adduct with nitriles, 397
 reaction in acidic and basic medium, 247
 reaction with alcohols, 247
 reaction with aldehydes, 95
 reaction with tert-alkyl halides, 247
 reaction with alkyl sulfates, 247
 reaction with 3,3-dimethyl-2,4-pentanedione, 8, 97
 reaction with ketones, 95
Hydroperoxides, primary and secondary, 151
 tertiary, 151
 tert-alkyl, 398
 tert-butyl, 397, 398
 cumyl, 396
 α-hydroxy-, 96, 399
o-(Hydroxyamino)phenylglyoxylic acid, 263
3-Hydroxycyclobutanone, 356
α-Hydroxyhydroperoxides, 96
Hydroxyisatin, 263
Hydroxy ketone, 395
Hydroxylamine, 167, 210
 N-methyl-, 167
 N-phenyl-, 167
 trisubstituted, 325
Hydroxyl cation, 130, 247
Hydroxyl radicals, 247
(+)-3-Hydroxymethyl-4-carene, 24, 208
Hypobromite formation, 152
Hypochlorite, tert-butyl, 390

$-I$ effect, 115
Imenes, 146
Imidazolines, aryl-, 211
Imidic esters, 353
Imidoyl chlorides, 121
Imine formation from α-haloacetamides, 315
Imine, C-phenyl-N-thiobenzoylnitrile-, 211
Imines, 70
Iminium ion in Beckmann rearrangement, 147
Incubation period in radical reactions, 389
Indane-1,3-diol, 2,2-dimethyl-1,3-diphenyl-, 28, 232
1,3-Indanedione, 2-methylsulfinyl-, 434
Indanone oxime, 4-bromo-7-tert-butyl-, 147
Indene, 3, 60
 1-chloro-2,3-diphenyl-, 112
Indole, 2,3-dimethyl-, 424, 425
Indole, 2-methyl-3-propyl-, 425
Indoline, 2,3-dimethyl-, 424
Indoline, 2-methyl-, 423, 424, 425
Inductive effect, 204

Inductive effect (*Continued*)
  of the ammonium group in aromatic substitution, 340
  of the methyl group, 78, 206
    in saturated hydrocarbons, 390
    in the Smiles rearrangement, 386
Inductive stabilization of carbanions, 329
Inhibitors, 389
Insertion, of carbenes, 134, 138
  of methylene, 14, 134, 135
  of nitrenes, 15, 146
  of positive nitrogen, 16, 147
  of positive oxygen, 16, 150
  transannular, of carbenes, 141
Intermediate, bicyclic, 300
Internal conversion, 409
Internal return, 246
Intersystem crossing, 414
  in methylene, 134
Intramolecular addition of carbenes, 14, 136
Inversion of configuration, 295
  in S_N1 substitution, 236
  in S_E1 substitution, 318
Inversion, Walden, 287
Ion pair formation in S_N2 substitution, 285
Ion pair, tight, 3, 61, 63, 84, 97, 111, 113, 151, 228, 246, 285, 308, 310, 311, 314, 317, 319, 320, 323, 325
Ions, non-classical, 59, 61, 265
Irradiation, in chlorination, 389
  of ketones, 414
  of olefins, 419
Isobenzofuran, 1,3-diphenyl-, 393
Isobutene, 295
Isobutylbenzene, 323
Isocyanate, phenyl, 9, 105
Isocyanic esters, 307
Isocyanide, phenyl, 15, 143
Isocyanides, 15, 142, 229
Isomerization, 7, 87
  of aluminum alkyls
  of boron alkyls, 88
  of olefins, 337
  photo-, 419
  thermal, of thujone, 21, 193
Isomers in Diels–Alder reactions, 170, 172
Isonitriles, 142
Isooctane, polychloro-, 223
Isophthalic acid, 5-chloro-, 175
Isoprene, 153, 155
Isoxazole, 216

Isoxazolium chloride, 5-methyl-5-phenyl-, 25, 216

Ketene, derivatives, 195
  dimerization of, 159
Ketimine, 290
Keto carbenes, 165
β-Keto carbonyl compounds, 350
α-Keto carboxylic acids, 234
β-Keto esters, 304
Keto glycols
α-Keto ketene imine, 216
α-Ketols, 75
Ketone, 2-chlorovinyl methyl, 348
  diethyl, 232
  diphenyl, 340
Ketones, cleavage of, 321
  α-acyl, 98
  α,β-dibromo, 294
  β,γ-dibromo, 294
  γ,δ-dibromo, 293, 294
  α-halo, 296, 315, 316
Ketoximes, 226, 228, 429
  α-amino, 28, 226

Labeling, of nitrogen, 169
  of oxygen in acetic anhydride, 383
  studies, 431
α-Lactam, 364
Lactones, 96, 303
Ladenburg prism, 184, 185
Lead tetraacetate, 152, 232, 399–402
  reaction by a radical mechanism, 399
1-Lithiopyrrole, 139
Lithium aluminum hydride, 123, 198
  addition to double bonds, 6, 81, 82
  addition to tropolone methyl ether, 8, 98
  addition to tropone, 6, 81
  reaction with the carbonyl bond, 8, 98
  reaction with norbornadien-7-ol, 6, 82
  reduction of esters, 81
Lithium azide, 18, 169
Lithium diethylamide, 141
Lithium-organic compounds, butyl-, 28, 231, 335, 346
  cyclopropyl-, optically active, 326
  methyl-, 25, 214
  1-methylheptyl-, 326
  phenyl-, 25, 213, 335
Lithium salt effect in Wittig reaction, 343

Magnesium, cyclopropyl-, 326
Maleic anhydride, 19, 175
Malonic ester, (4-bromo-2-butenyl)-, 303
Malononitrile, 7, 89
Maleonitrile, bis(trifluoromethyl)-, 157
Mannich synthesis, 162
Markownikoff's rule, 60, 63, 86
Mass law effect in S_N1 substitution, 240
McEwen–Streitwieser–Applequist–Dessy scale of carbon acids, 331
Medium-sized rings, 141
Meerwein–Ponndorf–Verley–Oppenauer reaction, 114
Meisenheimer complex, 383
Meisenheimer rearrangement, 325
Mercuric acetate, phenyl-, 328
Mercuric chloride, *trans*-4-methylcyclohexyl-, 326
  neophyl-, 318
Mercury-organic compounds, of alkanes, 330
  alkyl mercury, 327
  dialkyl mercury, 327
  diphenyl mercury, 327, 328
  optically active, 326
o-Mesitylbenzoic acid, 436
Mesityl oxide, 349
Mesityl oxide oxime, 430
Metallation, of the aromatic nucleus, 333
  of aromatic side-chains, 333
Methane, 414
Methanes, polyhalo-, 137
1-*p*-Methoxyphenyl-3-*p*-nitrophenylacetone, 339
(Methoxycarbonylmethylene)triphenylphosphorane, 341
Methylacetylene, 164
Methyl (*o*-{*N*-benzoyl-*N*-[*o*-(methoxycarbonyl)phenyl]amino}phenyl)-acetate, 351
α-Methylbenzyl chloride, 236
2-Methylbutane, 109
3-Methyl-2-butanone, 109
2-Methyl-2-butene, 247
1-Methylbutyl halides, 204
o-(2-Methylbutyl)phenyl azide, 146
Methyl chloride, 219
Methyl chloroacetate, 306
Methyl chlorosulfite, 438
Methyl crotonate, 18, 174
1-Methylcyclobutanol, 414

trans-4-Methylcyclohexylmercuric chloride, 326
1-Methylcyclopentyl chloride, 277
Methyl dichloroacetate, 298
2-Methyl-1,2-diphenyl-1-butanone, 321
Methylene, 13, 134
  electronic states of, 134
  generation of, 134
  "hot" singlet, 135
Methylene chloride, 139
1,6-Methylenecyclodecapentaene, 199
Methylenecyclononatetraene, 244
Methylene diiodide, 136
1-(α-Methylenefurfuryl)pyrrolidine, 4, 72
Methylenesulfene, 16, 159
4-Methylheptadien-2-one, 124
4-Methyl-2,6-heptanedione, 124
1-Methylheptyllithium, 326
2-Methylindoline, 423, 424, 425
Methyllithium, 25, 139, 214
2-Methylnorbornyl chloride
8a-Methyl-1,2,3,4,6,7,8,8a-octahydro-1,6-naphthalenedione, 305
(1-Methyl-2-oxocyclohexyl)methyl *p*-toluenesulfonate, 350
3-Methyl-2,4-pentanedione, 98
2-Methyl-4-pentenal, 213
Methyl phenethyl ether, 376
2-Methyl-3-phenyl-2-cyclopentenecarboxylate, 300
2-Methyl-5-phenylisoxazolium chloride, 25, 216
Methyl 3-phenylpropyl ether, 376
Methylphosphonate, di-*tert*-butyl, 295
2-Methyl-3-propylindole, 425
Methylsulfinylacetophenone, 433
2-Methylsulfinyl-1,3-indanedione, 434
Methyl *p*-toluenesulfonate, (2,3-diphenyl-2-cyclopropenyl)-, 32, 270
  diphenylcyclopropyl-, 271
  (1-methyl-2-oxocyclohexyl)-, 350
4-Methyl-4-(trichloromethyl)-2,5-cyclohexadienone, 10, 112
Methyltriphenylphosphonium bromide, 340
Michael addition, 90
Microscopic reversibility, principle of, 177
Migration, of alkyl groups, 108, 109, 110, 431
  of aryl groups, 108
  of the ethoxy group, 169

Migration (*Continued*)
  of halide ions, 246
  of the hydride ion, 115, 149, 150
  of hydrogen, 141
  of the methoxy group, 101
  of the methyl group, 112, 418
  of the phenyl group, 103, 219
  of protons, 70, 73, 84, 207, 208, 209, 210
  of the trichloromethyl group, 113
Molecular refraction, 285
Moloxides, 393
Monovinylcyclobutanes, 189, 190
Monovinylcyclopropane, 21, 190
Morpholine, 291
Multi-centre mechanism, 165, 170, 190
Multiplicity, 410

Naphthalene, 178
  9,10-dihydro-, 179
  1,4,6,7-tetrakis(trifluoromethyl)-, 179
  2,3,6,7-tetrakis(trifluoromethyl)-, 179
1,6-Naphthalenedione, 8a-methyl-1,2,3,4,6,7,8,8a-octahydro-, 305
Neber rearrangement, 337, 338, 339, 364
Neighboring carbon participation, 265
Neopentyl derivatives, 291
Neophylmercury chloride, 318
Neophyl rearrangement, 149
Nickel tetracarbonyl, 143
Nicotinic acid oxide, 382
Nitration, aromatic, 376
Nitrenes, 146
  ground state of, 146
  insertion into C–H bonds, 146
Nitriles, 11, 120
Nitrilium salts, 11, 119, 120, 228
Nitrimine, 431
  diphenyl-, 166
o-Nitrobenzaldehyde, 105
p-Nitrobenzaldehyde, 345
p-Nitrobenzene derivatives
o-Nitrobenzoyldiazomethane, 31, 261
Nitro compounds in 1,3-cycloadditions, 165
Nitrogen, insertion of positive, 147
Nitrogen pentoxide, 376
Nitrogen ylides, 340
Nitroimine, 431
Nitrones, 17, 18, 167
Nitrosation of oximes, 429
Nitrosonitrone, 430, 431

o-Nitrosophenylglyoxal, 262
5-Nitroso-2-phenylpyrimidine-4,6-diamine, 28, 230
Nitrosyl cation, 430
m-Nitro(trifluoromethyl)benzene, 372
Nitrous acid, reaction with oximes, 430
Nitrylium fluoroborate, 371
NMR spectroscopy, detection of valence isomers by, 190
Non-classical anions, 266
Non-classical cations, characteristics of, 266
  equilibrating, 267
Non-classical ion theory, 269
Non-classical norbornyl cation, 276
Non-classical radical, 266
Norbornadiene, 281
Norbornadien-7-ol, 6, 82
Norbornene epoxide, 131
Norbornene-*syn*-7-methanol, 82
Norbornen-7-ol, 82
Norbornyl cation, 276
  2-*p*-methoxyphenyl-7,7-dimethyl-, 279
Norbornyl chloride, 2-methyl-, 277
endo-Norbornyl derivatives, 276, 277
exo-Norbornyl derivatives, 276, 277
2-Norbornyl derivatives, 1,2-bis-(*p*-methoxyphenyl)-, 279
Norbornyl system, 276
Norcaradiene, 98
Nortricyclanone, 322
Nortricyclene derivative, 281
Nucleofugous, 224
Nucleophiles, phosphorus triamides as, 99
  reaction with the C=O bond, 95
  reaction with the C=C bond, 78
  reaction with the C≡C bond, 88
  reaction with the furan system, 72
  trialkylphosphines as, 99
  trialkyl phosphites as, 99
Nucleophilic β-carbon atom, 362, 363
Nucleophilicity, 285, 286

Octafluorocyclobutane, 153
trans-Octatetraene derivatives, 197, 198
Olefins, addition of aluminum alkyl to, 86, 87
  synthesis by Wittig reaction, 340
  β-hydroxy-, 122
Oligomerization, 87
Onium bases, 213

Orbital, antibonding, 61, 266
  benzyl, 420
  d-, 92
  d-, interaction, 335
  d-, stabilization, 330
  non-bonding, 411
  s-, effects, 329
Organo-alkali compounds, 332, 334
Organo-lithium compounds, 326, 331, 335
Organometallic compounds, optically active, 326
Orientation rule of thermal 1,2-additions, 153
Overvoltage in Clemmensen reduction, 108
Oxalyl chloride, nucleophilic attack in reaction with amides, 306
Oxazine, dihydro-, 118, 119
Oxetanes, 161
Oxetenes, 161
Oxidation, with lead tetraacetate, 399
  of sulfides with H_2O_2, 247
Oxime esters, 338
Oximes, reaction with HNO_2, 429
  syn- and $anti$-, 229
Oxirane, 2-benzoyl-3,3-dimethyl-, 417
  $trans$-1,2-dimethyl-, 288
  trichloromethyl-, 25, 214
3-Oxobutanal, 356
4-Oxocyclohexyl benzoate, 302
Oxonium salts, 120
6-Oxo-5-phenylheptanoic acid, 300
3-Oxo-5-phenylpentanol, 350
β-Oxo sulfoxides, 433
Oxy-Cope rearrangement, 133
Oxygen, addition to 1,3-dienes, 392
  addition to thiabenzene, 393
  diradical character of, 392
  electronic state of, 392
  insertion of positive, 16, 150
  positively charged, 65, 97, 150
Ozone, 64
Ozonides, mol-, 64
  normal, 65
  primary, 64
Ozonization, 64
  of aromatic double bonds, 64
  of the C=C bond, 64
  of indene in liquid NH_3, 4, 67
  of 2-pentene, 65

Partial rate factors, 377
Pauli principle, 406
1,1,2,3,4-Pentaalkyl-1,3-butadiene, 94
1,3-Pentadiene, 412
Pentamethyl(trichloromethyl)benzene, 26, 220
Pentane, 2,4-dimethyl-, 395
2,4-Pentanediol, 395
2,4-Pentanedione, 10, 109
  3,3-dimethyl-, 8, 97
  3-methyl-, 98
Pentanol, 3-oxo-5-phenyl-, 350
2-Pentanone, 414
  4-hydroxy-, 395
  5-phenyl-, 111
Pentaphenylcyclopentadiene, 422
Pentazole, 169
  p-(dimethylamino)phenyl-, 170
  p-ethoxyphenyl-, 170
Pentazoles, substituted, 169
Pentenal, 234
4-Pentenal, 2-methyl-, 213
1-Pentene, 2-chloro-5-phenyl-, 111
2-Pentene, 3-ethyl-, 232
Pentenediol monoenol, 347
3-Pentenoic acid lactone, 3-hydroxy-2,2,4-trimethyl-, 101
1-Penten-4-yne, 7, 88
Pentyl halide, 1,1,4-trimethyl-, 224
Pentyl isobutyrate, 3,4-epoxy-2,2,4-trimethyl-, 30, 250
Peracids, epoxidation of olefins with, 130, 131
Perchlorofulvalene, 316
Peresters, formation of, 97
  heterolysis of the O–O bond of, 151
Perfluoro-2,3-dimethyl-1,3-butadiene, 140
Perhydroazulene, 432
Periodic acid, 232
Perkow reaction, 296, 315
Pernitroso compounds, 429, 430
Peroxide, dibenzoyl, 311, 392
Peroxide anion, 397
Peroxides, reaction with phosphines, 311
  aroyl, 311
  cyclic, 97
  dimeric, trimeric, and polymeric, 65
Peroxyoxalate, di-$tert$-butyl, 397
Peroxy radical, 395
Phenacylidenetriphenylphosphorane, 7, 91

Phenanthrene, dihydro-, 319
  9,10-dimethyl-, 109
Phenanthrone, 109
Phenethyl cation, 284
Phenethyl chloride, 11, 119
Phenol, o- and p-amino-, 380
  o-pyrrolidino-, 72
  2,4,6-trimethoxy-, 380
Phenols, 422
Phenoxide, 2,4,6-trimethyl-, 380
4-Phenoxy-4-phenyl-3-butenoate, 416
Phenylacetylenes, 162, 195
10-Phenylanthrone, 244
Phenyl azide, 169
3-Phenyl-2H-azirine, 339
Phenylcyclobutadienediol, 195
4-Phenyl-2-butanone, 111
2-Phenyl-2-butene, 233
3-Phenyl-3-cyclobutene-1,2-dione, 22, 195
trans-2-Phenylcyclohexyl p-toluene-
  sulfonate, 23, 207
4-Phenyl-1,2-diazabicyclo[3.2.0]hept-2-en-
  6-one, 25, 215
Phenyldiazomethane, 441, 442
5-Phenyldibenzaluminol, 95
2-Phenyl-1,3-dioxolane, 28, 231
Phenethyl bromide, 354
Phenethyl chloride, 11, 119
Phenylglyoxal, 433
6-Phenyl-1-hexene, 332
N^2-Phenylhydrazonoyl chloride, 17, 166
Phenyl isocyanate, 9, 105
Phenyl isocyanide, 15, 143
Phenyllithium, 25, 94, 213, 219
Phenylmagnesium bromide, 197
Phenylmercuric acetate, 328
5-Phenyl-2-pentanone, 111
N-Phenylphthalimide, 143
Phenylpropiolic acid ester, 17, 166
C-Phenyl-N-thiobenzoylnitrile imine, 211
1-[(Phenylthio)carbonyl]aziridine, 290
Phenyl vinyl ether, 335
Phenyl vinyl sulfide, 335
Phosgene route to prepare isocyanides, 142
Phosphine, tertiary, 311
  trialkyl-, 99, 315
  tri-n-butyl-, 327, 328
  triphenyl-, 313, 328
Phosphine oxide, benzyldiphenyl, 219
  triphenyl-, 219

Phosphines, 218
  trialkyl-, 99
Phosphite, tri-$tert$-butyl, 295
  triethyl, 295
  trimethyl, 9, 101
Phosphite esters, 99
Phosphites, reaction with alkyl halides, 295
  trialkyl, 99
3-Phospholin 1-oxide, 1,3-diphenyl-, 9, 105
Phosphine sulfide, triphenyl-, 313
Phosphonium bromide, methyltriphenyl-, 340
Phosphonium chloride, (chloromethyl)tri-
  phenyl-, 219
Phosphonium hydroxide, (2,2-diphenyl-
  ethyl)triethyl-, 218
Phosphonium salts, 295
  reaction with base, 26, 218
  hydroxymethyl-, 219
  trialkoxyalkyl-, 99
  trialkyl-(β-oxoalkyl)-, 296
Phosphonium ylides, 340
Phosphoranes, hybridization in, 92
  alkylidenetriphenyl-, 341
Phosphorane, benzylidenetriphenyl-, 341
  ethylidenetriphenyl-, 342
  (methoxycarbonylmethylene)triphenyl-, 341
  phenacylidenetriphenyl-, 7, 91
Phosphorescence, 410
Phosphonic esters, 295
Phosphorimide, 106
Phosphorus compounds, trivalent, nucleophilic vs. electrophilic character of, 99
  quinquevalent, 218
Phosphorus pentabromide, 113
Phosphorus pentachloride, 72, 110
  reaction with amides, 316
  reaction with the C=O bond, 10, 110, 112, 113
  reaction with cyclohexadienones, 10, 113
  reaction with dibenzoylethylene, 10, 113
  reaction with furans, 72
Phosphorus tribromide, 114
Phosphorus tris(dimethylamide), 8, 99
Phosphorus ylides, 99, 340, 341
  addition to triple bonds, 7, 91
  preparation, 340
Photochemical process, secondary, 412
Photochemistry, theory of, 51
Photodecomposition, 408

Photoexcitation, 393, 405
Photoisomerization, 413, 419
Photolysis, of diazomethane, 134
  in the gas phase, 414
  of ketene, 134
  in the liquid phase, 414
  of ylides, 347
Photons, energy of, 406
Photoreaction, of dibenzoylethylene, 416
  of 2-pentanone, 414
  of stilbene, 419
Photorearrangement of epoxy ketones, 417
Phthalimide, N-phenyl-, 143
o-Phthalonitrile, 397
Pinacol, 108
Pinacol rearrangement, 105
α-Pinene, 122
α-Pinene oxide, rearrangement of, 13, 133
Piperidine oxide, 210
Piperidinomethyl thioacetate, 299
Pivalic acid, 97
Planck's constant, 405
Polarizability, 285
Polarization in thioketones, 122
Polychloroisooctane, 223
Polyenes, conjugated, 191
Polymerization of ethylene with Ziegler catalysts, 86
Polyphosphoric acid as catalyst in Beckmann rearrangement, 16, 147
Positive nitrogen, 16, 147
Positive oxygen, 16, 150
Potassium $tert$-butoxide, 199, 322
Potassium, butyl-, 332
Potassium hydride, 334
Potassium N-methylanilide, 323
Potassium methyl xanthate, 288
Potassium thiocyanate, 286
Potential energy, 237
  curves, 407, 408, 409
Predissociation, 409
Prins reaction, 209
Propane, 163
  2,2-dibromo-, 414
2-Propanol, 3-(1-cyclopentenyl)-2-methyl-, 11, 118
  3-(3-cyclopentenyl)-2-methyl-, 119
  2-(2-hydroxycyclopentyl)-2-methyl-, 11, 119

2-Propanone, 3-diazo-1-(1,2,3-triphenylcyclopropenyl)-, 136
Propene, 86, 87
  2-bromo-, 163, 414
4a-Propargyl-$trans$-decalin-2,7-dione, 361
Propiolic acid, 332
  ester, 17, 166
  phenyl-, 17, 166
Propyl cis-propenyl ether, 157
Protonation of sulfoxides, 434, 435
Proton donors in Birch reduction, 358
Pseudobases, 125
Pummerer rearrangement, 433
$4H$-Pyran, 124
Pyran, dihydro-, 72
$3H$-Pyrazole dioxide, 430
Pyridine, 2-acetoxy-, 382, 383
  triphenyl-, 125
2-Pyridinecarbaldehyde, 99
Pyridine oxide, 382, 383
Pyridinium cation, 1-acetyl-, 349
Pyridinium acetate, 90
  1-acetoxy-, 383
Pyridinium salts, 123
2-Pyridone, 383
Pyrimidine-4,6-diamine, 5-nitroso-2-phenyl-, 28, 230
Pyrolysis of thiocarbonates, 309
2-Pyrone, 19, 177
  3,5-dichloro-, 19, 175
Pyrrole, 393
  1-lithio-, 139
Pyrrolidine, 1-(α-methylenefurfuryl)-, 4, 72
Pyrrolidinium bromide, 1,1-dimethyl-, 25, 213
o-Pyrrolidinophenol, 72
Pyrylium perchlorate, 2,4,6-trimethyl-, 12, 123
  2,4,6-triphenyl-, 12, 124

Quantization of light, 405
Quantum yield, 412, 419
Quenching of energy, 412
Quinazolines, 120
Quinoline, 139
Quinolizinium salts, 123

Racemization, in S_E1 substitution, 317
  in S_N1 substitution, 240
Radical, $tert$-butoxy, 390, 398
  triphenylmethyl, 388

Radical addition, of HCl and HI, 391
  of polyhalomethanes, aldehydes, alcohols, amines, mercaptans, 391
Radical anion, 361, 362
Radical chain, 388
  mechanisms, 388, 389, 391
Radical complex, 391
Radical formation in Clemmensen reduction, 108
Radical initiators, 388
Radical polymerization, 391
Radical reactions, 388
Radical rearrangement, 391
Radical substitution, mechanism, 389
Radicals, electronic states of, 389
  generation of, 389
  stabilization of, 389, 398
  structure of, 389
Ramberg–Bäcklund reaction, 336, 364
Rate of addition, of hydrogen halides to C=C bonds, 60
  of bromine to the C=C bond, 62
  to the C=O bond, 95, 96
Rate theory, absolute, 237
Rearrangement, of acetylfuran, 73
  of acylcyclopropanes, 79
  of 1-(α-methylenefurfuryl)pyrrolidine, 4, 72
  of α-pinene oxide, 13, 133
  Baeyer–Villiger, 97, 151
  Baker–Venkataraman, 302
  benzilic acid, 79, 433
  Chapman, abnormal, 353
  Claisen, 424
  Cope, 133, 189, 191, 192
  Criegee, 97, 98, 151, 247
  Favorsky, 79
  Meisenheimer, 325
  Neber, 338, 364
  neophyl, 149
  oxy-Cope, 133
  photo-, 417
  Pummerer, 433
  radical, 391
  Ramberg–Bäcklund, 336, 364
  v. Richter, 262
  Smiles, 386
  Sommelet–Hauser, 346
  Stevens, 325
  Wagner–Meerwein, 256
  Wittig, 323, 324, 325

Wolff, 169, 263
Recarbonylation, 259
Recombination of radicals, 414
Reduction with $LiAlH_4$, 6, 81, 82
Relaxation, vibrational, 409, 410, 412
Resonance effects in aromatic substitution, 374
Resonance stabilization, of carbanions, 340
  in phosphorus ylides, 329, 330, 340
Resonance structures of ozone, 64
Retention of configuration, in carbene insertion, 135
  in nitrene insertion, 146
  in S_E2 substitution, 317, 318, 326
Retroaldol condensation, 256
Retrodiene cleavage, 179
Retro-Prins reaction, 208
v. Richter rearrangement, 262
Ring contraction, 242, 243, 244, 245
Ring opening of epoxides, 289, 417
Rho-values of the Hammett equation, 105, 378
Rotations in molecules, 405, 406

Salicylamide, o-glycyl-, 439
Sandmeyer reaction, 170
Saytzeff's rule, 202, 203, 204, 205
Schiff bases, 210
Schlenk equilibrium, 327
Sensitizers, 135, 393, 416, 419
Sigma-electron participation, 266
Sigma-values of Hammett equation, 105, 373, 376
Singlet state, 134, 135, 146, 413
  first excited, 135, 406
  of methylene, 134, 135
Six-membered rings, formation by Diels–Alder reaction, 170
Smiles rearrangement, 386
Sodium in liquid ammonia, 358
  pentyl-, 332
Sodium aluminum hydride, 81
Sodium o-benzoylbenzoate, 438
Sodium borohydride, 81, 123, 273, 274
Sodium cyanide, 276
Sodium cyclopentadienecarboxylate, 332
Sodium disulfide, 294, 295
Sodium hydride, 334, 364
Sodium iodide, 290, 291
Solvated electrons, 358
Solvation of nucleophiles, 286

Solvent cage effects, 383, 414
Solvent effects, in S_E1 substitution, 319
  in S_N1 substitution, 240
  in S_N2 substitution, 286
Sommelet–Hauser rearrangement, 346
Specific acid catalysis, 247
Spin conservation, 410
Spin flip, 414
Stability of primary vs. sec- and tert-carbanions, 329
Stability of cations, 329, 440
Stereochemistry, of aluminum hydride addition, 82
  of Beckmann rearrangement, 147, 228
  of carbene addition, 135, 138
  of 1,2-cycloaddition, 153, 154, 155
  of 1,3-cycloaddition, 165, 166
  of cyclobutane formation, 74
  of α-dicyclopentadien-1-ol rearrangement, 176
  of Diels–Alder reaction, 170, 171, 172, 173, 174
  of enamine alkylation, 71
  of epoxidation, 130, 131
  of halocarbene addition, 138
  of halogen addition, 61, 88
  of hydrogen halide addition, 59, 60, 61
  of lithium aluminum hydride addition, 82, 83
  of methylene addition, 135
  of ozone addition, 64, 65, 66
Stereospecificity of reactions of the norbornenyl cation, 280
Steric effects, in elimination reactions, 202, 203, 204, 205, 206, 207, 208
  in fragmentation reactions, 225
  in S_N1 substitution, 239
  at small rings, 246
  in valence isomerization, 190, 191, 192
Steric hindrance in cyclobutanone reactions, 356
Stevens rearrangement, 325
Stilbene, 346
  photoreactions of, 419
  trans-, 6, 85
Styrene, 398
  α-azido-, 339
  β-methyl-, 6, 84, 342
Substituent effects, in aromatic substitution, 373, 376
  in excited aromatic systems, 420

Substituents, electron donating, 374, 421
Substitution, electrophilic, in aliphatic systems, 372
  S_E1 substitution
    with carbon as leaving group, 317 318
    with deuterium, hydrogen, and nitrogen as leaving group, 321
    with oxygen as leaving group, 321, 323
  S_E2 substitution at asymmetric carbon, 326
  in aromatic systems, 371
    with OH$^+$, 379
  nucleophilic, in aliphatic systems, 236
  S_N1 substitution, 236
    leaving group in, 240
    rate equation of, 240
    transition state of, 240
  S_N2 substitution, 237, 284
    at the carbonyl group, 240, 284
    at chlorine, 315
    intramolecular, 291, 292
    at oxygen, 311
    rate-determining step, 284
    at divalent sulfur, 313
  S_N2' substitution, 246
  S_Ni substitution, 308, 314
  in aromatic systems, 383
  radical, 389
Succinimide, N-bromo-, 390
  N-chloro-, 389
Sulfenes, 16, 159
Sulfide, p-chlorobenzhydryl methyl, 310
  dimethyl, 345
  diphenyl, 346
  phenyl vinyl, 335
  vinyl 1,1-dimethyldecyl, 335
Sulfone, 1-bromoethyl ethyl, 336
  2-hydroxy-5-methylphenyl 2-nitrophenyl, 386
Sulfones, chlorinated cyclic, 337
  cyclic, 220, 337
  effect of the SO_2 group, 336
  vinylogous chloroalkyl, 337
Sulfonium salts, benzyldiphenyl-, 346
  trialkyl-, 213
Sulfonyl chlorides, 161
Sulfoxide, dimethyl, 322
  β-oxo-, 433
Sulfur, elimination of, 26, 220

Sulfur ylides, 344, 345, 346
Swain equation, 285
Symmetrization of alkylmercury salts, 327

Taft sigma-values, 310, 374
Telomerization, 391
Termination, chain-, 388
α-Terpinene, 393
Tetrachlorocyclopentadienone, 90
Tetracyanoethylene, 131, 175
Tetrafluoroethylene, 153
Tetrahedrane, 136
Tetrahydrofuran, 346
1,2,4,5-Tetrakis(trifluoromethyl)benzene, 179
1,4,6,7-Tetrakis(trifluoromethyl)naphthalene, 179
2,3,6,7-Tetrakis(trifluoromethyl)naphthalene, 179
2,2,4,4-Tetramethyl-1,3-cyclobutanedione, 101
4,4,5,5-Tetramethyl-1,3-dioxolan-2-one, 287
Tetramethylethylene, 12, 121
1,1′,3,3′-Tetraphenylbi-(2-imidazolidinylidene), 144
1,2,3,4-Tetraphenyl-1,3-butadiene, 4, 68
Tetraphenylcyclobutane, 420
Tetraphenylcyclopentadienone, 90
Tetraphenyl-1,2,4,6-thiabenzene, 393, 394
Tetrasulfides, alkyl, aryl, and aralkyl, 313
Tetrazole, 211
   2-acyl-, 24, 211
   2-thiobenzoyl-, 211
Thiabenzene, 1,2,4,6-tetraphenyl-, 393, 394
1,3,4-Thiadiazole, 2,5-diphenyl-, 211
1,2,3,4-Thiatriazole, 5-ethoxy-, 307
2-Thiazoline-4-carboxylic acid, 2-amino-, 428
Thietanes, 161
Thietenes, 161
Thiirane, cis-dimethyl-, 287
   dioxide, cis-2,3-dimethyl-, 337
Thioacetone, hexafluoro-, 12, 121, 161
Thiocarbonates, 309
   p-chlorobenzhydryl S-methyl, 310
   decomposition of, 309, 310
Thiocyanate, potassium, 286
3,3′-Thiodipropionate, 2,2′-dichloro-, 294
Thioformate, O-ethyl chloro-, 307
Thiohemiacetals, 248

Thiolactone, 269
Thiols, 248
Thionyl chloride, 31, 69, 264, 436
Thiopyran, 1,1-dichlorocyclopropa[b][1]-benzo-, 26, 220
Thiourea, 287
Third-order reaction, 218
Three-centre bond, 266
   mechanism, 135
Three-membered rings by carbene addition, 135
α-Thujene, 193
Thujone, 21, 193
Tight ion pairs, see also Ion pair, tight, 97, 111, 113, 151, 246
Tischtschenko reaction, 117
Titanium tetrachloride, 86
Tolane, 94
Toluene, chlorination of, 377
5-(p-Toluenesulfonyloxy)-1-decalone, 28, 230
Torsional effects in norbornyl derivatives, 277
Transannular, hydride shift, 438
   reactions, 15, 141
Transition, forbidden, 410
   radiative, 134
Transition-metal salts as catalysts, 86, 88
Transitions, electronic, 406
   rotational, 406
   vibrational, 406
Transition state, of alkylcarbonium ion formation, 265
   of allylic carbonium ion formation, 265
   of base-catalyzed olefin isomerization, 337
   bicyclo[3.1.0], 440
   bicyclo[3.1.1], 439
   bicyclo[3.2.1], 436
   cyclic five-membered, 146, 165, 207
   cyclic four-membered, 73, 106
   cyclic seven-membered, 300
   cyclic six-membered, 71, 73, 146, 207, 269, 336
   cyclic three-membered, 314
   of 1,3-cycloaddition, 165
   of Diels–Alder addition, 170, 174
   of ene-addition, 122
   of electrophilic aromatic substitution, 374
   of E1 elimination, 202

Transition state (*Continued*)
  of *E*2 elimination, 203
  of enamine alkylation, 71
  highly strained, 80
  of $S_N 2$ substitution, 285
transoid-Conformation of dienes, 170
Triacetylbenzene, 356
Trialkoxyalkylphosphonium salts, 99
Trialkylphosphines, 99, 315
Trialkyl phosphites, 99
Trialkyl-(β-oxoalkyl)phosphonium salts, 296
Trialkylphosphorus ylides, 345
Trialkylsulfonium salts, 213
1,2,4-Triazine, 427
  3-amino-5,6-diphenyl-, 427
  3-chloro-5,6-diphenyl-, 427
1,2,4-Tri-*tert*-butylbenzene, 185
1,3,5-Tri-*tert*-butylbenzene, 185
Tributylphosphine, 327, 328
Tri-*tert*-butyl phosphite, 295
Trichloroacetamide, 315
Trichloroacetate, ethyl, 138, 139
Trichloroacetic acid, 138
Trichloromethyl carbanions, 137, 138, 139
Trichloromethyloxirane, 25, 214
Tricyclic hydrocarbons, 272, 274, 276
Triethylaluminum, 94
Triethylphosphine, 218
Triethyl phosphite, 295
(Trifluoromethyl)benzene, 371
1-(Trifluoromethyl)cycloheptatrienyl cyanide, 200
Trihalide ion, 62, 89
Trimerization of acetylenes, 20, 182
Trimerization of dichloroacetylene, 182
1,3,5-Trimethoxybenzene, 379
Trimethoxybenzoic acid, 358, 360
Trimethylamine, 286
1,5,5-Trimethylbicyclo[2.1.1]hexane-6-carboxylic acid, 31, 258
3,5,5-Trimethyl-1,2-cyclohexanedione 2,3-monoenol, 5, 75
1,3,3-Trimethylcyclohexyl hydroperoxide, 150
2-(2,2,3-Trimethyl-3-cyclopentenyl)-acetaldehyde, 167
Trimethylene oxide, 232
1,1,4-Trimethylpentyl halide, 224
2,4,6-Trimethylphenoxide, 380
Trimethyl phosphite, 9, 101

2,4,6-Trimethylpyrylium perchlorate, 12, 123
2,4,6-Trinitroanisole, 383
Triplet state, see Photochemistry, 406
  of alkyl nitrenes, 146
  of cyclobutadiene, 183
  of 1,3-dienes, 392, 393
  of methylene, 134, 135
Triplet–triplet transfer, 410
Triphenylaluminum, 8, 94
2,2,3-Triphenylcyclobutanone, 357
Triphenylcyclopropenylium cation, 441
Triphenylmethyl cation, 441
  radical, 388
Triphenylphenol, 137
Triphenylphosphine, 313
  oxide, 219
  sulfide, 313
Triphenylpyridine, 125
2,4,6-Triphenylpyrylium perchlorate, 12, 124
Tripropylcyclopropenylium cation, 441
Tropolone, 269
  methyl ether, 8, 98
Tropone, 6, 81
Tropylium ion, 242

Urea, *N*-alkyl-*N*-nitroso-, 260

Valence electrons, 406
Valence isomerization, 189
  of bicyclo[3.2.0]hept-2-en-6-yl acetate, 21, 192
  of conjugated polyenes, 22, 195
  of cyclooctatriene derivatives, 22, 195
  of diallylic systems, 21, 189, 192
  of 1,1-dicyclopropylethylene, 21, 192
  of monovinylcyclobutanes, 21, 189, 192
  of monovinylcyclopropanes, 21, 189, 192
  of 3-phenyl-3-cyclobutene-1,2-dione, 22, 195
  by radical mechanisms, 190, 194
  of thujone, 21, 193
  of vinylcyclobutanes, 21, 189, 190, 192
  of vinylcyclopropanes, 21, 189, 192
Valence isomers, 189
  of benzene, 182, 183
van der Waals interaction, 337
Vibrations in molecules, 405
Vinyl acetate, 193
Vinylacetylenes, 162

Vinylcarbenes, 165
Vinyl compounds, 153, 157
Vinylcyclopropane, 1,2-cycloaddition of, 209
2-Vinylcyclopropane-1,1-dicarboxylic acid, 303
cis- and trans-2-Vinylcyclopropylformaldehyde, 133
Vinyl 1,1-dimethyldecyl sulfide, 335
Vinyl ether, phenyl, 335
Vinyl ethers, 335
  oxetanes from, 161
Vinyl hydrogen, 332
Vinylogs, 347
Vinylogous acid chlorides, 347
Vinyl radical, 392
Vinyl sulfides, 335

Wagner–Meerwein rearrangement, 256, 284
Walden inversion, 287
Wheland intermediate, 371
Wittig reaction, 340
  stereochemistry of, 341
Wittig rearrangement, 323, 324
Wolff rearrangement, 169, 263
Woodward–Hoffmann rules, 153, 158, 173, 191, 193

Xanthate, potassium methyl, 288

Ylene, 340
Ylides, 7, 91, 213, 218
  moderately reactive, 342
  of nitrogen, 340
  of phosphorus, 340, 341
  photolysis of, 347
  preparation of phosphorus, 340
  reactive, 341, 342, 345
  resonance stabilization in, 344, 345
  stable, 340, 341
  sulfur, 345, 346
  sulfonium, 325
  trialkylphosphorus, 345
Yukawa–Tsuno equation, 378

Ziegler catalysts, 86
Zinc, reaction with epoxides, 79
  reduction with, 79
Zinc, amalgamated, in Clemmensen reduction, 107, 109
Zinc chloride as catalyst, 67, 423
Zinc–copper couple, 136
Zwitterions, 65, 131, 132, 144, 220, 340, 394, 414, 428

COLLEGE OF MARIN LIBRARY
3 2555 00021304 6

DATE DUE   19939

MAY 4 72	JUN 3 0 '83		
APR 12 73	MAR 0 8 1990		
MAY 3 73	NOV 1 4 1996		
JAN 3 74	DEC 1 9 1997		
MAY 15 1976	NOV 0 2 2007		
DEC 15 1977			
MAR 8 1978			
APR 1 9 1979			
DEC 1 1 1980			
MAY 14 '81			
MAR 3 '83			

QD 257 .H64   Höver, Hermann.
   Problems in organic reaction mechanisms

C13752